Handbook of Research on Emerging Innovations in Rail Transportation Engineering

B. Umesh Rai
Chennai Metro Rail Limited, India

A volume in the Advances in Civil and Industrial Engineering (ACIE) Book Series

Published in the United States of America by
Engineering Science Reference (an imprint of IGI Global)
701 E. Chocolate Avenue
Hershey PA, USA 17033
Tel: 717-533-8845
Fax: 717-533-8661
E-mail: cust@igi-global.com
Web site: http://www.igi-global.com

Copyright © 2016 by IGI Global. All rights reserved. No part of this publication may be reproduced, stored or distributed in any form or by any means, electronic or mechanical, including photocopying, without written permission from the publisher. Product or company names used in this set are for identification purposes only. Inclusion of the names of the products or companies does not indicate a claim of ownership by IGI Global of the trademark or registered trademark.

Library of Congress Cataloging-in-Publication Data

Names: Rai, B. Umesh, 1959- editor.
Title: Handbook of research on emerging innovations in rail transportation
   engineering / B. Umesh Rai, editor.
Description: Hershey : Engineering Science Reference, 2016. | Includes
   bibliographical references and index.
Identifiers: LCCN 2016002410| ISBN 9781522500841 (hardcover) | ISBN
   9781522500858 (ebook)
Subjects: LCSH: Railroads--Planning. | Transportation--Forecasting.
Classification: LCC HE1031 .H36 2016 | DDC 385--dc23 LC record available at http://lccn.loc.gov/2016002410

This book is published in the IGI Global book series Advances in Civil and Industrial Engineering (ACIE) (ISSN: 2326-6139; eISSN: 2326-6155)

British Cataloguing in Publication Data
A Cataloguing in Publication record for this book is available from the British Library.

All work contributed to this book is new, previously-unpublished material. The views expressed in this book are those of the authors, but not necessarily of the publisher.

For electronic access to this publication, please contact: eresources@igi-global.com.

Advances in Civil and Industrial Engineering (ACIE) Book Series

ISSN: 2326-6139
EISSN: 2326-6155

Mission

Private and public sector infrastructures begin to age, or require change in the face of developing technologies, the fields of civil and industrial engineering have become increasingly important as a method to mitigate and manage these changes. As governments and the public at large begin to grapple with climate change and growing populations, civil engineering has become more interdisciplinary and the need for publications that discuss the rapid changes and advancements in the field have become more in-demand. Additionally, private corporations and companies are facing similar changes and challenges, with the pressure for new and innovative methods being placed on those involved in industrial engineering.

The **Advances in Civil and Industrial Engineering (ACIE) Book Series** aims to present research and methodology that will provide solutions and discussions to meet such needs. The latest methodologies, applications, tools, and analysis will be published through the books included in **ACIE** in order to keep the available research in civil and industrial engineering as current and timely as possible.

Coverage
- Ergonomics
- Quality Engineering
- Engineering Economics
- Coastal Engineering
- Optimization Techniques
- Production Planning and Control
- Materials Management
- Structural Engineering
- Operations Research
- Hydraulic Engineering

IGI Global is currently accepting manuscripts for publication within this series. To submit a proposal for a volume in this series, please contact our Acquisition Editors at Acquisitions@igi-global.com or visit: http://www.igi-global.com/publish/.

The Advances in Civil and Industrial Engineering (ACIE) Book Series (ISSN 2326-6139) is published by IGI Global, 701 E. Chocolate Avenue, Hershey, PA 17033-1240, USA, www.igi-global.com. This series is composed of titles available for purchase individually; each title is edited to be contextually exclusive from any other title within the series. For pricing and ordering information please visit http://www.igi-global.com/book-series/advances-civil-industrial-engineering/73673. Postmaster: Send all address changes to above address. Copyright © 2016 IGI Global. All rights, including translation in other languages reserved by the publisher. No part of this series may be reproduced or used in any form or by any means – graphics, electronic, or mechanical, including photocopying, recording, taping, or information and retrieval systems – without written permission from the publisher, except for non commercial, educational use, including classroom teaching purposes. The views expressed in this series are those of the authors, but not necessarily of IGI Global.

Titles in this Series

For a list of additional titles in this series, please visit: www.igi-global.com

Emerging Challenges and Opportunities of High Speed Rail Development on Business and Society
Raj Selladurai (Indiana University Northwest, USA) Peggy Daniels Lee (Kelley School of Business, Indiana University, USA) and George VandeWerken (Providence Bank, USA)
Engineering Science Reference • copyright 2016 • 289pp • H/C (ISBN: 9781522501022) • US $210.00 (our price)

Advanced Manufacturing Techniques Using Laser Material Processing
Esther Titilayo Akinlabi (Univeristy of Johannesburg, South Africa) Rasheedat Modupe Mahamood (University of Johannesburg, South Africa & University of Ilorin, Nigeria) and Stephen Akinwale Akinlabi (University of Johannesburg, South Africa)
Engineering Science Reference • copyright 2016 • 288pp • H/C (ISBN: 9781522503293) • US $175.00 (our price)

Handbook of Research on Applied E-Learning in Engineering and Architecture Education
David Fonseca (La Salle Campus Barcelona, Universitat Ramon Llull, Spain) and Ernest Redondo (Universitat Politècnica de Catalunya, BarcelonaTech, Spain)
Engineering Science Reference • copyright 2016 • 569pp • H/C (ISBN: 9781466688032) • US $310.00 (our price)

Emerging Design Solutions in Structural Health Monitoring Systems
Diego Alexander Tibaduiza Burgos (Universidad Santo Tomás, Colombia) Luis Eduardo Mujica (Universitat Politecnica de Catalunya, Spain) and Jose Rodellar (Universitat Politecnica de Catalunya, Spain)
Engineering Science Reference • copyright 2015 • 337pp • H/C (ISBN: 9781466684904) • US $235.00 (our price)

Robotics, Automation, and Control in Industrial and Service Settings
Zongwei Luo (South University of Science and Technology of China, China)
Engineering Science Reference • copyright 2015 • 337pp • H/C (ISBN: 9781466686939) • US $215.00 (our price)

Using Decision Support Systems for Transportation Planning Efficiency
Ebru V. Ocalir-Akunal (Gazi University, Turkey)
Engineering Science Reference • copyright 2016 • 475pp • H/C (ISBN: 9781466686489) • US $215.00 (our price)

Contemporary Ethical Issues in Engineering
Satya Sundar Sethy (Indian Institute of Technology Madras, India)
Engineering Science Reference • copyright 2015 • 343pp • H/C (ISBN: 9781466681309) • US $215.00 (our price)

Emerging Issues, Challenges, and Opportunities in Urban E-Planning
Carlos Nunes Silva (University of Lisbon, Portugal)
Engineering Science Reference • copyright 2015 • 380pp • H/C (ISBN: 9781466681507) • US $205.00 (our price)

www.igi-global.com

701 E. Chocolate Ave., Hershey, PA 17033
Order online at www.igi-global.com or call 717-533-8845 x100
To place a standing order for titles released in this series, contact: cust@igi-global.com
Mon-Fri 8:00 am - 5:00 pm (est) or fax 24 hours a day 717-533-8661

Dedicated to the two women in my life, one who gave me life, other who made life worth living.

Editorial Advisory Board

Paul Allen, *University of Huddersfield, Huddersfield, UK*
Björn Birgisson, *Kungliga Tekniska högskolan, Stockholm, Sweden*
Zhibin Jiang, *Tongji University, Shanghai, China*
William Powrie, *University of Southampton, UK*
L. Umanand, *Indian Institute of Science, Bangalore, India*

List of Contributors

An, Min / *University of Birmingham, UK* 173
Arboleya, Pablo / *University of Oviedo, Spain* 452
Asbach, Lennart / *German Aerospace Center (DLR), Germany* 250
Bardhan, Ronita / *Indian Institute of Technology Bombay, India* 40
Basu, Devasish / *Indian Railways, India* 489
Böhm, Thomas / *German Aerospace Center (DLR), Germany* 405
Bojović, Nebojša / *University of Belgrade, Serbia* 67,100
Butaud, Christophe / *Egis Tunnels, France* 580
Chaudhuri, Aritra / *Indian Institute of Technology Kharagpur, India* 273
Dasgupta, Anirban / *Indian Institute of Technology Kharagpur, India* 273
Dasgupta, Pallab / *Indian Institute of Technology Kharagpur, India* 212
Deb, Alok Kanti / *Indian Institute of Technology Kharagpur, India* 489
Galviz, Carlos Lopez / *Lancaster University, UK* 1
George, Anjith / *Indian Institute of Technology Kharagpur, India* 273
Hassannayebi, Erfan / *Tarbiat Modares University, Iran* 420
Hungar, Hardi / *German Aerospace Center (DLR), Germany* 250
Hwang, Jong-Gyu / *Korea Railroad Research Institute, Korea* 232
Jana, Arnab / *Indian Institute of Technology Bombay, India* 40
Jillella, Satya Sai Kumar / *Curtin University, Australia* 130
Jo, Hyun-Jeong / *Korea Railroad Research Institute, Korea* 232
Kabi, Bibek / *Indian Institute of Technology Kharagpur, India* 273
Lackhove, Christoph / *German Aerospace Center (DLR), Germany* 405
Mangal, Mahesh / *Indian Railways, India* 212
Mardani, Soheil / *Tarbiat Modares University, Iran* 420
Matan, Anne / *Curtin University, Australia* 130
Melichar, Vlastimil / *University of Pardubice, Czech Republic* 67
Meyer zu Hörste, Michael / *German Aerospace Center (DLR), Germany* 250,405
Milenković, Miloš / *Zaragoza Logistic Center, Spain & University of Belgrade, Serbia* 67,100
Naha, Arunava / *Indian Institute of Technology Kharagpur, India* 489
Narayanaswami, Sundaravalli / *IIM Ahmedabad, India* 313,387
Newman, Peter / *Curtin University, Australia* 130
Pradhan, Smitirupa / *Indian Institute of Technology Kharagpur, India* 524
Qin, Yong / *Beijing Jiaotong University, China* 173
Raturi, Varun / *Indian Institute of Science, India* 146
Routray, Aurobinda / *Indian Institute of Technology Kharagpur, India* 273,489

Sajedinejad, Arman / *Research Institute for Information Science and Technology (IRANDOC), Iran* 420
Samanta, Anik Kumar / *Indian Institute of Technology Kharagpur, India* 489
Samantaray, Arun K. / *Indian Institute of Technology Kharagpur, India* 524
Schimmelpfennig, Jörg / *Ruhr-Universität Bochum, Germany* 161
Sengupta, Anwesha / *Indian Institute of Technology Kharagpur, India* 273
Srinivasan, Sumeeta / *Tufts University, USA* 23
Švadlenka, Libor / *University of Pardubice, Czech Republic* 67
T G, Sitharam / *Indian Institute of Science, India* 130
Verma, Ashish / *Indian Institute of Science, India* 146
Waymel, Frederic / *Egis Tunnels, France* 580
Yalçınkaya, Özgür / *Dokuz Eylül University, Turkey* 335

Table of Contents

Preface xxii

Section 1
General

Chapter 1
Past Futures: Innovation and the Railways of Nineteenth-Century London and Paris 1
    Carlos Lopez Galviz, Lancaster University, UK

Chapter 2
The Potential for Rail Transit as a Way to Mitigate Accident Risk: A Case Study in Chennai 23
    Sumeeta Srinivasan, Tufts University, USA

Chapter 3
Planning Transit System for Indian Cities: Opportunities and Challenges 40
    Arnab Jana, Indian Institute of Technology Bombay, India
    Ronita Bardhan, Indian Institute of Technology Bombay, India

Section 2
Finance

Chapter 4
Railway Investment Appraisal Techniques 67
    Miloš Milenković, Zaragoza Logistic Center, Spain & University of Belgrade, Serbia
    Libor Švadlenka, University of Pardubice, Czech Republic
    Nebojša Bojović, University of Belgrade, Serbia
    Vlastimil Melichar, University of Pardubice, Czech Republic

Chapter 5
Railway Demand Forecasting 100
    Miloš Milenković, Zaragoza Logistic Center, Spain & University of Belgrade, Serbia
    Nebojša Bojović, University of Belgrade, Serbia

Chapter 6
Emerging Value Capture Innovative Urban Rail Funding and Financing: A Framework 130
    Satya Sai Kumar Jillella, Curtin University, Australia
    Sitharam T G, Indian Institute of Science, India
    Anne Matan, Curtin University, Australia
    Peter Newman, Curtin University, Australia

Chapter 7
Analyzing Intercity Modal Choice and Competition Between High Speed Rail (HSR) and Other Transport Modes in Indian Context 146
    Ashish Verma, Indian Institute of Science, India
    Varun Raturi, Indian Institute of Science, India

Chapter 8
Welfare Economic Principles of (Urban) Rail Network Pricing: First-Best and Second-Best Solutions 161
    Jörg Schimmelpfennig, Ruhr-Universität Bochum, Germany

Section 3
Safety

Chapter 9
Challenges of Railway Safety Risk Assessment and Maintenance Decision Making 173
    Min An, University of Birmingham, UK
    Yong Qin, Beijing Jiaotong University, China

Chapter 10
Formal Assurance of Signaling Safety: A Railways Perspective 212
    Pallab Dasgupta, Indian Institute of Technology Kharagpur, India
    Mahesh Mangal, Indian Railways, India

Chapter 11
Automatic Static Software Testing Technology for Railway Signaling System 232
    Jong-Gyu Hwang, Korea Railroad Research Institute, Korea
    Hyun-Jeong Jo, Korea Railroad Research Institute, Korea

Chapter 12
Automated Testing: Higher Efficiency and Improved Quality of Testing Command, Control and Signaling Systems by Automation 250
    Lennart Asbach, German Aerospace Center (DLR), Germany
    Hardi Hungar, German Aerospace Center (DLR), Germany
    Michael Meyer zu Hörste, German Aerospace Center (DLR), Germany

Chapter 13
Alertness Monitoring System for Vehicle Drivers using Physiological Signals 273
    Anwesha Sengupta, Indian Institute of Technology Kharagpur, India
    Anjith George, Indian Institute of Technology Kharagpur, India
    Anirban Dasgupta, Indian Institute of Technology Kharagpur, India
    Aritra Chaudhuri, Indian Institute of Technology Kharagpur, India
    Bibek Kabi, Indian Institute of Technology Kharagpur, India
    Aurobinda Routray, Indian Institute of Technology Kharagpur, India

Section 4
Operation

Chapter 14
Railway Operations Models: The OR Approach 313
    Sundaravalli Narayanaswami, IIM Ahmedabad, India

Chapter 15
A General Simulation Modelling Framework for Train Timetabling Problem 335
    Özgür Yalçınkaya, Dokuz Eylül University, Turkey

Chapter 16
Intelligent Transportation Systems: The State of the Art in Railways 387
    Sundaravalli Narayanaswami, Indian Institute of Management, India

Chapter 17
Integrated Traffic Management using Data from Traffic, Asset Conditions, Energy and
Emissions 405
    Thomas Böhm, German Aerospace Center (DLR), Germany
    Christoph Lackhove, German Aerospace Center (DLR), Germany
    Michael Meyer zu Hörste, German Aerospace Center (DLR), Germany

Chapter 18
Disruption Management in Urban Rail Transit System: A Simulation Based Optimization
Approach 420
    Erfan Hassannayebi, Tarbiat Modares University, Iran
    Arman Sajedinejad, Research Institute for Information Science and Technology (IRANDOC), Iran
    Soheil Mardani, Tarbiat Modares University, Iran

Section 5
Engineering

Chapter 19
Steady State Modeling of Electric Railway Power Supply Systems for Planning and Operation
Purposes 452
    Pablo Arboleya, University of Oviedo, Spain

Chapter 20
Online Condition Monitoring of Traction Motor 489
    Anik Kumar Samanta, Indian Institute of Technology Kharagpur, India
    Arunava Naha, Indian Institute of Technology Kharagpur, India
    Devasish Basu, Indian Railways, India
    Aurobinda Routray, Indian Institute of Technology Kharagpur, India
    Alok Kanti Deb, Indian Institute of Technology Kharagpur, India

Chapter 21
Dynamic Analysis of Steering Bogies 524
    Arun K. Samantaray, Indian Institute of Technology Kharagpur, India
    Smitirupa Pradhan, Indian Institute of Technology Kharagpur, India

Chapter 22
Ventilation and Air Conditioning in Tunnels and Underground Stations 580
    Frederic Waymel, Egis Tunnels, France
    Christophe Butaud, Egis Tunnels, France

Compilation of References 605

About the Contributors 651

Index 661

Detailed Table of Contents

Preface xxii

Section 1
General

Chapter 1
Past Futures: Innovation and the Railways of Nineteenth-Century London and Paris 1
　Carlos Lopez Galviz, Lancaster University, UK

Innovation was central to developments in urban railway transport in nineteenth-century London and Paris. Innovation was often political, the result of an encounter between and across a range of actors, including railway entrepreneurs and their companies, railway engineers, civil engineers, architects, intellectuals, a range of authorities –local, municipal, metropolitan, regional and national –, and the rich mix of people affected by the opening of a new railway line. The chapter opens up the notion of innovation to issues that cover three different dimensions: the politics, the culture and the social concerns behind the building of railways. It shows how London and Paris coped, but also dealt with one of the most transformative forces of nineteenth-century Britain and France. An important part of that story relates to the different futures that were envisioned in the two cities, in response to specific concerns and determined by a particular set of conditions. This approach highlights the process of how innovations took place rather than the end result.

Chapter 2
The Potential for Rail Transit as a Way to Mitigate Accident Risk: A Case Study in Chennai 23
　Sumeeta Srinivasan, Tufts University, USA

The city of Chennai has made road accident data available with the address location of road accidents and the total numbers of persons and pedestrians affected in the accident in 2009. These data were geocoded to locate the accidents with respect to the census wards within the Chennai Corporation area. Both the total number of persons as well as pedestrians in accidents as well as the rate of accidents normalized by population in the ward were modeled as dependent variables using Poisson based regression models to see the effect of location characteristics such as road length, vehicle traffic, proximity to existing and proposed transit infrastructure and the percentage of the land developed between 1991-2009. The results from the models suggest that location does indeed affect the risk for accidents in Chennai and that planners in the city may need to better understand the implications of roads, urban development, transit access and the built environment for traffic safety.

Chapter 3
Planning Transit System for Indian Cities: Opportunities and Challenges.. 40
Arnab Jana, Indian Institute of Technology Bombay, India
Ronita Bardhan, Indian Institute of Technology Bombay, India

Indian cities are currently in a phase of transition. Continuous urbanization and seamless connectivity is the paradigm. Proliferating bourgeois class is extending the demand for private automobiles. With limited opportunity to increment land use allocated to transportation and rapid shift towards automobile ownership, importance of transit system is being sensed. City managers believe that public transit could be an alternative in providing solution to ever increasing problem of traffic congestion, parking demand, accidents and fatalities, and global environmental adversities. This chapter examines the critical planning issues that need to be addressed. It highlights the opportunities and challenges these cities are poised towards transit system planning. The experiences from cities worldwide that have adopted transit systems to create compact city forms fostering mixed land use development are exemplified here. A '3P' developmental framework of 'provide', 'promote' and 'progress' has been proposed to harness the opportunity.

Section 2
Finance

Chapter 4
Railway Investment Appraisal Techniques 67
Miloš Milenković, Zaragoza Logistic Center, Spain & University of Belgrade, Serbia
Libor Švadlenka, University of Pardubice, Czech Republic
Nebojša Bojović, University of Belgrade, Serbia
Vlastimil Melichar, University of Pardubice, Czech Republic

Railway transport involves the expenditure of resources on a combination of investment in capital items (e.g. stations, tracks, equipment) and/or in operations (e.g. subsidies). Concerning the fact that there are limited amounts of resources, it is necessary to maximize the returns obtained from the investments of those resources. The best way to do this is to ensure that the resources will be allocated on those projects that maximize their return. Railway appraisals therefore represent a way of thinking about all the costs and benefits of different railway related spending projects in a systematic manner so that, the projects can be compared and investments made in those which are going to provide the maximum possible return on the investment. This chapter provides a review of the main analytical tools that should be used in the process of railway investments appraisal. Namely, a detailed description of discounting, Net Present Value (NPV), Internal Rate of Return (IRR) and Cost Benefit Analysis (CBA) is covered by this chapter.

Chapter 5
Railway Demand Forecasting 100
Miloš Milenković, Zaragoza Logistic Center, Spain & University of Belgrade, Serbia
Nebojša Bojović, University of Belgrade, Serbia

Forecasting represents an indispensable activity in railway transportation planning. Forecasting of demand levels is vital to the railway company as a whole as it provides the basic input for the planning and control of all functional areas including railway transport operations planning, marketing and finance. Demand levels and the timing of their appearance (on a day, week, month or seasonal basis) greatly effects capacity levels, financial needs and general structure of the business. Forecasting employs historical data and uses various forecasting methods to make accurate estimates of future demands. Forecasting approaches can be generally divided into two categories: econometric or causal and time series techniques. In this chapter a comprehensive review of methods belonging to these two broad classes will be made. Special emphasis will be given to the application of these techniques to railway demand modeling.

Chapter 6
Emerging Value Capture Innovative Urban Rail Funding and Financing: A Framework 130
    Satya Sai Kumar Jillella, Curtin University, Australia
    Sitharam T G, Indian Institute of Science, India
    Anne Matan, Curtin University, Australia
    Peter Newman, Curtin University, Australia

Urban rail transit is emerging around the world as a catalyzing developmental solution to enable 21st century sustainable cities. However, these transit systems are capital intensive and cities worldwide are seeking innovative funding and financing mechanisms. Recently, land based value capture (VC) mechanisms have emerged as a pioneering solution to fund urban rail projects. This chapter introduces the VC concept and provides global best practice. The chapter aims to help enhance the understanding and rationale behind VC approaches through assessing the transit impacted accessibility value proposition and various VC mechanisms to capture the value created. A six-step Strategic Value Capture (SVC) framework is proposed which offers a step-by-step guidance to help define the VC based urban rail transit funding and financing processes from VC planning to VC operations.

Chapter 7
Analyzing Intercity Modal Choice and Competition Between High Speed Rail (HSR) and Other Transport Modes in Indian Context 146
    Ashish Verma, Indian Institute of Science, India
    Varun Raturi, Indian Institute of Science, India

In this study, a theoretical framework is developed in order to assess the viability of transport infrastructure investment in the form of High Speed Rail (HSR) by assessing, the mode choice behaviour of the passengers and the strategies of the operators, in the hypothetical scenario. Discrete choice modelling (DCM) integrated with a game theoretic approach is used to model this dynamic market scenario. DCM is incorporated to predict the mode choice behaviour of the passengers in the new scenario and the change in the existing market equilibrium and strategies of the operators due to the entry of the new mode is analysed using the game theoretic approach. The outcome of this market game will describe the strategies for operators corresponding to Nash equilibrium. In conclusion, the impact of introduction of HSR is assessed in terms of social welfare by analysing the mode choice behaviour and strategic decision making of the operators, thus reflecting on the economic viability of the transport infrastructure investment.

Chapter 8
Welfare Economic Principles of (Urban) Rail Network Pricing: First-Best and Second-Best
Solutions 161
    Jörg Schimmelpfennig, Ruhr-Universität Bochum, Germany

The purpose of this chapter is to rectify the at best unprofessional intermingling of objectives and constraints and present a proper theory of first-best and second-best pricing in urban rail networks. First, in view of the flaws of both Dupuit's – though nevertheless ingenious idea of – consumer surplus as well its cannibalized version found in most of today's economics textbooks, a proper definition of economic welfare resting on Hicks'sian variations instead is provided. It is used to derive efficient pricing rules that are subsequently applied to specific questions arising from running an urban railway network such as overcrowding, short-run versus long-run capacity or competing modes of transport like the private motor car. At the same time, another look is taken at economic costs, and in particular economic marginal costs, differing from commercial or accounting costs. Among other things, it is shown that even with commercial marginal costs being constant first-best pricing might not necessarily be incompatible with a zero-profit budget.

Section 3
Safety

Chapter 9
Challenges of Railway Safety Risk Assessment and Maintenance Decision Making 173
Min An, University of Birmingham, UK
Yong Qin, Beijing Jiaotong University, China

Railway safety is a very complicated subject, which is determined by numerous aspects. Many of qualitative and quantitative railway safety and risk analysis techniques and methods are used in the industry. But, however, the railway industry faces problems and challenges on how to apply these techniques and methods effectively and efficiently, particularly in the circumstances where the risk data are incomplete or there is a high level of uncertainty involved in the risk data. This chapter approaches these subjects to discuss the problems and challenges of railway safety and risk analysis methods in dealing with uncertainties, and those growing needs of the industry. A well-established technique is also introduced in this chapter which can be used to identify major hazards and evaluate both qualitative and quantitative risk data, and information associated with railway operation efficiently and effectively in an acceptable way in various environments.

Chapter 10
Formal Assurance of Signaling Safety: A Railways Perspective 212
Pallab Dasgupta, Indian Institute of Technology Kharagpur, India
Mahesh Mangal, Indian Railways, India

The EN50128 guidelines recommend the use of formal methods for proving the correctness of railway signaling and interlocking systems. The potential benefit of formal safety assurance is of unquestionable importance, but the path towards implementing the recommendations is far from clear. The EN50128 document does not specify how formal assurance of railway interlocking may be achieved in practice. Moreover, the task of setting up an electronic interlocking (EI) equipment involves multiple parties, including the EI equipment vendor, the certification agency which certifies the resident EI software to be correct, and the end user (namely the railway service provider) who must configure the EI equipment. Considering the distributed nature of the development process, a feasible approach towards formal certification of the end product (post configuration) is not obvious. This chapter outlines the basics of formal verification technology and presents, from the perspective of the railways, a pragmatic roadmap for the use of formal methods in safety assurance of its signaling systems.

Chapter 11
Automatic Static Software Testing Technology for Railway Signaling System 232
    Jong-Gyu Hwang, Korea Railroad Research Institute, Korea
    Hyun-Jeong Jo, Korea Railroad Research Institute, Korea

In accordance with the development of recent computer technology, the railway system is advancing to be flexible, automatic and intelligent. In addition, many functions of railway signaling which are cores to the railway system are being operated by computer software. Recently, the dependency of railway signaling systems on computer software is increasing. The testing to validate the safety of the railway signaling system software is becoming more important, and related international standards for inspections on the static analysis based source code and dynamic test are a highly recommended (HR) level. For this purpose, studies in relation to the development of source code analysis tools were started several years ago in Korea. To verify the applicability of validation tools developed as a part of these studies, the applicability test was performed for the railway signaling system being applied to the Korean domestic railway. This automated testing tool for railway signaling systems can also be utilized at the assessment stage for railway signaling system software, and it is anticipated that it can also be utilized usefully at the software development stage. This chapter drew the result of the application test for this actual source code of the railway signaling system being applied to railway sites and analyzed its result.

Chapter 12
Automated Testing: Higher Efficiency and Improved Quality of Testing Command, Control and Signaling Systems by Automation 250
    Lennart Asbach, German Aerospace Center (DLR), Germany
    Hardi Hungar, German Aerospace Center (DLR), Germany
    Michael Meyer zu Hörste, German Aerospace Center (DLR), Germany

The need for time- and cost-efficient tests is highly relevant for state-of-the-art safety-related train control and rail traffic management systems. Those systems get increasingly more complex and so testing becomes a more and more and important cost factor. This chapter discusses some approaches to relocate tests from the field to the lab, reduce cost and duration while improving quality of lab tests. The European Train Control System (ETCS) is used as an example, but the approaches and results can be applied to other systems as well, for instance interlocking.

Chapter 13
Alertness Monitoring System for Vehicle Drivers using Physiological Signals 273
    Anwesha Sengupta, Indian Institute of Technology Kharagpur, India
    Anjith George, Indian Institute of Technology Kharagpur, India
    Anirban Dasgupta, Indian Institute of Technology Kharagpur, India
    Aritra Chaudhuri, Indian Institute of Technology Kharagpur, India
    Bibek Kabi, Indian Institute of Technology Kharagpur, India
    Aurobinda Routray, Indian Institute of Technology Kharagpur, India

The present chapter deals with the development of a robust real-time embedded system which can detect the level of drowsiness in automotive and locomotive drivers based on ocular images and speech signals of the driver. The system has been cross-validated using Electroencephalogram (EEG) as well as Psychomotor response tests. A ratio based on eyelid closure rates called PERcentage of eyelid CLOSure (PERCLOS) using Principal Component Analysis (PCA) and Support Vector Machine (SVM) is employed to determine the state of drowsiness. Besides, the voiced-to-unvoiced speech ratio has also been used. Source localization and synchronization of EEG signals have been employed for detection of various brain stages during various stages of fatigue and cross-validating the algorithms based in image and speech data. The synchronization has been represented in terms of a complex network and the parameters of the network have been used to trace the change in fatigue of sleep-deprived subjects. In addition, subjective feedback has also been obtained.

Section 4
Operation

Chapter 14
Railway Operations Models: The OR Approach 313
Sundaravalli Narayanaswami, IIM Ahmedabad, India

This chapter is intended as an exposure to OR based methods, particularly the analytical approach to modelling railway operations. An overview of several planned operations in railway transportation is provided in an academic context. Some of the applications and the associated models are applied in realistic settings in the transportation industry, and also have demonstrated evidence of acceptance over a long number of years. Primary coverage is on transportation scheduling and the concise discussions are on planning phases, various operations that can be deterministically modeled and analysed, model development, few exercises and real-world stories, wherever appropriate. All sections are adequately provided with the list of references and an interested reader can benefit from a conceptual understanding to model development and to implement and deploy, under some prior knowledge on the basics and programming experience.

Chapter 15
A General Simulation Modelling Framework for Train Timetabling Problem 335
Özgür Yalçınkaya, Dokuz Eylül University, Turkey

One of the most important problems encountered and needed to be solved in railway systems is train timetabling (scheduling) problem. This is the problem of determining a feasible timetable for sets of trains which does not violate track capacities and additionally satisfies some operational constraints of the railway system. In this chapter, a feasible timetable generator framework for stochastic simulation modelling is introduced. The objective is to obtain a feasible train timetable for all trains in the railway system, which includes train arrival and departure times at all visited stations and calculated average train travel time. Although this chapter focuses on train timetabling (scheduling) problem, the developed general framework can also be used for train dispatching (rescheduling) problem if the model can be fed by the real-time data. Since, the developed simulation model includes stochastic events, and it can easily cope with the disturbances that occur in the railway systems, it can be used for dispatching.

Chapter 16
Intelligent Transportation Systems: The State of the Art in Railways.. 387
Sundaravalli Narayanaswami, Indian Institute of Management, India

"Intelligent Transportation systems" is what everyone wants to know about, and about which very little is available as know-how. ITS technologies and monitoring systems are quite popular and reasonably well deployed in developed countries, particularly the roadways and airways. ITS holds a greater promise than ever before, as both availability of niche technologies and demand for more efficient transportation systems have increased multi-fold in recent years. Of late, there are huge railway projects all over the world that spans through several techniques, such as light / heavy rails, monorails etc. Apart from the social benefits that can be envisaged, these projects are genuine examples of public-private partnerships along with global business operations. Many of these projects demonstrate a classy trend of moving towards automation of operations of very large scales. Few agent architectures are discussed in brief in this chapter.

Chapter 17
Integrated Traffic Management using Data from Traffic, Asset Conditions, Energy and Emissions 405
    Thomas Böhm, German Aerospace Center (DLR), Germany
    Christoph Lackhove, German Aerospace Center (DLR), Germany
    Michael Meyer zu Hörste, German Aerospace Center (DLR), Germany

The traffic management is the core of the railway operations control technology. It receives the timetable information as a target definition and advises the command control and signaling systems to execute the rail traffic. Hence the traffic management system (TMS) has to take into account many sources of requests towards the traffic operation e.g. coming from the maintenance planning or the power supply system and to optimize the operation with respect to many criteria as e.g. punctuality, energy consumption, capacity and infrastructure wear. This chapter shows the sources of information for the TMS as well the resulting criteria. The final approach to configure a specific has to be done with respect to a specific application.

Chapter 18
Disruption Management in Urban Rail Transit System: A Simulation Based Optimization Approach 420
    Erfan Hassannayebi, Tarbiat Modares University, Iran
    Arman Sajedinejad, Research Institute for Information Science and Technology (IRANDOC),
      Iran
    Soheil Mardani, Tarbiat Modares University, Iran

The process of disruption management in rail transit systems faces challenging issues such as the unpredictable occurrence time, the consequences and the uncertain duration of disturbance or recovery time. The objective of this chapter is to adopt a discrete-event object-oriented simulation system, which applies the optimization algorithms in order to compensate the system performance after disruption. A line blockage disruption is investigated. The uncertainty associated with blockage recovery time is considered with several probabilistic scenarios. The disruption management model presented here combines short-turning and station-skipping control strategies with the objective to decrease the average passengers' waiting time. A variable neighborhood search (VNS) algorithm is proposed to minimize the average waiting time. The computational experiments on real instances derived from Tehran Metropolitan Railway are applied in the proposed model and the advantages of the implementing the optimized single and combined short-turning and stop-skipping strategies are listed.

Section 5
Engineering

Chapter 19
Steady State Modeling of Electric Railway Power Supply Systems for Planning and Operation Purposes 452
    Pablo Arboleya, University of Oviedo, Spain

This chapter describes the different electric railway power supply systems and their main characteristics from the point of view of the power flow modeling and simulation. It considers the DC traction systems and also the AC ones, explaining the different elements embedded in the network and proposing steady state electrical models of each element. The basic methods for modeling the train behavior in terms of demanded/regenerated power are also detailed. Finally, the procedures to simulate the trains motion into the electrical network and how their motion and power demand affect to the electrical variables will be unraveled, explaining how to merge all the models in a system of equations.

Chapter 20
Online Condition Monitoring of Traction Motor 489
    Anik Kumar Samanta, Indian Institute of Technology Kharagpur, India
    Arunava Naha, Indian Institute of Technology Kharagpur, India
    Devasish Basu, Indian Railways, India
    Aurobinda Routray, Indian Institute of Technology Kharagpur, India
    Alok Kanti Deb, Indian Institute of Technology Kharagpur, India

Squirrel Cage Induction Motors (SCIMs) are major workhorse of Indian Railways. Continuous online condition monitoring of the SCIMs like Traction Motor (TM) are essential to prevent unnecessary stoppage time in case of a complete failure. Before a complete failure, the TMs generally develop incipient or weak faults. Weak faults have minute influence on the motor performance but eventually leads to complete failure of the motor. If these weak faults are identified at the earliest then, a scheduled maintenance can be planned which will prevent any unplanned stoppage. The signals used for SCIM fault detection are motor current, voltage, vibration, temperature, voltage induced in search coil, etc. The most popular fault detection technology is based on Motor Current Signature Analysis (MCSA). MCSA based online and onboard TM condition monitoring system can be very useful for Indian railways to reduce the cost of operation and unplanned delay by shifting from unnecessary scheduled maintenance to condition-based maintenance of TM and other auxiliary SCIMs.

Chapter 21
Dynamic Analysis of Steering Bogies 524
    Arun K. Samantaray, Indian Institute of Technology Kharagpur, India
    Smitirupa Pradhan, Indian Institute of Technology Kharagpur, India

Running times of high-speed rolling stock can be reduced by increasing running speed on curved portions of the track. During curving, flange contact causes large lateral force, high frequency noises, flange wears and wheel load fluctuation at transition curves. To avoid derailment and hunting, and to improve ride comfort, i.e., to improve the curving performances at high speed, forced/active steering bogie design is studied in this chapter. The actively steered bogie is able to negotiate cant excess and deficiency. The bogie performance is studied on flexible irregular track with various levels of cant and wheel wear. The bogie and coach assembly models are developed in Adams VI-Rail software. This design can achieve operating speed up to 360 km/h on standard gauge ballasted track with 150mm super-elevation, 4km turning radius and 460m clothoid type entry curve design. The key features of the designed bogie are the graded circular wheel profiles, air-spring secondary suspension, chevron springs in the primary suspension, anti-yaw and lateral dampers, and the steering linkages.

Chapter 22
Ventilation and Air Conditioning in Tunnels and Underground Stations 580
    Frederic Waymel, Egis Tunnels, France
    Christophe Butaud, Egis Tunnels, France

This chapter is an overview of the state of the art and advanced principles in the field of ventilation and air conditioning (AC) in tunnels and underground stations. The first part is dedicated to the background which deals with the design objectives that are generally retained for normal and emergency operation of underground rail projects. The second part provides solutions and recommendations of ventilation and AC strategies that can be used in metro and rail projects. Advantages and drawbacks of the proposed solutions are also discussed. The main parameters that can influence the design are introduced in this section. The possibility of using draught relief shaft is detailed. Advantages of Platform Screen doors and heat sink effects are also described. Various cooling technologies of station air conditioning systems are presented. Critical issues when designing longitudinal ventilation system for tunnel emergency situations are also discussed. The last part is a short list of future research directions in the field of cooling / heating production for air conditioning systems.

Compilation of References 605

About the Contributors 651

Index 661

Preface

Railway has fascinated most of us with its capacity to transport a large mass of passengers and goods on land. Importantly, it does this with the most efficient use of energy. This is a common knowledge and policy planners are well aware of it, unfortunately, one less discussed aspect of Railway, in the technical literature is that the Railway has always been a cradle for innovation. Right from its birth, at the beginning of the Industrial Revolution, when the wheels were put on the boiler and tracks of steel laid to present, Railways has always been a pioneer in 'innovation' (Tony Judt, 2010). Recently automobiles have taken over the major share of short distance transport and air travel has similarly captured the long distance travel, but despite its relative decline, the spirit of innovation is still alive and kicking in Railway. It is therefore very apt, that a book is written that catches this spirit of innovation in Railway, while pointing to the direction where the innovations are most active.

Railway has used innovation as a tool for improving itself. It has also created a bunch of new services during its growth spurts. It gave birth to the travel and tourism industry (Tony Judt, 2010). It started the concept of circulating areas wherein the travelling public could relax, shop or dine while waiting for their next connecting train. From the Railway's desire to give last mile connectivity, the idea of common ticketing for multimodal transport was born.

The study of Railway is a multidisciplinary task. Railway has a rich history to discuss and the study, the building and the operation of the railway needs a complex finance study and analysis, operating and optimising the railway network is a management domain problem and finally, maintaining the asset has challenges in all spheres of engineering study. A comprehensive book covering all these railway functional areas is difficult to find. And even more difficult is finding a book totally dedicated to Railway research. Frequently, we may find literature on the application in a Railway field in a particular research field but not the other way around. Therefore, there is a felt need for a book on Railway related research to cater to a wider audience of researchers, policy maker, railway lovers and investors.

This book attempts to be an easy read and technically challenging too, a difficult task to accomplish!

Multidisciplinary Approach

The chapters in the book can be broadly grouped into five subgroups - General, Finance, Safety, Operation and Engineering. The general section discusses the history of railways in two historically important world cities (Ellis, C. H., 1954). In the study of cities of emerging economies, the Railway network plays an important role in increasing the productivity of Urban agglomerates (Ellis, P., & Roberts, M., 2015). This has led to city planners trying to build a railway network into the city transit system, and this is also

Preface

covered in the general section. It also has a chapter on the desirability of a more extensive railway network to overcome accident fatalities in a city (Covey, J., Robinson, A., Jones-Lee, M., & Loomes, G., 2010).

The next section of the book covers the very important topic of finance. Railway is a means of 'mass' transportation, be it passenger or freight. The scale associated with Railway Transportation precludes all small to medium-size investors. More often than not, it has to be built by government investments only. But the limited-size government funds are also needed by other competing sectors. This includes sectors of high importance like Defence, Healthcare, Education, etc. Therefore, Railway finds it difficult to earn its fair share of investment. It is more often than not, taken as a commercial investment and not as a welfare investment. This approach causes the investment to come in with the burden of expectation of a good rate of return (Perkins, S., 2005, September). This is unfair, as Railway has a much wider impact on welfare, productivity, environmental and other happiness indexes (Diener, E., 2000). This is usually not quantified in monetary terms. If these impacts are financially measured by tools (Gordon, L. A. 1974), which are already available, then the attractiveness of Railway investment is increased manifold. This is the central theme of the finance section.

The third section is on Safety. Railways always have had its right of way, therefore, it is safer than other modes of transport. Railways have the lowest fatality per passenger kilometre of any land-based transport system, but the right of way also leads to high speeds and most of the railway accident will be catastrophic, involving large-scale loss of life and material. Therefore, built-in safety while designing the system is of paramount importance in Railways. The contributing factors to accidents may range from human, to equipment and to weather. As the frequency of operation increases, with headways coming down to 90 seconds, maintenance windows are shrinking. This calls for automation in running and carrying out maintenance with predictive tools (Hall, S., 2003). The section concentrates on the modern day challenges on safety.

Once the decision for having the Railway network is made, investments are made and the tracks and rolling stock are in place, the next big challenge is to operate the network efficiently. The goal is to transport a maximum load in the minimum possible time. This 'Operation' which is significantly different from road traffic is covered in the next section (Landex, A., et.al, 2006). The Railway works with a constraint of being guided traffic. A train has to necessarily be at the track and it can overtake the front-running train only at predetermined places only. Commands for undertaking this operation are issued by the central controller. Thus, timetable management for normal running and its disruption management for abnormal running have to be carefully analysed. An overview of the tools available is covered in this section.

In the initial days Railway was a boiler on wheels and now it has evolved into a 10 Megawatt substation on wheels. The scale of the engineering challenge can be imagined, wherein not only these huge energy sources are to be moved around, but are moving at such breakneck speeds that it can challenge other transport competitors on speed, comfort and safety (Profillidis, V. A., 2000). The last section deals with engineering innovations for addressing these challenges. A train is a multidisciplinary system and incorporates the latest advances in the fields of computers, communications, electrical motors, vehicle dynamics, comfort engineering, etc. Therefore, the engineering section also follows a multidisciplinary approach.

Sustainable Growth

With increasing awareness of climate change and its consequence, there is a stress on using greener solutions to our transport problems. There is a renewed interest in all policy makers and city planners across the world to strengthen the public transport infrastructure. Railways have emerged as the primary choice due to its sustainability and for causing minimum damage to the environment (Schafer, A., et. al, 1999). There is a need for more financial tools which capture these social impacts. These tools will help the policy makers to justify their decision for committing finances to relatively costly railway infrastructure.

In the developing world where the railway has never lost its primacy. A new awareness has now dawned to policy planners of emerging countries that increasing Railway network is adding numbers to their GDP growth. The mainly agrarian economy of the developing world is becoming manufacturing and service oriented, resulting in new urban agglomeration. High Speed Railways and Metro Rail are the transport of choice for bringing in and moving people around in this new urban environment.

In a typical emerging economy, the governance structure consists of an executive comprising of an entrenched bureaucracy and a legislative body consisting of politicians. Politicians exist from one election to another but the bureaucracy has typically three decades of service in government. Because of longer tenures, the bureaucracy will look at the financial evaluation of a project mostly on discounted rates as the effects are long term while a political party with their ears to ground will look at the economic evaluation. The voting public will get a sense that the governance structure 'cares' for their wellbeing if a beneficial project springs up in their neighbourhood. This leads to a situation wherein the ruling party announces a railway project in the neighbourhood but the bureaucracy, tasked with implementing the decision, stalls it, trying to wait out the issue till the next election.

A Railway project may never be justified when compared to other means of transport if evaluated at Net Present Value (NPV) because of its high capital cost and longer gestation period. But if evaluated on economic appraisal basis (Grant-Muller, et. al., 2001), Railway projects will have an advantage as they have a much wider positive impact. Extensive goodwill for Railway projects have a reason which is not wholly intuitive. This book written with less emphasis on the complex mathematical equations makes for an easy reading and is thus recommended for policy makers to understand Railway impact better.

Organization of the Book

Chapter 1 reveals that developments in urban railway transport in nineteenth century London and Paris was driven by innovation. The author argues that a careful study of these developments will help emerging cities across the world when they face questions that are similar to those faced by London and Paris during that age.

Chapter 2 reports a high rate of traffic fatality in the city of Chennai, India. The analysis shows that pedestrian accidents contribute to a high percentage of the accident statistic in densely populated areas of the city. Modelling this problem reveals that a proximity to high carriage public transport, most likely railway, will reduce the risk of an accident.

Chapter 3 brings out the planning challenges for building a transit system in a major city of an emerging economy. The author argues for a developing a framework for building a transit system based on other cities across the world. He suggests that the primary aim of the planners should be to achieve cross-sector integration of transit systems with urban development.

Preface

Chapter 4 studies the different Railway investment appraisal techniques. The author admits that there is plenty of competition for scarce investment resources which are available with the government. For deciding where to invest, both economic and non-economic factors—political, social, etc.—have to be considered. The author discusses the tools for the same, namely Net Present Value (NPV), Internal Rate of Return (IRR) and Cost Benefit Analysis (CBA).

Chapter 5 discusses the technique for Railway demand forecasting. It suggests how to use the historical data together with various forecasting methods to make accurate estimates of future demand. Qualitative and various quantitative methods are studied and discussed, with particular emphasis on SARIMA model.

Chapter 6 addresses the common dilemma of local government to find funds for railway infrastructure. The author states that the Urban Rail infrastructure is coming up in a big way around the world and funding these projects is a big challenge. An innovative option of 'Land-based Value Capture' as one of the sources for funding is discussed in this chapter.

Chapter 7 analyses the intercity modal choice and argues for high-speed rail as the preferred choice, but a High-Speed Rail infrastructure needs a huge investment. The chapter uses gaming theory in capturing the modal decision of the travelling public to facilitate the investor's decision-making process.

Chapter 8 urges the administrators and the decision makers for pricing the public transport, treat budget constraints as constraints and not the goal of arriving at pricing. Intangible social benefits should be taken into consideration before fixing the public transport tariff.

Chapter 9 recognises the challenges of Railway maintenance managers in estimating the safety risk. FRA and FAHP modelling techniques are suggested in this chapter. This technique absorbs the domain expert knowledge in formulating railway safety and risk analysis. This is recommended when enough risk data is unavailable and there is a high level of uncertainty.

Chapter 10 lays out a roadmap for assurance of signaling safety. It outlines the formal model of a signaling system for safety assurance. It advocates the benefits of initial investment in formal capture and validation of signaling properties.

Chapter 11 is on automatic signaling software testing. For bringing in automation, Railway is becoming more and more dependent on signaling software. Development of automated software testing tools to achieve higher safety levels in software code is discussed in this chapter.

Chapter 12 is again on automated testing. The chapter argues that the modern safety systems in the trains are increasing complexity by the day. Testing these functions comprehensively in the field is becoming difficult. Therefore, it advocates model based testing to address these issue.

Chapter 13 proposes an alertness monitoring system for vehicle drivers. This is based on using physiological signals. It recommends the usage of PERCLOS, a drowsiness measuring tool to monitor the train driver alertness during a run. Speech is another variable monitored in conjunction.

Chapter 14 suggests an operation research model for railway operations. It argues that railway operations involve many subsets of operation problem. Therefore, while taking a holistic view of the operational problem the author develops an extensible basic model.

Chapter 15 attempts to develop an algorithm for solving a train timetabling problem. By definition timetabling is, running trains within the track capacity to satisfy operational needs. A stochastic simulation model for generating a feasible timetable is developed in the chapter.

Chapter 16 is on the emerging trends of intelligent transportation systems in Railways. Using the increasing availability of computing power and extensive networking, the author advocates the use of Intelligent Transportation Systems to achieve higher speeds, an increase of carrying capacity and more precise real-time information to travelers.

Chapter 17 develops an Integrated Traffic Management system using Data from Traffic, Asset Conditions, Energy and Emissions. The Traffic Management System (TMS) takes into account many sources of information and tries to optimize conflicting objectives. Data exchanges between different sources and decision-making is discussed in detail in this chapter.

Chapter 18 discusses the after effects of a disruption in the urban rail transit system. Recovery techniques after disruption are discussed in this chapter. A VNS algorithm, based on station skipping and short turning is developed and has proven to be the most efficient in recovery after a disruption.

Chapter 19 models the energy supply to the railway. There are different systems adopted worldwide for Railway Electric Traction. The system is complex as there is a moving load which may be consuming or regenerating. Moreover, it will transit from one feeding station to another during its journey. These complexities are discussed in the chapter.

Chapter 20 discusses the traction motor maintenance issue as this is the most critical equipment in the rolling stock. Maintenance hour reduction by condition-based monitoring in place of pre-scheduled maintenance is advised. Online algorithms to detect weak faults, together with a predictive algorithm are used for condition monitoring.

Chapter 21 is on another critical component of rolling stock - the bogies. The bogie in a train is critical safety, riding comfort and speed. In this chapter simulation model of bogie with passive and forced/active steering mechanisms is developed.

Chapter 22 is on safety and comfort issues of the railway infrastructure. This is a study on air-conditioning of the underground station and emergencies in tunnels. The authors discuss the objectives in the modern system and the tools available to achieve it. They also point to future research directions.

REFERENCES

Covey, J., Robinson, A., Jones-Lee, M., & Loomes, G. (2010). Responsibility, scale and the valuation of rail safety. *Journal of Risk and Uncertainty, 40*(1), 85–108. doi:10.1007/s11166-009-9082-0

Diener, E. (2000). Subjective well-being: The science of happiness and a proposal for a national index. *The American Psychologist, 55*(1), 34–43. doi:10.1037/0003-066X.55.1.34 PMID:11392863

Ellis, P., & Roberts, M. (2015). *Leveraging Urbanization in South Asia: Managing Spatial Transformation for Prosperity and Livability*. World Bank Publications. doi:10.1596/978-1-4648-0662-9

Gordon, L. A. (1974). Accounting rate of return vs. economic rate of return. *Journal of Business Finance & Accounting, 1*(3), 343–356. doi:10.1111/j.1468-5957.1974.tb00867.x

Grant-Muller, S. M., Mackie, P., Nellthorp, J., & Pearman, A. (2001). Economic appraisal of European transport projects: The state-of-the-art revisited. *Transport Reviews, 21*(2), 237–261. doi:10.1080/01441640119423

Hall, S. (2003). *Beyond Hidden Dangers-Railway Safety into The 21st Century*. Ian Allan Publishing Limited.

Judt, T. (2010). The Glory of the Rails. In The New York Review of Books (p. 23). Ellis, C. H. (1954). British railway history (Vol. 1). Allen and Unwin.

Preface

Landex, A., Kaas, A. H., & Hansen, S. (2006). *Railway operation*. Technical University of Denmark, Centre for Traffic and Transport.

Perkins, S. (2005, September). The role of government in European railway investment and funding. *Proceedings of the European Conference of Ministers of Transport, China Railway Investment & Finance Reform Forum*, Beijing, China (Vol. 20).

Profillidis, V. A. (2000). Railway engineering. Iwnicki, S. (Ed.), (2006). Handbook of railway vehicle dynamics. CRC press.

Schafer, A., & Victor, D. G. (1999). Global passenger travel: Implications for carbon dioxide emissions. *Energy*, *24*(8), 657–679. doi:10.1016/S0360-5442(99)00019-5

Section 1
General

Chapter 1
Past Futures:
Innovation and the Railways of Nineteenth-Century London and Paris

Carlos Lopez Galviz
Lancaster University, UK

ABSTRACT

Innovation was central to developments in urban railway transport in nineteenth-century London and Paris. Innovation was often political, the result of an encounter between and across a range of actors, including railway entrepreneurs and their companies, railway engineers, civil engineers, architects, intellectuals, a range of authorities –local, municipal, metropolitan, regional and national –, and the rich mix of people affected by the opening of a new railway line. The chapter opens up the notion of innovation to issues that cover three different dimensions: the politics, the culture and the social concerns behind the building of railways. It shows how London and Paris coped, but also dealt with one of the most transformative forces of nineteenth-century Britain and France. An important part of that story relates to the different futures that were envisioned in the two cities, in response to specific concerns and determined by a particular set of conditions. This approach highlights the process of how innovations took place rather than the end result.

INTRODUCTION

Innovation was central to developments in urban railway transport in nineteenth-century London and Paris. Innovation was often political, the result of an encounter between and across a range of actors, including railway entrepreneurs and their companies, railway engineers, civil engineers, architects, intellectuals, a range of authorities –local, municipal, metropolitan, regional and national –, and the rich mix of people affected by the opening of a new railway line: shopkeepers whose business would be affected by the scale of the works; landlords who were forced to deal with the noise, the pollution, and the viaducts across their properties; tenants displaced without recourse to much else beyond their own means, the largest majority consisting of the poor.

DOI: 10.4018/978-1-5225-0084-1.ch001

This rich and diverse mix of people and interests is very important. When we think about railway innovation, a common tendency is to think about which new technologies are now at our disposal: lighter and more spacious cars, new signaling instruments, automated doors, improved tracks, faster trains, and so on and so forth. Important as they are, however, these are only part of the kinds of innovation that are prompted by the very conceiving, planning, designing and building of railways. I believe this is a reality that is felt most acutely in cities. In nineteenth-century London and Paris, for example, innovations involved a range of topics:

- Using the underground spaces and cellars of market buildings
- Defining an area that railways would not cross
- Early trains for the working and poorer classes
- Collaboration between private companies and metropolitan authorities so that the building of a new railway line might be linked to street improvements
- New forms of governance, especially in terms of the degree to which London and Paris might use railways to direct their growth
- Concessionary fares for excursion trains on Sundays
- Conditions of employment in the context of municipal socialism, characteristic of developments in cities in Europe and North America at the turn of the 20th century

My aim in this contribution is twofold. I want to open up the very notion of innovation to issues that cover at least three different and inter-related dimensions: the politics, the culture and the social concerns behind the opening of new railway lines in nineteenth-century London and Paris. Secondly, I wish to show how the two cities coped, but also dealt with one of the most transformative forces of nineteenth-century Britain and France. An important part of that story relates to the different futures that were envisioned in the two cities, in response to specific concerns and determined by a particular set of conditions. This approach highlights the process of how innovations took place rather than the end result. My concern is, therefore, with the debates, ideas and challenges of getting to the object or point we call innovation, not the ready-packed model that we know circulates, widely and far.

LONDON AND PARIS IN THE NINETEENTH CENTURY: A BRIEF OVERVIEW

There are important similarities and differences in relation to the transformation that London and Paris experienced during the nineteenth century. Key among them are population growth, in both cities largely fueled by immigration; changes in their administration which present us with a sharp contrast between, on the one hand, the City of London and the metropolitan-wide authorities, the first of which was the Metropolitan Board of Works, created in 1855, succeeded in 1889 by the London County Council; and, on the other, the Paris municipal council, appointed by the Seine Prefect, in turn accountable to the national authorities which, throughout the nineteenth century, changed a number of times with three republics, two empires and an eighteen-year long monarchy (Porter 2000; Jones 2004; White 2008; Marchand 1993). Administration was directly concerned with the limits and extent of the two cities' built-up areas, in other words, up to which point did London and Paris extend. The contrast is again illuminating: Paris was a walled city up to after the First World War; an area called *intra-muros* was contained for a period of nearly two decades in-between the late-eighteenth-century wall of the *Fermiers Généraux* or Farmers

Past Futures

General and the outer fortifications built in the 1840s. The wall, first that of the Fermier Generaux and, since 1860, the Thiers fortifications, performed an important function for Paris's finances, namely the *octroi* or the tax levied on any products entering the city (Picon 1994; López Galviz 2013b). Walls had become something of a relic in London since at least the Great Fire in 1666. At the same time, the City of London retained full control of its jurisdiction, as it still does today, within what is often called the 'old square-mile'. Growth in and around the West End, Westminster, Southwark and eastwards by the river docks was largely the result of private initiative and brought under one administration only gradually from the mid nineteenth century onwards, following the creation of the Metropolitan Board of Works.

Between 1801 and 1901(Table 1), the rate of London's population increase was both more consistent and generally higher than that of Paris. At the same time, the populations in the two cities grew exponentially, about five times during this period. This growth would have significant consequences on the provision of public utilities such as water, food, sanitation, housing and transport, but also in terms of public order, education and health.

As for the way people travelled during the second half of the nineteenth century there were, again, important similarities and differences between the two cities. Tables 2 and 3 provide an overview of the key tendencies.

In London, the number of travellers had increased by ten times between the first half of the nineteenth century and the mid-1890s. There were ninety-three journeys per head in 1894, an increase of more than five times compared with 1864. The share of the Metropolitan Railway and the Metropolitan District Railway companies, the operators of the first two 'Underground' lines, was limited to approximately four journeys in 1864, subsequently increased to thirty in 1894. In Paris, the number of travelers increased nearly an eightfold between 1855 and 1890. While thirty-six journeys were made per head in

Table 1. Population in London and Paris, 1801 – 1901

London (a)		Paris (b)	
Year	Population	Year	Population
1801	959,130	1801	547,756
1811	1,139,355	1807	580,609
1821	1,379,543	1817	713,966
1831	1,655,582	1831	785,862
1841	1,949,277	1841	936,261
1851	2,363,341	1851	1,053,261
1861	2,808,494	1861	1,696,141
1871	3,261,396	1872	1,851,792
1881	3,830,297	1881	2,269,023
1891	4,227,954	1891	2,477,957
1901	4,536,267	1901	2,714,068

Notes: (a) The population of Greater London in 1899 was approximately 6,528,000. I have considered the jurisdiction of the Metropolitan Board of Works (smaller in area) as applicable to the entire period based on the Census figures (Ball & Sunderland 2001, p. 42). (b) The population of the Département de la Seine (inclusive of the arrondissements of Saint Denis and Sceaux was approximately 3,670,000 in 1901 (Chevalier 1973, pp. 182 – 83). The figures from 1861 to 1901 include the suburbs annexed in 1860. The comparison between the London and Paris figures is based on their respective administrative areas which leaves out significant sections of Greater London and the Département de la Seine, respectively.

Table 2. London passenger traffic per operating company 1864 – 1894

	1864	1874	1884	1894
LGOC (a)	42,650,000	48,340,000	75,110,000	133,132,000
Metropolitan	11,720,000	44,120,000	75,930,000	88,514,000
District		20,770,000	38,520,000	42,097,000
Tramways		41,930,000	119,260,000	231,522,000
Road Car Co			3,060,000	44,610,000
CSL (b)				6,959,000
Total (T)	54,370,000	155,160,000	311,880,000	546,834,000
Population (P)	2,940,000	3,420,000	4,010,000	5,900,000 (c)
Ratio T to P	18 to 1	45 to 1	78 to 1	93 to 1

Notes: The unity of measurement is the number of journeys as recorded by the operating companies: (a) London General Omnibus Company; (b) City and South London Railway; (c) estimate for the Greater London area whose population in 1891 was 5,572,012 and in 1901 was 6,506,954. Source: J. Greathead 1896.

Table 3. Paris passenger traffic per operating company 1855 – 1890 (1)

	1855	1865	1875	1885	1890
CGO (a)	40,000,000	107,358,111	125,061,957	191,218,501	198,228,364
Ceinture (b)	2,407,039	4,902,554	13,883,681	31,007,212	34,032,588
Riverboats (c)		3,567,010	9,578,631	18,820,922	23,591,967
Mainlines (d)			9,383,128	13,619,324	18,010,272
Tramways			5,723,882	50,578,734	51,858,179
Total (T)	42,407,039	115,827,675	163,631,279	305,244,693	325,721,370
Population (P)	1,174,346	1,825,274	1,988,800	2,344,550	2,477,957
Ratio T to P	36 to 1	63 to 1	82 to 1	130 to 1	131 to 1

(1) Population figures are for the years immediately after, namely, 1856, 1866, 1876, 1886, and 1891. (a) Compagnie Générale des Omnibus; figures comprise urban and suburban omnibus routes; the railroad service to Saint Cloud (including the service from the Louvre to Versailles since 1881); and tramway services since 1875; (b) the 1855 figure corresponds to 1856 and is only from the Auteuil line; (c) the 1865 figure corresponds to 1867 when riverboat services started operation on the occasion of the *Exposition Universelle* of the same year; as an indication of the substantial increase in passenger traffic during the exhibitions the figure of 1889 was 52,885,104. Figures incorporate suburban and urban passengers. (d) Main line railways, the figure of 1875 does not include the services of Paris-Charenton (Vincennes line). Source: A. Martin 1894.

1855 (the vast majority of them by omnibus), there were one hundred and thirty-one in 1890 distributed across omnibuses, tramways, riverboats, main line railways, and a suburban railway ring (Ceinture). It is important to bear in mind that no underground or metropolitan railway was opened before 1900 in the French capital. Paris was able to learn from the London experience for a period of over thirty-five years.

Figures for passenger traffic in the two cities are only estimates and should be treated cautiously. There was nothing like a consolidated way of recording traffic, let alone one that was consistent. In the case of railway companies, receipts were a generally accurate indication, but every company would account for a journey differently: was it a return or a single journey? Or, were different times of the day

Past Futures

when different fares were charged accounted for? At the same time, what can be said with a degree of certainty is that more people travelled more frequently and, here the comparison is again illuminating: in London people travelled longer distances than they did in Paris, which is not surprising, given that London's built-up area was over three times that of Paris. If there was anything like a general tendency in terms of urban and suburban travel, we might say that there was a correspondence between a higher number of modes of transport and ever growing numbers of passengers. Which side drove the other is a far more contentious matter and one that I will not be discussing here.

The idea of taking trains beneath and above the streets of London and Paris emerged, therefore, in a context where population growth, shifts in but also changing regimes interacting with metropolitan administration, and ever newer –allegedly better – modes of transport combined to transform the two cities into new entities that would develop a close relationship with railways, for better and for worse. The encounter between railways and the two cities was fraught with challenges, but also opportunities. Innovations sprung up, specifically in response to the conditions that London and Paris posed: connectivity to the central food market, the post office, the river docks; the necessity to think about railway building and street improvements as part of one and the same vision in the interest of a more cohesive urban form; the different ways in which new technologies –electricity – might prompt the emergence of a system suited for the specific needs of the metropolis.

These are questions that planners, authorities, architects, engineers and others face in cities across the world today: in India, Latin America, the Middle East, China. The contexts are different, of course, but I think that some valuable insights and parallels across space and over time might be drawn; the question is how and using which criteria. The notion of 'past futures' is of service here.

Past futures concern the futures that have been envisioned in the past, using a range of media –textual, visual, oral, performative – and as a result of a combination of circumstances. An important part of that process relates to who took part in envisioning which future and in response to what kind of motivations. This, of course, raises the question of whether or not there are or there have been conflicting futures in the past; futures that were chosen over others; futures that were obliterated, ignored, sidelined; futures that, by contrast, were celebrated, however unrealised.

In the specific context of cities, we may talk about locating the imagining of their future at different times in their history, in other words, were problems such as traffic congestion and housing overcrowding, or, discriminating the traffic of goods and people in relation to a changing urban geography understood in similar ways in medieval European cities than in classical Rome, or in the megalopolises of the Global South in the twenty-first century and, if so, do the solutions that we are able to reconstruct reflect those similarities, or, on the contrary, are solutions specific to the contexts where they are produced and, therefore, contingent?

Thinking about past futures invites us to move away from the idea of 'models' that can be imported and exported –as, indeed, they do – and reflect on issues such as innovation as historically contingent, the result of a long process that is riddled with unforeseen conditions and unintended consequences and through which a range of actors, institutions, beliefs and perceptions collide, mix and diverge. It is the process that counts: innovating where, how, and with whose involvement.

Past futures is useful in another important respect. One of the notions that is central to German historian Reinhard Koselleck's exploration of the relationship between the past and the future is what he called a horizon of expectation (*Erwartungs-horizont*), that is, the imagined domain that structures action in the present in the interest of a vision of the future over which a monopoly, religious, political, cultural or otherwise, keeps a tight hold (Koselleck 2004). An important part of the 'moments of innovation' that I discuss here forecast the future, structuring a horizon of expectations that gives purpose to action not by the consistency of how real the forecast is but by the commonality of moving in a certain direction. That direction often involves a future that is measurable, manageable so that concerns about the present become subordinate to the vision of a future that might never materialize. Both what did not happen, the routes not taken, and the actual outcomes of innovation are therefore constitutive of what past futures entail. They supplement each other and qualify how we understand the kinds of futures that have been envisioned in the past.

In what follows, I concentrate on three moments of innovation and what each tells us about the transformation that London and Paris experienced during the nineteenth century: the centrality of the food market and the extent to which railways provided an alternative to, but also a nuisance against the 'circulatory' needs of the two metropolises; the completion of the inner circle in London, an important part of which involved joining railway building and street improvements; and the refining of electric traction as the technology that was suited for the specific needs of the metropolis, namely, lighter, speedier and more frequent trains for passengers and their luggage rather than steam locomotives skirting out of town in a growing network that with time and some direction might become a system.

1. Connecting to the Heart of Metropolis

The London and Greenwich was the first railway line built in London reaching, first, Spa Road, Bermondsey –from Deptford – open on 8 February 1836, followed by an extension to London Bridge on 14 December 1836. The Gare St Lazare was the first railway terminal built in Paris, open in 1837 by the Chemin de Fer Paris-St-Germain. By the mid nineteenth century, there would be seven railway lines with a terminus in London, five on the north bank of the Thames, at Paddington, Euston, King's Cross, Bishopsgate and Fenchurch Street (in the City), and two on the south bank at London Bridge and Waterloo. Seven termini would serve Paris around the same time, four on the right bank of the Seine: Gare St Lazare, Gare du Nord, Gare de l'Est, Gare de Lyon; and three on the left bank: Gare d'Orléans, Sceaux, and Montparnasse also known as Gare de l'Ouest rive gauche (Kellett 1979, Freeman and Aldcroft 1985, Caron 1997, Bowie and Texier 2003).

Railways attracted higher numbers of traffic, both of goods and passengers. Existing streets became visibly insufficient accommodating a growing number of vehicles, horses, animals en route to the market, hackney cabs, pedestrians, carters and porters. The extent to which railways could take some of that traffic off the streets raised at least two different questions: whether or not to allow railways further into the city centre, and how to connect the lines that had been built as well as those that were planned. Railway connectivity took a number of forms: through a link line connecting the railway termini, via junctions in the periphery, or, through a central station. Whatever the option, the idea was to allow traffic to connect to, cross or bypass the central districts of the two cities. Markets at Smithfield in London and Les Halles in Paris posed some of the most important challenges and they did so against the background of their very transformation from cattle to meat markets, with abattoirs being relocated outside the centre, to the periphery in areas like Copenhagen Fields and La Villette. The traffic related to the Post Office and the river docks was also important.

The debates around the redesigning of Les Halles in Paris followed by the findings of a royal commission on railways termini in London will show precisely how innovation was a part of railway developments in both cities and, perhaps more importantly, the degree to which the new transport technology prompted readings of the city that were both original and an important precedent for future growth.

Past Futures

Paris

By the mid 1830s, the *Quartier des Marchés* in Paris consisted of four different buildings: the Marché aux Poissons, the Halle à la Viande, the Marché aux Oeufs, and the Halle aux Draps, catering for fish, meat, eggs and drapery, respectively (Fleury and Pronteau 1987). In 1842, a commission decided to preserve the location on the right bank of the Seine, after considering plans for its transfer to a site next to the Halles aux Vins on the left bank. The planning and design of a new market was the beginning of the larger transformation that Paris would experience as a result of the imperial 'politics of airing', a key aim of which was to sanitise the centre (Boudon 1977).

One of the plans produced on the occasion of the 1842 commission was by Hippolyte Meynadier, one of the many intellectuals involved in thinking about the Paris of the future and who placed the market building in a larger and more ambitious 'general plan of grand circulation' consisting of the opening of wide thoroughfares and parks; the redistribution of public buildings of both national and municipal significance; and the sale of those properties owned by the city, with little or no artistic and historic value (Meynadier 1843; Perreymond 1844). Interestingly, the basis for his proposal was Meynadier's own close examination of everyday life in the city: 'It is by journeying Paris in every sense; travelling its sites at different times of the day; observing the oscillations of the plebeian masses on the public road; penetrating the corners and nooks of all the old streets […] that one can estimate the need for large thoroughfares in Paris and that one can indicate assuredly the points that [people] use most' whether it is 'for their departures, destinations [or] crossings.' (Meynadier 1843, Bourillon 2001, pp. 150-51). Improving circulation required wide thoroughfares that would help regulate the dithering movement of the plebeian masses. Similarly, railways provided a suitable alternative for the kind and scale of circulation required in a new and larger central market.

A new public enquiry took place in 1845, specifically concerned with the new building of the central market or Les Halles Centrales (Lemoine 1980, Lavedan 1969). The project conceived by Victor Baltard and Félix Callet would receive a direct commission from the municipal administration in August the same year (Baltard and Callet 1863, pp. 13-14). In response to the decision, the architect Hector Horeau published an *Examen critique du projet d'agrandissement et de construction des Halles Centrales d'Approvisionnement pour la Ville de Paris* in October the same year, and in which he outlined the disadvantages of the commissioned project contrasting these with the benefits of his own scheme. The publication was followed soon after by a new pamphlet *Nouvelles Observations*, which focused on how other European markets functioned. This was a critique of the report of the official visit to markets in England, Belgium, Holland and Prussia, also published in 1846. (Horeau 1845, Horeau 1846, Lemoine 1980, p. 84). Baltard was part of the official mission. Of particular interest to him were the markets at Liverpool and Birkenhead due to the full enclosure of the market area and the use of cellars for the preservation of foodstuffs (Baltard and Callet 1863, pp. 12-13; Lemoine 1980, pp. 79, 82).

The importance of the Halles lay in the provisioning of the capital to which effective connectivity with regional centres and lines of distribution across France was central. Baltard's project was primarily concerned with the functional apparatus that would concentrate all the market activities in one building complex and was focused on the engineering, architectural and decorative features of its structure. Horeau's ideas were different, as his plan devised connections to the main streets of the immediate surroundings and the river so as to consolidate an effective movement within and without the market, a vision that resonated with Meynadier's general plan of grand circulation (Papayanis 2004). Horeau's scheme consisted of six different pavilions and was served by railways connecting the central market to the eastern and western districts of Paris (Horeau 1845, p. 7). Connection to the main line railways was underground using the cellars for the storage and circulation of classified produce, which occupied the entire area beneath the proposed pavilions. Horeau's concern was how the waste and produce would circulate through the building complex without interfering with the acute traffic congestion of the area in and around Les Halles.

After a renewed attempt in 1849, Horeau's plans were deferred and Baltard's scheme prepared for their execution. Construction works for the new market building started in 1851. Two large wings, east and west, accommodating six pavilions were complete in 1858 (Lemoine 1980, pp. 94-103, 164). Access and connectivity to the market remained dependent on existing roads.

In 1869, *Le Figaro* would accuse Baltard of plagiarising Horeau's design. Despite their support to Horeau's plans, the *Revue Générale d'Architecture et Travaux Publics* – a key publication for contemporary trends and developments in Paris during the second half of the nineteenth century – gave Baltard the opportunity to publish a letter dismissing the claims and reasserting the originality of his building (Revue Générale d'Architecture et Travaux Publics 1869, col. 206 ; Saboya 1991, p. 258). Differences between the two architects aside, what was characteristic of Horeau's vision was the way in which he subordinated the idea of improving an existing building to accommodating a special function, namely, the reception, storage, and distribution of foodstuffs into, across and out of the city. The circulatory needs of the capital became central to his project as a result of this, more specifically, the centrality of effective connections to the market for which railways provided an increasingly sound alternative. Horeau's ideas provided an innovative reading of the city's needs alluding to the importance of thinking about the market in relation to the new kinds of connectivity that the combination of railways and underground spaces beneath buildings might accommodate. His was a vision of urban circulation rather than an isolated architectural statement.

London

Connections to the market at Smithfield were also central to the initial plans of what would become the Metropolitan Railway in London. In the 1840s, dozens, if not hundreds, of plans put forward by railway companies proposed to connect London to the rest of Britain in what became known as the 'railway mania'. A good number of proposals included connections to Smithfield market. The 'mania' had its peak in 1847, by which time the fortunes of businessmen and laymen alike had bulged, shrunk and disappeared in vast numbers (Lambert 1984, Lewin 1936, Barker and Hyde 1982).

At the same time, traffic congestion in London was turning distance into a function of time more so than of space: 'You may find it takes you as long to go from Kensington to the London-bridge terminus of the Brighton Railway, as from London-bridge to Brighton. Nay, of two friends taking leave at London-bridge, one for Brighton by rail, and one for Kensington by omnibus, the traveler to Brighton might reach his destination first.' London, the commentary on *The Times* went on to assert, 'will speedily find the means of balancing these disparities; and when that has been done by an internal system of railways, the long-lined railways will obtain the means of using the internal system as an extension of their own.' (The Times, 13 October 1845, p. 6).

Plans for London railways before Parliament soared in 1846. A royal commission was appointed in response to the situation. Its remit: assessing the role that railways should play in metropolitan communications. Nineteen different schemes were examined, fifteen north of the Thames, four in the south.

Past Futures

With the exception of two lines, both south of the river, the commission advised against all the schemes. The commission's report was to provide an assessment of the benefits and problems of crossing the city centre. Their suggestion was to connect the railway termini by means of 'branches skirting the Town, and terminating at points without the line described in our Instructions [rather] than by penetrating within it.' (Houses of Commons and Lords, 1846, p. 6; Bradley 2006, p. 33) The 'line' introduced a new geography of the city in an attempt to limit the further extension of railways into London's central and inner districts and covered both riverbanks.

Four other aspects were central to the commission's report: the choice of a central terminus, the handling of goods traffic outside the central districts, the perceived and real impact of railway plans on property, and the relationship between new railway lines and street improvements.

As the names of several proposals indicated, Farringdon Street was the site that companies favoured for the sitting of their termini in the City, given its proximity to Smithfield Market. But having a central station raised the question of who was to benefit: 'If the convenience of passengers does not call for the prolongation of railways into the heart of the Metropolis', the commissioners reported, 'still less does it require the establishment of any one Central Terminus, at which the railways from different parts of the country should unite.' The evidence proved sufficiently just how complicated the arrangements were when several different competing companies shared the use of a central terminal rendering the suggestion both impractical and unwelcome (Houses of Commons and Lords, 1846, p. 6).

In terms of goods traffic the commission recommended 'a line which should pass outside the Metropolis on the North, at such a distance as to avoid interference with populous districts and the thronged thoroughfares, and so connect the goods' stations of the various railways from West to East with each other, terminating at some convenient point on the Thames or within the Docks.' A junction crossing the river west of Vauxhall Bridge would connect the northern and southern lines avoiding concentrating traffic at a central station. A different yet related benefit of this 'circuitous communication' was the need 'to establish an unbroken connection between the railways of the North, South, and West' across Britain, also in the interest of strengthening national defense (Houses of Commons and Lords, 1846, p. 6, 21).

As for property the commissioners affirmed that they were 'not disposed to attach any weight to the assumption which [they] found to be a common one, that districts thickly inhabited by a population of the lowest class, and where vice and destitution prevail, are sensibly improved by the passage of a railway through them.' Often the evidence showed the opposite, for the railway 'does not open the streets to any new traffic, nor does it lead to the improvement of dwellings on either side of the line; and where the improvements which such districts most require, viz., the formation of new streets, with better built houses, better ventilation, and better drainage, are in contemplation, or are likely to be effected, it tends in most cases to obstruct, rather than to facilitate them.' Examples of these were the viaducts of the London and Blackwall and the Eastern Counties railways on the eastern end of the City (Houses of Commons and Lords, 1846, pp. 7, 9; Kellett 1979, pp. 36-37).

The disruption to public works prompted most proponents 'to combine, more or less with their own works, and at their own expense, the improvement of existing, or the formation of new Thoroughfares for the benefit of the public.' But as the commissioners explained, the situation, though specific to every scheme, had developed into equalling railway building to works that were planned and carried out in the public interest, with important differences in terms of who absorbed the costs and under which conditions (Houses of Commons and Lords, 1846, pp. 7-8). Any collaboration between railway companies and public authorities, their report advised, 'should be planned and prescribed to the companies [that agreed to these conditions], and finally carried out under the authority of some Department of Your Majesty's Government, in conjunction with the Corporation of the City of London, or with the local Authorities of the District in which the works are to take place.'

The debates around and the evidence provided in favour or against a central railway terminus, the bypassing of goods traffic around the city, the effects of railways on property, and the challenges of connecting railway plans to street improvements contributed to recognising the innovative role that railways might play in the transformation that London was experiencing. The significance of these debates, in Parliament, but also in a whole array of specialist circles, lay in their providing the arena where the question of metropolitan communication was addressed. Moreover, through the work of commissions such as that of 1846, ideas about the public rendered the benefits of railways in a light that subordinated the disjointed and conflicting efforts behind railway operation to the goals that were shared by citizens: 'under no circumstances should the Thoroughfares of the Metropolis, and the property and comfort of its inhabitants, be surrendered to separate schemes, brought forward at different times, and without reference to each other.' (Houses of Commons and Lords 1846, p. 21; Kellett 1979, p. 43).

These were the debates, however. The reality looked very different. Railways continued to cover London's geography with their viaducts and trenches, their river bridges and steam locomotives. The innovative dimension of the debates rests on their challenging of existing practices and the further refining of the capitalist spirit, forcibly heralded in the culture of laissez faire. A significant part of what the 1846 royal commission advocated was how to direct capital so that it met the interests of the many and not the few.

2. London's Inner Circle

The contrast between railway developments in London and Paris would take a distinct turn in the second half of the nineteenth century. For one thing, the first section of the Metropolitan Railway in London would open to services in January 1863, whereas the first line of the Métropolitain in Paris opened in 1900. Innovation in London concerned learning about specifically metropolitan needs and the degree to which railways were a means to address them. Similar questions were raised in Paris, but in this section, I wish to concentrate on one specific challenge that led to a distinct form of innovation closely related to the transformation of Paris's central market, but also recognized by the royal commission in London, namely, whether and how to combine railway building and street improvements (López Galviz 2013b).

The first section of the Metropolitan Railway represented a clear benefit for a population consisting primarily of businessmen who travelled between Paddington and Bank, the busiest omnibus route ever since the 1830s. As the chairman of the Metropolitan, William Malins, said soon after the opening: the time saved by trains running beneath the congested London streets was well received and supported by 'gentlemen engaged in commercial and business avocations' (quoted in Bradley 2006, p. 53). It was for this particular segment of the population that the journey between home and work might represent a daily pattern.

Trains of several main line railway companies used the new line. Separate agreements between the Metropolitan and the companies included tolls charged to trains running on Metropolitan tracks, tolls the Metropolitan paid to other companies for using sections of their tracks, or, indeed, reciprocal agreements (Huet 1878, p. 45). One example was the convention of 2 September 1867 that gave the Midland Railway rights to run their trains, machines, and use their own staff on the new 'widened' lines between King's Cross and Moorgate. The Midland had also a separate platform for their passengers and storage premises at Moorgate Station. In return, the Metropolitan received a percentage of the Midland revenues for the operation of this section of the line. (Huet 1878, pp. 45-47; Jackson 1986).

Although, overall, these arrangements helped Metropolitan services, the regular and intensive use of their tracks by other main line railways (the Great Western, the Midland, the Great Northern and the London Chatham and Dover) constituted a challenge for a vision of railway development that was based on outer and inner circles, each specializing in either local or through traffic, an idea that emerged out of the debates connected to a new parliamentary committee on metropolitan communications in 1863. Things would change little after the opening in December 1868 of the first part of the southern section of the inner circle, between South Kensington and Westminster, built and operated by a separate company, the Metropolitan District Railway. The opening was reported as part of the works on the 'Metropolitan Inner Circle' or the 'Metropolitan underground system', terms that seemed 'a little perplexing to the uninitiated' (*Illustrated London News*, 2 January 1869, quoted in Halliday 2001, p. 28; *The Builder*, 2 January 1869, p. 2).

Agreements like those between the Metropolitan and the main line railways were planned at Victoria Station which, according to a report in *The Builder* was 'exceptional in its arrangement', in that 'a mezzanine floor [was] introduced for the booking-office, which is on a level, about half-height between the platforms of the Brighton and Chatham companies and the street, for the more easy access of passengers from these lines.' (*The Builder*, 2 January 1869, p. 3) Similarly, in one of the shareholders' annual meetings, the District's chairman, James Staats Forbes, would mention that the company had made an agreement with the London and North Western Railway whereby the latter would run its trains from its own lines, north of Euston, on to the District's' via the West London to Earl's Court, using 'the District's tracks to run into Mansion House, thus giving it access to the City.' The London and North Western paid for the agreement along with the 'tolls for the two trains an hour using the facility' (Halliday 2001, p. 31).

The District extension from Westminster to Mansion House opened in 1871, where the company's terminal would remain for over thirteen years. Competition, financial difficulties, but also coordination between the Metropolitan and the Metropolitan District, and the metropolitan and City authorities were among the factors contributing to the time that completing the inner circle would take. In 1874, a new company, the Inner Circle Completion Railway, was formed precisely to that effect. But eventual opposition from the Metropolitan Board of Works, the City Commissioners of Sewers and the Lord Mayor and Aldermen halted the initiative.

In 1878, the Metropolitan and District approached John Hawkshaw, a prominent civil and railway engineer responsible for several projects across Britain and overseas, including the Severn railway tunnel, which he would take charge of a year later in 1879 (Chrimes 2013). Hawkshaw 'recommended that the circle should be completed by extending the railway southward from Aldgate to Tower Hill, and thence westward along Great Tower Street, Eastcheap and Cannon Street to join the District Company's railway at Mansion House'. Hawkshaw's idea was, essentially, joining the completion of the inner circle to street improvements, namely, 'the widening of Eastcheap and Great Tower Street, and the construction of a new street between Mark Lane and Trinity Square', projects that were under the jurisdiction of the Metropolitan Board of Works and the City (Barry 1885, p. 35). Hawkshaw and John Wolfe Barry were appointed the same year to produce the plans for and execute the works. Their plan included an eastern extension to join the East London railway, which would give both the District and the Metropolitan, as well as the main line railways on both riverbanks, access to the more easterly and south-easterly districts. The extension ran under Whitechapel Road and included a terminus for the exclusive use of the District adjoining the East London's Whitechapel Station (Barry 1885, pp. 35-36).

The street improvements, inner circle and the Whitechapel extension were complete in October 1884. The Metropolitan Board of Works and the City Commissioners of Sewers together contributed £800,000 to cover the costs of the street improvements (total of £929,412). The Metropolitan and District, also responsible for the purchase of property and the execution of all works, met the rest (Barry 1885, pp. 36-38). As Barry asserted: 'The completion of the ring of railway has been rather the joining together of two parallel lines than the completion of a circle.' Moreover, the arrangement of trains per hour consisted of six different types of services, reflecting the degree to which the so-called circle was divided into lines, rings, loops and circuits:

1. Eight Inner Circle trains going 'completely round the circle'
2. Six District trains running 'from Ealing, Richmond and Fulham, by way of Earl's Court, South Kensington, and Mansion House to Whitechapel or (via the Thames Tunnel) to New Cross'
3. Two Middle Circle trains from Aldgate by King's Cross, Bishop's Road, Paddington, via the Hammersmith Branch to Latimer Road onto the West London railway to Earl's Court, South Kensington and Mansion House
4. Two Outer Circle trains from the London and North Western terminal at Broad Street, Bishopsgate, via the North London railway through Dalston, Camden Town, Willesden and onto the West London to Mansion House
5. Two Metropolitan trains from Aldgate to New Cross via the Whitechapel extension
6. Eleven Metropolitan trains between Moorgate and Edgware Road during the busiest hours, excluding 'the traffic of foreign companies' that used the widened lines. (Barry 1885, p. 50; López Galviz 2013b)

This amounted to up to thirty-one steam-operated trains per hour in each direction during the busiest times running through different sections that extended far beyond the inner circle. Not surprisingly, there were severe problems: 'Despite the four days of experimental working which preceded the opening [...] services were thrown into chaos by the new schedules [...] Dislocation was so severe that traffic sometimes came to a complete standstill for hours on end, and there is at least one well-authenticated case of exasperated passengers having to get out of their train in the tunnel and walk to the nearest station' (Barker and Robbins 1963, pp. 232-33).

The difference between through, long-distance, and local traffic in London was the result of the agreements between private railway companies. The interest of the Metropolitan and the Metropolitan District, in particular, was suburban not local traffic as the extensions of their lines illustrated. Their extensions were coupled to suburban growth, notably north and west of London, where the main lines that used the inner circle originated (Jahn 1982). At the same time and despite the problems concerning the operation of outer, middle and inner circles, there was a sound awareness of the complexity and scale of railway communication in London and the extent to which its development might be used in the interest of metropolitan growth. This qualifies further the assertion that London seemed to be 'rearranging its built fabric around a completely new means of transport, but in a way which did not immediately suggest that the metropolis was being consciously and centrally planned' (Sutcliffe 1983, p. 10). The centrality was somewhat manifest in the role that the Metropolitan Board of Works and the City played. True, the building of any system depended on the actual financial model behind railway operation. But rather than being the long-overdue outcome of the battle between the chairmen of the two railway companies, as we have learned from previous historiography (Halliday 2001, Wolmar 2004, Jackson

Past Futures

1986, Barker and Robbins 1963), the completion of the inner circle might also be seen as the history of a joint innovative effort: between the metropolitan and the City authorities, on the one hand, and the Metropolitan and the District, on the other. Steps were taken, indeed, to direct the centrifugal forces of railways in favor of a more orchestrated development of London, one that was representative of different publics: private companies encouraging and competing for suburban traffic; authorities with interests that were both complementary and mutually exclusive; and passengers for whom time was money and the difference between home and work a question of daily travel. Not one dominated; nor could either account for the benefit of all.

3. Shield Tunneling and Electricity

Using electricity in railway operation became a real option towards the end of the nineteenth century. In November 1890, *The Engineer*, a specialist periodical, would produce a concise summary of the state of electric traction in connection with means of urban transport: 'A number of small tramways, both on the Continent and in the United Kingdom, have been worked electrically, and in the United States many of the street tramways are worked in this way; but it has not hitherto been applied on any large scale to the working of a railway of the usual gauge for passengers.' (*The Engineer*, 7 November 1890, quoted in Barker and Robbins 1963, pp. 309-10). 1890 was the year when the City and South London, the first of the electric 'Tube' lines in London opened to passenger services shortly before Christmas on 18 December. As a pioneer, the operation of the City and South London set the pace of developments that were yet to come, but it also showed the challenges and limitations of using a new technology. The combination of tunnel, rail, car and electricity proved to be one of the most significant innovations in urban railways ever since. What is more, electric traction prompted the conception of a system that was based on circuits, a significant precedent that would supersede the allure of steam locomotives, particularly in cities.

London

In the early 1880s, the Siemens brothers obtained the concession to build an electric railway connecting Trafalgar Square to the area in the immediate surroundings of Waterloo Station, crossing the river with a bridge: 'the scheme did not get beyond the building of 20 yards of tunnel under Northumberland Avenue and the Embankment', and was subsequently abandoned (Barker and Robbins 1963, p. 304). A second attempt was the London Central Electric Railway (1884), also by the Siemens, which proposed to link this time Charing Cross (a few yards from Trafalgar Square) and Cheapside, in the City. The officer of the Metropolitan Board of Works, William R. Selway, objected to the scheme, calling it 'speculative and experimental' (quoted in Barker and Robbins 1963, p. 305). Like their first scheme it was also abandoned.

The same year, in 1884, the City of London and Southwark Subway received an Act of Parliament to build a line between King William Street in the City and the Elephant and Castle, south of the Thames. The line was to be built by a 'tube' system of tunneling with trains operated by cable traction (Sekon 1899). Tunneling was devised with an entirely different construction technique than the 'cut-and-cover' of the Metropolitan and District lines. The idea was to take the line deeper into the city's soil using a system that incorporated both traction and tunneling as had been conceived in the mid-1860s by Peter William Barlow.

In the patent of his invention 'Improvements in Constructing and Working Railways, and in Constructing Railway Tunnels', Barlow explained the benefits of a new system for working underground and other railways. This consisted in 'employing local in contradistinction to constant power to propel trains of carriages on railways', whether underground lines, tunnel crossings, or standard over ground 'where trains are required to start and stop at short intervals.' (Barlow 1864, p. 2). The trains would be attached to and hauled by a rope connected to 'cylinders worked by hydraulic power'. Tunnelling would follow the cylinder principle (in effect, a tube), devising gradients on either end of every station as to alleviate the pressure on the brakes of arriving trains and facilitating their departure by gravity and inclination (Barlow 1864, pp. 1-3).

In terms of its operation this was a system 'consisting of iron tunnels 8 feet in diameter, in which single steel omnibuses, [would] seat twelve passengers each.' There were no stations, at least not in the ordinary sense, passengers paying their fare directly in the omnibuses. The arrangement became more elaborate as the topography required it with 'three series of subways at different levels, the carriages as well as the passengers being lifted in passing from one to the other' at intersections with steep gradients. (Barlow 1867, Barlow 1871, Greathead 1896). The Tower Subway, opened in August 1870, would be a test version of Barlow's system. The subway was fitted with lifts on both riverbanks and a tramcar hauled by a cable which in turn was propelled by stationary engines. Barlow's initial plan was for the subway to be 'manumotive', relying on the strength of 'two and a half men, if the journey was made in one minute' in a carriage with twelve passengers covering a distance of 1320 ft (402 m), or, alternatively and as was originally proposed, 'one manpower constantly applied' which increased the journey time to approximately two and a half minutes (López Galviz 2013a, pp. 71-73). The subway later became a footpath, following the bankruptcy of the Tower Subway Company in November 1870, closing to the public in 1894 when Tower Bridge opened offering the same connection above (Thornbury 1878, pp. 122-28; Dennis 2008; Lascelles 1987; Lee 1973).

James H. Greathead, one of Barlow's collaborators during the construction of the subway, was appointed chief engineer of the City of London and Southwark Subway (CLSS) while John Fowler, engineer of the Metropolitan and District, accepted a role as consulting engineer (City of London & Southwark Subway Company 1884-1889, pp. 8, 29). By 1890, the largest part of the tunnelling work was ready and the electric locomotives tested (Greathead 1896). The Lord Mayor and other gentlemen were taken on a trial journey from the City to Elephant and Castle on 5 March; the results seemed satisfactory. Lighting remained a problem and so the company's 'solicitor was instructed to communicate to Mss. Mather and Platt [contractors for the electric equipment] that as they were unable to efficiently light the stations with electricity under the terms of the contract it had been decided to substitute [electric lamps with] the use of gas.' Prior to the official opening, an agreement was reached for a five-minute frequency and a service restricted to weekdays (no Sundays nor Christmas day), starting at 8 am. (City of London & Southwark Subway Company 1889-1892, pp. 29, 44, 68, 103).

The names of the six stations were King William Street, Borough, Elephant and Castle, Kennington, The Oval, and Stockwell, where the car sheds and generating plant had been built. Regular passenger services started on 18 December 1890, though the company, now called the City and South London, experienced the problems and difficulties associated with the use of a new technology (Lascelles 1987; Barker and Robbins 1963, p. 310). The combination of the insufficient power generated from a plant situated at the southern end of the line and the weight of the locomotives produced questionable results. The locomotives used were 'ponderous, noisy, and slow' (Simmons 1995, p. 94). Stations were eventually lit by gas, while 'on the trains themselves [the electricity supplied] gave only a feeble glimmer whenever a number of locomotives all accelerated at the same time.' (Barker and Robbins 1963, p. 312-13). The state of lighting on the trains was still unsatisfactory by 1895 (Greathead 1896, pp. 17-18). Furthermore, the decision to follow the pattern of streets forced an awkward arrangement on the City end station (King William Street) whereby the two tunnels that for the largest part of the route ran more or less parallel were built one above the other (Greathead 1896, p. 6). The station was reorganised in 1895, with two pairs of tracks and an 'island platform', replacing a single line and platforms on either side as it had been built in the first place (Barker and Robbins 1963, p. 314). But it was not until the opening of the northern extension first to Moorgate and Bank stations, in 1900 and then to Angel, Islington in 1901, that the company could solve the problems at King William Street, by closing the station on 24 February 1900, and adjusting the line's route.

As stated by the general manager of the City and South London, Thomas Chellew Jenkin, the first 'north-to-south railway' incorporated a number of different features that distinguished it from the Metropolitan and the District (Sekon 1899, p. 8). Stations were fitted with lifts to cover the considerably larger distance between platforms and streets and, perhaps more visibly, the power and smoke of steam engines had been replaced by the pale light and traction of electric locomotives. By contrast, there was also something that the City and South London, the Metropolitan and the District shared, as the Prince of Wales, later Edward VII, stressed during his speech preceding the official opening ceremony on 4 November 1890:

'This railway today', the Prince affirmed, 'this first electric railway which has been started in England will, I hope, do much to alleviate the congestion of the traffic which now exists, so that business men who have a great distance to go will find easy means of getting away from this great city and enjoying the fresh air of the country and I hope that it will also be a great boon to working men who are obliged to work in an unpleasant atmosphere, and who by its means will be able to get away for a little fresh air.' (*Daily News*, 5 November 1890; quoted in Wolmar 2004, p. 136)

This was the trinity of congestion, travel, and health. Overcrowded streets and residences that might give way to comfortable travel in a growing London by opening up the healthy country to businessmen and workmen alike: this was the promise of the new railway, a promise that had been heard before, a number of times in rather similar terms. The royal comment encapsulated an understanding of urban life that was characteristic throughout the nineteenth century. What had changed by 1890 when the City and South London opened to services was the context in which these remarks were made, namely, London had a metropolitan-wide authority: the London County Council, created in 1889 and succeeding the Metropolitan Board of Works.

Without a doubt, the advent of electricity added a new dimension to urban transport, but the question was also whether transport might be a way to solve other problems, some pressing like the acute housing crisis that Parliamentary commissions examined from the mid-1880s onwards. Four decades had passed since some of the early plans to house artisans in North London were first voiced. The struggle to house the working classes and the poor turned into a political battle fought between moderates and progressives in the London County Council and in other circles, but also including figures such as Charles Booth, Sydney and Beatrice Webb, Octavia Hill and others (López Galviz 2012). These were issues beyond the choice of electricity or steam locomotion for the running of railway lines, old and new.

At the same time, now lines could be dug deeper, creating a system of tunnel, rail and carriage. This differed substantially from the cut and cover technique and shallow tunnels that had been employed before. Difficulties were all too evident: the feeble glimmer, the ponderous noise, the wanting pull. In hindsight, the City and South London differed, but only so much, from the Metropolitan and the District: trains were still pulled by a locomotive, electric, yes, but locomotive nonetheless. Trains ran to the same frequency: five minutes, no less, no more. The newness of the electric railway consisted in setting a precedent for future developments; the very precedent that continues to shape the London Underground today.

Paris

Which rolling stock to use and what kind of electric system should be in place became paramount to refining the Métropolitain as it would be built in Paris. Fulgence Bienvenüe, engineer of the Technical Service of the Métropolitain, explained each in detail in a report on urban railways operated by electric traction (Bienvenüe 1896). In it, Bienvenüe distinguished between two types of vehicle: coupling or trailing carriages (*voitures d'attelage*) and carriages fitted with their own motors (*voitures automobiles*). Broadly speaking, this recalls the difference between the electric locomotives of the City and South London, and the Sprague multiple unit system introduced in the later tube lines (Duffy 2003, pp. 23-33). In Bienvenüe's estimates, each carriage of the Métropolitain would seat forty-four passengers, while the length of platforms (seventy-five metres) would accommodate six carriages.

In terms of operation, the proposed network –in its mid-1890s version consisting of a circular line and two transversals – was divided into six different electric circuits: west, east, north, Porte Maillot (circular north), Porte Maillot (circular south), and north-south diameter (Bienvenüe 1896, p. 7). The circular line followed for the most part the external boulevards, while the two transversals ran, one, east to west (partly along the Rue Réaumur) and, the other, north to south crossing the river Seine. Four types of circulation resulted from this arrangement, namely, at one, two, three, or four circuits, based upon traffic sections of different intensities (*sections de trafic à intensités differentes*). In other words, the type of service that the Métropolitain would provide depended on the frequency of trains, their speed, the timing of boarding stops and, of course, the power needed to supply the system: eight trains per hour in each direction in the sections that were operated with one circuit; sixteen trains with two circuits; twenty four trains with three circuits; and thirty two trains with four circuits (Bienvenüe 1896, p. 8).

In its entirety, the network supported 'forty-five trains circulating simultaneously in each direction [at any one time] or ninety in total'. Vehicles were fitted with dynamos; the lighting of stations and the powering of other facilities were by electricity. Two depots housed the power plants, the rolling stock and yards: the depot at Vaurigard supplied the circuits west, Porte Maillot (south circular), and the north-south line. Charonne's, in turn, supplied the circuits east, north, and Porte Maillot (north circular). An intermediate substation was planned at Montmartre to break the distance between the two (Bienvenüe 1896, pp. 9-10). Bienvenüe's report was instrumental to clarifying the extent to which electricity was necessary to create a metropolitan railway system. By the end of the nineteenth century, there were little doubts that electricity was 'the only possible [type of traction] for the operation of a metropolitan network' (Huet 1896, annexe p. 2).

The agreement between the city and the concessionaire stipulated that the latter was to raise capital enough to operate the network, providing the rolling stock, including tracks, as well as the access points to stations. In its turn, the city was responsible for the infrastructure, including platforms. Further to the execution of the works, 'the concessionaire was required to begin the works of superstructure two months after the platforms of each section were delivered by the city' and have every such section ready for operation within ten months (Robert 1967, p. 26). Infrastructure consisted of the works relating to tunneling, the diversion of existing networks (water, sewerage, gas) whenever needed, and the handling

Past Futures

of disruption and changes to existing streets, buildings and public spaces where appropriate. Superstructure concerned the works needed for the final operation of the system including power supply, plants, substations and the layout of the electric system. The distinction between the two, infrastructure and superstructure, would prove to be a contentious point illustrated later by the disagreement between the eventual concessionaire and the municipal council about the construction of a new plant for the generation of electricity, a point that is, nonetheless, beyond the scope of our discussion (Beltran 1988, p. 118-20).

The initial contract was granted to the *Compagnie Générale de Traction*, which associated itself with the *Établissements Schneider du Creusot*, the well-known steel manufacturers and with firm interests in electricity under the leadership of Eugène, Henri's son. Henri Schneider was among the most prominent members of the family, having been régent of the Banque de France, vice-president of the Comité des Forges, administrator of two railway companies, the Paris-Orléans and the Midi, mayor of Creusot, and general councilor and deputy of Autun, Burgundy, eastern France (de la Broise and Torres, 1996). The association of the two companies –the *Compagnie Générale de Traction* and the *Établissements Schneider du Creusot* – represented an important shift in that the new industries concerned with the production and distribution of electricity would take a position normally occupied by main line railway companies (Larroque 2002, p. 78). Shifts in financing would be accentuated further with the agreement between the eventual concessionaire, the *Compagnie du Chemin de Fer Métropolitain de Paris* (CFMP), and the *Société d'Électricité de Paris* on the construction of the generating plant at St. Denis, in operation from 1906. Moreover, foreign capital, notably from the Belgian conglomerate of Général Baron Édouard Empain, would become increasingly central to the operation of the Métropolitain as the twentieth century progressed (Conseil Municipal de Paris 1898, *Procès verbal* and *Délibérations*, pp. 835, and 463-64, respectively; Beltran 1988, pp. 115-17).

Line 1 opened on 19 July 1900, three months after the opening of the Exposition Universelle on 14 April. The extension of the Compagnie de l'Ouest from their Gare St. Lazare to Champ de Mars would serve the exhibition grounds first, including the junction between the Champ de Mars and the Invalides, also operated by electric traction and in service the same day (14 April) (Robert 1967, p. 34). The new spaces underground of the Métro would be a much needed shelter from the heat of the Parisian summer by late July (*L'Illustration*, 14 July 1900, pp. 22-23; *The Builder*, 28 July 1900, p. 72). Eight stations were in service at Porte de Vincennes, Place de la Nation, Gare de Lyon, Place de la Bastille, Hôtel de Ville, Palais Royal, Champs-Élysées, and Porte Maillot (Robert 1967, p. 35). The sections between Étoile and Trocadéro and between Étoile and Porte Dauphine of line 2 opened in October and December the same year. At peak hours, trains ran every ten minutes; from 20 September the frequency was increased to six minutes between 5:30 am and 9:30 pm and back to ten between 9:30 pm and 12:30 am. From 30 January 1901, and largely as a response to passenger demand, trains would run to a three-minute frequency (Robert 1967, pp. 35-36, 38). Lines 3, 4, 5 and 6 would be complete by 1910, all operated by electricity.

Bienvenüe, and indeed Parisians, had reasons to celebrate. Their Métropolitain had left the drawing rooms and the meeting halls of national, regional and municipal assemblies where it had remained for nearly forty years. The map of Paris was now a map with a cohesive network that covered the city from east to west and north to south, although there were no connections between the Métropolitain and the main line railways, to a large extent the very instigators of the new railway in the first place. By contrast to how metropolitan railways evolved in London, Paris's own Métropolitan was based upon a vision that was limited to the city walls: it was contained, perhaps insular, more so than it was expansive. At the same time, it was the crystallization of a vision that had been refined for years. The city was able to benefit from the range of innovations that electricity provided, not least a system that worked like one rather than a network consisting of separate lines, connecting at certain points, reaching into an ever growing periphery. That was the service of steam. In the city, in Paris itself, metropolitan railway transport was electric.

PAST FUTURES

To an important degree the building of railways in nineteenth-century London and Paris was the result of innovations in a range of fields. They were technological, most exemplary in the combined system of shield tunneling, tube, rail and carriage, but also in connection to the emergence of an exclusively urban system separate from main line railways as it was conceived and built in Paris. Innovation also concerned governance in the greater or lesser degree of influence that authorities were able to exert over the planning and building of railways that were metropolitan, that is, railways for local or urban traffic rather than extending outwards to the suburbs and beyond. Envisioning the city as a circulatory system at the centre of which were key structures such as the food market was also innovative in the sense that scale –more produce for more people – prompted new kinds of structures, the market buildings themselves, but also using railways to connect to them, whether underground or outwards to their new locations in the periphery. Each was a field of innovation in its own right. Each was determined by the political cultures and the social concerns around the joining of a new railway line to street improvements, or, the challenging of existing practices so that the capitalist spirit was refined and directed in the interest of a range of publics.

At a time when cities across the world face questions that are similar to those London and Paris faced in the nineteenth century, it is important to remind ourselves of the historical context in which innovations emerged: should it be a line connecting main line railway termini, port facilities or the food market? Or should it be a network of two or more lines with transfer stations, each dedicated to passengers or goods? Or might it be a system that directs the growth of a city using a technology suited to specifically metropolitan needs? The answers will of course be local and contingent. At the same time, what is important to realize is that whatever the strategy, a line, a network or a system, a horizon of opportunities is formed. The character of that horizon is political: at its most basic it involves praxis, the very practice of debate, argumentation and disagreement; on the other hand, it draws on allegiances that move and change as do regimes, institutions and individuals.

The role that railways can play in imagining the future of cities must include learning about the processes that led to the very envisioning of those futures and the innovations they shaped: less the system of tunnel, rail and car than the debate and process of getting to that solution in the first place. Such an approach recuperates voices and visions that have not been seen let alone heard loudly enough.

REFERENCES

Ball, D., & Sunderland, M. (2001). *An economic history of London, 1800-1914*. New York: Routledge.

Baltard, V., & Callet, F. F. (1863). Monographie des Halles Centrales de Paris, construites sous le règne de Napoléon III et sous l'administration de M. le Baron Haussmann. Paris.

Barker, F., & Hyde, R. (1982). *London. As it might have been*. London: John Murray.

Barker, Th., & Robbins, M. (1963). *A History of London Transport. Passenger travel and the development of the metropolis* (Vol. 1). London: George Allen & Unwin Ltd.

Barlow, P. W. (1864, September 9). Patent No. 2207. London Transport Museum.

Barlow, P. W. (1867). *On the Relief of London Street Traffic, with a Description of the Tower Subway now Shortly to be Executed*. London.

Barlow, P. W. (1871). *The Relief of Street Traffic. Advantages of the City and Southwark Subway, with Reasons why the Proposed Connection of Street Tramways from the Elephant and Castle through the City is Unnecessary and Undesirable. Second Pamphlet*. London.

Barry, J. W. (1885). The City Lines and Extensions (Inner Circle Completion) of the Metropolitan and District Railways. *Minutes of Proceedings of the Institution of Civil Engineers, 81*, 34-51.

Beltran, A. (1988). Une Victoire Commune. L'alimentation en énergie électrique du Métropolitain (1re moitié du XXe siècle). In *Métropolitain. L'autre dimension de ville* (pp. 111–122). Paris: Mairie de Paris.

Bienvenüe, F. (1896). *Chemins de Fer Urbains à traction électrique. Devis descriptif et estimatif*. Paris: Régie Autonome des Transports Parisiens.

Boudon, F. et al. (1977). Système de l'architecture urbaine: Le Quartier des Halles de Paris. Paris.

Bourillon, F. (2001). A propos de la Commission des embellissements. In K. Bowie (Ed.), (textes réunis par) La Modernité avant Haussmann. Formes de l'espace urbain à Paris 1801 – 1853 (pp. 139–151). Paris: Éditions Recherches.

Bowie, K., & Texier, S. (Eds.), (2003). Paris et ses Chemins de Fer. Paris: Action Artistique de la Ville de Paris.

Bradley, K. (2006). The Development of the London Underground, 1840-1933: The Transformation of the London Metropolis and the Role of Laissez-Faire in Urban Growth [PhD Thesis]. Emory University.

Caron, F. (1997). *Histoire de chemins de fer en France* (Vol. I). Paris: Fayard.

Chevalier, L. (1973). *Labouring Classes and Dangerous Classes in Paris during the first half of the nineteenth century* (F. Jellinek, Trans.). London: Routledge & Kegan Paul. (Original work published 1958)

Chrimes, M. (2013). Hawkshaw, Sir John (1811–1891). Oxford Dictionary of National Biography. Retrieved from http://0-www.oxforddnb.com.catalogue.ulrls.lon.ac.uk/view/article/12690

City of London & Southwark Subway Company. (n. d.). Minute book 1, 1884 – 1889. London Metropolitan Archives.

City of London & Southwark Subway Company. (n. d.). Minute book 2, 1889 – 1892. London Metropolitan Archives.

Procès verbal. (1898, June 27). Conseil Municipal de Paris.

Délibérations. (1898, June 27). Conseil Municipal de Paris.

de la Broise, T., & Torres, F. (1996). *Schneider, histoire en force*. Paris: Éditions de Monza.

Dennis, R. (2008). Cities in Modernity Representations and Productions of Metropolitan Space, 1840-1930. Cambridge: Cambridge University Press.

Duffy, M. (2003). *Electric Railways 1880-1990*. London: Institution of Electrical Engineers.

Fleury, M., & Pronteau, J. (1987). *Petit Atlas Pittoresque des quarante-huit Quartiers de la ville de Paris (1834) par A.-M. Perrot, ingénieur*. Paris: Commission des Travaux Historiques.

Freeman, M. J., & Aldcroft, D. (1985). *The Atlas of British Railway History*. London: Croom Helm.

Greathead, J. (1896). The City and South London Railway; with some Remarks upon Subaqueous Tunnelling by Shield and Compressed Air. London.

Halliday, S. (2001). *Underground to Everywhere. London's underground railway in the life of the capital*. Sutton Publishing London Transport Museum.

Horeau, H. (1845). *Examen critique du projet d'agrandissement et de construction des Halles Centrales d'Approvisionnement pour la Ville de Paris*. Paris.

Horeau, H. (1846). *Nouvelles observations sur le projet d'agrandissement et de construction des Halles Centrales d'Approvisionnement pour la Ville de Paris*. Paris.

Report of the Commissioners appointed to investigate the Various Projects for establishing Railway Termini Within or in the immediate Vicinity of the Metropolis. (1846). Parliamentary Papers, Houses of Commons and Lords, London.

Huet, E. (1896). Métropolitain urbain à traction électrique: Rapport du Directeur Administratif des Travaux. Paris, 2 March 1896, plus Annexe au rapport, Paris, 23 March 1896. Régie Autonome des Transports Parisiens.

Huet, O. (1878). *Les Chemins de Fer Métropolitains de Londres. Étude d'un réseau de chemins de fer métropolitains pour la ville de Paris. Mission à Londres en mai, 1876*. Paris.

Jackson, A. (1986). *London's Metropolitan Railway*. Newton Abbot: David & Charles.

Jahn, M. (1982). Suburban development in Outer West London, 1850-1900. In F.M.L. Thompson (Ed.), *The rise of suburbia*. Leicester: Leicester University Press.

Jones, C. (2004). *Paris Biography of a City*. Penguin Books.

Kellett, J. R. (1979). Railways and Victorian Cities. London et al. Routledge.

Koselleck, R. (2004). Futures Past on the Semantics of Historical Time. Translated and with an Introduction by Keith Tribe. New York: Columbia University Press. originally published in German in 1979

Lambert, A. J. (1984). *Nineteenth Century Railway History through the Illustrated London News*. London: David and Charles.

Larroque, D. (2002). Le Métropolitain: histoire d'un projet. In D. Larroque, M. Margairaz, P. Zembri, Paris et ses Transports XIXe – XXe siècles. Deux siècles de décisions pour la ville et sa région. Paris: Éditions Recherches, 41-94.

Lascelles, T. S. (1987). *The City & South London Railway*. Oxford: The Oakwood Press.

Lavedan, P. (1969). *La question du déplacement de Paris et du transfert des Halles au Conseil municipal sous la monarchie de Juillet*. Paris: Commission des Travaux Historiques.

Lee, C. E. (1973). *The Northern Line. A Brief History*. London: London Transport.

Lemoine, B. (1980). Les Halles de Paris. L'histoire d'un lieu, les péripéties d'une reconstruction, la succession des projets, l'architecture d'un monument, l'enjeu d'une "Cité". Paris: L'Equerre.

Lewin, H. G. (1936). *Railway mania and its aftermath: 1845-1852*. London: Railway Gazette.

López Galviz, C. (2012). Converging Lines Dissecting Circles: Railways and the Socialist Ideal in London and Paris at the Turn of the Twentieth Century. In M. Davies & J. Galloway (Eds.), *London and Beyond: Essays in honour of Derek Keene* (pp. 317–337). London: Institute of Historical Research Series.

López Galviz, C. (2013a). Mobilities at a standstill: Regulating circulation in London c.1863-1870. *Journal of Historical Geography, 42*, 62–76. doi:10.1016/j.jhg.2013.04.019

López Galviz, C. (2013b). Metropolitan Railways: Urban Form and the Public Benefit in London and Paris 1850-1880. *The London Journal, 38*(2), 184–202. doi:10.1179/0305803413Z.00000000030

Marchand, B. (1993). *Paris, histoire d'une ville XIXe – XXe siècles*. Paris: Éditions du Seil.

Martin, A. (1894). *Étude historique et statistique sur les moyens de transport dans Paris avec plans, diagrammes et cartogrammes*. Paris: Ministère de l'Instruction Publique et des Beaux-Arts.

Martin, G. (2004). *Past Futures: The Impossible Necessity of History*. Toronto, Buffalo, London: Toronto University Press.

Meynadier, H. (1843). *Paris sous le point de vue pittoresque et monumental ou éléments d'un plan général d'ensemble de ses travaux d'art et d'utilité publique*. Paris.

Papayanis, N. (2004). *Planning Paris before Haussmann*. Baltimore, London: The Johns Hopkins University Press.

Perreymond. (1844). *Revue Générale d'Architecture et des Travaux Publics*, col. 184-188 and 232-235.

Picon, A. (1994). Les Fortifications de Paris. In B. Belhoste, F. Masson, & A. Picon (Eds.), Le Paris des Polytechniciens. Des ingénieurs dans la ville 1794 – 1994 (pp. 213 – 221). Paris: Délégation à l'action artistique de la ville de Paris.

Porter, R. (2000). *London: A Social History*. London: Penguin Books.

Robert, J. (1967). *Notre Métro*. Paris.

Saboya, M. (1991). *Presse et Architecture au XIXe siècle. César Daly et la Revue Générale de l'Architecture et des Travaux Publics*. Paris: Picard Éditeur.

Sekon, G. A. (1899). Illustrated Interviews, Mr. Thomas Chellew Jenkin. The Railway Magazine (Vol. 5, pp. 1–16). July to December.

Simmons, J. (1995). *The Victorian Railway*. London: Thames & Hudson.

Sutcliffe, A. (1983). London and Paris: Capitals of the Nineteenth Century. *The Fifth H.J. Dyos Memorial Lecture*. London, Paris: Victorian Studies Centre, University of Leicester.

Thornbury, W. (1878). The Tower Subway and London Docks. In *Old and New London:* (Vol. 2, pp. 122-128). London: Cassell, Petter, & Galpin. Retrieved from http://www.british-history.ac.uk/report.aspx?compid=45081

White, J. (2008). *London in the Nineteenth Century: A Human Awful Wonder of God*. London: Vintage.

Wolmar, C. (2004). *The Subterranean Railway. How the London underground railway was built and how it changed the city forever*. London: Atlantic Books.

Chapter 2
The Potential for Rail Transit as a Way to Mitigate Accident Risk:
A Case Study in Chennai

Sumeeta Srinivasan
Tufts University, USA

ABSTRACT

The city of Chennai has made road accident data available with the address location of road accidents and the total numbers of persons and pedestrians affected in the accident in 2009. These data were geocoded to locate the accidents with respect to the census wards within the Chennai Corporation area. Both the total number of persons as well as pedestrians in accidents as well as the rate of accidents normalized by population in the ward were modeled as dependent variables using Poisson based regression models to see the effect of location characteristics such as road length, vehicle traffic, proximity to existing and proposed transit infrastructure and the percentage of the land developed between 1991-2009. The results from the models suggest that location does indeed affect the risk for accidents in Chennai and that planners in the city may need to better understand the implications of roads, urban development, transit access and the built environment for traffic safety.

INTRODUCTION

The World Health Organization (WHO, 2009) estimated that 1.2 million people a year die and 50 million are injured in traffic accidents and that a majority of these accidents occur in low income countries. Mohan et al (2009) report that traffic fatalities in India increased by about 5% per year from 1980 to 2000, and have continued to increase by about 8% per year since then. After New Delhi (the capital of India), in 2006 Chennai had the highest fatality rate from accidents with over 1300 fatalities in that year alone (Mohan et al, 2009). There are significant geographical and socioeconomic disparities in the risk of traffic accidents related injuries. A national survey of road traffic injuries in India found that the age-adjusted mortality rate was greater in urban than in rural areas, and was notably higher than that estimated from national police records (Hsiao et al, 2013). The found that pedestrians, motorcyclists

DOI: 10.4018/978-1-5225-0084-1.ch002

and other vulnerable road users constituted 68% of the deaths due to road traffic accidents and 81% of pedestrian deaths were associated with less education and living in poorer neighborhoods. Hsiao et al also note that the state of Tamil Nadu (of which Chennai is the capital) had the third highest death rate in the country due to road traffic injuries. Patel et al (2011) found that rapid motorization, along with the heterogeneous composition of road traffic and infrastructural deficiencies can be directly linked to the increased number of road traffic injuries in India. They report that more than half of total road deaths in 2007 were in Andhra Pradesh, Maharashtra, Tamil Nadu, and Karnataka, which account for only 27% of India's population but about 37% of its motor vehicles. Research suggests that the overall fatality rates are likely to continue to increase as GDP continues to increase and motorization increases (Grimm and Treibech, 2012). These grim statistics also have implications for the economy. Yearly losses due to road traffic accidents have been estimated at 750 billion rupees (Sikdar and Bhavsar, 2009). This was nearly 3% of the gross domestic product (GDP). In the light of these grim statistics it is vital that we address traffic accidents as a major health burden especially in urban areas where residents face a higher risk.

Many studies in India have focused on traffic accident risk at the scale of states, regions and cities. Very few studies have looked at risk within the city. It is important to look at both macro level and micro level policies to be able to address the risk of accidents at different scales. While macro level policies are necessary for traffic safety micro-scale policies can be used by urban planners within cities. Among the few Indian studies that address this within cities – Tiwari and Jain (2014) estimate a 43% reduction in accidents along the Bus Rapid Transit (BRT) corridor in New Delhi suggesting that access to transit may a play a role in reducing the risk of accidents. De Andrade et al (2014) observed in Brazil that the built environment has a direct influence on the occurrence of accidents, as well as the specific behaviors that causes them. In their built environment analysis, the variables length of road in urban area, limited lighting, double lane roadways, and fewer auxiliary lanes were associated with a higher incidence of fatal accidents. Harvey et al (2015) estimated regression models for New York City that suggest that vehicle accidents on smaller, more enclosed streetscapes were less likely to result in injury or death compared with those on larger, more open streetscapes. Hanson et al (2013) also suggest that severity of pedestrian casualties is associated with the lack of sidewalks and buffers, high-speed roads, roads with six or more lanes and a median, and lack of traffic lighting. Clifton et al (2011) note that transit access and greater pedestrian connectivity, such as central city areas, are significant and negatively associated with injury severity. Ewing et al (2013) have found that for the US urban sprawl is both directly and indirectly a significant risk factor for traffic fatalities suggesting that the built environment may play a role in accident risk. Demographic factors may also play are role in accident risk: researchers in developed countries like the UK have found that find that economically deprived areas within urban areas tend to have higher levels of casualties due to road traffic (Noland and Quddus, 2005; Anderson, 2010). A study by Cottril and Thakuriah (2010) in the US notes that pedestrian-vehicle crashes are more common in locations with high low-income and minority populations (also called environmental justice or EJ areas in the United States). They find that pedestrian traffic accidents in EJ areas are related to variables of exposure (including the suitability of the area for walking and transit accessibility), transit availability, and general population demographics such as income and the presence of children.

As India continues to urbanize a greater proportion of the population will be exposed to the risk of accidents. Therefore it is useful to understand the risk of accidents within an urban context. In this chapter we study the geographic distribution of traffic accidents within the city of Chennai and estimate the probability of accidents while controlling for socioeconomic and built environment characteristics.

The Potential for Rail Transit as a Way to Mitigate Accident Risk

Background

Chennai is the sixth largest city in India with an estimated population of 4.6 million administered by the Chennai Metropolitan Corporation based on the 2011 Census. The larger urban metropolitan area had a population estimated as close to 9 million making it the fourth largest urban concentration in India. The Chennai Metropolitan Corporation covers approximately 460 sq km in 155 wards (now 200 wards). The Bay of Bengal lies on the eastern shore of the city. The Metropolitan Transport Corporation of Chennai has a fleet of about 3500 buses with 25 bus depots with an estimated 5 million passengers a day. Overloading is as high as 150% along certain routes. As a result, overcrowding at the bus stops and spillover on the carriageways has become common (CMDA, 2006). The Chennai Mass Rapid Transit System (MRTS) operates an elevated rail line within the city which carries about 16,000 passengers a day (Muthukannan and Thirumurthy, 2008) and along the same corridor buses carried about 65,000 passengers. The Chennai Metropolitan area also has three commuter rail lines which radiate from the city-centre and account for 300,000 commuter trips per day (CMDA, 2006). While the first two corridors carry intercity passengers on separate dedicated lines, the third corridor carries both commuters and intercity passengers on the same lines. There is also a dedicated rail project called the Chennai Metro. Phase I of the project has two corridors covering a length of 45.1 km (42 stations) is under construction and scheduled to be completed in 2015. In 2010 there were an estimated 5047 accidents in Chennai in which 621 persons were killed and 1749 "were grievously injured" (TNSA, 2015). This has almost doubled in 2014 to about 10,000 persons in accidents of which about 1046 were killed and 1341 were severely injured.

MAIN FOCUS OF THE CHAPTER

Data at the sub-city scales of accident locations and at the ward level were used to map the spatial distribution of accidents and locate clusters within the city of Chennai. Mapping the spatial clusters of accidents within Chennai may help policy makers understand the risk of accidents at a neighborhood level. Some of the questions that this chapter asks that arise from past research include: Do marginalized residents of the city have a larger burden of accidents? Are there certain locations in the city that are especially prone to accidents? If so, is it related to the exposure due to the presence of major roads and traffic? Does providing more transit access through the metro, bus routes and MRTS lower the risk of accidents? Do residents in peripheral locations that may have different built environment characteristics have a different risk burden for road accidents?

MODELING ACCIDENT RISK

Data and Software

The original data for the road accidents comes from the Transparent Chennai website which posted an excel file indicating every accident in 2009 recorded by the Chennai Traffic Police. The number of fatalities was not recorded but the spreadsheet lists the number of pedestrians and the total number of persons in the accident. Of the 3822 addresses recorded by the Chennai Traffic Police, 2784 were successfully geocoded using GPS visualizer, Google Earth and Open Street Maps base map. The addresses involved a total of 3164 persons and 835 pedestrians in accidents. These locations were aggregated by ward to assess the relationships between socioeconomic and demographic characteristics and accidents. The quality of the base map data available through the Open Street Map initiative is steadily improving but the ability to pinpoint addresses, especially on the peripheral locations of the city, is still not as reliable as it is in the central parts of the of the city. The ward data was from the Census 2001 (the latest available data at this resolution). The Census data included demographic data including population, and workers by occupation. Street data from Open Street Map were used to measure road length by ward. Data on buses and vehicle counts were downloaded from the website Geocommons.org. All the maps were created using Quantum GIS and Geoda which are both freely available. The statistical analysis was conducted using R, an open source software.

Analytic Methods

Local Moran's I (Anselin, 1995) with and without empirical Bayes (EB) smoothing statistics for the accidents by ward are shown in Figure 5 (Assuncao and Reis, 1999). A Local Moran's I statistic is an indicator of the extent to which the value of an observation is similar to or different from its neighbors. Monte Carlo simulations are used to evaluate the statistical significance of concentrations. The local Moran (Anselin, 1995) is evaluated as follows:

$$I = z_i \sum_j w_{ij} z_j \tag{1}$$

The observations z_i and z_j are deviations from the mean, and the summation over j is such that only neighboring values of ward j are included. For ease of interpretation, the weights w_{ij} are be in row standardized form and by convention, $w_{ii} = 0$. The results are mapped as "low–low" or "high–high" wards that contribute to positive autocorrelation and "low–high" or "high–low" cases that contribute to negative autocorrelation. For example, "high–high" cases are ones where the value of ward i is high and neighboring ward values are also high. While all wards are in one of the four classes only wards that are statistically significant at the 5% level are colored in the maps.

The presence of hot spots can be investigated with Getis-Ord statistics (Getis and Ord, 1992). Here the G* is as follows:

$$G^* = \frac{\sum_{i=1}^{n}\sum_{j=1}^{n} \omega_{ij} y_k^i y_k^j}{\sum_{i=1}^{n}\sum_{j=1}^{n} \omega_{ij}} \tag{2}$$

where, i is the ward to be measured and j is a ward defined as a neighbor by weight matrix, n is the total number of wards in the study area. y_k^i and y_k^j are accidents or other measures k, in ward i and j respectively, and ω_{ij} is spatial weight matrix.

As evident from Figures 1 and 2 accident data for Chennai, like most count data, are not normally distributed. Therefore, either a Poisson or negative binomial model has the correct distributional prop-

Figure 1.

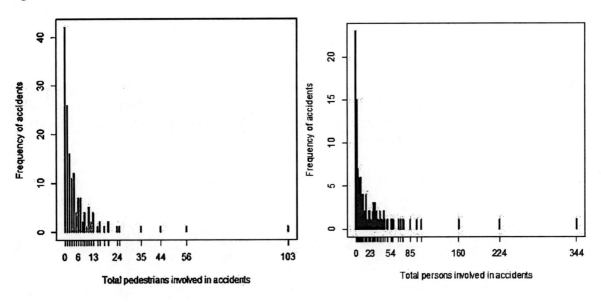

Figure 2.

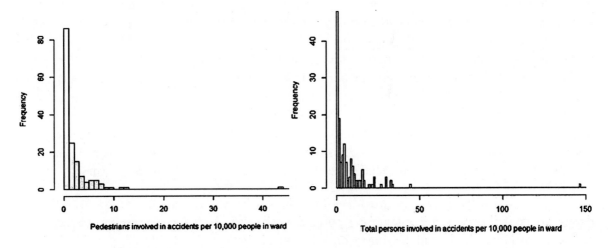

erties for model estimation. Poisson distributions assume the variance is equal to the mean, a condition that is violated in this dataset and is known as "over-dispersion". A Poisson distribution is given by

$$\Pr[Y = y] = \frac{e^{\lambda}\lambda^y}{y!}, y = 0,1,2.. \qquad (3)$$

where, λ is the mean number of accidents in a specified interval

The most common formulation is through a log linear specification and estimation is carried out through Maximum Likelihood Estimation. The expected number of accidents is.

$$E[y_i \mid x_i] = \lambda_i = e^{x_i'\beta} \qquad (4)$$

While the Poisson model is often useful for describing the mean it underestimates the variance in the data, rendering all model-based tests liberal. One way of dealing with this is to use the same estimating functions for the mean, but to base inference on the more robust sandwich covariance matrix estimator (Zeleis, et al, 2007). The authors note that another way of dealing with over-dispersion is to use the mean regression function and the variance function from the Poisson GLM but to leave the dispersion parameter unrestricted which leads to similar coefficient estimates as the standard Poisson model but inference is adjusted for over-dispersion. A third way of modeling over-dispersed count data is to assume a negative binomial (NB) distribution which can arise as a gamma mixture of Poisson distributions:

$$P(Y = y) = \left(\frac{\nu}{\nu + \lambda}\right)^\nu \frac{\Gamma(r + y)}{\Gamma(y + 1)\Gamma(r)}\left|\frac{\lambda}{\nu + \lambda}\right|^y \qquad (5)$$

Γ is the gamma function; λ is the mean, and $(\lambda + \lambda^2/\nu)$ is the variance, so as ν is smaller, then variance is larger; ν is sometimes referred to as the dispersion parameter. Zeileis et al note that in addition to over-dispersion, many empirical count data sets exhibit more zero observations than would be allowed for by the Poisson model. They suggest the use of two-component models: a truncated count component, such as Poisson, or negative binomial, is employed for positive counts, and a hurdle component models zero versus larger counts.

The accidents data for Chennai were both over-dispersed and had a many zero observation wards. Of the 155 wards in the city there were 23 wards (15%) with no persons involved in accidents and 42 wards (27%) with no pedestrians involved in accidents. Thus the zero observation issue was more severe for pedestrian accidents. Table 3 shows the coefficients estimated for Poisson and quasi-Poisson models (with sandwich variance derived significance values) and a hurdle model for total accidents and a negative binomial model for pedestrian counts (models with the lowest Log likelihood and significant Likelihood Ratio tests are reported). Models results are further described in the next section.

DESCRIPTIVE RESULTS OF LOCATIONS ACCIDENTS AND CORRELATES

Not surprisingly, the ward with the highest number of persons in accidents (344) also had the most pedestrian accidents (103). This is a densely populated central ward in the Thousand lights area which includes the main traffic artery through Chennai –Anna Salai. This location also had the highest rate of persons involved in accidents per 10,000 people (147) and the highest rate of pedestrians involved in accidents per 10,000 persons (44). As Figures 1 and 2 indicate the frequency distribution of persons in accidents as counts and rates are quite similar. Note that the rate in India was 42.5 per 10,000 in 2010 (MoRTH, 2010) and the mean rate per 10,000 population in Chennai was 11.8 per 10,000 for 2010.

The spatial distribution of both persons and pedestrians in accidents (Figure 3) appear to concentrate in the southern and western periphery and along major roads within the city – mainly Anna Salai and Poonamali High Road and at the intersections with state highways such as Old Mahabalipuram Road and

Figure 3.

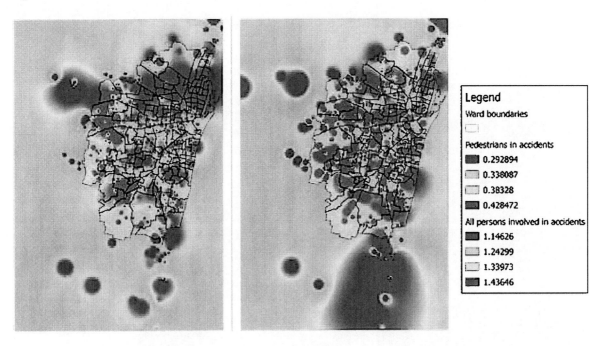

East Coast Road on the southern periphery and highways leading to Bangalore and Thiruvallur on the western periphery. Fatality data was only available for the 25 worst locations. Figure 4 suggests that the fatality patterns appear to echo the overall patterns in Figure 3 with high values along western and the southern periphery and in central commercial locations in Egmore, Thousand lights along major arteries.

The Local Moran's I maps (Figure 5) are similar for both the counts as well as the rates showing clustering of accidents in the central locations mentioned as high-high in Figure 3 and 4. The peripheral locations appear not to have clusters. Unfortunately, Census data at the ward level is not available beyond the city periphery though accidents do occur at high concentrations in these locations just outside the corporation area limits. Getis-Ord (G) statistic clustering (Figure 6) indicates a similar pattern of clustering in accidents as seen in the Local Moran's I maps– the locations shaded red in the central areas and some peripheral locations in the west have the highest share of accidents and northern locations show lowest shares of accidents. Again, G statistic cluster maps of rates (accidents per 10,000 people) do not pick up high accident locations on the periphery. However, clear patterns of clustering in accidents do emerge in several locations.

Tables 1 and 2 include descriptive statistics and correlations between various variables describing the wards. The Pearson correlation coefficients suggest that accident counts and rates are highly correlated with each other. The correlations for rates and for counts are identical and for this reason only counts are reported in the next section. Not surprisingly, road length is positively correlated with accidents indicating that more exposure leads to more accidents. However, road density is negatively correlated with the count of accidents suggesting perhaps that locations with high densities of roads per sq. km tend to have lower speeds and possibly fewer accidents. As can be expected traffic counts (interpolated) were also correlated with higher accidents and accident rates. The highest traffic counts were in the central locations and southern periphery.

Figure 4.

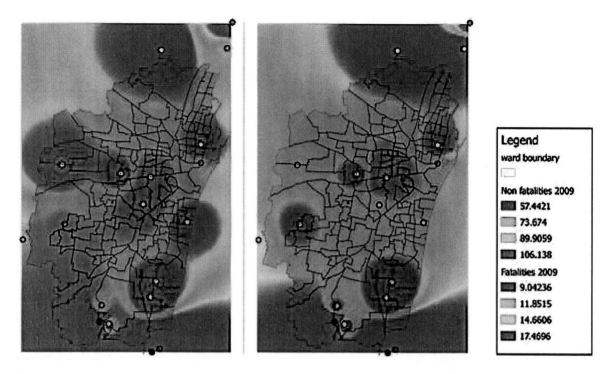

Table 1. Descriptive variables by ward for Chennai

Variable (by ward)	Mean	Median	SD
Total pedestrians in accidents	5.4	2.0	11.0
Total in accidents	20.4	8.0	38.8
Total in accidents per 10K persons	7.3	3.2	13.9
Pedestrians in accidents per 10K persons	2.0	0.8	4.1
Road length (m)	31785.0	20594.0	41399.1
Road density (m/ sq km)	31525.0	29856.0	14446.4
Percentage of the ward developed between 1991-2009	31.8	31.8	24.7
Percentage developed before 1991	65.5	67.8	24.1
Marginal worker percentage	7.0	5.8	4.4
Low income workers	10.2	9.1	4.7
Agricultural workers percentage	1.4	1.2	0.9
Illiteracy percentage	23.1	22.0	59.0
Children below 6 percentage	9.8	9.9	1.4
SC ST percentage	14.0	10.8	1.2
Interpolated mean vehicle counts	70948	68761	18872
Number of metro stops	0.2	0	0.6
Number of bus stops	16	10	21
Number of MRTS stops	0.3	0	0.8
Distance to metro stop (km)	1.6	1.5	1.1
Total number wards	155		

Figure 5.

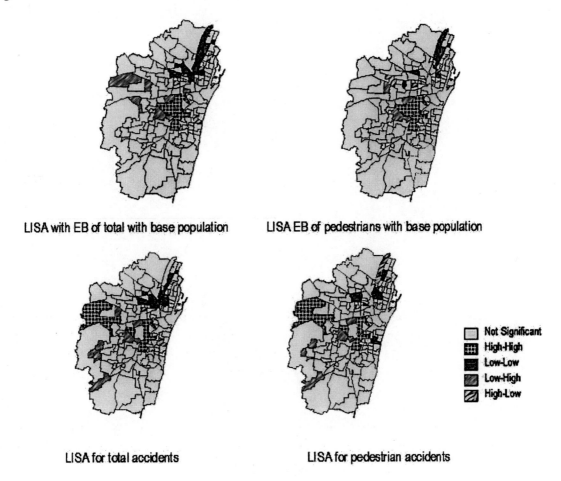

Demographic variables indicating high illiteracy, large percentages of children, marginal workers, and low income workers and high percentages of SC/ ST (scheduled caste and tribes) are all correlated positively with each other suggesting spatial clustering within Chennai of marginalized low income and high illiteracy locations. Surprisingly, accident rates and counts seem to be lower in these locations (which are concentrated in the northern parts of Chennai except for a narrow strip along the northern coast). This appears to suggest that unlike the developed countries locations with predominantly low income residents (as measured by Census data proxies such as literacy and SC/ST percentage) are at lower risk for accidents. This does not however mean that the risk of accidents is lower for the poor as they are likely to be employed in higher income areas. Since our data did not include socioeconomic details about the persons involved in the accidents we were unable to test if they were at higher risk individually.

Two variables describing the percentage of ward developed between 1991 and 2009 and the percentage of ward that was developed by 1991 were also highly correlated with accidents but with opposite signs (See Figure 7). As the map shows, wards that had a higher percentage of land urbanized between

Figure 6.

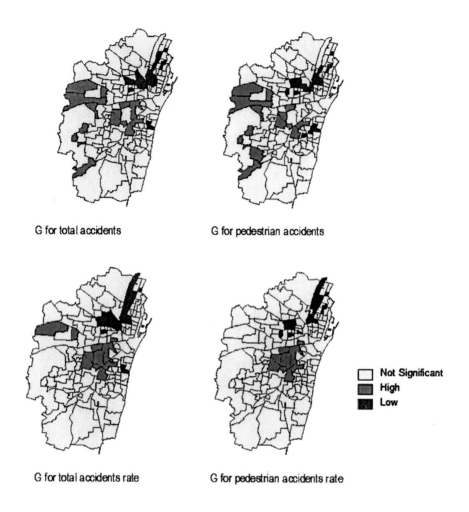

1991 and 2009 are mostly peripheral locations which tend to be less dense and have poorer transit access to employment which is concentrated in central locations or along highway corridors. Such peripheral locations tended to have more accidents. Conversely locations that were highly developed by 1991 had fewer accidents. In the next section regression models are used to predict accident counts to estimate the effect of location characteristics.

MODEL RESULTS: PREDICTING ACCIDENTS

The variables that were significant in most of the models included traffic, road density, road length and the percentage of urbanization that occurred in the ward between 1991 and 2009. The coefficient on road length in both the models suggested an increase of 10 km of road within a ward would increase the number of accidents by about 10%. In other words, if a ward with below average road length (the mean road length was 32 km) built an additional 10 km of roads the count is likely to rise by 2 accidents and about 1 pedestrian accident on average in the ward. If a ward was to build a road length approaching

Table 2. Pearson correlation coefficients between variables used in the model ($p < 0.001$ ***, $p < 0.01$ **, $p < 0.5$ *)

Variables (by ward)	Pedestrian Acc	Total Acc	Total Acc per 10K	Pedestrian Acc per 10K	Road length	Road density	Marginal Worker	Low income worker	Illiteracy	Children
Total in accidents	0.97***									
Total per 10K persons	0.91***	0.91***								
Pedestrians per 10K persons	0.93***	0.88***	0.97***							
Road length	0.47***	0.54***	0.28***	0.23**						
Road density	-0.22**	-0.25**	-0.18*		-0.24**					
Percentage of ward developed between 1991-2009	0.25**	0.28**	0.19*	0.17*	0.26**					
Percentage developed by 1991	-0.24**	-0.30***	-0.25**	-0.19*	-0.29***	0.59***				
Illiteracy		-0.14*	-0.16*				0.33***	0.37***		
Children below 6			-0.20*				0.28***	0.32***	0.77***	
SC ST percentage							0.40***	0.40***	0.55***	0.39***
Total bus fleet					0.20*	-0.17*				
Mean interpolated traffic		0.20*	0.20*	0.16*				-0.20*	-0.39***	-0.47***
Agriculture labor								0.21*		

Note: Not all coefficients are reported and only significant coefficients are included

Figure 7.

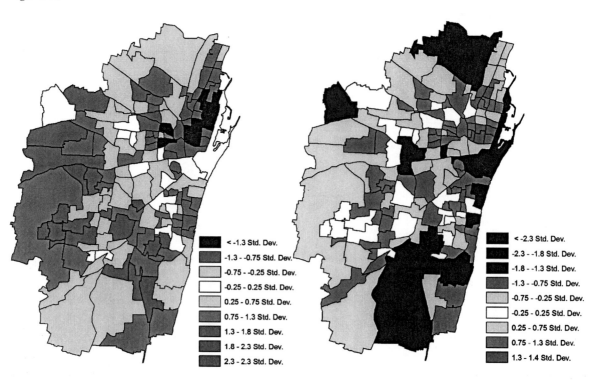

Table 3. Regression results (significance codes: p < 0.001 '***'; p < 0.01 '**'; p < 0.05 '*'; p < 0.1 'x')

Variables (by ward)	Count of all accidents			Count of pedestrian accidents		
	Poisson (sandwich)	Quasi-Poisson (sandwich)	Hurdle (Negative Binomial)	Poisson (sandwich)	Quasi-Poisson (sandwich)	Negative Binomial
Intercept	**3.9*****	**3.9****	**3.7*****	**1.8*****	1.8	1.9
Mean interpolated traffic in ward	**7.1e-06*****	7.1e-06	-	**8.7e-06*****	8.7e-06	-
Road length	**5.4e-06*****	**5.4e-06*****	9.7e-06.	**5.2e-06*****	**5.2e-06****	8.6e-06
Road density	**-2.5e-05*****	**-2.5e-05***	-	**-2.7e-06x**	**-2.7e-06***	-
Percentage of ward developed 1991-2009	**0.012*****	**0.012****	**0.012***	**0.012*****	**0.012***	0.006
Percentage developed 1991	**0.011****	0.011.	**-0.014****	**-0.006****	-0.006	-0.009
Low income worker percentage	**0.015****	0.015	0.02	**0.019***	0.019	0.03
Agriculture labor percentage	-0.017	-0.0162	0.004	0.052	0.052	-0.008
Illiteracy percentage	**-0.038*****	**-0.037x**	-0.03	**-0.023***	-0.023	-0.05
Distance to Metro stop (m)	**-1.00e-04*****	-1.00e-04	-4.9e-05	-7.7e-05	-7.7e-05	2.7e-05
Number of MRTS stops	**-0.075****	-0.075	-0.05	**-0.098***	-0.098	-0.04
Number of bus stops	1.5e-03	1.5e-03	0.004	1.5e-03	1.5e-03	4.3e-03
Log Likelihood	-1910	-	-556	-722	-	-376

The Potential for Rail Transit as a Way to Mitigate Accident Risk

New Delhi's (the capital) average of about 180 km per ward then the number of accidents is likely to rise by about 45 total accidents. Road density was also significant in predicting pedestrian accidents but the effect was the opposite sign of road length. An increase of 10 km/ sq km of ward would decrease the pedestrian in accidents by about 10%. This is probably a function of increased traffic congestion leading to lower speeds. The effect of illiteracy was also similar in that a ward with about 10% more illiteracy (these are locations in the northern part of Chennai which are also more densely populated) there would be about 10 fewer accidents overall. Illiteracy percentage was not significant in predicting pedestrian accidents.

The effect of urbanization variables was very interesting as it could be interpreted as a measure of the built environment. Dense locations that were fully urbanized in 1991 were in the older, northern and central parts of Chennai (see Figure 7) and so did not change much during the period till 2009. Such locations are also likely to be fully urbanized by 1991 as seen in the map on the right. Locations that developed during this time period (1991-2009) were more likely to have a higher count of accidents and were mostly along the peripheries. Thus if the percentage of the ward that was urbanized in this time period was increased by 10% it would have 2.5 more accidents and 1 more pedestrian involved in an accident. On the other hand if the ward was 10% more urbanized by 1991 it had 2.5 fewer accidents. This suggests that peripheral locations which developed in the last 20 years had a higher risk of accidents.

The effect of proximity to transit (proposed Metro stops) and the number of MRTS (commuter rail) and bus stops in the ward was also tested in the model. These variables are a good measure of the accessibility of a ward since a central location had more bus stops, MRTS stops and was close to a metro stop. Some of these variables were significant in the Poisson models and the signs were negative suggesting that proximity to the Metro or the presence of MRTS stops is likely to reduce the risk of accidents. The Poisson models suggest that increased access to transit is likely to reduce the risk of accidents. However after taking account of the relatively high numbers of zero accident wards these variables were not significant in the negative binomial and hurdle models. It should be noted here that these are relatively coarse measures of access to transit. If we had data for employment that could be used to estimate finer grain measures of transit access across Chennai it is likely that these measures would show more variation. Furthermore the overall transit access at the peripheries is quite low as measured by distance to metro lines, transit stops and the number of bus stops. Measures of demand that also include route frequencies would better indicate how connected wards are to employment through transit.

Figure 8 shows that the models predict higher counts of accidents along the peripheries and lower counts in the north and along the coast on the east but that they over-predict on the periphery and under-predict in the central commercial locations. It is likely that the model is missing other variables that might be significant in predicting accidents. These include the type of vehicles involved in the accident, speeds on the roads, the presence of flyovers and other physical indicators of the roads. It is evident from Figure 8 that the proposed Metro stops are in locations that have relatively high predicted accident risk.

CONCLUSION AND POLICY IMPLICATIONS

Poisson and negative binomial based regression models allow for a fine grained spatial analysis of accident risk in Chennai by ward. The model results suggest that two phenomena that are closely linked– roads and urbanization are likely to affect the risk of accidents at the ward level. As long as highways continue to link Indian cities and per capita incomes continue to rise, urban development of the peripheries is likely

Figure 8.

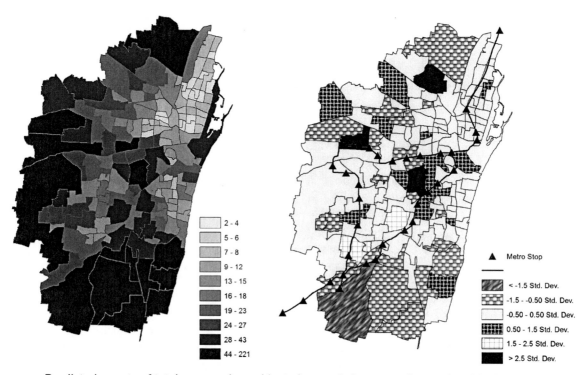

Predicted counts of total persons in accidents by wards in percentiles and residuals

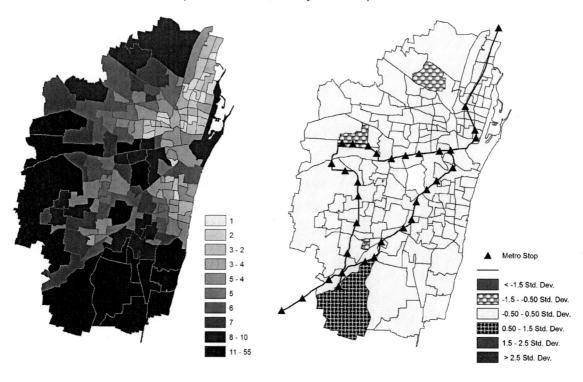

to be accompanied by increased vehicle ownership. Based on the models estimated for Chennai, this will result in increased traffic accidents unless they are also accompanied by urban planning that addresses the built environment and transit based accessibility. The risk for accidents could be addressed by diverting traffic on the roads to rail or other forms of transit such as bus only lanes (BRT). Future lines for the Metro in Chennai could also be planned along the wards of the periphery that have the highest risk for accidents. The CMDA plan (2006) cautions that bus and rail are developing as competing modes rather than being complementary to each other and that sprawling suburban development without adequate transport facilities has placed considerable demand in favor of private vehicles. Clearly for improved transit access route plans for bus, and all kinds of rail have to be integrated.

Tamil Nadu has been on the forefront of improved accident databases in India with the implementation of the Road Accident Data Management System (RADMS) since 2008 (Raban et al, 2014) and this is likely to provide better tools for improving safety. However, Spoerria et al (2011) note that prevention of accidents has focused on improvements to road infrastructure but that prevention should also consider social context and aim to reduce inequities across socio-demographic groups and across geographical areas. The data currently provided by the Chennai traffic police do not allow us to predict based on individual characteristics such as gender, income and age of those involved in the accidents. Our models do not indicate that the risk is higher for low income groups based on the ward level data we do not have sufficient data to predict if this is true at the individual scale. It is possible that most pedestrians are from low income households and are affected at or on the way to their workplace and not at their home ward. It is also likely that the accidents mostly affect bus riders, bicyclists and two wheelers who are also likely to be less wealthy, less educated and younger.

The link between peripheral urbanization and roads is worth exploring in future research. At the very least planners need to link land use policy for the new Indian suburbs with the effects of regional road building. Patel et al (2011) note that population-wide interventions for reduction of number of road traffic injuries (through enforcement of drink-driving laws and speed limits) are the most feasible and cost effective and could be implemented first. They also note that the main focus of road safety efforts in India has been to change the behavior of road users through isolated, sporadic, and non-systematic approaches whereas globally the approach has shifted to building safe vehicles and safe road environments through engineering, enforcement, and education. The results of models estimated in this paper suggest that beyond these larger scale policies, micro level policies that integrate access to public transit and encourage smart growth of newly developing urban areas should be integrated into planning for Indian cities like Chennai in order to mitigate the risk of road traffic accidents (Tables 1, 2, & 3).

ACKNOWLEDGMENT

The author is grateful to Transparent Chennai for making traffic accident and other data related to transportation available on their website.

REFERENCES

Anderson, T. K. (2010). Using geodemographics to measure and explain social and environment differences in road traffic accident risk. *Environment & Planning A*, *42*(9), 2186–2200. doi:10.1068/a43157

Anselin, L. (1995). The Local Indicators of Spatial Association LISA. *Geographical Analysis*, *27*(2), 93–115. doi:10.1111/j.1538-4632.1995.tb00338.x

Assuncao, R., & Reis, E. A. (1999). A new proposal to adjust Moran's I for population density. *Statistics in Medicine*, *18*(16), 2147–2162. doi:10.1002/(SICI)1097-0258(19990830)18:16<2147::AID-SIM179>3.0.CO;2-I PMID:10441770

Clifton, K. J., Burnier, C. V., & Akar, G. (2009). Severity of injury resulting from pedestrian–vehicle crashes: What can we learn from examining the built environment? *Transportation Research Part D, Transport and Environment*, *14*(6), 425–436. doi:10.1016/j.trd.2009.01.001

Development Plan for Chennai Metropolitan Area. (2006). CMDA.

Cottrill, C. D., & Thakuriah, P. (2010). Evaluating pedestrian crashes in areas with high low-income or minority populations. *Accident; Analysis and Prevention*, *42*(6), 1718–1728. doi:10.1016/j.aap.2010.04.012 PMID:20728622

de Andrade, L., Vissoci, J. R. N., Rodrigues, C. G., Finato, K., Carvalho, E., Pietrobon, R., & de Barros Carvalho, M. D. et al. (2014). Brazilian road traffic fatalities: A spatial and environmental analysis. *PLoS ONE*, *9*(1), e87244. doi:10.1371/journal.pone.0087244 PMID:24498051

Ewing, R., Hamidi, S., & Grace, J. B. (2014). Urban sprawl as a risk factor in motor vehicle crashes. *Urban Studies*.

Getis, A., & Ord, J. K. (1992). The analysis of spatial association by use of distance statistics. *Geographical Analysis*, *24*(3), 189–206. doi:10.1111/j.1538-4632.1992.tb00261.x

Global Status Report on Road Safety: Time for Action. (2009). World Health Organization, Geneva, Switzerland.

Grimm, M., & Treibech, C. (2012). Determinants of road traffic crash fatalities across Indian States. *Institute of Social Studies Working Paper Series/General Series*, Working Paper 531. The Hague, Netherlands.

Hanson, C. S., Noland, R. B., & Brown, C. (2013). The severity of pedestrian crashes: An analysis using Google Street View imagery. *Journal of Transport Geography*, *33*, 42–53. doi:10.1016/j.jtrangeo.2013.09.002

Harvey, C., & Aultman-Hall, L. (2015). Urban Streetscape Design and Crash Severity. *Proceedings of the Transportation Research Board 94th Annual Meeting* (No. 15-2942).

Hsiao, M., Malhotra, A., Thakur, J. S., Sheth, J. K., Nathens, A. B., Dhingra, N., & Jha, P.Million Death Study Collaborators. (2013). Road traffic injury mortality and its mechanisms in India: Nationally representative mortality survey of 1.1 million homes. *BMJ Open*, *8*, 1–9. PMID:23959748

Road Accidents in India. (2010). *Ministry of Roads Transport and Highways (MoRTH)*. Retrieved from http://morth.nic.in/writereaddata/mainlinkFile/File761.pdf

Mohan, D., Tsimhoni, O., Sivak, M., & Flannagan, M. J. (2009). Road safety in India: challenges and opportunities (Report UMTRI-2009-1). University of Michigan, Transportation Research Institute.

Muthukannan, M., & Thirumurthy, A. M. (2008). Modeling for optimization of urban transit system utility: A case study. *Journal of Engineering and Applied Sciences (Asian Research Publishing Network)*, *3*, 71–74.

Noland, R. B., & Quddus, M. A. (2004). A spatially disaggregate analysis of road casualties in England. *Accident; Analysis and Prevention*, *36*(6), 973–984. doi:10.1016/j.aap.2003.11.001 PMID:15350875

Patel, V., Chatterji, S., Chisholm, D., Ebrahim, S., Gopalakrishna, G., Mathers, C., & Reddy, K. S. et al. (2011). Chronic diseases and injuries in India. *Lancet*, *377*(9763), 413–428. doi:10.1016/S0140-6736(10)61188-9 PMID:21227486

Raban, M. Z., Dandona, L., & Dandona, R. (2014). The quality of police data on RTC fatalities in India. *Injury Prevention*, *20*(5), 293–301. doi:10.1136/injuryprev-2013-041011 PMID:24737796

Sikdar, P. K., & Bhavsar, J. N. (2009). *Road Safety Scenario in India and Proposed Action Plan. Transport and Communications Bulletin for Asia, 79*. Road Safety.

Spoerri, A., Egger, M., & von Elm, E. (2011). Mortality from road traffic accidents in Switzerland: Longitudinal and spatial analyses. *Accident; Analysis and Prevention*, *43*(1), 40–48. doi:10.1016/j.aap.2010.06.009 PMID:21094295

Tamil Nadu State Transport Authority. (2010). Retrieved from http://www.tn.gov.in/sta/ra2.pdf

Tiwari, G. (2011). Key Mobility Challenges in Indian Cities. International Transport Forum Discussion Papers, Discussion Paper No. 2011-18.

Tiwari, G., and D. Jain. (2012). Accessibility and safety indicators for all road users: case study Delhi BRT. *Journal of Transport Geography*, *22*, 87-95.

Zeileis, A., Kleiber, C., & Jackman, S. (2007). Regression models for count data in R. Research report series. *Vienna Univ. of Economics and Business Administration*.

Chapter 3
Planning Transit System for Indian Cities:
Opportunities and Challenges

Arnab Jana
Indian Institute of Technology Bombay, India

Ronita Bardhan
Indian Institute of Technology Bombay, India

ABSTRACT

Indian cities are currently in a phase of transition. Continuous urbanization and seamless connectivity is the paradigm. Proliferating bourgeois class is extending the demand for private automobiles. With limited opportunity to increment land use allocated to transportation and rapid shift towards automobile ownership, importance of transit system is being sensed. City managers believe that public transit could be an alternative in providing solution to ever increasing problem of traffic congestion, parking demand, accidents and fatalities, and global environmental adversities. This chapter examines the critical planning issues that need to be addressed. It highlights the opportunities and challenges these cities are poised towards transit system planning. The experiences from cities worldwide that have adopted transit systems to create compact city forms fostering mixed land use development are exemplified here. A '3P' developmental framework of 'provide', 'promote' and 'progress' has been proposed to harness the opportunity.

INTRODUCTION

Importance of transportation and its allied infrastructure to economic growth and productivity is undeniable. Competiveness and vibrancy of any urban areas is dependent on the ease of connectivity and seamless mobility of desired resources and manpower. As the economic activity increased, per capita income levels rose significantly. With more employment opportunity, cities attracted and retained talents. However, majority of the Indian cities could not cope up with the increased demand for infrastructure. Similar has been the case with public transportation service. It has been reported in 2005 that public

Planning Transit System for Indian Cities

bus service for intra city transportation was available in 17 cities and rail transit existed in only four cities, out of 35 million plus cities in India (Singh, 2005). As an alternative, the nouveau riche opted for personal mobility. If compared globally India still lags behind in ownership pattern with respect to developed nations (see Table 1). With increasing economic affluence, ownership pattern is evident to increase. Currently the roads in CBDs in different cities across the country have exceeded the capacity, causing congestion, delays, accidents and pollution. Unreliability and deteriorated service delivery of the public transport system further added on to the agony of the working middle class, who as well opted for personalized vehicle as an alternative. There is a dire need for planning an integrated transit system to cater to the growing demand.

Under this purview, this chapter examines the planning challenges that need address at this wake of change the Indian cities are currently witnessing. It highlights the opportunities and challenges of transit planning. The experiences from cities worldwide have been assimilated to develop a developmental framework that might aid in policy decisions and planning transit systems for Indian cities, fostering efficient public transportation.

Subsequently the chapter will discuss:

- Growth of Indian cities and planning challenges
- Adaptive transit planning
- Policy Initiative in India
- Harnessing the opportunities

Table 1. Vehicular penetration in select developed & developing countries

	Country	GNI per Capita, (current US$)	Passenger Cars (per 1,000 people)*	Motor Vehicles (per 1,000 people)†	Road Density (km of road per 100 sq. km of land area)‡
Developed Nation	USA	48040	440	26	68
	UK	41130	456	21	172
	Germany	42550	510	46	180
	Japan	37610	452	28	88
Developing Nations	Brazil	8140	179	68	19
	India	1170	11	84	136
	China	3610	34	72	40
	South Africa	5630	110	7	NA
Other Asian Nations	Sri Lanka	1970	19	115	NA
	Philippines	2480	9	35	NA
	Malaysia	7590	308	325	41
	Korea	21090	265	37	105

* Passenger cars refer to road motor vehicles, other than two-wheelers, intended for the carriage of passengers and designed to seat no more than nine people (including the driver).

† Motor vehicles include cars, buses, and freight vehicles but do not include two-wheelers. Population refers to midyear population in the year for which data are available.

‡ Road density is the ratio of the length of the country's total road network to the country's land area. The road network includes all roads in the country: motorways, highways, main or national roads, secondary or regional roads, and other urban and rural roads.

Source: World Bank; http://data.worldbank.org

NA: Data not available

BACKGROUND

Post-independence, growth of Indian cities can be phased into three phases:

1. **1947 to 1981:** The formative stage of cities and vis-a-vis building of public transportation system.
2. **1981 to 1990:** The development of the first metro railway being built in Kolkata.
3. **Post 1990:** The phase of rapid growth and impact of globalization.

From the agrarian based society, India passed through phases of industrialization and currently service sectors are driving the high tides. As the need for travel is an induced demand, generated as an outcome of purposes, perhaps internalizing the patterns of purpose becomes prudent. With further upsurge of service sector, majority of cities would generate more employment and therefore need for mobility and demand for services. Transportation sector contributes about 6.4% of nation's Gross Domestic Product (see Table 2).

Indian Cities are currently in the phase of transition. As per Census 2011, million plus cities increased to 53 in 2011 (see Table 3). With the current rate of urbanization, there will be further upsurge in the demand for infrastructure and mobility, thereby prompting to plan for effective and efficient provision of transportation services. Analyzing the current situation, it is observed that non-motorized transport still occupies the larger share of the pie; however, with the current state of public transportation and increasing risk to pedestrians and bicyclist, the share of personalized vehicle is expected to grow phenomenally. The current modal split of Greater Mumbai is ("Development Plan for Greater Mumbai 2014-2034: Preparatory Studies," 2013):

- 51% by walking
- 25% by train
- 12% by bus
- 5% by taxi/auto rickshaws
- 2% by private vehicle and rest uses other modes

This majorly indicates that current travel behavior is oriented towards public transportation system.

Table 2. Share of different modes of transport in GDP (TRW, 2013)

Sector	2003-04	2004-05	2005-06	2006-07	2007-08	2008-09	2009-10	2010-11	2011-12
As percentage of GDP (at factor cost and constant prices)									
Railways	1.0	1.0	1.0	1.0	1.0	1.0	1.0	1.0	1.0
Road Transport	4.6	4.8	4.8	4.8	4.7	4.8	4.7	4.6	4.8
Water Transport	0.2	0.2	0.2	0.2	0.2	0.2	0.2	0.2	0.2
Air Transport	0.2	0.2	0.2	0.2	0.2	0.2	0.2	0.3	0.3
Services*	0.5	0.5	0.5	0.5	0.5	0.4	0.4	0.4	0.4
Transport Sector	6.5	6.7	6.7	6.7	6.6	6.6	6.5	6.5	6.7

Note: * Unadjusted Financial Intermediation Services Indirectly Measured (FISIM).

Planning Transit System for Indian Cities

Table 3. Urbanization pattern of Indian cities (Census of India 2011)

Class	Population Size	No. of UAs/Towns 2001	No. of UAs/Towns 2011
Class I	100,000 and above	393	465
	Out of above, Million plus	35	53
All classes		5161	7935
Statutory Towns		3799	4041
Census Towns		1362	3894

With the aspiration to remain competitive and foster economic growth megapolis such as Mumbai, Delhi and Kolkata UA[1] tend to promise policies and infrastructure to attract employment and therefore talents. It has been felt by city managers that linkages and connectivity alone cannot drive the image of the city rather ease of accessibility and timeliness holds the key.

Behavioral studies on MIG's consumerism depict that willingness of the middle class consumer to pay a little extra might add on substantially (Murphy, Shleifer, & Vishny, 1989). Kharas (2010) argued that "several Asian countries, in particular China and India, have reached a tipping point where large numbers of people will enter the middle class and drive consumption."

Megapolis like Mumbai, Delhi, Kolkata, Chennai, or Bangalore face considerable challenges in integration of transit system under the purview of existing land use patterns. Unavailability of land and limited scope of acquisitions are major hindrances. While under the current stress of traffic congestion and constraints of limited carrying capacities of networks, the demand for timely transport system is ever increasing. In the face of rapid growth of population and urbanization, low incomes, and extreme inequality, the supply of transport infrastructure and services are lagging behind, more so as public sector financing for transport sector remains inadequate (Pucher, Korattyswaropam, Mittal, & Ittyerah, 2005).

Growth of economic activity and increase in average income led to phenomenal growth in vehicular traffic in last three decades, from about 0.3 million on 31st March, 1951 to about 159.5 million as on 31st March 2012[2] (refer to Figure 1). On the overall context, share of buses have reduced to 1% (2012) from 11.1% (1951). Among the million plus cities, in 2011 Delhi had the highest registered motor vehicles, while cites like Pune, Hyderabad and Chennai witnessed a CAGR of more than 10%. It can be argued that the bicycle users are shifting towards two wheelers which have an annual growth rate of 13% reaching to a statistic of 115.42 million in 2012.

To cope up with the increased demand of mobility and with the objective to improve the bottlenecks, state and national governments envisaged several strategies ranging from conceiving and constructing new infrastructure to management of existing ones. In last few decades several projects ranging from construction of new roads, flyovers and mass rapid transit systems (MRTS) in various cities of India, has been implemented. Most of these initiatives are piecemeal solutions without defining the holistic goal of a given urban area. Often these projects were conceptualized based on emerging issues and rarely conformed to the existing master plan of the region. This lack of integration often leads to speculative land market and attracted urban sprawl. In many cases the opportunity to develop these mega projects as 'iconic development, while strengthening the prima fascia was lost. For example, in case of Kolkata subway system, at the conception phase there was no vision to develop the subway stations as potential commercial and retail hub (especially in areas where authority had access to land resources). Positive

Figure 1. Number of registered vehicles in India (in million) (TRW, 2013)
Notes:*: Two-wheelers include auto-rickshaws for the years ending 31st March 1959, 1960, 1962, 1963, 1964, 1965, 1967, 1968 and 1969. For the remaining years, auto-rickshaws are included in Others;**: Others include tractors, trailers, three wheelers (passenger vehicles)/LMV and other miscellaneous vehicles which are not classified separately; #: Includes Omni buses since 2001; Totals may not tally due to rounding off of data.

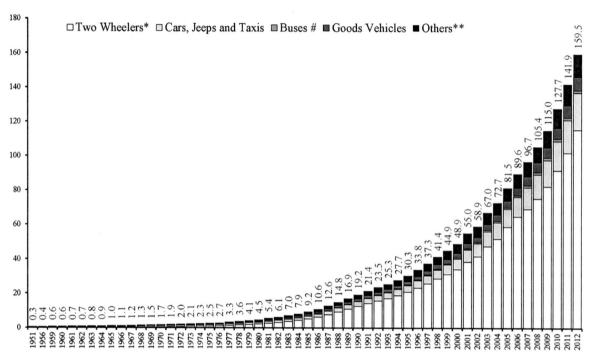

impacts of simultaneous developments might have significant impact on ridership, average office rents, average building densities, and economic growth rate (Cervero, 1994).

Not only was the lack of vision but disconnect between several stake holders has spoiled the opportunity. Transit systems can catalyze development only if appropriate integration of several other dimensions is looked into simultaneously. As illustrated in Table 3, India currently has tremendous opportunity to develop the emerging cities into transit oriented compact development, serviced by Internet and Communication Technology (ICT). The upcoming projects should therefore leverage on the lessons learnt from the 'lost opportunity' and build on it.

From planning perspective, there is an evident disintegration between land use pattern and transportation provision. Under the dynamism of rapid rate of urbanization, the cities have expanded and several planned and unplanned satellite towns have emerged. Considering the case of Mumbai, the scarcity of housing has led to the development of satellite towns of Thane, Kalyan – Dombivili, Navi Mumbai and all along the road and railway corridor. Similar is the case of NCR and Kolkata. Additionally geographical constraints often determined the shape and form of the city. In Mumbai, as the residential locations are being pushed northwards, the demand of trips to the southern CBD necessitates the working class to undertake trips catered by sub urban railway and the public bus service (BEST[3]). There are reportedly 7 million railway and 5.5 million bus trips per day. The existing infrastructure fails to satisfy the demand seamlessly and lead to bottleneck formation. As governmental initiative, monorail and metro system were conceptualized to satisfy the growing demand.

DYNAMICS OF TRANSIT PLANNING IN INDIA

Fouracre, Dunkerley, and Gardner (2003) argued "Selection of the most appropriate mass transit mode can be difficult, and there are many pressures on civic leaders to favor a particular system." The randomness of the choice of the MRTS has been criticized by several researchers, especially considering the outcome verses the cost of developing and maintaining such infrastructure (Gomez-ibanez, 1985; Pickrell, 1992). While Bus rapid transit systems (BRTS) remains cost effective and demands lesser investment and can be easily incorporated within the existing infrastructure with/without dedicated lanes than other forms of MRTS. It is being observed that there is a bias towards selection of BRTS in several metropolitan as well as Class I cities of India.

Transit systems affect stake holders like, users, transport operators, non-users, and shape the urban development and form of the city. Therefore, planning, constructing and operating transit systems have a multifaceted impact over the urban area. At the scale of the users, it has been found that users usually gain time as well as convenience, once the multi-modality of the system is sorted out and appropriate feeder networks align with the demand nodes. Reduction of fuels consumption, improvement on the environmental quality and better experience of travelling itself adds on to the enhancement of the quality of urban life of the residents. From the perspective of the city managers, authorities and operators, transit systems often reduces the stress of congestion and eases mobility. In developing countries context, the fares are heavily subsidized with the aim to reduce social seclusion and improve ridership.

In Indian cities, in spite of growing economic affluence, there is a considerable proportion of disadvantaged population living in city core, in shanties as well as in suburban and peri-urban areas. Having no access to own vehicle, dependency on the public transit as main mode of transportation still persists. Although these transit dependents are the most steadiest customers, transit policies fails to focus on them, hence, proving to be more expensive and marginally effective (Garrett & Taylor, 1999). The relevance of transit systems with the rate of labor market participation is evident (Sanchez, 1999). Garrett & Taylor, further argued that poor or mediocre transit systems in areas which are public transport dependents often leading to environmentally costly modal shift. Krumholz (1982) defined equity in planning as "to provide a wider range of choices for … residents who have few, if any, choices."

On this note, it becomes prudent to enumerate the different typologies of MRT systems:

- **BRTS:** Usually include dedicated lanes with operational features to increase mobility and passenger capacity. The systems are laid with well-designed bus stops, organized operations, and efficient ticketing methods.
- **Light Tram Services:** Electrically powered cars, on tracks which are partially or completely shared with other traffic.
- **Light Rail Transit (LRT):** A fully segregated and often grade-separated ROW, with advanced control and signaling systems.
- **Metros:** Fully segregated, and grade-separated, track (track may be elevated or underground); the metros are equipped with very advanced control systems allowing high-frequency operations. The trains are made up of multiple units of high-capacity cars.
- **Suburban Rail:** Part of a larger rail network, often at grade but separated from road traffic operate within the context of the wider network demands, and are characterized by higher headways and longer station spacing (as compared with both metros and LRT).

LEARNING FROM PAST

MRTS is accepted as a means of providing public transportation, while minimizing the effects of the travel delay, congestion and accidents. Various countries have inducted transit system as main mode of public transport system.

Rail Based Transit System: Metro, LRTS and Suburban Railway

The major drawback of the rail based system is their high operating and maintenance cost. Winston and Maheshri (2007) reported that "In 1980, two million Americans got to work by rail transit. ..., in spite of an increase in urban jobs and transit coverage, fewer than one million U.S. workers commute by rail, causing its share of work trips to drop from 5 percent to 1 percent." Therefore the revenue generated is insufficient to cater to the operating and capital cost. On the other hand Tokyo and Hong Kong exemplifies the governmental initiatives to privatize rail transit systems. Furthermore these transit providers in their continuous efforts to improve ridership reduced labor and capital costs while improving the railway coaches and introducing user friendly ticketing and payment options using ICT. Nelson, Baglino, Harrington, Safirova, and Lipman (2007) argued that benefits from congestion reduction in Washington DC exceeds the subsidies and the benefits of the transit system as a whole comprising of bus and rail exceeds transit subsidies. Further, Litman (2007) argued that rail based transit is an effective investment if benefits and impacts are considered comprehensively. On this note the urban development opportunity at the station areas should be harnessed and assessed, pedestrian malls as part of downtown redevelopment has been undertaken in several North American cities (Table 4) (Cervero, 1984).

On the overall aspect, it can be ascertained that transit does improve the property value (Bowes & Ihlanfeldt, 2001) and vibrancy of the neighborhood that houses the stations, and it also affects the travel behavior of "few" people, thereby reducing congestion and have a positive impact on the environment (Baum-Snow & Kahn, 2000).

One of the major criteria for undertaking a mega project such as rail based transit is often noted to be 'image' of the city rather than what the city actually needs (Bollinger & Ihlanfeldt, 1997). Moreover the malice of actualizing these projects has several undesired consequences as well. Siemiatycki (2006) argued, "The development of a metro has been accompanied by land appropriations, slum clearances and a broad range of political and economic opportunism that threaten to exacerbate the rift between the wealthy and the poor."

Table 4. Benefits of transit systems

Benefits of Transit Systems	Effects
Reduce travel time	Improve transportation option and choice, especially for non-drivers
Grade-separated transit reduces delays on parallel roadways	Reduction in congestion; Improvement of road safety
Rail transit can stimulate transit oriented development (TODs)—compact, mixed-use urban development	Efficient feeder network might further improve accessibility; Reduction in parking demand; reduction in consumer transportation cost; High density development & reduction of sprawl;
Improvement of environmental quality	Reduced emission;
Might have positive impact on public health	Promoting walking and cycling

Kolkata Metro[4]

Kolkata subway system is the first metro system in India. Currently it transports about half a million passengers a day[5]. The city of Kolkata saw the first MRTS in form of metro in 1984. The impact of the subway was evident while reducing congestion and easing traffic. However, being the first of its kind in India, the project faced several barriers such as non-availability of sufficient funds, shifting of underground utilities, court injunctions, irregular supply of construction materials and components. The project seen various phases of operational commencement and by 1995 the project started operation between 'Belgachia' (northern parts of Kolkata) to 'Tollygunj' (South Kolkata) covering 11 stations and 9.79 kilometers.

The experience of existing metro corridor and the success of the Delhi metro network contributed to the sanctioning of the second phase of the Kolkata metro, namely, the East – West corridor connecting Howrah Railway Station (Western part of KMA[6]) and 'Bidhannagar' (eastern part of KMA) via the CBD in 2008. This project aimed to connect 'Howrah' and 'Sealdah' Railway Stations, two of the busiest railway stations in the world and which would pass through under River Hooghly.

However, several barriers lead to delay and imposed a cumulative loss, increasing cost of the project. It was reported that 1.5 km of road blockage with a diversion of 0.9-1.09 km incurred a substantial loss in terms of extra fuel, subsidy, lost man-hour, pollution etc. (Das, 2014).

Bangalore Metro[7]

The city of Bangalore witnessed a massive population growth between 2001 and 2010. The city saw a huge immigration due to the job opportunities in service and allied sectors leading to excessive increase in automobile usage. It is reported that around 70% of the total vehicular split of Bangalore is two-wheelers followed by cars. However, considering modal split by trips, 42% of the trips are made by public buses, followed by 38% by two wheelers. The Bangalore metropolitan transport corporation has applied ICT based framework[8] for operation of public bus services, owning 6,472 buses that ply on 2,398 routes carrying 4.9 million passengers daily.

To augment public transport, a Special Purpose Vehicle named Bangalore Metro Rail Corporation Limited (BMRCL), a joint venture of Government of India and Government of Karnataka, was created to entrust the responsibility of implementation of Bangalore Metro Rail Project. Bangalore metro (also known as "Namma Metro") have two lines: the east-west (18.1 km) and north-south (24.2 km) corridor, out of which both are operational partially. The commercial operations of the lines have started recently. Operation between 'MG Road' to 'Baiyappanahalli station' commenced on 2011 and operation between 'Peenya Industry' to 'Mantri Square Sampige Road Station' started from 2014. With the commencement of the MRTS, it became necessary to integrate the service with the existing public transport system. Bangalore has an extensive public bus services, integration might aid in utilization of the existing infrastructure, together with development of new services to minimize travel time and congestion while improving mobility and accessibility. The policy to augment the transit service with feeder bus service might not only augment ridership but also might lead to lower use of private vehicle while improving environmental quality as well.

Delhi Metro[9]

Delhi witnessed a considerable construction of road infrastructure in last three decades; the total road length (km. lane) was 14,316 km in 1981 which increased to 28,508 km in 2001 and 31,373 km in 2009. However, the number of vehicles increased from 0.56 million in 1981 to 3.45 million in 2001 and 6.45 million in 2010.

Considering the necessity to decongest the NCR, metro railway started operational in 2002 between 'Shahdara' and 'Tis Hazari'. After a decade of its operation and further augmentation, currently there are 193 operational kilometers with 140 stations along with six stations of the Airport Express Link. Delhi Metro Rail Corporation Limited (DMRC) network currently connects NOIDA and Ghaziabad in Uttar Pradesh and Gurgaon in Haryana. According to a study conducted by the Central Road Research Institute (CRRI), in the year 2011, Delhi Metro has helped in removing about 1.17 lakh vehicles from the streets of Delhi. It is reported that during the financial year 2012-13, there was a total ridership of 702.9 million, which is an increase of 15.68% over previous financial year (2011-12), of 607.6 million. The daily peak was reported to be 2.3 million (DMRC, 2013).

Monorail[10]

One of the early proponents of monorail is Japan and its operation dates back to 1964, when the country introduced straddle type monorail system in Tokyo. Since then more lines have been introduced in Kitakyushu, Osaka, Tama, and Okinawa. Advantages of monorail over other forms of MRTS are that a) occupies comparatively lesser space on the ground, b) can negotiate sharp turns very easily, c) can climb and descend steep gradient with ease, d) improve environment, (e) shorter construction period, and (f) lower costs of construction (Sugita et al., 2005).

In 2014 Mumbai adopted monorail to decongest the growing stress of congestion from the Mumbai suburban to the CBD. It has conceptualized that the monorail will connect 20 km[11] stretch from 'Chembur' via 'Wadala' to 'Sant Gadge Maharaj Chowk'. The first phase covering a distance of eight kilometers is currently operational. It was believed that monorail could efficiently connect unattended demand nodes with the existing suburban railway, while acting as a feeder network to the mass transit systems.

Bus Rapid Transit System (BRTS)

Levinson et al. (2003) defined BRT as a "fully integrated system of facilities, services, and amenities that are designed to improve the speed, reliability, and identity of bus transit. In many respects, it is rubber-tired LRT, but with greater operating flexibility and potentially lower capital and operating costs. Often, a relatively small investment in dedicated guide ways can provide regional rapid transit."

There are three kinds of BRTS:

- **Open System:** Buses can enter or leave the BRTS unrestricted, therefore there is a little need for the feeder system,
- **Closed System:** The BRT fleets cannot leave the system, and
- **Hybrid System:** A combination of open and closed systems. In case of India, all three of the above mentioned are operational.

Planning Transit System for Indian Cities

There are sound economic and environmental factors which make it imperative to support the BRTS to meet the rapidly growing demand. On an average, energy consumption per passenger km by bus is the least and that by car highest amongst road based modes of passenger transport. Bus transport makes the most optimum use of the available road space and fossil fuel by transporting the maximum number of people per unit of road space and passenger km/litre. On an average, a car consumes nearly 6 times more energy than an average bus, while two wheelers consume about 2.5 times and three-wheelers 4.7 times more energy in terms of per passenger km.

It is reported that in most of the Chinese cities the BRT system is successful. Analyses of the different BRT systems are elaborated in Table 2. Guangzhou BRT has an average ridership of 805,000 passengers per day; however, in general other Chinese cities have daily passenger volumes between 200,000 and 350,000 per day. In Beijing as metro is the main mode of public transportation, BRT plays secondary role catering to 1.4% (Zhang, Liu, & Wang, 2013). Dedicated lanes as well as elevated road ways often tend to increase the speed of BRT, in Xiamen city the average speed is 42 km/hr. Moreover the operational windows of the BRT system in majority of Chinese cities vary between 14-17h per day. It is therefore noteworthy that BRT systems play a pivotal role of public transportation system especially in high density cities of developing nations. Additionally, ridership have a strong correlation with the service frequency, inter-modal connectivity, population density and employment density (Cervero, Murakami, & Miller, 2010).

It is argued that for universal acceptability i.e. provisioning for the disabled, elderly mobility and ease of ingress and egress, BRTS should employ specification such as:

- Low Floor buses,
- Fare Collection,
- Smart card or cash,
- Intelligent transportation System,
- Signal priority at intersection,
- Real-time vehicle information,
- Automated on board stop announcements,
- Electric route map in bus, and
- Real time monitoring system.

TransMilenio

Columbia, with a high urbanization rate of more than 75% replaced their failed attempt in rail-based transit networks with a co-financed BRTS. The aim was to develop a system that could cater to the yearly 100 per cent of city's public transport demand. In 2000, a special purpose vehicle (SPV) named *TransMilenio* (Bogota's BRTS) started their operations in Bogota with a 3:1 public-private partnership. This was coupled with a program of access to transit through pedestrian ways and walk able networks which were integrated into the system by free cycle parking at stations. Restrictions were also imposed on the private car usage by the "*pica y placa*" system which banned car use in the city center at the peak hours on a number plate basis. *TransMilenio* was a successful venture and saw a magnificent shift in transit ridership. This lead the national government to re-negotiate the rail based system in Cali to a BRTS system. The initiative of BRTS did not completely replace the traditional bus operations but existed as a parallel service, augmenting the market.

However, the system could not be fully realized due to major challenges like:

- Continued excessive capacity in the traditional bus market led to a predatory behavior
- Competition between the traditional market and the BRTS due to lack of proper integration. BRTS buses and traditional buses continued to ply on same routes leading to more competition.
- Reluctance of the traditional buses to be integrated with BRTS.
- Continued rise in crowding on the BRT lines and there by fall in the satisfaction levels of the user - by the first three years of operation of *Transmilenio*, it catered to 50% more demand in compared to traditional bus operators. However the overcrowding in the peak hours lead to drop in the survey ratings by passengers in compared to the traditional counterparts.

Despite these shortcomings the government continued to lay stress on the BRTS and started to see it as the model for development of a disciplined and effective public transport system in large and medium sized cities. The Government of Colombia has now adopted a comprehensive policy platform and introduced BRTS in the highest demand corridors of nine largest cities. Twelve medium-sized cities (between 250,000 and 600,000 inhabitants), are also in the purview of adopting BRTS although the challenges in these cities are more complex with the existing multi-modal ridership.

In order to implement an efficient public transit system the government has adopted a policy platform to incorporate the challenges. Arrays of national and local policies under National Urban Transport program (NUTP) were established and are in their way for implementation. The aim of program was to provide a sustainable efficient and improved safety of public urban transport services. NUTP provisioned reliable transport accessibility for the poor with enhanced private sector involvement in service provision. Financial vehicles like fiscal transfers ("*vigencias futuras*") were undertaken to pay for the Government's share of project's costs in each participating city. Subsequently complimentary policy measures were created to support the context of BRTS: like reducing the oversupply of obsolete bus fleet, investing in the renewal of public space and bicycle paths to promote non-motorized transit, strengthening the institutional and regulatory capacity of the local transport authority, etc.

Delhi BRTS[12]

Delhi constructed High Capacity Bus System (HCBS). The system constituted exclusive bus lanes operating in the median with parallel bike lanes and pedestrian access. However, it has been critiqued that in the mixed lanes the speed slowed down during the peak hours. Moreover, creating an open system was reported to be an issue because it resulted in lane intrusions by existing buses and consequently, lowering the speed (13 km/hour) (Infrastructure Development Finance Company Limited, 2012). Additionally it has been appraised that as the operation was created for only a stretch of six km, it offered little advantage to riders. Another critical issue was the intermediate bus stops created at crossroads and junctions, which led to clogging of buses.

Ahmedabad BRTS

It was devised as a response to the city's traffic troubles after the population increased to six million. However, the city was served by public bus services operated by Ahmedabad Municipal Transport Services (AMTS). Old public buses were replaced by user friendly buses with smart cards and providing greater comfort and convenience. The image of the system was boosted by bus stops that provided seating and waiting space with off board ticketing facility and other features like air-conditioning in buses, dedicated fast lanes, public information system, Global Positioning System and centralized control room. To minimize the interference from the traffic, Ahmedabad BRTS deployed central dedicated (closed system) lanes, with bus stops about 400 meters beyond crossroads. Together, these measures led to increased bus frequencies and reduced the waiting period to two minutes during peak hours and 8-10 minutes during off-peak hours. As on 2014 the daily ridership is around 1.28 million a day. The success of Ahmedabad BRTS' has caught the attention of other Indian cities. Metros such as Chennai, Kolkata, Bangalore and other tier-1 cities such as Hyderabad, Pune, Indore, along with smaller towns such as Pimpri, Chinchwad and Hubli are either implementing or are looking to launch a similar transport system.

Identified Barriers in Implementation of Transit Systems

In spite of success of transit systems, it still has major challenges and barriers that make it a critical system to adapt. Overlapping jurisdiction and lack of consensus and coordination among various stakeholders with coinciding responsibility is a challenge. In India special purpose vehicles (SPVs) are often created but it lack technical capacity. The strong promotions of competing modes and opposition from other existing players reduce the acceptability of transits. Underestimating the implementation effort, discontinuities due to political cycles and lack of national policies supporting public transit development often hinders the implementation. Insufficient funding, rushed inauguration without realizing the environmental/land availability issues and expropriation often delays and hinders the success of public transit system in India.

INSTITUTIONAL INITIATIVES TOWARD MASS TRANSIT PLANNING IN INDIA

Importance to transit planning in India has been a recent phenomenon. To support the prospects of efficient transit planning government of India has initiated a number of supporting policies in the recent past. Although the shift of focus from road infrastructure augmentation to efficient mobility was first realized in the 7th five year plan through exclusive budgetary allocation to urban transport, concentrated effort to urban public transport is apprehended in the recent past. A number of enabling policies were framed in pursuit of reliable and efficient public transport. Since most of these policies fulfilled the larger agenda of sustainability, low carbon development countering climate change, most of the projects perceived within these policies were related to public transit planning and development (Doll, Dreyfus, Ahmad, & Balaban, 2013). These policies were often operationalized simultaneously to realize an urban transit project. Table 5 lists some of the key enabling policies.

Working Group on Urban Transport for the 11th Five-Year Plan (2007-2012)

The growing demand for passenger transit infrastructure was formally recognized by India in its 11th Five Year Plan. It essentially realized the need for a colossal investment to meet the demands. The plan objectively laid out an extensive list of investments needed in the different tiers of cities and sectors. It estimated a budget of 132,590 Cr (see Table 6) and about 435,000 Cr to close the gaps, over next 20

Table 5. Policy towards urban transportation planning

Name of the Policy	Year of Initiation/ Operation
Working Group on Urban Transport for the 11th Five-Year Plan	2007 - 2012
Jawaharlal Nehru National Urban Renewal Mission (JNNURM)	2005
National Urban Transport Policy	2006
National Sustainable Urban Transport Program: Sustainable Urban Transport Project (SUTP)- World Bank – UNDP – Global Environment Facility (GEF)	2009 - 2015
National Mission on Sustainable Habitat (NMSH) - under the ambit of the National Action Plan on Climate Change (NAPCC)	2009 - 2017

Table 6. Estimated breakup of investment as per 11th Five Year plan, Government of India

Items for Investment Allocation	Estimated Investment in '00,000,000
Capacity building and urban transport planning	350
0.1 – 0.5 million cities	7400
0.5 – 1.0 million cities	7800
1.0 – 4.0 million cities	26040
Above 4.0 million cities	21000
Mass Rail Transit for megacities	32000
Bus Transit systems/ modern buses for urban transport	38000
Total	1,32,590

Source: Lohia (2011)

years (Lohia, 2011). It was first time that a national plan pointed out the bias of transport plans favoring motorized vehicles. The plan suggested a specific allocation of 25% of road budget towards promotion, design, standard formulations and guidelines for non-motorized vehicles. The need and importance of MRTS was identified as an important specific action to be undertaken within the purview of the plan.

Jawaharlal Nehru National Urban Renewal Mission (JnNURM): 2005

In 2005, Indian Ministry of Urban Development and Ministry of Poverty Alleviation launched the Jawaharlal Nehru National Urban Renewal Mission (JnNURM) to support state and local government for investment in urban development. The total tenure of the mission was decided as seven years commencing with an estimated budgetary allocation of 1,20,536 Cr for investment in 63 million plus identified "mission cities" across India. The number of cities was later increased to 65. The objective of the Mission was to *create economically productive, efficient, equitable and responsive cities* (JNNURM, 2005). It was introduced with the sole purpose of encouraging urban reforms and fast tracking planning & development for identified "mission cities" with a sustained focus on efficiency in urban infrastructure and service delivery mechanisms, community participation and accountability of all urban local bodies and para state agencies. It offered a financial support for infrastructure development related projects under a proportioned cost sharing basis with the local and state governments. A structured governance model which included both central assistance and mandatory and optional reforms, was adopted to mobilize the funds.

Most of the state governments/ ULB's and para state agencies which had large projects stalled due to insufficient financial assistance were immediately revived through major revamping of the projects' objective in line with the mission goals. While infrastructure sectors like water, sanitation, solid waste management, governance got a large share from the mission, urban transport got special attention. Many mission cities which had stalled "road widening" projects were overnight reinvigorated as public transit planning projects like BRTS, LRTS, pedestrian safe transport planning and created special "urban transport funds". This mission provided such boost that by September, 2009 the MoUD had already released 1,041 crore as the first installment for procurement of 15,260 buses in its first 61 mission cities (MoUD, 2012).

National Urban Transport Policy

National Urban Transport Policy (NUTP) was launched by government of India in 2006, about the same time when JnNURM was initiated (NUTP, 2006). The aim of the policy was to *move people – not vehicles*. It was probably the first time in India when a policy laid the importance to people oriented approach in urban transport planning. The policy objectively tried to ensure safe, affordable, quick, comfortable, reliable and sustainable access for the growing urbanites. It was also first time that 'urban public transport' was emphasized through an institutional framework. It encouraged the establishing of effective regulatory, institutional and enforcement mechanisms for augmenting and providing reliable and efficient public transit in Indian cities. Modern technologies like intelligent transport systems, cleaner fuel, and green technologies were identified under the gamut of NUTP. Overall NUTP expected to improve urban mobility and consequently quality of life in Indian cities.

The timely coincidence of JnNURM and NUTP provided a major boost to urban transit planning in India. Neglected sectors like non-motorized vehicles, equitable allocation of road space, integrated land use and transport planning, strategies for parking, innovative financing mechanisms to raise resources got their due attention. JnNURM and NUTP was also tied up together to achieve a holistic approach. It was mandated that all urban transport projects receiving financial assistance from the JnNURM program were to conform to the rules of NUTP mission.

National Sustainable Urban Transport Program (SUTP)- World Bank – UNDP – Global Environment Facility (GEF)

This was the flagship project initiated by government of India in partnership with Global Environment Facility (GEF), The World Bank and United Nations Development Programme (UNDP), to internalize NUTP's objectives[13]. The program aimed at two strategic objectives of capacity building in the field of Urban Transport and creating best practices of 'green' and/or 'sustainable' transport planning through actual implementation of the projects in the selected cities. Realization of the aims was done through a three-component framework. While component-one dealt with capacity building through education and training at individual and institutional levels in the field of planning, implementing, financing, operating and managing sustainable urban transport systems.; Component-two and three emphasized on building demonstration projects and strengthening of project management capabilities. SUTP had 1400 Cr for achieving the components within a time frame of six years commencing in 2009.

National Mission on Sustainable Habitat (NMSH)

Conceived under the ambit of India's first National Action Plan on Climate Change (NAPCC), the National Mission on Sustainable Habitat (NMSH) 2010 had energy efficient transport as one of its prime agenda. Specifically it focused on improved urban planning through promotion of modal shift to public transits and low carbon transport options (Sharma & Tomar, 2010). Under NMSH, cities could seek assistance for long term urban transport plans that are re-oriented in light of climate change. Mainly projects that looked into parameters like walkability, high quality transit planning, last mile connectivity to transit, behavioral and physical modal shift to public transits were more favored.

Recent Initiatives Envisaged

Currently, Government of India has envisaged two new initiatives for augmenting mass rail transit in India (MoUD-GoI, 2014):

1. The High-Speed Rail (HSR) Corridor project
2. Smart Mobility under the aegis of Smart City Framework

The details of the two initiatives are outlined below.

High-Speed Rail (HSR) Corridor Project

In India, while introduction of HSR system has been matter of discussion since 1980s. It was first objectively spelt in the Vision 2020 plan of Indian Railways in 2009. With changing of the political scenario, the current government has embarked on an ambitious project of "10,000 kms long Diamond Quadrilateral Network of High Speed Rail". This would connect major metros and growth centers of the country. Although there is no existing standard definition of High-Speed rail globally, but as per International Union of Railways any system which has a speed exceeding 250 kmph is generally considered to be high speed. In India the HSR is envisioned to have a speed variation of 160-200 kmph connecting identified growth centres like Mumbai, Delhi, Chennai and Hyderabad along with many other Tier II economic hubs. A special purpose vehicle named High Speed Rail Corporation of India Limited (HSRC) has been set up in 2012 under Rail Vikas Nigam Limited. HSRC is entrusted to carry out for feasibility studies, detail project formulation to implantation of the project. Currently, feasibility studies of eight HSR projects namely Pune-Mumbai-Ahmedabad (650 kms), Delhi-Agra-Lucknow-Varanasi-Patna (991 kms), Howrah-Haldia (135 kms), Hyderabad-Dornakal-Vijaywada-Chennai (664 kms), Chennai-Bangalore-Coimbatore-Ernakulam-Thiruvananthapuram (850 kms), Delhi-Chandigarh-Amritsar (450 kms), and Delhi-Jaipur-Ajmer-Jodhpur (591 kms) are being undertaken. A provisionary budget of INR 100 crore has also been allocated for its implementation. It is believed that the High-Speed Rail project will significantly reduce travel time between metro centers. The HSR initiative is still in the planning phase and its impact is yet to be realized.

Smart City

In July 2014, the Central government rolled out the vision of developing 'one hundred Smart Cities' to harness the potential of emerging India. With ICT as its backbone the Smart cities are envisioned to provide the cities a competitive edge while remaining coherent in terms of environment sustainability thereby creating a better quality of life (MoUD, 2014). One of the essential part of Smart City is Smart Infrastructure which encompasses Urban Mobility as its central component. This will be based on a "three pronged approach" which will include:

- Public transport improvements by introducing Metro Rail, BRT, LRT, Monorail, Trams etc.
- Augmentation and improvement of existing carriageways
- Provisioning for promotion of walkability in urban centres

This framework is still in conceptualization phase and a clear definition of Smart mobility for Indian cities is yet to be defined.

ASSIMILATING THE CHALLENGES AND INITIATIVES

Current traffic and transportation in India is maligned with traffic accidents causing fatalities and injuries, increasing rate of emission of GHGs and degradation of level of service. These adversely affect the urban quality of life. With the current growth rate of private vehicles in major Indian cities, transits could aptly reduce the growing stress. Several studies revealed that the existing modal share in urban areas of India is in favor of non-motorized and public transport, for example (see Tiwari, 2011). Taking the example of Mumbai, 11 million trips are generated per day, of which 88% of the trips are catered by sub urban railway and public bus service provided by BEST (MCGM, 2005). The share of passenger km travelled is 22.15 kilometers for railway and 4.67 kilometers for bus.

As a component of a public transportation system, BRT may play different roles in different cities. In medium cities, BRT is a very important travel mode that bears a large proportion of passengers. In some large cities, BRT can be a substitute for a metro system, since its capacity can be made to be as high as that of a metro system. In megacities, BRT can be a major traffic corridor in suburban areas, feeding passengers into the metro network. Under the given condition of sub optimal usage of public infrastructure, transit systems can play a vital role to reduce social seclusion.

Physical Challenges in Public Transit

- Despite low car ownership (150/1000 person) the flow of traffic is significantly slowed down at peak hours. In Kolkata, for example, the average speed during peak hours in CBD area (Padam & Singh, 2001) goes down as low as 7 km per hour.
- High competition of transit systems from very high motorization rate has led to decline in the share of public transport over the past few decades. Currently the ownership rate of motorbikes stands at a high rate of 300-500/1000p. The cumulative percentage of two-wheelers and cars together constitute more than 91% in Kanpur, 88% in Hyderabad, and 86% in Nagpur whereas buses constitute 0.5, 0.5, and 0.4 percent respectively.

- Reluctance of modal shift behavior from non-motorized and para-transit is still persistent.
- Incommensurate spending on transit in comparison to economic growth is witnessed. Spending on transport is too often influenced by a notion of political prestige than by rational calculations of economic growth. Most Indian cities instead of making modest labor-intensive road improvements, extending city streets, and promoting low-cost bus operations, spend too much on politically attractive costly facilities, such as elevated roadways and MRTS.
- Low percentage of transport infrastructure space: area occupied by roads and streets in Class – I cities in India is only 16.1 per cent of the total developed area while the corresponding statistic for US is 28.19 per cent.
- The present suburban rail services in India are limited. Three metropolitan cities i.e., Mumbai, Kolkata, and Chennai are served by suburban rail systems carrying a major share of daily commuters.
- Traffic diet plans generally favor cars and lead to car obesity. There always exists a huge latent demand for cars as the demand is tied up with people's aspiration. Only 'planned structure for cars' – which in turn 'induces car'.
- As illustrated in preceding sections, policy should be people centric not vehicle centric. In 2006 NUTP spelt out that plans should be made for people not vehicles, whereas prior to this initiative, investment was more of building transport infrastructure that would be inevitably for the cars.
- Choice of buses or metro is only feasible for medium trip length (i.e. above 12km), for shorter trip inefficient methods of access to transit dissuades people to choose public transits like BRTS and LRTS. The average trip length in Indian cities is less than 6 km. Hence people might generally opt for para-transit, bicycle or walking under constrained conditions.
- Lack of policies facilitating pedestrianisation; however, UTIPEC, MoUD and IRC103 first drafted pedestrian guidelines in 2009.

Apart from the challenges there are also multiple hurdles in implementing the current policies. Some of the critical barriers in implementation are as follows:

- **Unavailability of Land:** Transit project especially rail and metro systems are land intensive infrastructure. They require vast parcels of uninterrupted land, which seldom exists. Although there exists mechanisms in government to acquire lands for public projects using compensatory instruments or incentive based land-pooling mechanisms, but these often do not provide sufficient avenues to Government to acquire sufficient land. Seldom these mechanisms provide market driven incentives for the private landowners. Acquisition of private lands is often interjected due to unsatisfactory compensatory packages or divergent self-interests of the local democratic-politicos.
- **Lack of Funding Resources:** Implementation of transit projects has specific funding requirements at various stages of development. Most of the time funding is available as per budgetary value of the project. However, improper project planning, unscheduled delays due to legal injunctions from various stakeholders, unavailability of adequately skilled labor often stretches the planned tenure of the sub-phases of these projects. Thereby overshooting budgetary allocations.
- **Unclear or Ambitious Policy Goals:** It has been often observed that the policies are driven by socio-political sentiments without undertaking a thorough feasibility of the overall framework. Although project specific feasibility studies are carried out at the projectisation stage, but conceptualization phase often remains ambitious.

Planning Transit System for Indian Cities

- **Prolonged Political Commitment:** Political commitment from the leadership play a critical role in success of mass transit policies. Instability of the government and unsure tenure of the leadership due to constant political turmoil often put ambitious or large infrastructure projects at risk of incompleteness. There have several instances where governments have failed to provide the needed political support for sustaining a conceptual policy initiative. Many projects of preceding governments are often discontinued by its successors.
- **Lack of Cooperation:** Among different departments of government and their mutual mistrust & disrespect create several deadlocks that hamper the implementation.
- **Heavy Top-Down Approach:** Mis-specification of the *real* stakeholders and inadequacy of involving them in policy implementation is a major drawback of centralized top-down approach.

STRATEGIZING THE OPPORTUNITIES: A 3P FRAMEWORK

To achieve an efficient transit planning in India, a systematic alternative developmental approach of 3P framework is conceptualized here. Spurring from the inherent concepts of sustainability, this alternative approach seeks to achieve a milieu for modal shift towards transits, energy efficient planning, reduce congestion and articulate incorporation of advanced technologies for improving brand value and usability of public transits in India. The primary objective of 3P is to integrate transit planning with the urban development vision of a city, thereby generating a sustainable path.

The 3P framework is a systematic cyclic process comprising of three main components (see Figure 2):

- Provide,
- Promote, and
- Progress.

Provide

The framework initiates with the 'provide' component which implies that in order to create transit systems that benefit the society, one must first provide the avenues. By 'providing' it is aimed to improve the efficiency of transport system. The avenues for provide are as follows:

Integrated Land Use Transport Plan

India is currently in transition with cities planning their future strategies for development. If land use and transport are planned as an integrated system then it would be easier to internalize the benefits of each other in a structured manner (Bardhan, 2013). Land use-transport integration calls for designing neighborhoods that literally follow public transport systems. This would not only ensure shorter trip lengths but improve the system efficiency through improvement in accessibility. It allows for dedicated public transport route ways for proposed large developments. Social exclusion which is a growing vice with rising disparity of incomes, can be reduced considerably by providing the ease of access to public transit. The major lacuna that leads to failure of public transit systems is the inability to provide reliable and cost effective solution for last mile connection. Integrated system will automatically cater to such gaps as neighborhoods will be lying within a perceivable distance from the transit stations.

Figure 2. The 3P framework

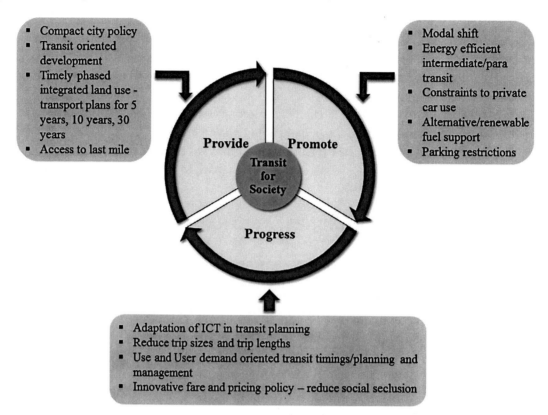

Some of the other benefits of an integrated system are the capability of incorporation of multiple stakeholders in planning, which is essential in generating sense of ownership; simultaneous demand and network management, ensuring safety and inducing behavioral change leading to modal shifts. As an externality, researchers also unanimously agree on the economic gains that an integrated land use and transport system brings.

Achievable concepts like transit oriented development (TOD), compact city policy, walkability come under the purview of integrated land use-transport system. While TOD allows for bringing the residents closer to the transit thereby improving accessibility, policies like compact city, walkability ensure energy efficiency, low carbon and reduction of trip lengths through provisioning of land-use mix and bringing the utilities of a city closer to the residents. Suzuki, Cervero, and Iuchi (2013) outlined the major pre-conditions required for successful integration of transit and land use. Apart from the existing threshold demand for transit systems, there must be sufficient government commitment in actualizing the integrated system. They emphasize that integration is only possible when public sector reforms and land re-development projects are coupled with transit planning.

Phased Transit Plans

Phasing of transit system planning should be such that it allows for latent demand based capacity augmentation, ease of interconnection to future feeder networks and ability to absorb advanced technology.

Planning Transit System for Indian Cities

Promote

The 'promote' instrument seeks to improve travel efficiency through

- Endorsement of public transit as a choice mode by prioritizing trips and improving transfers during peak hours
- Stimulating modal shifts towards non-motorized transport (NMT) by connecting transits through safe non-motorized pathways and proving NMT parking facilities at inter-changes
- Institutionalizing the intermediate public transits (IPT) or para transits – recognizing the importance of IPT's in context of India is important as they cater to large demand for shirt trips especially in tier 2 and tier 3 cities. Inclusivity and potential livelihood support for the socio-economically weaker segment makes IPT more important in developing situation like India. Currently the IPT has an ad hoc modus operandi. Outdated regulations and lack of any organized provision to this mode of transit is leading to the deterioration of the level of service of IPTs. Although it can be easily acknowledged that organized IPTs have the potential to improve mass transit efficiency significantly. Especially they play a pivotal role in connecting transit to destinations.
- Shifting to alternative renewable fuel
- Limiting car use through dynamic road usage pricing
- Enabling market forces to set and control parking price policies
- Creating mechanism for service level benchmarking and comparing the statistics across cities to create competition and branding – generating a brand value for public transit can provide a city its competitive edge as witnessed in the case of Bogota in Columbia and Ahmedabad in India.

Progress

This component provides the capability to the framework to incorporate advanced technology in transit planning through adoption of ICT or ITS based strategies for operation and pricing. Progress gives the avenue to reconstruct traditional planning in line with modern pathways. ICT in transit planning can provide an alternative route to solve micro-spatial to regional traffic and transport problems. It has the capability to eliminate certain travel demands completely through activity-place substitution (Timmermans & Zhang, 2009). As ICT has the potential to tap the 'city knowledge' and change the dimension of space, place, distance and time and probably transform them also (Houghton, Miller, & Foth, 2013). The impact of ICT on transit planning would be immense. Directly it can improve productivity through integrating fare collection, vehicle location routing and real time monitoring. Indirect impacts would include creating new economies of place, discreet choice of facility location and support for continuous integrated planning process (Höjer, 2000). This auto-regeneration of new dimensions to place will need a significant support from ICT in transit planning. This will become more exigent once the society starts to age. Although in India this crossover will not to happen soon but it is inevitable. Advantages of using transit fare smart cards to integrate activity and travel has been exemplified by Japan, Singapore and Hong Kong through PASMO or Suica, EZ-Link and Octopus respectively. In Tokyo, these cards act as electronic money which pays for the travel in any modes of public transit- rail based or bus. These also double as mobile wallets for convenience purchases by commuters like drinks, books and food, that they might require during the travel (Kusakabe, Iryo, & Asakura, 2010). Hence improve quality and convenience in urban living.

To realize the strength of the proposed 3P framework and make it implementable, incorporation within the master/ perspective plans of the cities can be an appropriate mechanism. This could be further strategized through detailed action plan with appropriate phasing for incorporation within existing urban governance system. Especially the 'provide' and the 'promote' phase enables opportunities for incorporation of multiple stake holders, cooperation among different organs of government, while the 'progress' phase would enable transition to smarter cities with higher transparency and with smoother functioning.

CONCLUDING REMARKS AND DISCUSSION

This chapter tried to identify the challenges and opportunities of public transit planning in India. It builds on the historical practices and policy initiatives of public transport planning in India and sketches the path for achieving cross-sector integration of transit systems with urban development. Multicity best practices are taken as anecdotes to showcase the multi-variant forms and utility of transit systems and planning. Emphasis has been given on integrated transit-land use plans as a promising strategy for economic profitability, social stability and environmental sustainability.

Infrastructure development Strategy in India has been sequential rather than simultaneous, primarily due to financial constraints and lack of political will. Under the current purview of rapid urbanization, growing economic affluence of MIG, and growth of service sector industries, need and importance of seamless mobility has emerged to be critical for sustaining growth rate and remaining competitive. Exiting public transportation system in majority of the megapolis of India failed to satisfy the growing demand for timely transportation services. In last few decades the sub urban areas of majority of these UAs have densified and resulted in increase in travel demand. Additionally the fast changing travel behavior and car ownership pattern is adding on to the stress on the infrastructure.

As a reactive measure, several kilometers of road length has been added, yet the registered vehicles per 100 kilometers of road length have increased significantly (from 328,000 in 2000 to 389,000 in 2010), similar is the observation in case of registered vehicle per 1000 population (see Figure 3).

At this juncture, it is prudent to relook the strategies and reframe based on the occupational pattern, demand and affordability of the citizens. This chapter elaborately discussed several issue affecting the accessibility. Several case studies have been discussed, elaborating on their success and failures.

Need for augmenting the public transport system in Indian cities has emerged critical. With limited opportunity to increment land use allocated to transportation, implementing MRTS projects seems to be an alternative. As a transit system may provide reliability, safety, convenience and comfort to the daily commuters it is believed that transits might be an appropriate alternative to ever increasing problem of traffic congestion, parking demand, accidents and fatalities, and global environmental adversities. However, green mobility, bicycling and walking needs to be promoted and adequate infrastructure is required to be created and augmented. Currently, the production of bicycles (see Figure 4) in India has fallen significantly post 2000, depicting the emergence of the MIG and consequent stress on road and allied infrastructure. A 3P framework has been forwarded here that enables in achieving the strategy.

Planning Transit System for Indian Cities

Figure 3. Growth of motor vehicles in India
Source: TRW (2013)

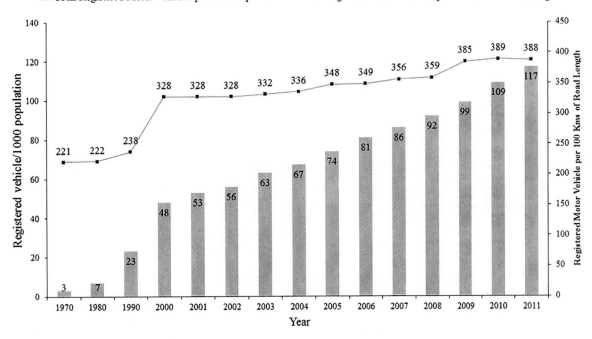

Figure 4. Production of bicycles in the country during 1950-51 to 2006-07
Source: 1) Economic Survey 2002-03 for the period 1950-51 to1980-81 and 2)CSO for the period 1990-91 and onward

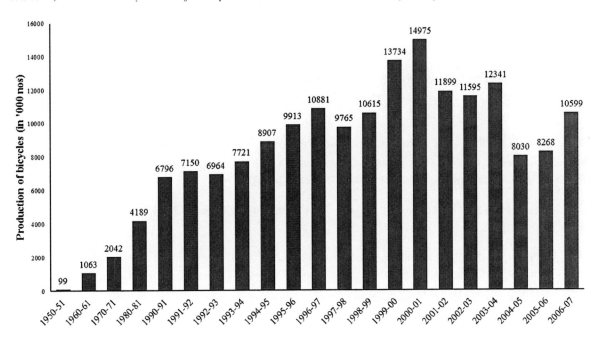

Although, no framework can be foolproof, the 3P framework provides ample scope to overcome the existing barriers of policy implementation in India. In the "provide" and "promote" phase the framework is capable of absorbing the views of the real stakeholders. Hence enable an inclusionary participatory planning approach. If implemented well, the 3P can bring about an integrated spatial development that not only reduces the current public transit challenges but can also provide meaningful outcomes through absorption of advanced technologies.

REFERENCES

Annual Report 2012-2013. (2013). *Delhi Metro Rail Corporation Ltd.*, New Delhi.

Bardhan, R. (2013). Simulation of Land Use Consequences of Urban Corridor in Kolkata: An Integrated Spatial and Expert System Model. *Environment and Urbanization Asia*, *4*(2), 267–286. doi:10.1177/0975425313510767

Baum-Snow, N., & Kahn, M. E. (2000). The effects of new public projects to expand urban rail transit. *Journal of Public Economics*, *77*(2), 241–263. doi:10.1016/S0047-2727(99)00085-7

Bollinger, C. R., & Ihlanfeldt, K. R. (1997). The Impact of Rapid Rail Transit on Economic Development: The Case of Atlanta's MARTA. *Journal of Urban Economics*, *42*(2), 179–204. doi:10.1006/juec.1996.2020

Bowes, D. R., & Ihlanfeldt, K. R. (2001). Identifying the Impacts of Rail Transit Stations on Residential Property Values. *Journal of Urban Economics*, *50*(1), 1–25. doi:10.1006/juec.2001.2214

Cervero, R. (1984). Journal Report: Light Rail Transit and Urban Development. *Journal of the American Planning Association*, *50*(2), 133–147. doi:10.1080/01944368408977170

Cervero, R. (1994). Rail Transit and Joint Development: Land Market Impacts in Washington, D.C. and Atlanta. *Journal of the American Planning Association*, *60*(1), 83–94. doi:10.1080/01944369408975554

Cervero, R., Murakami, J., & Miller, M. (2010). Direct Ridership Model of Bus Rapid Transit in Los Angeles County, California. *Transportation Research Record: Journal of the Transportation Research Board*, *2145*, 1–7. doi:10.3141/2145-01

Das, S. (2014). Hidden Cost in Public Infrastructure Project: A Case Study of Kolkata East–West Metro. In K. S. Sridhar & G. Wan (Eds.), *Urbanization in Asia: Governance, Infrastructure and the Environment* (pp. 149–164). Springer. doi:10.1007/978-81-322-1638-4_9

Development Plan for Greater Mumbai 2014-2034: Preparatory Studies. (2013). Retrieved from http://www.mcgm.gov.in/

Doll, C. N. H., Dreyfus, M., Ahmad, S., & Balaban, O. (2013). Institutional framework for urban development with co-benefits: The Indian experience. *Journal of Cleaner Production*, *58*(0), 121–129. doi:10.1016/j.jclepro.2013.07.029

Fouracre, P., Dunkerley, C., & Gardner, G. (2003). Mass rapid transit systems for cities in the developing world. *Transport Reviews*, *23*(3), 299–310. doi:10.1080/0144164032000083095

Garrett, M., & Taylor, B. (1999). Reconsidering Social Equity in Public Transit. *Berkeley Planning Journal, 13*(1).

Gomez-ibanez, J. A. (1985). A Dark Side to Light Rail? The Experience of Three New Transit Systems. *Journal of the American Planning Association, 51*(3), 337–351. doi:10.1080/01944368508976421

Höjer, M. (2000). What is the Point of IT? Backcasting Urban Transport and Land-use Futures.Royal Institute of Technology, Stockholm (Trita-IP. FR; 00-72)

Houghton, K., Miller, E., & Foth, M. (2013). Integrating ICT into the planning process: Impacts, opportunities and challenges. *Australian Planner, 51*(1), 24–33. doi:10.1080/07293682.2013.770771

Evolving Perspectives in the Development of Indian Infrastructure (Vol. 2). *Infrastructure Development Finance Company Limited*. (2012). Orient Blackswan Private Limited, Greater Noida.

Jawaharlal Nehru National Urban Renewal Mission. (2005). New Delhi: Overview.

Kharas, H. (2010). The emerging middle class in developing countries *Global Development Outlook* (Working Paper No. 285). OECD Development Centre.

Krumholz, N. (1982). A Retrospective View of Equity Planning Cleveland 1969–1979. *Journal of the American Planning Association, 48*(2), 163–174. doi:10.1080/01944368208976535

Kusakabe, T., Iryo, T., & Asakura, Y. (2010). Estimation method for railway passengers' train choice behavior with smart card transaction data. *Transportation, 37*(5), 731–749. doi:10.1007/s11116-010-9290-0

Levinson, H., Zimmerman, S., Clinger, J., Rutherford, S., Smith, R. L., Cracknell, J., & Soberman, R. (2003). Case Studies in Bus Rapid Transit Bus Rapid Transit (TCRP Report 90). Washington, D.C.

Litman, T. (2007). Evaluating rail transit benefits: A comment. *Transport Policy, 14*(1), 94–97. doi:10.1016/j.tranpol.2006.09.003

Lohia, S. K. (2011). Sustainable Urban Transport Sustainable Urban Transport; Initiatives by Govt. of India. *Paper presented at theThird Biennial Conference of the Indian Heritage Cities Network Karnataka*, India.

Mumbai City Development Plan 2005-2025. (2005). *MCGM*. Retrieved from http://www.mcgm.gov.in/irj/go/km/docs/documents/MCGM%20Department%20List/City%20Engineer/Deputy%20City%20Engineer%20(Planning%20and%20Design)/City%20Development%20Plan/Strategy%20for%20transportation.pdf

Transforming City Bus Transport in India through Financial Assistance for Bus Procurement under JnNURM. (2012). MoUD. New Delhi: Retrieved from http://jnnurm.nic.in/wp-content/uploads/2012/02/booklet-on-transforming-City-Bus-Transport-in-India.pdf

Draft Concept Note on Smart City Scheme. (2014). MoUD-GoI.

Murphy, K. M., Shleifer, A., & Vishny, R. W. (1989). Industrialization and the Big Push. *Journal of Political Economy, 97*(5), 1003–1026. doi:10.1086/261641

Nelson, P., Baglino, A., Harrington, W., Safirova, E., & Lipman, A. (2007). Transit in Washington, DC: Current benefits and optimal level of provision. *Journal of Urban Economics, 62*(2), 231–251. doi:10.1016/j.jue.2007.02.001

NUTP. (2006). *National Urban Transport Policy*. New Delhi.

Padam, S., & Singh, S. K. (2001). Urbanization and Urban transport in India: the sketch for a policy. *Paper presented at theTransport Asia project workshop*, Pune, India.

Pickrell, D. H. (1992). A Desire Named Streetcar Fantasy and Fact in Rail Transit Planning. *Journal of the American Planning Association, 58*(2), 158–176. doi:10.1080/01944369208975791

Pucher, J., Korattyswaropam, N., Mittal, N., & Ittyerah, N. (2005). Urban transport crisis in India. *Transport Policy, 12*(3), 185–198. doi:10.1016/j.tranpol.2005.02.008

Sanchez, T. W. (1999). The Connection Between Public Transit and Employment. *Journal of the American Planning Association, 65*(3), 284–296. doi:10.1080/01944369908976058

Sharma, D., & Tomar, S. (2010). Mainstreaming climate change adaptation in Indian cities. *Environment and Urbanization, 22*(2), 451–465. doi:10.1177/0956247810377390

Siemiatycki, M. (2006). Message in a Metro: Building Urban Rail Infrastructure and Image in Delhi, India. *International Journal of Urban and Regional Research, 30*(2), 277–292. doi:10.1111/j.1468-2427.2006.00664.x

Singh, S. K. (2005). Review of urban transportation in India. *Journal of Public Transportation, 8*(1), 79–97. doi:10.5038/2375-0901.8.1.5

Sugita, Y., Okamoto, S., Kuwabara, T., Hiraishi, M., Goda, K., & Ito, A. (2005). *New Solution for Urban Traffic: Small-Type Monorail System Automated People Movers 2005*. Orlando, Florida, United States: American Society of Civil Engineers.

Suzuki, H., Cervero, R., & Iuchi, K. (2013). *Transforming Cities with Transit*. Washington, DC: The World Bank. doi:10.1596/978-0-8213-9745-9

Timmermans, H. J. P., & Zhang, J. (2009). Modeling household activity travel behavior: Examples of state of the art modeling approaches and research agenda. *Transportation Research Part B: Methodological, 43*(2), 187–190. doi:10.1016/j.trb.2008.06.004

Tiwari, G. (2011). Key Mobility Challenges in Indian Cities. *International Transport Forum*, Germany. doi:10.1787/5kg9mq4m1gwl-en

Road transport Year Book (2011-12). (2013). Transport Research Wing. New Delhi.

Winston, C., & Maheshri, V. (2007). On the social desirability of urban rail transit systems. *Journal of Urban Economics, 62*(2), 362–382. doi:10.1016/j.jue.2006.07.002

Zhang, X., Liu, Z., & Wang, H. (2013). Lessons of Bus Rapid Transit from Nine Cities in China. *Transportation Research Record: Journal of the Transportation Research Board, 2394*, 45–54. doi:10.3141/2394-06

ENDNOTES

1. As per Census of India, Urban Agglomeration (UA) is defined as follows: An urban agglomeration is a continuous urban spread constituting a town and its adjoining outgrowths (OGs), or two or more physically contiguous towns together with or without outgrowths of such towns. An Urban Agglomeration must consist of at least a statutory town and its total population (i.e. all the constituents put together) should not be less than 20,000.
2. 2013, Road Transport Year Book (2011-12), Transport Research Wing, Ministry Of Road Transport & Highways, Government Of India, New Delhi (available at http://morth.nic.in/showfile.asp?lid=1131, accessed on 15th October 2014)
3. For more details, see http://www.bestundertaking.com/
4. For more details, see http://www.kmrc.in/
5. For more details, see http://www.kmrc.in/overview.php
6. Kolkata Metropolitan Area
7. For more details, see http://bmrc.co.in/
8. For more details, see http://btis.in/bus
9. For more details, see http://www.delhimetrorail.com/
10. For more details, see https://mmrda.maharashtra.gov.in/mumbai-monorail-project
11. Passengers Capacity: 562 Max Passengers; Design Speed: 80 km/h; Design Headway: 3 minutes; Total number of stations: 17; Stations
12. For more details, see http://www.dimts.in/Projects_Bus-Rapid-Transit.aspx
13. For more details, see http://www.sutpindia.com/

Section 2
Finance

Chapter 4
Railway Investment Appraisal Techniques

Miloš Milenković
Zaragoza Logistic Center, Spain & University of Belgrade, Serbia

Libor Švadlenka
University of Pardubice, Czech Republic

Nebojša Bojović
University of Belgrade, Serbia

Vlastimil Melichar
University of Pardubice, Czech Republic

ABSTRACT

Railway transport involves the expenditure of resources on a combination of investment in capital items (e.g. stations, tracks, equipment) and/or in operations (e.g. subsidies). Concerning the fact that there are limited amounts of resources, it is necessary to maximize the returns obtained from the investments of those resources. The best way to do this is to ensure that the resources will be allocated on those projects that maximize their return. Railway appraisals therefore represent a way of thinking about all the costs and benefits of different railway related spending projects in a systematic manner so that, the projects can be compared and investments made in those which are going to provide the maximum possible return on the investment. This chapter provides a review of the main analytical tools that should be used in the process of railway investments appraisal. Namely, a detailed description of discounting, Net Present Value (NPV), Internal Rate of Return (IRR) and Cost Benefit Analysis (CBA) is covered by this chapter.

1. INTRODUCTION

To accomplish a continuous development process a railway company must invest its own and borrowed assets, to postpone possible consumption today, in order to provide new consumption and new investments in the future. On this way a railway company is forced to accumulate capital and to invest it, because investing represents the only way of achieving development objectives. Therefore, investments represent the necessity because the future development of every railway company merely depends on good planning and efficient management of investment process (Jovanović, 1991).

An investment represents a very complex process that involves a number of activities and participants. The most part of financial assets for planning and development is spent within the investing process as

DOI: 10.4018/978-1-5225-0084-1.ch004

a main mean for achieving planned development. The importance and complexity of investing process implies the need for its management in order to realize it on the best way. Total process of investing and gaining effects from an investment is characterized by a single- and a multi- period investments performed today, and in most cases a sequence of effects expected in the future. In order to get a real picture of this process and to manage it, especially to evaluate the benefits of an investment, it is necessary to analyze and determine the effects made by realization of an investment project.

There are different classifications of investment effects. These effects are mostly categorized on economic and non-economic. Economic effects represent a certain volume of services in railway passenger or freight transport and they are the most often expression of results of some investment. Non-economic effects – political, social, etc., may in certain cases be more important than economic and therefore it is necessary to include them in evaluation of some investment. Railway line across some area may be economically inefficient, but non-economic effects from this investment (environment, employment, etc.) justify its realization. Also, effects from an investment, as a result of investment realization, may be expressed through a specific volume of railway service production or some production expressed in certain value terms. Measuring of total effects gained from an investment and their quantitative expressing through some indicators or criterion, enable us to evaluate if the estimated effects will exceed the total investments. This procedure represents the estimate of efficiency of an investment program and it serves for making investment decision.

In practice, validity of realization an investment program is performed through analysis and evaluation of effects obtained by investment realization. Considering that economic effects of an investment may be measured and quantitatively expressed, analysis is in most cases focused on economic effects which will be delivered by investment realization and an estimate if that effects are enough with respect to needed investments. This procedure is known as analysis and evaluation of efficiency and profitability of an investment program.

Effects of an investment may be measured by calculation of certain indicators or criterion by which certain effects of an investment may be expressed. There are static and dynamic approaches for profitability evaluation of an investment proposal. Static approach does not involve entire time horizon of investing and exploitation of certain investment, just one time period. In contrast to static approach, dynamic estimate takes the whole time period in analysis and evaluation of an investment project covering its whole period of investing and exploitation. By the use of discounting method all positive and negative effects from every year of investing and exploitation period are included and dynamic criterion are calculated. Dynamic criterion represent complex indicators that provide more realistic analysis of different aspects of some investment project and evaluation of its profitability. A number of dynamic criterion is proposed in literature and practice for investment proposals evaluating. Therefore, in this chapter, the Present Worth and Internal Rate of Return are introduced as the most important dynamic criterion. In the case of investments that have substantial indirect effects (non-economic), like investments in transport infrastructure, Cost Benefit analysis will be presented. The use of these techniques in the context of railway related investments will be reviewed.

The chapter is organized as follows. In Section 2. the discounting cash flow technique is explained, as the main prerequisite for dynamic evaluation of investment proposals. Section 3. presents an explanation and classification of investment alternatives. Section 4. contains Present Worth, its definition and main features. In Section 5. Internal Rate of Return is explained. Section 6. contains basic prerequisites for applying Cost Benefit approach in evaluation of public projects. Section 7. contains a summary of presented methods. In Section 8. a review of railway related applications of appraisal techniques is given.

Railway Investment Appraisal Techniques

A case study that demonstrates usability and importance of considered methods is given in Section 9. Concluding remarks are given in the last section.

2. DISCOUNTED CASH FLOW CALCULATION

In order to account for the time dynamics in the evaluation process of investment proposals it is necessary to use the cash flow discounting method. This method covers effects of an investment during the whole exploitation period. Yearly amounts of revenues and costs expected from an investment are translated to the time of evaluation, in that case appears a possibility of comparison and evaluation of investment projects.

Cash flow discounting represents a well-known approach by which estimated future cash flows are discounted at an interest rate. Involved interest rate reflects the perceived riskiness of cash flows (Yao et al., 2005). Discount rate usually includes the time value of the money and the risk of the investment. Time value of the money is based on the fact that investors would rather have cash immediately than having to wait and must therefore be compensated by paying for the delay. Risk of the investment represents the extra return investors demand because they want to be compensated for the risk that the cash flow might not materialize after all.

Discounted cash flow approach (Figure 1) translates a cash flow composed from yearly revenues and costs of an investment during its economic life on present value.

It can be seen from Figure 1 that the present value at time t for each time $t+k$, $k=1,2,...,n$ can be expressed as:

$$P_{t+1} = \frac{F_{t+1}}{(1+i_{t+1})} \qquad (1)$$

$$P_{t+2} = \frac{F_{t+2}}{(1+i_{t+1})(1+i_{t+2})} \qquad (2)$$

Figure 1. Cash flow and present value at time t

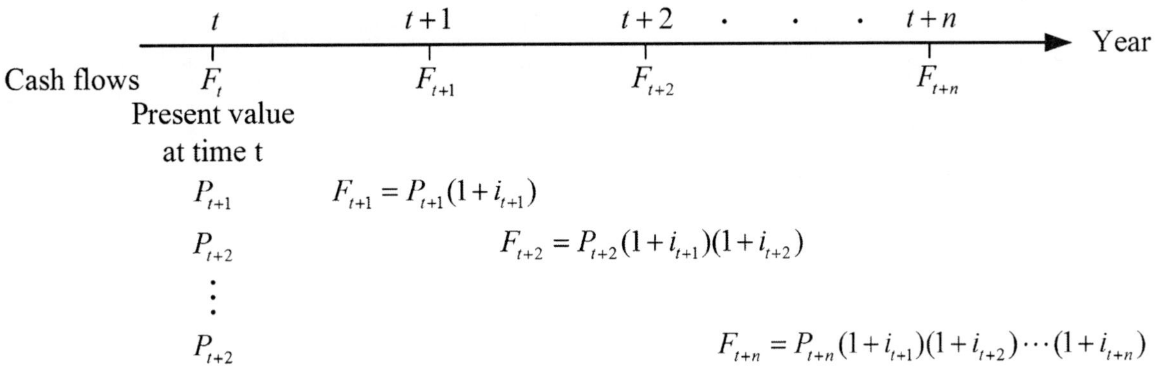

$$
\begin{array}{l}
\text{Cash flows} \quad F_t \quad F_{t+1} \quad F_{t+2} \quad \cdots \quad F_{t+n} \quad \text{Year} \\
\text{Present value at time t} \\
\quad P_{t+1} \quad F_{t+1} = P_{t+1}(1+i_{t+1}) \\
\quad P_{t+2} \quad F_{t+2} = P_{t+2}(1+i_{t+1})(1+i_{t+2}) \\
\quad \vdots \\
\quad P_{t+n} \quad F_{t+n} = P_{t+n}(1+i_{t+1})(1+i_{t+2})\cdots(1+i_{t+n})
\end{array}
$$

$$P_{t+n} = \frac{F_{t+n}}{(1+i_{t+1})(1+i_{t+2})\cdots(1+i_{t+n})} \tag{3}$$

Discounted cash flow can now be expressed as the present value of estimated future cash flows at time t as follows:

$$S_t = \sum_{k=1}^{n}\left[\prod_{j=1}^{k}\frac{1}{(1+i_{t+j})}\right]F_{t+k}, \ t=0,1,2...,n \tag{4}$$

n: economic life of an investment;
P_t: discounted value of a cash flow at time t;
F_{t+k}: the cash flow in period $t+k$;
i_{t+j}: discount rate at time $t+j$;

The amount that represents present value of one money unit available for $t=1,2,\ldots,n$ years $\dfrac{1}{(1+i_{t+j})}$

is known as the discounting factor.

Example 1: (Figure 2) demonstrates the process of discounting.

For given discount rate $i=15\%$ by discounting all future values for their equivalent present values they are made directly comparable. For example, what is the PW of $6000, 2 years from now?

Figure 2. Discounting a project cash flow

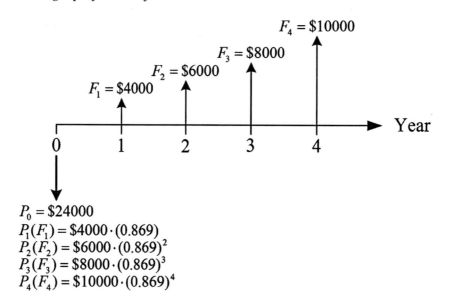

$P_0 = \$24000$
$P_1(F_1) = \$4000 \cdot (0.869)$
$P_2(F_2) = \$6000 \cdot (0.869)^2$
$P_3(F_3) = \$8000 \cdot (0.869)^3$
$P_4(F_4) = \$10000 \cdot (0.869)^4$

Railway Investment Appraisal Techniques

P in year 1= $6000(0.869) = $5214
P in year 0 = $5214(0.869) = $4530.1

In summary, discounting factor determines a single present worth *P* from the amount of money *F* accumulated after N years (or periods), with interest compounded one time per year (or period) for discount rate *i*. It represents a fundamental factor in engineering economy. However, cash flows occur in many configurations and amounts isolated single values (P, present; F, future value), series that are uniform (A, annuities), and series that increase or decrease by constant amounts or constant percentages (G, gradients). Table 1 contains mathematical forms and a standard notation format for all the commonly used engineering economy factors that take the time value of money into account (Blank and Tarquin, 2002).

These factors do not have to be analytically calculated each time in order a calculation to be performed, they are usually available in compound interest factor tables, which give the interest factors for different discount rates (interest rates) and years.

Example 2: Example 2 illustrates the use of interest factors. Suppose that a railway company just installed a new software for rail freight car management for a price of $70000 and with estimated yearly costs of software maintaining of $5000 for 6 years beginning from 3rd year. What is the present value of these outflows under the yearly discount rate of 8%? Cost flow diagram is presented on Figure 3.

Table 1. Interest factors for discrete cash flow with end-of-period compounding

Factor	To Find	Given	Symbol	Mathematical Expression
Compound amount	Future worth, F	Present amount, P	(F/P, i%, N)	$(1+i)^N$
Present worth	Present worth, P	Future amount, F	(P/F, i%, N)	$\dfrac{1}{(1+i)^N}$
Sinking fund	Annuity amounts, A	Future amount, F	(A/F, i%, N)	$\dfrac{i}{(1+i)^N - 1}$
Series compound amount	Future worth, F	Annuity amounts, A	(F/A, i%, N)	$\dfrac{(1+i)^N - 1}{i}$
Capital recovery	Annuity amounts, A	Present amount, P	(A/P, i%, N)	$\dfrac{i(1+i)^N}{(1+i)^N - 1}$
Series present worth	Present worth, P	Annuity amounts, A	(P/A, i%, N)	$\dfrac{(1+i)^N - 1}{i(1+i)^N}$
Arithmetic gradient conversion	Annuity amounts, A	Uniform increase in amount, G	(A/G, i%, N)	$\dfrac{1}{i} - \dfrac{N}{(1+i)^N - 1}$

(Riggs and West, 1986)

Figure 3. Cash flow diagram of a railway project

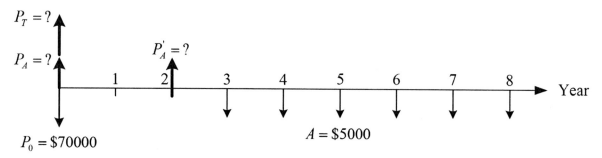

Symbol P_A is used for present value of regular yearly flows A, and P_A' for representing a present value in period which is not period 0. As it can be seen from the cash flow diagram P_A' is located in 2nd year, not in 3rd. Let first calculate P_A' value of shifted cash flow:

$$P_A' = \$5000(P/A, 8\%, 6) \tag{5}$$

As P_A' is located in 2nd year, P_A in year 0 is:

$$P_A = P_A'(P/F, 8\%, 2) \tag{6}$$

Final present value is determined by adding P_A and starting investment in 0th year:

$$\begin{aligned} P_t &= P_0 + P_A \\ &= \$70000 + \$5000(P/A, 8\%, 6)(P/F, 8\%, 2) \\ &= \$70000 + \$5000(4.6229)(0.8573) \\ &= \$898816.1 \end{aligned} \tag{7}$$

3. INVESTMENT ALTERNATIVES

Investment alternatives are developed from project proposals to accomplish a stated purpose (Blank and Tarquin, 2002). Some projects are economically and technologically viable, and others are not. After the viable projects are defined it is possible to formulate the alternatives. For example, assume an investment initiative for decreasing transport costs and travelling time between two cities in a region. Three projects have been proposed: a high speed railway line, highway and a river ferry service. Economically (and technologically) only the first two project proposals can be accepted, considering the hydrological characteristics of rivers in a given region. Therefore, project proposals represent precursors to economic alternatives. Projects can be:

Railway Investment Appraisal Techniques

- Mutually exclusive where only one of the viable projects can be selected by the economic analysis and each viable project is an alternative;
- Independent where more than one viable project may be selected by the economic analysis (there may be dependent projects requiring a particular project to be selected before another and contingent projects where one project may be substituted for another).

The do-nothing (DN) option is usually understood to be an alternative when the evaluation is performed. No new costs, revenues or savings are generated by the DN alternative.

A mutually exclusive alternative selection takes place, for example, when a railway operator must select the one best locomotive from several competing models.

Independent projects do not compete with one another in the evaluation. Each project is evaluated separately, and thus the comparison is between one project at a time and the do-nothing alternative.

4. PRESENT WORTH ANALYSIS

For the case when the investment alternative is assessed over a number of years, the predictive model used will normally calculate the benefits and costs for each year of the project (Cowie, 2010). However, consider that the model gives an estimated net inflow for the year 2015. of $15.000 and for 2020., also $15.000. Disregarding any inflation that might exists it is not appropriate to add these values together along with net inflows for all the other years in order to derive the total benefit of investment. The reason lies in the fact that even if they may serve to buy the same amount of goods with $15.000 in 2020. as in 2015., the two sums would be worth different amounts from the point of view of the present. This is because it is needed to wait longer before the benefit arising from the investment in 2020. would be enjoined, consequently there is a cost involved of having to wait that additional time before deriving the benefit. Therefore, before costs and benefits that are predicted to arise in different years are added together, they must be subject to a process that converts them to a common unit known as their (Net) Present Value (NPV) or Present Worth (PW).

It is already clear that PW belongs to a class of dynamic criterion which are determined by the use of discounting technique. A future amount of money converted to its equivalent value has a present value that is always less than that of the actual cash flow because for any interest rate greater than zero, all discounting factors have a value less than one. For this reason, PW values are often referred to as discounted cash flows. Similarly, used interest rate is referred to as the discount rate.

In present worth analysis, the P value, now called PW is calculated at the Minimum Attractive Rate of Return (MARR) for each alternative. The Minimum Attractive Rate of Return (MARR) is a reasonable rate of return established for the evaluation and selection of alternatives. It represents either the organization's minimum attractive rate of return or the cut-off rate of return. By the PW criterion a sum of discounted net inflows earned within the economic life of an investment is considered. Mathematically,

$$PW = \frac{NI_1}{(1+i)} + \frac{NI_2}{(1+i)^2} + \cdots + \frac{NI_n}{(1+i)^n} \tag{8}$$

$$PW = \sum_{k=1}^{n} \frac{NI_k}{(1+i)^k} = \sum_{k=1}^{n} NI_k b_k \qquad (9)$$

where

NI_k: Net inflow, the difference between inflow and outflow of an investment in k-th year of project exploitation.
i: Discount rate or MARR
b: Discount factor
n: Economic life of an investment

Figure 4 shows the present worth in relation to the discount rate.

The PW comparison of alternatives with equal lives is straightforward. The following guidelines are applied to select one alternative:

One Alternative: Calculate PW at the MARR. If $PW \geq 0$, the requested MARR is met or exceeded and the alternative is acceptable;

Two or More Alternatives: Calculate the PW of each alternative at the MARR.

Example 3: This example demonstrates the use of PW criterion in railway appraisal applications. Suppose that there are two alternatives for investing in renewal of the railway equipment (Table 2). All values are hypothetical in order to demonstrate the use of evaluating criterion.

By using the PW criterion follows that:

$$\begin{aligned} PW_A &= -\$70000 + \$15000(P/A,15,10) + \$8000(P/F,15,10) \\ &= -\$70000 + \$15000 \cdot 5.019 + \$8000 \cdot 0.247 \\ &= \$7260 \end{aligned} \qquad (10)$$

Figure 4. Present worth in relation to the discount rate

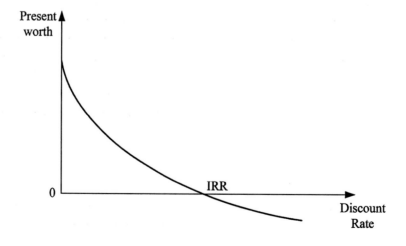

Table 2. Cost parameters for alternatives A and B

	A	B
Investment	$70000	$10000
Net inflow	$15000	$5000
Salvage value	$8000	$2000
Economic life	10	10
MARR [%]	15	15

$$PW_B = -\$10000 + \$5000(P/A, 15, 10) + \$2000(P/F, 15, 10)$$
$$= -\$10000 + \$5000 \cdot 5.019 + \$2000 \cdot 0.247 \quad (11)$$
$$= \$15588$$

So it may be concluded that by the PW criterion and within the assumption that there is no budget constraint, investing in Alternative B returns higher discount net effect and therefore it is more preferable alternative.

In this example it is assumed a 15% discount rate. What happens to the PW if the discount rate is increased to 25% or the discount rate is decreased to 10%.

Consider investment in Alternative A at a discount rate of 20%:

$$PW(A)_{0.2} = -\$70000 + \$15000(P/A, 20, 10) + \$8000(P/F, 20, 10)$$
$$= -\$70000 + \$15000 \cdot 4.192 + \$8000 \cdot 0.161 \quad (12)$$
$$= -\$5832$$

Note that *PW(A)* has decreased and even become negative. Therefore, if the discount rate is 20%, reject alternative A.

In case of decreased discount rate to 10% following results are obtained:

$$PW(A)_{0.1} = \$25255 \quad (13)$$

$$PW(B)_{0.1} = \$21495 \quad (14)$$

Therefore, at the decreased discount rate both alternatives are acceptable and considering PW criterion alternative A is preferred. It should be noted that the PW at 15% is much lower than the PW at 10%. The reason for this lies in the fact that as the projects net benefits occur in the future whereas the net cost are all at the beginning. The higher the rate of discount, the lower will be the present value of the future benefits and, therefore, the PW.

When the present worth method is used to compare mutually exclusive alternatives that have different lives, the procedure of the previous section is followed with one exception. The PW of the alternatives must be compared over the same number of years and end at the same time. This is necessary, since a present worth comparison involves calculating the equivalent present value of all future cash flows for each alternative.

The equal-service requirement can be satisfied by either of two approaches:

- Compare the alternatives over a period of time equal to the least common multiple (LCM) of their lives;
- Compare the alternatives using a study period of length n years, which does not necessarily take into consideration the useful lives of the alternatives.

This is also called the planning horizon approach. In either case, the PW of each alternative is calculated at the MARR, and the selection guideline is the same as that for equal-life alternatives.

Example 4: Assume that there are two alternative investments in improvement of distribution channels of a railway company. Alternative 2 (A2) has initial cost of $3200 and estimated salvage value of $400 at the end of its 4 year long economic life. Alternative 1 (A1) has $900 less initial expenses, with economic life 1 year shorter then A2, with zero salvage value and its yearly operating costs higher for $250. If MARR is 15%, let us try to find better alternative in case of:
1. The repeated projects method. Alternatives are joined by selecting an analysis period that spans a common multiple of the lives of involved alternatives.
2. A 2-year study period. This analysis is based on a specified duration that corresponds to the length of a period of time the alternatives are expected to be in service.

In case of 1) follows that:

$$PW(A1) = -\$2300 - \$2300(P/F,15,3) - \$2300(P/F,15,6) - \$2300(P/F,15,9) - \$250(P/A,15,12)$$
$$= -\$2300 - \$2300 \cdot 0.657 - \$2300 \cdot 0.432 - \$2300 \cdot 0.284 - \$250 \cdot 5.421 \quad (15)$$
$$= -\$6816$$

$$PW(A2) = -\$3200 - \$3200(P/F,15,4) - \$2800(P/F,15,8) + \$400(P/F,15,12)$$
$$= -\$3200 - \$2800 \cdot 0.572 - \$2800 \cdot 0.327 + \$400 \cdot 0.187 \quad (16)$$
$$= -\$5642$$

Considering that each alternative has only cost cash flow estimates, according to the PW criterion A2 is preferred option.

In case 2) a service period comparison is utilized which means that a limited period of ownership of assets is set by specific operational requirements. A 2-year study period for A1 and A2 indicates that the service required from either asset will be needed only 2 years and will be disposed of at that time. In this case it is needed for estimates of the worth of assets at the end of the study period to be secured. These salvage values may be quite large when the service period is only a small fraction of the economic life which is the case with railway related facilities and equipment. When it is difficult to secure reliable estimates of worth after use during the ownership period, minimum resale levels (S) can be calculated to make alternatives equivalent. Then only a judgment is needed about whether the market value will be above or below the minimum level (Riggs and West, 1986). For instance, assuming S=0 for alternatives A1 and A2 after 2 years of service, follows that

$$PW(A1) = -\$2300 - \$250(P/A, 15, 2) = -\$2300 - \$250 \cdot 1.626 = -\$2707 \quad (17)$$

$$PW(A2)=\$3200 \tag{18}$$

Which show that A1 has the lower present worth of costs for 2-year service period. The salvage value for A2 that would make PW(A2) equal to PW(A1) is

$$PW(A2) + S(P/F, 15, 2) - PW(A1) \text{ or } \$2707 = \$3200 - S(P/F, 15, 2) \tag{19}$$

$$S = \frac{\$3200 - \$2707}{(P/F, 15, 2)} = \frac{\$493}{0.756} = \$652 \tag{20}$$

Which means that A2 is preferred to A1 when the resale value of A2 at the end of 2 years is more than 652 greater than the resale value of A1 at the same time. The advantage of calculating this aspiration level it to avoid making an estimate of S. Only a judgment is required about whether S will exceed a certain amount, the aspiration value which is 652 in this case.

5. INTERNAL RATE OF RETURN

It can be seen from the Figure 4 that the PW curve slopes downwards from left to right. At some point the curve intersects the horizontal axis. That is, it becomes "0". The discount rate at which the PW becomes "0" is called the Internal Rate of Return (IRR). In other words, IRR is the discount rate which equates the present value of the future cash flows of an investment with the initial investment. Once IRR of an investment is known it can be compared with the cost of project financing (MARR). If the rate of return, the IRR, is greater than the cost of financing the project, the investment should be accepted. When the IRR is less than the cost of finance, the project should be rejected. This decision rule can be summarized as follows:

When $IRR \geq MARR$, accept
When $IRR < MARR$, reject

Mathematical expression of IRR criterion is:

$$PW = \sum_{k=1}^{n} NI_k b_k = 0 \tag{21}$$

From this equation, by solving for unknown i, internal rate of return can be determined. IRR represents that discount rate for which an investment does not generates benefits nor loses, or in other words, investment represents "empty business". IRR shows the minimum discount rate by which the realization of an investment project is still acceptable.

There are two ways of solving the equation (21). Which of the two is used depends on the type of cash flow the project has (Campbell and Brown, 2003):

- When the cash flow is not regular in the sense that there is not an identical value every year after the initial (year 0) investment, it must be estimated by trial and error – by iteration and interpolation.
- When the cash flow is regular the easiest method is to use the compound interest factor tables and then by interpolation find unknown internal rate.

For the case of interpolation as an approach to IRR calculation let us consider again the Example 3. The $PW(A)$ is positive at MARR=15% ($PW(A)_{0.15}$=$7260), but not at MARR=20% ($PW(A)_{0.2}$=$–5832). Therefore, it may be concluded that the IRR must lie somewhere between these two rates. Then, the actual IRR by interpolation may be calculated:

$$IRR = 15 + 5\left[\frac{\$7260}{\$7260 + \$5832}\right] = 17.8 \qquad (22)$$

Therefore, the rule for interpolation is as follows:

$$IRR = \begin{bmatrix}lower\\discount\\rate\end{bmatrix} + \begin{bmatrix}difference\\between\\the\ two\ discount\\rates\end{bmatrix} \times \left[\frac{PW\ at\ the\ lower\ disount\ rate}{sum\ of\ the\ absolute\ values\ of\ the\ PWs}\right] \qquad (23)$$

However, if the absence of salvage value at the end of economic life of Alternative A is assumed, there is a constant or regular cash flow. In this case it becomes very easy to calculate the IRR using the compound interest factor tables. Therefore, the expression for $PW=0$ from which the IRR will be found as follows:

$$PW_A = -\$70000 + \$15000(P/A, i, 10) = 0 \qquad (24)$$

Or,

$$(P/A, i, 10) = 4.67 \qquad (25)$$

Now it is only needed to find out at what discount rate the annuity factor for year 10 has a value equal to 4.67. To determine this compound interest factor tables are used. Namely, it is necessary to search along the row for year 10 until (P/A, i, 10) that is approximately equal to 4.67 is found. This occurs when the discount rate is between 16% ((P/A, 16, 10)=4.83) and 18% ((P/A, 18, 10)=4.49). The exact value can be obtained by interpolation as follows:

$$IRR(A) = 16 + (18-16)\frac{4.83-4.67}{4.83-4.49} = 16.94 \qquad (26)$$

As it is already seen in previous examples investment alternatives are examined for situations in which a decision must be made between a set of independent alternatives. In these situations, each alternative is evaluated separately from the others, and more than one can be selected. PW method and IRR method give the same results. However, in case two or more mutually exclusive alternatives are evaluated, or in other words, in case where decision maker have to select one of two or more alternatives, the PW and IRR methods can yield conflicting results.

Example 5: Let us consider two mutually exclusive railway projects, A and B (Table 3).

Using the PW criterion, discounted net effect of each investment is:

$$PW(A)=\$280192 \tag{27}$$

$$PW(B)=\$39946.6 \tag{28}$$

However, using the IRR criterion appear that Project B is preferable to Project A given that the

$$IRR(A)=27.78\% \tag{29}$$

$$IRR(B)=49.34\% \tag{30}$$

As it can be seen, this is the reverse of the choice using the PW method. Solving of this paradox and the explanation of why Project A is correct choice lies in understanding of the meaning of MARR rate. In either case the derivation of the rate is directed at maximizing the organization's profits for all of its invested capital (Sprague and Whittaker, 1986). Therefore, assume that there is 100000 to invest and that the choice must be made between projects A and B.

For project A: The rate of return on the $100000 invested in Project A is 27.78%.

For Project B: The rate of return on the $20000 invested in Project B is 49.34%. The remaining 80000 will be invested elsewhere at the MARR=10% and the average rate of return is (20/100)49.34% + (80/100)10% = 17.86%. 17.86%<27.78% therefore, invest $100000 in Project A. Thus, although Project B has the higher rate of return, Project A is the correct choice. Project A maximizes profit.

However, there is a way in which the IRR rule could still be used to choose between mutually exclusive projects. Consider the incremental project where this is defined as the difference between cash flows of Projects A and B. The cash flow of a hypothetical project A-B is given in next Table 4.

Table 3. Cost parameters for alternatives A and B

	A	B
Investment	$100000	$20000
Net inflow	$30000	$10000
Salvage value	$10000	$3000
Economic life	10	10
MARR [%]	10	10

Table 4. Cost parameters for alternative A-B

	A-B
Investment	$80000
Net inflow	$20000
Salvage value	$7000
Economic life	10
MARR [%]	10

The IRR of this cash flow have to be calculated now. If the IRR of the incremental project is equal or greater than the cost of capital, Project A is selected. In contrary, Project B will be selected. In effect, we try to analyze if it makes sense to invest additional $80000 in Project A. After calculation, we obtain that IRR_{A-B}=21.41 so the IRR is greater than the cost of capital and it does make sense to invest the extra amount in Project A.

6. COST BENEFIT ANALYSIS

When we consider projects which generate valuable effects, not only for investor but also for a state in whole, these projects can be properly evaluated if we include total effects (economic as well as non-economic) obtained by the investment.

Cost-Benefit analysis (CBA) provides this kind of evaluation. CBA considers all social costs and benefits generated by an investment. CBA analysis is a method used in evaluation of investments that have influence on development of social community – certain region, economy and the state in whole. Therefore, CBA is not used for projects that generate explicit commercial effects that can be measured and quantitatively expressed, but for projects which generate significant indirect and immeasurable effects. These projects are mainly focused on investments in transport infrastructure (rail, road, air, water), investments in big energetic facilities as well as investments in agriculture development.

CBA analysis lies on an idea that the same effect can't be positive for both, the industry as well as the state in whole. An investment can generate significant positive effects to an investor, but at the same time, because of environmental impacts, to be harmful for state in whole. Because of this possible difference in contribution to particular and total social objectives, CBA insist on social effects, or on considering and evaluating the effects from the society's point of view, and this represents the main feature of this method. If we consider, for example, a project of a new railway line construction, this project brings various social effects which have to be included. Besides the benefits from decreased costs of exploitation of transport organizations, there are more significant possibilities for faster development of a particular region, shorter traveling time, increased comfort and other benefits which have to be included along with total costs.

6.1. General Framework for Cost Benefit Analysis

CBA of major infrastructure projects includes following steps (European Commission, 2008):

- **Context Analysis and Project Objectives:** The first step of the project appraisal aims to understand the social, economic and institutional context in which the project will be implemented. In fact, the possibility of achieving credible forecasts of benefits and costs often relies on the accuracy in the assessment of the macro-economic and social conditions of the region.
- **Project Identification:** A clear statement of the project's objectives is an essential step in order to understand if the investment has social value. The broad question any investment appraisal should answer is 'what are the net benefits that can be attained by the project in its socio-economic environment?' The benefits considered should not be just physical indicators (km of railways) but socio-economic variables that are quantitatively measurable.

Railway Investment Appraisal Techniques

- **Feasibility and Option Analysis:** Once the socio-economic context and the potential demand for the project output have been analyzed, then the next step consists of identifying the range of options that can ensure the achievement of the objectives of the project. Feasibility analysis aims to identify the potential constraints and related solutions with respect to technical, economic, regulatory and managerial aspects. Moreover, as mentioned, several project options may be feasible. The main result of feasibility and option analysis is to identify the most promising option on which detailed CBA should be carried out.
- **Financial Analysis:** The main purpose of the financial analysis is to use the project cash flow forecasts to calculate suitable net return indicators. A particular emphasis is placed on two financial indicators: the Financial Net Present Value (FNPV) and the Financial Internal Rate of Return (FRR), respectively in terms of return on the investment cost, FNPV(C) and FRR(C), and return on national capital, FNPV(K) and FRR(K).
- **Economic Analysis:** The economic analysis appraises the project's contribution to the economic welfare of the region or country. It is made on behalf of the whole of society instead of just the owners of the infrastructure, as in the financial analysis. The key concept is the use of accounting shadow prices, based on the social opportunity cost, instead of observed distorted prices. Observed prices of inputs and outputs may not mirror their social value (i.e. their social opportunity cost) because some markets are socially inefficient or do not exist at all.
- **Risk Assessment:** Project appraisal is a forecasting exercise rather than the formulation of an opinion. However, no forecast is without problems. For example, one may know that because of data limitations the forecasts for the demand for drinking water are affected by estimates that are prone to considerable errors. One may also have doubts about some parameters crucial to the calculation of the return, such as the shadow wage. A risk assessment consists of studying the probability that a project will achieve a satisfactory performance (in terms of some threshold value of the IRR or the NPV).

6.2. Benefit Cost Ratio

The benefit/cost ratio represents a fundamental analysis method for public sector projects, was developed to introduce more objectivity into public sector economics (Blank and Tarquin, 2008). It was developed in response to the U.S. Flood Control Act of 1936. There are several variations of the B/C ratio; however, the basic approach is the same. All cost and benefit estimates must be converted to a common equivalent monetary unit at the discount rate. The B/C ratio is then calculated using one of these relations:

$$B/C = \frac{PW \text{ of benefits}}{PW \text{ of costs}} \tag{31}$$

In case of a single project, if this ratio is equal or greater than unity then accept the project. If it is less than unity then reject the project.

Salvage values, when they are estimated, are subtracted from costs. Disbenefits are considered in different ways depending upon the model used. Most commonly, disbenefits are subtracted from benefits and placed in the numerator. According to these assumptions there are two different forms of B/C ratio:

- The conventional B/C ratio in which disbenefits are subtracted from benefits

$$B/C_C = \frac{\text{benefits - disbenefits}}{\text{costs}} \qquad (32)$$

- The modified B/C ratio places benefits (including income and savings), disbenefits and operation costs in the numerator. The denominator includes only the equivalent PW of the initial investment. Salvage value is included in the denominator with a negative sign.

$$B/C_M = \frac{\text{benefits - disbenefits - operation costs}}{\text{initial investment}} \qquad (33)$$

Example 6: Consider three alternative investments in a railway infrastructure. Initial costs are 100000, and estimated economic life is 15 years. MARR=10% is used for efficiency evaluation (table 5).

According to the conventional B/C form we have for alternative A:

$$B/C_C(A) = \frac{\text{PW of net benefits for user}}{\text{PW of capital + operating costs of investor}} = \frac{(\$35000 - \$3200)(P/A,10,15)}{\$100000 + \$13000(P/A,10,15)} = 1.22 \qquad (34)$$

By Equation (33) the modified B/C ratio treats the operating cost as a reduction to benefits:

$$B/C_M(A) = \frac{\text{PW of net total benefits}}{\text{PW of total capital invesmtents}} = \frac{(\$35000 - \$3200 - \$13000)(P/A,10,15)}{\$100000} = 1.43 \qquad (35)$$

The benefit-cost ratio, if properly applied, should lead to the same accept/reject decisions as the other criterion (PW, IRR). Example shows that the order of alternatives by preference is not proper by using the conventional B/C. Therefore, to maintain consistency between various methods, it is necessary to calculate B/C ratio as follows (Sprague and Whittaker, 1986):

Table 5. Cost parameters for alternative investments

Alternative	User Annual Benefits [10^3]	User Annual Costs [10^3]	Supplier Annual Costs [10^3]	B/C	PW [10^3]
A	$35 000	$3 200	$13000	$B/C_C(A)=1.22$ $B/C_M(A)=1.43$	$42992.8
B	$25 000	$4 100	$1300	$B/C_C(B)=1.45$ $B/C_M(B)=1.49$	$49077.6
C	$18 000	$1000	$600	$B/C_C(C)=1.24$ $B/C_M(C)=1.25$	$24.738.4

- Determine the PW of the net annual cash flows associated with project. They represent the difference between annual operating revenues (benefits to the user) and annual operating costs (operating costs to the user + operating costs to the agency supplying the service).
- Determine the PW of all capital investment costs.
- Calculate the ratio of these two values using B/C_M.

The B/C_M treats all annual cash outflows (regardless they originate with the user (citizen) or supplier (government agency) as an annual operating cost. The benefits that result are usually benefits to the user. For example, railway ticket fares represent a cost to the user and thereby a cost to the system. One may prefer to think of the user and supplier as being the same group of people. That is, the funds supplied by the government agency to cover capital and operating costs are actually supplied by the user through service charges.

7. SUMMARY

This section aims to present a synthetic overview of the methods described above. Table 6 contains the main strengths and weaknesses of methods seen above.

8. REVIEW OF APPRAISAL TECHNIQUES IN RAILWAY TRANSPORT RELATED APPLICATIONS

Deciding where to focus investment is a key part of building successful railway transportation system. Presented appraisal techniques provide a tool for evaluating the effect a railway investment will have on cash flow of investor. This section contains a brief review of some of a number of applications of appraisal techniques within the field of railway transportation.

Table 6. Main features of presented appraisal tools

Appraisal Techniques	Strengths	Weaknesses
Present Worth Analysis	• Absolute measure – it represents the dollar amount of value added or lost by undertaking an investment; • Ability to notch the discount rate up and down to allow for different risk level of projects.	• The project size is not measured; • Exact estimations are needed for discount rate, size of each cash flow and time it its occurrence; • Opportunity cost of making an investment is not built in PW calculation.
Internal Rate of Return	Relative measure – it represents rate of return an investment offers over its lifespan.	• It can give conflicting answers when compared to NPV for mutually exclusive projects; • A multiple IRR problem occurs when cash flows during the project lifetime is negative.
Cost-Benefit Analysis	Simplicity - measuring the dollar amount of the benefits and the costs involved in a project, the cost benefit is very easy to see;	• Exact estimations are needed for costs and benefits; • Common measurement of costs and benefits – difficult in presence of both quantitative and qualitative costs and benefits; • Accuracy with regard to benefits and costs must be closely monitored because benefits are easy to double count.

Tao et al. (2011) conducted a cost-benefit analysis of high speed rail (HSR) link between Hong Kong and Mainland China. The most controversial part of the HSR investment is whether its cost could be compensated by the social benefits. In their study, authors first defined all the direct and indirect costs, and social benefits. Then, they assigned monetary equivalents to these elements. Third, all the future values are discounted into present values and aggregated. The results show that the project has a positive net present value (NPV) up to USD$2,068.49 million, which proves that the investment is worth. Wiegmans (2008) used various financial modeling methods, such as present and annual worth analysis and internal rate of return, in combination with analysis of the political arguments in order to assess financial performances of the Betuweline. Namely, Betuweline is a freight railway line across the Netherlands that provides easier and more environment-friendly transport options into the port of Rotterdam. Judged from an economic point of view, the Betuweline is not a profitable investment. However, besides the financial analysis, various political arguments have also been taken into account. Therefore, to invest in the Betuweline could be a good decision because:

- There were no alternatives for increasing the rail freight transport over the existing networks;
- The Betuweline can relieve the mixed passenger-freight rail network;
- The transport of dangerous goods can be redirected to the Betuweline; and
- The Betuweline is important for international rail freight transport network.

Political arguments against the Betuweline were:

- The lack of public support;
- The lack of a capacity link with Germany; and
- Negative external effects.

The government has worked hard to compensate for or solve these issues. Ibraheem (2011) studied the development of a high-speed passenger rail link between airports and cities centers in Iraq. Author develops the evaluation process for selecting a most preferable railway transportation project. Within this approach net present worth, annual worth and benefit cost ratio are applied to perform economic evaluations of various alternatives. Olsson et al. (2012) presented a cost-benefit methodology for the appraisal of railway infrastructure in Norway, Sweden, Denmark, United Kingdom, France, Germany and Switzerland. The consequences of differences in methodology are illustrated by a case study undertaken with the methodology from each of the seven countries. Even though the basic principles are the same in the CBA in the seven countries, the methodological differences means that the result for the CBA is depending on which country's methodology that is used. Wang et al. (2014) conducted a CBA of three possible alternatives along with the existing scenario on the Dhaka-Narayanganj (Bangladesh) route. The first alternative is to reduce headway by improving existing railway tracks and signaling systems. The second alternative is to increase train frequency by introducing Diesel Electric Multiple Units (DEMU). The third alternative is to attain expected headway by double tracking with DEMU. The benefits considered are user's benefits, accident reduction savings, vehicle operating costs and emission reduction savings. The costs included capital and construction costs, operating and maintenance costs. The analyses utilized 3%, 5% and 10% Annual Average Ridership Growth Rates (AARGR) depending on passenger trends on the route. Two analysis tools such as Net Present Values (NPV) and Benefit-Cost

Ratios (BCR) are used and compared among the alternatives to decide which alternatives are economically feasible on the basis of short terms and long terms perspectives. Hunt (2005) made a report in which author addresses how to demonstrate what the public obtains in terms of benefits from its investment in rail capacity improvement(s) by:

- Exploring the current practice of evaluating benefits attributable to public investments in freight rail projects though a set of 11 case studies;
- Describing the methods and software models that have been developed and adapted to freight rail projects;
- Discussing the potential funding mechanisms for public investment in freight rail;
- Combining current practice and methods with future funding requirements to develop a framework for establishing public benefits accruing from investments in freight rail capacity.

In this report methods such as CBA and IRR are used to evaluate and help determine the best allocation of public investments. Kolonko and Engelhardt-Funke (2001) developed a tool for CBA of future investments into a railway network. Authors presented a method to obtain a cost benefit curve that shows the effect of investments (cost) on the quality of the network measured by the waiting time of passengers (benefit). This curve is obtained from the solutions of multi-criterion timetable optimization problem. Casares and Coto-Millan (2011) examined several high speed lines in Spain, by using the CBA, in order to obtain more information and determine the social costs and benefits linked to each project. Also, an IRR per project is determined, in order to classify investments with positive CBA. de Rus (2012) made an analysis of whether investing in the construction of HSR infrastructure in a standard medium-distance corridor is socially desirable. In his study, author described the construction and maintenance costs of HSR, discussing the cost structure of a standard medium-distance HSR line and highlighting the main direct and indirect benefits of a new HSR line. Then, author presented a new model for economic evaluation of three HSR lines. Namely, an ex-post CBA of two similar lines in Spain (Madrid-Seville and Madrid-Barcelona) as well as a CBA of the Stockholm-Gothenburg project is presented. Raju (2008) estimated the NPV of the Heartland Light Rail project in Kansas City (USA). Using best estimates of construction costs, operating expenses and federal funding, resulted net present value of the project was negative $343 million. Therefore, from a standard NPV perspective the Kansas City light rail transit (LRT) system is unlikely to break even. However, in case if the negative externalities of auto travel and the positive externalities associated with light rail are properly accounted for in a comprehensive social cost benefit framework, investment in the Kansas City LRT system becomes an increasingly feasible option. Hossain and Rahman (2007) analyzed the present condition of rail transportation in Bangladesh especially between Dhaka-Chittagong by comparative analysis with other modes of transportation (road and inland water transport). The present railway link between Dhaka and Chittagong is not straight one, rather it has a huge rounding loop between Dhaka and Laksam. Presented study deals with the necessity of a direct railway connection between the two major cities for the transport related economic benefits of the country. Using the discounted net benefit and cost, the BC ratio of the proposed direct link has been estimated for different conditions. A sensitivity analysis of BC ratio with different project evaluation period and discount rate has also been performed in this paper.

Assessing the feasibility and profitability of large investment projects requires the consideration of various aspects and procedures. The projected outcomes of feasibility and profitability are frequently subject to a partially or even fully undeterminable future, encompassing uncertainty and various types of risk. Current approaches mainly consider sensitivity analysis as a way to treat uncertainty in input parameters. Therefore, it is necessary to further strengthen uncertainty in railway related investment appraisal applications through stochastic, fuzzy or fuzzy stochastic approach.

9. CASE STUDY: ECONOMIC ANALYSIS OF RAILWAY LINE RENEWAL

Let us now analyze the possibility of railway line renewal from the aspect of economic analysis. Namely, in this decision problem there are two mutually exclusive alternatives:

Alternative A: Without renewal of a railway line. This alternative is known as "do minimum" scenario - this option includes a certain amount of investments for necessary expanding of fleet, in order to cope with increasing intensity of traffic services.

Alternative B: With renewal of a railway line. This alternative is known as "do-something" scenario and it includes complete reconstruction of railway line. This investing process is a multi period process and it is planned to be realized with following dynamics: 30% of investments in first and second year, and remaining 40% in third year. Total amount of investment is:
   - $58.260 million, in substructure;
   - $29.130 million, in superstructure;
   - $19.42 million, in contact line;
   - $24.23 million, in signal safety devices;

Renewed railway line will be capable for maximum speed of 200 km/h. For purposes of this analysis, it is assumed that passenger trains will operate by average speed of 200 km/h, and freight trains by 120 km/h. Cost savings relative to „do minimum" alternative (A) are calculated for 1., 2., 3., 5., 10., 15., 20., i 25. year. Therefore, it is needed to estimate the effects of investment on time horizon of 25 years. Realized volume of passenger and freight traffic is given in Table 7.

In order to compare these two alternatives, it is needed to estimate future flows on considered railway line. Using some of the forecasting methods (in this case exponential smoothing) it is necessary to make a one year ahead forecast of volume of passenger and freight traffic. Then, for remaining periods, perspective volume of passenger traffic is predicted to increase for 2.25% in first 6 years, 2.5% in next 5 years and 3% in rest of the planning period. In freight traffic, perspective volume is assumed to increase for 2.75% in first 6 years and 2.5% in remaining periods.

Other input data:

- Average mass of passenger trains $m_p=400(t)$;
- Average mass of passenger trains $m_t=955+5k(t)$, $0<k<24$;
  - Average number of wagons in passenger trains is 10;
  - Average number of cars in freight trains is 32;
  - Immobilization coefficient for passenger wagons, freight cars and locomotives is 1.15;

Table 7. Volume of passenger and freight traffic in previous period

Year	Gross Tones of Passenger Trains [10³]	Weight of Commodities in Tones [10³]
1	4426,63	2805
2	4763,84	2890
3	4777,81	3241
4	4872,84	3421
5	4264,27	3988
6	4226,23	3158
7	3978,53	2573
8	3970,61	2880
9	4520,81	3331
10	4503,11	3500
11	4958,82	3886
12	5178,65	3882
13	4843,29	4144
14	5016,33	4713
15	5223,98	5064
16	5136,12	5060

- Purchasing price of locomotives is $2.95 mil., passenger wagons $1.3 mil., and freight cars $0.07 mil.;
- Gross work coefficient is 2.22;
- Monetary value of one hour of train crew in passenger traffic is $121.80, and in freight traffic $36.13;
- The length of railway line is 97.1 km.

In the first step it is necessary to make a one year ahead forecast of future intensity in passenger and freight traffic on considered railway line. For that purpose exponential smoothing or exponentially weighted moving average method is selected (Table 8). Exponential smoothing model predicts the value of a variable in next period (F_{n+1}) as sum of forecast for the last period (F_n) and a part of error made in that period ($Y_n - F_n$). In other words, forecast in $n+1$, is equal to:

$$F_{n+1} = F_n + a(Y_n - F_n) \tag{36}$$

where a represents exponential smoothing coefficient ($0<a<1$). For any a between 0 and 1, the weights attached to the observations decrease exponentially as we go back in time, hence the name "exponential smoothing". If a is small, more weight is given to observations from the more distant past. If a is large, more weight is given to the more recent observations. At the extreme case where $a=1$, $F_{n+1}=Y_n$ and forecasts are equal to the naive forecasts.

Table 8. Forecasting of the volume of railway passenger traffic in next period

Godina	Y_i	a=0,2		a=0,5		a=0,8	
		F_i	e_i	F_i	e_i	F_i	e_i
1	4426.63						
2	4763.84	4426.63	337.21	4426.63	337.21	4426.63	337.21
3	4777.81	4494.07	283.74	4595.24	182.58	4696.40	81.41
4	4872.84	4550.82	322.02	4686.52	186.32	4761.53	111.31
5	4264.27	4615.22	350.95	4779.68	515.41	4850.58	586.31
6	4226.23	4545.03	318.80	4521.98	295.75	4381.53	155.30
7	3978.53	4481.27	502.74	4374.10	395.57	4257.29	278.76
8	3970.61	4380.72	410.11	4176.32	205.71	4034.28	63.67
9	4520.81	4298.70	222.11	4073.46	447.35	3983.34	537.47
10	4503.11	4343.12	159.99	4297.14	205.97	4413.32	89.79
11	4958.82	4375.12	583.70	4400.12	558.70	4485.15	473.67
12	5178.65	4491.86	686.79	4679.47	499.18	4864.09	314.56
13	4843.29	4629.22	214.07	4929.06	85.77	5115.74	272.45
14	5016.33	4672.03	344.30	4886.18	130.15	4897.78	118.55
15	5223.98	4740.89	483.09	4951.25	272.73	4992.62	231.36
16	5136.12	4837.51	298.61	5087.62	48.50	5177.71	41.59
17		4897.23		5111.87		**5144.44**	
Σ			5518.23		4366.89		3693.41

For a set of three values of smoothing coefficients (0.2, 0.5 and 0.8), time series is fitted and appropriate forecasts calculated successively according to (36). The first observation is used as the forecast for the second period ($F_2=Y_1$). The last row of the table gives the total error for each a. The forecast made for value of a equal to 0.8 is selected. Namely, for this value of smoothing coefficient the forecast error ($e_i=|F_i-Y_i|= 3693.41$) is minimized. The same procedure is conducted for railway freight case in Table 9.

Table 10 contains future estimates of flows respecting to input assumptions about projections of increasing of the volume of passenger on considered railway line. Therefore, starting with one year ahead forecast of passenger trains in previous step, appropriate growth rates are applied and gross tones of passenger trains are calculated (1). Gross tone-km (2) are obtained by multiplying (1) with the length of railway line (97.1). Yearly number of trains (4) is a result of division (2)/(3). Daily number of trains (5) is equal to division of (4) with number of days in a year.

The same procedure is applied for estimating the perspective volume of freight traffic (Table 11.), except that there is a need to transform net tones of freight trains in gross tones. Therefore, (3) is obtained as a product of net tones (1) and gross work coefficient (2.22).

Table 12 contains travelling time data in case of "do minimum" alternative (current situation, speed of passenger trains is 80 km/h, and freight trains 60 km/h), and renewal alternative which assumes a 3

Table 9. Forecasting of the volume of railway freight traffic in next period

Year	Y_i	a=0.2		a=0.5		a=0.8	
		F_i	e_i	F_i	e_i	F_i	e_i
1	2805						
2	2890	2805.00	85.00	2805.00	85.00	2805.00	85.00
3	3241	2822.00	419.00	2847.50	393.50	2873.00	368.00
4	3421	2905.80	515.20	3044.25	376.75	3167.40	253.60
5	3988	3008.84	979.16	3232.63	755.38	3370.28	617.72
6	3158	3204.67	46.67	3610.31	452.31	3864.46	706.46
7	2573	3195.34	622.34	3384.16	811.16	3299.29	726.29
8	2880	3070.87	190.87	2978.58	98.58	2718.26	161.74
9	3331	3032.70	298.30	2929.29	401.71	2847.65	483.35
10	3500	3092.36	407.64	3130.14	369.86	3234.33	265.67
11	3886	3173.89	712.11	3315.07	570.93	3446.87	439.13
12	3882	3316.31	565.69	3600.54	281.46	3798.17	83.83
13	4144	3429.45	714.55	3741.27	402.73	3865.23	278.77
14	4713	3572.36	1140.64	3942.63	770.37	4088.25	624.75
15	5064	3800.49	1263.51	4327.82	736.18	4588.05	475.95
16	5060	4053.19	1006.81	4695.91	364.09	4968.81	91.19
17		4254.55		4877.95		5041.76	
Σ			8967.51		6870.00		5661.45

year process of reconstructing (in the 1. year 30% of the total length will be reconstructed, in the 2. year 60% and in the third year the whole line will have projected speeds).

In Tables 13 and 14 relative savings between two alternatives are calculated for passenger and freight cars on the base of decreasing of travelling time on railway line. Number of passenger/freight cars in a train (3) is obtained as a product of yearly number of trains and average number of cars in a train. Decreasing of travelling time will imply savings in car-hours (carh) proportional to total yearly number of passenger/freight cars in a train ((5)=(3)•(4)). Savings in number of cars (6) are obtained as a division of savings in carh with total number of hours in a year (8760h). These savings are increased due to immobilization (7). According to purchasing price of passenger/freight cars total savings can be calculated (9). These total savings due to investment ("renewal" alternative) are considered as investments in "do minimum" alternative (10). The same procedure is applied for savings in locomotives (Table 15).

Tables 16 and 17 contain an assessment of total savings in train crew costs due to decreased travelling time for passenger as well as for freight traffic, respectively.

Plan of investment is given in Table 18. With respect to alternative B (Renewal), alternative A (do minimum) will contain costs for expanding the fleet (passenger wagons, freight cars and locomotives) due to increased intensity of traffic. Investments in B are scheduled in first three years of planning horizon.

Table 10. Perspective volume of passenger traffic on railway line A-B

Year	Estimated Growth Rate (%)	Gross Tones of Passenger Trains (1)	Gross Tone-KM of Passenger Trains (2)	Mass of Trains (3)	Yearly Number of Trains (4)	Daily Number of Trains (5)
1	2.15	5144.44	499524.93	400.00	12861	35
2		5255.04	510264.72	400.00	13138	36
3		5368.03	521235.41	400.00	13420	37
4		5483.44	532441.97	400.00	13709	38
5		5601.33	543889.47	400.00	14003	38
6		5721.76	555583.09	400.00	14304	39
7	2.45	5861.95	569194.88	400.00	14655	40
8		6005.56	583140.15	400.00	15014	41
9		6152.70	597427.09	400.00	15382	42
10		6303.44	612064.05	400.00	15759	43
11		6457.87	627059.62	400.00	16145	44
12		6616.09	642422.58	400.00	16540	45
13		6778.19	658161.94	400.00	16945	46
14		6944.25	674286.90	400.00	17361	48
15		7114.39	690806.93	400.00	17786	49
16		7288.69	707731.70	400.00	18222	50
17		7467.26	725071.13	400.00	18668	51
18		7650.21	742835.37	400.00	19126	52
19		7837.64	761034.84	400.00	19594	54
20		8029.66	779680.19	400.00	20074	55
21		8226.39	798782.36	400.00	20566	56
22		8427.94	818352.52	400.00	21070	58
23		8634.42	838402.16	400.00	21586	59
24		8845.96	858943.01	400.00	22115	61
25		9062.69	879987.12	400.00	22657	62

In Table 19, structure and dynamic of investments is given. For A there are necessary expenses in passenger cars, freight cars and locos due to increased volume of traffic (from Table 13-15.). Alternative B generates planned investments in renewal of railway line.

On the base of calculated parameters total costs for both alternatives are calculated in Table 20. These costs are composed from investment and operational costs. Investment cost of A are expenses for increasing the fleet of wagons, cars and locos whereas for B these costs are based on railway line renewal process. Operational costs include crew costs and costs for current and investment maintenance for A, and only the second component of these costs in case of B.

Table 11. Perspective volume of freight traffic on railway line A-B

Year	Estimated Growth Rate (%)	Net Tones of Freight Trains (1)	Net Tone-KM of Freight Trains (2)	Gross Tones of Freight Trains (3)	Gross Tone-KM of Freight Trains (4)	Mass of Trains (5)	Yearly Number of Trains (6)	Daily Number of Trains (7)
1	3.00	5041,76	489555,09	11192,71	1086812,30	955	11720	32
2		5193,01	504241,74	11528,49	1119416,67	960	12009	33
3		5348,81	519369,00	11874,35	1152999,17	965	12305	34
4		5509,27	534950,07	12230,58	1187589,14	970	12609	35
5		5674,55	550998,57	12597,50	1223216,82	975	12921	35
6		5844,78	567528,52	12975,42	1259913,32	980	13240	36
7	2.75	6005,52	583135,56	13332,24	1294560,94	985	13535	37
8		6170,67	599171,79	13698,88	1330161,37	990	13837	38
9		6340,36	615649,01	14075,60	1366740,80	995	14146	39
10		6514,72	632579,36	14462,68	1404326,18	1000	14463	40
11		6693,88	649975,29	14860,40	1442945,14	1005	14786	41
12	2.35	6851,18	665249,71	15209,62	1476854,36	1010	15059	41
13		7012,18	680883,08	15567,05	1511560,43	1015	15337	42
14		7176,97	696883,83	15932,87	1547082,10	1020	15620	43
15		7345,63	713260,60	16307,30	1583438,53	1025	15910	44
16		7518,25	730022,22	16690,52	1620649,34	1030	16204	44
17		7694,93	747177,75	17082,75	1658734,60	1035	16505	45
18		7875,76	764736,42	17484,19	1697714,86	1040	16812	46
19		8060,84	782707,73	17895,07	1737611,16	1045	17124	47
20		8250,27	801101,36	18315,60	1778445,02	1050	17443	48
21		8444,15	819927,24	18746,02	1820238,48	1055	17769	49
22		8642,59	839195,53	19186,55	1863014,08	1060	18101	50
23		8845,69	858916,63	19637,43	1906794,92	1065	18439	51
24		9053,57	879101,17	20098,91	1951604,60	1070	18784	51
25		9266,32	899760,05	20571,24	1997467,30	1075	19136	52

In order to find internal rate of return, relative effect of this two investments (T1-T2) is calculated and discounted for a set of discount rate values. Results of discounting are given in Table 21.

Various facilities have lifetime which is often different than discounting period, so it is necessary to calculate the salvage value and its value. Then, total investments have to be decreased for calculated salvage value. Salvage value in this case study is calculated as follows:

$$V_m = (I-V)\frac{(1+i)^n - (1+i)^m}{(1+i)^n - 1} + V \qquad (37)$$

Table 12. Average travelling time of passenger and freight trains

Type of Train	Average Travelling Time of a Train (h)				Decreasing of Travelling Time		
	Without Renewal	With Renewal					
		1	2	3	1	2	3
Passenger	1.214	0.995	0.777	0.486	0.218	0.437	0.728
Freight	1.618	1.376	1.133	0.809	0.243	0.486	0.809

Table 13. Savings in passenger wagons

No.	Description	Year							
		1	2	3	5	10	15	20	25
1	Daily number of passenger trains	35	36	37	38	43	49	55	62
2	Yearly number of passenger trains	12861	13138	13420	14003	15759	17786	20074	22657
3	Number of passenger cars in a train	128610	131380	134200	140030	157590	177860	200740	226570
4	Decreasing of travelling time	0.218	0.437	0.728	0.728	0.728	0.728	0.728	0.728
5	Savings (carh)	28098.07	57406.49	97731.15	101976.85	114764.92	129526.54	146188.90	164999.60
6	Savings (car)	3.208	6.553	11.157	11.641	13.101	14.786	16.688	18.836
7	Increasing due to immobilization	3.689	7.536	12.830	13.387	15.066	17.004	19.191	21.661
8	Purchasing price of a wagon [$]	1.300	1.300	1.300	1.300	1.300	1.300	1.300	1.300
9	Total savings (10^6) [$]	4.795	9.797	16.679	17.404	19.586	22.105	24.949	28.159
10	Investments (10^6) [$]	4.795	5.002	6.882	0.725	2.182	2.519	2.844	3.210

Table 14. Savings in freight cars

No.	Description	Year							
		1	2	3	5	10	15	20	25
1	Daily number of freight trains	32	33	34	35	40	44	48	52
2	Yearly number of freight trains	11720	12009	12305	12921	14463	15910	17443	19136
3	Number of freight cars in a train	375040	384288	393760	413472	462816	509120	558176	612352
4	Decreasing of travelling time	0.121	0.243	0.405	0.405	0.405	0.405	0.405	0.405
5	Savings (carh)	45379.8	93381.9	159472.8	167456.2	187440.4	206193.6	226061.2	248002.5
6	Savings (car)	5.180	10.660	18.205	19.116	21.397	23.538	25.806	28.311
7	Increasing due to immobilization	5.957	12.259	20.935	21.983	24.607	27.069	29.677	32.557
8	Purchasing price of a freight car [$]	0.070	0.070	0.070	0.070	0.070	0.070	0.070	0.070
9	Total savings (10^6) [$]	0.417	0.858	1.465	1.539	1.722	1.895	2.077	2.279
10	Investments (10^6) [$]	0.417	0.441	0.607	0.073	0.184	0.172	0.183	0.202

Table 15. Savings in locomotives

No.	Description	Year							
		1	2	3	5	10	15	20	25
1	Daily number of locos	67	69	71	73	83	93	103	114
2	Yearly number of locos	11720	12009	12305	12921	14463	15910	17443	19136
		12861	13138	13420	14003	15759	17786	20074	22657
3	Number of locos in a train	1	1	1	1	1	1	1	1
4	Decreasing of travelling time	0.218475	0.43695	0.72825	0.72825	0.72825	0.72825	0.72825	0.72825
		0.24275	0.4855	0.809167	0.809167	0.809167	0.809167	0.809167	0.809167
5	Savings (locoh)	5682.535	11625.832	19820.133	20740.479	23284.337	25978.296	28946.076	32269.081
6	Savings (locos)	0.649	1.327	2.263	2.368	2.658	2.966	3.304	3.684
7	Increasing due to immobilization	0.746	1.526	2.602	2.723	3.057	3.410	3.800	4.236
8	Purchasing price of a locomotive [$]	2.950	2.950	2.950	2.950	2.950	2.950	2.950	2.950
9	Total savings (10^6) [$]	2.201	4.502	7.676	8.032	9.017	10.061	11.210	12.497
10	Investments (10^6) [$]	2.201	2.302	3.173	0.356	0.985	1.043	1.149	1.287

Table 16. Savings in train crew for passenger traffic

No.	Description	Year							
		1	2	3	5	10	15	20	25
1	Yearly number of passenger trains	12861	13138	13420	14003	15759	17786	20074	22657
2	Decreasing of travelling time	0.218	0.436	0.728	0.728	0.728	0.728	0.728	0.728
3	Savings (trainh)	2809.8	5740.6	9773.1	10197.6	11476.4	12952.6	14618.8	16499.9
4	Value of 1h of train crew [$]	121.8	121.8	121.8	121.8	121.8	121.8	121.8	121.8
5	Savings in crew (10^6) [$]	0.342	0.699	1.190	1.242	1.398	1.578	1.781	2.010

Table 17. Savings in train crew freight traffic

No.	Description	Year							
		1	2	3	5	10	15	20	25
1	Yearly number of freight trains	11720	12009	12305	12921	14463	15910	17443	19136
2	Decreasing of travelling time	0.242	0.485	0.809	0.809	0.809	0.809	0.809	0.809
3	Savings (trainh)	2845.0	5830.3	9956.7	10455.2	11702.9	12873.84	14114.29	15484.21
4	Value of 1h of train crew [$]	36.13	36.13	36.13	36.13	36.13	36.13	36.13	36.13
5	Savings in crew (10^6) [$]	0.103	0.211	0.360	0.378	0.423	0.465	0.510	0.559

Table 18. Plan of investments

Year		1	2	3	5	10	15	20	25
Investments [$]	A	7.412978	7.744623	10.66265	1.154367	3.351249	3.734876	4.175545	4.698808
	B	39.312	39.312	52.416	0	0	0	0	0

Table 19. Structure and dynamic of investments

No.	Structure	Dynamics														
		1		2		3		5		10		15		20		25	
		A	B	A	B	A	B	A	B	A	B	A	B	A	B	A	B
1	Substructure [$]		17.471		17.471		23.304										
2	Superstructure [$]		8.742		8.742		11.652										
3	Contact line [$]		5.824		5.824		5.826										
4	Signal-safety devices [$]		7.269		7.269		9.692										
5	Locos [$]	2.200		2.301		3.173		0.356		0.985		1.043		1.149		1.286	
6	Passenger wagons [$]	4.795		5.001		6.881		0.724		2.182		2.519		2.843		3.210	
7	Freight cars [$]	0.417		0.441		0.607		0.073		0.183		0.172		0.182		0.201	
8	Total [$]	7.412	39.31	7.74	39.31	10.662	50.474	1.154	0	3.351	0	3.734	0	4.175	0	4.698	0

where

V_m: Salvage value after *m* years of exploitation;
I: Purchasing price of an investment;
V: Salvage value at the end of lifetime of assets (after *n* years);
n: Lifetime of asset;
m: Age of use of an asset at the end of the last year of discounting period;

This expression gives discounted value of an asset which is expected after the period of use, by selling of that asset. Results of salvage value calculation are given in Tables 22 and 23.

At the end, in Table 24, incremental of investments and salvage values are summed up and a total discounted effect is obtained.

Therefore, on the base of comparison of alternatives A (without renewal) and B (with renewal) profitability of investment of renewal a railway line from the aspect of investor is calculated. Results of calculation give the IRR equal to 26.1% which is larger than MARR=12% (Figure 5). So, from the aspect of direct economic effects for investor this investment is reasonable. However, in order to perform a cost benefit analysis it is needed to include all indirect or social effects of considered investment.

Table 20. Plan of investments and operational costs (mil.)

| Year | Without Renewal ||||||| With Renewal ||| |
| | Investments for Increased Number ||| Operational Cost |||| | | | |
	Locos [$]	Passenger Wagons [$]	Freight cars [$]	Crew [$]	Current and Investment Maintenance [$]	Investments [$]	Operational Costs [$]	Total T1 [$]	Investments in Railway Line Renewal [$]	Current and Investment Maintenance [$]	Total T2 [$]
1	2.200	4.795	0.417	0.445	20.0	7.412	20.4	27.9	39.312	15	54.312
2	2.301	5.001	0.441	0.909	32.0	7.744	32.9	40.7	39.312	16	55.312
3	3.173	6.881	0.607	1.550	30.0	10.662	31.6	42.2	50.474	10	60.474
4	0	0	0	0	20.0	0	20.0	20.0	0	10	10
5	0.356	0.724	0.073	1.619	25.0	1.154	26.6	27.8	0	13	13
6	0	0	0	0	30.0	0	30.0	30.0	0	12	12
7	0	0	0	0	40.0	0	40.0	40.0	0	11	11
8	0	0	0	0	30.0	0	30.0	30.0	0	12	12
9	0	0	0	0	55.0	0	55.0	55.0	0	17	17
10	0.985	2.182	0.183	1.820	55.0	3.351	56.8	60.2	0	17	17
11	0	0	0	0	40.0	0	40.0	40.0	0	17	17
12	0	0	0	0	30.0	0	30.0	30.0	0	17	17
13	0	0	0	0	30.0	0	30.0	30.0	0	17	17
14	0	0	0	0	30.0	0	30.0	30.0	0	17	17
15	1.043	2.519	0.172	2.042	30.0	3.734	32.0	35.8	0	17	17
16	0	0	0	0	30.0	0	30.0	30.0	0	17	17
17	0	0	0	0	30.0	0	30.0	30.0	0	17	17
18	0	0	0	0	30.0	0	30.0	30.0	0	17	17
19	0	0	0	0	30.0	0	30.0	30.0	0	17	17
20	1.149	2.843	0.182	2.290	30.0	4.175	32.3	36.5	0	17	17
21	0	0	0	0	30.0	0	30.0	30.0	0	17	17
22	0	0	0	0	30.0	0	30.0	30.0	0	17	17
23	0	0	0	0	30.0	0	30.0	30.0	0	17	17
24	0	0	0	0	30.0	0	30.0	30.0	0	17	17
25	1.286	3.210	0.201	2.569	30.0	4.698	32.6	37.3	0	17	17

Table 21. Discounted effect

Year	T1-T2 [$]	Discounted Value				
		15%	20%	25%	30%	35%
1	-26.454	-23.003	-22.045	-21.163	-20.349	-19.596
2	-14.658	-11.083	-10.179	-9.381	-8.673	-8.043
3	-18.261	-12.007	-10.568	-9.350	-8.312	-7.422
4	10.000	5.718	4.823	4.096	3.501	3.011
5	14.774	7.345	5.937	4.841	3.979	3.295
6	18.000	7.782	6.028	4.719	3.729	2.974
7	29.000	10.902	8.093	6.082	4.622	3.549
8	18.000	5.884	4.186	3.020	2.207	1.632
9	38.000	10.802	7.365	5.100	3.583	2.551
10	43.172	10.671	6.973	4.636	3.132	2.147
11	23.000	4.944	3.096	1.976	1.283	0.847
12	13.000	2.430	1.458	0.893	0.558	0.355
13	13.000	2.113	1.215	0.715	0.429	0.263
14	13.000	1.837	1.013	0.572	0.330	0.195
15	18.778	2.308	1.219	0.661	0.367	0.208
16	13.000	1.389	0.703	0.366	0.195	0.107
17	13.000	1.208	0.586	0.293	0.150	0.079
18	13.000	1.050	0.488	0.234	0.116	0.059
19	13.000	0.913	0.407	0.187	0.089	0.043
20	19.466	1.189	0.508	0.224	0.102	0.048
21	13.000	0.691	0.283	0.120	0.053	0.024
22	13.000	0.601	0.235	0.096	0.040	0.018
23	13.000	0.522	0.196	0.077	0.031	0.013
24	13.000	0.454	0.164	0.061	0.024	0.010
25	20.268	0.616	0.212	0.077	0.029	0.011
Total	35.276	12.395	-0.849	-8.784	-13.623	-33128.083

Figure 5. IRR of alternative B

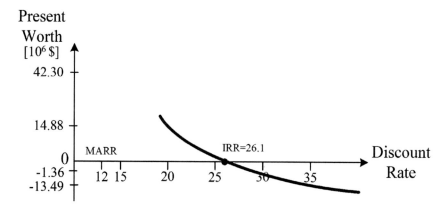

Table 22. Salvage value for Alternative A (without renewal)

No.	Name of asset	Value of asset [$]	Salvage Value [$]	Lifetime	Years of Use	Salvage Value [$]				
						15%	20%	25%	30%	35%
0	Locos	2.95	0.59	40	25	2.67	2.80	2.87	2.90	2.92
1		2.95	0.59	40	24	2.71	2.82	2.88	2.91	2.93
2		2.95	0.59	40	23	2.74	2.85	2.90	2.92	2.94
3		2.95	0.59	40	20	2.81	2.89	2.92	2.94	2.94
4		2.95	0.59	40	15	2.89	2.93	2.94	2.95	2.95
5		2.95	0.59	40	10	2.92	2.94	2.95	2.95	2.95
6		2.95	0.59	40	5	2.94	2.95	2.95	2.95	2.95
7		2.95	0.59	40	0	2.95	2.95	2.95	2.95	2.95
0	Passenger wagons	1.3	0.195	35	25	1.03	1.12	1.18	1.22	1.25
1		1.3	0.195	35	24	1.07	1.15	1.21	1.24	1.26
2		1.3	0.195	35	23	1.10	1.18	1.22	1.25	1.27
3		1.3	0.195	35	20	1.17	1.23	1.26	1.28	1.29
4		1.3	0.195	35	15	1.24	1.27	1.29	1.29	1.30
5		1.3	0.195	35	10	1.27	1.29	1.30	1.30	1.30
6		1.3	0.195	35	5	1.29	1.30	1.30	1.30	1.30
7		1.3	0.195	35	0	1.30	1.30	1.30	1.30	1.30
0	Freight cars	0.07	0.007	40	25	0.06	0.07	0.07	0.07	0.07
1		0.07	0.007	40	24	0.06	0.07	0.07	0.07	0.07
2		0.07	0.007	40	23	0.06	0.07	0.07	0.07	0.07
3		0.07	0.007	40	20	0.07	0.07	0.07	0.07	0.07
4		0.07	0.007	40	15	0.07	0.07	0.07	0.07	0.07
5		0.07	0.007	40	10	0.07	0.07	0.07	0.07	0.07
6		0.07	0.007	40	5	0.07	0.07	0.07	0.07	0.07
7		0.07	0.007	40	0	0.07	0.07	0.07	0.07	0.07
	Total					32.64	33.52	33.97	34.21	34.35

10. CONCLUSION

In today's increasingly competitive business climate for a railway company, there is a growing requirement for stronger cost control and a demand for higher returns while minimizing risks of investments. Recognition of the potential impact of railway services on the strategic power of a railway company makes the evaluation, justification and control of such investments a critically important issue. The primary purpose of this chapter was to present the basic principles and applications of the most important investment appraisal techniques in a clearly written fashion, supported by a number of railway related examples. Additionally, chapter contains a review of applications of presented methods within the do-

Table 23. Salvage value for Alternative B (with renewal)

No	Name of Asset	Value of Asset [$]	Salvage Value [$]	Lifetime	Years of Use	Salvage Value [$]				
						15%	20%	25%	30%	35%
1	Substructure	58.26	0	60	25	57.836	58.1624	58.23646	58.25402	58.2584
2		58.26	0	60	24	57.893	58.17884	58.24118	58.2554	58.25882
3		58.26	0	60	23	57.942	58.19254	58.24496	58.25646	58.25912
1	Superstructure	29.13	2.913	25	25	2.913	2.913	2.913	2.913	2.913
2		29.13	2.913	25	24	6.440	7.328789	8.176284	8.971663	9.713752
3		29.13	2.913	25	23	9.506	11.00861	12.38691	13.63217	14.75135
1	Contact line	19.42	2.913	50	25	18.933	19.24876	19.35787	19.39664	19.4109
2		19.42	2.913	50	24	18.999	19.2776	19.37035	19.40204	19.41326
3		19.42	2.913	50	23	19.056	19.30164	19.38032	19.40619	19.41501
1	Signal-safety devices	24.23	1.938	25	25	1.938	1.9384	1.9384	1.9384	1.9384
2		24.23	1.938	25	24	4.937	5.693025	6.413627	7.089916	7.720894
3		24.23	1.938	25	23	7.545	8.821879	9.993808	11.05262	12.00422
	Total					263.938	270.0655	274.6532	278.5685	282.0571

Table 24. Present value of total costs in function with discount rate

	Discount Rate				
	15%	20%	25%	30%	35%
Discounted costs [10⁶ $]	35.28	12.39	-0.84	-8.78	-13.62
Discounted salvage value [10⁶ $]]	7.03	2.48	0.91	7.42	0.14
Total [10⁶ $]]	42.30	14.88	0.06	-1.36	-13.49

main of railway engineering. This material will serve reader to gain valuable insight into how this tools work, how they can be applied and what difficulties decision makers on railway face when implementing presented methods. At the same time, readers will have a good base for considering further developments of presented methods in the way of undeterminable future, uncertainty and various types of risk.

REFERENCES

Blank, L., & Tarquin, A. (2002). *Engineering Economy* (6th ed.). USA: McGraw-Hill Higher Education.

Blank, L., & Tarquin, A. (2008). *Basics of Engineering Economy*. USA: McGraw-Hill Higher Education.

Campbell, H. F., & Brown, R. P. (2003). Investment Appraisal: Decision-Rules. In Benefit-Cost Analysis (pp. 36-61). United Kingdom: Cambridge University Press.

Casares, P., & Coto-Millan, P. (2011). Passenger transport planning. A Benefit-Cost Analysis of the High Speed Railway: The case of Spain. *Atlantic Review of Economics, 2*, 1–12.

Cowie, J. (2010). *The Economics of Transport – A theoretical and applied perspective*. USA: Routledge.

de Rus, G. (2012). Economic evaluation of the high speed rail. Expert Group on Environmental Studies. Sweden: Ministry of Finance. Retrieved from http://www.ems.expertgrupp.se/Default.aspx?pageID=3

Guide to Cost-Benefit Analysis of Investment Projects. (2008). *European Commission*. Retrieved from http://ec.europa.eu/regional_policy/sources/docgener/guides/cost/guide2008_en.pdf

Hossain, M., & Rahman, M. S. (2007). Economic feasibility of Dhaka-Laksam direct railway link. *Journal of Civil Engineering*, 35(1), 47–58.

Hunt, D. (2005). *Return on Investment on Freight Rail Capacity Improvement*. United Kingdom: Cambridge Systematics. Inc.

Ibraheem, A. Th. (2011). Evaluating light-rail transit alternatives using the rating and ranking method. *Journal of Engineering and Applied Sciences*, 6(10), 93–104.

Jovanović, P. (2005). Upravljanje investicijama. Faculty of Organizational Sciences, Serbia.

Kolonko, M., & Engelhardt-Funke, O. (2001). Cost-Benefit Analysis of Investments into Railway Networks with Periodically Timed Schedules. *Computer-Aided Scheduling of Public Transport Lecture Notes in Economics and Mathematical Systems*, 505, 443–459. doi:10.1007/978-3-642-56423-9_25

Olsson, N., Okland, A., & Halvorsen, S. (2012). Consequences of differences in cost-benefit methodology in railway infrastructure appraisal - A comparison between selected countries. *Transport Policy*, 22, 29–35. doi:10.1016/j.tranpol.2012.03.005

Raju, S. (2008). Project NPV, positive externalities, social cost-benefit analysis—the Kansas City light rail project. *Journal of Public Transportation*, 11(4), 59–88. doi:10.5038/2375-0901.11.4.4

Riggs, J., & West, T. (1986). *Engineering Economics*. USA: McGraw-Hill.

Sprague, J. C., & Whittaker, J. D. (1986). *Economic Analysis for Engineers and Managers*. USA: Prentice Hall.

Tao, R., Liu, S., Huang, C., & Tam, C. M. (2011). Cost-Benefit Analysis of High-Speed Rail Link between Hong Kong and Mainland China. *Journal of Engineering*, 1(1), 36–45.

Wang, R., Kudrot-E-Khuda, M., Nakamura, F., & Tanaka, S. (2014). A Cost-Benefit Analysis of Commuter Train Improvement in the Dhaka Metropolitan Area. Bangladesh. *Procedia: Social and Behavioral Sciences*, 138, 819–829. doi:10.1016/j.sbspro.2014.07.231

Wiegmans, B. W. (2008). *The economics of a new rail freight line: The case of the Betuweline in the Netherlands*. Association for European Transport and contributors.

Yao, J.-S., Chen, M.-S., & Lin, H.-W. (2005). Valuation by using a fuzzy discounted cash flow model. *Expert Systems with Applications*, 28(2), 209–222. doi:10.1016/j.eswa.2004.10.003

Chapter 5
Railway Demand Forecasting

Miloš Milenković
Zaragoza Logistic Center, Spain & University of Belgrade, Serbia

Nebojša Bojović
University of Belgrade, Serbia

ABSTRACT

Forecasting represents an indispensable activity in railway transportation planning. Forecasting of demand levels is vital to the railway company as a whole as it provides the basic input for the planning and control of all functional areas including railway transport operations planning, marketing and finance. Demand levels and the timing of their appearance (on a day, week, month or seasonal basis) greatly effects capacity levels, financial needs and general structure of the business.Forecasting employs historical data and uses various forecasting methods to make accurate estimates of future demands. Forecasting approaches can be generally divided into two categories: econometric or causal and time series techniques. In this chapter a comprehensive review of methods belonging to these two broad classes will be made. Special emphasis will be given to the application of these techniques to railway demand modeling.

1. THE NATURE AND USES OF FORECASTS IN RAILWAY FIELD

Forecasting represents a process of predicting or estimating the future. It provides information about the potential future events and their consequences for an organization. It may not reduce the complications and uncertainty of the future but it increases the confidence of the management to make important decisions.

Railway companies use forecasting methods in order to anticipate potential issues and results for the business in the upcoming months and years. The essence of railway transportation planning and management is to match transport supply with railway demand. A thorough understanding of existing pattern of railway customers is the key for identifying and analyzing existing railway traffic related problems. Detailed data on current pattern and railway traffic volumes are needed also for developing demand forecasting/prediction models. The prediction of future demand is an essential task of the long-range railway transportation planning process for determining strategies for accommodating future needs. These strategies may include land use policies, pricing programs, and expansion of transportation supply – high speed railway lines and express services.

DOI: 10.4018/978-1-5225-0084-1.ch005

The demand model is a base of railway transport forecasts. Via this model a possible causal relationship can be found between the subject of forecasting process (number of passenger trains on a railway line) and the factors influencing on it (GDP, quality of service, travel times, prices, etc.). After determining the causalities and checking for statistical and logical validity, model can be used for forecast of railway demand in future.

Railway transport demand may be divided into passenger and freight demand. Passenger demand consists of intercity demand and commuting demand. Intercity demand is based on business activities or leisure. The same holds for commuting, but all components of commuting demand has similar characteristics and for this reason there is no difference between business and leisure. Freight demand is almost always part of the industrial process. It is under the great influence of type of goods, geographical and socio-economic parameters, pricing policy, seasonality, etc. (Profillidis, 2006).

Forecasting methods used for railway demand modeling can include both quantitative data and qualitative observations. The chief advantage of qualitative methods is that the main source of data derives from the experiences of qualified executives and employees. On the other hand, projections of quantitative forecasting methods rely on the strength of past data.

However, any demand forecasting has one primary disadvantage as that of any other method of predicting the future – there is no absolute certainty about the future. Any unforeseen factors may influence on forecast usability, regardless of the quality of its data.

This chapter is organized as follows. In Section 2, a classification of forecasting methods is presented. In Section 3, qualitative forecasting techniques are described, and a review of their application in the field of railway demand forecasting is given. Section 4, contain a review of quantitative forecasting techniques and their applications in the domain of railway demand forecasting. A survey of other forecasting methods with railway related applications is presented in Section 5.

2. METHODS OF FORECASTING

Considering railway demand forecasting, one classification of forecasting methods concerns the time in the future they cover. Long-term forecasts look ahead 5 to 10 years. Medium-term forecasts extend from 2 to 5 years into the future and short-term forecasts involve predictive intervals within 6 to 18 months.

Short and medium-term forecasts are required for activities that range from operations management to budgeting and selecting new research and development projects in railway transport. Long term forecasts impact issues such as strategic planning in the domain of railway infrastructure and mobile capacities.

The time horizon affects the choice of forecasting method because of the availability and relevance of historical data, time available to make the forecast, cost involved, seriousness of errors, effort considered worthwhile (Waters, 2008).

Despite the wide range of problem situations that arise in the field of railway transportation (railway operations management, marketing, finance and risk management) there are only two types of forecasting techniques – qualitative and quantitative methods.

Qualitative forecasting techniques are often subjective in nature and require judgment on the part of experts. These techniques are often used in situations where there is little or no historical data (Figure 1).

Quantitative methods for forecasting the future railway demand are used in case a company has records of past sales and knows the factors that affect them. Quantitative methods may be classified into two broad categories (Figure 2).

Figure 1. Qualitative forecasting techniques

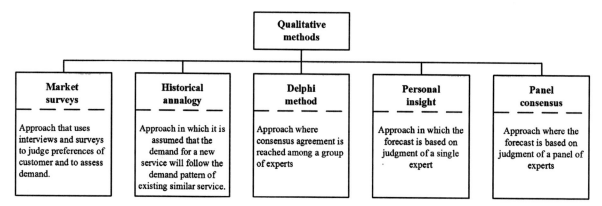

Figure 2. Quantitative forecasting techniques

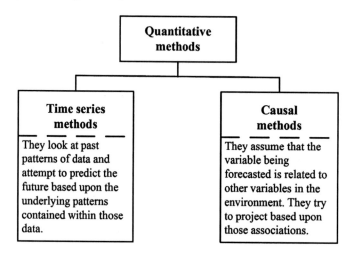

Both of these methods need accurate quantitative data. However, suppose the case of introduction a new railway passenger service for which there is no relevant history. So, there are no historical data for the operator to project forward, and also there is no information about the factors that affect demand for a causal forecast. In this case the use of qualitative techniques is the only choice.

3. QUALITATIVE FORECASTING METHODS

Consider a case where a railway operator is about to market an entirely new service, or a board of managers is considering plans for next 20 years. There is no historical data for a quantitative forecast. In this case, railway operator might use the expert opinion of sales and marketing personnel to subjectively estimate passenger service sales during new service introduction phase in its life cycle. The essence of qualitative forecasts is that they use subjective opinions from informed people.

3.1. Market Surveys

Market surveys are very often used qualitative methods. They however require a significant amount of time (2-5 months) and have a high cost. The market survey is the only method that can be applied when there are no statistical data (opening of a new railway station or construction of a new railway line) or when it is attempted to identify the reactions of customers to certain changes or to the supply of new rail services (Cahantone et al., 1987). Transport market surveys, besides the determination of passenger characteristics, can (by means of appropriate questions) identify passenger intentions (Profillidis, 2006). Next table gives an example of a questionnaire that was used in a market survey of passenger satisfaction with rail services.

3.2. Historical Analogy

When a railway operator introduces a new service, it may have a similar service that it launched recently, and can assume that demand will follow the same pattern. To use historical analogy, managers must have a service that is similar enough to the new one, that was launched recently and for which there is reliable information.

3.3. Delphi Method

Delphi method (Dalkey and Helmer, 1963; Brown, 1968; Sackman, 1974; Linstone and Turoff, 1975) represents the most formal of the judgmental methods with a well defined procedure. A number of experts are contacted by post and each is given a questionnaire to complete and data is collected from a group of experts without the problems of face to face discussions. The replies are analysed and summaries passed back to the experts – with everything done anonymously to avoid undue influences of status. Then each expert is asked to reconsider their original reply in the light of the replies from others and perhaps to adjust their responses. This process of modifying responses in the light of replies made by the rest of the group is repeated usually from three to six times.

Makitalo and Hilmola (2010) made an assessment on the direction in which the Finish railway freight competition will develop and analyzed different views on railway transport policy. This research work is based on a Delphi questionnaire directed at 52 Finnish experts in this branch. Responses on the questionnaire were gathered during year 2005 (competition in Finland in railway freight started 2007) within two rounds with appropriate amount of response rate. Respondents were from the public and private sectors, actors working closely with railway transports and logistics. With an expert profiling matrix, three different railway transport policy viewpoint groups are identified, and character descriptions for these are constructed.

In order to obtain objective and reasonable results on a smaller survey sample some researchers use an integration of the fuzzy concept and the Delphi method, known as the fuzzy Delphi method. With this method, time and costs of collecting questionnaires can be reduced, and experts' opinions can be kept as they are without being twisted (Maskeliunaite et al., 2009).

Wu (2011) explored the indexes for service quality after re-opening the Old Mountain Line Railway in Taiwan from the viewpoints of some tourism experts, tour guides and scholars. Through literature review and interviews with experts, preliminary indexes were determined. Then the fuzzy Delphi method was applied to select proper ones from them, to discuss the key factors needed to be improved based on the current service quality.

3.4. Personal Insight

This method relies entirely on one person, a single expert familiar with the situation producing a forecast based on his own judgment. The method sometimes gives good results, but more often it gives very bad ones and there are countless examples of experts being totally wrong (Waters, 2008).

3.5. Panel Consensus

In case where the forecast is based on the opinion of a number of experts then the approach is called a panel consensus forecasting. This method may be sometimes unfavorably affected by the force of personality of one or few key individuals despite of the fact that it usually results in forecasts that embody the collective wisdom of all consulted experts.

4. QUANTITATIVE FORECASTING METHODS

Quantitative forecasting methods make formal use of historical data and a forecasting model (Montgomery et al., 2008). The model formally summarizes patterns in the data and expresses a statistical relationship between previous and current values of the variable. Then the model is used to project the patterns in the data into the future. In other words, the forecasting model is use to extrapolate past and current behavior into the future. There are two types of forecasting models in general use – econometric and general time series models. Econometric models make use of relationships between the variable of interest and one or more related predictor variables. Sometimes econometric models are called causal forecasting models, because the predictor variables are assumed to describe the forces that cause or drive the observed values of the variable of interest. General time series models employ the statistical properties of the historical data to specify a formal model and then estimate the unknown parameters of this model (usually) by least squares. In subsequent chapters, both types of quantitative forecasting models will be analyzed and their applications to railway demand modeling will be reviewed.

4.1. Econometric Models

The implementation of econometric models has become increasingly fashionable in transport demand modeling. The main reason for this is that nowadays transport demand can involve the analysis of large amounts of data on revealed preferences, such as service quality, ticket price, car ownership, GDP per capita etc., and stated preferences such as opinions, attitudes and intentions for a particular. Many transport companies collect data on these performance measures for their current and their prospective customers, and they usually try to relate these measures with individual-specific characteristics and marketing-mix efforts. The main reason for considering econometric models is that in many cases the number of data points and the number of variables is rather large, and hence simply performing a range of bivariate analyses seems impractical (Fok, 2002). The econometric analysis of a certain model for the above mentioned measures usually involves a range of steps. The first step amounts to specifying a model given the available data, the relevant explanatory variables, and the transport demand problem at hand. Once the model has been specified, one needs to estimate the parameters and their associated confidence regions. Third, one usually considers the empirical validity of the model by performing diagnostic tests on its adequacy, where one typically focuses on the properties of the unexplained part of the model. Given the potential availability of two or more adequate rival models, one seeks to compare these models either on within-sample fit or on out-of-sample forecasting performance. Finally, one can use the ultimately obtained model for forecasting or for policy analysis. It should be noted that the focus in econometric textbooks tends to be on parameter estimation, but it is by no means the single most important issue. Indeed, in practice it is often difficult to specify the model and to compare it with alternatives. In the remaining part of this section, the most important econometric approaches to railway demand modeling are described beginning with the linear regression as a basic tool for econometric.

4.1.1. Univariate and Multivariate Regression Analysis of Passenger Railway Demand

Odgers and Schijndel (2011) looked at passenger rail demand in the Melbourne metropolitan area, i.e., the specific area of interest to the present study, over a twenty-seven year period 1983/84 to 2009/10. A series of univariate linear regression analyses was performed, followed by several multivariate regression analyses of various combinations of the independent variables which showed the highest explanatory power.

The dependent variable investigated in this study is the annual passenger boardings (millions) per year on Melbourne's trains. Selected independent variables for this study are:

Table 2 presents selected characteristics of the univariate regression analysis for the independent variables studied. The first characteristic selected is adjusted R^2. The adjusted R^2 for each of the independent variables apart from variables X1 and X2 to X2b indicates a medium to high level of statistical association the dependent variable and these explanatory variables. The three independent variables with the highest explanatory power based on the $adj\ R^2$ value are in descending order the annual average percentage of total interest payments to household income (X5a, X5b, X5), the estimated resident population in the Melbourne Statistical Division (X6), and the estimated number of persons employed (both full and part time) in the Melbourne Statistical Division (X3) (Table 3). The Standard Error of Estimates (SEE) in Table 1 range from 13.91 for variables X5a to 24.546 for variable X2. In relation to the average value of the dependent variable, annual train boardings, these SEE equal standard errors of estimate of 10.8% to 18.8% of that mean value. Use of this convention indicates that variable X5a offers the lowest standard error of estimate. The t-values presented in Table 1 of each of the ten independent variables are statistically significant when compared with the critical t value for 25 degrees of freedom at the five percent level of significance of 2.06.

Multivariate regression analysis was then performed to investigate the extent of the change in predictive power that resulted from combining two or more of the independent variables into a multiple linear regression function. Forty five sample multiple linear regression functions were produced and examined. Twenty four of these were rejected because one or more of the independent variables t-value(s) failed the critical value significance test (at the 5% significance level), or because the extent of multicollinearity exceeded acceptable levels (more will be noted on this point shortly).

Table 1. Questionnaire for a market survey on passengers' satisfaction with rail services

1. Have you travelled by train in the past 12 months, i.e. from [MONTH OF INTERVIEWING IN 2010] till [MONTH OF INTERVIEWING IN 2011] in [YOUR COUNTRY]? Please don't include those travels that you made by a sub-urban train, or within the city limit or to / from the airport. Yes [CONTINUE] No [THANK AND TERMINATE] [DK/NA] [THANK AND TERMINATE]
2. How often do you travel by train [IN YOUR COUNTRY]? Most days 1-3 times per week 1-3 times per month Less than once a month [DK/NA]
3. What is the most frequent purpose of your rail trip [IN YOUR COUNTRY]? [ONLY ONE ANSWER IS POSSIBLE] Travelling to work/school/university Business trips Leisure Other [DK/NA]
4. Are you very satisfied, rather satisfied, rather dissatisfied or very dissatisfied with the following features of the train stations [IN YOUR COUNTRY]? [READ OUT- ROTATE-ONE ANSWER PER LINE] Very satisfied Rather satisfied Rather dissatisfied Very dissatisfied [Not applicable] [DK/NA] *A. Connections with other modes of public transport* *B. Facilities for car parking* *C. Quality of the facilities and services (e.g. toilets, shops, cafes, etc.)* *D. Provision of information about train schedules/platforms* *E. Ease of buying tickets* *F. Easy and accessible complaint handling mechanism put in place* *G. Cleanliness / good maintenance of station facilities* *H. Your personal security in the station*
5. Are you very satisfied, rather satisfied, rather dissatisfied or very dissatisfied with the following features of the trains [IN YOUR COUNTRY]? [READ OUT- ROTATE-ONE ANSWER PER LINE] Very satisfied Rather satisfied Rather dissatisfied Very dissatisfied [Not applicable] [DK/NA] *A. Frequency of the trains* *B. Length of time the journey was scheduled to take (commercial speed/ the travelling speed of the trains)* *C. Punctuality/reliability (i.e. departing and arriving on time)* *D. Your personal security whilst on board* *E. Cleanliness and good maintenance of rail cars, including the toilet on the train* *F. The provision of information during the journey, in particular in case of delay* *G. Sufficient capacity for passengers in rail cars* *H. The comfort of the seating area* *I. Connections with other train services* *J. Availability of staff on trains* *K. Assistance and information for disabled or elderly people in station and in rail cars*

continued on following page

Table 1. Continued

6. Gender [DO NOT ASK - MARK APPROPRIATE] Male Female
7. How old are you? [_][_] years old [00] [REFUSAL/NO ANSWER]
8. How old were you when you stopped full-time education? [WRITE IN THE AGE WHEN EDUCATION WAS TERMINATED] [_][_] years old [STILL IN FULL TIME EDUCATION] [NEVER BEEN IN FULL TIME EDUCATION] [REFUSAL/NO ANSWER]
9. As far as your current occupation is concerned, would you say you are self-employed, an employee, a manual worker or would you say that you are without a professional activity? Does it mean that you are: **Self-employed** farmer, forester, fisherman owner of a shop, craftsman professional (lawyer, medical practitioner, accountant, architect,...) manager of a company other **Employee** professional (employed doctor, lawyer, accountant, architect) general management, director or top management middle management civil servant office clerk other employee (salesman, nurse, etc...) other **Manual worker** supervisor / foreman (team manager, etc...) manual worker unskilled manual worker other **Without a professional activity** looking after the home student (full time) retired seeking a job other [Refusal]
10. Would you say you live in a...? metropolitan zone other town/urban centre rural zone [Refusal]

(Gallup organization, 2011)

Table 4 provides a summary of the five multiple regression functions multiple regression functions that produced the statistically strongest results in respect of adjusted R2, standard error of estimate (SEE). Each of the sample regression functions shown in Table 4. yield adjusted R2 of more than 0.9. Each function presented in Table 4. also generates a reasonably low proportional value of Standard Error of Estimate (SEE) relative to the actual average annual patronage levels for train boardings over the twenty

Table 2. Independent variables used

X1	Real average annual price of a full fare weekly Zone 1 public transport ticket;
X2	Real average annual price/litre of unleaded petrol;
X2a	Real average annual price per litre of unleaded petrol lagged 3 months;
X2b	Real average annual price per litre of unleaded petrol lagged 6 months;
X3	Estimated number of persons employed in the Melbourne;
X4	Total weekly earnings of persons;
X5	Average annual housing interest paid as a percentage of household disposable income;
X5a	Average annual housing interest paid as a percentage of household disposable income lagged 3 months;
X5b	Average annual housing interest paid as a percentage of household disposable income lagged 6 months;
X6	Estimated resident population in the Melbourne Statistical Division.

Table 3. Univariate regression results

Variable	Adjusted R^2	SEE	t value	Coefficients
X1	0.539	24.209	5.606	20.244
X2	0.526	24.546	5.468	490.122
X2a	0.586	22.953	6.146	521.788
X2b	0.576	23.221	6.028	508.646
X3	0.830	14.725	11.293	0.137
X4	0.789	16.371	9.920	0.181
X5	0.836	14.428	11.570	17.279
X5a	0.848	13.910	12.080	17.681
X5b	0.845	14.043	11.947	17.651
X6	0.831	14.647	11.365	0.099

Table 4. Multivariate analysis results with regression equation

Variable	Adjusted R^2	SEE	t Statistics
X_1, X_{2a}, X_6	0.955	7.56	-4.81; 7.575; 11.258
X_1, X_{2b}, X_6	0.954	7.668	4.693; 7.424; 11.389
X_1, X_2, X_6	0.941	8.667	-4.265; 6.717; 10.469
X_{2b}, X_3	0.918	10.193	5.308; 10.283
X_{2a}, X_3	0.917	10.258	5.246; 10.059

seven years studied of 128.7 million: the lowest is 6.7% (using variables X1, X2 and X6) and the highest is 8.2% (for variables X2a and X3). The value of the t statistics for each of the explanatory variables is greater than the critical value, indicating that there exists a significant linear statistical relationship between the dependent and each of these independent variables.

Railway Demand Forecasting

The forecasts of train patronage (T) for the years 2010-11 to 2012-13 presented in this section are based on the following sample regression Equation (1) drawn from the results of multivariate analysis:

$$T = -258.59 - 12.095X1 + 266.919X2a + 0.116X6 \tag{1}$$

Using this formula to forecast the number of train boardings calls in the first instance for the the creation of a forecast each of its independent variables. The forecast of the real average annual price of full fare Zone 1 ticket is based on the result of a series of time series analyses of the behaviour of the variable over the whole twenty seven year time period of this study and of its annual value over the last fifteen years. A simple three year moving average was also computed for comparative purposes. The polynomial formula set down as (2) is the one used to compute the values of variable X1:

$$y = -0.0023x^3 + 0.0565x^2 - 0.2182x + 9.7826 \; (R^2 = 0.967) \tag{2}$$

Proposed polynomial function in Equation (3) produce a usable forecasting equation for second independent variable – unleaded petrol prices:

$$y = -0.0003x^3 + 0.00215x^2 - 0.01808x + 0.41673 \; (R^2 = 0.946) \tag{3}$$

The task of forecasting X6, the final explanatory variables in Equation (1) is less daunting given the quite consistent and virtually linear nature of the changes in the time horizon of this study. Forecasting formula for estimated resident population in the Melbourne Statistical Division is:

$$y = -0.0003x^3 + 0.00215x^2 - 0.01808x + 0.41673 \; (R^2 = 0.946) \tag{4}$$

The forecasts of all independent variables are presented in Table 5 and they are incorporated into (1) to forecast the value of train boardings over the same five year period. The results are given in Table 6.

Table 5. Forecasts of selected explanatory variables

Year	Real Price Zone 1 Weekly Full Fare	Real Price/Litre of Unleaded Petrol	Estimated Resident Population
2010-11	11.34	0.555	4105
2011-12	11.11	0.583	4195
2012-13	10.75	0.593	4291

Table 6. Forecasts of train boardings (millions) based on multivariate regression function

Year	Point Estimate Forecast	95% Confidence Intervals
2010-11	229.4	214.3 to 244.5
2011-12	250.1	235.0 to 265.2
2012-13	268.2	253.1 to 283.3

4.1.2. Cointegration and Error Correction Approach for Passenger Rail Demand Modeling

Wijeweera et al. (2014) conducted a multi-city railway passenger demand study in case of Australia. Namely, four cities are included in this study, Sidney, Perth, Adelaide and Melbourne. Passenger rail demand (Y) is measured by the number of boarding passengers at time t. There are seven explanatory (X) variables (determinants of demand) (Table 7).

This study uses the cointegration and error correction techniques to estimate the long- and short-run passenger rail demand responses respectively for the selected Australian cities. Authors estimated a passenger rail demand model by a single equation conintegration method. Namely, they used the two-step procedure proposed by Engle and Granger (1987). The Engle and Granger cointegration method (1987) consists of two main steps. First, the best possible linear model for the passenger rail demand model needs to be estimated. Second, the residuals of the estimated model for possible unit roots require testing. If the residuals are stationary, the variables are regarded as cointegrated. This done, the results can be used for the purposes of analysis.

The model used to estimate the long-run passenger rail demand elasticities is given as follows:

$$\log BOARDING_t = \beta_0 + \beta_1 \log FARE_t + \beta_2 \log PCI_t + \beta_3 \log FUEL_t \\ + \beta_4 \log POPULATION_t + \beta_5 \log KMRUN_t + \beta_6 \log FATALITIES \\ + \beta_7 \log VEHICLE_t + e_t \quad (5)$$

After estimating the long-run elasticities, an error correction model (ECM) was employed to obtain short-run elasticities. It was also used to validate the cointegration results of the estimation of (5).

The Engle–Granger approach is used in this approach because the error correction term can be constructed easily using the already estimated long-run results. Granger representation theorem states that, if variable X and Y are generated by error correction models, they are cointegrated. The dependent variable (BOARDING), together with its explanatory variables (FARE, PCI, POPULATION, FUEL, VEHICLE, KMRUN, FATALITIES) are I(1).

Furthermore, the first difference of these variables (ΔBOARDING, ΔFARE, ΔCPI, ΔPOPULATION, ΔFUEL, ΔVEHICLE, ΔKMRUN, ΔFATALITIES) is I(0). The error correction model in terms of I(0)

Table 7. Independent variables used

FARE	This represents the control variable for price in the demand function. Authors used revenue per kilometer run as the fare variable.
PCI	Australian Per Capita Income (PCI) is used as a proxy for the per capita income of rail users in the selected cities being investigated in this study.
FUEL	Fuel price index over the period being studied.
POPULATION	It might be assumed that the demand for rail will increase as a city grows in population size.
KMRUN	Number of kilometers run annually. It is stated that a change in the number of kilometers run might result from any increase in the frequency of rail services run, or as a result of the rail network's expansion
FATALITIES	Therefore the number of accidental deaths relating to the Australian rail sector
VEHICLE	Australian vehicle price index. This is included as a possible substitute for other transport modes.

variables is given in (6). The ECM contains variables in first differences and an error correction term (ECT). The ECT is the one period lag residuals obtained from the cointegrating model.

$$\Delta \log BOARDING_t = \alpha_0 + \alpha_1 \Delta \log FARE_t + \alpha_2 \Delta \log PCI_t + \alpha_3 \Delta \log FUEL_t \\ + \alpha_4 \Delta \log POPULATION_t + \alpha_5 \Delta \log KMRUN_t + \alpha_6 \Delta \log FATALITIES \quad (6) \\ + \alpha_7 \Delta \log VEHICLE_t + \lambda ECT + e_t$$

Here, the parameter α_1 is the short-run elasticity of passenger rail demand with respect to FARE, α_2 is the short-run income elasticity of demand, and α_3 is the short-run cross-price elasticity of demand. Other parameters can be interpreted similarly. After the own-price, income and cross-price elasticity, the most important other parameter is λ, which represents the disequilibrium error. If the cointegrating relationship is correct, the estimate on the parameter λ must be negative and statistically significant. The parameter λ is called the adjustment parameter because it reveals the degree to which the disequilibrium is corrected within one period.

Cointegration (long-run) results are summarized as follows:

- FARE exerts a negative and statistically significant effect on passenger rail demand in only Sydney and Melbourne, i.e., in two of the four cities studied.
- With regard to the other variables, the coefficient of income (PCI) has a negative sign in all cities. It is only statistically significant at the 5% level for Sydney and Adelaide.
- Fuel price (FUEL) and passenger rail demand are positively related in the case of Melbourne. POPULATION and passenger rail demand are positively related. The coefficient was highly statistically significant for Sydney and Melbourne.
- The number of kilometres run (KMRUN) is highly statistically significant at a 1% level of significance in Sydney, Perth and Adelaide.
- The estimate on the FATALITIES coefficient is not statistically significant at conventional levels of significance, with the exception of Adelaide.

Authors estimated the short-run elasticity via an Engle and Granger ECM. The fare elasticity of demand has the expected sign and is statistically significant at a 1% and 10% level of significance for Sydney and Melbourne respectively. Although the fare elasticity of demand is statistically significant, it has the unexpected positive sign in Adelaide. The short-run fare elasticity is statistically insignificant in the case of Perth. The most important findings in this phase are:

- In general, short-run relationships are quite similar to long-run relationships for the case of Sydney, while they are quite different in the other cities investigated;
- The long-run income elasticity of demand is not statistically significant in Melbourne and Perth, while the short-run elasticity seems significant and sizable;
- In Melbourne, short-run POPULATION elasticity is only significant at a lower significance level (i.e., 10%);
- While long-run FUEL elasticity exhibits the expected sign and is statistically significant, short-run FUEL elasticity is not statistically significant;

- In Perth, VEHICLE is not statistically significant in the short-run—a different result from the long-run;
- For Adelaide, FARE, PCI, POPULATION, KMRUN and FATALITES are significant variables of the passenger rail demand function in the long-run, while only FARE and KMRUN are statistically significant in the short-run.

Table 8 presents a summary of results for long and short run demand responses in case of Melbourne.

4.1.3. Other Econometric Approaches for Railway Demand Forecasting

Doi and Allen (1986) presented two time series regression models, one in linear form and the other in logarithmic form, to estimate the monthly ridership of a single urban rail rapid transit line. The model was calibrated for a time period of about six and a half years (from 1978–1984) based on ridership data provided by a transit authority, gasoline prices provided by a state energy department, and other data. The major findings from these models are: (1) seasonal variations of ridership are −6.26%, or −6.20% for the summer period, and 4.77%, or 4.62% for the October period; (2) ridership loss due to a station closure is 2.46% or 2.41%; and (3) elasticities of monthly ridership are −0.233 or −0.245 with respect to real fare, 0.113 or 0.112 with respect to real gasoline price, and 0.167 or 0.185 with respect to real bridge tolls for the competing automobile trips. Such route specific application results of this inexpensive approach provide significant implications for policy making of individual programs in pricing, train operation, budgeting, system changes, etc., as they are in the case reported herein and would be in many other cities.

Kulshreshtha et al. (2001) proposed a cointegrated VAR methodology to draw inferences about the responsiveness of freight transport demand for Indian Railways at an aggregate level. Authors estimated long run structural relationships between demand for freight transport and its influencing economic variables using annual time series data for the period 1960±1995. The analysis was done in a multivariate cointegrating VAR framework to avoid the shortcomings associated with single-equation approaches to cointegration. To check the stability of cointegrating relationships, the analysis was performed over two sub-samples as well as the full sample. Evidence of at most one cointegrating relationship between GDP, TKM and P was found for all samples. The long-run structural relationship between these variables suggested that economic growth was the major determinant of freight transport demand and vice-versa. The price elasticity was found to be low, declining further in the last two decades indicating low price

Table 8. Summary of Long and Short Run Demand Responses for a Case of Melbourne

Long-Run Responses			Short-Run Responses			
Statistically significant variables (estimated coefficients)	Statistically not significant variables	R-squared	Statistically significant variables (estimated coefficients)	Statistically not significant variables	Adjustment parameter (λ)	R-squared
FARE (-0.26204); POPULATION (4.035545); FUEL (0.223924); VEHICLE (-0.30525).	PCI; KMRUN; FATALITIES.	97%	FARE (-0.15847); POPULATION (3.182886); PCI (0.776549); VEHICLE (-0.32331).	FUEL; KMRUN; FATALITIES.	Statistically significant (-0.36081)	63%

elasticity of freight transport demand. This result may indicate that more and more incremental demand is contributed by commodities like coal which are required to be transported in bulk and cannot be transported economically over long distances by alternative means and hence have lower sensitivity to price. The freight demand system appears to be fairly stable, and the major effect of a typical system-wide shock dissipates within a period of around three years with both GDP and TKM adjusting to correct the disequilibrium. Evidence of strong contemporaneous correlation was found between GDP and TKM. To the extent that growing demand for freight transport in general seems to be directly related to growth in the economy and rapid industrialization, the freight transport demand has implications for large investments in wagons, locomotives and tracks to augment the transport capacity. A related and critical issue concerns the freight rate structure and price elasticities and whether revenues and earnings will rise if rates increase. Both price and income elasticities could be important policy parameter in rationalizing the freight rates.

Within the British context, econometric demand methods are used for modeling and forecasting demand and they are prescribed by the Passenger Demand Forecasting Handbook (ATOC, 2005) which is unique amongst railway administrations and has for over 20 years recommended a forecasting framework and set of demand parameters with a firm basis in empirical evidence. forecasting performances.

Profillidis and Botzoris (2006) developed a set of three econometric models for the forecast of passenger demand in Greece, one for total demand, one for rail demand and one for private car demand. The validity of each model is tested by means of statistical and diagnostic tests. Through calculation of the U-theil Statistics for the proposed models it is concluded that all models have good forecasting performances.

Shen et al. (2009) applied six econometric time series models to modeling and forecasting the road plus rail freight demand in Great Britain, based on annual time series data for the period 1974-2006. These models comprise: the traditional OLS (Ordinary Least Squares) regression model, the PA (Partial Adjustment) model, the reADLM (reduced Autoregressive Distributed Lag Model), the unrestricted VAR (Vector Autoregressive) model, the TVP (Time-Varying Parameter) model and the STSM (Structural Time Series Model). The relative forecasting accuracy of alternative models has been evaluated based on MAPE in the context of freight demand. The estimation results show that industrial production generally offers a good explanation of road plus rail freight demand in GB. However, the sensitivity of road plus rail freight demand to the change in the industrial production varies across different commodity groups, as different commodities have different transport requirements and each estimate reflects particular circumstances for each commodity group. The actual magnitudes of income elasticity estimates also vary due to the different models estimated. The ranges of estimated income elasticity for different sectors have been provided. The forecasting performance comparison results show that no single model outperforms the others in all situations. Overall, it can be concluded that for short-term (one-year-ahead) forecasting, the STSM is the best forecasting model, followed by the TVP model. For medium-term (three-year-ahead) forecasting the STSM is superior to its competing models, followed by the PA model. For relatively longer horizons (five-year-ahead in this study), the PA model and reduced ADLM seem to perform best although the STSM is not far behind. Forecasting horizons do seem to have an effect on the forecasting performance of different models. The STSM seems to perform better for short to medium term horizons, whereas the PA model outperforms others for longer term horizons. The TVP model generally performs better in the short term (one-year-ahead) than for longer-term forecasting.

Wijeweera et al. (2013) examined the impacts of exchange rate, freight rate and economic activities on the growth rate of non-bulk freight demand in Australia. The paper uses a simple but robust econometrics method to estimate the demand growth function and utilizes a relatively large annual data set encompassing over four decades (1970-2011). The findings provide convincing evidence that the volatility of the Australian dollar has a substantial impact on freight rail demand within Australia. Furthermore, the study finds that, although freight rate and macroeconomic activities exhibit the expected relationship with freight rail demand, the relationships are not strong enough to make valid statistical inferences.

Xia et al. (2014) forecasted short-term passenger flow by means of both support vector regression method and BP (Back Propagation) neural network method, and the results show that the support vector regression model has such a theoretical superiority as minimized structural risk, thus having a higher forecasting accuracy under small sample conditions for short-term urban rail transit passenger flow, which predicts a promising forecasting performance the method has.

4.2. Time Series Models for Railway Demand Forecasting

Time series forecasting methods produce forecasts based solely on historical values. They are widely used in situations where short term forecasts are required. Time series methods have the advantage of relative simplicity, but certain factors need to be considered:

- Better suitability to short term forecasts;
- Sufficient past data and data of high quality are required;
- The most appropriate for relatively stable situations. In case of substantial fluctuations time series methods may give poor results.

In the rest of this section time series approaches for railway demand forecasting are reviewed.

4.2.1. Use of SARIMA Models to Assess Rail Passenger Flows

Milenkovic et al. (2013) analyzed SARIMA models in the forecasting of the passenger flows on Serbian railways on a base of a time series of a total monthly number of passengers travelled by the railway. Data set covers the period from 01 January 2006 to 31 December 2012 (84 monthly values). The time series is presented on Figure 1. Authors used first 75 months to fit the SARIMA models and the last 9 months as a hold-out period to evaluate forecasting performance.

As it can be seen from the Figure 3, time series data (x_t, $t=1,...,75$) for rail passenger traffic in Serbia have strong seasonality pattern with a trend which is not constant throughout the whole study period, rather slightly decreasing from 2006 to 2010, and increasing in period from 2011 to 2012. Variance-mean plots indicated an increase in variance with the series mean. Therefore, authors used $\log_{10}$ transformation to stabilize the variance of the series. They log-transformed the data and then used the log transformed data set (z_t, $t=1,...,75$) as input to the SARIMA analysis.

Selection of the candidate model set was carried out by first analyzing sample estimates of autocorrelation function (ACF) and partial autocorrelation function (PACF) in order to select three major orders of the SARIMA models: d, D, and S. Large autocorrelations were recorded for lags 1, 2, 3, 11, 12, 13, 14, 23, 24 and 25 with values 0.69, 0.47, 0.21, 0.38, 0.49, 0.38, 0.28, 0.22, 0.30 and 0.23, respectively (Figure 4). Gradual decrease in autocorrelation values on first three lags indicated a long term trend. Consequently, there was a need to include a first-lag difference term in the SARIMA model structure (d=1). Also, large auto correlation S=12, D=1 values were registered at annual lags (and its

Figure 3. Time series of monthly rail passenger flows (in thousands) in Serbia (January 2006 to December 2012) – Raw data

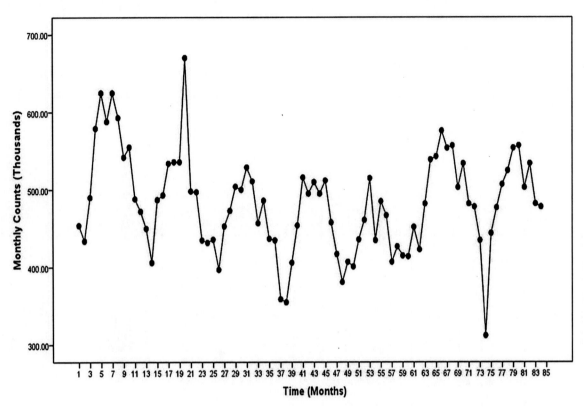

multiplies) which indicated the need to include a 12-month difference term in models (). The ACF and PACF plots of the differenced time series further support for these conclusions (Figure 4). Therefore, a SARIMA(p,1,q)×(P,1,Q)$_{12}$ was selected as the basic structure of the alternative SARIMA models.

Among the statistical models, SARIMA(0,1,1)×(0,1,1)$_{12}$ was selected as the best model, with the lowest normalized BIC of 6.780 and a Mean Absolute Percent Error (MAPE) of 3.567. The model explained 84.9% of the variance of the series (stationary R-squared). The Ljung-Box statistic provides an indication of whether the model is correctly specified. The value of 0.804 shown here is not significant, so we can be confident that the model is correctly specified. This model has the following equation:

$$\log_{10} Z_t = \log_{10} Z_{t-1} + \log_{10} Z_{t-12} - \log_{10} Z_{t-13} + \varepsilon_t - 0.199\varepsilon_{t-1} - 0.642\varepsilon_{t-12} + 0.128\varepsilon_{t-13} \qquad (7)$$

Diagnostic checks indicated that the SARIMA model was stationary and invertible and did not have redundant parameters. The residuals were white noise (Ljung-Box Q = 11.085, P-value>0.05), so there is no significant autocorrelation between residuals at different lag times (Figure 5.). All tests were performed at a significance level of α=0.05.

We evaluated 9 months of model forecasts, using the first month (April 2012) after the fitting part of time series as the forecast origin. Forecasts were obtained in the original scale of the data ($\hat{x}_l, l = 1, ..., 9$).

Figure 4. Sample autocorrelation function (acf) and partial autocorrelation function (PACF) of the transformed rail passenger flow data. ACF/PACF plots for $\log_{10}$- transformed data (z_t, far left), lag-1 differenced series ($\nabla_1^1 z_t$), lag-12 differenced series ($\nabla_{12}^1 z_t$), and lag-1 and lag-12 differenced ($\nabla_1^1 \nabla_{12}^1 z_t$, far right) are displayed. Horizontal dashed lines represent the 95% confidence limits valid under the null hypothesis of white noise error structure

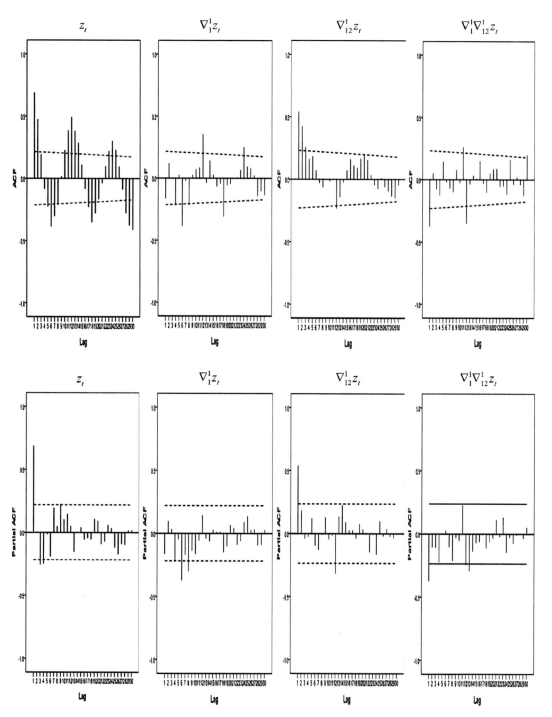

Figure 5. Autocorrelation function (ACF) and partial autocorrelation function (PACF) of residuals

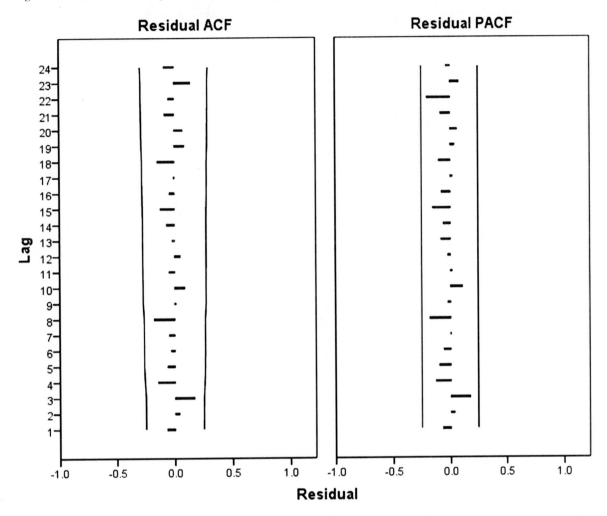

SARIMA model performance was assessed by comparing *l*-step forecasts with monthly number of passengers travelled in period from April 2012 to December 2013. This was done by evaluating monthly forecast errors and then considering a set of accuracy measures: 1) Root mean square error (RMSE); 2) mean error (ME); 3) absolute percent error (APE_l); 4) mean absolute percent error (MAPE). Additionally, we compared the forecasting performance of the SARIMA model against simple seasonal exponential smoothing measures (Table 9). For this purpose trial version of IBM SPSS (Version 19) software is used.

Observed passenger counts (x_l), forecasted passenger counts ($\hat{x}_l$), monthly forecast errors (e_l) and monthly absolute percent error (APE_l) are displayed for simple seasonal exponential smoothing model (SSES) and SARIMA model (SAR).

Figure 6 displays observed values and SARIMA model fit and forecast values. Model gives slightly lower forecasts then observed values, but pattern of model forecasts almost matched the one in observed passenger counts except for period May - June 2012. The main reason for this is a slightly increasing trend for the last two years of the study period.

Table 9. Forecasts of railway passenger flows in thousands (April 2012 – December 2012)

Month	Step (l)	Obs(x_i)	Forecasts (x_i)		Forecast Errors (e_i)		APE_i	
			SSES	SAR	SSES	SAR	SSES	SAR
Apr-12	1	477.00	479.61	488.01	2.61	11.01	0.55	2.31
May-12	2	507.00	519.27	521.82	12.27	14.82	2.42	2.92
Jun-12	3	525.00	501.44	522.55	-23.56	-2.45	4.49	0.47
Jul-12	4	554.00	519.61	520.79	-34.39	-33.21	6.21	5.99
Aug-12	5	557.00	528.61	511.39	-28.39	-45.61	5.10	8.19
Sep-12	6	503.00	466.27	458.42	-36.73	-44.58	7.30	8.86
Oct-12	7	534.00	472.61	480.53	-61.39	-53.47	11.50	10.01
Nov-12	8	482.00	425.44	440.25	-56.56	-41.75	11.73	8.66
Dec-12	9	478.00	415.11	431.49	-62.89	-46.51	13.16	9.73
Mean	1:9	513.00	480.89	486.14	-32.11	-26.86	6.94	6.35
Sum	1:9	4617.00	4327.97	4375.25	-289.03	-241.75		

Figure 6. Results of SARIMA forecasting for the passenger demand on Serbian railways

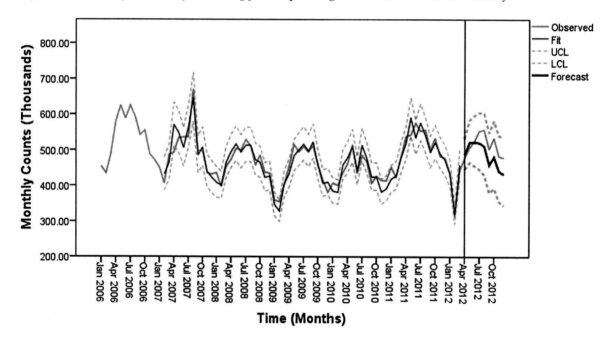

RMSE during the prediction period (RMSE=36.94) was 1.6 times the RMSE of the fitting period (RMSE=22.74). Seven of the eight forecasts registered negative errors but the low ME (ME=26.86) indicated that underestimation was minor in global terms. However, it has to be noted that this conclusion is valid for the case of random error, but this will not be the case if the error is systematic (system changes –transportation, activity or flow patterns). MAPE was 6.35% reflecting the slightly lower forecasts during the period June-December 2012. The SARIMA model forecasts also registered better performances with respect to SSES, resulting in 9.4% reduction in RMSE, 16.3% reduction in ME, and 8.4% reduction in MAPE.

4.2.2. Other Approaches of Time Series Methods for Railway Demand Forecasting

Babcock et al. (1999) concluded that the economic process generating quarterly railroad grain car loadings is quite complex and very difficult to model with regression techniques. Therefore, the authors developed a time series model to make a short-run forecast of quarterly railroad grain car loadings. An AR(4) model was estimated using the Maximum Likelihood estimation procedure for the 1987:4±1997:4 period. The actual railroad grain car loadings for this period were compared to the forecast car loadings generated by the time series model. For 92% of the 37 quarters the percentage difference between the actual and forecast values was 10% or less. Of the 9 annual observations, the per cent difference between the actual and forecast value was less than 2.6% for 8 of the 9 years.

Guo et al. (2010) analyzed trend and seasonal fluctuation of China's monthly railway freight and applied ARIMA and Holt-Winters models to forecast freight flows. After comparison of forecasting results, a final freight result from January to December of 2010 is generated. After verification, the predict results of ARIMA model and Holt-Winters model, with errors lower than 4%, are preferable to be used of railway freight forecast.

Jiuran and Bingfeng (2013) combined the ARIMA model and RBF (Radial Basis Function) neural network model to formulate the ARIMA-RBF model by analyzing passenger flow's temporal characteristics, the mechanism of ARIMA model with RBF model. The authors used proposed model to forecast Beijing urban rail transit passenger flow. According to the mean absolute percentage error as the evaluation indicator, they concluded that the developed ARIMA-RBF model is much better then ARIMA and RBF models.

4.3. Review of Other Forecasting Methods

Besides these two broad classes of forecasting techniques there is a number of other approaches suitable to handle non linear data like Kalman filtering, neural networks or fuzzy logic. In this section, an introduction and review of some alternative but very popular forecasting tools has been made.

4.3.1. State-Space Models

State space time series analysis began with the path breaking paper of Kalman (1960) and early developments of the subject took place in the field of engineering. The term "state space" came from engineering, and although it does not strike a natural rapport in statistics and econometrics, it's have been used because it is strongly established. The distinguishing feature of state space time series models is that observations are regarded as made up of distinct components such as trend, seasonal, regression elements and disturbance terms, each of which is modeled separately. The models for the components are put together to form a single model called a state space model which provides the basis for analysis. The techniques that emerge from this approach are very flexible and are capable of handling a much wider range of problems than the main analytical system for time series analysis, the Box-Jenkins ARIMA system (Durbin and Koopman, 2001). In the rest of this section, a recursive Kalman filter approach is described for railway passenger flow forecasting.

Milenkovic et al. (2013) and Milenkovic and Bojovic (2014) presented a recursive Kalman filter approach to railway passenger flow forecasting for the case of Serbian Railways JSC. Identified SARIMA$(0,1,1)(0,1,1)_{12}$ model as the most appropriate (among the selected alternative SARIMA models) for modelling the railway passenger demand is embedded into the state space framework for the sake of applying the Kalman recursions for forecasting the rail passenger flow.

The Gaussian state space form consists of a transition equation for the $m \times 1$ state vector α_t and a measurement equation for the $N \times 1$ observation vector y_t for $t=1,\ldots,n$. The model can be formulated as in (Dubrin and Koopman, 2001):

$$\alpha_{t+1} = T_t \alpha_t + R_t \eta_t, \quad \eta_t \sim N(0, Q_t), \quad t = 1, \ldots, n \tag{8}$$

$$y_t = Z_t \alpha_t + \varepsilon_t, \quad \varepsilon_t \sim N(0, H_t), \quad \alpha_1 \sim N(a_1, P_1) \tag{9}$$

where η_t and ε_t represent the error terms assumed to be serially independent and independent of each other at all time points. The matrices T_t, R_t, Z_t, Q_t, and H_t are referred to as the state space system matrices. The initial state vector is α_1 with mean vector α_1 and variance matrix P_1. Model (8)-(9) is linear and driven by Gaussian disturbances. Therefore, the state space model can be treated by standard time series methods based on the Kalman filter (Durbin and Koopman, 2001; Anderson and Moore, 1979).

SARIMA models can be dealt with by constructing ARMA models for the stationary differenced series $y_t^* = (1-B)^d (1-B^s)^D y_t$ and placing the non-stationary variables such as y_{t-i} and $(1-B)^d y_{t-1}$ in the state vector. y_t^* is a seasonal $ARMA(p^*, q^*)$ process with $p^* = p + SP$ and $q^* = q + SQ$. Appropriately constructed state vector of this ARMA process for $d=D=0$ can be defined as α_t^*, so that $y_t^* = y_t$. In this case there is

$$\begin{aligned}\alpha_t^* = (&y_t^*, \phi_2 y_{t-1}^* + \cdots + \phi_p \Phi_P y_{t-p^*+1} + \theta_1 \varepsilon_t + \cdots \theta_q \Theta_Q \varepsilon_{t-q^*+1}, \\ &\phi_3 y_{t-1}^* + \cdots + \phi_p \Phi_P y_{t-p^*+2} + \theta_2 \varepsilon_t + \cdots \theta_q \Theta_Q \varepsilon_{t-q^*+2}, \ldots, \phi_p \Phi_P y_{t-1} + \theta_q \Theta_Q \varepsilon_t)\end{aligned} \tag{10}$$

With the dimension of α_t^* equal to $m = \max(p^*, q^*+1)$. The complete state vector α_t has dimension $SD+d+m$, and for the case of $d=1$ and $D=1$ can be written a

$$\alpha_t = (y_{t-1}, (1-B) y_{t-1}, \ldots, (1-B) y_{t-S}, \alpha_t^*)^T \tag{11}$$

where the term y_t^* in the state vector α_t^* changes according to the orders of d and D, but the structure of α_t^* stays the same (Hindrayanto, 2010). The MA parameters are included in the disturbance vector, which is given by

$$H_t \varepsilon_t = (0_{1\times(SD+d)}, \varepsilon_{t+1}, \theta_1 \varepsilon_{t+1}, ..., \theta_{m-1} \varepsilon_{t+1})^T \tag{12}$$

The transition matrix T_t has dimension $(SD+d+m)\times(SD+d+m)$ and Z_t is a row vector of dimension $1\times(SD+d+m)$. For seasonal models with $d=D=1$, the T_t and Z_t matrices can be defined as follows:

$$T_t = \begin{pmatrix} 1 & 0_{1\times(S-1)} & 1 & 1 & 0 & 0 & ... & 0 \\ 0 & 0_{1\times(S-1)} & 1 & 1 & 0 & 0 & ... & 0 \\ 0 & I_{S-1} & 0 & 0 & 0 & 0 & ... & 0 \\ 0 & 0_{1\times(S-1)} & 0 & \phi_1 & 1 & 0 & ... & 0 \\ 0 & 0_{1\times(S-1)} & 0 & \phi_2 & 0 & 1 & ... & 0 \\ . & . & . & . & . & . & & . \\ . & . & . & . & . & . & & . \\ . & . & . & . & . & . & & . \\ 0 & 0_{1\times(S-1)} & 0 & \phi_{m-1} & 0 & 0 & ... & 1 \\ 0 & 0_{1\times(S-1)} & 0 & \phi_m & 0 & 0 & ... & 0 \end{pmatrix} \tag{13}$$

$$Z_t = (1, 0_{1\times(S-1)}, 1, 1, 0, 0, ..., 0) \tag{14}$$

The $r \times r$ identity matrix is denoted by I_r and an $r \times c$ matrix of zeros is denoted by $0_{r \times c}$.

The Kalman filter represents a set of mathematical equations that provides an efficient computational (recursive) means to estimate the state of a process, in a way that minimizes the mean squared error.

The classical Kalman recursions were introduced by Rudolph E. Kalman in 1960 (Kalman, 1960). The objective is to obtain the conditional distribution of α_{t+1} based on the observations $Y_t = \{y_1, y_2, ..., y_t\}$. Since all distributions are normal, conditional distributions of subsets of variables given other subsets of variables are also normal. The required distribution is therefore determined by a knowledge of $a_{t+1} = E(\alpha_{t+1} | Y_t)$ and $P_{t+1} = Var(\alpha_{t+1} | Y_t)$. It is assumed that α_t given Y_{t-1} is $N(a_t, P_t)$. Recursive procedure for determining α_{t+1} and P_{t+1} from α_t and P_{t+1} is as follows:

Since $\alpha_{t+1} = T_t \alpha_t + R_t \eta_t$, there is

$$a_{t+1} = E(T_t \alpha_t + R_t \eta_t | Y_t) = T_t E(\alpha_t | Y_t) \tag{15}$$

$$P_{t+1} = Var(T_t \alpha_t + R_t \eta_t | Y_t) = T_t Var(\alpha_t | Y_t) T_t' + R_t Q_t R_t' \tag{16}$$

for $t=1,...,n$. Let

$$v_t = y_t - E(y_t | Y_{t-1}) = y_t - E(Z_t \alpha_t + \varepsilon_t | Y_{t-1}) = y_t - Z_t a_t \tag{17}$$

where v_t represents the one-step forecast error of y_t given Y_{t-1}.

$$\begin{aligned} E(\alpha_t | Y_t) &= E(\alpha_t | Y_{t-1}, v_t) \\ &= E(\alpha_t | Y_{t-1}) + Cov(\alpha_t, v_t)[Var(v_t)]^{-1} v_t \\ &= a_t + M_t F_t^{-1} v_t \end{aligned} \quad (18)$$

where $M_t = Cov(\alpha_t, v_t) = P_t Z_t'$, $F_t = Var(v_t) = Z_t P_t Z_t'$, and $E(\alpha_t | Y_{t-1}) = a_t$.
Substituting in (16) and (19) gives:

$$a_{t+1} = T_t a_t + T_t M_t F_t^{-1} v_t = T_t a_t + K_t v_t \quad (19)$$

where

$$K_t = T_t M_t F_t^{-1} = T_t P_t Z_t' F_t^{-1} \quad (20)$$

It is clear that α_{t+1} has been obtained as a linear function of the previous value a_t and v_t, the forecast error of y_t given Y_{t-1}.
Since

$$Var(\alpha_t | Y_t) = P_t - P_t Z_t' F_t^{-1} Z_t P_t \quad (21)$$

Substituting it in (17) gives

$$P_{t+1} = T_t P_t L_t' + R_t Q_t R_t' \quad (22)$$

with $L_t = T_t - K_t Z_t$.

The recursions (19)-(22) constitute the Kalman filter for the model (9)-(10). They enable to update the knowledge of the system each time a new observation comes in (Durbin and Koopman, 2001).

Due to partially diffuse initial state vector of considered SARIMA model, minor extensions to the classical Kalman filter equations are necessary (Peng and Aston, 2006; Gomez, 2012). In general, h-step ahead forecasts of future state values are recursively obtained as $a_{t+h} = T_{t+h-1} a_{t+h-1}$ with covariance matrix $P_{t+1} = T_{t+h-1} P_{t+h-1} T_{t+h-1}' + R_{t+h-1} Q_{t+h-1} R_{t+h-1}'$.

As in the case of SARIMA modeling the same data set is used. It covers the period from 01 January 2006 to 31 December 2012 (84 monthly values). In this case, first 78 months are used to fit the SARIMA models and the last 6 months as a hold-out period to evaluate forecasting performance. Identified SARIMA$(0,1,1)(0,1,1)_{12}$ model has a slightly changed equation with respect to Equation (7):

$$\log_{10} Y_t = \log_{10} Y_{t-1} + \log_{10} Y_{t-12} - \log_{10} Y_{t-13} + \varepsilon_t + 0.195 \varepsilon_{t-1} + 0.606 \varepsilon_{t-12} + 0.118 \varepsilon_{t-13} \quad (23)$$

Railway Demand Forecasting

This model is then incorporated into the state space framework for the purpose of forecasting. SARIMA model is presented to initial the measurement and state Equations (8)-(9) for a Kalman model.

For $s=1,\ldots,12$, the state α_t vector of SARIMA$(0,1,1)(0,1,1)_{12}$ model is defined as

$$\alpha_t = (y_{t-1},(1-B)y_{t-1},\ldots,y_{t-12},(1-B)(1-B)^{12}y_t, 0.195\varepsilon_t + 0.606\varepsilon_{t-11} \\ +0.118\varepsilon_{t-12}, 0.606\varepsilon_{t-10}+0.118\varepsilon_{t-11}, 0.118\varepsilon_{t-10},\ldots,0.118\varepsilon_{t-1},0.118\varepsilon_t) \quad (24)$$

with corresponding disturbance vector

$$H_t\varepsilon_t = (O_{1\times 13}, \varepsilon_{t+1}, 0.153\varepsilon_{t+1}, O_{1\times 11}, 0.111\varepsilon_{t+1}) \quad (25)$$

The transition matrix T_t is therefore 27×27 and T_t and Z_t are given by:

$$T_t = \begin{bmatrix} T_a & T_b \\ T_c & T_d \\ T_e & T_f \end{bmatrix}, \quad T_a = \begin{bmatrix} 1 & O_{1\times 11} & 1 & 1 \\ 0 & O_{1\times 11} & 1 & 1 \\ O_{11\times 1} & I_{11} & 0 & 0 \end{bmatrix}, \quad T_b = [O_{13\times 13}], \quad T_c = [O_{13\times 14}] \quad (26)$$

$$T_d = [I_{13}], T_e = [O_{1\times 14}], T_f = [O_{1\times 13}], Z_t = [1 \ O_{1\times 11} \ 1 \ 1 \ O_{1\times 13}]$$

The variance-covariance matrix of the state disturbances is given by

$$H_t H_t' = \begin{bmatrix} O_{13\times 13} & O_{13\times 14} \\ O_{14\times 13} & H_t^* H_t^{*'} \end{bmatrix} \quad (27)$$

Where the 14×14 stationary part of this variance matrix is given by $H_t^* H_t^{*'}$,

$$H_t^* H_t^{*'} = \begin{bmatrix} 1 & 0.195 & O_{1\times 11} & 0.606 \\ 0.195 & 0.038 & O_{1\times 11} & 0.118 \\ 0 & 0 & O_{11\times 11} & 0 \\ 0.606 & 0.118 & O_{1\times 11} & 0.367 \end{bmatrix} \quad (28)$$

The mean of the initial state vector is given by $a = E(\alpha_1) = O_{27\times 1}$ and the corresponding variance matrix is given by

$$P = \begin{bmatrix} kI_{13} & O_{13\times 14} \\ O_{14\times 13} & P_{14\times 14}^* \end{bmatrix} \text{ with } k \to \infty \quad (29)$$

Matrix $P^*_{14 \times 14}$ is the unconditional variance matrix for the stationary part of the state vector.

The forecasting process with SARIMA-Kalman hybrid model was performed in Matlab using the SSM Toolbox (Gomez, 2012), where appropriate diffuse initialization methods for the state covariance matrix and Kalman filter recursions are already implemented.

Authors evaluated 6 months of model forecasts, using the first month (July 2012) after the fitting part of time series as the forecast origin. Forecasts were obtained in the original scale of the data $(\hat{x}_l, l = 1,...,6)$. SARIMA-Kalman model performance was assessed by comparing l-step forecasts with monthly number of passengers travelled in period from July 2012 to December 2012. This was done by evaluating monthly forecast errors and then considering a set of accuracy measures: 1) Root mean square error (RMSE); 2) mean error (ME); 3) absolute percent error (APE_l); 4) mean absolute percent error (MAPE). Additionally, we compared the forecasting performance of the SARIMA-Kalman model against simple seasonal exponential smoothing measures (Table 10.).

Observed passenger counts (x_l), forecasted passenger counts ($\hat{x}_l$), monthly forecast errors (e_l) and monthly absolute percent errors (APE_l) are displayed for simple seasonal exponential smoothing model (SSES) and SARIMA-Kalman model (SARK)

Figure 7 displays observed values, SARIMA-Kalman and SSES forecast values. For the first three months developed SARIMA-Kalman model gives slightly higher forecasts then observed values, but pattern of model forecasts almost matched the one in observed passenger counts. For period October – December 2012 forecast is very near to the realized values which can be seen from the average APE_l=1.46% for these three months. RMSE during the prediction period (RMSE=21.33) was 0.95 times the RMSE of the fitting period (RMSE=22.28). Three of the six forecasts registered negative errors but the low ME (ME=10.53) indicated that underestimation was minor in global terms. MAPE was 3.38% reflecting the slightly lower forecasts during the period July-September 2012. The SARIMA-Kalman model forecasts also registered better performances with respect to SSES, resulting in 50.1% reduction in RMSE, 56.5% reduction in MAE, and 3.72% reduction in MAPE.

4.3.2. Neural-Networks for Railway Demand Forecasting

Neural networks are mathematical models inspired by the functioning of biological neurons. There are many neural network models. In some cases, these models correspond closely to biological neurons,

Table 10. Forecasts of railway passenger flows in thousands (July 2012 – December 2012)

Month	Step (*l*)	Obs(x_l)	Forecasts (x_l)		Forecast errors (e_l)		APE_l	
			SSES	SARK	SSES	SARK	SSES	SARK
Jul-12	1	554.00	528.98	583.33	-25.02	29.33	4.5	5.3
Aug-12	2	557.00	535.80	591.32	-21.20	34.32	3.8	6.2
Sep-12	3	503.00	476.72	524.98	-57.28	21.98	5.2	4.4
Oct-12	4	534.00	483.34	527.27	-50.66	-6.73	9.5	1.3
Nov-12	5	482.00	438.09	478.64	-43.91	-3.36	9.1	0.6
Dec-12	6	478.00	427.79	465.64	-50.21	-12.36	10.5	2.5
Mean	1:6	513.00	490.78	528.53	-41.38	10.53		
Sum	1:6	4617.00	2944.72	3171.18	-248.28	63.18		

Figure 7. Forecasting results for hold-out period (July 2012-December 2012)

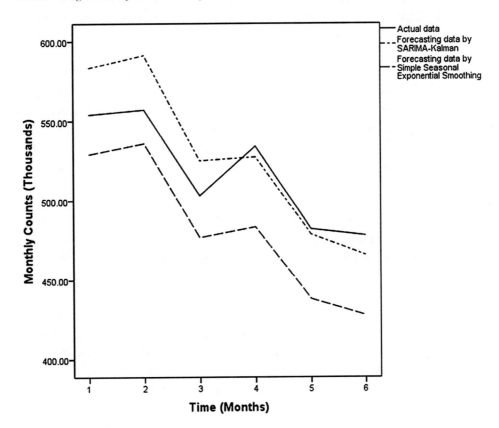

and in other cases, the models depart from biological functioning in significant ways. The most prominent, back propagation, is estimated to be used in over 80 percent of the applications of neural networks (Kaastra and Boyd, 1996). Rumelhart and McClelland (1986) discuss most of the neural network models in detail. Given sufficient data, neural networks are well suited to the task of forecasting. They excel at pattern recognition and forecasting from pattern clusters. The key issue to be addressed is in which situations do neural networks perform better than traditional models. Researchers suggest that neural networks have several advantages over traditional statistical methods:

- Neural networks may be as accurate or more accurate than traditional forecasting methods for monthly and quarterly time series;
- Neural networks may be better than traditional extrapolative forecasting methods for discontinuous series and often are as good as traditional forecasting methods in other situations;
- Neural networks are better than traditional extrapolative forecasting methods for long-term forecast horizons but are often no better than traditional forecasting methods for shorter forecast horizons;
- To estimate the parameters characterizing neural networks, many observations may be required. Thus, simpler traditional models (e.g., exponential smoothing) may be preferred for small data sets.

Tsai et al. (2005) developed two dynamic neural network structures to forecast short-term railway passenger demand. The first neural network structure follows the idea of autoregressive model in time series forecasting and forms a nonlinear autoregressive model. In addition, two experiments are tested to eliminate redundant inputs and training samples. The second neural network structure extends the first model and integrates internal recurrent to pursue a parsimonious structure. The result of the first model shows the proposed nonlinear autoregressive model can attain promising performance and most cases are fewer than 20% of Mean Absolute Percentage Error. The result of the second model shows the proposed internal recurrent neural network can perform as well as the first model does and keep the model parsimonious. Short-term forecasting is essential for short-term operational planning, such as seat allocation. The proposed network structures can be applied to solve this issue with promising performance and parsimonious structures.

Tsai et al. (2009) addressed two novel neural network structures for short-term railway passenger demand forecasting. An idea to render information at suitable places rather than mixing all available information at the beginning in neural network operations is proposed. The first proposed network structure is multiple temporal units neural network (MTUNN), which deals with distinctive input information via designated connections in the network. The second proposed network structure is parallel ensemble neural network (PENN), which deals with different input information in several individual models. The outputs of the individual models are then integrated to obtain final forecasts. Conventional multi-layer perceptron (MLP) is also constructed for comparison purposes. The results show that both MTUNN and PENN outperform conventional MLP in the study. On average, MTUNN can obtain 8.1% improvement of MSE and 4.4% improvement of MAPE in comparison with MLP. PENN can achieve 10.5% improvement of MSE and 3.3% improvement of MAPE in comparison with MLP.

4.3.3. Fuzzy Models

Mukaidono (2002) concluded, "It is a big task to exactly define, formalize and model complicated systems", and it is at precisely at this task that fuzzy logic has excelled. In fact, fuzzy logic has routinely been shown to outperform classical mathematical and statistical modeling techniques for many applications involving the modeling of real world data. Researches shown that fuzzy logic methods are as accurate as the more sophisticated methods and simpler to utilize.

Stefanis et al. (2001) proposed fuzzy regression method for forecasting railway demand. The appropriate explanatory variables are determined and an econometric and a fuzzy model are suggested to establish relationships important for the future demand of railway services. Using the similarity ratio authors compared models that come from different set of data. Application of the method in the case of the intercity rail demand in Greece is conducted. Based on the historical data of demand, several combinations of data are tested with fuzzy regression analysis. Similarity ratios are calculated and the more appropriate set of data is concluded.

Milenkovic et al. (2012) studied the urban railway passenger demand in the city of Belgrade. On the base of available historical data about rail passenger flows two forecasting techniques are proposed and compared, a hybrid of Neural Network and Fuzzy Logic, known as Adaptive Network-based Fuzzy Inference System (ANFIS) and Autoregressive Integrated Moving Average (ARIMA) model. The presented models are used to provide one year ahead rail passenger demand forecast.

REFERENCES

Anderson, B. D., & Moore, J. B. (1979). *Optimal Filtering*. Englewood Cliffs: Prentice-Hall.

ATOC. (2005). *Passenger Demand Forecasting Handbook*. London: Association of Train Operating Companies.

Babcock, M., Lu, X., & Norton, J. (1999). Time series forecasting of quarterly railroad grain carloadings. *Transportation Research Part E, Logistics and Transportation Review, 35*(1), 43–57. doi:10.1016/S1366-5545(98)00024-6

Brown, B. (1968). Delphi Process: A Methodology Used for the Elicitation of Opinions of Experts. An earlier paper published by RAND (Document No: P-3925, 1968).

Dalkey, N., & Helmer, O. (1963). An Experimental Application of the Delphi Method to the use of experts. *Management Science, 9*(3), 458–467. doi:10.1287/mnsc.9.3.458

Doi, M., & Allen, W. B. (1986). A Time Series Analysis of Monthly Ridership for an Urban Rail Rapid Transit Line. *Transportation, 13*(3), 257–269. doi:10.1007/BF00148619

Durbin, J., & Koopman, S. (2001). *Time series analysis by state space methods*. Oxford University Press.

Engle, R. F., & Granger, C. J. (1987). Cointegration and error correction: Representation, estimation and testing. *Econometrica, 55*(2), 251–276. doi:10.2307/1913236

Fok, D., Frances, H., & Paap, R. (2002). Econometric Analysis of the Market Share Attraction Model. In P. H. Frances & A. L. Montgomery (Eds.), *Econometric Models in Marketing* (pp. 223–256). New York: JAI/Elsevier. doi:10.1016/S0731-9053(02)16010-5

Gomez, V. (2012). SSMMATLAB, a Set of MATLAB Programs for the Statistical Analysis of State-Space Models. Retrieved from http://www.sepg.pap.minhap.gob.es/sitios/sgpg/en-GB/Presupuestos/Documentacion/Paginas/SSMMATLAB.aspx

Guo, Y. N., Shi, X. P., & Zhang, X. D. (2010). A study of short term forecasting of the railway freight volume in China using ARIMA and Holt-Winters models. *Proceedings of the 8th International Conference on Supply Chain Management on Information Systems*.

Harold, L., & Turoff, M. (1975). *The Delphi Method: Techniques and Applications*. Reading, Mass.: Addison-Wesley.

Hindrayanto, I., Koopman, S. J., & Ooms, M. (2010). Exact maximum likelihood estimation for nonstationary periodic time series models. *Computational Statistics & Data Analysis, 54*(11), 2641–2654. doi:10.1016/j.csda.2010.04.010

Jiuran, H., & Bingfeng, S. (2013). The application of ARIMA-RBF model in urban rail traffic volume forecast.*Proceedings of the 2nd International Conference on Computer Science and Electronic Engineering* (pp. 1662-1665).

Kalman, R. (1960). A new approach to linear filtering and prediction problems. *Journal of Basic Engineering, 82*(1), 35–45. doi:10.1115/1.3662552

Kulshreshtha, M., Nag, B., & Kulshreshtha, M. (2001). A Multivariate Cointegrating Vector Auto Regressive Model of Freight Transport Demand: Evidence from Indian Railways. *Transport Research Part A*, *35*, 29–45.

Makitalo, M., & Hilmola, O. P. (2010). Analysing the future of railway freight competition: A Delphi study in Finland. *Foresight*, *12*(6), 20–37. doi:10.1108/14636681011089961

Maskeliunaite, L., Sivilevičius, H., & Podvezko, V. (2009). Research on the quality of passenger transportation by railway. *Transport*, *24*(2), 100–112. doi:10.3846/1648-4142.2009.24.100-112

Milenković, M., & Bojović, N. (2014). A Recursive Kalman Filter Approach to Forecasting Railway Passenger Flows. *International Journal of Railway Technology*, *3*(2), 39–57. doi:10.4203/ijrt.3.2.3

Milenković, M., Bojović, N., Glisovic, N., & Nuhodžić, R. (2013). Use of SARIMA models to assess rail passenger flows: a case study of Serbian Railways. *Proceedings of the 2nd International Symposium & 24th National Conference on Operational Research*, Athens, Greece (pp. 296-302).

Milenković, M., Bojović, N., Glišović, N., & Nuhodžić, R. (2014). Comparison of Sarima-Ann and Sarima-Kalman Methods for Railway Passenger Flow Forecasting. *Proceedings of the Second International Conference on Railway Technology: Research, Development and Maintenance*, Ajaccio, France. Stirlingshire, UK: Civil-Comp Press.doi:10.4203/ccp.104.193

Milenković, M., Bojović, N., & Nuhodžić, R. (2012). A Comparative Analysis of Neuro-Fuzzy and Arima Models for Urban Rail Passenger Demand Forecasting. *Proceedings of International Conference on Traffic and Transport Engineering*, Belgrade, Serbia (pp. 569-577).

Montgomery, D. C., Jennings, C. L., & Kulahci, M. (2008). *Introduction to Time Series Analysis and Forecasting*. Wiley-Interscience.

Mukaidono, M. (2002). *Fuzzy Logic for Beginners*. Singapore, New Jersey, London, Hong Kong: World Scientific Publishing Co.

Odgers, J. F., & Schijndel, A. V. (2011). Forecasting Annual Train Boardings in Melbourne Using Time Series Data. *Proceedings of the 34th Australian Transport Research Forum*. Retrieved from http://www.atrf11.unisa.edu.au/Assets/Papers/ATRF11_0109_final.pdf

Peng, J., & Aston, J. (2006). The State Space Models Toolbox for MATLAB. *Journal of Statistical Software*, *41*(6), 1–26.

Profillidis, V. A. (2006). *Railway Management and Engineering*. Ashgate Publishing Limited.

Profillidis, V. A., & Botzoris, G. N. (2006). Econometric models for the forecast of passenger demand in Greece. *Journal of Statistics & Management Systems*, *9*(1), 37–54. doi:10.1080/09720510.2006.10701192

Sackman, H. (1974). Delphi Assessment: Expert Opinion, Forecasting and Group Process. Brown, Thomas, An Experiment in Probabilistic Forecasting. (R-944-ARPA, 1972).

Shen, S., Fowkes, T., Whiteing, T., & Johnson, D. (2009). Econometric modelling and forecasting of freight transport demand in Great Britain.*Proceedings of the European Transport Conference*, Noordwijkerhout, The Netherlands.

Stefanis, V., Profillidis, V., Papadopoulos, B., & Botzoris, G. (2001). Analysis and Forecasting of Intercity Rail Passenger Demand by Econometric and Fuzzy Regression Models. *Proceedings of the 8th SIGEF Congress*, Italy.

Waters, D. (2008). *Quantitative methods for business*. Prentice Hall.

Wijeweera, A., To, H., & Charles, M. B. (2013). An Econometrics Analysis of Freight Rail Demand Growth in Australia. *Proceedings of the 42nd Australian Conference of Economists Conference Proceedings beyond the Frontiers: New Directions in Economics*, Murdoch University, Perth, Western Australia.

Wijeweera, A., To, H., Charles, M. B., & Sloan, K. (2014). A time series analysis of passenger rail demand in major Australian cities. *Economic Analysis and Policy*, *44*(3), 301–309. doi:10.1016/j.eap.2014.08.003

Wu, K. Y. (2011). Applying the fuzzy Delphi method to analyze the evaluation indexes for service quality after railway re-opening: using the old mountain line railway as an example. *Proceedings of the 15th WSEAS international conference on Systems* (pp. 474-479).

Chapter 6
Emerging Value Capture Innovative Urban Rail Funding and Financing:
A Framework

Satya Sai Kumar Jillella
Curtin University, Australia

Sitharam T G
Indian Institute of Science, India

Anne Matan
Curtin University, Australia

Peter Newman
Curtin University, Australia

ABSTRACT

Urban rail transit is emerging around the world as a catalyzing developmental solution to enable 21st century sustainable cities. However, these transit systems are capital intensive and cities worldwide are seeking innovative funding and financing mechanisms. Recently, land based value capture (VC) mechanisms have emerged as a pioneering solution to fund urban rail projects. This chapter introduces the VC concept and provides global best practice. The chapter aims to help enhance the understanding and rationale behind VC approaches through assessing the transit impacted accessibility value proposition and various VC mechanisms to capture the value created. A six-step Strategic Value Capture (SVC) framework is proposed which offers a step-by-step guidance to help define the VC based urban rail transit funding and financing processes from VC planning to VC operations.

INTRODUCTION

Cities across the globe have pressures to build urban rail transit systems as a solution to a range of urban mobility issues driven by rapid urbanization challenges. These urban rail transit systems include metro (or subway systems), mono rail and light rail systems. There is a growing recognition among cities across developed and developing nations that urban rail transit helps maintain a city's economic competitiveness and also helps to enable livable and sustainable communities around station areas. There has been a dramatic turnaround in rail's fortunes globally as well as an increased awareness of rail's value to enable

DOI: 10.4018/978-1-5225-0084-1.ch006

21st century cities to achieve their sustainability goals (Newman et al., 2013). For example, in India currently urban rail (metro) is operational in 9 Indian cities, with another 7 cities currently constructing metros and a further 14 cities with rail transit in the planning stage. A further 16 cities have rail transit under initiation and eight regional rail corridors from Delhi are being planned. However, despite this move towards rail transit, none of the operational urban rail systems in India are financially viable and they are heavily dependent on government subsidies and grants. This is a problem worldwide (Gadgil, 2011) leading to the question *how can urban rail transit systems be funded and financed*?

Traditional funding sources for urban rail transit that include a mixture of federal and state aid grant programs, local taxes and fees, are grossly inadequate to meet the demand for new urban rail infrastructure. In the United States these traditional resources are typically combined to fund projects on a "pay-as-you-go" basis, meaning that projects have often been built in phases or increments as funds become available over a period of years (Chen, 2012). The scenario is not much different in other countries, even in developing economies. Many cities in developing countries depend on scarce grants from Federal or State governments or on loans from donor agencies with limitations leading to funding shortfalls to create any new urban rail transit infrastructure. Urban rail systems are being constrained as a social sector project as it is difficult to increase the fares beyond a point. Turning to the private sector for financing only works with urban rail if the necessary funding is provided to enable private financing to achieve their necessary return. A confluence of all these factors has prompted an urgent need to look for innovative funding and financing mechanisms to build such projects and enable them to be sustained.

In this context, land based value capture (VC) mechanisms, through the monetization of urban land values, are gaining attention as an innovative solution to fund urban rail projects and enable the involvement of private financing. Many studies have established the relationship between urban rail services, accessibility and residential and commercial property values (see below) and this is providing the basis for establishing mechanisms that can capture some of this value as alternative funding. Though it is not new to recognize the value of integrating transport and land use, it is new to integrate these with funding and financing, and is conceptually more challenging (Cervero, 1994; Newman & Kenworthy, 1999; Newman et al., 2013).

This chapter first introduces the concept and rationale of a value capture based rail transit funding and financing approach, introduces various VC mechanisms, and then shares some of the global experiences from cities across North America, Australia, Africa and Asia regions in utilizing VC to fund urban rail. The chapter further elaborates on the various successful VC mechanisms suggested for value capture implementation towards achieving sustainable urban mobility goals, and also identifies future research requirements in this important area. There is limited research available with regards to VC implementation, VC fund governance and VC strategic deliverables, especially for urban rail transit in an emerging cities context but also in many places like Europe and Australia where the mechanism is not used yet. In this chapter, the authors have tried to address these limitations by introducing a six-step *Strategic Value Capture* (SVC) framework. The SVC framework offers step-by-step methodical guidance to help define the VC process from VC planning to VC operationalization within the context of an urban rail transit financing project.

The topic of this chapter is of interest to policy makers, treasuries, city and transport planners, developers, economists, government agencies, mass transit organizations, academics and infrastructure banks.

BACKGROUND

Historically, it is evident that transportation infrastructure, especially urban rail transit infrastructure, effectively creates time utility, and space utility, thereby, adding value to people, goods, and markets by improving accessibility and through catalyzing land development opportunities. Local government initiated community infrastructure improvements have been shown to lead to increased local land values through streets, sidewalks, water and energy resources, schools, hospitals, businesses, up-zoning, parks, and mass transit stops/nodes,. Generally these infrastructure investments are sourced through tax revenue and/or grants. However, the local land owners and property developers gain the benefits and profits from these increased in land values as unearned income by their proximity. Land and property values vary spatially due to differing characteristics of properties' location, land use type, neighborhoods and accessibility amongst other characteristics. The real challenge for VC in regards to enabling capital for transit investment is to establish what portion of the increase in property values is due to the impact of the transit investment proposed versus the other many contributing factors and then seek to capture part of this value as the basis of a funding and financing mechanism.

Construction of a new highway will also spur development along the road network and thereby increase the value of adjacent land and properties. However, unlike rail transit, the accessibility gains found from the highway investment only hold for the initial few years. Then the accessibility gains are reduced due to congestion through this investment encouraging an increase in vehicle population. By contrast, investment in urban rail transit systems not only enable the initial accessibility gains and increases in demand but also enable increased accessibility on a continuous basis (Levinson & Istrate, 2011). The underlying success factor is that the beneficiary people, namely the land owners, residents, business communities near the transit stations, are willing to pay for their accessibility and hence may be willing to help with raising the funds to pay for the rail, otherwise they do not get the accessibility and land value changes (McIntosh et al, 2014a)..

VC CONCEPTS AND RATIONALE

VC is the process by which all, or a portion of, increments in land value attributed to the public investments independent of land or property owner interactions are 'captured' by the public sector to recover the full, or a portion of, the public investments made. Medda (2012) defines VC as it refers to a type of innovative public financing in which, increases in private land values generated by a new public transportation infrastructure investment are all or in part "captured" through a land or property related tax or any other innovative mechanism, to pay for that investment or other public projects. Smolka(2013) explains that the notion of VC is to mobilize for the benefit of the community at large some or all of the land value increments (windfall or unearned income) generated by actions other than the landowner's such as public investments in infrastructure or administrative changes in land use norms and regulations.

In recent times, a few cities worldwide have tried out value capture mechanisms to fund urban rail transit systems by tapping increases in urban land values due to the rail transit proximity as an alternate funding source. Bahl and Linn (2013) examined VC practices in over 50 cities and concluded that all cities (except Hong Kong and Singapore) had a deficient experience in the implementation of VC, and in the VC mechanisms applied and had varying levels of success. Thus although the theory seems to be accepted the practical implementation may have some way to go.

As a first step in defining the VC based funding and financing strategies for transit projects, it is important to understand who are the key stakeholder groups to be involved or focused in defining the VC based transit financing process.

VC Stakeholders

Jillella et al., (2014) identified and classified the VC stakeholders in a VC based rail transit project context into three groups namely: investors, beneficiaries and the community (page 4-5). Table 1, adapted from Jillella et al. (2014), illustrates the various VC stakeholders and their aspirations across these three groups in the context of rail transit project funding and financing.

Firstly, the *investor group* is the stakeholder group that provides the capital or investment for the project and constructs the project. Based on the nature of the project contract, the investor group may broadly include public agencies, private sector companies, local bodies, railway agencies and donor organizations such as banks. This group is primarily responsible for building the infrastructure, and creating a public good based value proposition, and also generally requires a return on the investments made. The second group, the *beneficiary group*, is the stakeholder group who benefits from the delivery of the project, generally through their proximity to the project. These stakeholders include those within the project catchment area, and generally include land owners, real-estate developers, businesses and vacant public land owned government agencies. Primarily these stakeholders accrue the benefits due to the implementation of the rail project as windfall gains such as increased accessibility, increased property values, increased rents, agglomeration of new economies, and/or land use changes, amongst other benefits. These stakeholders play a key role in defining the VC process from the perspective of the proposed project value proposition and also their willingness to pay for such benefits and participate in the VC process. Lastly, the *community* group is the stakeholder group which includes the local community members with direct access to the project. Particularly important in this group are the low income

Table 1. VC based urban rail funding and financing project stakeholders and their aspiration (adapted from Jillella et al., 2014)

Groups	Stakeholders	Aspirations
Investors	• Public agencies • Private sector • Local bodies • Railway agencies • Donor agencies	• Build infrastructure • Improved productivity • Return on investment • Regulations & budget • Sustaining operations
Beneficiary	• Land owners • Real-estate • Developers • Businesses • Government with public land	• Land values appreciation • Windfall gain • Speculation • Private developments • Land banking
Community	• Local residents • Local businesses • Low-income group • Civic societies • Precinct community	• Improved accessibility • Improved commuting • Improved business • Value shareholders • Sustainable mobility

(adapted from Jillella et al., 2014)

community groups within the vicinity of the project, civic agencies, precinct community members and groups and the city tax payer community at large. Primarily this group will aspire for sustainable mobility, accessibility improvements and sustainable land development along the project.

Understanding the aspirations and cross-sectoral objectives of these three stakeholder groups provides a solid platform to define the appropriate VC process to be undertaken. By involving these groups in the early stages of the project planning can also help validate the best transit alignment options and also elicit value proposition potentials through elevated or underground or at surface based transit operations. Further the authors strongly recommend that a participatory stakeholder engagement route is undertaken to plan the VC process from the planning stage to the implementation stage as this is a key success factor to enable a sustained VC based project funding and financing process that can integrate community aspirations along with the project objectives (Jillella et al., 2014). The subsequent section details the transit impact of accessibility benefits on land values.

Transit Accessibility Impact Assessment on Property and Land Values

The key principle associated with VC is the unlocking of the increased land and property values based on increased accessibility value. The hypothesis is that urban rail transit driven accessibility improvements lead to gains in proximate land, property and rental values. In case of a subway it opens up new urban spaces with subway transit shopping or business places including public spaces below the busy streets and business districts. For all rail systems the reductions in traffic increase the opportunities for other urban activities (especially knowledge economy jobs, see Newman and Kenworthy, 2015) and the ability to attract new urban development to make the most of improved accessibility. The gains are reflected in the generalized cost of travel as well as agglomeration benefits depending on a number of factors such as location, land use and density factors that are used to generate the economic value proposition for the transit system (McIntosh et al, 2011; 2014b).

The improved economic value of accessibility is internalized by businesses and residents so the relationship between the distance to a station and the property value is inverse and the value decreases as distance from a station increases. This has been validated through a nationwide survey conducted in the United Kingdom in April 2012 which showed that property prices within 500m of a railway station were 9% higher than similar properties away from the line. Further a similar study in Montreal in 2011 showed that property value had a 13% increase within 500m of a metro station, 10% within 1 km and 5% within 1.5km. A San Francisco Bay Area study found that for every meter a single-family home was closer to a Bay Area Rapid Transit station in 1990, its sales price increased by $2.29 (Landis et al., 1994).

Several research studies have demonstrated the impact of urban rail transit investments on property values (for example; Cervero & Duncan, 2001; Cervero & Landis, 1997). This assessment is, however, difficult when it comes to assessing the true value accrued due to the transit investments. The value of the uplift can vary due to a number of factors such as location, multi-modal presence, development concentration, density, property type, land use type and, of course, the assessment methodology used (Ewing & Cervero, 2010; Duncan 2008, Pan and Zhang, 2008). Furthermore, the variances in land value uplift can also be impacted by other issues related to the operation of the service and its surrounds, namely other issues such as noise, pollution and crime levels within the close proximity to the station (Diaz, 1999; Hui and Ho, 2004). In addition, Mohammad et al. (2013) observed lower land value uplift premiums in the car dependent North American and Australian cities compared to a higher uplift premium in the transit focused East Asian and European cities which have more patronage for public transport services. Research in Perth, Western Australia found an increase of 17% in residential land and over 50% in commercial land values around a suburban rail network but a new fast rail service to the south increased residential land values 42% in a five year period beyond those in similar areas (McIntosh et al 2014a; Newman et al., 2013). Another study conducted in 1993 on residential properties adjacent to the 14.5 mile urban rail in Philadelphia, USA, using hedonic price models, recommended that access to rail created an average housing value premium of 6.4% (Voith, 1993). In an Indian study it was noted that a decline in accessibility during the construction phases of several Metro projects due to increased dust, noise, congestion and air pollution, etc., had a huge impact on local businesses and residents adjacent to these project construction areas (ref?). These studies all illustrate the challenge to estimate the true uplift value attributable to improved accessibility due to transit investments made among other local factors and the nature of the project.

There are several approaches have been used to estimate the transit impact on property values. The simple comparison method and hedonic price models are popularly used in a number of studies (for example; Cervero & Landis, 1993). The simple comparison method examines the relationship between land value or property price and transport accessibility by isolating transport accessibility from other factors through comparisons of land value/property price. The hedonic pricing method examines the relationship between land value and transport accessibility by standardizing a number of attributes in a multiple regression model with the dependent variable of land value. Martinez et al. (2012) further used a Monte Carlo simulation procedure to estimate synthetic population of residential and non-residential properties while evaluating the value capture potential of the Lisbon subway. In this study two simulation procedures were developed: one for the residential market and another for the non-residential market (commercial, office, industry, etc.). In another interesting study by Mohammed, et al (2012) a meta-analysis of the impact of rail projects on land and property values was conducted. Meta-analysis is a regression based approach that can be used to distinguish the main factors characterizing a range of studies.

The theory of how accessibility can impact on land and property values comes from original work by Alonso (1964). In more recent times it is explained by the bid-rent theory, which views the price that the consumer is willing to pay for a particular property as a decreasing function of distance to a certain attraction due to certain benefits realized from the attraction (McIntosh et al, 2014a). Banister and Goodwin (2011) identified three conditions needed for transport investment to spur economic growth: a buoyant economic environment, supportive political conditions, and sound decisions relating to the nature of the investment itself. The other value assessment methods include geographically weighted regression; direct differencing of land values; ratio and bench marking analysis, etc., based on the data availability.

McIntosh, et al (2014a) provided a compilation of the transit induced value uplift academic studies. The majority of these are based on hedonic methods on land and property prices with respect to light rail transit, metro and commuter rail. These are given in Table 2.

Typically to estimate the impact of rail on land values the explanatory data parameters considered for the value assessment include: Land, House or Property, Neighborhood, Proximity to Transit, and Time based variables. The most popular data variables to measure the premium value include:

- **Land:** Lot size, land ownership or available public land;
- **Property/House:** Focused on structural variables include built up area, number of bed rooms, bath rooms, and car parks, age of the building, building coverage ratio, floor area ratio, etc.;

Table 2. Compilation of transit induced value uplift academic studies

	Author	Transit/Location	Dependent variable	Proximity Variable	Premium Rate
Light rail impact	Golub, et al., (2012)	Phoenix, USA Phoenix LRT	Land Adjusted Sale Price	200ft	25%
	Du and Mulley (2007)	England, UK Tyne & Wear light rail	House Price	200m	17.1%
	Cervero & Duncan (2002)	San Diego, USA LRT	Sale Price	400m	3.8% to 17.3%
	Hess and Almeida, (2007)	Buffalo, NY, USA LRT	Assessed Property Value	1/4 mile	2 to 5%
	Garrett, (2004)	Missouri, USA St Louis Metrolink LRT	House Price	700m	32%
Metro/ heavy rail	Banister (2007)	London, UK, London Metro Jubilee Line	Land & Property Value	2000m	75%
	Laakso (1992)	Helsinki, Finland Helsinki Metro	Land Sale price	250m	3.5% to 6%
	Bae et al. (2003)	Seoul, South Korea Heavy Rail KoRail	Land (Sales Price)	400m	0.3% to 2.6%
	Yankaya and Celik (2004)	Izmir, Turkey Izmir Metro	Property Sale Price	500m	0.7% to 13.7%
	Medda (2011)	Warsaw, Poland Warsaw Metro	Property Sale Price	1000m	6.7%-7.13%
Commuter rail	Cervero and Duncan, (2002)	San Diego, USA Commuter Rail	Property Sale Price	1/2 mile	-7.1% to 46.1%
	Gruen, 1997	Chicago, USA METRA Commuter Rail	Property Value	400m	20%
	Armstrong, 1994	Boston, USA Commuter Rail	Property Value	400m	6.7%
	Voith (1991)	Pennsylvania & NJ, USA Commuter Rail	Property Value	400m	3.8% to 10%

(adapted from McIntosh et al., 2014a)

- Neighborhood variables include socio-economic parameters of the area, number of open spaces or parks, population density, employment density, percentage of differential land uses, road network within the area, crime rate, etc.; and
- **Proximity to Transit:** Typically proximity is measured from the perspective of walkability, or accessibility through cycling distance as measured less than 200m, 200m-400m, 400m-800m, 800m-1600m, etc. scales widely used. Adjacent properties tend to be for commercial and the rest predominantly for residential, employment or schools/hospital related facilities.
- Time based variables include land value price impacts during the announcement of alignment, pre-transit condition, during construction, post construction and transit operations phases indicate a dynamic land value changes impacted by transit. In most cases it was noted that the initial stages show a good jump from the pre-transit conditions generally due to speculation and marketing excitement from the developer community. Similar trends were observed from various studies post construction and initial operational phases.

Emerging Value Capture Innovative Urban Rail Funding and Financing

This section discussed the accessibility driven value proposition due to the transit investment on land values. The next section will discuss how to 'capture' the value created along with the various capturing mechanisms.

VC MECHANISMS TO CAPTURE THE VALUE

The main idea behind value capture is that urban rail will increase land values when it is built; this must be beyond what would happen anyway due to rising incomes and other economic activity. These land value increases can be captured by various government mechanisms and put into a Transit Fund that can then be used to raise finance for building and operating the rail system. Thus it can involve private sector financing (for building, owning and operating) as well as government sources of financing, but in all cases it will require a government funding mechanism to capture the value as the first step in unlocking the finance.

Capturing the value in an acceptable, transparent and equitable way involves multiple methods and complex mechanisms. There is no *one size fits all* solution possible. The applicability of a specific VC mechanism may or may not be applicable in another project due to a number of factors such as location, legislation, project type, willingness-to-pay, ease of adaption, administration, duplicity, etc., and these are all areas that need to be looked into on a case-by-case basis.

The most important way to categorize VC is into passive and active VC mechanisms. Active VC revenue sources are mostly revenue flows from active intervention such as buying property or creating a special levy on the station precinct; passive VC revenue flows are more asset value dependent so funds come from tax based revenue flows without intervention to actively pursue the value directly.

Active VC mechanisms can be applied to government owned property or vacant public lands which benefit from the transit accessibility driven increased land values. The said revenues can be accrued if governments either sell their land holdings or sell the development rights to the land holdings. In case of deep cut subway transit projects, the newly created underground space holdings around transit station area can yield more revenues through shopping or business activities as revealed in the case of subway projects of Hong Kong, SAR China, Japan and London. The public transport financing practiced by Mass Transit Railway Corporation (MTR) in Hong Kong SAR, China and the Japan Railway Construction Agency (JRCA), a public corporation of Japan Railway are good examples of this active, development based approach. Hong Kong's MTR co-developed the sites along the transit corridors and above the transit station rather than selling those sites. In 1993, the corporation financed about 22% of operating cost of their transit system through property rental income. Similarly the greater Tokyo's private railways have practiced transit value capture through development on an even grander scale, building massive new towns along rail-served corridors and cashing in on construction, retail and household service opportunities created by these investments (Suzuki et al., 2013, page 183).

Active VC mechanisms can include betterment tax, benefit area levies, infrastructure levies, special assessment districts, developer contributions, density bonuses or sale of air rights. In Australia the Gold Coast Transport Levy, which is collected across the whole of the Gold Coast municipal area, was used to help fund and operate a new light rail. The Transport Levy was able to provide the on-going costs of operation and was used to induce state and federal capital for building the system. A PPP was able to be used based on this active intervention to create a fund suitable to be used for raising the finance.

Passive VC mechanisms are mostly on private land where the revenue flow is focused through ad valorem tax instruments namely capital gains tax, stamp duty tax, land tax, GST on land sales and any other land-based taxes. These will rise due to the increased accessibility from the urban rail service and will flow into various levels of government. If scientifically estimated the increased flow of funding can be hypothecated into a Transit Fund and used to attract financing from banks involving various combinations of the private sector and government. Passive VC mechanisms still require government actions but not directly in the market place; they are therefore more politically acceptable. They do require Treasury Departments to hypothecate revenues.

Both active and passive value capture can enable more significant private involvement in the urban rail projects. If private financing is being used it is possible to involve private consortia in a PPP where not only do they bid to build, own and operate the rail system but they can also do entrepreneurial land development as part of their bid. In the case of private properties a number of active VC mechanisms are available such as: development of private property or government land that has been specified to be part of the bid process; joint development of government land with the private sector; leasing the property for parking or development as fee based revenues; rental returns on government property; and leasing the space for other revenue. Thus a combination of active and passive mechanisms could all be used to create the Transit Fund and hence create the financing opportunities for the rail project. Table 3 summarizes various VC mechanisms implementation as compiled in McIntosh, et al (2014, page 6) from various academic studies, and relevant secondary sources.

A combination of active and passive VC mechanisms may work as a better approach to providing the best potential value proposition. The underlying success factor is that stakeholders are willing to pay provided they are convinced about the value proposition. Prior to the implementation of the VC mechanisms listed, each mechanism should therefore be required to be evaluated against an existing policy evaluation framework. This would include factors such as administrative ease of collection; legislation related challenges; socio-economic-demographic preferences; and political priorities.

The next section of this chapter describes a strategic value capture (SVC) framework approach to VC based rail transit project financing that would be possible to be undertaken across the transit project life cycle.

Six-Step SVC Framework for VC funded Urban Rail Transit Projects

It is a potentially more beneficial approach if the VC based funding and financing strategy is decided by the decision makers at the early stages of the project life cycle itself as a key objective. The general findings from the various value capture experiences suggest one thing in common: that the VC based funding and financing process is a staged process. The proposed SVC framework described in this section offers a platform defining a VC based rail transit funding and financing development plan covering the planning, implementation and operationalization stages. This SVC framework offers a six step strategic process across each stage of a rail transit project life cycle. These stages include: initiation; planning; design; funding strategy; execution; and operations. The six steps identified for the proposed SVC Framework across the six project stages with key VC processes identified are:

Step 1: VC Initiation: VC concept due diligence
Step 2: VC Planning: VC value proposition analysis
Step 3: VC Design: VC revenue capturing mechanisms

Table 3. Compilation of VC mechanisms implementation from academic studies, and related websites

	VC Mechanism	Implementation & Transit	Comments
Passive Government Property	Sale of surplus property/development rights/air rights	• Hong Kong SAR, China (Metro) • Washington DC, USA (Metro) • Sydney, Australia (Heavy Rail)	These VC instruments can be used when vacant public land or government property is available and got transit proximity asset value increase.
	Sale of naming rights to stations	• New York, USA • Philadelphia, USA	
Active Government Property	Direct development of government property	• Hong Kong SAR, China (Metro)	These are more induced VC mechanisms and maximize the returns on the government land or property available and also own the asset value.
	Joint development	• Hong Kong SAR, China (Metro) • Tokyo, Japan (Metro) • London, UK (Metro)	
	Returns on public parking fee	• Portland, USA (Street car/LRT)	
	Government property leasing	• Philadelphia, USA	
	Advertising revenue at station areas	• Popular international practice	
Passive Non-Government Property	Tax increment financing	Widely used in USA, UK & Australia	Primarily focussed on additional portion accrued due to increase in land values to the existing ad valorem taxes
	State transfer duty/sales tax	• Atlanta, USA (Heavy Rail) • Dallas, USA (LRT)	
	State land/property tax	• Dallas, USA (LRT) • Portland, USA (Street car/LRT)	
	Local government taxes	• Portland, USA (Street car/LRT)	
Active Non-Governmental Property	Special assessment districts	• London UK (Metro) • Seattle, USA (Streetcar/LRT) • Portland, USA (Streetcar/LRT)	These are integrated transit and land use development oriented VC mechanisms.
	Special area rates/service charges	• Atlanta, USA (Heavy Rail) • Dallas, USA (LRT)	
	Infrastructure tax hypothecation	• London UK (Metro) • Portland, USA (Streetcar/LRT)	
	Developer contributions	• Popular practice	
	Density bonuses	• New York, USA (Metro) • Curtiba, Brazil	
	Local parking levy	• San Francisco, USA	

(adapted from McIntosh et al. 2014)

Step 4: VC Funding Strategy: VC fund redistribution plan
Step 5: VC Implementation: VC Governance
Step 6: VC Operations: VC Performance evaluation and monitoring

The remainder of this section details each of these steps.

Step 1: VC Initiation: VC Concept due Diligence

This stage is more like a due diligence phase on VC and builds on the normal transport planning processes but adds in a much more defined land use element. Firstly, the need for such investment and the problem it addresses needs to be determined. This will help justify the investment requirement and expected accessibility improvement. Secondly, a study is needed on the proposed corridor network with anticipated VC opportunities such as current densification, developmental opportunities, availability of vacant public land and also other accrued benefits from the development and also improved quality of life in the adjacent neighborhood. This will help from the perspectives of validation of VC potentials, and project impact zone. Primarily, the proposed VC catchment area is selected from considerations of transit accessibility and from the walkability and cycling perspectives, i.e. it is unlikely to go beyond 500m from the station area. In a few cases, where alternate transit alignments are available, this will aid in determination of the transit network based on anticipated VC potentials. Further project stakeholders including the beneficiary community are identified and the objectives of the proposed project outlined. A macro-level qualification of the project for VC based funding and financing would be done during this step. VC due diligence can be included as a pre-project VC assessment study along with project feasibility or detailed project report preparation stages.

At the end of this Step 1, the following outcomes are expected:

- Need for transit established through problem definition.
- The best alignment with maximum anticipated VC potential is determined.
- VC catchment area as proposed transit project impact area identified as influence zone.
- VC stakeholders are identified.
- VC macro level goals defined.

Step 2: VC Planning: VC Value Proposition Analysis

This step involves defining the value proposition potential. Several studies need to be conducted to analyze the improved accessibility-driven transit impacts on land use and land values. Validation of the stakeholder opinion on the value proposition and assessment of their willingness-to-share the accrued unearned benefits anticipated through accessibility improvement need to be determined. During this step, the transit value proposition in terms of increased land and property values are analyzed, stakeholders are contacted and engaged in the process, and development strategies around the transit station areas are identified from the land use change or densification perspectives and also support infrastructure requirements from the sustainability and accessibility improvement perspective. This is a key step in the VC project life cycle which validates the value proposition of the transit project.

At the end of this Step 2, the following outcomes are expected:

- Assessment of transit project impact on land and property values (a willingness to pay assessment).
- Assessment of active and passive VC potentials (assessed against criteria outlined above).
- Validation of stakeholder participation.
- Review the support infrastructure and other sustainability priorities around stations to support land development.

Step 3: VC Design: VC Revenue Capturing Mechanisms

Step 3 is an important phase in the process which focuses on how to translate the value created into monetary terms through appropriate VC mechanisms. This step identifies the revenue flows through various combinations of VC mechanisms. Revenue flows with induced land use strategies through integration of transit and land use are estimated and documented. Stakeholders need to be actively engaged in the process of selecting innovative alternate revenue flows. Care needs to be taken that sustainable development goals are kept in mind while finalizing the revenue streams. During this step, legislative measures and modalities as required are identified for implementing the recommended VC mechanisms as appropriate. The full set of funding potential raised needs to be specified. There will be some revenues that will flow immediately and some revenues will flow through the project life cycle as such as passive value capture from taxation revenues due to the on-going accessibility advantages of the rail system. The possible total amount that could go into a Transit VC Fund would be established. Such fund valuation potentials can act as security collateral to attract private sector financing through PPP or Joint Development methods as part of procurement decision.

At the end of this Step 3, the following outcomes are expected:

- Passive and active revenue mechanisms identified and revenue generation plan finalized.
- Legislation challenges and ease of revenue capturing for identified VC mechanisms evaluated.
- Stakeholders engaged in the finalization of the VC mechanisms and modalities.
- VC fund prospects and revenue generation plan finalized.

Step 4: VC Fund Strategy - VC Fund Redistribution Plan

This is a very interesting phase for planners, stakeholders and decision makers as the VC fund redistribution strategies with stakeholder gain share model will be determined. The key strategic decision during this stage is focused on the captured fund redistribution strategies with proportionate share and ensuring the equity based revenue gain share strategies are finalized. The redistribution of the captured value will be primarily through three complimentary strategies. First strategy is to give a major share to finance the investments made to build the rail transit project. This may be with private or government approaches or a joint PPP approach but the key is to see how much of the VC Fund would be needed for the transit system financing. Second strategy is to share some revenues to boost the rail impact through local infrastructure. For example: improvement of access roads, pedestrian and bicycle parking facilities around the station areas covering the transit influence zone will bring more patronage to the transit and also sustainable infrastructure improvements will be made. This amount would need to be worked out with the local council and with the proponents of the rail system as well as the local stakeholders such as local businesses. Third strategy is to extend a partial benefit to the city community at large catering to the needs of their accessibility requirements. For example equity issues may need to be considered to ensure those pushed out by the redevelopment process may be adequately compensated or provided with access to the station precinct by bus. At this stage, it is recommended to involve key stakeholders including community participation in finalizing these VC redistribution strategies. Participatory budgeting practices can be explored to ensure more accountability and transparency in the decision making. Once the strategies are finalized then a detailed activity based fund allocation plan could be determined.

At the end of this Step 4, the following outcomes are expected:

- Fund allocation strategies finalized and target beneficiaries identified.
- Eligible projects plan with funding stages finalized.
- Overall fund redistribution plan with multiple projects, targets and milestones detailed.
- Participatory budgeting options explored.

Step 5: VC Implementation: VC Governance

Now that the value proposition has been assessed and the VC mechanisms to capture the revenues are identified, this step is about translating plans and strategies into action. Step 5 mainly focuses on VC implementation mechanisms through establishment of institutional and administration set-up and strengthening with staff for activities to be undertaken. A core function will be establishing a Transit VC Fund within an institutional framework that can deliver the project. Such an institutional set-up can function as part of a project management authority but with a focus on fund collection and fund management as it may be best to deliver the project through a PPP.

The procurement process needs to be specified to generate the kind of involvement from the private sector that is seen to be preferred for building, owning, operating and doing land development around the transit system.

Governance structures are also needed for stakeholder participation and empowerment to steer the VC process to achieve the set objectives and goals. This can be required as part of a PPP. Execution of various VC mechanisms can be done on a collaborative basis through various participating agencies. Against each of the VC mechanisms a detailed implementation plan with administrative mechanisms and protocols is to be established. It is also essential to formulate transparency in the actions of fund collection and distribution.

At the end of this Step 5, the following outcomes are expected:

- VC institutional and administrative setup established.
- Procurement process specified to enable private sector involvement.
- Executive body with stakeholders representation formed including transparent budget allocation plan.
- Stakeholder engagement and community empowerment plan.

Step 6: VC Operations: VC Performance evaluation and monitoring

This is a VC sustaining phase. Once the various VC mechanisms, institutional and administrative setup are put in place, the focus turns to monitoring and evaluation. Step 6 focuses on periodic review and continuous monitoring of fund flows and the effectiveness of the VC set objectives. During this stage the VC yield revenues flow back to recover the investments made, and also ensures appropriate fund commitments to the community at large are fulfilled. Leveraging software technologies, it is easy to define a VC based balanced score card approach with dash board analytics to validate the VC governance strategy and performance on a continuous basis. The main emphasis is on understanding the performance of the various stakeholder groups against the set objectives and targets. From time to time it is advisable to have a check point to measure the effectiveness of the engagement performance and also evaluate any risk groups or dependent activities to be monitored. It is advisable to have a performance evaluation criteria and monitoring plan put in place in the initial stages of VC implementation for an objective evaluation. This is an ongoing process. As an outcome of Step 6, periodical performance metrics of VC implementation and fund management will be published and actions to review each of the Six Steps approach would be made.

The Six-Step SVC framework has been conceived on the principles of accountability and the utilization of a participatory approach. At the heart of this framework is the goal of achieving sustainable mobility and sustainable development along the proposed rail transit corridor.

FUTURE RESEARCH DIRECTIONS

Despite the good research progress made on VC based-financing approaches, there are many inadequacies in the current research in exploring the true potentials of VC based financing. In future more focused research needs to be conducted on the delivery and implementation aspects of VC based project financing as this is where there is currently only limited knowledge available. Furthermore, research is needed to look at the effective institutional set-ups necessary, along with how participatory governance models can be utilized for an effective delivery of VC based rail transit funding and financing project implementation. In addition, there is limited research on the efficient VC fund redistribution strategies beyond simply recovering the rail transit investments made, and also on how VC can contribute to the place-making around the transit stations.

CONCLUSION

Urban rail transit is a major part of future urban mobility as there is a growing recognition and demand for more sustainable urban mobility linked in to the building of more sustainable land use patterns. Cities worldwide are facing a daunting funding challenge to build urban rail transit systems and there is a great need for innovative funding and financing mechanisms to enable these systems to be built and operated. Land based VC mechanisms offer a potentially feasible approach for financing urban rail projects in a way that enables more integrated, sustainable land use patterns. The general findings of the chapter suggest that transit investments if planned strategically with VC based approaches should help catalyze development opportunities along transit corridors and hence create access for more people without the need for a car. Therefore, integrating transit, land use and finance is a workable approach to achieve transit oriented sustainable development. It is evident from the chapter that VC concepts are here to stay in the future as a sustainable public transportation funding and financing solution.

ACKNOWLEDGMENT

This chapter is one of a series of writings, as part of PhD on "Innovative Financing for Urban Rail in Indian Cities: Land based Strategic Value Capture Mechanisms" at Curtin University, Australia funded partly by AusAID and PATREC.

REFERENCES

Alonso, W. (1964). *Location and Land-use: Towards a General Theory of Land Rent*. Cambridge: Harvard University Press. doi:10.4159/harvard.9780674730854

Bahl, R., Linn, J.F., & Wetzel, D.L. (Eds.), (2013). Financing Metropolitan Governments in Developing Countries. Cambridge, MA, USA: Lincoln Institute of Land Policy.

Banister, D., & Goodwin, M. T. (2011). Quantification of the non-transport benefits resulting from rail investment. *Journal of Transport Geography, 19*(2), 212–223. doi:10.1016/j.jtrangeo.2010.05.001

Cervero, R. (1994). Rail transit and joint development: Land market impacts in Washington DC and Atlanta. *Journal of the American Planning Association, 60*(1), 83–94. doi:10.1080/01944369408975554

Cervero, R., & Duncan, M. (2001). *Rail transit's value added: effects of proximity to light and commuter rail transit on commercial land values in Santa Clara County, California*. Washington, DC, USA: Urban Land Institute, National Association of Realtors.

Cervero, R., & Landis, J. (1997). Twenty years of the Bay Area Rapid Transit system: Land use and development impacts. *Transportation Research Part A, Policy and Practice, 31*(4), 309–333. doi:10.1016/S0965-8564(96)00027-4

Chen, X. (2012). Managing transportation financing in an innovative way. *Management Research and Practice, 4*(3), 5–17.

Gadgil, V. B. (2011). *No metro project is financially viable on its own across the world*. Retrieved from http://www.reachouthyderabad.com/newsmaker/hw376.htm

Jillella, S., Newman, P., & Matan, A. (2014). Participatory Sustainability Approach to Value Capture Based Urban Rail Financing in India Through Deliberated Stakeholder Engagement.*Proceedings of the 4th World Sustainability Forum* (Vol. 4). doi:10.3390/wsf-4-b002

Landis, J., Guathakurta, S., & Zhang, M. (1994). *Capitalization of transportation investments into single-family home prices* (Working paper 619). Institute of Urban and Regional Development, University of California, Berkeley, USA.

Levinson, D. M., & Istrate, E. (2011). *Access for value: financing transportation through land value capture*. The Brookings Institute, Transportation Research Board.

Martinez, L.M.G., & Viegas, J.M. (2012). The value capture potential of the Lisbon subway. *The journal of transport and land use*, 5(1), 65-82.

McIntosh, J. (2014). *Framework to capture the value created by urban transit in car dependent cities* [Unpublished doctoral dissertation]. Curtin University, Perth, Australia.

McIntosh, J., Newman, P., Crane, T., & Mouritz, M. (2011). *Alternative Funding Mechanisms for Public Transport in Perth: the Potential Role of Value Capture, Committee for Perth*. Retrieved from http://www.committeeforperth.com.au/pdf/Advocacy/Report%20-%20AlternativeFundingforPublicTransportinPerthDecember2011.pdf

McIntosh, J., Trubka, R., & Newman, P. (2014a). Can Value Capture Work in Car Dependent Cities? Willingness to pay for transit access in Perth, Western Australia. *Transport Research – Part A, 67*(September), 320–339.

McIntosh, J., Trubka, R., & Newman, P. (2014b). Tax Increment Financing framework for integrated transit and urban renewal projects in car dependent cities. *Urban Planning and Research Journal Online*, *3*(December).doi:10.1080/08111146.2014.968246

Medda, F. R. (2012). Land value capture finance for transport accessibility: A review. *Journal of Transport Geography*, *25*, 154–161. doi:10.1016/j.jtrangeo.2012.07.013

Mohammed, S. I., Graham, D. J., Melo, P. C., & Anderson, R. J. (2013). A meta-analysis of the impact of rail projects on land and property values. *Journal of Transport Research Part A*, *50*, 158–170.

Newman, P., Glazebrook, G., & Kenworthy, J. (2013). Peak car use and the rise of global rail: Why this is happening and what it means for large and small cities. *Journal of Transportation Technologies*, *3*(04), 272–287. doi:10.4236/jtts.2013.34029

Newman, P., & Kenworthy, J. (1999). *Sustainability and cities overcoming automobile dependence*. USA: Island Press.

Newman, P., & Kenworthy, J. (2015). *The End of Automobile Dependence: How Cities are Moving Beyond Car-based Planning*. Washington, DC: Island Press. doi:10.5822/978-1-61091-613-4

Smolka, M. O. (2012). A New Look at Value Capture in Latin America. *Land Lines*, *24*(3), 10–15.

Suzuki, H., Cervero, R., & Luchi, K. (2013). *Transforming cities with transit*. Washington, DC, USA: The World Bank. doi:10.1596/978-0-8213-9745-9

Voith, R. (1993). Changing capitalization of CBD-oriented transportation systems: Evidence from Philadelphia 1970-1988. *Journal of Urban Economics*, *33*(3), 361–376. doi:10.1006/juec.1993.1021

Chapter 7
Analyzing Intercity Modal Choice and Competition Between High Speed Rail (HSR) and Other Transport Modes in Indian Context

Ashish Verma
Indian Institute of Science, India

Varun Raturi
Indian Institute of Science, India

ABSTRACT

In this study, a theoretical framework is developed in order to assess the viability of transport infrastructure investment in the form of High Speed Rail (HSR) by assessing, the mode choice behaviour of the passengers and the strategies of the operators, in the hypothetical scenario. Discrete choice modelling (DCM) integrated with a game theoretic approach is used to model this dynamic market scenario. DCM is incorporated to predict the mode choice behaviour of the passengers in the new scenario and the change in the existing market equilibrium and strategies of the operators due to the entry of the new mode is analysed using the game theoretic approach. The outcome of this market game will describe the strategies for operators corresponding to Nash equilibrium. In conclusion, the impact of introduction of HSR is assessed in terms of social welfare by analysing the mode choice behaviour and strategic decision making of the operators, thus reflecting on the economic viability of the transport infrastructure investment.

INTRODUCTION

This chapter focuses on analyzing the intercity modal choice of the passengers which eventually affects the intermodal competition between the modes. In the event of High speed rail mode entering the market, passenger's mode choice in the new scenario may differ from the mode choice in the current scenario. This change in the mode choice is examined using a Discrete choice model too get the mode share in

DOI: 10.4018/978-1-5225-0084-1.ch007

the new market scenario. The mode share thus obtained is used to study the resulting strategies played by the mode operators to maximize their utility using a game theoretic approach. Thus the primary objectives of this chapter are:

- Analyze intercity modal choice in the new hypothetical scenario.
- Determine the optimal mode attributes which maximizes the social welfare and the operator's profit.
- Determine the optimal strategies of the mode operators under different market scenarios using game theoretic approach.

BACKGROUND

In India, rapid urbanization, growing economy and increasing per capita income has caused increasing growth in intercity travel. High speed rail (HSR) system has been globally proven to be an efficient transportation mode to fulfill the demand gap for faster intercity movement of passenger traffic. According to International Union of Railways (UIC), passenger trains that travel at 250 km/h or more on a new track or 200 km/h for an upgraded track is a High Speed Rail. Thus, in order to cater to the ever increasing passenger traffic and demand for better services, the Government of India is exploring the option of introducing HSR system as a sustainable mode of transportation. Verma et al. (2010) described HSR as a sustainable mode while comparing it with the current mode scenario in Indian context. They compared HSR and conventional rail using parameters like per-passenger km, occupancy levels and electricity de-carbonization. They concluded that HSR is anticipated to produce lower GHG emissions than Conventional Rails.

India has one of the largest rail networks in the world. Currently, India does not have any high-speed rail lines capable of supporting speeds of 200 km/h (124 mph) or more. However, high-speed corridors have been proposed and are under prefeasibility studies. The Indian Railways' vision 2020 envisages the following on High Speed Corridors: "India is the only country among the major nations of the world which do not have a high speed rail corridor. In order to escalate the speed of the corridors Indian Railway will follow a two-step approach. Based on the feasibility of the passenger corridors, speed will be raised to either 160-200 kmph using conventional technology or up to 350 kmph by building state-of-the-art high-speed corridors through on PPP mode in partnerships with the State Governments. By 2020, at least four corridors of 2000 km would be developed and planning for 8 other corridors would be in different stages of progress". High Speed Rail Corporation of India Limited (HSRC) has been formed on the directions of Ministry of Railways, Government of India, for development and implementation of high speed rail projects. HSRC (2015) mentions the railway budget speech 2012-2013 which states the issue of capital intensiveness and innovative funding mechanism to make this project a reality. Six corridors have already been identified for technical studies on setting up of HSR as shown in Table 1.

Since Indian government perceives HSR as a feasible future mode of intercity transportation, this study is timely as current literature on HSR in Indian context is sparse. The Indian Railway budget 2014 stated some of the issues related to HSR thereby ascertaining the need for studies in the mentioned context.

HSR projects are highly capital intensive, requiring high passenger volumes and high tariff to justify investment (Railway Budget, 2014).

Table 1. Proposed corridors for technical studies

High-Speed Corridor	Route	Stations	Speed (kmph)	Length (km)
Howrah-Haldia	Howrah-Haldia	TBD	250-300	135
Delhi-Patna	Delhi-Agra-Kanpur-Lucknow-Varanasi-Patna	TBD	200-350	991
Delhi-Amritsar	Delhi-Chandigarh-Amritsar	TBD	TBD	450
Hyderabad-Chennai	Hyderabad-Dornakal-Vijayawada-chennai	TBD	TBD	664
Pune-Mumbai-Ahmedabad	Pune-Mumbai-Ahmedabad	7	300-350	650
Chennai-Bangalore-Coimbatore-Ernakulam	Chennai-Bangalore-Coimbatore-Ernakulam	TBD	350	649

Source: Vision 2020, Indian Railways

While there is a considerable amount of studies on HSR in developed nations especially the European countries, very little is available for the developing nation's scenario. Does the region where the study is done, matters? Rus and Nombela (2005) concluded that decisions to invest in high speed rail (HSR) infrastructure are not always based on sound economic analysis. A mix of arguments –strategic considerations, regional development, technology, congestion in competing modes, and so forth– usually makes the discussion on the economic rationality of investing in HSR vague and imprecise. To recover costs, HSR investments not only requires a high level of demand but also a willingness to pay (WTP) for the new technology and this, in turn, depends on the magnitude of time savings and its economic value (de Rus & Roman, 2006). Thus to take an informative decision about investing on the new infrastructure, the policy makers should comprehend the changes in the existing market with the introduction of the new mode. The entry of HSR in the market will lead to a change in the existing equilibrium of the market, therefore understanding the conditions that allow stable access of the new mode under existing scenario should form the basis of decision making.

Indian Railway provides the most important mode of public transport in India. This is the most commonly used and cost effective long distance transport system of the country. Indian Railway operated by Ministry of Railways is touching life of almost every people across India. Since its inception, the Indian Railways has served to integrate the fragmented markets and thereby, stimulating the emergence of a modern market economy. It connects industrial production centers with markets and with sources of raw materials and facilitates industrial development and link agricultural production centers with distant markets. It links places, enabling large-scale, rapid and low-cost movement of people across the length and breadth of the country. In India the cultural diversity, inequitable job distribution, economic disparity leads to a demand for an affordable and reliable mode of transportation to fulfill this need for travel. In the process, the Indian Railways has become a symbol of national integration and a strategic instrument for social welfare. Thus through this reasoning, profit maximization may not be the primary objective for the Railways. Thus this study tries to incorporate maximization of social welfare function as the primary objective which is more relevant for the Indian context. Majority of the competition studies in foreign context deals with profit maximization as the primary objective as the railways operators are private sectors. Maximizing social welfare directly relates to increasing the consumer surplus in the new scenario i.e. due to the introduction of the new mode and hence can be provided as a feasibility parameter for the new mode.

Literature discussing high-speed rail infrastructure, several interesting papers have reached different conclusions. Janic (1993) studied the competition between the two modes, concluding that high-speed rail can compete with air transport over a relatively large range of distances (from 400 to over 2000 km). However, the model assumes that all demand is met and that the aim is to minimize total system costs for both passengers and transport operators. Hansen (1990) developed a non-cooperative game in which the airlines choose the frequencies of service, assuming fixed airfares and inelastic demand. Mandel et al. (1997) developed a cox-box logit model to model the intercity travel in Germany and determine its implication on the high speed rail demand forecast. Roman et al. (2007) analyzed the potential competition of the high-speed train with the air transport between Madrid and Barcelona. They developed a disaggregate demand model using mixed RP/SP data and obtained different willingness to pay measures for improving service quality. Demand responses to various policy scenarios that consider the potential competition between high-speed train and air transport were examined and concluded that the HST market share would not exceed 35%. This implied that low modal share and generally low rate of return on HST projects, cast doubts on the competition that HSTs can exert in markets characterized by high frequency air services. Shyr and Hung (2010) estimated the modal shares using discrete choice modelling by conducting revealed preference and stated preference survey and stated that if Taiwan High speed rail continues to increase its service frequency and lower its promotion prices, the chances of airline market to compete with HSR is very slim. They incorporate co-operative game theory to study the coalition structures between the airlines predicting that to maintain profitability, airlines would have to unify as an alliance and cut their daily flights by 50%. For given payoff values, Shapley value is computed in order to solve the profit distribution problem. M Bierlaire (2001) used a RP/SP survey of long distance road and rail travelers to investigate the entry of Swiss metro in the market. He explored the stability of the modelling results with respect to different approaches of capturing the heterogeneity in the data. The results indicated a relative stability of the relative parameter values and nearly consistent results with regards to the assessment of parameter significance. Impact of deregulation of the bus market in terms of profits and social cost to the society by modelling the strategic interactions between the operators in the bus market was assessed by Wang and Yang (2005). Assuming value of time distribution as uniform they calculated the modal share for optimized fare and frequency which maximizes their profit. They analyzed the extensive form game between the incumbent and the entrant for various strategies. Incumbent has three strategies namely monopolist, accommodation and deterrence whereas the entrant has two options of entering the market or staying out of the market. They concluded that deterrence is the dominant strategy in many market scenarios which is best for the society whereas accommodation occurs when the sunk cost is low and demand is high. Natural monopoly arises when sunk cost is high and route density is low. Levinson et al. (1997) utilize an engineering, full-cost approach to argue that high-speed rail infrastructure is significantly more costly than expanding air services and should not be assumed to substitute air transport. Raturi et al. (2013) incorporated a game theoretic approach to investigate the competition between HSR and bus system for Bangalore-Mysore corridor. Value of time approach was used to compute the modal share in the hypothetical scenario. They used hypothetical data to illustrate the importance of sunk cost and different strategies in the decision of HSR to enter or stay out of the market. They modelled the competition scenario as an extensive form game and the sub-game perfect Nash equilibrium was found out by backward induction. Discrete choice model is better alternative to value of time approach to compute the modal share as it takes individual characteristics into account. Fu et al. (2012) investigated the effects of HSR services on Chinese airlines. They suggested that HSR services will be competitive in short to medium distance city pairs, diverting the traffic to rail.

They also indicated that to compete airlines must transform their point to point networks to effective hub and spoke networks. Adler et al. (2010) developed a methodology to assess infrastructure investments and their effects on transport equilibria. The operators maximized best response functions via prices, frequency and mode sizes. They applied the methodology to all 27 European Union countries and concluded that development of HSR should be encouraged across Europe if the objective is to maximize social welfare. Zito et al. (2011) used a game approach to model airline choices in a duopolistic market. They formulated a bi-level optimization program to maximize profit for the operator and maximizing consumer surplus for the users.

ISSUES AND PROBLEMS

The vast majority of the literature to date analyzes airline competition and relatively little has been published on the rail sector and there is no significant work on studying the impacts of HSR in Indian context. As Indian government is exploring the option of HSR as a sustainable mode, this study is timely appropriate. Many studies focus on the profit maximization for operators in a deregulated market scenario which may not be the case in India. Thus social welfare maximization by the introduction of the new mode is more suitable for Indian scenario. In a competitive environment, mode operators may vary their strategies which result in changes in travel variables. General practices while analyzing the intermodal competition do not take dynamic behavior into account thus this study tries to bridge the gap in the existing literature by analyzing the competition scenario in Indian context in a dynamically changing market while keeping the social welfare maximization as the basis. Currently the Indian government is in a dilemma of whether to go for commercial viability or social viability of the new project which may have direct impact on the engineering aspect of the new project (speed, fare, up gradation of track etc.). This study will help the policy makers to address this dilemma by analyzing modal behavior in new scenario and thereby determining the optimal strategies for the operators.

METHODOLOGY

The core of the research work is to model the intercity modal competition between HSR and other conventional transport modes in Indian context. In transportation field, mode share on a given corridor becomes the foundation of intermodal competition where each mode operators try to strategize their decision variables in order to gain a significant market share. In this study, discrete choice modeling is used to determine the modal share in the new scenario coupled with game theory to understand the strategic behavior.

Modal Share

Disaggregate demand analysis is one of the ways to determine the modal share. Disaggregate demand analysis has its theoretical basis on the microeconomic of discrete choices. Utility depends on consumption of continuous goods and on the characteristics of discrete alternatives. Discrete choice models assume utility-maximizing behavior by the decision maker (McFadden, 1974). It states that the utility of alternative j for individual n has the expression:

$$U_{nj} = V_{nj} + \varepsilon_{nj}$$

Where, V_{nj} is the representative or systematic utility observed by the analyst of individual n for alternative A_j and ε_{nj} is a random term that includes unobserved effects. V_{nj} depends on the observable attributes of alternative j, X_{nj}, as well as on the socioeconomic characteristics of individual n.

The dependent variable represents individual's choices. Estimation provides the probability distribution of the dependent variable for every individual observation. Hence, the probability that individual n chooses alternative j is given by:

$$P_{nj} = P\left(V_{nj} + \varepsilon_{nj} \geq V_{ni} + \varepsilon_{ni} \text{ for all } i \neq j\right)$$

This probability will depend on the hypotheses formulated about the distribution of the vector of random terms ε_{nj}.

By far the easiest and most widely used discrete choice model is logit. Its popularity is due to the fact that the formula for the choice probabilities takes a closed form and is readily interpretable (Train, 2009). McFadden (1974) completed the analysis by showing that the logit formula for the choice probabilities necessarily implies that unobserved utility is distributed extreme value. The logit model is obtained by assuming that each ε_{nj} is independently, identically distributed extreme value. The distribution is also called Gumbel and type I extreme value. The density for each unobserved component of utility is

$$f\left(\varepsilon_{nj}\right) = e^{-\varepsilon_{nj}} e^{-e^{-\varepsilon_{nj}}}$$

By assuming the variance is $\pi^2/6$, we are implicitly normalizing the scale of utility and the difference between two extreme value variables is distributed logistic (Akiva & Lerman, 1985). Hence the choice probabilities are given by:

$$P_{nj} = \frac{e^{V_{nj}}}{\Sigma e^{V_{nj}}}$$

which is the logit choice probability.

The Game

A game in the everyday sense is "a competitive activity in which players contends with each other according to a set of rules" (Osborne, 2003) and game theory is a way to mathematize these strategic interactions among players and finding the outcome of a game. In this study, High speed rail is entering the market thereby creating a competitive scenario for the incumbent (say Airlines) hence resulting in an entry game. Assuming in this scenario; the incumbent i.e. airlines has two strategies; Accommodation and Monopolist. In Accommodation strategy the airlines modes will try to accommodate the entry of HSR by varying their fare and frequency whereas in monopolist strategist they will not bother about the entry of the HSR. High speed rail will enter the market if the payoff function is greater than zero otherwise it will stay out of the market. This situation may be modeled as the following extensive game with perfect information given in Figure 1.

Where $\{\Phi_{air}, \Phi_{HSR}\}_{AE}$ is the payoff for airlines and HSR respectively in the first strategic scenario i.e. Accommodation – Entry. Similarly for other scenarios payoffs are mentioned.

Data Required

To model the mode share models information about the current behavior of the individuals is collected and their change in the behavior in the new scenario is assessed by conducting stated preference surveys. Revealed preference survey contains information about various socioeconomic factors for an individual like household income, age, gender, occupation and marital status, vehicle ownership status. Travel variables information i.e. in vehicle travel time, out vehicle travel time, frequency, comfort, safety is also collected. Stated preference survey will focus on individual's reaction to the introduction of the HSR. In this survey their willingness to pay for the new mode should be collected. The process of pooling RP and SP data and estimating a model from the pooled data is called data enrichment. This process originally was proposed by Akiva and Morikawa (1990), whose motivation was to use SP data to help identify parameters that RP data could not, and thereby improve the efficiency (*i.e.*, obtain more precise and stable estimates) of his model parameters.

The Model

After the data collection, the model is estimated using maximum likelihood estimation to determine the modal share in the new scenario. The logit probability is given by

Figure 1. Extensive form game for entry

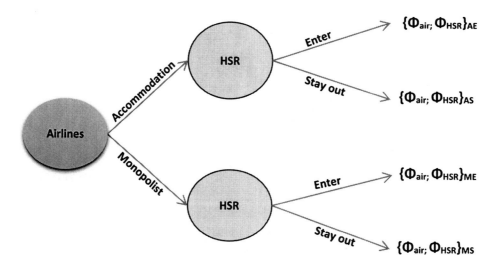

$$P_{nj} = \frac{e^{V_{nj}}}{\Sigma e^{V_{nj}}}$$

Where V_{nj} is the deterministic utility of an individual represented in the form of travel variables and socioeconomic variables of the individual.

$$V_{in} = ASC + \beta_1 \times X_{1ni} + \beta_2 \times X_{2ni} + \ldots + \beta_k \times X_{kni}$$

Where ASC is the Alternative Specific Constant

'β's are the parameters to be estimated

'X_{ni}' are the variables included in the model

'X_{ni}' variables can be generic, alternative specific or dummy variables. Generic Variables are the variables which have same marginal (dis)utility on all the modes equations. For e.g. Travel Time, Fare, Frequency etc. Alternative specific socioeconomic variables are the variables which reflect the differences in the preferences for the modes as functions of individual characteristics. For e.g. Income, age, vehicle ownership etc. Dummy Variables are the variables which take only 0 or 1 as values. For e.g. gender, marital status etc.

The model is estimated using a maximum likelihood estimation method. The likelihood function is given by

$$L^* = \prod_{n=1}^{N} \prod_{i \in C_n} P_n(i)^{y_{in}}$$

where y_{in} is an indicator variable

$y_{in} = 1 \quad \text{if the individual 'n' choses the alternative 'i'}$

$\quad\quad\; = 0 \quad \text{otherwise}$

Taking log of the likelihood function gives the log-likelihood function.

$$L = \sum_{n=1}^{N} \sum_{i \in C_n} y_{in} \left(\beta X_{in} - \ln \sum_{j \in C_n} e^{\beta X_{jn}} \right)$$

Differentiating the log-likelihood function with respect to each parameter and equating it to zero gives the estimates of the parameters which maximize the log-likelihood function.

$$\frac{\partial L}{\partial \beta_k} = 0$$

Following the estimation of the model, tests are carried out to check the significance of the parameters and validation of the model is carried out using a chi-square test.

$$\chi^2 = \sum_{i=1}^{m} \frac{(O_i - E_i)^2}{E_i}$$

where

Ei = Expected number of individuals choosing alternative 'i' (from the model)

Oi = Observed number of individuals choosing alternative 'I (from stated preference survey)

After the calibration and validation of the model, the model can be used to determine the mode share in the new scenario given by q_j.

Pay Off Functions

The incumbent mode i.e. airlines will have the motive to maximize their profit in different strategic scenarios. Thus the payoff function for the airlines will be the profit function and can be estimated by calculating the net revenue which is daily total revenue generated minus the daily operation cost for a mode.

$$\Phi_{air} = \pi_j = (P_j - b_j) \times q_j - a_j \times F_j$$

The daily operating cost of the operators can be determined by taking average direct operation cost per trip (a_j) and the average service cost (b_j).

$$DOC_j = a_j \times F_j + b_j \times q_j$$

$$F_j = Round\left\{ q_j \Big/ Seat_j \times L_j \right\}$$

where $Seat_j$ and L_j are the capacity and load factor for each mode respectively.

The entrant mode i.e. the HSR will be operated by the government and therefore maximizing the profits may not be the primary objective. While considering the approval of major projects, social welfare is one of major concern for the policy makers. Thus, maximizing social welfare by the introduction of the HSR service can be the primary objective of the government. Chang and Hsu (2001) used social welfare concept to estimate the mode variables for entering mode. They maximized the social welfare function with respect to fare and frequency. A major limitation of their work is that they varied the mode parameters of the entering mode by keeping the incumbent mode characteristics constant. In a competitive environment, a change in a mode characteristic may lead to a change in other modes variables. Jong et al. (2006) used logsum as an evaluation measure of the consumer surplus in the context of logit choice models. They used the change in consumer surplus as a measure in project appraisal and evaluation.

In the field of policy analysis, the researcher is mostly interested in measuring a change in consumer surplus that results from a particular policy. By definition, a person's consumer surplus is the utility (also taking account of the disutility of travel time and costs), in money terms, that a person receives in the choice situation. The decision-maker n chooses the alternative that provides the greatest utility, so that, provided that utility is linear in income, the consumer surplus (CS_n) can be calculated in money terms as

$$CS_n = \left(1/\alpha_n\right) max_n \left(U_{nj} \forall j\right)$$

where α_n is the marginal utility of income and equal to dU_{nj}/dY_n if j is chosen, Y_n is the income of person n, and U_n the overall utility for the person n.

Small and Rosen (1981) determined that if the model is MNL and utility is linear in income (that is α_n is constant with respect to income), then expected consumer surplus becomes;

$$E(CSn) = \left(1/\alpha_n\right) ln\left\{\sum_{j=1}^{J} e^{V_{nj}}\right\} + C$$

where C is an unknown constant that represents the fact that the absolute value of utility can never be measured. Under the usual interpretation of distribution of errors, $E(CS_n)$ is the average consumer surplus in the subpopulation of people who have the same representative utilities as person n. Total consumer surplus in the population can be calculated as the weighted sum of $E(CS_n)$ over a sample of decision-makers, with the weights reflecting the number of people in the population who face the same representative utilities as the sampled person.

The change in consumer surplus is calculated as the difference between the calculation of $E(CS_n)$ under the conditions before the change and the calculation of $E(CS_n)$ after the change (e.g. introduction of policy)

$$\Delta E(CSn) = \left(1/\alpha_n\right) ln\left\{\sum_{j=1}^{J^1} e^{(V_{nj}^1)}\right\} - \left(1/\alpha_n\right) ln\left\{\sum_{j=1}^{J^0} e^{(V_{nj}^0)}\right\}$$

Thus a social welfare (SW) can be defined as the sum of the profit of all the operators and the consumer surplus.

$$SW = \sum_{j=1}^{J} \pi_j(P) + E[CS_n(P)]$$

Solution Approach

The market share obtained by DCM method can be used to estimate the fare and frequency of each mode under different strategies played by the incumbent operator. These strategies can be monopolist and accommodation. In monopolist strategy incumbent operator tries to maximize the payoff without considering the entry of the new mode whereas in accommodation strategy the incumbent accommodate the entry of the new mode by fixing the fare and frequency of the mode based on the travel parameters of the new mode. The incumbent will fix the strategy where it achieves the maximum profit whereas the new mode can play two strategies i.e. whether to enter the market or stay out it.

Objective function:

Maximize $\pi_j(mon, acc \mid in, out)$

Such that Capacity constraints

The model can also be used to conduct sensitivity analysis by varying the sunk cost of the entering mode and thus observing the change in the strategy played by the incumbent mode.

The entrant mode will try to maximize the social welfare due to the entry of HSR. The upper bound of the fare price is estimated by first maximizing the social welfare function with respect to fare. This upper bound price is then used in the constraint of the profit maximization of the operators.

Objective Function:

$$\text{Maximize}_{\mathbf{P}}\, SW = \sum_{j=1}^{J} \pi_j(\mathbf{P}) + E[CS_n(\mathbf{P})]$$

Subject to $\mathbf{P} \geq 0$

After estimating the upper bound of the fare price that maximizes the social welfare function, we maximize the individual profit function.

Objective function:

$\text{Maximize}_{\mathbf{p}}\, \pi_j(\mathbf{p})$

Subject to $0 \leq \mathbf{p} \leq \mathbf{P}$

This procedure is followed until equilibrium is reached. This equilibrium state will define the optimum value of fare for the modes under the new market scenario. The policy sensitivity and associated changes in travel patterns of individuals can be investigated by comparing current scenario and the new

Analyzing Intercity Modal Choice and Competition

scenario. Policy should look into maximizing the social welfare but operational profitability of the new mode should not be neglected.

Game Outcome: Nash Equilibrium and Sub Game Equilibrium

After computing the payoff of the modes, the game outcome can be determined by solving the extensive form game for different choice scenarios. Nash equilibrium can be computed to determine the outcome of the game. Nash equilibrium is a strategy profile with the property that no player can induce a better outcome for himself by changing his strategy, given the other players' strategies. The notion of Nash equilibrium ignores the sequential structure of an extensive game; it treats strategies as choices made once and for all before play begins. In an equilibrium that corresponds to a perturbed steady state, the players' behavior must correspond to a steady state in every sub-game, not only in the whole game (Osborne, 2003). Thus the notion of Sub-game equilibrium comes into picture. A sub-game perfect equilibrium is a strategy profile with the property that in no sub-game can any player 'i' do better by choosing a different strategy, given that every other player j adheres to their strategies. A sub-game is represented in Figure 2.

A sub-game perfect equilibrium(s) of the game will be the outcome(s) of the extensive form game and the operators can fix their payoff function maximizing variables accordingly.

SOLUTIONS AND RECOMMENDATIONS

The decision to invest on a highly capital intensive project like HSR will depend on the respective corridor. Demand is an important factor but type of demand should be the underlying factor and the passenger's willingness to pay for the new service should be properly carried out. The modal share in the new scenario will depend on the competition between the modes and thus understanding the each player's

Figure 2. Sub-game in an extensive form game

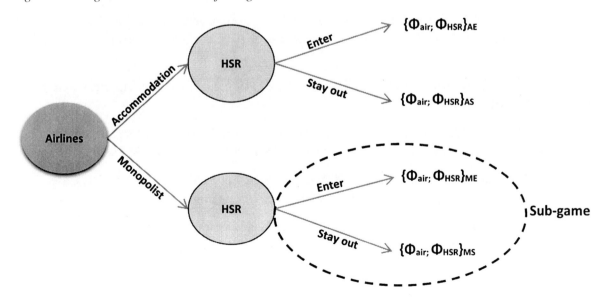

actions and strategies is important. After determining the modal share in the hypothetical scenario, the profit or the payoffs can be determined which in turn affect the engineering aspect of the service. The cost of the project can be reduced by going for up gradation of the tracks instead of new tracks where a little sacrifice can be done on the speed aspect of the HSR service. Frequency, capacity, schedule of the service etc. will depend on the modal share of the service and thus care should be taken while calibrating and estimating the model. Social viability of the project can be justified by maximizing the consumer surplus.

CONCLUSION

The Research will contribute to the knowledge by comprehensively analyzing and model the intermodal competition between HSR and other modes in Indian Context and thereby providing an insight in engineering aspect of the project. The mode choice behavior of individuals in the hypothetical scenario representing the introduction of the new mode is predicted to provide a scientific tool for decision making. This prediction of the mode choice behavior will assist in efficient judgment of the engineering aspect of the project. Considering social welfare as the primary objective for the Indian government will help in taking social viability of the project into account.

FUTURE RESEARCH DIRECTIONS

This work can be further improved by incorporating multiple player game and thus representing the market scenario more efficiently. Cooperation among the airlines and different operators can also be looked into using cooperative game theory.

REFRENCES

Adler, N., Pels, E., & Nash, C. (2010). High-speed rail and air transport competition: Game engineering as tool for cost-benefit analysis. *Transportation Research Part B: Methodological, 44*(7), 812–833. doi:10.1016/j.trb.2010.01.001

Ben-Akiva, M. E., & Lerman, S. R. (1985). *Discrete Choice Analysis: Theory and Application to Travel Demand*. MA: MIT Press.

Ben-Akiva, M. E., & Morikawa, T. (1990). Estimation of travel demand models from multiple data sources.*Proceedings 11th International Symposium on Transportation and Traffic Theory, Yokohama.*

Bierlaire, M., Axhausen, K., & Abbay, G. (2001). Acceptance of Model Innovation: The Case of the Swiss metro.*Proceedings of the 1st Swiss Transport Research Conference.*

Chang, S. K. J., & Hsu, C. L. (2001). Fare and Service Headway for a High Speed Rail system with private sector involvement. *Journal of the Eastern Asia Society for Transportation Studies, 4*(1), 2001.

Fu, X., Zhang, A., & Lei, Z. (2012). Will China's airline industry survive the entry of high-speed rail? *Research in Transportation Economics, 35*(1), 13–25. doi:10.1016/j.retrec.2011.11.006

Hansen, M. (1990). Airline competition in a hub-dominated environment: An application of non-cooperative game theory. *Transportation Research Part B: Methodological, 24*(1), 27–43. doi:10.1016/0191-2615(90)90030-3

HSRC. (n. d.). Retrieved from http://hsrc.in/backgroundBudgetSpeech.html

Jong, G., Daly, A., Pieters, M., & Hoorn, T. (2006). The logsum as an evaluation measure: Review of the literature and new results. *Transportation Research Part A: Policy and Practice, 41*(9), 874–889.

Levinson, D., Mathieu, J. M., Gillen, D., & Kanafani, A. (1997). The full cost of high speed rail: An engineering approach. *The Annals of Regional Science, 31*(2), 189–215. doi:10.1007/s001680050045

Mandel, B., Gaudry, M., & Rothengatter, W. (1997). A disaggregate Box-Cox logit mode choice model of intercity passenger travel in Germany and its implications for high-speed rail demand forecasts. *The Annals of Regional Science, 31*(2), 99–120. doi:10.1007/s001680050041

Mcfadden, D. (1974). Conditional logit analysis of qualitative choice behavior. In Frontiers of econometrics (pp. 105-142).

Milan, J. (1993). A model of competition between high speed rail and air transport. *Transportation Planning and Technology, 17*(1), 1–23. doi:10.1080/03081069308717496

Osbourne, M.J. (2003). *An Introduction to Game Theory*. Toronto, Canada: Oxford University Press.

Raturi, V., Srinivasan, K., Narulkar, G., Chandrashekharaiah, A., & Gupta, A. (2013). Analyzing intermodal competition between high speed rail and conventional transport systems: A game theoretic approach. *Proceedings of the 2nd Conference of Transportation Research Group of India. Procedia: Social and Behavioral Sciences, 104*, 904–913. doi:10.1016/j.sbspro.2013.11.185

Roman, C., Espino, R., & Martin, J. (2007). Competition of high-speed train with air transport: The case of Madrid–Barcelona. *Journal of Air Transport Management, 13*(5), 277–284. doi:10.1016/j.jairtraman.2007.04.009

Rus, G., & Gustavo, N. (2007). Is investment in High Speed Rail socially profitable? *Journal of Transport Economics and Policy, 41*(1), 3–23.

Shyr, O., & Hung, M. (2010). Intermodal Competition with High Speed Rail-A Game Theory Approach. *Journal of Marine Science and Technology, 18*(1), 32–40.

Small, K. A., & Rosen, H. S. (1981). Applied welfare economics with discrete choice models. *Econometrica, 49*(1), 105–129. doi:10.2307/1911129

Train, K. (2003). *Discrete Choice Methods with Simulation*. Cambridge: Cambridge University Press. doi:10.1017/CBO9780511753930

Verma, A., Sudhira, H., Rathi, S., King, R., & Dash, N. (2010). Sustainable urbanization using high speed rail (HSR) in Karnataka, India. *Research in Transportation Economics, 38*(1), 67–77.

Vision 2020 Board. Government of India - New Delhi: Vision 2020. (2009). *Indian Railways*. Retrieved from http://www.nwr.indianrailways.gov.in/uploads/files/1299058054467-englishvision.pdf

Wang, J. Y., & Yang, H. (2005). A game-theoretic analysis of competition in a deregulated bus market. *Transportation Research Part E, Logistics and Transportation Review*, *41*(4), 329–355. doi:10.1016/j.tre.2004.06.001

Zito, P., & Salvo, G., & LaFranca, L. (2011). Modelling Airlines Competition on Fares and Frequencies of Service by Bi-Level Optimization. *Procedia: Social and Behavioral Sciences*, *20*, 1080–1089. doi:10.1016/j.sbspro.2011.08.117

Chapter 8
Welfare Economic Principles of (Urban) Rail Network Pricing:
First-Best and Second-Best Solutions

Jörg Schimmelpfennig
Ruhr-Universität Bochum, Germany

ABSTRACT

The purpose of this chapter is to rectify the at best unprofessional intermingling of objectives and constraints and present a proper theory of first-best and second-best pricing in urban rail networks. First, in view of the flaws of both Dupuit's – though nevertheless ingenious idea of – consumer surplus as well its cannibalized version found in most of today's economics textbooks, a proper definition of economic welfare resting on Hicks'sian variations instead is provided. It is used to derive efficient pricing rules that are subsequently applied to specific questions arising from running an urban railway network such as overcrowding, short-run versus long-run capacity or competing modes of transport like the private motor car. At the same time, another look is taken at economic costs, and in particular economic marginal costs, differing from commercial or accounting costs. Among other things, it is shown that even with commercial marginal costs being constant first-best pricing might not necessarily be incompatible with a zero-profit budget.

1. INTRODUCTION

The problem of optimal pricing in transport systems does not just go back right to the beginnings of microeconomics, it actually was the first truly microeconomic question ever asked. The concept of utility is of course older. However, while, e.g., Bentham (1789) discussed the idea of social welfare as the sum of individual utility functions, utility was never used before as an instrument of microeconomics in the sense of individual utility maximisation.

When during the early 1840s the French government turned to Jules Dupuit, an engineer at what today would be called the French Ministry of Transport, and tasked him with devising some system on the basis of which one should be able to decide on whether or not to build a bridge and subsequently which

DOI: 10.4018/978-1-5225-0084-1.ch008

toll to charge for its use, microeconomics did not yet exist. One must not forget that Adam Smith, even though talking about equilibria – he called it "natural prices" – and demand and supply, did not visualize demand *functions* or supply *functions*, let alone profit maximisation or utility maximisation. Similarly, the only concept of welfare he used just looked at the labour which was required to produce some given aggregate output. Dupuit (1844) subsequently not only conceived the brilliant idea of consumer welfare, but before being able to do that he even had to "invent" the demand curve, or "courbe de consummation", as he called it, first. The resulting fundamental principle followed pretty straightforwardly. It is the very idea of first-best pricing, it is intuitively appealing, and it is still taught in microeconomics courses at every level all over the world: as long as there are individuals, or users, around who would be willing to pay more than the additional, i.e. marginal, cost that would arise if the good, or service, is to be provided to them, one should go ahead. The optimum quantity then would be the one for which the marginal willingness-to-pay equals the marginal cost. Unfortunately, though, Dupuit's concept not only rests on the assumption of a cardinal utility function which, at least from a 20th century microeconomic perspective, is somewhat difficult to comprehend. It violates the budget constraint; it breaks down once more than one good is considered due to a possible path dependence of the integral; and once one tries to make it compatible with the budget constraint, the concept would only work for demand functions with an income elasticity of one, i.e. straight-line Engel curves running through the origin.

The purpose of this chapter is two-fold: first, a proper and coherent framework for optimal pricing compatible with modern microeconomic thinking is provided. Formulae for first-best as well as second-best pricing schemes are derived. Second, the formulae are applied to rail networks, paying special attention to budget constraints, the competition between different modes of transport and externalities within and out of the network.

The paper is organised as follows. A summary of Dupuit's theory of consumer surplus, including its limitations, is presented in section 2. In section 3, a general measure, based solely on the concept of ordinal utility and building on Hicksian compensating and/or equivalent variations, is developed. It is subsequently used to derive a first-best pricing rule which, even though resulting from a completely different technique, turns out to be no different at all from the original result by Dupuit. While this admittedly might look like using the proverbial sledgehammer to crack a nut, the difference shows when turning to second-best pricing where a whole range of new applications, in particular with regard to transport systems, is offered. In section 4, the results are applied to rail transport networks. They include rules how to deal efficiently with budget constraints, short-run capacity constraints and/or externalities both outside of and within a given network. It will be shown that even with constant marginal costs, first-best marginal cost pricing may well be compatible with a subsidy-free budget.

2. DUPUIT'S CONSUMER SURPLUS

The starting point is Dupuit's idea of utility maximisation: as long as the utility to be gained from consuming one additional unit is greater than its price, the consumer would buy that additional unit. Thus, he would only stop once the two are equal, i.e. marginal utility is equal to the price: with x, p and U denoting quantity, price and utility,

$$\frac{dU}{dx} = p. \tag{1}$$

Using the same logic, (1) then would not only provide the first-order condition, but can at the same time be interpreted as the inverse demand functions: for any given price, a consumer would demand that amount for which the additional utility yielded by the very last unit is equal to its price.

The idea of consumer surplus immediately follows. Using the fundamental theorem of integration and assuming the utility to be gained from consuming nothing to be equal to zero,

$$\int_0^x \frac{dU}{dx}(t)dt = U(x) - U(0) = U(x), \tag{2}$$

i.e. for any given price, and, thus, the corresponding quantity, the area beneath a demand curve just gives the overall utility the individual would enjoy when consuming the good. All of this rests of course on the assumption of utility being cardinally interpretable as otherwise it could never be compared to a monetary price. Deducting the money spent on that good then gives the net utility, or consumer surplus CS, i.e.

$$CS(x) = \int_0^x \frac{dU}{dx}(t)dt - p(x) \cdot x. \tag{3}$$

Turning to the task originally given to Dupuit, and assuming x to be the number of crossings of any given bridge, maximising welfare W defined as the sum of consumer surplus and revenue minus the cost of building and maintaining the bridge, i.e.

$$W(x) = CS(x) + p(x) \cdot x - C(x), \tag{4}$$

would yield

$$\frac{dCS}{dx}(x^*) = \frac{dU}{dx}(x^*) = p(x^*) = \frac{dC}{dx}(x^*), \tag{5}$$

i.e. marginal utility, or the price or toll, should be equal to marginal cost. In a more general context, it is the earliest formal proof of Adam Smith's vision that a free market, even though every single one participant is merely interested in his own well-being, is the best that could happen to society, very much in line with Bernard de Mandeville's "Fable of the Bees": as profit maximisation by price-taking firms, i.e. firms acting in a perfectly competitive market, implies that prices should equal marginal costs, economic welfare will be maximised automatically, too. Note, though, that the welfare maximum can thus be defined by quantity alone rather than by price. Just imagine, that a smaller price would be charged but the number of crossings be limited by the same x as yielded by (5): consumer surplus would go up, revenue would go down by the same amount, and overall welfare would thus remain the same.

Further, when having to decide on whether a bridge should be built in the first place, one would only have to look at the welfare implied by (5): if and only if the sum of consumer surplus and revenue exceeds the cost, it should be built, which is in contrast to the business-only perspective requiring a non-zero profit. The difference becomes all the striking for the very case Dupuit had to deal with as a bridge is nothing but a textbook-style natural monopoly, featuring constant – and even close to zero – marginal costs and positive fixed cost. Marginal cost pricing would then result in revenue just paying for the variable cost, i.e. the variable profit, or producer surplus, would be zero. Total profit would be minus the fixed costs and would thus be negative. Still, if the resulting consumer surplus exceeds the fixed cost, welfare would be positive and the natural monopoly should thus be run at a loss.

A first drawback, though, is obvious: both demand, and, thus, willingness-to-pay, and consumer surplus, as defined by (1) and (2), respectively, refer to one individual only. For the result to hold for a society consisting of more than one individual, the very least one would have to do would be to look at the aggregate which would not only require utility to be cardinal but to be interpersonally comparable, too. While some followers of cardinal utility theory might well be willing to accept such an assumption, there is another and much more serious problem: the analysis so far ignores the budget constraint. Once this is added and the analysis subsequently extended to more than one good, the resulting Lagrange first-order condition, with λ being the Lagrange multiplier, yields

$$\int_0^{x_1} p_1(t)dt = \int_0^{x_1} \frac{1}{\lambda} \cdot \frac{\partial U}{\partial x_1} U(t, x_2, ..., x_n) dt \tag{6}$$

which, in order to be equal to at least a multiple of U, would require λ to be independent of prices. Unfortunately, though, as has been shown by, e.g., Ekelund and Hébert (1985), this is equivalent of assuming that the income elasticity of demand is equal to one which would render the theory next to useless as practically no such examples of goods or services exist.

Fast-forwarding to the cannibalized version of Dupuit's concept as it appears in practically all of today's microeconomic textbooks is anything but helpful as it reveals another two, and final, straws. Due to Marshall (1898), who not only detached consumer surplus from utility altogether by defining it as the area, or triangle – in the case of linear demand functions –, between his, i.e. the Marshallian, demand curve and the price, but, by dropping Dupuit's name from later editions of his "Principles" onwards, even effectively suggested that it was he, i.e. Marshall, rather than Dupuit, who had conceived the whole idea of consumer surplus in the first place. Unfortunately, though, by arguing that this area yields one individual's total willingness-to-pay, one wrongly assumes that inverted demand is equal to the marginal willingness-to-pay. At any given point inverted demand is but equal to the overall willingness-to-pay, implying that any attempt to add these up would only violate the budget constraint. Further, when looking at price changes and assuming that two prices are changed simultaneously, the integral may become path-dependent, i.e. the result would depend on the way the integral is computed, implying consumer surplus would no longer be uniquely defined and would thus not even constitute a measure any longer. The simplest example for this to happen would be a quasi-linear utility function.

3. CONSUMER SURPLUS BASED ON THE COMPENSATING VARIATION

Using Hicksian rather than Marshallian demand provides an escape route. Instead of defining consumer surplus as the total benefit that arises from any given equilibrium price just the way Dupuit's concept works, John Hicks only looked at the change of benefit due to the change of prices, say from some price (vector) p^0 to p^1, by defining the compensating variation as

$$CV(y, p^0, p^1) = e(U^0, p^1) - e(U^0, p^0). \tag{7}$$

It is the difference in income, given by the expenditure function e, that would be needed in order to leave utility unchanged at the original level U^0. Correspondingly, a similar measure can be defined by using the new utility level U^1, i.e. the utility achieved following the price change, but without any change in income, as a point of reference, which Hicks (1941) called the equivalent variation, or EV.

First, as CV is defined in money terms, neither would cardinal utility nor interpersonal comparability be needed. Second, a positive CV would signal a worsening, because income had to rise in order to compensate the individual, while a negative CV would signal an improvement. Third, in case only one price, say the ith price, changes, CV can even be written as an integral: as Silberberg (1990) has shown, due to the Envelope Theorem

$$\frac{\partial e}{\partial p_i} = h_i \tag{8}$$

which thus implies

$$\int_{p_i^0}^{p_i^1} h_i(U^0, p_1, \ldots, p_{i-1}, t, p_{i+1}, \ldots, p_n) dt = e(U^0, p^1) - e(U^0, p^0). \tag{9}$$

Then, as the Marshallian demand, because of the Slutzki Equation, runs somewhere between the two respective Hicksian demand functions, i.e. $h_i(U^0, p_i)$ and $h_i(U^1, p_i)$, Dupuit's consumer surplus would fall right between two very measures and, despite of all the shortcomings discussed in the previous section, does not look that wanting any longer.

Most interestingly, though, overall economic welfare can be properly defined using either of the two variations. As can be seen from (9) in conjunction with the Slutzki Equation, both would of course differ as long as the income effect is different from zero. In contrast to Dupuit's idea, only the change of welfare would be measured. Using, e.g., the compensating variation and denoting the welfare-maximising price by p^*, the change of welfare when using the welfare-maximising price as the point of reference would be given by the required change in total income on the side of the consumers and the change in profits on the side of the producers, i.e.

$$-\sum\left(e\left(U^{*},p\right)-e\left(U^{*},p^{*}\right)\right)+\left(p\cdot X(p)-C\left(X(p)\right)\right)-\left(p^{*}\cdot X\left(p^{*}\right)-C\left(X\left(p^{*}\right)\right)\right). \tag{10}$$

Note that (i) the negative sign in front of the first term is due to a higher CV indicating a worsening, (ii) the sum is taken over all consumers, (iii) X denotes respective aggregate demand, and (iv) the difference between the second and the third term is equal to the change in aggregate profits.

Taking the partial derivative of (10) with respect to any p_j yields

$$p_j^* = \frac{\partial C}{\partial X_j}. \tag{11}$$

The same result would follow if the equivalent variation, i.e.

$$-\sum\left(e\left(U(p),p\right)-e\left(U(p),p^{*}\right)\right)+\left(p\cdot X(p)-C\left(X(p)\right)\right)-\left(p^{*}\cdot X\left(p^{*}\right)-C\left(X\left(p^{*}\right)\right)\right) \tag{12}$$

is employed instead, implying that in spite of the differing measures the results, when solving for the welfare-maximising price, are identical, i.e. at the margin the differences just vanish.

To summarize, even though welfare theory has been placed on a consistent but rather different-looking footing perfectly compatible with standard microeconomics, the marginal cost principle holds true: Jules Dupuit was right after all.

When it comes to second-best pricing, two cases are considered. The first would be one that immediately springs to mind when talking, in the Dupuit context, about natural monopolies. With constant marginal costs, natural monopolies can quite often only be run at a loss because demand is too small in the sense that the demand curve is running below the average cost curve throughout. And even if demand intersects average costs at some point, first-best pricing would still dictate marginal cost pricing, implicating a loss even though a positive profit had been possible. Neither case may be politically sustainable and whatever pricing is chosen, a zero-profit restriction, or one just limiting the loss, has to be met, giving rise to a second-best pricing solution.

In order to get a clearer idea of the resulting solution, cross-elasticities of demand are assumed to be zero. The task would be to maximise (10), or, equivalently, (12), subject to the constraint that the revenue from the natural monopoly, denoted here by the index 1, plus the subsidy financed by raising a tax in some other market, denoted by the index 2, should cover the full cost of the natural monopoly, i.e.

$$p_1 \cdot X_1(p_1) + t \cdot X_2(p_2(t)) = C_1(X_1(p_1)). \tag{13}$$

Assuming that market 2 is perfectly competitive, the first-order condition implies

$$\frac{p_1^* - \frac{\partial C}{\partial X_1}}{p_1^*} = \frac{\varepsilon_2}{\varepsilon_1}, \tag{14}$$

with ε denoting the respective elasticities of demand. The result suggests that, in order to maximise welfare subject to a constraint, prices in the natural monopoly can be allowed to deviate the more from marginal costs the smaller the elasticity of demand for that service and/or the higher the elasticity of demand in the other market is.

Another example for the need of a second-best pricing scheme would be if there is any deviation from the marginal cost principle in some other market that is connected in the sense that the cross-derivatives of demand are non zero. Staying with just two goods or services in order to keep the argument as simple as possible, maximising (10), or (12), subject to

$$p_2 = \bar{p}_2 \neq \frac{\partial C}{\partial X_2} \tag{15}$$

yields

$$\left(p_1^* - \frac{\partial C}{\partial X_1}\right) \cdot \frac{\partial X_1}{\partial p_1} + \left(p_2 - \frac{\partial C}{\partial X_2}\right) \cdot \frac{\partial X_2}{\partial p_1}. \tag{16}$$

In the case of substitute goods, both bracket terms should have the same sign, i.e. if the second good, e.g., is sold too cheaply, so should the first, and vice versa. In the case of complementary goods, the signs should differ: if the second good is sold above marginal cost, in order to compensate that effect and keep the resulting welfare loss as small as possible, the first good should be sold below its marginal cost in order to keep the price of the "package" more or less unchanged.

4. APPLICATIONS

Railway networks exhibit a couple of special features. First, in many textbooks they are mentioned as a prime example for a natural monopoly, very much like Dupuit's bridge. And even if marginal costs were not constant – for respective estimates of the cost functions of transport systems and, in particular, railway networks, see, e.g. and in chronological order, Wedgwood (1909), Keeler (1974), Harris (1977) or Schimmelpfennig (2011) –, due to their very nature of being state-run, i.e. public, enterprises demand would still be too small in most cases in order for them to be able to earn a positive profit. Railway networks would therefore be prime examples for the application of the principles that have evolved so far. The purpose of this section, though, goes beyond that of merely conveying these principles to railway networks, but to identify railway-specific features that might lead to other, and sometimes unexpected, solutions.

First-Best Pricing

First-best pricing prescribes marginal cost pricing, and would thus, provided marginal costs are constant, ceteris paribus generate a loss as average costs would be higher than marginal costs and thus price, implying average costs to exceed average revenue. Provided this is politically sustainable, i.e. the government decides to provide the financial means for building the network and purchasing the necessary rolling stock and would thus require the railway to just pay for the variable costs, welfare would be at its maximum. In case the overall deficit, i.e. including capital costs, is not sustainable, though, not everything is lost as will be seen by having a closer look at marginal costs.

Marginal Costs and Long-Run Capacity

While funny ideas originating from creative accounting can give rise to instances where brand-new locomotives were classified as rebuilts in order to be able to ascribe the building costs as working expenditure rather than capital costs – it was, as has been pointed by, e.g., Reed (1996), a common practice for British railway companies from the late 19th century onwards right until nationalisation in 1948 –, at least on paper the distinction between fixed costs and marginal costs, seems to be non-contentious: any longer-term investment, or capital expenditure, should belong to the former category, all running costs to the latter; it should be obvious that whenever economists talk about costs, they should always refer to economic, i.e. opportunity costs. In transport networks, either assignment can be wrong, though. To give an example, if a railway is running at capacity, in order to satisfy additional demand and/or reduce overcrowding, more rolling stock would be needed. From an economic viewpoint, its annualised procurement costs should be counted as marginal costs as they originated only because of additional, i.e. marginal demand. According to (11), then, they should not only be taken into account when computing p^*, leading of course to higher, but still first-best prices. When deciding on whether to increase capacity, the same principle should rule: the (annualised) marginal cost of providing additional capacity must not exceed the willingness-to-pay of the marginal customer. On the other hand, provided there is spare rolling stock, introducing an additional train service would only require an additional train crew and the usual running expenses; while the latter would of course depend on the number of passengers and therefore to be counted as marginal costs, at least part of the former should be counted as fixed enabling the railway company to conduct a cost-benefit analysis just for the additional train.

Marginal Costs and Short-Run Capacity

If demand exceeds capacity, i.e. if the number of seats, or standing places, on a train is insufficient to accommodate all passengers, the marginal cost of providing space for an additional passenger would become infinite. The standard economic principles nevertheless would apply; in particular, prices have to be raised. While marginal costs would no longer be a function in the mathematical sense – and neither could (11) work –, they could still be drawn as a schedule, running parallel to the x-axis up to capacity and then having a ninety-degree kink and breaking off vertically. The principle of marginal cost pricing would remain the same: while (11), from a graphical perspective, defines the solution as being the point of intersection between demand and marginal cost, this point would merely shift to the vertical part as has been pointed out by Rees (1984). The price would be higher than constant marginal costs, generating additional revenue which may even suffice to cover all fixed costs. It should be remembered here that, as already mentioned in Section 2, every welfare maximum will be uniquely determined by quantity. Charging any other than the market-clearing price would lead to queuing and thus avoidable transaction costs.

It carries another implication. Traditionally, as the principle of first-best pricing prescribes marginal cost pricing, it would imply a negative profit in a natural monopoly situation because marginal costs are smaller than average costs. However, as costs should include all costs and, thus, all kind of external effects, too, charging for these implies that the price should exceed the commercial marginal costs and could thus even be sufficient in order to yield a zero profit.

Peak-Load Pricing

If demand regularly shifts over some time period, price differentiation becomes imperative in order to maximise overall economic welfare. Consider a simple case where every day can be divided into two time periods, one, called the peak period, with demand being high, and the other one, the off-peak period, with demand being low. As both periods are disjunct, the network infrastructure as well as all rolling stock can be interpreted as a public good: using it during one period does not preclude it from being used during the second period as well. Then two cases can be distinguished: if peak demand can be met with the existing capacity, off peak prices should equal the running costs only because the provision of capacity has already been justified by the willingness-to-pay of the peak users. If, however, charging marginal running costs for off-peak users would exceed peak capacity – the scenario was fittingly called "shifting peaks" by Rees (1984) –, Samuelson prices should be applied: the sum of net – i.e. without the proportion needed to cover marginal usage costs only – peak and off-peak prices should be equal to the marginal capacity costs. The latter case has been described as a "shifting peaks" by Rees (1984). Whatever the case, peak users should pay a higher price than off-peak users. It would be in line with the idea of second-best pricing, too: because of (14), if demand becomes less elastic, i.e. if the elasticity of demand becomes, in absolute terms, smaller, it should be charged more. By its very definition, peak demand would be the less elastic of the two.

Externalities within the Railway Network

As already mentioned in the previous section, a rail network may become congested, too. Rather than interpreting capacity as fixed as in the second last paragraph, assume instead that additional passengers can be accommodated. It would happen at a price, though, as it would give rise to both overcrowding as well as delays because of the additional time that would be needed for alighting from or boarding an overcrowded train. While neither would raise the commercial costs of providing the train service, from an economic viewpoint they would of course have to be classified as negative external effects, and thus economic costs, because they would lead to both inconveniences and lost time. Again, charging for these negative external effects might both make the rail network commercially viable as well as constitute a first-best solution. If capacity could only be increased by discrete intervals, but increasing it by such a minimum amount would not be justifiable in terms of a cost benefit analysis, the situation could even become a long-run optimum, i.e. even though marginal costs would still appear to be constant to the railway, a first-best solution could be attainable without any subsidisation.

Competing Modes of Transport

For any rail network, there is always at least one competing mode of transport which is the private motor car. Even though in many cities fuel duties and/or road user charges are levied in order to reduce congestion, it can be doubted that the levels take account of the whole range and extent of the negative external effects caused by using a car such as pollution, noise and congestion. As, obviously, both modes of transport are substitute services, (16) tells that as long as the marginal car user does not have to pay for all the costs brought about by him, the marginal rail user should not have to pay his marginal cost either. It would thus be a rare exception to marginal cost pricing, but, taking account of the externalities within the rail system discussed in the previous paragraph, rail prices might still exceed their commercial marginal costs.

Second-Best Pricing Subject to a Budget Constraint

It would of course be a rather standard case and there is only so much to add that politicians should understand that it is the least preferable of all the options and cases discussed so far and is but a last resort. However, there is of course an even worse alternative left which would be to charge on the basis of a budget constraint without paying any attention to economic welfare. Unfortunately, it is not extremely rare to find.

5. SUMMARY

The chapter outlined the principles of first-best and second-best pricing with special regard to characteristic features of an urban rail network. In particular, it showed possible paths that serve justice to both political realities and economic welfare. Admittedly, not all are feasible in every case. Still, a menu has been provided for politicians and their economic advisers from which a pricing scheme can be devised that makes at least some economic sense.

Whatever the case, though, there is one piece of advice that has to be heeded come what may: constraints, and in particular budget constraints, must never ever be mistaken as objectives.

REFERENCES

Bentham, J. (1789). *An Introduction to the Principles of Morals and Legislation*. Oxford: Clarendon Press.

Dupuit, J. (1844). De la mesure de l'utilité des travaux publics. *Annales des Ponts et Chaussées*, 8.

Ekelund, R., & Hébert, R. (1985). Consumer's surplus - the first hundred years. *History of Political Economy*, *17*(3), 419–454. doi:10.1215/00182702-17-3-419

Harris, R. G. (1977). Economies of traffic density in the rail freight industry. *The Bell Journal of Economics*, *8*(2), 556–564. doi:10.2307/3003304

Hicks, J. (1941). The rehabilitation of consumers' surplus. *The Review of Economic Studies*, *8*(2), 108–116. doi:10.2307/2967467

Keeler, T. E. (1974). Railroad costs, returns to scale, and excess capacity. *The Review of Economics and Statistics*, *56*(2), 201–208. doi:10.2307/1924440

Marshall, A. (1898). *Principles of Economics* (4th ed.). London: Macmillan.

Reed, M. E. (1996). *The London & North Western Railway*. Penryn: Atlantic Transport Publishers.

Rees, R. (1984). *Public Enterprise Economics* (2nd ed.). Oxford: Weidenfeld and Nicolson.

Schimmelpfennig, J. (2011). The South Eastern and Chatham Railways Managing Committee: A case for vertically-integrated regional duopolies? *International Journal of Strategic Decision Sciences, 2*, 95–103. doi:10.4018/jsds.2011040106

Wedgwood, R. L. (1909). Statistics of railway costs. *The Economic Journal, 19*(73), 13–31. doi:10.2307/2220505

Section 3
Safety

Chapter 9
Challenges of Railway Safety Risk Assessment and Maintenance Decision Making

Min An
University of Birmingham, UK

Yong Qin
Beijing Jiaotong University, China

ABSTRACT

Railway safety is a very complicated subject, which is determined by numerous aspects. Many of qualitative and quantitative railway safety and risk analysis techniques and methods are used in the industry. But, however, the railway industry faces problems and challenges on how to apply these techniques and methods effectively and efficiently, particularly in the circumstances where the risk data are incomplete or there is a high level of uncertainty involved in the risk data. This chapter approaches these subjects to discuss the problems and challenges of railway safety and risk analysis methods in dealing with uncertainties, and those growing needs of the industry. A well-established technique is also introduced in this chapter which can be used to identify major hazards and evaluate both qualitative and quantitative risk data, and information associated with railway operation efficiently and effectively in an acceptable way in various environments.

1. INTRODUCTION

Railways are by far one of the safest means of ground transportation, especially for their passengers and employees. However, there are serious issues involved in both maintaining this position in reality and sustaining the public perception of railway safety excellence. The railway industry now finds itself in a situation where actual and perceived safeties are real issues, to be dealt with in a new public culture of rapid change, short-term pressures, and instant communications.

Railways are a traditional industry, whose history extends for at least two centuries. Much of their safety record depends upon concepts developed many, many years ago and established practices over

DOI: 10.4018/978-1-5225-0084-1.ch009

the whole of their history. After half a century of decline and public/political isolation, railways have started to expand rapidly again, but in a situation where parallel but newer industries have moved on. The parallel aerospace, nuclear and oil exploration industries in particular have developed approaches to safeguarding their assets, customers and employees which reflect their different traditions, their shorter histories and their different generic cultures. There are many possible causes of risk through operation and design of vehicles and rail infrastructure, and also from outside the railway such as vandalism and road incidents. Specifically, in the design, modification, and maintenance of plain line, the largest number of serious incidences are from derailments and vehicles fouling infrastructure such as station platforms. There are many combinations of potential causes, each involving several disciplines and work groups.

Risk management is being used increasingly to support decision making in the railway industry. While risk management is used to justify and prioritize investment, its main function and the one being considered here is in demonstrating safety. Any safety and risk information produced must be processed for decision making purposes. If risks are high, risk reduction measures must be applied or the design, operation and maintenance have to be reconsidered to reduce the occurrence probabilities or to control the possible consequences. If risks are negligible, no actions are required but the information produced needs to be recorded for certification purpose. However, the acceptable and unacceptable regions are usually divided by a transition region. Risks that fall in this transition region need to be reduced to as low as reasonably practicable (ALARP). This chapter aims to introduce a systemic bottom-up safety and risk analysis approach, which addresses the problems and challenges of railway safety and risk analysis and methods of these growing needs of the industry, and discusses standards and current practice of safety and risk management in the railway industry. A well-established technique will also be introduced in this chapter which can be used to identify major hazards and evaluate both qualitative and quantitative risk data, and information associated with railway operation efficiently and effectively in an acceptable way in various environments. A case study is used to demonstrate the application of application of railway safety and risk assessment methodology.

2. BACKGROUND

As stated in section 1, risk management is being increasingly for the railway industry in order to improve safety to safeguard their passengers and employees. The principal risks in the railway industry appear to be to people and property as a result of collision, derailment, and fire. Recent structured hazard identification work within the industry has confirmed the high-risk scenarios of these types of accidents (HSE, 2000; Railway Safety, 2003; LUL, 2001; An et al, 2006, 2007 & 2011). In the railway industry, many people had severe injuries and even were killed in relation to this industry over the past years (LUL, 2001; Gadd et al, 2003; Kennedy, 2003; Railway Safety, 2002). The figures of accidents and incidents include not only workers, but also a significant number of people not employed in the industry, including children and members of the public. In addition, more than a hundred people had their quality of life affected by railway maintenance work in the past ten years, brought about through either a major injury or an injury that rendered them unable to work for over 3 days. This shows the dangerous nature of the railway industry and demonstrates the need for increased awareness and better safety management (Railway Safety, 2003; Gadd eta al, 2003; An et al, 2007 & 2011). There are many accidents and incidents occurred in the railway industry over the years, demanding improvement of safety management. To assess how this can be effectively achieved knowledge on the nature and causes of these accidents are fundamental. Therefore, risk analysis plays a central role in the railway safety and health management framework. For example, the most common hazards in the railway depots identified by the railway industry over the years (Railway Safety, 2002 & 2003; Metronet SSL, 2005a & b; Tube Lines, 2004) provide very useful information for risk analysis, such as derailment hazards, collision hazards, fire hazards, electrocution hazards, falls hazards, train strike hazards, slip/trip hazards, platform train interface hazards. The Railway Safety Management has to achieve a goal where the occurrence likelihood of events causing train collisions or derailment, results in an accidental equivalent fatality rate needs to be achieved a certain standard (Railway Safety, 2003) and the associated risks are reduced to ALARP. Therefore, railway safety analysts need to develop and employ safety assessment approaches for their safety document preparation. Additionally, the accident statistics also present not only human tragedy but also substantial economic cost. These costs can be incurred through, for example, damage to equipment and plants, damage to work already completed, delay completion of time, increased insurance premiums, legal costs, fines, compensation and even loss of reputation of the companies. Fines from governing authorities play an influential role in deterring companies who are prepared to cut back on health and safety to increase profits. Insurers can show that the modest extra cost of a safe and orderly railway depot is well invested because the risk and cost of serious accidents are so high. The economic expense of both these incidents in fines and costs far outweighs the additional costs of a safer depot which could have prevented the incidents. This is a fact that needs to be made more aware to the railway industry to improve their safety management performance.

Risk assessment is a process to determine the risk magnitude to assist with decision-making. Many of the railway risk assessment techniques currently used are comparatively mature tools. The results of using these tools highly rely on the availability and accuracy of the risk data (Railway Safety, 2002 & 2003; An et al, 2005, 2006, 2007 & 2011). However, railway safety analysts often face the circumstances where the risk data are incomplete or there is a high level of uncertainty involved in the risk data. Additionally, there are numerous variables interacting in a complex manner which, due to the large amount of data available, cannot be explicitly described by an algorithm, a set of equations or a set of rules. There may be both a shortage of key information and an excess of other information. In many circumstances, it may be extremely difficult to conduct probabilistic risk assessment to assess the occurrence likelihood of hazards and the magnitudes of their possible consequences because of the uncertainty with risk data. Therefore, it is essential to develop new risk analysis methods to identify major hazards and assess the associated risks in an acceptable way in various environments where such mature tools cannot be effectively or efficiently applied (An et al, 2005 & 2006). The railway system safety problem is appropriate for examination by fuzzy reasoning approach (FRA) combined with fuzzy analytical hierarchy process (FAHP). The FRA allows imprecision or approximate information in risk assessment process. In this method:

- A membership function is regarded as a possibility distribution based on a proposed theory, and
- An apparent possibility distribution expressed by fuzzy set theory is transferred into a possibility measure distribution.

The FRA method has been widely used in engineering applications. Bowles and Pelaes (1995a & b) developed a FRA model to analyse risk and reliability of failures of a water supply system. Bojadziev and Bojadziev (1997) introduced FRA method to business, finance and management for market risk analysis. Sii et al (2001) applied a fuzzy logic based approach to qualitative safety modelling for offshore platform systems. Ghiaus (2001) proposed a FRA model for control of fan-coil systems in Energy and Buildings industry. Ngai et al (2003) Designed and developed a fuzzy expert system for hotel selection. The FRA method has also been used in the assessment of the financial conditions of public schools (Ammar, 2004). As Sadiq and Husain (2005) indicated that FRA technique is a useful method in risk analysis, and they also proposed a fuzzy based methodology for environmental risk assessment. Peters and Peters (2006) applied a FRA model to analyse human errors in the design and manufacturing industry. More recently, An et al (2007 & 2011) developed a fuzzy based risk analysis system for railway system safety and risk assessment. The FRA method provides a useful tool for modelling risks and other risk parameters for risk analysis where the risks with incomplete or redundant safety information. Because the contribution of each hazardous event to the safety of a railway depot is different, the weight of the contribution of each hazardous event should be taken into consideration in order to represent its relative contribution to the risk level (RL) of the railway depot.

The FAHP is an important extension of the typical AHP method which was firstly introduced by Laarhoven and Pedrycz (1983). The FAHP is a very useful technique that has been applied in many fields of, for example, design and maintenance planning, reliability analysis, selecting a best alternative and resource allocations, etc. Ramanathan & Ganesh (1995) applied FAHP to analyse resource allocation problems in manufacturing industry. Lootsma (1996) proposed a model based on multiplicative FAHP for analysis of the relative importance of the criteria in smart systems. Hauser & Tadikamalla (1996) developed a simulation approach based on FAHP method to assess uncertainties in environment science. Cheng (1997) also applied FAHP and developed the grade value of membership function to evaluate naval tactical missile systems. Yu et al (2002) proposed a FAHP decision making model to solve group decision making problems in construction projects. Wang et al (2007) developed a decision making model based on a FAHP for selection of optimum maintenance strategies of production economics. Cheng et al (1999) also applied FAHP to produce linguistic variable weight for the evaluation of attack helicopters. Tang et al (2000) integrated FAHP method with multi criteria decision making approach to study electronic marketing strategies in the information service industry. Kuo et al (1999) proposed a decision support system for locating convenience store through FAHP approach, and late the decision support system was developed further to integrate artificial neural network in, which was used for selecting convenience store location more effectively (Kuo et al, 2004). Lee et al (2008) applied FAHP method to the evaluating performance of IT department in the manufacturing industry. Kahraman et al (2004) also applied FAHP method to a case study of multi-attribute comparison of catering service companies. Therefore, a FAHP technique needs to be incorporated into the risk model to use its advantage in determining the relative importance of the risk factors so that the risk assessment can be progressed from hazardous event level to the identified hazard group level and finally to depot level. An advantage of the FAHP is its flexibility to be integrated with different techniques (Huang et al, 2005; Andriantiatsaholiniaina et al, 2004; Saaty, 1980; Laarhoven & Pedryez, 1983; Buckley, 1985; Fahmy, 2001, Mikhailov, 2004; Mon et al, 1994). In this case, it is FRA technique. Therefore, the application of FRA in risk assessment may involve the use of FAHP technique. The application of FAHP may solve the problems of risk information loss in the hierarchical process so that risk assessment can be carried out from hazardous event level to a railway depot level. Both of these processes result in a set of probability distributions which can be used not only to predict RLs but also to design safety maintenance intervals. The use of these techniques is especially appropriate in the railway environment because of the volume of experience, which is still available from long-term employees.

Challenges of Railway Safety Risk Assessment and Maintenance Decision Making

As stated earlier in this chapter, in many circumstances, the application of probabilistic risk analysis (PRA) tools may not give satisfactory results because the risk data are incomplete or there is a high level of uncertainty involved in the risk data. This chapter introduce a risk assessment methodology for railway safety and risk analysis using FRA and FAHP techniques, in which FRA is employed to estimate the risk level (RL) of each hazardous event (/component failure event) in terms of frequency of occurrence (FO) and consequence severity (CS). This allows imprecision or approximate information in the risk analysis process. Fuzzy-AHP technique is then incorporated into the risk assessment to use its advantage in determining the relative importance of the risk contributions so that the risk assessment can be progressed from hazardous event (/component failure event) level to hazard group (sub-system) level and finally to railway system level. The outcomes of the described risk assessment method are represented as the risk degrees and the defined risk categories with a belief of percentage, which provide very useful risk information to safety analysts, managers, engineers and decision makers. The structure of this chapter is as follows. After the introduction, a bottom-up railway safety assessment approach is described. The application of qualitative descriptors and corresponding membership functions (MF) to represent risk factors and RLs are discussed and the development of fuzzy rule base is also addressed to describe the relationship among risk factors and RL expressions. The application of FRA to risk assessment of the RLs of hazardous events and the identified hazard groups is also described. Then, the FAHP method is described which is employed to synthesis the risk information produced at the group level to obtain the overall RL of risk assessment at railway depot level. After that, a case study on risk assessment of shunting at Waterloo depot is presented to demonstrate the application of the risk assessment methodology. Finally, it is conclusions and a summary of the main benefits of using the methodology in risk analysis.

3. A BOTTOM UP RAILWAY SAFETY AND RISK ASSESSMENT APPROACH

A risk assessment is a process that can be divided into five phases: problem definition phase; data and information collection and analysis phase; hazard identification phase; risk estimation phase and risk response phase. The process provides a systematic approach to the identification and control of high-risk areas. Figure 1 depicts the typical steps in the bottom up risk assessment process of railway depots (/systems). This framework is considered to be generally applicable to most risk analysis processes of railway depots (/systems) but it should be noted that different depot (/system) might require variation of the process, including total elimination of some steps.

Risk assessment begins with an identified need. Problem definition involves identifying the need for safety, i.e. specific safety requirements. The requirements regarding safety should be specified and they may have to be made at different level, e.g. hazardous event (/component failure event) level, hazard group (/sub-system) level and the railway depot (/system) level. The following typical items may need to be specified in the problem definition (An et al, 2005, 2006, 2007 & 2011; An,2005; Sii et al, 2001; Huang et al, 2005):

- Sets of rules and regulations made by the national authorities and classification societies, etc.
- Deterministic requirements for safety, reliability, availability, maintainability, etc.
- Criteria referring to probability of occurrence of serious hazardous events and the possible consequences.

Figure 1. The bottom up risk assessment process

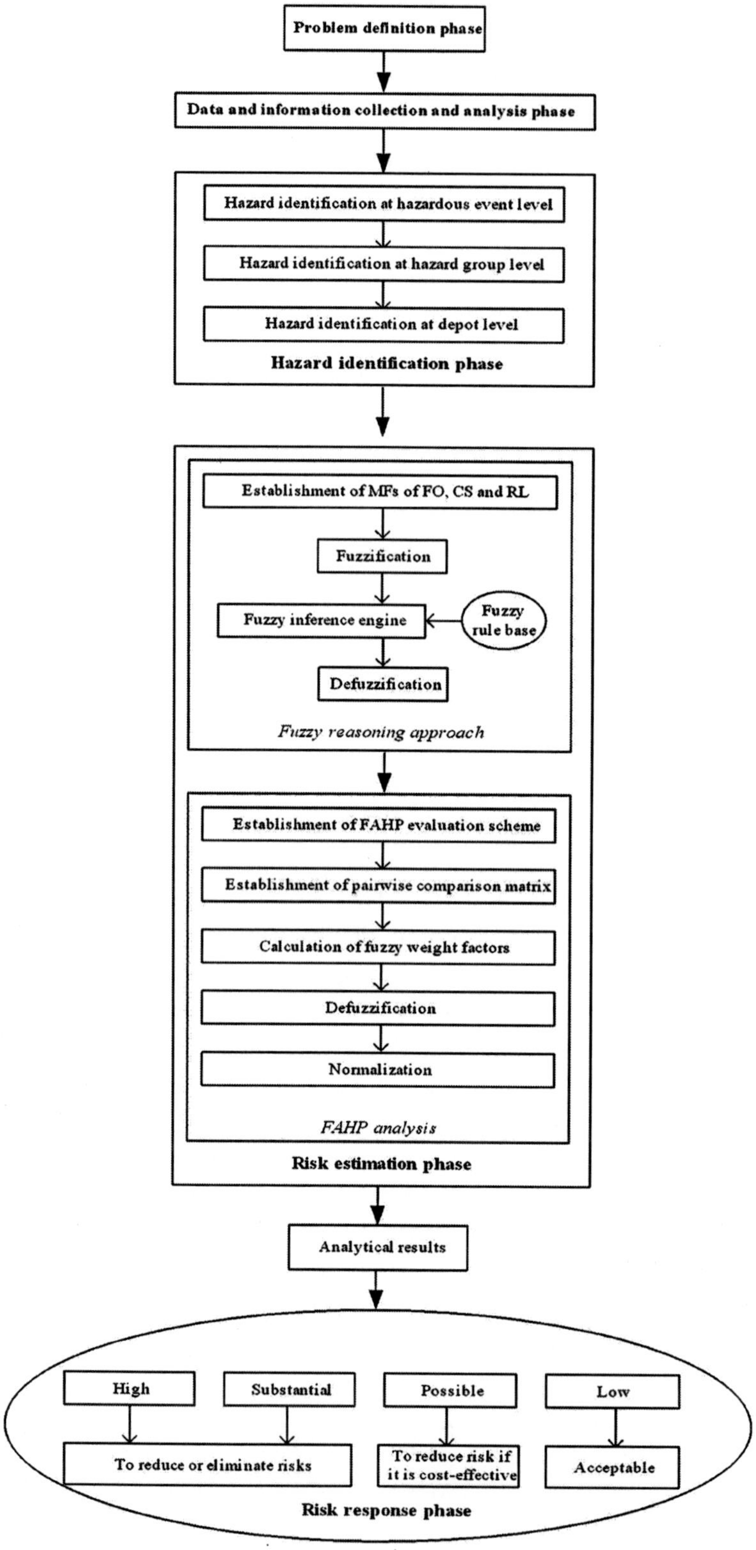

Once the need for safety is established, the risk assessment moves from the problem identification phase to the data and information collection and analysis phase. The aims of data and information collection and analysis are to develop a good understanding of what serious accidents and incidents occurred in a particular railway depot (/system) over the years and generate a body of information. If the statistic data does not exist, expert and engineering judgements should be applied. The information gained from data and information collection will then be used to develop qualitative descriptors and associated MFs. A number of the most commonly used techniques can be used to gather information and knowledge such as statistical data and information analysis, domain human experience and engineering knowledge analysis and concept mapping. These techniques are not mutually exclusive and a combination of them is often the most effective way to derive the risk information (An et al, 2006, 2007 & 2011; Sii et al, 2001; Ammar, 2004).

The purpose of hazard identification phase is to systematically identify all potential hazardous events associated with a railway depot(/system) at each required level, e.g. hazardous event (/component failure event) level, hazard group (/sub-system) level with a view to assessing their effects on railway depot (/system) safety. Various hazard identification methods such as brainstorming approach, check-list, 'what if?', HAZOP (Hazard and Operability), and failure mode and effect analysis (FMEA), may be used individually or in combination to identify the potential hazardous events of a railway depot (/system) (An, 2005; Huang et al, 2005). The hazard identification can be initially carried out to identify hazardous events at event (/component failure event) level, then progressed up to hazard group (/sub-system) level and finally to the depot (/system) level.

Risk estimation phase aims to assess RLs of hazardous event (/component failure event), hazard group (/sub-system) and the railway depot (/system), which can be carried out on either a qualitative or a quantitative basis. Various risk analysis techniques such as fault tree analysis, event tree analysis, Monte Carlo simulation analysis, FMEA, programme evaluation and review technique that are applicable across sectors, are currently used in the railway industry. As stated earlier in this chapter, in many circumstances, it may be extremely difficult to conduct a quantitative risk assessment due to the great uncertainty involved in the risk data. When the uncertainty in the risk data is very high, subjective risk analysis incorporating FRA and FAHP modelling may prove to be more suitable in risk estimation.

The results produced from the risk estimation phase may be used through the risk response phase and may also be used to assist risk analysts, engineers and managers in developing maintenance and operation policies. If risks are high, risk reduction measures must be applied or the depot (/system) operation has to be reconsidered to reduce the occurrence probabilities or to control the possible consequences. If risks are negligible, no actions are required but the information produced needs to be recorded for audit purpose (An et al, 2006; An, 2005). However, the acceptable and unacceptable regions are usually divided by a transition region. Risks that fall in this transition region need to be reduced to ALARP (HSE, 2000; Railway Safety, 2003; LUL, 2001). In this study, the RLs are characterized into four regions, i.e. 'High', 'Substantial', 'Possible' and 'Low'.

As risk assessment is not a one-off activity it is essential that the assessment needs to be reviewed at the appropriate intervals in order to update the risk assessment and provide risk information for responses. Obviously, all risk analyses may involve uncertain or incomplete risk information, which have been proved by the industry (Railway Safety, 2003; LUL, 2001; Metronet SSL, 2005a & b; Tube Lines, 2004). It is particular true when:

- The railway operates in a very changeable environment;
- Human error is a major contributor to possible accidents;
- There is a lack of detailed risk information;
- There is insufficient existing database.

Therefore, in many circumstances, a railway risk analyst may have to describe a given event in vague and imprecise terms such as 'likely' and 'impossible'. Such judgements are obviously subjective and hence a FRA analysis may be more appropriate to analyse the risks of a railway depot (/system) with incomplete risk information. It should be noted that FRA has its unique characteristics which are different from those associated with qualitative risk analysis. FRA usually involves the use of fuzzy set modelling. Additionally, risk analysis is also a hierarchical process where risk information obtained at lower levels may be used for risk assessment at higher levels. This is also true for the synthesis of other parameters. Therefore, a hierarchical procedure, e.g. FAHP method, is usually required for synthesising the information produced at lower levels to obtain the results of risk assessments at higher levels.

4. DEVELOPMENT OF FUZZY MFs FOR REPRESENTING RLs

Three risk parameters can be used to assess RLs of hazardous events (/component failure events) and hazard groups (/sub-systems), and an overall RL of the railway depot (/system), i.e. frequency of occurrence (FO), consequence severity (CS) and weight factor (WF). The FO defines the number of times an event occurs over a specified period, e.g. number of events/year. CS represents the number of fatalities, major injuries and minor injuries resulting from the occurrence of a particular hazardous event (/component failure event). The risk assessment methodology employs FRA method that allows the incomplete risk information, imprecise knowledge and subjective information to be used in the risk assessment process. As stated in section 3, risk data and information can be obtained from a number of available sources such as previous accident and incident reports, historical data, engineering knowledge and expert experience to conduct the risk assessment. Risk identification can be carried out to identify potential hazardous events (/component failure events), then grouping these events (/failures) into a number of category hazard groups (/sub-systems) based on their contributions to the safety of a railway depot (/a system). For example, hazardous events(/component failure events) in a railway depot(/system) can usually be grouped into derailment hazards, fire hazards, electrocution hazards, falls hazards, train strike hazards, slip/trip hazards, platform train interface hazards, vehicle collision hazards, health hazards (/sub-system failure hazards) etc. Each hazard group (/sub-system) contains a number of hazardous events, e.g. the hazard group of falls from height consists of minor injury events, major injury events and fatal events. Risk assessment can be carried out from hazardous event (/component failure event) level, to hazard group (/sub-system) level and finally to railway depot (system) level. The following two steps are used in risk analysis:

Step 1: Application of FRA for risk analysis at the hazardous event (/component failure event) level and hazard group(/sub-system) level

Step 2: Application of FAHP to synthesis the information produced at the group (/sub-system) level to obtain the overall RL of risk assessment at railway depot(/system) level

The outcomes of risk assessment using FRA at the hazardous event (/component failure event) level are represented as the risk degrees of hazardous events (/component failure events) and the defined risk categories with a belief of percentage. Then fuzzy aggregation operation is applied to calculate the RLs of hazard groups (/sub-system). Risk analysis at the depot (/system) level can then be conducted by introducing WF in the risk assessment process. FAHP is employed to calculate WF of each hazard group(/sub-system), which represents the relative importance of the identified hazard groups(/sub-systems) to the safety. The results of the RLs of hazard groups(/sub-system) and corresponding WFs are then synthesised to obtain overall RL of a railway depot (/system). Assume a railway depot (/system) has n identified hazard groups (/sub-systems). The overall RL of the railway depot (/system) is defined by:

$$RL = \sum_{i=1}^{n} RL_i w_i \qquad (1)$$

where RL_i is the RL of the i-th hazard group(/sub-system), w_i stands for the WF of the i-th hazard group(/sub-system) and RL is the overall RL of a railway depot(/system).

4.1 Fundamental of the Fuzzy Reasoning Approach

Fuzzy reasoning approaches possess the ability to mimic the human mind to effectively employ modes of reasoning that are approximate rather than exact. It enables to specify mapping rules in terms of words rather than numbers and approximate function rather than exact reasoning. In other words, fuzzy reasoning approaches are knowledge-based or rule-based ones constructed from human knowledge in the form of fuzzy If-then rules (An et al, 2006, 2007 & 2011; Ghiaus, 2001; Homnan & Benjapolakul, 2004). An important contribution of fuzzy reasoning theory is that it provides a systematic procedure for transforming a knowledge base into a non-liner mapping. A fuzzy If-then rule is an If-then statement in which some words are characterised by continuous MFs. An introduction to fuzzy reasoning theory is given in the following sections.

4.1.1 Background of Fuzzy Reasoning Approach

A fuzzy set A on a universe of discourse U is defined as a set of ordered pairs (Saaty, 1990; An et al, 2006):

$$A = \{(x, \mu_A(x)) \mid x \in U\} \qquad (2)$$

where $\mu_A(x)$ is called the MF of x in A that takes values in the interval (0, 1). The element x is characterised by linguistic values, e.g. in railway risk assessment, the FO is defined as 'Remote', 'Infrequent', 'Frequent', 'Common', 'Rare', 'Occasional' and 'Regular'; the CS is defined as 'Minor', 'Marginal', 'Moderate', 'Severe' and 'Catastrophic'; and the RL is defined as 'low', 'possible', 'substantial' and 'high' (An et al, 2006, 2007 and 2011). Various types of MFs can be used, including triangular, trapezoidal, generalized bell shaped and Gaussian functions (Saaty, 1980 & 1990). However, triangular and trapezoidal MFs are the most frequently used in risk analysis practice (Bowles & Pelaez, 1995a & b; An et al, 2006, 2007 & 2011).

The most elementary fuzzy set operations are union and intersection, which essentially correspond to OR and AND operators, respectively. Let A and B be two fuzzy sets.

Union. The union of A and B, denoted by $A \cup B$ or A OR B, contains all elements in either A or B which is calculated by the maximum operation and its MF is defined as

$$\mu_{A \cup B}(x) = \max\{\mu_A(x), \mu_B(x)\} \tag{3}$$

Intersection. The intersection of A and B, denoted by $A \cap B$ or A AND B, contains all the elements that are simultaneously in A and B, which is obtained by the minimum operation and its MF is defined as

$$\mu_{A \cap B}(x) = \min\{\mu_A(x), \mu_B(x)\} \tag{4}$$

Fuzzy reasoning approaches are rule-based methodologies constructed from human knowledge in the form of fuzzy If-then rules. As stated earlier in section 3, a fuzzy If-then rule is a statement in which some words are characterised by continuous MFs; e.g. the following is a frequently used fuzzy If-then rule in railway risk assessment:

If FO is frequent AND CS is critical, then RL of the failure event is high. where FO, CS and RL are qualitative descriptor. frequent, critical and high are qualitative descriptor values characterised by MFs.

A fuzzy rule base consists of a set of fuzzy If-then rules. Consider A_1^i the input space $U = U_1 \times U_2 \times \cdots \times U_n \subset R^n$ and output space $V \subset R$. Only the multi-input-single-output case is considered here, as a multi-output system can always be decomposed into a collection of single-output systems. Specifically, the fuzzy rule base comprises the follow fuzzy If-then rules:

$$R_i: \text{If } x_1 \text{ is } A_1^i \text{ and } \ldots \text{ and } x_n \text{ is } A_n^i, \text{ then } y \text{ is } B^i \tag{5}$$

where and B^i are the fuzzy sets in $U \subset R$ and $V \subset R$, respectively, and $x = (x_1, x_2, \ldots, x_x)^T \in U$ and $y \in V$ are the input and output qualitative descriptors of the fuzzy reasoning system, respectively. Owing to their concise form, fuzzy If-then rules are often employed to capture the imprecise modes of reasoning that play an essential role in the human ability to make decisions in an environment of uncertainty and imprecision. Therefore, in the fuzzy reasoning system, human knowledge has to be represented in the form of the fuzzy If-then rules (equation (5)). There are three major properties of fuzzy rules that are outlined as follows (Saaty, 1980 & 1990; An et al, 2006, 2007 & 2011; Sadiq & Husain, 2005):

1. A set of fuzzy If-then rules is complete if for any $x \subset U$, there exists at least one rule in the fuzzy rule base, say rule R_i in the form of equation (5), thus

$$\mu_{A_1^i}(x_i) \neq 0 \tag{6}$$

Challenges of Railway Safety Risk Assessment and Maintenance Decision Making

2. For all $i = 1, 2, \ldots, n$. Intuitively, the completeness of a set of rules means that at any point in the input space there is at least one rule that 'fires', i.e., the membership value of the If part of the rule at this point is non-zero.
3. A set of fuzzy If-then rules is consistent if there are no rules with the same If parts but different then parts.
4. A set of fuzzy If-then rules is continuous if there do not exist such neighbouring rules whose then part fuzzy sets have empty intersection, i.e. they do not intersect.

4.1.2 Fuzzy Inference System

The railway risk assessment system consists of two subsystems: fuzzy inference system (FIS) and user interface system. A generic fuzzy reasoning risk analysis system is shown in Figure 2.

The FIS consists of four components: the fuzzy rule base, fuzzification, fuzzy inference engine and defuzzification. The development of the rule base involves various knowledge acquisition techniques to generate a body of information that could be useful in developing qualitative descriptors and their associated MFs to qualify RLs. For many practical situations, several approaches can be used to gather information and knowledge required in deriving fuzzy rules. The knowledge acquisition methodologies used may include:

a. Historical data analysis
b. Failure analysis
c. Concept mapping
d. Domain human expert experience and engineering knowledge analysis

These techniques are not mutually exclusive and a combination of them is often the most effective way to determine the rule base (An et al, 2006, 2007 and 2011; Sii et al, 2001).

The fuzzification converts input qualitative descriptor values into the corresponding fuzzy MF values. It determines the degrees of input qualitative descriptor values belonging to each of the appropriate fuzzy sets by MFs.

In a fuzzy inference engine, fuzzy logic principles are used to combine the fuzzy If-then rules in the fuzzy rule base into a mapping from a fuzzy set A in U to a fuzzy set B in V. Once inputs have been fuzzified, these fuzzified values are employed to each rule to find out whether or not the rule will be fired. If a rule has true value in its premise, it will be fired and then contribute to the conclusion part. If the premise of a given rule has more than one part, the fuzzy operator is applied to evaluate the composite firing strength of the rule. Considering the i-th rule has two parts in the premise

$$R_i : \text{if } x_1 \text{ is } A_1^i \text{ and } x_2 \text{ is } A_2^i, \text{ then } y \text{ is } B^i \qquad (7)$$
$$i = 1, 2, \cdots, r$$

The two parts in the premise are connected with 'and' and the firing strength α_i can be obtained using fuzzy intersection (minimum) operation

Figure 2. A generic fuzzy reasoning risk analysis system

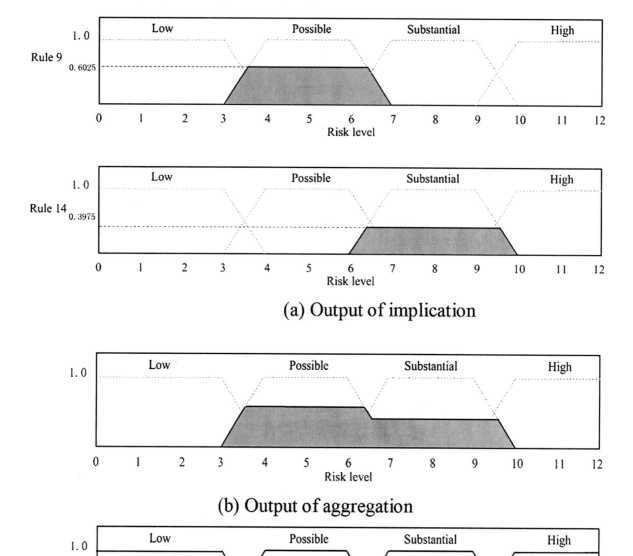

$$\alpha_i = \min\{\mu_{A_1^i}(x_1), \mu_{A_2^i}(x_2)\} \tag{8}$$

where $\mu_{A_1^i}(x_1)$ and $\mu_{A_2^i}(x_2)$ are the MFs of fuzzy sets A_1^i and A_2^i. The firing strength is implicated with the value of the conclusion MF and the output is a truncated fuzzy set. The implication using fuzzy intersection (minimum) operation is given by

$$\mu_{imp^i}(y) = \min\{\alpha_i, \mu_{B^i}(y)\} \tag{9}$$

where $\mu_{B^i}(y)$ is the MF of the conclusion part of a fuzzy rule and $\mu_{imp^i}(y)$ is the MF of the truncated fuzzy set after implication. The truncated fuzzy sets that represent the implication outputs of each rule are aggregated into a single fuzzy set. The aggregation using fuzzy union (maximum) operation is denoted by

$$\mu_{agg}(y) = \max\{\mu_{imp^1}(y), \mu_{imp^2}(y), \cdots, \mu_{imp^r}(y)\} \tag{10}$$

where $\mu_{agg}(y)$ is the MF of the fuzzy set after aggregation.

On the basis of the aggregated fuzzy set, defuzzification calculates the defuzzified value, which is a crisp value, standing for the final result of the fuzzy inference. The centroid of area method, which determines the centre of gravity of an aggregated fuzzy set, is the most frequently used method in fuzzy reasoning systems (Saaty, 1990; Sii et al, 2001; Bowles & Pelaez, 1995a & b; An et al, 2006, 2007 & 2011), defined as

$$y_{def} = \frac{\int_y \mu_{agg}(y) y \, dy}{\int_y \mu_{agg}(y) \, dy} \tag{10}$$

where $\mu_{agg}(y)$ is the aggregated output MF. The process of fuzzy inference is depicted in Figure 3.

4.2 Application of FRA

As described earlier in this chapter, in many circumstances, it may be extremely difficult to conduct PRA to assess the occurrence likelihood of hazards and the magnitudes of their possible consequences because of the uncertainty in the risk data. Therefore, the application of FRA in risk assessment may have the following advantages (An et al, 2008; Sii et al, 2001; An, 2003; Ghiaus, 2001; Bow;es & Pelaez, 1995a & b):

1. The risk can be evaluated directly by using qualitative descriptors;
2. It is tolerant of imprecise data and ambiguous information;
3. It gives a more flexible structure for combining qualitative as well as quantitative information.

Figure 3. Fuzzy inference process

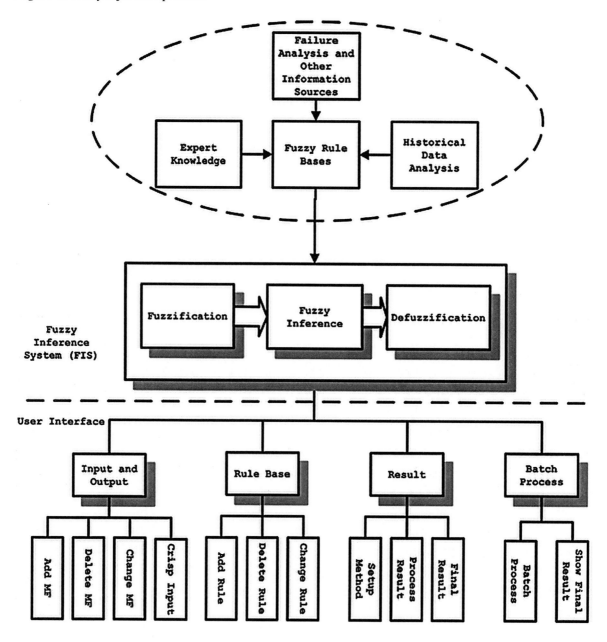

FRA focuses on qualitative descriptors in natural language and aims to provide fundamentals for approximate reasoning with imprecise propositions. Qualitative descriptors can be used to represent the condition of a risk factor at a given interval. Therefore, the risk assessment methodology using FRA includes the development of fuzzy qualitative descriptors and MFs for describing FO, CS and RL expressions, which is summarised as follows.

Table 1 describes the range of the FO. To estimate the FO, one may often use such qualitative descriptors as 'Remote', 'Infrequent', 'Frequent', 'Common', 'Rare', 'Occasional' and 'Regular' are suggested to

Table 1. Frequency of Occurrence

Qualitative Descriptors	Description	Mid-point of the Estimated Frequency	Approximate Numerical Value (event/yr)	Parameters of MFs
Remote	< 1 in 175 years	1 in 500 years	0.002	0, 0, 0.002, 0.01 (trapezoid)
Rare	1 in 35 years to 1 in 175 years	1 in 100 years	0.01	0.002, 0.01, 0.05 (triangle)
Infrequent	1 in 7 years to 1 in 35 years	1 in 20 years	0.05	0.01, 0.05, 0.25 (triangle)
Occasional	1 in 1 ¼ years to 1 in 7 years	1 in 4 years	0.25	0.05, 0.25, 1.25 (triangle)
Frequent	1 in 3 months to 1 in 1 ¼ years	1 in 9 months	1.25	0.25, 1.25, 6.25 (triangle)
Regular	1 in 20 days to 1 in 3 months	1 in 2 months	6.25	1.25, 6.25, 31.25 (triangle)
Common	1 in 4 days to 1 in 20 days	1 in 12 days	31.25	6.25, 31.25, 91.25 (trapezoid)

be between 1 in 35 years and 1 in 175 years, between 1 in 1/4 years to 1 in 7 years, and between 1 in 20 days and 1 in 3 months, respectively (An, 2005; Metronet SSL, 2005a). The triangular and trapezoidal MFs are assigned to describe these qualitative descriptors as shown in Figure 4. As can be seen, each qualitative descriptor of FO has a range to describe the FO. A mid-point of the estimated frequency is

Figure 4. MFs of frequency of occurrence (X-axis in logarithmic format)

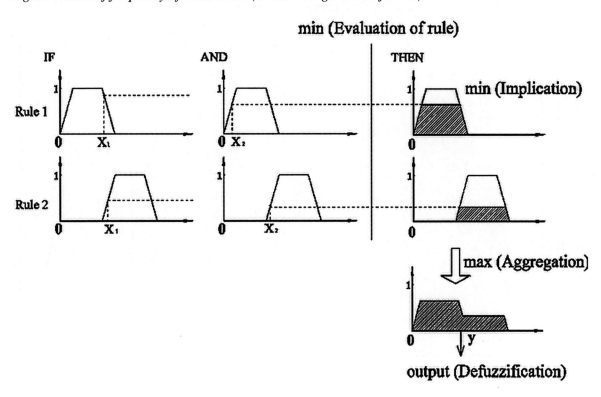

used in each category so that approximate numerical value can be obtained. For example, qualitative descriptor 'Rare' is defined to cover the range of FO between non-occurrence and 1 in 175 years. A mid-point between them is 1 in 500 years, and then approximate numerical value can be calculated as 0.002 events per year.

The CS describes the magnitude of possible consequences. One may often use such qualitative descriptors as 'Minor', 'Marginal', 'Moderate', 'Severe' and 'Catastrophic'. Table 2 shows the criteria used to rank the CS of hazardous events. The MFs of CS are shown in Figure 5. It should be noted that seven and five qualitative descriptors to describe FO and CS in this study, respectively, but this does not have to be the case. The number of qualitative descriptors used to describe FO and CS is flexible and depends on particular cases.

Table 2. Consequence Severity

Qualitative Descriptors	Description	Approximate Numerical Value	Parameters of MFs
Minor	Minor injury	0.005	0, 0, 0.005, 0.025 (trapezoid)
Marginal	More serious injury/multiple minor injuries	0.025	0.005, 0.025, 0.125 (triangle)
Moderate	Major injury	0.125	0.025, 0.125, 0.625 (triangle)
Severe	Multiple major	0.625	0.125, 0.625, 3.125 (triangle)
Catastrophic	Fatality/multiple fatalities	3.125	0.625, 3.125, 5, 5 (trapezoid)

Figure 5. MFs of consequence severity (X-axis in logarithmic format)

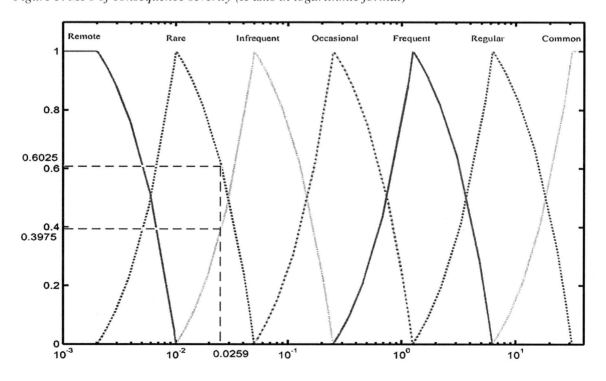

Challenges of Railway Safety Risk Assessment and Maintenance Decision Making

It is commonly understood that risk can be expressed by degrees to which it belongs with such qualitative descriptors, 'Low', 'Possible', 'Substantial' and 'High', that are referred to as risk expressions(An et al, 2006 & 2007; Sii et al, 2001; Ammar, 2004; Bowles & Pelaez, 1995a & b; Ghiaus, 2001; Sadiq and Husain, 2005). Table 3 shows the qualitative descriptor categories of RL. Trapezoidal MFs are employed to describe each qualitative descriptor of RL as shown in Figure 6.

Table 3. Risk Level

Qualitative Descriptors	Description	Risk Scores	Parameters of Membership Functions
Low	Risk is acceptable	0 – 4	0, 0, 3, 4 (trapezoid)
Possible	Risk is tolerable but should be further reduced if it is cost-effective to do so	4 – 7	3, 4, 6, 7 (trapezoid)
Substantial	Risk must be reduced if it is reasonably practicable to do so	7 – 10	6, 7, 9, 10 (trapezoid)
High	Risk must be reduced safe in exceptional circumstances	10-12	9, 10, 12, 12 (trapezoid)

Figure 6.

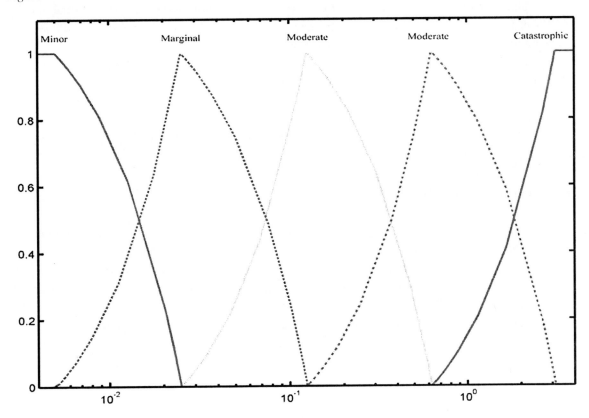

4.3 Development of Fuzzy Rule Base

Fuzzy rule base consists of a set of fuzzy If-then rules. It is the core of a fuzzy logic system in the sense that all other components are used to implement these rules in a reasonable and efficient manner, which is comprised of the following fuzzy If-then rules:

If x_1 is A_1 and x_2 is A_2 ... x_n is A_n, then y_n is B_n where A_n and B_n are qualitative descriptor values defined by the fuzzy sets on universes of discourse U. "If x_1 is A_1 and x_2 is A_2 ... x_n is A_n" is called the precondition while "then y_n is B_n" is called the conclusion. For example, in railway risk analysis, the following rule is commonly used

If FO is High and CS is Critical, then RL is High where 'High', 'Critical' and 'High' are qualitative descriptors characterised by MFs.

Several approaches such as data collection, expert judgement and engineering knowledge can be used to derive the fuzzy rules. These approaches are not mutually exclusive and a combination of them is often the most effective way to determine the rule base. As fuzzy rules are expressed in qualitative descriptors rather than numerical values, they present a natural platform to deliver information based on expert judgments and engineering knowledge. Therefore, railway engineering experts often find that it is a convenient way to express their knowledge in the risk assessment. Other factors also need to be considered in constructing the fuzzy bases. The fuzzy rule base should be:

1. **Completeness:** The fuzzy rule base must cover all matches between inputs and outputs;
2. **The Number of Rules:** Although there is no general procedure for deciding the optimal number of rules, the decision is important when performance, efficiency of computations and choice of qualitative descriptors are important considerations;
3. **Consistency and Correctness:** The choice of fuzzy rule should minimize the possibility of contradiction, and unwanted interactions between the rules.

The number of fuzzy rules in the fuzzy rule base depends on the number of qualitative descriptors adopted for representing FO and CS. For example, if there are 7 qualitative descriptors of FO and 5 of CS, the fuzzy rule base therefore consists of ($7 \times 5 =$) 35 fuzzy rules.

4.4 Fuzzy Inference System

A fuzzy inference system consists of four components: fuzzificaiton, fuzzy rule base, fuzzy inference engine and defuzzification as shown in Figures 1 and 2. Fuzzification is the interface between the input and the fuzzy inference engine. It converts inputs into fuzzy qualitative descriptors and determines the degrees to which belong each of the appropriate fuzzy sets via MFs. For example, the FO of hazardous event of Electrocution is 0.0259 per year. As a result of fuzzificiation, qualitative descriptor FO has qualitative descriptors of 'Rare' with a membership degree of 0.6025, and 'Infrequent' of 0.3975. The result of fuzzificiation is $F_i =$ {(Remote, 0), (Rare, 0.6025), (Infrequent, 0.3975), (Occasional, 0), (Frequent, 0), (Regular, 0), (Common, 0)} as shown in Figure 4. However, in many circumstances, because the process of risk analysis is very complex and data available may be incomplete for risk assessment. For example, if risk information of a hazardous event (\failure event) is incomplete or uncertainty in the risk data, a subjective judgement may be provided by the expert on the basis of his experience under consideration. The expert may provide a range of numerical values, a qualitative term, or a fuzzy number. In this case, a flexible method for expressing expert judgement and engineering knowledge is proposed based on fuzzy reasoning theory as described earlier in this section. The triangular fuzzy numbers can be defined as

$$\tilde{M} = (t^l, t^m, t^u) \tag{11}$$

where t^l, t^m and t^u are real numbers with satisfaction of the relationship $t^l \leq t^m \leq t^u$. t^l and t^u stand for the lower and higher values, respectively. t^m is defined as the most likely value. For example, an expert is required to make his judgement on FO of a hazardous event (failure event), the expert may provide a range of value such as (0.1, 0.2, 0.3), where 0.2 is the most likely value, 0.1 and 0.3 represent lower and higher values respectively as shown in Figures 6 and 7, which shows the results of fuzzification, F_i = {(Remote, 0), (Rare, 0.6025), (Infrequent, 0.8), (Occasional, 0.8), (Frequent, 0.05), (Regular, 0), (Common, 0)}.

In a fuzzy inference engine, fuzzy logic principles are used to combine the fuzzy If-then rules in the fuzzy rule base into a mapping from input fuzzy sets, i.e. FO and CS to an output set, i.e. RL expression. It consists of evaluation of fuzzy rules, implication and aggregation. Evaluation of fuzzy rules is to determine which rule in the rule base is fired or not. The purpose is to apply the fuzzy logic principles

Figure 7. Fuzzification for fuzzy input (X-axis in logarithmic format)

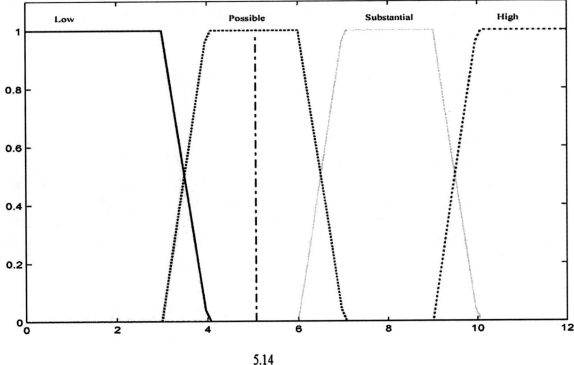

5.14

to combine the fuzzy If-then rules in the fuzzy rule base into a mapping from a fuzzy set A in U to a fuzzy set B in V. Once inputs have been fuzzified, these fuzzified values are employed to each rule to find out whether the rule will be fired. If a rule has true value in its precondition part, it will be fired and contribute to the conclusion part. The fuzzy operations are then adopted to evaluate the composite firing strength of the precondition part (An et al, 2006, 2007 & 2011). In railway safety risk assessment, FO and CS are in the precondition part. Let A, B and C be three fuzzy sets, a fuzzy rule can be represented as

$$R_q: \text{IF } x \text{ is } A_i \text{ and } y \text{ is } B_j, \text{THEN } z \text{ is } C_k$$

where i, j and k represent numbers of qualitative descriptors for A, B and C. The firing strength α_q is implicated with the value of the conclusion MF, which can be calculated by using fuzzy minimum operation

$$\alpha_q = \min(\mu_{(A_i)}(x), \mu_{(B_j)}(y)) \tag{12}$$

where α_q are defined as the output of precondition of the qth rule.

The implication is to shape the conclusion part of fired rule from precondition part by the firing strengths. The fuzzy intersection operation is employed to fulfil implication.

$$\beta_q = \min(\alpha_q, \mu_{C_q}(z)) \tag{13}$$

where β_q is the output of the qth rule after implication, $\mu_{C_q}(z)$ stands for the MF of the fuzzy set of conclusion part of qth rule, i.e. RL expression.

Aggregation is the process to synthesise the fuzzy sets which represent the outputs of all fired rules into a single fuzzy set, which can be calculated by fuzzy maximum operation

$$\mu_{agg}(z) = \max(\beta_1, \beta_2, \ldots, \beta_n) \tag{14}$$

where $\mu_{agg}(z)$ is the output of aggregation.

Defuzzification is to convert the aggregated result to a crisp number that represents the final result of the fuzzy inference. The centroid of area method, which determines the centre of gravity of an aggregated fuzzy set, is a popular method in fuzzy reasoning techniques(An et al, 2006; Sii et al, 2001; Bowles & Pelaez,1995a & b; Huang et al, 2005). Suppose an aggregated fuzzy set $\mu(z)$, $z \in [z_0, z_n]$ as shown in Figure 8, defuzzification can be calculated

$$z_c = \frac{\sum_{q=1}^{n-1} z_q \mu(z_q)}{\sum_{q=1}^{n-1} \mu(z_q)} \tag{15}$$

Figure 8. Defuzzification

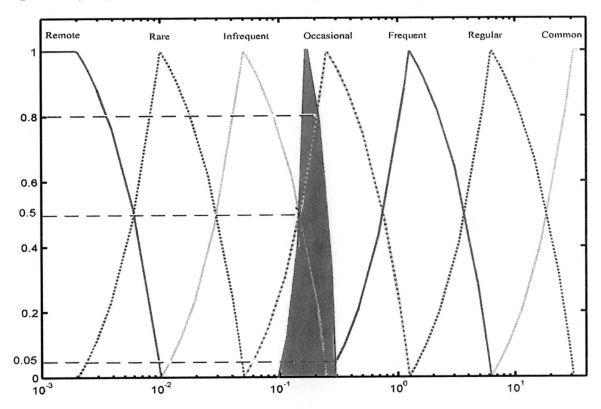

where z_c is the aggregated output MF which indicates the position in the RL expression and also shows the extent to which RL belongs and the corresponding membership degrees.

5. FAHP ANALYSIS

As stated earlier in this chapter, because the contribution of each risk factor to the overall RL is different, the weight of the contribution of each risk factor should be taken into consideration in order to represent its relative contribution to the RL of a railway depot (/system). The application of FAHP may solve the problems of risk information loss in the hierarchical process in determining the relative importance of the risk factors in the decision making process so that risk assessment can be progressed from hazardous event (/component failure event) level to hazard group (/sub-system) level and finally to a railway depot (/system) level. An advantage of the FAHP is its flexibility to be integrated with different techniques, for instance, FRA techniques in risk analysis. Therefore, a FAHP analysis leads to the generation of WFs for representing the primary risk factors within each category.

As the expert experience and engineering knowledge are often expressed in nature of language that describe the risks associated with a system, there are many FAHP techniques have been proposed and are different from the traditional AHP (Tang et al, 2000; Huang et al, 2005, Laarhoven & Pedrycz, 1983; Buckley, 1985; Fahmy, 2001; Mikhailov, 2004; Mon et al, 1994). The traditional AHP may have the following problems in risk analysis:

- The AHP is mainly applied to nearly crisp (non-fuzzy) decision by a standardized estimation scheme which adopts crisp number from 1 to 9 to represent the relative importance between alternatives.
- The AHP does not take into account the experts' imprecise subjective judgments associated with uncertainty. In practice, experts often feel more confident to give imprecise judgments by using qualitative descriptors.

FAHP uses a similar framework of AHP to conduct risk analysis but fuzzy ratios of relative importance replaces crisp ratios to allow the existence of uncertainty in risk assessment.

5.1 FAHP Estimation Scheme

The FAHP determines WFs by conducting pairwise comparison. The comparison is based on an estimation scheme which lists intensity of importance using qualitative descriptors. Each qualitative descriptor has a corresponding triangular fuzzy number that is employed to transfer experts' judgments into a comparison matrix as

$$\tilde{a}_k = (t_k^l, t_k^m, t_k^u) \tag{16}$$

where t_k^l, t_k^m, t_k^u are the numbers from 1 to 9 with satisfaction of the relationship $t_k^l \leq t_k^m \leq t_k^u$, t_k^l and t_k^u correspond to the lower and upper values of a range to describe kth qualitative descriptor, t_k^m stands for the most likely value to represent kth qualitative descriptor. For example, Table 4 describes qualitative descriptors and their corresponding triangular fuzzy numbers for shunting risk assessment at railway depots. Each grade is described by an importance expression and a general intensity number. When two events are equal importance, it is considered $(1,1,2)$. Fuzzy number of $(8,9,9)$ describes that one event is absolute important than the other one. Figure 7 shows triangular MFs (solid lines) with equal importance – $(1,1,2)$, weak importance – $(2,3,4)$, strong importance – $(4,5,6)$, very strong importance – $(6,7,8)$ and absolute importance – $(8,9,9)$, respectively. The other triangular MFs (dash lines) describe the corresponding intermediate descriptors between them.

5.2 Construction of Fuzzy Pairwise Comparison Matrix

The purpose of application of the estimation scheme is to construct a fuzzy pairwise comparison matrix. Assume two event A and B, if A is very strong importance than B, a fuzzy number of $(6,7,8)$ is then assigned to event A. Obviously, event B has a fuzzy number of $(1/8,1/7,1/6)$. Let $\tilde{a}_p (t_p^l, t_p^m, t_p^u)$ and $\tilde{a}_q (t_q^l, t_q^m, t_q^u)$ be two triangular fuzzy numbers. The arithmetic operations on fuzzy number are defined as follows

$$\tilde{a}_p \oplus \tilde{a}_q = (t_p^l + t_q^l, t_p^m + t_q^m, t_p^u + t_q^u) \tag{17}$$

Table 4. The Estimation Scheme

Intensity of Importance in Qualitative Descriptors	Explanation	Triangular Fuzzy Numbers
Equal importance	Two hazard groups contribute equally to the shunting event	(1,1,2)
Between equal and weak importance	When compromise is needed	(1,2,3)
Weak importance	Experience and judgment slightly favor one hazard group over another	(2,3,4)
Between weak and strong importance	When compromise is needed	(3,4,5)
Strong importance	Experience and judgment strongly favor one hazard group over another	(4,5,6)
Between strong and very strong importance	When compromise is needed	(5,6,7)
Very strong importance	An hazard group is favored very strongly over another	(6,7,8)
Between very strong and absolute importance	When compromise is needed	(7,8,9)
Absolute importance	The evidence favoring one hazard group over another is of the highest possible order of affirmation	(8,9,9)

Figure 9. The MFs of the triangular fuzzy numbers

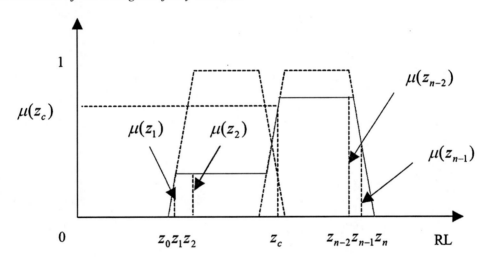

$$\tilde{a}_p \otimes \tilde{a}_q = (t_p^l \times t_q^l, t_p^m \times t_q^m, t_p^u \times t_q^u) \tag{18}$$

$$\tilde{a}_p \oslash \tilde{a}_q = (t_p^l / t_q^u, t_p^m / t_q^m, t_p^u / t_q^l) \tag{19}$$

$$\exp(\tilde{a}_p) = (\exp(t_p^l), \exp(t_p^m), \exp(t_p^u)) \tag{20}$$

where $\oplus$, $\otimes$ and $\oslash$ stand for fuzzy logic addition, multiplication and division operations, respectively, symbol 'exp' presents exponential operation.

Assume m experts in the risk assessment group, the element in a fuzzy pairwise comparison matrix can then be calculated by

$$\tilde{a}_{i,j} = \left(\frac{1}{m}\right) \otimes (e^1_{i,j} \oplus e^2_{i,j} \oplus \cdots e^k_{i,j} \cdots \oplus e^m_{i,j})$$
$$\tilde{a}_{j,i} = \frac{1}{\tilde{a}_{i,j}} \tag{21}$$

where $\tilde{a}_{i,j}$ is the relative importance by comparing event i with event j and $e^k_{i,j}$ stands for the kth expert's judgment in triangular fuzzy number format.

A $n \times n$ fuzzy pairwise comparison matrix $\tilde{A}$ can be obtained

$$\tilde{A} = \begin{pmatrix} \tilde{a}_{1,1} & \tilde{a}_{1,2} & \cdots & \tilde{a}_{1,n} \\ \tilde{a}_{2,1} & \tilde{a}_{2,2} & \cdots & \tilde{a}_{2,n} \\ \vdots & \vdots & \ddots \tilde{a}_{i,j} & \vdots \\ \tilde{a}_{n,1} & \tilde{a}_{n,2} & \cdots & \tilde{a}_{n,n} \end{pmatrix} \tag{22}$$

5.3 Calculation of Fuzzy WFs

The WFs can be calculated by using geometric mean technique (Tang et al, 2000, Huang et al, 2005; Saaty, 1980; Mikhailov, 2001).

$$\tilde{f}_i = (\tilde{a}_{i,1} \otimes \tilde{a}_{i,2} \otimes \cdots \tilde{a}_{i,j} \cdots \otimes \tilde{a}_{i,n})^{1/n}$$
$$= ((t^l_{i,1} \times t^l_{i,2} \times \cdots t^l_{i,j} \cdots \times t^l_{i,n})^{1/n}, (t^m_{i,1} \times t^m_{i,2} \times \cdots t^m_{i,j} \cdots \times t^m_{i,n})^{1/n}, (t^u_{i,1} \times t^u_{i,2} \times \cdots t^u_{i,j} \cdots \times t^u_{i,n})^{1/n}) \tag{23}$$

$$\tilde{w}_i = \frac{\tilde{f}_i}{\tilde{f}_1 \oplus \tilde{f}_2 \oplus \cdots \tilde{f}_j \cdots \oplus \tilde{f}_n} \tag{24}$$

where $\tilde{f}_i$ is the geometric mean of the ith row in the fuzzy pairwise comparison matrix and $\tilde{w}_i$ is the fuzzy WF of the ith event.

5.4 Defuzzification

As the outputs of geometric mean methods are triangular fuzzy WFs, a defuzzifiction is adopted to convert triangular fuzzy WF to the corresponding crisp WF in which the FAHP employs a defuzzifica-

Challenges of Railway Safety Risk Assessment and Maintenance Decision Making

tion approach proposed (Tang et al, 2000; An et al, 2007 & 2011). Assume a triangular fuzzy WF of $\tilde{w}_i(t_i^l, t_i^m, t_i^u)$

$$DF_{\tilde{w}_i} = \frac{[(t_i^u - t_i^l) + (t_i^m - t_i^l)]}{3 + t_i^l} \tag{25}$$

where $DF_{\tilde{w}_i}$ is defuzzified mean value of fuzzy WF. w_j can be calculated by

$$w_i = \frac{DF_{\tilde{w}_i}}{\sum DF_{\tilde{w}_i}} \tag{26}$$

On the basis of RLs of hazard groups and corresponding WFs being obtained, the overall RL at the railway depots can be calculated by equation (1).

6. CASE STUDY: RISK ASSESSMENT OF SHUNTING AT WATERLOO DEPOT

An illustrated case example on risk assessment of shunting at Waterloo depot is used to demonstrate the proposed risk assessment methodology. The case study materials have been collected from the industry (Metronet SSL, 2005a & b). The input parameters are the FO and CS of hazardous events. The outputs of risk assessment are RLs of hazardous events, hazard groups, and the overall RL of shunting at Waterloo depot with risk scores located from 0 to 10 and risk categorised as 'Low', 'Possible', 'Substantial' and 'High' with a percentage belief. The RLs of hazard groups are calculated using the fuzzy reasoning approach based on the aggregation results of each hazardous event belonging to the particular hazard group. The overall RL of shunting at Waterloo depot is obtained based on the aggregation of the RLs of each hazard group contribution weighted by using FAHP methods.

Waterloo depot is the one of largest depots in London Underground. The historical data of accidents and incidents have been recorded over the past 10 years. In this case, the historical accident and incident databases have been reviewed in the Waterloo depot. Ten hazard groups have been identified and defined, and each hazard group consists of a number of hazardous events (An, 2005; Metronet SSL, 2005a & b; Tube Lines, 2004) as shown in Table 6:

1. *Derailment* hazard group includes two sub-hazard groups, i.e. typical outcome (minor injury) and worst-case scenario (major injury) which have been identified based on the pervious accidents and incidents. Each sub-hazard group consists of a number hazardous events such as track related faults including mechanical failure of track e.g. broken rail and fishplates, signaling related faults including mechanical failure of signals and points, rolling stock faults including mechanical failure of rolling stock e.g. brakes, axles and bogies, structure failure including collapsed drain or civil structure beneath track leading to derailment, object from train including object falls from train (e.g. motor) leading to derailment, human errors including human error causing derailment e.g. overspeeding, incorrect routing, etc.

2. *Collision* hazard group consists of four sub-hazard groups i.e. collision between trains of worst case scenario (fatality), collision between trains of typical outcome (multiple minor injuries), collision hazard of worst case scenario (fatality) and collision hazard of typical outcome (minor injury). Collision between trains involves a number of scenarios, for example, MR moving train from 7 road (no train stop) and LU train moving from arrivals platform to 5 or 6 road etc. Collision hazards include, for example, collision with object on track, collision with terminal e.g. overrunning at end of any of the depot roads, collision with platforms involving both the track and/or the train being out of gauge without anybody noticing it, collision with other civil structures involving track/train being out of gauge and nobody noticing.
3. *Train fire* hazard group includes arcing from conductor rail causing train fire, and electrical, oil or hydraulic failure leading to train fire.
4. *Electrocution* hazard group has two sub-hazard group, typical outcome (minor injury) and worst case scenario (major injury) in which cover a number of hazardous events, for example, contacting with conductor rail whilst entering/leaving cab, contacting with conductor rail whilst walking to train, plugging in gap jumper leads if train is stalled/gapped, flooding e.g. sumps or pumps leading to surface water and conducting electricity from conductor rails, etc.
5. *Slips/trips* hazard group includes three sub-hazard groups i.e. minor injury, major injury and fatality. The hazardous events includes, for example, instances when shunter is required to leave train, risks to other persons involved in move and instances when person is required to approach train when it is stalled/gapped.
6. *Falls from height* hazard group consists of three sub-hazard groups i.e. minor injury, major injury and fatality which cover the events of falls from height agreed as when shunter leaves train cab.
7. *Train strikes person* hazard group has been identified based on the record in the past 10 years into two sub-hazard groups – major injury and fatality. The events in these two sub-hazard group include, for example, train strikes authorized person including other depot workers (e.g. ground shunter) or track side staff and train strikes unauthorized person e.g. trespassers etc.
8. *Platform train interface* hazard group includes major injury and fatality sub-hazard groups which covers train hitting person on platform. For example, train moves will not take place with passengers present (either outside of passenger hours or closing platform for move). Persons are considered at risk including station staff and contractors.
9. *Structural failure* hazard group has two sub-hazard groups, minor injury and major injury due to structural failure. The hazardous events include in these two group covering scenarios of partial or catastrophic collapse of structures hitting train e.g. wall collapse, train wash collapse, ceiling collapse and cables/pipes becoming loose etc.
10. *Health* hazard group includes the hazardous event such as failure of pumps and sumps leading to flooding, health hazard posed by mercury and arsenic in ballast which would be washed to surface.

6.1 Development of Qualitative Descriptors and Fuzzy Rule Base

The FO in term of qualitative descriptors, and their trapezoidal and triangular MFs defined in Section 3.2 have been used to assess RL of each hazardous event in different hazard groups. The qualitative descriptors of FO are defined as 'Remote', 'Infrequent', 'Frequent' and 'Common'. 'Rare', 'Occasional' and 'Regular' as shown in Figure 4 and their meanings are presented in Table 1. The CS is described as 'Minor', 'Marginal', 'Moderate', 'Severe' and 'Catastrophic' characterised by triangular and trapezoidal

MFs as shown in Figure 5. The definition of qualitative descriptors of CS is given in terms of the number of fatalities, major and minor injuries resulting from the occurrence of a particular hazardous event. Table 4 shows the definition of these qualitative descriptors where major and minor injuries are calculated in terms of equivalent fatalities. Ten major injuries or 200 minor injuries are considered equal to one equivalent fatality. For example, qualitative descriptor 'Minor' is defined to describe the consequence level of minor injury with an approximate numerical value of 0.005. The fuzzy set of RL in terms of qualitative descriptors is defined as 'Low', 'Possible', 'Substantial' and 'High' (An et al, 2006 & 2007; Huang et al, 2005). Their definitions, which are generally similar to those described in EN50126, EN50129, GE/GN8561 and MIL-STD-882D (Railway Safety, 2002; European Standard, 1999; MIL-STD-882D, 2000), are listed in Table 3. The risk score is defined in a manner that the lowest score is 0, whereas the highest score is 12. For example, qualitative descriptor, 'Low', is defined on the basis of the risk score ranging from 0 to 4. Similar to the input qualitative descriptors of FO and CS, the trapezoidal MFs are used to describe the RL as shown in Figure 6. The result of RLs can be expressed either as risk score located in the range from 0 to 12 or as risk category with a belief of percentage.

The 35 rules in the rule base that are used in this study are listed as follows:

R_1 : If FO is Remote and CS is Minor, then RL is Low

R_2 : If FO is Remote and CS is Marginal, then RL is Low

R_3 : If FO is Remote and CS is Moderate, then RL is Possible

R_4 : If FO is Remote and CS is Severe, then RL is Possible

R_5 : If FO is Remote and CS is Catastrophic, then RL is Possible

R_6 : If FO is Rare and CS is Minor, then RL is Low

R_7 : If FO is Rare and CS is Marginal, then RL is Possible

R_8 : If FO is Rare and CS is Moderate, then RL is Possible

R_9 : If FO is Rare and CS is Severe, then RL is Possible

R_{10} : If FO is Rare and CS is Catastrophic, then RL is Substantial

R_{11} : If FO is Infrequent and CS is Minor, then RL is Possible

R_{12} : If FO is Infrequent and CS is Marginal, then RL is Possible

R_{13} : If FO is Infrequent and CS is Moderate, then RL is Possible

R_{14} : If FO is Infrequent and CS is Severe, then RL is Substantial

R_{15} : If FO is Infrequent and CS is Catastrophic, then RL is Substantial

R_{16} : If FO is Occasional and CS is Minor, then RL is Possible

R_{17} : If FO is Occasional and CS is Marginal, then RL is Possible

R_{18} : If FO is Occasional and CS is Moderate, then RL is Substantial

R_{19} : If FO is Occasional and CS is Severe, then RL is Substantial

R_{20} : If FO is Occasional and CS is Catastrophic, then RL is Substantial

R_{21} : If FO is Frequent and CS is Minor, then RL is Possible

R_{22} : If FO is Frequent and CS is Marginal, then RL is Substantial

R_{23} : If FO is Frequent and CS is Moderate, then RL is Substantial

R_{24} : If FO is Frequent and CS is Severe, then RL is Substantial

R_{25} : If FO is Frequent and CS is Catastrophic, then RL is High

R_{26}: If FO is Regular and CS is Minor, then RL is Substantial
R_{27}: If FO is Regular and CS is Marginal, then RL is Substantial
R_{28}: If FO is Regular and CS is Moderate, then RL is Substantial
R_{29}: If FO is Regular and CS is Severe, then RL is High
R_{30}: If FO is Regular and CS is Catastrophic, then RL is High
R_{31}: If FO is Common and CS is Minor, then RL is Substantial
R_{32}: If FO is Common and CS is Marginal, then RL is Substantial
R_{33}: If FO is Common and CS is Moderate, then RL is High
R_{34}: If FO is Common and CS is Severe, then RL is High
R_{35}: If FO is Common and CS is Catastrophic, then RL is High

6.2 Risk Assessment

Risk assessment is carried out from hazardous event level to sub-hazard group levels and then progress to hazard group level. The results of RLs of sub-hazard groups and hazard groups are shown in Tables 5 and 6, respectively. For example, it can be seen that Electrocution (Typical outcome) sub-hazard group in Table 6 has risk scores of 6.22 and risk categories of possible with a belief of 78 per cent and substantial with a belief of 22 per cent and Electrocution (Typical outcome) sub-hazard group has risk scores of 5 belonging to possible with a belief of 100 per cent. Whilst Electrocution hazard group in Table 6 has risk scores of 6.22 and risk categories of possible with a belief of 78 per cent and substantial with a belief of 22 per cent. The process of risk assessment from sub-hazard group level to hazard group level can be demonstrated as follows.

As can be seen in Table 6 that Electrocution (Typical outcome) sub-hazard group has 0.0259 of FO and 0.625 of CS. FO is evaluated as 'Rare' with a MF value of 0.6025 and 'Infrequent' of 0.3975 as shown in Figure 4. The CS is evaluated as 'Severe' with a MF value of 1.0000 as shown in Figure 5. In this case two rules are fired in the fuzzy inference system.

R_9: If FO is Rare and CS is Severe, then RL is Possible
R_{14}: IF FO is Infrequent and CS is Severe, then RL is Substantial

The firing strength can be calculated by equation (15). R_9 has a firing strength of 0.6025 with risk category of possible and R_{14} has a firing strength of 0.3975 with risk category of substantial. The process of fuzzy inference is depicted in Figure 10. The shadowed areas as shown in Fig 10(a) indicate the output of implication of each fired rule. Figure 10(b) shows the implication outputs of each rule are aggregated into a single fuzzy set by using equation (14). On the basis of the aggregated fuzzy set, defuzzification calculates the defuzzified value, which is a crisp value, standing for the final result of the fuzzy inference. The RL of Electrocution hazard group can then be calculated by equation (15) as shown in Figure 10(c), which indicates the risk scores of 6.22 and risk categories of possible with a belief of 78 per cent and substantial of 22 per cent. Similarly, the RLs of other hazard groups can be calculated following this approach. The summary of RLs of other hazard groups is shown in Table 6.

Table 5. Risk Levels of Hazardous Events

Operation	Hazard Groups	Ref. No.	Hazardous Events	Failure Likelihood	Consequence Severity	Risk Score	Risk Category
Shunting at Waterloo depot	Derailment	1	Derailment (Typical outcome)	6.43E-03	0.005	2.84	Low: 100%
		2	Derailment (Worst case scenario)	7.14E-04	0.125	5.00	Possible:100%
	Collision	3	Collision between trains (Worst case scenario)	2.27E-04	0.625	5.00	Possible:100%
		4	Collision between trains (Typical outcome)	4.32E-03	0.025	3.88	Low: 12%, Possible: 88%
		5	Collision hazard (Typical outcome)	2.05E-01	0.005	5.00	Possible:100%
		6	Collision hazard (Worst case scenario)	2.07E-03	0.625	5.00	Possible:100%
	Train fire	7	Train fire (Typical outcome)	2.50E+00	0.005	5.79	Possible:100%
	Electrocution	8	Electrocution (Typical outcome)	2.59E-02	0.625	6.22	Possible: 78%, Substantial: 22%
		9	Electrocution (Worst case scenario)	2.59E-02	0.125	5.00	Possible:100%
	Slips/trips	10	Slips / trips (Minor injury)	6.80E-01	0.005	5.00	Possible:100%
		11	Slips / trips (Major injury)	1.04E-01	0.125	5.79	Possible:100%
		12	Slips / trips (Fatality)	1.60E-02	0.625	5.49	Possible:100%
	Falls from height	13	Falls from height (Minor injury)	9.29E-04	0.025	2.75	Low: 100%
		14	Falls from height (Major injury)	6.07E-03	0.125	5.00	Possible:100%
		15	Falls from height (Fatality)	1.43E-04	0.625	5.00	Possible:100%
	Train strikes person	16	Train strikes person (Major injury)	5.00E-02	0.125	5.00	Possible:100%
		17	Train strikes person (Fatality)	5.00E-02	0.625	8.00	Substantial:100%
	Platform train interface	18	Platform train interface (Major injury)	2.50E-02	0.125	5.00	Possible:100%
		19	Platform train interface (Fatality)	2.50E-02	0.625	6.15	Possible: 85%, Substantial: 15%
		20	Structural failure (Minor injury)	3.86E-03	0.025	3.71	Low: 29%, Possible: 71%
	Structural failure	21	Structural failure (Major injury)	5.91E-04	0.125	5.00	Possible:100%
		22	Structural failure (Fatality)	9.09E-05	0.625	5.00	Possible:100%
	Health hazard	23	Health hazard (Most likely)	5.00E-02	0.125	5.00	Possible:100%

6.3 FAHP Analysis: Assessment of Priority Weights for Different Risk Categories

In the FAHP process, the weights in the additive utility function in the mathematical model can be evaluated by using FAHP analysis. As stated earlier in section 4, Table 4 shows pairwise comparisons of the criteria. The case of ten hazard groups is shown in Table 6. Three experts with high qualification

Table 6. Risk Levels of Hazard Groups

Operation	Index	Hazard Groups	Risk Score	Risk Category
Shunting at Waterloo depot	1	Derailment	4.53	Possible: 100%
	2	Collision	4.43	Possible: 100%
	3	Train fire	5.79	Possible: 100%
	4	Electrocution	6.22	Possible: 78% Substantial: 22%
	5	Slips/trips	5.78	Possible: 100%
	6	Falls from height	4.28	Possible: 100%
	7	Train strikes person	6.50	Possible: 50% Substantial: 50%
	8	Platform train interface	6.15	Possible: 85% Substantial: 15%
	9	Structural failure	4.39	Possible: 100%
	10	Health hazard	5.00	Possible: 100%
Overall RL			5.14	Possible: 100%

Figure 10. The process of implication, aggregation and defuzzification

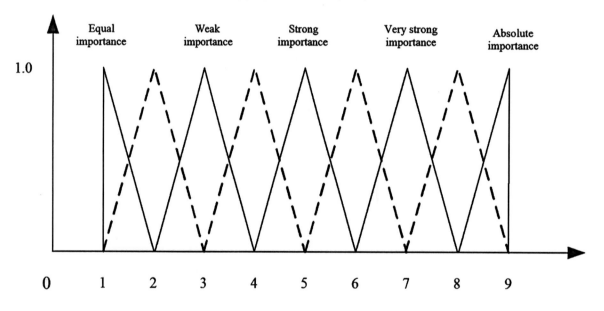

regarding this subject are selected to form a risk assessment group for undertaking the risk assessment by using such a risk assessment methodology. Comparison of two criteria is made at a time by comparing the relative importance of going from the worst to the best value of two criteria. An allocation of a total of nine points to the two criteria reflects the judgement made on the relative importance of each, and then these judgements are transferred into fuzzy number. For example, three safety analysts made these comparisons of derailment hazard group (index 1) with electrocution hazard group (index 4). One estimated "strong importance" and his judgement has then transferred into fuzzy numbers of (4,5,6). One evaluated "between strong and very strong importance" which corresponds to triangular fuzzy number of (5,6,7). The third safety analyst's judgement was "between strong and very strong importance and most likely to be 6 in the range of 4 to 8'. His evaluation is therefore transferred to a triangular fuzzy number of (4,6,8). By using equation (24), the elements in the pairwise comparison can be calculated as

$$\tilde{a}_{1,4} = \frac{1}{3}((4,5,4) \oplus (5,6,7) \oplus (5,6,8)) = ((\frac{4+5+4}{3}),(\frac{5+6+6}{3}),(\frac{6+7+8}{3})) = (4.33, 5.67, 7.00)$$

$$\tilde{a}_{4,1} = \frac{1}{\tilde{a}_{1,4}} = (0.14, 0.18, 0.23)$$

Similarly, when the safety analysts made judgements, they implicitly assigned values to each criterion, in the case, risk category. Those values can be recovered until all elements in the pairwise comparison matrix is obtained by calculating the ratios of reciprocal cells of the pairwise comparisons using equation (21) as shown in Table 7, where $\tilde{a}_{j,i} = 1/\tilde{a}_{i,j}$ for all i, j except for $i=j=1.00$. It should be noted that each row of the pairwise comparison matrix contains the ratios of the judgemental values assigned by the safety analysts to the various criteria as shown in Table 7.

On the basis of the geometric mean technique, each hazard group's fuzzy weight can be calculated by equations (25) and (26).

$$\tilde{f}_1 = (\tilde{a}_{1,1} \otimes \tilde{a}_{1,2} \otimes \cdots \otimes \tilde{a}_{1,10})^{1/10}$$

$$= ((t_{1,1}^l \times t_{1,2}^l \times \ldots \times t_{1,10}^l)^{1/10}, (t_{1,1}^m \times t_{1,2}^m \times \ldots \times t_{1,10}^m)^{1/10}, (t_{1,1}^u \times t_{1,2}^u \times \ldots \times t_{1,10}^u)^{1/10})$$

$$= ((1.00 \times 1.00 \times \cdots \times 7.00)^{1/10}, (1.00 \times 1.00 \times \cdots \times 8.00)^{1/10}, (1.00 \times 2.00 \times \cdots \times 9.00)^{1/10})$$

$$= (3.50, 4.15, 5.03) = (3.50, 4.15, 5.03)$$

Similarly,

$$\tilde{f}_2 = (3.17, 4.03, 4.55)$$
$$\tilde{f}_3 = (1.36, 1.86, 2.35)$$
$$\tilde{f}_4 = (0.73, 0.98, 1.40)$$
$$\tilde{f}_5 = (0.30, 0.37, 0.51)$$
$$\tilde{f}_6 = (0.28, 0.36, 0.46)$$
$$\tilde{f}_7 = (1.15, 1.69, 2.36)$$
$$\tilde{f}_8 = (0.36, 0.51, 0.76)$$
$$\tilde{f}_9 = (0.71, 0.95, 1.19)$$
$$\tilde{f}_{10} = (0.28, 0.38, 0.53)$$

Table 7. Pairwise Comparison Matrix

Index	1	2	3	4	5	6	7	8	9	10
1	1.00,1.00,1.00	1.00,1.00,2.00	3.00,4.00,5.00	4.33,5.67,7.00	7.33,8.33,9.00	8.00,9.00,9.00	2.00,3.00,4.00	4.33,5.33,6.33	6.00,7.00,8.00	7.00,8.00,9.00
2	0.50,1.00,1.00	1.00,1.00, 1.00	3.00,4.00,5.00	3.67,4.67,5.67	8.00,9.00,9.00	7.33,8.33,9.00	2.00,3.00,4.00	5.67,6.67,7.67	4.00,5.00,6.00	7.00,8.00,9.00
3	0.20,0.25,0.33	0.20,0.25,0.33	1.00,1.00, 1.00	1.00,2.00,3.00	4.67,5.67,6.67	4.33,5.33,6.33	1.00,2.00,3.00	3.00,4.00,5.00	3.00,4.00,5.00	3.00,4.00,5.00
4	0.14,0.18,0.23	0.18,0.21,0.27	0.33,0.50,1.00	1.00,1.00, 1.00	3.00,4.00,5.00	3.00,4.00,5.00	0.30,0.43,0.75	1.00,2.00,3.00	1.00,1.00,2.00	2.00,3.00,4.00
5	0.11,0.12,0.14	0.11,0.11,0.13	0.15,0.18,0.21	0.20,0.25,0.33	1.00,1.00, 1.00	1.00,1.00,2.00	0.25,0.33,0.50	0.33,0.50,1.00	0.20,0.25,0.33	1.00,2.00,3.00
6	0.11,,0.11,0.13	0.11,0.12,0.14	0.16,0.19,0.23	0.20,0.25,0.33	0.50,1.00,1.00	1.00,1.00, 1.00	0.20,0.25,0.33	0.33,0.50,1.00	0.20,0.25,0.33	1.00,2.00,3.00
7	2.00,3.00,4.00	0.25,0.33,0.50	0.33,0.50,1.00	1.33,2.33,3.33	2.00,3.00,4.00	3.00,4.00,5.00	1.00,1.00, 1.00	1.00,2.00,3.00	1.00,1.67,2.67	3.00,4.00,5.00
8	0.16,0.19,0.23	0.13,0.15,0.18	0.20,0.25,0.33	0.33,0.50,1.00	1.00,2.00,3.00	1.00,2.00,3.00	0.33,0.50,1.00	1.00,1.00, 1.00	0.25,0.33,0.50	0.33,0.50,1.00
9	0.13,0.14,0.17	0.17,0.20,0.25	0.20,0.25,0.33	0.50,1.00,1.00	3.00,4.00,5.00	3.00,4.00,5.00	0.37,0.60,1.00	2.00,3.00,4.00	1.00,1.00, 1.00	2.00,3.00,4.00
10	0.11,0.13,0.14	0.11,0.13,0.14	0.20,0.25,0.33	0.25,0.33,0.50	0.33,0.50,1.00	0.33,0.50,1.00	0.20,0.25,0.33	1.00,2.00,3.00	0.25,0.33,0.50	1.00,1.00, 1.00

Then,

$$\tilde{w}_1 = \frac{\tilde{f}_1}{\tilde{f}_1 \oplus \tilde{f}_2 \oplus \cdots \oplus \tilde{f}_{10}} = \frac{(3.50, 4.15, 5.03)}{(3.50, 4.13, 5.03) \oplus (3.17, 4.03, 4.55) \oplus \ldots \oplus (0.28, 0.38, 0.53)}$$

$$= (0.18, 0.27, 0.42)$$

Similarly,

$$\tilde{w}_2 = (0.17, 0.26, 0.38)$$

$$\tilde{w}_3 = (0.07, 0.12, 0.20)$$

$$\tilde{w}_4 = (0.04, 0.06, 0.12)$$

$$\tilde{w}_5 = (0.02, 0.02, 0.04)$$

$$\tilde{w}_6 = (0.01, 0.02, 0.04)$$

$$\tilde{w}_7 = (0.06, 0.11, 0.20)$$

$$\tilde{w}_8 = (0.02, 0.03, 0.06)$$

$$\tilde{w}_9 = (0.04, 0.06, 0.10)$$

$$\tilde{w}_{10} = (0.01, 0.02, 0.04)$$

As all fuzzy weights are obtained which are represented as triangular fuzzy numbers, these fuzzy numbers have to be defuzzified into crisp weights by equation (26). For example, the result of defuzzification of derailment hazard group can be calculated

$$DF_{\tilde{w}_1} = \frac{[(t_1^u - t_1^l) + (t_1^m - t_1^l)]}{3 + t_1^l} = \frac{[(0.42 - 0.18) + (0.27 - 0.18)]}{3 + 0.18} = 0.10$$

Similarly,

$$DF_{\tilde{w}_2} = 0.09$$

$$DF_{\tilde{w}_3} = 0.06$$

$$DF_{\tilde{w}_4} = 0.03$$

$$DF_{\tilde{w}_5} = 0.01$$

$$DF_{\tilde{w}_6} = 0.01$$

$$DF_{\tilde{w}_7} = 0.06$$

$$DF_{\tilde{w}_8} = 0.02$$

$$DF_{\tilde{w}_9} = 0.03$$

$$DF_{\tilde{w}_{10}} = 0.01$$

When all FWs are defuzzified into crisp weights, those crisp weights have to be normalized by equation (26).

$$w_1 = \frac{DF_{\tilde{w}_1}}{\sum_{i=1}^{10} DF_{\tilde{w}_i}} = \frac{0.1}{0.1+0.09+0.06+0.03+0.01+0.01+0.06+0.02+0.03+0.01} = 0.24$$

Similarly,

$$w_2 = 0.21$$

$$w_3 = 0.14$$

$$w_4 = 0.07$$

$$w_5 = 0.03$$

$$w_6 = 0.03$$

$$w_7 = 0.14$$

$w_8 = 0.05$

$w_9 = 0.07$

$w_{10} = 0.02$

The sum: $\sum = 1.00$

The overall RL estimation of shunting at Waterloo depot can be evaluated by equation (1)

$$RL_{Waterloo} = \sum_{i=1}^{10} RL_i w_i = RL_1 w_1 + RL_2 w_2 + ... + RL_9 w_9 + RL_{10} w_{10}$$

$= 4.53 \times 0.24 + 4.43 \times 0.21 + 5.79 \times 0.14 + 6.22 \times 0.07 + 5.78 \times 0.03$

$+ 4.28 \times 0.03 + 6.15 \times 0.05 + 4.39 \times 0.07 + 5.00 \times 0.02$

$= 5.14$

which indicates that the overall RL of shunting at Waterloo depot is 5.14 belonging to 'Possible' with a belief of 100 percent as shown in Figure 6. The result provides very useful information to risk managers and engineers for railway maintenance decision making.

6.4 Discussions

The overall RL of shunting at Waterloo depot is 5.14 that belongs to risk category of possible with a belief of 100 per cent. Ten identified hazard groups effect the overall RL estimation at the Waterloo depot. It should be noted that each hazard group contributes a different weight value to the overall RL of depot. The major contributions are from the hazard groups of derailment, collision, train fire, and train hits person which contribute 21, 18, 16 and 18 per cent, respectively, to the overall RL of shunting at Waterloo depot. Each of these groups consists of a number of hazardous events. For example, there are six main hazardous events in derailment hazard group such as track related faults, signal related faults, rolling stock faults, structural failures, objects from trains, human errors and sabotage/malicious acts, which result in derailment. Based on the accident and incident reports and statistics, majority of derailment risk (92 per cent) is put down to human error such as overspeeding and incorrect routing. Therefore, in order to reduce RLs of derailment, staff training should be provided to shunters and signallers, also to limit speed at depot, to improve liaison between MR shunters and signalers, to provide reference manual procedures, to use plunger on 2 and 3 roads and use of 2 people in cab/shunt panel during degraded modes. The other potential control measures to reduce derailment risks include track maintenance, inspection training of track staff for reducing track related faults; signal/points maintenance and inspection engineering controls, e.g. interlocking, training and competence of signal maintenance staff for reducing signalling related faults; fleet maintenance and inspection training and competence of fleet staff and brake test before moving for reducing rolling stock related faults; civil maintenance and inspection training and competence of civil staff for reducing structural failures; fleet maintenance and inspection training and competence of fleet staff for reducing objects from trains risk; depot security measures, e.g. controlled access/egress CCTV on platforms, platform end barriers, and station assistants for reducing sabotage/malicious acts. The potential additional control measures to reduce other hazardous events in other hazard groups and the proposed strategy for each additional control measure have been recommended to industry.

The hazard groups of electrocution, slips/trips, falls from height, platform train interface, structural failure and health hazard contribute less than the above hazard groups with 9, 2, 2, 6, 6 and 2 per cent, respectively. Although these hazard groups have minor contribution to the overall RL of shunting at the depot, the control measures are still done to reduce those hazardous events. For example, the hazard group of health hazard contributes 2 per cent to the overall RL estimation. As the major health related hazardous events include failure of pumps/sumps leading to flooding, and health hazard posed by mercury and arsenic in ballast which would be washed to surface, therefore, the suggested control measures are to provide: maintenance and inspection of pump assets; training depot staff and six-monthly environmental monitoring.

Case study of shunting at Waterloo depot has demonstrated that there are potential benefits of the described risk assessment methodology. The risk analysis method described in this chapter can handle the expert knowledge, engineering judgments, and historical data for the railway safety and risk assessment in a consistent manner. This risk analysis method can assess risks directly using the qualitative descriptors that are more expressive and natural to describe the risk issues. The application of FAHP may solve the problems of risk information loss in the hierarchical process so that risk assessment can be carried out from hazardous event level to a railway depot (/system) level.

7. CONCLUSION

This chapter presents a systematic risk assessment methodology for railway safety and risk analysis using FRA and FAHP. Various knowledge acquisition techniques that could be used in developing fuzzy qualitative descriptors and their MFs to qualify RLs are discussed. The prototype method using FRA and FAHP is presented in this chapter with a case study of risk assessment of shunting at Waterloo depot for the application of the described risk analysis methodology.

As FRAs offered a great potential in risk assessment modelling of railway systems, especially in the circumstances where the risk data are incomplete or there is a high level of uncertainty involved. Risk analysis using FRA and FAHP approaches can formulate domain human experts' experience and risk management knowledge. In addition, information from various sources can be transformed to be the knowledge base such as qualitative descriptors, MFs, fuzzy rules as used in the fuzzy inference process. The result from case study has demonstrated that the described risk analysis method based on FRA and FAHP approaches provides risk analysts, engineers and managers with a convenient tool that can be used in various circumstances in performing risk analysis.

REFERENCES

Ammar, S., Duncombe, W., Jump, B., & Wright, R. (2004). Constructing a fuzzy-knowledge-based-system: An application for assessing the financial condition of public schools. *Expert Systems with Applications*, 27(3), 349–364. doi:10.1016/j.eswa.2004.05.004

An, M. (2003, July 6-7). Application of a knowledge-based intelligent safety prediction system to railway infrastructure maintenance. *Proceedings of International Railway Engineering Conference*, London. Edinburgh: Engineering Technics Press.

An, M. (2005). A review of design and maintenance for railway safety – the current status and future aspects in the UK railway industry. *World J. Eng.*, 2(3), 10–23.

An, M., Chen, Y., & Baker, C. J. (2011). A fuzzy reasoning and fuzzy-analytical hierarchy process based approach to the process of railway risk information: A railway risk management System. *Information Sciences*, 181(18), 3946–3966. doi:10.1016/j.ins.2011.04.051

An, M., Huang, S., & Baker, C. J. (2007). Railway risk assessment – the fuzzy reasoning approach and fuzzy analytic hierarchy process approaches: a case study of shunting at Waterloo depot. *Proc. Instn Mech. Engrs, Part F: J. Rail and Rapid Transit*, 221(3), 365–383. doi:10.1243/09544097JRRT106

An, M., Lin, W., & Stirling, A. (2006). Fuzzy-based-approach to qualitative railway risk assessment. *Proc. Instn Mech. Engrs, Part F: J. Rail and Rapid Transit*, 220, 153–167. doi:10.1243/09544097JRRT34

Andriantiatsaholiniaina, L. A., Kouikoglou, V. S., & Phillis, Y. A. (2004). Evaluating strategies for sustainable development: Fuzzy logic reasoning and sensitivity analysis. *Ecological Economics*, 48(2), 149–172. doi:10.1016/j.ecolecon.2003.08.009

Bojadziev, G., & Bojacziev, M. (1997). *Fuzzy logic for business, finance, and management*. Singapore: World Scientific.

Bowles, J. B., & Pelaez, C. E. (1995a). Fuzzy logic prioritisation of failure in a system failure mode, effects and criticality analysis. *Reliability Engineering & System Safety*, 50(2), 203–213. doi:10.1016/0951-8320(95)00068-D

Bowles, J. B., & Pelaez, C. E. (1995b). Application of fuzzy logic to reliability engineering. *Proceedings of the IEEE*, 83(3), 435–449. doi:10.1109/5.364489

Buckley, J. J. (1985). Fuzzy hierarchical analysis. *Fuzzy Sets and Systems*, 17(3), 233–247. doi:10.1016/0165-0114(85)90090-9

Cheng, C. H. (1997). Evaluating naval tactical missile systems by fuzzy AHP based on the grade value of membership function. *European Journal of Operational Research*, 96(2), 343–350. doi:10.1016/S0377-2217(96)00026-4

Cheng, C. H., Yang, K. L., & Hwang, C. L. (1999). Evaluating attack helicopters by AHP based on linguistic variable weight. *European Journal of Operational Research*, 116(2), 423–435. doi:10.1016/S0377-2217(98)00156-8

EN50126 Railway applications-the specification and demonstration of reliability, availability, maintainability and safety (RAMS). (1999). *European Standard*, CENELEC.

Fahmy, H. M. A. (2001). Reliability evaluation in distributed computing environments using the AHP. *Computer Networks, 36*(5-6), 597–615. doi:10.1016/S1389-1286(01)00175-X

Gadd, D. S., Keeley, D. D., & Balmforth, D. H. (2003). Good practice and pitfalls in risk assessment. Health & Safety Laboratory, HMSO, HSE Book, London.

Ghiaus, C. (2001). Fuzzy model and control of a fan-coil. *Energy and Building, 33*(6), 545–551. doi:10.1016/S0378-7788(00)00097-9

Hauser, D., & Tadikamalla, P. (1996). The analytic hierarchy process in an uncertain environment: A simulation approach. *European Journal of Operational Research, 91*(1), 27–37. doi:10.1016/0377-2217(95)00002-X

Railways (safety case) regulations 2000. (2000). Health & Safety Executive (HSE).

Homnan, B., & Benjapolakul, W. (2004). Application of fuzzy inference to CDMA soft handoff in mobile communication systems. *Fuzzy Sets and Systems, 144*(2), 345–363. doi:10.1016/S0165-0114(03)00206-9

Huang, S., An, M., & Baker, C. J. (2005). A fuzzy based approach to risk assessment for track maintenance incorporated with AHP. Proceedings of International Railway Engineering Conference, London, Edinburgh: Engineering Technical Press.

Kennedy, A. (1997). Risk management and assessment for rolling stock safety cases. *Proc. Instn Mech. Engrs, Part F: J. Rail and Rapid Transit, 211*(1), 67–72. doi:10.1243/0954409971530914

Kuo, R. J., Chi, S. C., & Kao, S. S. (1999). A decision support system for locating convenience store through fuzzy AHP. *Computers & Industrial Engineering, 37*(1-2), 323–326. doi:10.1016/S0360-8352(99)00084-4

Kuo, R. J., Chi, S. C., & Kao, S. S. (2002). A decision support system for selecting convenience store location through integration of fuzzy AHP and artificial neural network. *Computers in Industry, 47*(2), 199–214. doi:10.1016/S0166-3615(01)00147-6

vanLaarhoven, P.J.M., & Pedrycz, W. (1983). A fuzzy extension of Saaty's priority theory. *Fuzzy Sets and Systems, 11*(1), 229–241. doi:10.1016/S0165-0114(83)80082-7

Lee, A. H. I., Chen, W. C., & Chang, C. J. (2008). A fuzzy AHP and BSC approach for evaluating performance of IT department in the manufacturing industry in Taiwan. *Expert Systems with Applications, 34*(1), 96–107. doi:10.1016/j.eswa.2006.08.022

Lootsma, F. A. (1996). A model for the relative importance of the criteria in the multiplicative AHP and SMART. *European Journal of Operational Research, 94*(3), 467–476. doi:10.1016/0377-2217(95)00129-8

LUL. (2001). London Underground Limited Quantified Risk. *Assessment Update, 2001*, 1.

Metronet SSL, (2005a). *Development of Metronet staff risk model* (First Metronet SSL Interim Report).

Metronet SSL (2005b). *Framework for the assessment of HS&E risks* (Second Metronet SSL Interim Report).

Mikhailov, L. (2004). A fuzzy approach to deriving priorities from interval pairwise comparison judgements. *European Journal of Operational Research, 159*(3), 687–707. doi:10.1016/S0377-2217(03)00432-6

MIL-STD-882D Standard practice for system safety. (2000). USA Department of Defense.

Mon, D. L., Cheng, C. H., & Lin, J. C. (1994). Evaluating weapon system using fuzzy analytic hierarchy process based on entropy weight. *Fuzzy Sets and Systems, 62*(2), 127–134. doi:10.1016/0165-0114(94)90052-3

Ngai, E.W.T., & Wat, F.K.T. (2003). Design and development of a fuzzy expert system for hotel selection. *International Journal of Management Sciences, 31*(4), 275–286.

Peters, G.A., & Peters, B.J. (2006). *Human error: causes and control.* CRC Press. doi:10.1201/9781420008111

Guidance on the preparation of risk assessment within railway safety cases. (2002). Railway Safety Railway Group Guidance Note – GE/GN8561, 1.

Profile of safety risk on the UK mainline railway. (2003). Railway Safety SP-RSK-3.1.3.11, 3.

Ramanathan, R., & Ganesh, L. S. (1995). Using AHP for resource allocation problems. *European Journal of Operational Research, 80*(2), 410–417. doi:10.1016/0377-2217(93)E0240-X

Rolling stock asset strategic safety risk model. (2004). *Tube Lines*, Interim Report.

Saaty, T. L. (1980). *Analytical Hierarchy Process.* New York: McGraw-Hill.

Saaty, T. L. (1990). *The analytic hierarchy process.* Pittsburgh: RWS Publications.

Sadiq, R., & Husain, T. (2005). A fuzzy-based methodology for an aggregative environmental risk assessment: A case study of drilling waste. *Environmental Modelling & Software, 20*(1), 33–46. doi:10.1016/j.envsoft.2003.12.007

Sii, H. S., Ruxton, T., & Wang, J. (2001). A fuzzy-logic-based approach to qualitative safety modelling for marine systems. *Reliability Engineering & System Safety, 73*(1), 19–34. doi:10.1016/S0951-8320(01)00023-0

Tang, M. T., Tzeng, G. H., & Wang, S. W. (2000). A hierarchy fuzzy MCDM method for studying electronic marketing strategies in the information service industry. *J. Inter. Info. Man., 8*(1), 1–22.

Wang, L., Chu, J., & Wu, J. (2007). Selection of optimum maintenance strategies based on a fuzzy analytic hierarchy process. *International Journal of Production Economics, 107*(1), 151–163. doi:10.1016/j.ijpe.2006.08.005

Yadaiah, N., Kumar, A. G. D., & Bhattacharya, J. L. (2004). Fuzzy based coordinated controller for power system stability and voltage regulation. *Electric Power Systems Research, 69*(2-3), 169–177. doi:10.1016/j.epsr.2003.08.008

Yu, C. S. (2002). A GP-AHP method for solving group decision-making fuzzy AHP problems. *Computers & Operations Research, 29*(14), 1969–2001. doi:10.1016/S0305-0548(01)00068-5

Chapter 10
Formal Assurance of Signaling Safety:
A Railways Perspective

Pallab Dasgupta
Indian Institute of Technology Kharagpur, India

Mahesh Mangal
Indian Railways, India

ABSTRACT

The EN50128 guidelines recommend the use of formal methods for proving the correctness of railway signaling and interlocking systems. The potential benefit of formal safety assurance is of unquestionable importance, but the path towards implementing the recommendations is far from clear. The EN50128 document does not specify how formal assurance of railway interlocking may be achieved in practice. Moreover, the task of setting up an electronic interlocking (EI) equipment involves multiple parties, including the EI equipment vendor, the certification agency which certifies the resident EI software to be correct, and the end user (namely the railway service provider) who must configure the EI equipment. Considering the distributed nature of the development process, a feasible approach towards formal certification of the end product (post configuration) is not obvious. This chapter outlines the basics of formal verification technology and presents, from the perspective of the railways, a pragmatic roadmap for the use of formal methods in safety assurance of its signaling systems.

INTRODUCTION

Railway signaling has been one of the most well studied safety critical systems for nearly two centuries. During this period, the notion of railway signaling has evolved in various ways, including the protocol for signaling, the technology used for implementing the signaling system, and most importantly the way in which safety guarantees are assured. The very early form of signaling relied on *temporal separation* of trains, which was primarily implemented by setting up time tables that ensured that two trains never shared a track at the same time. With the increase in railway traffic, it became necessary to divide the

DOI: 10.4018/978-1-5225-0084-1.ch010

Formal Assurance of Signaling Safety

tracks into segments (or blocks), thereby giving birth to the notion of *block signaling*, where one or more trains can be on the same track, but on different blocks – which effectively means that the trains are *spatially separated*. Signals guard the entry of the block and implement the spatial separation.

Railway yards also have *points* where two tracks intersect. The *point setting* determines whether the train will continue to move on the same track or whether it will move into the intersecting track. A complex yard may contain many *points*, with the tracks crisscrossing each other, which significantly increases the complexity of ensuring that a train can safely move from one track to another possibly passing through several intermediate tracks. In signaling parlance such complex passages are called *routes*. A railway yard can have hundreds of routes, and the signaling system must ensure that conflicting routes are kept free when a train is passing through a route. This is achieved by a mechanism called *interlocking*.

In the past railway interlocking was primarily manual. The hand-operated point levers shown in Figure 1 are reminiscent of the times when point positions were changed manually using such levers. The interlocking system, implemented using electrical relays, would ensure that signal aspects change only when points are in proper position.

In recent times, railways use *electronic interlocking*, where the signals and point positions are controlled by software running on an *electronic interlocking* (EI) system. Such computer controlled systems are designed to adhere to the signaling standards. Once configured properly with respect to a specific railway yard, they continue to work reliably over time and are not subject to human errors.

The primary concern with EI equipment is in determining with absolute certainty that the system configuration is correct, and that given a correct system configuration, the software is guaranteed to never err in signaling decisions. In current practice this verification task is performed using a high level of certification of the software and through rigorous testing of the configured equipment.

In spite of the best practices followed in configuring and testing of EI equipment, there are known instances of failures resulting out of human errors in the design of the logic for the EI equipment. The following two incidents may be highlighted:

- [The Milton Keynes Incident, Dec 29, 2008] As reported in (Railway Accident Investigation Branch Report, 2010), on December 29, 2008, a serious signaling error was detected at the Milton Keynes Central station on the West Coast Main Line in UK. The driver of a train observed a signal change from red to green, although the track beyond the signal was occupied by another train.

Figure 1. A point (left) and hand-operated point lever (right)

An accident was averted by the driver, but the signaling error raised serious questions about the safety of modern EI systems. Investigations revealed that the error was due to incorrect configuration data for the EI equipment, namely that a track segment was missed in a specific part of the control logic. An even more intriguing fact which came out of the investigation was that interlocking data errors had in fact caused several other incidents in the past at Rugby, Glasgow Central, Peterborough, and Shenfield. In all of these cases, data errors had not been detected before the interlocking was commissioned.

- [The Cootamundra Incident, Nov 12, 2009] As reported in (Australian Transport Safety Bureau Report RO-2009-009, 2009), possible collision was averted by the driver of a train approaching No.1 Platform Road at Cootamundra, New South Wales, who noticed that the last wagon of a freight train was obstructing his approach. Investigations revealed that the track segment used for clearing the signal which allowed the train to approach was incorrect, namely that the data used to configure the EI system defined a wrong track segment for clearing the signal.

Such incidents reveal the uncomfortable fact that human errors in preparing the interlocking data may not be detected at the time of commissioning a EI equipment. The problem is that post-configuration testing may not sensitize the data error, particularly if the error is manifested only in very few corner case scenarios, say involving a very specific pattern of train movements in the yard. The commissioned equipment may then work perfectly for as long as providence does not create the scenario in which the error is manifested.

The problem of exhaustive testing hits what is known as the *complexity barrier*, that is, the number of possible test cases multiplies with every signal, every point, and every track segment. This combinatorial growth causes *state explosion*, that is, the number of states of the signals and points exceeds the capacity of any computer, and exhaustive testing becomes infeasible beyond that point. A deadline for commissioning the EI equipment further limits the number of tests that can be run. Therefore, most railways have their own well defined lists of tests, which are performed and then the equipment is commissioned. The above incidents show that such test plans are not adequate to cover all relevant scenarios.

The EN50128 guidelines (European Committee for Electrotechnical Standardization, 2001), issued by Comité Européen de Normalisation Électrotechnique (CENELEC), is the main reference for railway signaling equipment manufacturers in Europe, and also a widely referred standard in other railways around the world. The EN50128 document uses the concept of *Software Safety Integrity Level* (SSIL) to define the criticality of the components of a signaling system – specifically, for the software for railway control and protection systems. The SILs are: 4 (very high), 3 (high), 2 (medium), 1 (low), and 0 (not safety-related). While the guidelines do not contain any clear prescription of the software development methodology, they classify some of the commonly used techniques as *forbidden*, *highly recommended*, and *mandatory*.

One of the remarkable aspects of the EN50128 guidelines is in classifying a class of techniques, loosely called *formal methods*, as *highly recommended* for components with higher levels of SIL. Intuitively, formal methods are techniques for *mathematically proving* the correctness of software or a system, where the correctness requirement is *formally specified* using formalisms like *mathematical logic*. The basis of this recommendation lies in the realization that formal methods can comprehensively prove the correctness of a system using an arsenal of mathematical techniques, whereas conventional testing fails to provide the desired level of assurance on those systems within feasible time. The EN50128 guidelines are however silent on how formal methods may be applied to arrive at better safety assurance for the

Formal Assurance of Signaling Safety

high integrity software/system components, thereby posing a very pertinent research problem before the signaling and formal methods communities.

Any study on formal assurance of signaling safety must not only dwell upon the key technical challenges in verifying a configured signaling system, such as EI equipment, but must also focus on a deeper understanding on the commissioning process of such equipment. EI equipment is a sophisticated system, manufactured by only a handful of companies around the world. The software residing in such systems is a proprietary property of the EI equipment vendor. On the other hand, the yard specific data for configuring EI equipment is developed and provided by the signal engineers of the railway company. The vendor configures the equipment on the basis of the data received from the customer and then the configured equipment is subjected to rigorous testing by the end user. The software residing in the EI equipment is of the highest SIL, and may have been verified formally (to be compliant with railway signaling principles), but if the yard specific data has human induced errors, then the configured system will have errors.

The railway company has no means for formally verifying the configured equipment without having code-level access to the vendor's software. The vendor, on the other hand, has to rely on the data provided by the railway company and therefore all its assurances (formal or otherwise) are under the assumption that the data is correct. Therefore, the best intent of the railway company should be to ensure that the data provided to the vendor is correct, and that too with formal assurance, as recommended by the EN50128 guidelines. This chapter outlines the suggested steps towards achieving this goal, with important ramifications in ruling out future incidents similar to the Milton Keynes and Cootamundra incidents.

It is important to point out that various types of technologies are available today for automatic train control with/without signals. Notable among these is the ETCS standards (European Train Control System) which are used at various levels by different railways. Driverless metro rails are also operational in some cities. Formal methods are being used for validation of such systems as well. Though we touch upon the verification issues in such systems, our primary focus in this chapter shall be on signaling and interlocking systems, which is still the most widely used safety critical technology among railways worldwide.

A PRIMER ON RAILWAY INTERLOCKING

This section outlines the traditional development approach for a railway interlocking system, and prepares the reader for the main focus of this article, namely the formal assurance of signaling safety. We begin with the definition of the key components of a signaling system, and then walk through the main steps in developing the signaling logic.

Signaling Terminologies

The components of a signaling system and the terminologies used to define them are readily available from various sources. In this section we outline a small fragment of these definitions – only those that are used to explain the formal assurance approach.

Figure 2 shows a typical control panel in a railway station. Observe the layout of the various lines and their intersections (points). The operator of this panel can issue *route requests* to the EI equipment,

Figure 2. Section of a control panel

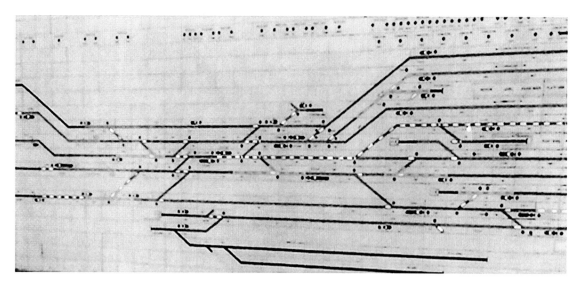

as we shall explain shortly, and the EI equipment allows the route to be *locked* only if *conflicting routes* are not locked at that time. In order to understand these key terms, let us walk through the steps in setting up a EI equipment.

We begin with a brief explanation of the layout of a simple railway yard from a remote village in India. Figure 3 shows the layout diagram of the yard. Trains may approach this station from both di-

Figure 3. Layout diagram of a railway yard

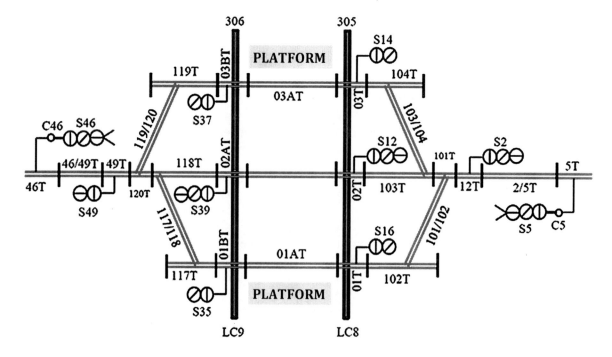

Formal Assurance of Signaling Safety

rections (left or right) along a single (bidirectional) line. At the line approaches the station from either side, it branches into three lines as shown in red. The upper and lower lines are called *loop lines*, which are adjacent to platforms shown in yellow. The middle line is called the *main line*. The two blue lines represent two level crossings located at the two sides of the station.

The main components of the layout from a signaling perspective are as follows:

1. **Track segments:** The tracks are divided into segments, and each segment has a label. In Figure 3 the labels ending with T represent track segments. The track segments have circuits to sense whether it is occupied by a train. Various types of track circuits are used in practice, but the purpose of all of them is to sense the presence of a train on the track. Note that a train may occupy more than one track segment at a time, for example, when it is passing from one segment to another. Track circuits provide inputs to the signaling system.

2. **Points:** The points are located at the intersections of two tracks. The position of a point determines the direction of the passage of the train. When the point is in *normal position*, the train continues on the same track; when it is in *reverse position*, the train moves to the adjacent track. For example, consider the point connecting segments 119T and 120T in Figure 3. When this point is in the normal position, a train approaching from 49T will pass towards 118T. On the other hand if the point is in reverse position, then a train approaching from 49T will change track from 120T to 119T along the point 119/120. Point positions are controlled by the signaling equipment.

3. **Signals:** The signals are asserted by the signaling system. In Figure 3 the signals are shown beside the tracks. The names of the main signals start with S (for example, S35). Since these are bidirectional tracks, there are signals on both sides of each track in Figure 3, but for a train moving in a particular direction, only the ones on its left hand side of the track are applicable. For example, the signal S35 is applicable only for trains leaving the platform segment 01AT towards 117T, and not for trains approaching the platform from the point 117/118. The signal C46 shown with signal S46 in Figure 3 is a special type of signal called *calling-on signal*. The calling-on signal asks an approaching train to pass a red signal of the main signal at a low speed. For example, when C46 may be used for calling-on a train even when S46 is red. Calling-on signals are used for various purposes, for example, to attach a coach at the end of an existing train.

4. **Routes:** Routes are sequences of consecutive track segments that define the path of a train from one part of the yard to another. For example in Figure 3, if a train approaching from the right has to be admitted on the upper platform, we need to clear the path from 5T to 03BT consisting of the track segments 2/5T, 12T, 101T, 103T, 104T, 03T, 03AT, 03BT. The point 101/102 has to be in normal position, and point 103/104 has to be in reverse position. The route is characterized by the starting and ending signals, which in this case are S5 and S37 respectively. Point positions may exclude one route when the other is locked. For example, when the route from S5 to S37 is locked, the point 103/104 is in reverse position, which excludes the route from S5 to S39, since it requires the same point to be in normal position. If two routes share one or more track segments, then they are said to be *conflicting*. For example, the route from S5 to S39 and the route from S12 to S2 are conflicting. The request for locking a route is made from the control panel of the yard, and it is the responsibility of the interlocking equipment to guarantee that conflicting routes are not asserted at the same time.

The locking of routes in a yard must adhere to the international railway signaling principles, which consists of a set of universally adopted rules. Any signaling system that follows these rules under all situations is guaranteed to be safe from signaling errors. In the EI equipment, adherence is ensured by the correct development of the *application logic* and the resident software that interprets this logic. We shall explain the development of this logic in the next section.

The Route Locking Logic

The basis for locking a route for the passage of a train is to guarantee that all track segments in that route and overlap are free, all points in that route and overlap are in proper position, and most importantly, the guarantee remains valid *until the train has exited the route*. In order to develop the logic for locking and releasing routes in a systematic way, a table called a *control table* or a *route locking table* is created. Table 1 shows a portion of the control table of the yard shown in Figure 3. Each row of the table corresponds to a route in the yard. Each route is defined by the starting and ending signals of the route. For each route the control table specifies the points that must be locked in normal position, the points that must be locked in reverse position, and the track circuits that must be clear. The last column of the table highlights the conflicting routes and whether the level crossing gates need to be closed. There are various other aspects in the control table which are not shown here for simplicity.

It is important to introduce the notion of overlap at this stage. When a train approaches a signal, it may not be able to stop *before* the signal, because the distance between the signal post and the location at which the driver sees the signal may be too less for the driver to stop the train before the signal post. Therefore, as a safety precaution, the railway signaling principles require the signaling system to clear the routes up to the next signal before clearing (making green/yellow) a signal. It also requires the point positions in the routes ahead to be locked in a way that no other train can come into those track segments. For example, consider the route R46(a) in Table 1, which starts from signal S46 and ends in signal S14 (see Figure 3). The *overlap* requires the point, 103/104, to be in reverse position so that the train can roll forward past signal S14. The track segments 104T, 103T, 101T and 12T must be free. Also the point, 101/102, must be in normal position to prevent a train from rolling past signal, S16, towards this direction. The overlap requirement does not apply for *calling on* signals (such as C5 and C46 in Figure 3) and this is useful for allowing more than one train to approach this specific yard. A *called-on* train must stop *before* the red signal (that is, no overlap is allowed for the called-on train).

The development of the control table is a highly safety critical activity, since this table forms the basis for developing the application logic. A busy railway station has more than a thousand routes, and it is close to impossible to manually guarantee that all entries in the control table are correct. Also the task of validating the application logic generated corresponding to a control table is a non-trivial task.

The software in typical EI equipment consists of the *executive software* which is factory installed, and the *application software* which is programmed for every yard. The application software represents the yard specific logic, which is interpreted by the executive software while regulating the signals and points. The application data prepared by the end user (namely, the railway company) is used to define the application software.

The application data consists of a set of Boolean equations that define the signaling logic for the yard. For historical reasons, most types of EI equipment continue to accept a form of logic called *relay ladder logic*. In order to ensure that the failure of a signal or a point always leaves the system in a fail-safe state, traditionally railways around the world have been using different types of relays which guarantee

Formal Assurance of Signaling Safety

Table 1. A portion of the control table

| RT | MOVEMENT | | IN THE ROUTE | | OVERLAP | | | REMARKS |
| | FROM | TO | POINTS | | POINTS | | TRACK CIRCUITS | |
			NOR	REV	TRACK CIRCUITS	NOR	REV		
R2	S2	-	-	-	2/5T	-	-	-	LOCKS S5
R5	S5	S39	101/102, 103/104	-	2/5T, 12T, 101T, 103T, 02/02AT	117/118, 119/120	-	118T, 120T, 39T	LOCKS S2, S12, S46, GATES 305, 306
R5b	S5	S37	101/102	103/104	2/5T, 12T, 101T, 103T, 104T, 03/03A/03BT		119/120	119T, 120T, 39T	LOCKS S2, S14, S46 GATES 305, 306
R12	S12	S2	101/102, 103/104		103T, 101T, 12T	-	-	-	LOCKS S5, S39
R14	S14	S2	101/102	103/104	104T, 103T, 101T, 12T	-	-	-	LOCKS S5, S37
R37	S37	S49	-	119/120	119T, 120T, 39T	-	-	-	LOCKS S14, S46
R39	S39	S49	117/118, 119/120	-	118T, 120T, 39T	-	-	-	LOCKS S12, S46
R46	S46	S12	117/118, 119/120	-	46/49T, 39T, 120T, 118T, 02/02AT	101/102, 103/104	-	103T, 101T, 12T	LOCKS S5, S39, S49 GATES 305, 306
R46 (a)	S46	S14	-	119/120	46/49T, 39T, 120T, 119T, 03/03A/03BT	101/102	103/104	104T, 103T, 101T, 12T	LOCKS S5, S37, S49 GATES 305, 306
R49	S49	-	-	-	46/49T	-	-	-	LOCKS S46

such fail safe operations. The logic which determines when a relay will be energized (or de-energized) is developed in the form of *ladder logic*, which became popular because of its visual simplicity and verifiability through visual inspection.

Figure 4 illustrates the notion of ladder logic in the context of the yard of Figure 3. ASR relays are used to lock routes. Here the logic is shown for the relay, 14ASR, which is responsible for locking the route from the signal S14 to the signal S2. The route is locked when the ASR relay is de-energized. The logic shows that the 14ASR relay can be energized by a combination of the relays shown in the upper

Figure 4. An illustration of ladder logic

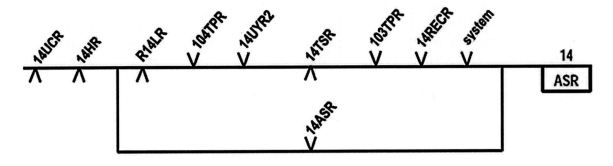

rung of the ladder. The relays shown with the ∨ symbol must be energized, while the relays shown with the ∧ symbol must be de-energized for the 14ASR relay to be energized. Some of these relays may have ladder logics of their own. For example, the logic of 14UCR guarantees that the points 103/104 and 101/102 are locked in their appropriate positions, and this is necessary for the route represented by 14ASR to be locked.

The lower rung of the ladder has the 14ASR relay itself, which allows the relay to hold its state once it is energized. The ability of a relay to hold its state is an important feature of relay logic, since it shows that relay logic describes *sequential circuits* with an underlying state transition system. In a real yard, the ladder has many more rungs, which allows the relay to be energized in other ways. It may be noted that 14ASR can be de-energized (say) by energizing the relay14UCR.

The ladder logic can be expressed in Boolean logic as well. For example, the logic of Figure 4 is equivalent to the following Boolean function for the next state of 14ASR:

~14UCR * ~14HR *
        (14ASR +
            (~R14LR * 104TPR * 14UYR2 *
              ~14TSR * 103TPR * 14RECR * system))

The primary focus of this chapter is to outline how the application logic developed by the signal engineers can be proven to be correct using formal methods, so that data errors that can potentially escape testing are ruled out before the commissioning of the logic. It may be prudent to point out that formal verification of the application data should not be viewed as a measure for replacing the testing of the EI equipment after it is configured. This is because, even if both the executive software and the application software are proven to be correct, we still have to make sure that the EI equipment has been correctly integrated into the yard, namely that the relays have been correctly connected to the EI equipment. The goal of formal methods in this case is to rule out data errors prior to commissioning, and understanding that in its entirety is our goal in this chapter.

FORMAL VERIFICATION

The notion of *formal verification* has its origin in proving the correctness of programs. In general the domain of the inputs to a program need not be finite, and hence a software program is typically tested using only a subset of the possible input combinations. One of the most fundamental results in Computer Science establishes that in general the task of deciding whether a program is correct for all input scenarios is *undecidable*, that is, it is mathematically proven that no algorithm can read a program and decide whether it is correct for all inputs. This is not true for the special class of programs where all variables have finite domains, because finiteness allows us to enumerate all input combinations and prove the correctness in each of them. For most systems of moderate complexity however, the theoretical possibility of enumerating all input combinations is not practically possible within feasible limits of time, and hence the objective is to prove the correctness of the system *without explicit enumeration*. Formal verification techniques achieve this objective using an arsenal of smart abstractions and intelligent decision procedures that work on succinct representations of large state spaces. Many languages and formalisms exist for developing formal specifications, including *temporal logic* (Pnueli, 1977), Z

Formal Assurance of Signaling Safety

(O'Regan, 2012), B (Lano, 1996), and VDM (Bjorner & Jones, 1978). In this chapter we shall restrict ourselves to properties expressed in temporal logic.

Formal verification encompasses several types of approaches – in terms of its objective, and in terms of the methodology used to achieve that objective. For example, *equivalence checking* and *model checking* are two popular notions associated with formal verification. An *equivalence checking* problem consists of the design-under-test and an abstract golden model, and the task is to determine whether for every input sequence the output of the design-under-test is identical to that of the golden model. A *model checking* problem consists of the design-under-test and a set of formal properties, and the task is to determine whether the properties are satisfied for *every* execution of the design-under-test.

The *railway signaling principles* define the safety properties that a signaling logic must satisfy at all reachable states of the signaling system. This verification task has been attempted in the past using several types of formal methods as discussed in the next section, which work with various degrees of user intervention. The approach that we advocate in this chapter is a fully automated one, namely one that uses automated *model checking* techniques (Dasgupta, 2006).

Figure 5 illustrates the model checking approach. The design-under-test (DUT) represents the logic that we wish to verify. In the context of the signaling validation problem, our DUT is the application logic developed by the signal engineer. A model checking tool has a front-end pre-processor that extracts a finite state machine representation from the DUT. The formal properties represent the *specification* – we shall explain how temporal logic can be used for this purpose. In our case, formal properties are gleaned from the railway signaling principles as interpreted on the specific yard. The core model checking engine receives the formal properties and finite state representation of the DUT as inputs and determines whether the formal properties are guaranteed by the DUT. If any property fails in the DUT under some scenario, then the model checker shows a specific scenario under which the DUT fails to satisfy that property, namely, it finds a counter-example scenario.

Let us consider a principle that says: *A signal is not cleared (made green/yellow) until it is proven that the track circuits in the route up to the next signal and its overlap are clear*. If we interpret this principle for the signal S14 in Figure 3, then we wish to prove that the signaling logic never clears S14 until tracks 104T, 103T, 101T, 12T are unoccupied. This property can be formally specified in terms of the relays involved as:

LTLSPEC G(X 14HR ⇒ 104TPR & 103TPR & 101TPR & 12TPR)

Figure 5. Model checking

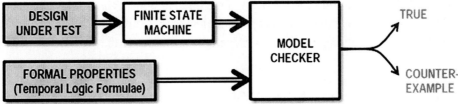

The relay, 14HR, is the relay controlling the signal S14. The relay, 104TPR, indicates whether the track segment 104T is unoccupied (the other TPRs are similarly defined). It may be noted from Table 1, that the route R14 does not have any overlap tracks.

The above property is expressed in *Linear Temporal Logic*, which extends Boolean logic with temporal operators. This property uses the standard Boolean operators like **&** (logical AND) and **~** (negation), and two temporal operators, namely the **G** operator and the **X** operator. The meanings of these operators are intuitively described as follows:

- A property **G**φ is true in a finite state machine, if and only if the property φ is true at all reachable states of the machine. Thus, the **G** operator (also called the *always* operator) is used to define properties that are *globally* true.
- A property **X**φ is true at a state, if and only if the property φ is true at all next states of the state. Thus, the **X** operator (also called the *next-time* operator) is used to specify what must hold in the next state of the system.

The above property specifies that a pre-requisite for 14HR to be high at a time (that is, for the signal S14 is clear) is that the TPR relays for 104T, 103T, 101T, and 12T must all be high. The property is violated if we reach a state where the following holds (where the symbol, **|**, represents the logical OR operator):

(**X** 14HR) & (~104TPR | ~103TPR | ~101TPR | ~12TPR)

The above expression can become true in broadly two ways:

1. 14HR goes high in some time point, even though the prerequisite condition (namely the conjunction of the TPR relays) is not true. This is in direct violation of the signaling principle stated earlier.
2. 14HR and the TPR relays are high, but one (or more) TPR relay goes low before 14HR becomes low (that is, a track circuit fails before S14 becomes red). This is also in violation of a signaling principle.

A formal proof must guarantee that the above expression is not satisfied in *all* reachable behaviors of the signaling system.

Traditional testing will typically examine whether 14HR is energized when the track segments represented by the above TPR relays are unoccupied. On the other hand, formal verification aims to prove that the 14HR relay is *not* energized *in any other circumstance*. It may be noted that the relay logic for 14HR is not directly defined in terms of these relays, but indirectly through several steps involving other relays (such as 14ASR, which was defined earlier). If the logic for any of those relays has an error, it may be manifested through a failure of the above property under some peculiar sequence of train movements. Formal verification helps in finding such counter-example scenarios if they exist.

FORMAL VERIFICATION IN RAILWAY SIGNALING

Railway signaling has been a problem of interest to the formal verification community for many years, because of its safety critical nature and its legacy of being treated as a logic design problem. However the

Formal Assurance of Signaling Safety

adoption of formal methods in actual practice has been scarce, even after the publication of the EN50128 guidelines which has a clear recommendation for the use of formal methods in SIL4 components. This is primarily because the path towards adoption of formal methods in not well understood by the railway companies and their vendors, and secondly because the investments needed to set up the tools and best practices for this purpose have not been taken up.

Formal verification methods have been researched in the railways context broadly for *interlocking systems* and *train control systems*. While *interlocking* systems control the movement of trains by operating the signals and points along railway tracks, train control systems aim to control the movement of trains directly based on the knowledge of the position, velocity and other parameters of the train. For example, in driverless metro railways, signals have no purpose and trains have to be directly controlled. Train control systems and interlocking systems may coexist, and in such cases the driver may sometimes be overridden by the train control system (for example, a train may brake automatically when an approaching signal is red and the train is travelling too fast). One of the primary references for train control systems is the ETCS standards (European Train Control System).

Several modeling and formal verification techniques have been studied for train control protocols and subsystems. This includes the use of formalisms like Petri nets (Meyer zu Horste & Schnieder, 1999), (Zimmermann & Hommel, 2005) and state charts (Damm & Klose, 2001).

The most primitive form of train control systems (ETCS Level 1) use trackside balises for issuing the movement authority. When the train passes over a balise the train control system receives a telegram from the balise warning it about the movement authority in its way forward. For example, a balise located in the approach to a signal will issue the appropriate authority to the train control system to warn the driver of the approaching signal, and to apply the brakes automatically if the driver does not respond. Similar control may be exercised before bends in the track, in the approach to bridges or tunnels, in specific blocks, and in general, wherever speed restrictions are to be imposed.

When a train control system receives a telegram from a balise, it has to apply the brakes in a manner that the speed is restricted within some specified distance. This requirement can be formally modeled by a *braking curve* which shows speed as a function of distance. Methods for formally proving that a train with a given inertial model will always be able to obey the movement authority have been studied by researchers (Platzer & Quesel, 2009).

More advanced levels of ETCS allow movement authorities to be imposed dynamically and also allow moving blocks. In moving blocks, a safe zone is defined around each train as it moves. As opposed to traditional signaling systems where only one train is allowed in a block section between two signals, a moving block system allows trains to move with less spacing between themselves. As a train moves into a yard, it is important to prove that the movement authorities to different trains are issued in a way that they are safe with respect to each other. Formal methods have been studied for such systems as well (Cimatti & et. al., 2012).

Formal methods have also been applied to prove the correctness of communication protocols used in railways (Lee, Hwang, & Park, 2005) and signaling protocols (Lee & et. al., 2007).

Railway interlocking, which is the main focus of this chapter, has been extensively studied by the formal verification community. A recent survey on the use of formal methods in railway signaling can be found in (Fantechi, Fokkink, & Morzenti, 2012). For example, researchers have studied the problem of gleaning the precise and complete set of yard specific rules from the railway interlocking principles (Fokkink, 1996).

As discussed earlier, EI equipment has *executive software,* which is the factory set software for driving the signaling system, and *application software* which is configured in a yard specific manner from the *application data/ logic* provided by the signaling engineers. For a given *application data*, the guarantee that the EI equipment will faithfully drive the signals as per the specified logic, must come from the EI equipment vendor. For this purpose, manufacturers of EI equipment have considered a wide variety of formal approaches, some of which can be found in (Fringuelli, Lamma, Mello, & Santocchia, 1992), (LeGoff, 1996), (Bernardeschi, Fantechi, & et. al., 1998). Model checking techniques have also been attempted, for example, (Cimatti & et. al., 1998), (Eisner, 2002). An approach based on theorem proving for verification of signaling data can be found in (Morley, 1993).

Formal Verification of Application Data/ Logic

Research on formal verification of railway signaling has primarily focused on the verification task of the EI equipment vendor. It is important to understand the benefits of the formal verification techniques for the end user, namely the railway company. We summarize the benefits under two broad categories:

1. The model can be validated against yard specific properties generated from layout based on railway signaling principles. This helps in eliminating data errors before the logic is passed on to the EI equipment.
2. The model can be used as the scientific basis for generating test scenarios for the configured equipment. When a yard is modified, the EI equipment has to be reconfigured. In a fully operational yard, the downtime resulting out of the reconfiguration has to be within tolerable limits, and hence the level of testing done prior to commissioning EI equipment cannot be repeated for every modification. Formal analysis can help in prioritizing the test scenarios.

The norm of every verification problem in engineering design is to ensure that the task of verification is carried out independent of the design team, so that the logical bias of the design team (and the logical errors thereof) does not get carried over to the verification team. Figure 6 shows the design tasks and the verification tasks in the mandated flow for commissioning EI equipment. The tasks shown inside blue clouds are the ones that are performed today under the following responsibilities:

1. **Rigorous Logic Development**: This task is performed by the signal engineers in a manual/semi-automated way. The kinds of practices followed vary from one company to another. The logic so developed may have redundancies, may be divided into multiple EI equipment, and may be represented in various functional forms. The internal steps of this task, which may involve the development of control table by the signal engineers, can be found in the manuals of individual companies and are not discussed here. We shall refer to this task as the *design phase*.
2. **EI Equipment Configuration:** This task is typically done by the EI equipment vendor, possibly with participation from the end user. This step may involve verification of various forms (including formal methods) which are performed by the EI equipment vendor. Since the executive software is typically proprietary in nature, the railway company has no role in the verification performed under this task.

Formal Assurance of Signaling Safety

Figure 6. Formal assurance of application data: Railway's perspective

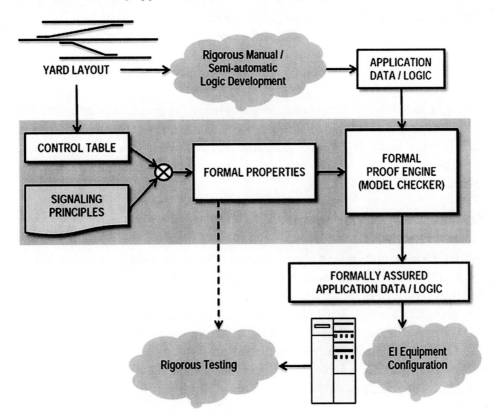

3. **EI Equipment Testing:** The testing of the EI equipment is the responsibility of the end user, namely the railway company. The railway company has well defined procedures for creating the yard specific test plan and executing the tests on the configured equipment. This phase is a very critical phase, since this is the last step before commissioning the equipment. We shall refer to this task as the *testing phase*.

In some cases, the application data may be simulated for verifying its logical correctness. It is infeasible to achieve 100% coverage of scenarios in simulation or testing, and this is the reason why there are numerous instances where data errors have made their way into the commissioned equipment.

Our prescription for railway companies is to introduce the tasks shown on the saffron banner between the design and testing phases. This is a fully automated flow and can achieve 100% coverage in validating the application data in a matter of few minutes. The main tasks in enabling this flow are:

1. Generating a complete sound and complete suite of yard specific formal properties using the governing railway signaling principles. We shall explain this step in the next subsection. We shall refer to this task as *specification development*.
2. Reading the application data / logic developed by the signal engineer and extracting a *finite state machine* (FSM) model from the design. In keeping with traditional practices, many railways con-

225

tinue to use legacy formalisms for expressing the logic (such as ladder logic), and it is fairly straight forward to extract the FSM from such representations.
3. Developing a model checker that can scale to the size of logic entailed by large complex yards. This step requires some insights into the nature of the properties to be proven, and tailoring the model checker for such properties. Elaborating this step is beyond the technical scope of this article.

Before we elaborate the task of *specification development*, it is important to point out that the properties generated in this step can also be used to ascertain whether relevant test scenarios have been exercised during testing. In other words the property suite also acts as a metric for coverage analysis during the testing phase. This is a very important when partial test coverage is achieved following logic updates after incremental changes in the yard.

Formal Specification Development

Specification development is the task of interpreting the railway signaling principles on a given yard and capturing the resulting requirements through formal properties. The signaling principles are almost the same in all railways and govern the general set of rules that must be followed by any signaling system. The signaling principles are independently published by different countries – for example, see (Paul Woolford, 2003) for the British standards, (IRISET, 2009) for the Indian standards, and (Warwick Allison, 2013) for the principles followed by RailCorp in Australia.

The signaling principles are comprehensive documents covering all aspects of signaling and safe movement of trains. The principles start with a set of broad requirements, and then present further details on how these requirements are expected to be achieved. For example, the following is an excerpt from one of the earliest versions of the Indian Railways Signaling Principles:

1. Lever frames and other apparatus installed for the operation and control of signals, points etc. must be so interlocked and arranged as to comply with the following essentials:
   a. It must not be possible to take 'OFF' a running signal unless the following are ensured not only in the actual portion of the route on which the train has to travel, but also in the overlap:
      i. All the points are correctly set.
      ii. All the facing points are locked (at site).
      iii. All the interlocked level crossing gates are closed and locked against road traffic.
      iv. The isolation is ensured.
   b. Once the signal has been taken OFF, it must not be possible to do any of the following unless the signal has first been put back to the 'ON' position:
      i. Alter the position of the relevant points.
      ii. Unlock the relevant facing points.
      iii. Unlock and open the relevant level crossing gate.
      iv. Disturb the relevant isolation.
   c. It must not be possible to take 'OFF' at the same time any two fixed signals, which may lead to any conflicting movements.
   d. Where feasible, the points shall be so interlocked as to avoid any conflicting movement.

Formal Assurance of Signaling Safety

The above statements provide the guidelines for *route locking*, *route holding* as well as *route release*. Each of these subtasks has more intricate requirements in terms of how the above guarantees are met. In this chapter we consider the signaling principles for *route locking* to illustrate the task of formal specification development.

Let us consider the route R14 (that is, the route from signal S14 to S2) of the yard shown in Figure 3. Note from Table 1, that the conflicting routes for R14 are R5, R5b and R37. We highlight some of the properties that are relevant for this route in Table 2.

In reality, a yard has many other types of routes and objects, including *shunt routes, calling-on routes, level crossings,* etc., and there are specific signaling principles for each of these.

Most railway companies have their versions of signaling principles, all of which are broadly in agreement in terms of safety requirements. Following the EN50128 guidelines, it becomes the responsibility of the company to develop formal safety rules governing the application logic and proving that the application logic is bug-free. Indian Railways in collaboration with the Indian Institute of Technology Kharagpur has established the broad outline and proof-of-concept for the approach presented in this chapter. It is envisaged that initial investment in this area will translate to more structured verification practices and compliment the testing procedures for the configured EI equipment.

The properties illustrated in this section are called *safety properties* in formal verification parlance. Intuitively, safety properties are those that assert that the *bad behaviors* of the system will never happen. For example, consider the following property:

G(X14HR ⇒ 104TPR & 103TPR & 101TPR & 12TPR)

Table 2. Some route-locking properties

Prove that interlocking is free	We need to prove that all conflicting routes with the same point setting are free. At the logic level this translates to checking that the LR relays for these routes are in the de-energized state. The following temporal properties are generated when this is interpreted for route R14: G(X R14LR ⇒ ~R37LR) G(X R14LR ⇒ ~R5bLR)
Prove that all points in the route, isolation and overlap are operated to the required position, locked, and detected	One way to verify this requirement is to check whether the point position is supported by some route's LR relay. For example, we have the following properties for the points 101/102 and 103/104: G(X 101_102NCR ⇒ R14LR | R5bLR | R12LR | R5LR) G(X 103_104RCR ⇒ R14LR | R5bLR) The NCR relay for a point is energized when the LR relay of any route that requires the point in the normal position is energized. Likewise, the RCR relay for a point is energized when the LR rely of any route that requires the point in the reverse position is energized. Note that we have only described some of the routes in Table 1, and therefore, the above properties are only partially complete.
Prove that the selected route, overlap and isolation is locked	A route is locked when its ASR relay is de-energized. The UCR relay is energized when all points in the route and overlap are detected to be in the required position. This requirement may be verified for route R14 by checking the following property: G (X~14ASR ⇒ 14UCR)
Prove that the track circuits in the route up to the next signal and its overlap are clear	This may be verified by checking whether track occupancy has been correctly considered in the HR logic. For example, for route R14, we have the following property: G(X14HR ⇒ 104TPR & 103TPR & 101TPR & 12TPR)
Clearing of signals	Before clearing the signals (that is, making them green/yellow), we check whether the points have been locked. For route R14, we have the following property: G(X14HR ⇒ ~101_102WLR & ~103_104WLR)

This property states that a state where 14HR is high (energized) will never be reached from a state where one or more among 104TPR, 103TPR, 101TPR, 12TPR are low. It does not say that 14HR *has to be asserted* when all the above TPRs are high.

Safety properties are only responsible for assuring the safety of the signaling system. The signaling principles, on the other hand, are not only concerned with safety, but also with *progress*. For example, a signaling system which keeps all signals red at all times is safe, but it is not useful. Signaling principles also require that *progress* in passage of trains happen under enabling circumstances. When trains are operated manually, the guarantee of progress involves a human being, and therefore the *guarantee of progress* must make some *assumptions* about the admissible behaviors of the driver. In formal verification parlance, such assumptions are also captured using formal properties which are called *fairness constraints* or *assume properties*.

On the other hand, the train control system must guarantee the *progress* when trains are operated under automatic control (as in driver-less metros). As various degrees of automated control gets adopted by various railway companies around the world, what is needed is a unified standard for railway signaling and train control, and formally laid out rules governing the safety and progress in such unified systems. This will help in better integration of formal verification practices in the next generation signaling system design flow.

CONCLUSION

Safety and efficient utilization of resources are the primary concerns in railway signaling. Efficient utilization aims towards facilitating movement of trains in and out of a yard with minimum waiting time. Safety criteria aim to introduce appropriate constraints on the movement of trains so that accidents are avoided *with adequate margin of error* from the human elements involved in the system. As we move into the modern era of automated control, some of the archaic signaling principles will have to be revised and made more specific and non-ambiguous, so that the ideal balance between efficiency and safety is achieved. Formal methods have an important role in this direction, since both safety and progress criteria can be proven using formal methods. Eventually, the end users, namely the railway companies will have to assume leadership on this subject.

In the present system, guaranteeing the correctness of the application data is an important task for the signaling engineers. For large and complex yards it is very difficult to conceive all possible sequences of train movements in the yard, and the task of achieving adequate coverage of scenarios through simulation and testing has become infeasible in practice. As a consequence, in spite of the best practices of the railway companies and their vendors, incidents like Milton-Keynes and Cootamundra have taken place in not too distant past. Formal verification of the application data, as professed in this chapter, is not only feasible in practice, but is of great value in ruling out data errors.

Having a formal model of the signaling system also opens up other interesting possibilities. We already mentioned the benefits towards formally assisted identification of relevant test scenarios for incremental testing following (minor) changes in the yard. Formal properties may also be monitored online over configured EI equipment for enhanced debugging capabilities. In conjunction with data-loggers (also called *historians*) which record the signal aspects and movement of trains, monitors can be used to catch violations of *assume properties* (for example, abnormal behavior of the driver or the train) and provide real-time warnings to the entities on the approach path of the train. All these point to the benefits of the initial investments in developing a tool flow for formal capture and validation of signaling properties.

REFERENCES

Allison, W. (2013). *Signal Design Principles*. NSW, Australia: RailCorp.

Australian Transport Safety Bureau Report RO-2009-009. (2009). *Australian Safety Bureau*.

Bernardeschi, C., Fantechi, S., Gnesi, S., Larosa, S., Mongardi, G., & Romano, D. (1998). A Formal Verification environment for railway signaling system design. *Formal Methods in System Design*, *12*(2), 139–161. doi:10.1023/A:1008645826258

Bjorner, D., & Jones, C. B. (1978). *The Vienna Development Method: The Meta-Language, LNCS* (Vol. 61). Springer. doi:10.1007/3-540-08766-4

Cimatti, A., Corvino, R., Lazzaro, A., Narasamdya, I., …, Tchaltsev, A. (2012). Formal Verification and Validation of ERTMS Industrial Railway Train Spacing System. Computer Aided Verification, LNCS (Vol. 7358, pp. 378–393). Springer.

Cimatti, A., Giunchiglia, F., Mongardi, G., Romano, D., Torielli, F., & Traverso, P. (1998). Formal Verification of a railway interlocking system using model checking. *Formal Aspects of Computing*, *10*(4), 361–380. doi:10.1007/s001650050022

Damm, W., & Klose, J. (2001). Verification of a radio-based signaling system using the STATEMATE verification environment. *Formal Methods in System Design*, *19*(2), 121–141. doi:10.1023/A:1011279932612

Dasgupta, P. (2006). *A Roadmap for Formal Property Verification*. Springer. doi:10.1007/978-1-4020-4758-9_8

Eisner, C. (2002). Using symbolic CTL model checking to verify the railway stations of Hoorn-Kersenboogerd. *Software Tools for Technology Transfer*, *4*(1), 107–124. doi:10.1007/s100090100063

EN 50128, Railway Applications Communications, Signaling and Processing Systems Software for Railway Control and Protection Systems. (2001European Committee for Electrotechnical Standardization. CENELEC.

ETCS / ERTMS Specifications. (n. d.). *European Train Control System*. http://www.era.europa.eu/Core-Activities/ERTMS/Pages/Set-of-specifications-2.aspx

Fantechi, A., Fokkink, W., & Morzenti, A. (2012). Some trends in Formal Methods applications to Railway Signaling. In S. Gnesi, & T. Margaria (Eds.), *Formal Methods for Industrial Critical Systems: A survey of applications*. John Wiley and Sons, Inc. doi:10.1002/9781118459898.ch4

Fokkink, W. (1996). Safety criteria for the vital processor interlocking at Hoorn-Kersenboogerd. *Proceedings of the 5th Conference on Computers in Railways*. Computational Mechanics Publications.

Fringuelli, B., Lamma, E., Mello, P., & Santocchia, G. (1992). Knowledge-based technology for controlling railway stations. *IEEE Intelligent Systems*, 7(6), 45–52.

Lano, K. (1996). *The B Language and Method: A Guide to Practical Formal Development*. Springer. doi:10.1007/978-1-4471-1494-9

Lee, J.-D., Jung, J.-I., Lee, J.-H., Hwang, J.-G., Hwang, J.-H., & Kim, S.-U. (2007). Verification and conformance test generation of communication protocol for railway signaling systems. *Computer Standards & Interfaces*, 29(2), 143–151. doi:10.1016/j.csi.2006.03.001

Lee, J.-H., Hwang, J.-G., & Park, G.-T. (2005). Performance evaluation and verification of communication protocol for railway signaling systems. *Computer Standards & Interfaces*, 27(3), 207–219. doi:10.1016/S0920-5489(04)00097-2

LeGoff, G. (1996). Using synchronous languages for interlocking. *First Int. Conf. on Computer Application in Transportation Systems*.

Meyer zu Horste, M., & Schnieder, E. (1999). Formal Modeling and Simulation of Train Control Systems using Petri nets. *Proceedings of the World Congress on Formal Methods in the development of Computing Systems*. Toulouse, LNCS (*Vol. 1709*). Springer.

Morley, M. (1993). Safety in Railway Signaling Data: A behavioral analysis. *Proceedings of the 6th Workshop on Higher Order Logic Theorem Proving and its Applications, LNCS(Vol. 740)*. Springer.

O'Regan, G. (2012). Z Formal Specification Language. In *Mathematics in Computing* (pp. 109–122). Springer.

Platzer, A., & Quesel, J.-D. (2009). European Train Control System: A Case Study in Formal Verification. *Proceedings of the 11th International Conference on Formal Engineering Methods ICFEM 2009* (pp. 246-265). Springer. doi:10.1007/978-3-642-10373-5_13

Pnueli, A. (1977). The Temporal Logic of Programs. In Foundations of Computer Science (pp. 46-57).

Principles of Interlocking. (2009). *Indian Railways Institute of Signal Engineering and Telecommunications*.

Report, R. A. I. B. (2010). *RAIB review of the railway industry's investigation of an irregular signal sequence at Milton Keynes, 29 Dec 2008. Railway Accident Investigation Branch*, Department for Transport, UK.

Woolford, P. (2003). *Interlocking Principles*. London, UK: Railway Safety and Standards Board Ltd.

Zimmermann, A., & Hommel, G. (2005). Towards modeling and evaluation of ETCS real-time communication and operation. *Journal of Systems and Software*, 77(1), 47–54. doi:10.1016/j.jss.2003.12.039

KEY TERMS AND DEFINITIONS

Formal Verification: A verification methodology for mathematically proving the compliance of a design against a formally defined logical specification.

Interlocking: A procedure for ensuring separation of trains in a railway yard through acquiring locks on signals, points, and track segments on a route.

Model Checking: A formal decision procedure for proving a given formal specification on a design implemented as a state machine.

Railway Signaling Principles: Internationally adopted guidelines governing signaling rules and regulations

Relay Ladder Logic: A logic for specifying sequential control implemented using electrical relays.

Chapter 11
Automatic Static Software Testing Technology for Railway Signaling System

Jong-Gyu Hwang
Korea Railroad Research Institute, Korea

Hyun-Jeong Jo
Korea Railroad Research Institute, Korea

ABSTRACT

In accordance with the development of recent computer technology, the railway system is advancing to be flexible, automatic and intelligent. In addition, many functions of railway signaling which are cores to the railway system are being operated by computer software. Recently, the dependency of railway signaling systems on computer software is increasing. The testing to validate the safety of the railway signaling system software is becoming more important, and related international standards for inspections on the static analysis based source code and dynamic test are a highly recommended (HR) level. For this purpose, studies in relation to the development of source code analysis tools were started several years ago in Korea. To verify the applicability of validation tools developed as a part of these studies, the applicability test was performed for the railway signaling system being applied to the Korean domestic railway. This automated testing tool for railway signaling systems can also be utilized at the assessment stage for railway signaling system software, and it is anticipated that it can also be utilized usefully at the software development stage. This chapter drew the result of the application test for this actual source code of the railway signaling system being applied to railway sites and analyzed its result.

1. INTRODUCTION

The railway signaling system, which is in charge of the most core function in a railway system, is changing from the existing electrical device to a computer-based control system in accordance with the development of the recent computer technology. Accordingly, the software operates many parts of the main functions of the signaling system, and the dependence on software has been continuously increasing. Especially,

DOI: 10.4018/978-1-5225-0084-1.ch011

there are many major signaling functions has been operated by software, such as the distance control function between two running trains, safety route control function in trackside, signaling aspects control, etc. As the railway signaling system software becomes more complex, the importance of the software being utilized more within the railway signaling system is increasing (Signor et al., 2014), (Zhu et al., 2014), (Rao et al., 2010), (Lawrence, 2000), and (Fewstar & Gramham, 1999)

Safety requirements for railway signaling system software was recently internationally standardized by IEC 61508, IEC 62279 and IEC 62425 like Figure 1. Due to the enactment of the Korean Railway Safety Act, the rules regarding safety standards of railway facilities, etc., various software testing and validation activities required by international standards, in relation to the railway signaling system, are also beginning to be required in Korea. There are several efforts to learn more information about software testing tools, but there are not suitable for the railway signaling system software on vital one required high level safety like Robyn et al. (1999), Fewstar and Graham (1999), Yeom et al. (2009), and Kadry (2011). The software verification is become main issues in safety assessment of railway signaling system based on related international standards. So the software testing and verification very important. In addition in Korea, the validation on software is mainly dependent on documents needed for the development process, and the quantitative analysis tests are completed for only for a fraction of it. Accordingly, the study for testing and validation of railway signaling system software according to the related international standards has progressed from several years ago like Hwang et al. (2008, 2009, 2010, 2013) in Korea (Figure 1).

Situations when concrete technical development is highly required not only to validate documents on safety activities required for international standards in relation to the system software, but also to cope with the analysis and assessment through software testing. In the "A.4 software design and development" of IEC 62279, it is especially required that the "design & coding standards" and "dynamic testing" be observed in the design and development of SW-SIL(Software Safety Integrity Level) 3 or 4 grade

Figure 1. Railway RAMS related international standards

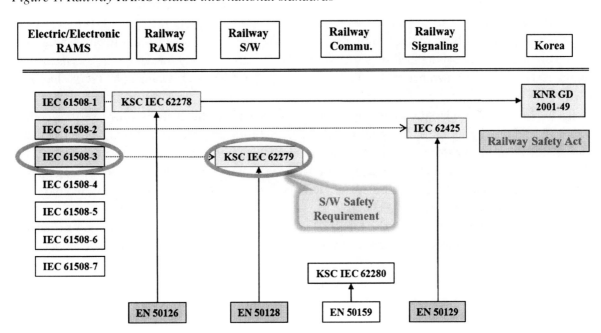

software(HR). Although international standards IEC 62279 require SWSIL 3 or 4 grade software for which a high safety grade is required to comply with the coding standard, they do not mention the rule to be observed concretely. Provided, however, guidelines for coding standards are presented within the inside of document for international standards in a qualitative form, and it takes the form of compliance to determine coding standards by actual project.

Accordingly, the coding standard with which the railway signaling system software must comply was drawn in Korea through the analysis on related international standards. The international standards were in relation to software safety in Hwang et al. (2009, 2010, 2013), and the studies on the drawn coding standard, MISRA-C which is the software coding standard for embedded control systems being applied to the motor industry, and the automated validation tool to support the drawn coding standards performed in Hwang et al. (2013). The studies performed tests through the application on the railway signaling system being applied to the existing actual domestic railway sites through automated testing tools for coding standards of the railway signaling system software, and analyzed their results in Hwang et al. (2013). The existing system, which was the target for the application, is the system having been applied before international standards were introduced in Korea. This chapter summarizes a collection of several previously published papers on related software testing studies by the authors. We would like to check the effectiveness of the validation tool being developed through the test and analysis results for the software of this system, and to consider the effect of learning more information on static testing technology for software on the railway industry in respect to its validation on safety.

2. CODING RULES FOR RAILWAY SIGNALING SYSTEM

As existing railway signaling systems us mechanical or electric logic, related safety standards of software for railway systems were internationally standardized. The demand for development of technology in relation to the software testing of railway systems, such as the enactment of standard requiring the validation of railway signaling system software, etc. is also increasing in Korea. In addition, studies in relation to the testing for railway signaling system software have been on going for several years in Korea.

Generally, there are three parts of embedded software testing, and although the software testing parts can be divided into several detailed stages it can also be classified largely into three stages as shown in Figure 2. This work of Lawrence et al. (2000), Fewstar et al. (1999), and Hwang et al. (2008) has been used in this research.

That is, they can be classified into the stages of source code static analysis which verifies coding standards and metrics for the source code of the developed software, white box testing which analyzes the test coverage, etc. of software, and black box testing which tests functional safety of software. This classification categorized and shows the testing stages, the coding standard to be applied among them is determined in the stage of the software module design. Among of them, this chapter covered the static testing of software testing (Figure 2).

A railway signaling system software for which a high safety grade is required shall be verified and validated to check that it was developed suitable for the coding standard presented in. In relation to that standard, international standards require the compliance with HR coding standards in IEC 61508, IEC 62279, IEC 62425. However, though it is required to check such coding standards, it is not presented in the corresponding standard, and international organizations in relation to the railway, etc. also does

Figure 2. Classification of software testing

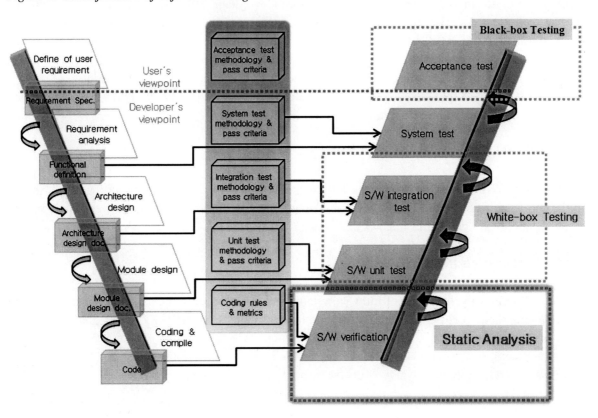

not define the relevant coding standard. However, overseas railway system software assessment agencies generally apply the MISRA-C standard, which also happens to be applied to the automobile field.

In the automobile field, the representative embedded systems are standardized in the form of "MISRA-C", which defined the coding rules and standards for control units; such programed software must be observed in C language. Like MISRA coding standards, the automobile field is making their own list and applying internally authorized coding standards, but in the aviation or railway field, etc., such the formalized coding rules are not defined. Therefore, currently most of the industrial software including those for aviation and railway are using MISRA. The safety requirements necessary for the software for railway systems, including railway signaling system, are being standardized by IEC 62279, and the activities that must be performed by the system development life cycle stage to secure the safety of vital software, such as the railway signaling system, are being explained in this standard. Besides, according to this standard, the software for the embedded system of railway signaling systems are developed in such a way that the source code is suitable for proper coding standards by safety grade and validation.

Studies have drawn the coding standard suitable for the railway field from regulations presented by the IEC 61508 and IEC 62279. These international standards must be observed when developing railway system software like the MISRA-C coding standard, which is widely used in the automobile industry. Such standards must be observed when developing the embedded software of railway signaling system being performed in Korea. Inspection Tool for Design & Coding Standards largely made MISRA-C coding standards, which are being widely used in the automobile industry, as the standard to be observed at the time of built-in software development by using C language. IEC 61508 and IEC 62279 are the international standards to be observed at time of software development for railway systems as their basis. Especially, in the cases of IEC 61508 and IEC 62279, their levels are conceptual and, comprehensive, and they do not define any concrete design or coding standard. Therefore, Hwang et al. (2009, 2010, 2013) drew and implemented the coding rules through analyzing relevant standards. Through these studies, the coding standard to apply the domestic railway signaling system software was drawn, and it is being prepared as the Korean domestic standard.

2.1 MISRA-C Coding Rules

MISRA-C, which is the coding rule for the automobile field, presents a total of 141 coding rules. Among all the coding rules presented by this MISRA-C, the items that could be automated were implemented in this development tool. Table 1 shows the coding rules of MISRA-C.

2.2 Coding Rules from IEC 61508

IEC 61508 defines the life cycle of railway systems and RAMS requirements, and consists of total 7 parts. Among them, part 7 deals with the method to measure systems (HW/SW). The configuration of part 7 is shows like Table 2.

Among them, the Annex C includes the explanation on the measurement of safety and technology of software, and the following Table 3 rules were analyzed and drawn on this basis.

Table 1. MISRA-C Coding rules

MISRA-C Item	Contents of Rules
1.1	To check ISO 9989:1990
2.1	Whether assembly languages were encapsulated or not
2.2	Permits C comment style only
2.3	No use of nested comment (C style)
3.4	To inspect whether any non-standard #pragma directive was used
4.1	Permissible escape character
4.2	Not permissible string pattern
5.1	Limitation to the identifier length
5.2	Not to sort out identifier on the scope
5.3	To inspect whether any typedef or tag name is an unique identifier
...	...
15.1	To inspect whether most closely-enclosing sentence of switch label is a switch door
15.2	If a switch is composed of more than one sentence, to inspect whether there is any break
15.3	To inspect whether there are any defaults existing in the switch door
15.4	No use of (in)equality formula in the conditional formula of switch door
15.5	To inspect whether a switch door has at least one case door
...	...

Table 2. Configuration of IEC 61508-Part 7

Table of Contents	Remarks
1. Scope	
2. Normative reference	
3. Definitions and abbreviations	
Annex A Overview of techniques and measures for E/E/PES: control of random hardware failure	Viewpoint from hardware
Annex B (informative) Overview of techniques and measures for E/E/PES: avoidance of systematic failures	Centered on concept tonal explanation
Annex C (informative) Overview of techniques and measures for achieving software safety integrity	Explanation on the measurement of safety and technology of software

Table 3. IEC 61508 Supported Rules

61508 Item	Contents of Rules (Relevant Sentence)	Detailed Contents of Rules
C.2.6.3	Prohibition function("If dynamic variables or objects are not used, these faults are avoided.")	Dynamic memory variables/memory creation function are prohibited
C.2.6.4	Memory return (""the whole memory which was allocated to it must be freed.")	To check if memory allocation/return function couple is matched
C.2.6.4	Error handling routine ("if allocation is not allowed, appropriate action must be taken.")	
C.2.6.6	Limited use of pointers ("pointer arithmetic may be used at source code level only if pointer data type and value range are checked before access.")	
C.2.6.7	No direct, indirect recursive call ("Limited use of recursion")	
C.2.7	Limitation to the cyclomatic complexity ("keep the number of possible paths through a software module small, and the relation between the input and output parameters as simple as possible.")	Designated value: 10
C.2.7	No uses of certain kind of sentence ("avoid complicated branching and, in particular, avoid unconditional jumps (goto) in higher level languages.")	
C.2.7	To inspect whether initialization of 'for' sentence, control, increase/decrease expression are all related with loop control ("where possible, relate loop constraints and branching to input parameters.")	
C.2.8	No definition on the global variables (""The key data structures are "hidden" and can only be manipulated through a defined set of access procedure. This allows the internal structures to be modified or further procedures to be added without affecting the functional behavior of the remaining software")	
C.2.9	Limitation to the function size(LOC) ("subprogram sizes should be restricted to some specified value, typically two to four screen sizes")	Designated value: 100
C.2.9	To inspect if the function has one exit point ("subprograms should have a single entry and a single exit only.")	
C.2.9	To inspect unused factors ("any software module's interface should contain only those parameters necessary for its function.")	

2.3 Coding Rules from IEC 62279

IEC 62279 consists of a total 17 chapters, and among them, the following Table 4 rules were drawn from '10. Software design and implementation', '11. Software verification and testing', '14. Software assessment.'

3. DEVELOPMENT OF SOFTWARE TESTING TOOL

Inspection technology on source code coding rules for railway signaling system software is used to test the software to judge whether it is suitable for the required standard. To a certain extent, by checking whether the software complies with coding standards, required by relevant standards, etc., inspection results can validate the risk of malfunctioning systems in advance. Due to software errors, violations against standards and potential errors embedded in the source code can be detected.

For the software validation of railway signaling systems, there are direct static analyses which analyzes coding rules of source codes and various metrics through source codes, dynamic testing which performs various coverage analyses with source codes as its target, and the black-box testing method which performs validation on functional safety with embedded railway signaling systems. Among them, tools for supporting static testing for railway signaling system software have been designed and developed.

There is no automated testing tool for coding rules to evaluate railway signaling system software yet overseas, as well as within our country, and there are some automatic tools commercialized to inspect coding rules for general industrial embedded software. However, most of these inspection tools for industrial embedded software coding rules are made to inspect MISRA-C coding rules only, or made

Table 4. IEC 62279 Supported Rules

62279 Item	Contents of Rules (Relevant Sentence)	Detailed Contents of Rules
Table 1. 10 Design and Coding Standards	Prohibition function("3. No Dynamic Objects 4. No Dynamic Variables")	Dynamic memory variables/memory creation function are prohibited
	Limited use of pointers ("5. Limited use of pointers")	
	No direct, indirect recursive call ("6. Limited use of recursion")	
	No use of certain kind of sentence ("7. No unconditional jumps")	No use of goto
Table 9. 10 Modular Approach	Limitation to the file size (LOC) ("1. Module Size Limited")	Designated value: 320
	No use of global variables ("2. Information Hiding/ Encapsulation")	
	Limitation to the parameter number of functions ("3. Parameter Number Limit")	Designated value: 5
	To inspect if the function has one exit point ("4. One Entry/One Exit Point in Subroutines and Functions")	
B.61 Structured Programming	Limitation to the cyclomatic complexity	Designated value: 10
	To inspect whether initialization of 'for' sentence, control, increase/decrease expression are all related with loop control	

Automatic Static Software Testing Technology for Railway Signaling System

possible to inspect several rules designated randomly by some testing tool developing companies only. Therefore, the automated testing tool for coding rules for vital software such as, railway signaling systems, has not been developed yet.

Coding rules testing tools are modules which enable a railway signaling system software to judge if it is suitable for the required rule, to a certain extent, by checking whether it observes coding rules required by relevant standards, etc. such as IEC 61508 and IEC 62279. Its basic purpose is to be utilized at the assessment stage for railway signaling system software, but the architecture was designed so that it can also be utilized at the software development stage. This section describes the architecture of Inspection Tool for coding rules for the automated source code analysis tool of railway signaling systems and showed in Hwang et al. (2008, 2009). Inspection tools for coding rules is the module that makes railway signaling system software judge if it is suitable for the required rule, to a certain extent, by checking whether it observes coding rules required by relevant standards, etc. Therefore, this module performs the testing for source code itself. Besides, the coding rules being applied was applied to the testing tool that wants to develop coding rules described in the previous section. Extensibility tools are also to be developed, such as the case where additional coding rules, etc. are required, it was made to enable coding rules to be edited with commands such as, select, add, etc.

Figure 3 is the one showing the configuration the diagram of developed static analysis tools for railway signaling system software suggested through this study. As shown in the figure, it was designed to present results through a report generator after analyzing whether the object source code input through a prescribed coding rule test kit is suitable for the required coding standard by analyzing the input source code, or if the source code targeted to be tested is being input into the automated testing tool.

On the basis of the system and functional architecture, screen design and design contents of coding standards of the Inspection Tool for Design & Coding Standard for the railway signaling system software explained so far, the automated testing tool was implemented. Figure 4 shows the new rule-creating window in the inspection module developed through this study. As explained in the previous section, the coding rules for railway signaling system software was not prescribed yet, and it is allowed to apply MISRA-C rules, otherwise to apply safety requirements of IEC 61508 or IEC 62279.

Figure 3. Outline of the automated testing tool for software coding standard

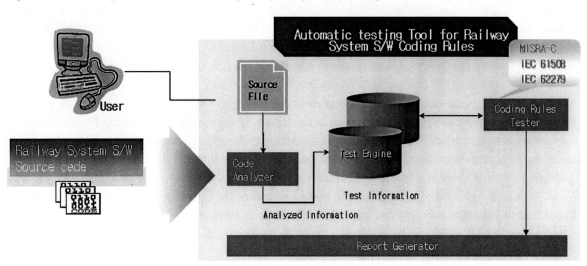

Figure 4. Coding rules setting windows

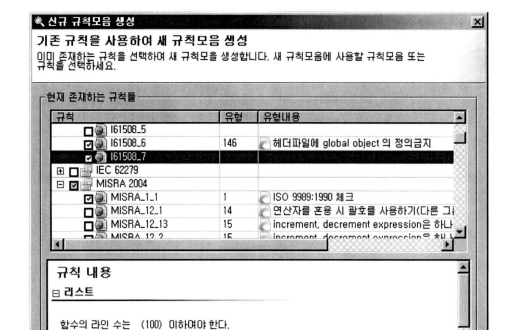

Accordingly, this study drew and implemented coding rules from MISRA-C and IEC 61508 & IEC 62279. Since there is no stipulated coding standard yet, it was made to create new rules by adopting or rejecting some parts from existing rule lists (see Figure 4), in order to be selected properly at the testing process. That is, it was made to enable coding rules being required by the source code, which becomes a testing target to be selected through a coding rule setting windows, as shown in the figure. It also enables the new rules set by project to be edited using the rule set of coding rules, which was data based previously. Besides, the developed tool is able to implement the function possible by inputting new rules, also in consideration of the extensibility for coding rules that would be prescribed in the future. Figure 5 shows the coding rule setting windows implemented through this study, and it is made to show the explanation on corresponding rules as shown in the figure, if any coding rule is selected.

Figure 6 shows the window that performs actual testing with random source codes as its target through a developed inspection module. Violated rules are shown in the upper right side of the window in the figure, and if one of them was selected, it is made to point to the violated source code part, like the window in the lower right end of the figure. The lower left end of the figure is shows the control flowchart of source code through analyzing the input source codes, and it was made to enable whole architectures of target source code testing to be checked. Due to these functions, this tool is expected to be utilized usefully even at the software debugging stage.

Figure 5. Coding Rules Editing Windows

Figure 6. Screen of the static test result

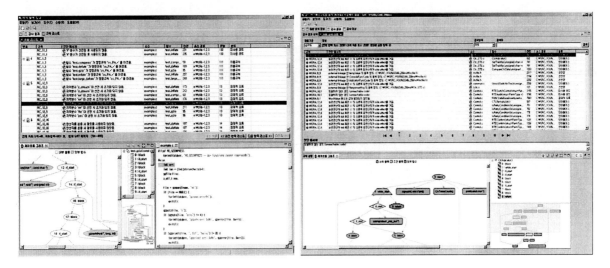

4. RESULTS OF APPLYING SOFTWARE TESTING FOR TARGET SYSTEM

The applicability test was performed to validate its applicability by focusing on the domestic railway signaling system software after developing an automated static testing tool for source codes, to inspect coding standards of railway signaling system software (Hwang et al, 2013). First of all, target software to validate the applicability of developed tools were aimed at software for railway signaling systems (Target system A: hereinafter referred to as 'A target system') that were developed prior to the introduction of international standards, in relation to the validation of railway system software in Korea, and applied to the actual railway sites. Although this target software is being applied to actual railway sites, and since the obstacle to train operation has been caused due to the recent occurrence of software failure, the static testing tool developed through this study was checked to be see if it could detect source code parts where the obstacle had occurred. This failed software was the obstacle that was not detected during the validation test through the simulator, or during the operation at the actual railway sites for quite some time. The corresponding part was modified since the obstacle had recently occurred. In the case where the obstacle was drawn while operating the part for a long time that has not been filtered in the stages of existing conventional software development, validation and operation.

Railway signaling system software being recently developed was selected as the target for applicability test as another target (hereinafter referred to as 'B target system'). Development of this software started after studies on technology for testing railway signaling system software began in Korea. Although the results of the studies were in relation to coding standard, etc. (the preceding study was not applied directly), it was selected as an application target because it could find out whether results of the study were spreading in Korea to domestic industrial sites through demonstration, exhibition and paper presentation, etc. That is, we would like to check the effectiveness of the validation tool being developed and want to consider the effect of relevant studies on the railway industry through tests and analyses of the results from the software for these two systems.

Basic information on the target system is as shown in Table 5, and both of the two systems are VxWorks-based embedded software. Figure 6 shows the window under analysis after performing the static software test. As shown in this Figure, it can also be utilized for source code debugging since the list of violated coding standards and its content and location are indicated within the file. It is also made to indicate source codes violations, if a corresponding coding list is selected.

Table 5. Basic information on applicable target systems

	A Target System	B Target System
Language	C/C++	C/C++
Total number	22 (14 input files/8 include files)	5 (4 input files/1 include file)
Total number of lines	48,024 lines	3,626 lines
Number of comment line (%)	10,778 lines (22.44%)	654 lines (18.04%)
Number of function	517	42
Main function	Route control function	Route control function

4.1 Results of Applying a Target System

Like B target system, SWSIL 2 level was selected as the criteria of analysis on the result of application to tests, and if summarizing results of applying the test, they are as shown in the following Table 6.

In Table 6, the source code complexity among applied standards was based on the CC (Cyclomatic Complexity) presented by software engineering in Fewstar et al. (1999). The complexity of the function among source codes of railway signaling system software must be simple. In relation to it, Signore et al. (2014) and Zhu et al. (2014) are require to analyze the complexity of software, and the preceding study presents the complexity of 1~50 on the basis of satisfaction to SIL 2. Total number of functions among functions corresponding to test targets whose complexity exceeds 50, was 16 each, of which the biggest complexity was shown to be 221. With this, we could see that every function was implemented in a very complex manner. The software was developed by orienting it to the implementation of function, without consideration on the safety aspect of the software. The function that has a very high complexity like this, can implement its function sufficiently, but the complexity by function is limited to 50 in SIL 2 because due to its complexity there may be a potential error, which the developer does not know. Thus, although no obstacle has occurred for now, it is necessary to divide corresponding functions into several functions, or to implement them by changing them to the simple logic that has the identical function.

As a result of performing tests, it was analyzed that 2,310 items requiring improvements were detected; 86 coding standards each needed necessary improvements. The coding standard compliance rate was 39.4%, and the violation density against coding standard was 0.68. As a result of the analysis, there were many violated coding standards and a number of violations, and accordingly, the coding compliance rate was less than 50%. It seems that the software had been developed and used in Korea prior to the beginning of study on static testing technology for software in the aspect of safety. Especially, the part modified recently after occurrences of the obstacle that were detected by the static testing tool developed through this study, during the operation of the system.

This is the coding standard for 'difference in function declaration and definition'. Although it may not be a problem in a functional aspect in general, if the function declaration and definition of the corresponding function are different, malfunctions could occur due to the error in memory during the execution, according to the circumstance. Problems such as an obstacle in train operation, etc. caused by this part did occur while operating the actual system. It was confirmed that the obstacle could have been

Table 6. Result of applying static testing to A target system software

Information on Test Performance	Applied Standard	IEC 61508, IEC 62279, MISRA 2004, Complexity (CC)
Information on summary of result	Total number of items required to be improved	2,310(severity 'very high': 223, 'high': 208)
	Number of files including improved items	20 / 22
	Number of standards for which items to be improved were identified / Total number of standards	86 / 142 (severity 'very high': 7, 'high': 15, corresponding standards: 22)
	*Standard compliance rate	39.44%
	**Violation density against standard	0.68

* Standard compliance rate= ((Total number of standards - Number of violated standards) / Total number of standards) X 100 ** Violation density against standard = Number of violated standards / Valid line

sufficiently prevented if it was detected through the static testing process in advance. The part actually violated and the improved form shown in Table 7. As you see in this Table, an explicit return value must exist for the function whose return value is not 'void', and this is to prevent occurrence of return of the unintended value if an explicit return value is not existed. For example, when calling a 'calc' function, and not 'execute', the sentence that was intended by the result of return after calling the 'func' function, but the 'error' syntax will be performed.

This is the coding standard for 'difference in function declaration and definition'. Although it may not be a problem in a functional aspect in general, if the function declaration and definition of the corresponding function are different, malfunctions could occur due to the error in memory during the execution, according to the circumstance. Problems such as an obstacle in train operation, etc. caused by this part did occur while operating the actual system. It was confirmed that the obstacle could have been sufficiently prevented if it was detected through the static testing process in advance. The part actually violated and the improved form shown in Table 8. As you see in this Table, an explicit return value must exist for the function whose return value is not 'void', and this is to prevent occurrence of return of the unintended value if an explicit return value is not existed. For example, when calling a 'calc' function, and not 'execute', the sentence that was intended by the result of return after calling the 'func' function, but the 'error' syntax will be performed (Figure 7).

4.2 Results of Applying B Target System

SWSIL 2 level was selected as the criterion of analysis on the result of application to tests that are identical to that for A target system, and results of application to software tests are summarized Table 9.

Table 7. Violated form due to the absence of return value of function and improved form

Violated Form	Improved Form

int g_var = 0;
int func(int var) {
int ret = 1;
ret += var;
g_var = var;
}
int calc(void) {
int local=0;
local = func(0);
if(local > 0)
execute;
else
error;
}
	int g_var = 0;
	int func(int var) {
	int ret = 1;
	ret += var;
	g_var = var;
	return ret;
	}

Table 8. Violated form due to the inability of performing intended operations and improved form

Violated Form	Improved Form

while ((var == 0)&&(x != i++)) {
execute;
if(i > 1000)
break;
}
	while((var==0)&&(x!=i) {
	execute;
	i++;
	if (i >10000) break;
	}

Figure 7. Results of Screen of the static test result
(a) Severity ratio of violation (b) Violation rules ratio

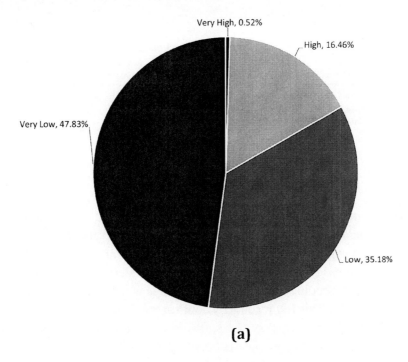

(a)

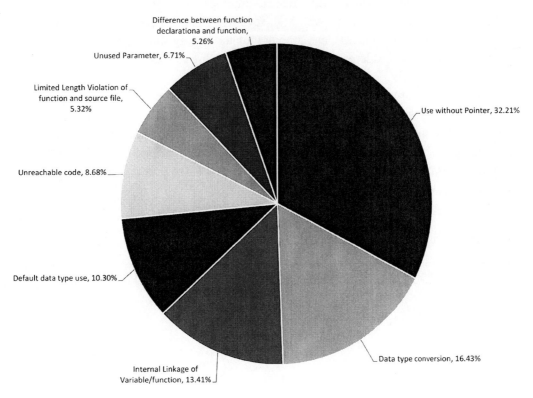

(b)

Table 9. Result of applying static testing to target B system software

Information on Test Performance	Applied Standard	IEC 61508, IEC 62279, MISRA 2004, Complexity(CC)
Information on summary of result	Total number of items required to be improved	98 ea. (severity 'very high': 3 ea. 'high': 95 ea.)
	Number of files including improved items	2 ea. (dynamic generation: 3 ea.)
	Number of standards for which items to be improved were identified / Total number of standards	33 / 146 ea.
	Standard compliance rate	77.40%
	Violation density against standard	0.19

CC, which is the same criterion applied to A target system, was selected as the complexity of source code among application standards in Table 8, and no item existed that exceeded the criteria since the biggest complexity was 41, among functions corresponding to the test target. As a result of performing tests, it was analyzed that 98 items requiring improvements were detected; coding standards necessary to be improved were 33 each. The coding standard compliance rate was 77.4%, and the violation density against coding standard was 0.19. This test result may confirm that B target system software complies with the coding standard better than A target system software as a whole, and the complexity of source code also satisfies the criteria required by the international standards.

Figure 8 shows the ratio of each severity level among improved items detected, and 4 violated items that take high importance among items with high severity. The difference in function declaration and definition, unreachable code standards, etc. among violated coding standards is identical to those in A target system. That is, it is considered that these parts are coding standards to which the software developer tends to make a mistake.

The violated contents of 'unreachable code' are MISRA 14.1; among these violated coding standards, and the case where corresponding parts were improved, are shown in Table 10. That is, in cases of code analyzed, corresponding coding standards were violated; the return sentence cannot be performed since the infinite loop is operated. In addition, since 'OK' will be returned unconditionally in the corresponding code, in case of exiting from the operation of infinite loop, it is the part that may become the obstacle factor of system.

Table 10. Violated form due to the unreachable code and improved form

Violated Form	Improved Form

// FOREVER: for(;;)
FOREVER {
switch(...) {
case ...:
case ...:
}
}
return(OK);
}
	// FOREVER: for(;;)
	FOREVER {
	switch(...) {
	case ...:
	case ...:
	}
	}
	}

Figure 8. Result of static testing to B target system
(b) Severity ratio of violation (b) Violation rules ratio

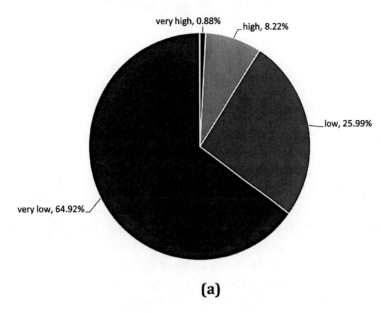

(a)

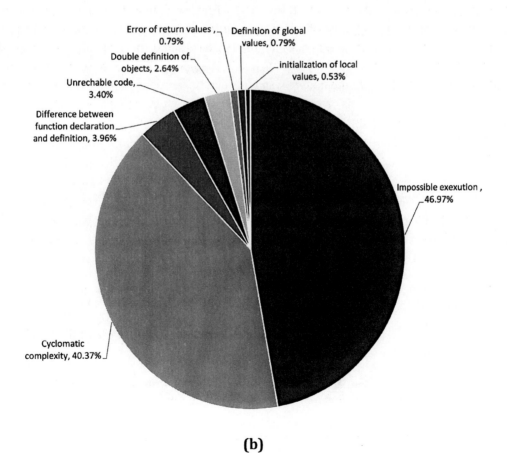

(b)

As a result of the application, several cases of violation to coding standards were detected, and as a result of checking these parts, it was confirmed that they were parts in violation of the coding standard. Figure 8 shows the main detection cases, and it shows an example that the pointer was used without inspection.

5. CONCLUSION

Recently, according to the development of computer technology, the dependency of railway signaling systems on computer software is being increased rapidly, and in accordance with this technical development, high levels of safety and reliability in the railway signaling system software is required. The necessity of automated testing tools for software coding rules for the railway signaling system is required to secure the safety of software, but the study in relation to the validation on safety of software is in the beginning stage. Related to this is the study of static testing tools, including derivation of the coding standards and validation on coding standards in connection with the validation of safe railway signaling system software, which began several years ago in Korea. As a part of this study, the applicability test through static testing tools that were developed for the railway signaling system software being operated in the domestic railway sites was performed. This chapter explained contents of development for automated testing tools in accordance with the coding rules for railway signaling system software. Basically from the standpoint of the assessor, the Automated Inspection Tool for coding rules for the railway signaling system software presented through this thesis may be utilized at the software development process to check if the software had observed coding rules required to the target software at the software assessment stage. That is, if this automated testing tool is used in the debugging process at the software development stage, the source code suitable for the coding standards can be developed, and it is anticipated that the software having higher safety levels can be secured through it.

REFERENCES

MISRA Coding Standard. (2004). MISRA (Motor Industry Software Reliability Association).

Del Signore, E., Giuliano, R., Mazzenga, F., & Petracca, M. (2014). On the suitability of public mobile networks for supporting train control/management systems. *Proceedings of the IEEE WCNC/14 conference* (pp. 3302 – 3307).

Fewstar, M., & Graham, D. (1999). *Software Testing Automation: Effective use of test execution tools.* ACM Press, Addison Wesley.

Fewstar, M., & Graham, D. (1999). *Software Testing Automation: Effective use of test execution tools.* ACM Press, Addison Wesley.

Hwang, J.-G., Cho, H.-J., & Kim, B.-H. (2013, May). Results of coding rules testing of train control system software. *Int'l Journal of Software Engineering and Its Applications, 7,* 249–259.

Hwang, J.-G., & Jo, H.-J. (June 2008). Development of automatic testing tool for design & coding standard for railway signal system software. *Proceedings of ICCMSE* 2008.

Hwang, J.-G., & Jo, H.-J. (2009). Development of automated testing tool for coding standard to check the safety of train control system software. *Journal of the Korean Society for Railway, 12*, 81–87.

Hwang, J.-G., & Jo, H.-J. (2009). Development of the automated metric analysis tool for train control system software. *Journal of the Korean Society for Railway, 12*, 450–456.

Hwang, J.-G., & Jo, H.-J. (2010). Deduction of Coding Rules and Implementation of Automatic Inspection Tool for Railway Signaling Software. *Proceedings of theInternational Congress on Railtransport Technology* (pp. 81-85).

Hwang, J.-G., Jo, H.-J., Kim, B.-H., & Baek, J.-H. (2013, April). Applicability analysis of software testing for actual operating railway software. *Proceedings of SoftTech* (Vol. *2013*, pp. 257–278).

Hwang, J.-G., Jo, H.-J., & Kim, H.-S. (2008). Design of the safety assessment tool for train control system software. *Journal of the Korean Society for Railway, 11*, 139–144.

IEC 61508-3. (1998). IEC.

IEC 62279. (2002). Railway Applications – software for Railway Control and Protection Systems.

IEC 62425. (2005). Railway Application: Communications, signaling and processing systems - Safety related electronic system for signaling.

Kadry, S. (2011). A New Proposed Technique to Improve Software Regression Testing Cost. *International Journal of Security and Its Applications, 5*, 45–48.

Lawrence, J. D. (2000). Software qualification in safety applications. *Reliability Engineering & System Safety, 70*(2), 167–184. doi:10.1016/S0951-8320(00)00055-7

Li Zhu, Yu F.R., Bing Ning, & Tao Tang. (2014). Communication-based train control (CBTC) systems with cooperative relaying: design and performance analysis. *IEEE Vehicular Technology, 63*, 2162-2172.

Rules regarding safety standards of railway facilities. (June 2011). Ordinance of the Ministry of Land, Transport Affairs.

Lutz, R.R. & Woodhouse, R.M. (1999). Bi-directional Analysis for Certification of Safety-Critical Software. Proceedings of 1st International Software Assurance Certification Conference. *Proceedings of the International Software Assurance Certification Conference* (pp. 1-9).

Railway Safety Act. (2006). *Act of the Ministry of Land, Transport Affairs.*

Rao; D.S., Disha Handa, Gaurav Bagga, Ajay Kumar Rangra, & Nandini Nayar. (2010). Extra-Organizational Systems: A Challenge to the Software Engineering Paradigm. *International Journal of Advanced Science and Technology, 20*, 25–42.

Yeom, H.-G., & Hwang, S.-M. (2009). A Study on Tool for supporting the Software Process Improvement based on ISO /IEC 15504. *International Journal of Software Engineering and Its Applications, 3*, 1–8.

Chapter 12
Automated Testing:
Higher Efficiency and Improved Quality of Testing Command, Control and Signaling Systems by Automation

Lennart Asbach
German Aerospace Center (DLR), Germany

Hardi Hungar
German Aerospace Center (DLR), Germany

Michael Meyer zu Hörste
German Aerospace Center (DLR), Germany

ABSTRACT

The need for time- and cost-efficient tests is highly relevant for state-of-the-art safety-related train control and rail traffic management systems. Those systems get increasingly more complex and so testing becomes a more and more and important cost factor. This chapter discusses some approaches to relocate tests from the field to the lab, reduce cost and duration while improving quality of lab tests. The European Train Control System (ETCS) is used as an example, but the approaches and results can be applied to other systems as well, for instance interlocking.

INTRODUCTION

Railway operation is the procedure to run trains on railway tracks. While the engine driver cannot oversee the full braking distance at higher speed those trains must be protected against passing a signal at danger or overspeeding by an automatic train protection system (ATP). Those safety-related systems in railways must be highly reliable and safe, which means that they cause hazards at most at an extremely low rate. The relevant standards mandate that safety and functionality of systems and components have to be proven. Therefore, they have to be tested comprehensively before being taken into operation. Tests are performed on a railway line or in a specifically equipped laboratory. Increasing the functionality

DOI: 10.4018/978-1-5225-0084-1.ch012

Automated Testing

of the systems while keeping the rate of hazards low leads to an increasing number and complexity of those tests. On the other hand the open, worldwide market requires to lower costs and to reduce the time to market. So the challenge is to improve the quality of the tests and reduce redundancy in them while speeding up and reducing cost. There are tests with different aims: they can be used to show that a system fulfils the relevant specification, the foreseen operational tasks and/or safety requirements. All these different tests need to be described and specified to be performed in the field or to be formalized to be executed in a lab. Field and lab testing require different levels of formal definition and description. The ideas presented here aim to use basically the same test case formalization and lab automation for different kinds of lab tests, and to relocate some kinds of field tests to the laboratory.

The specific objectives of this chapter are approaches for automation of different steps in the testing process: specifically tailored approaches can be applied to automate the generation, execution and evaluation of different kinds of lab tests. The generation can be done by e.g. software tools, and checking by web-based evaluations which implement suitable checking automations. The test automation requires real-time interaction, so the suitable approach is based on real-time event triggering and logging software as well as the use of tailored robots for manual inputs. The analysis can be automated by using software for automatic comparison and analysis of the logged data as well as report generators.

This chapter presents ways to improve the test process from automations of test campaigns to conceptual approaches which are still in an experimental stage. First, as an example of automation, it is shown how tests involving a graphical human machine interface can be performed without manual interaction, and how test results can be evaluated mechanically even in complicated cases. These two techniques have already been applied successfully. Following that, a prototype tool for constructing ETCS test sequences is presented, which addresses the problem of parameterizing generic test cases. And lastly, it is discussed how tests can be generated systematically and partly automatic from system specifications given in the form of semi-formal models. A full-scale application of the latter still constitutes a scientific challenge.

The structure of this chapter follows a shell-like approach. The innermost idea to improve the testing effort and duration is to execute tests as far as possible automatically. More or less in the same shell is the automation of the analysis of the test results. The next section discussing this aspect is presenting results on high readiness level. The next outer shell is the automation of the specification and documentation of the test spec. In the associated section a method and a tool are shown, which are available as a prototype. The outermost shell is a complete model-based system and test specification, which is still a scientific challenge. Current results and further work are discussed in the corresponding section.

BACKGROUND

Train Control Systems

The movement of trains needs to be supervised by technical systems, because the driver cannot oversee the full braking distance. So-called train control systems are used for this purpose. Historically they were designed to stop the train when passing a signal at danger. Today the functionality is much more complex and supervises at least the direction of travel, speed and stopping location, cf. - (Schön, Larraufie, Moens, & Poré, 2013). Typically the information needed to supervise the movement is transmitted to an on-board unit at least at the leading vehicle. State-of-the-art systems are using wireless communication for the transmission of the current control data to the trains. Hence those systems are called

"Communications-based train control systems - CBTC" (Ning, 2010). Typically those systems are using either spot transmission as e.g. transponders, beacons or continuous transmission as e.g. radio, coded track circuits or loop transmission. Some examples are listed in Table 1 and Table 2 (Meyer zu Hörste, 2004). The System which has the largest spread of the state of the art CBTC Systems is the European Train Control System ETCS. Hence this shall be used as the case study in the main part of this chapter.

While ETCS is the most widely used system in many countries in Europe as Spain, Switzerland, Italy, Hungary, Austria, The Netherlands, Denmark and Germany as well as outside of Europe as e.g. in Turkey and Mexico it should be used as case study in the further part of this chapter.

Case Study: The European Train Control System ETCS

The European Train Control System (ETCS) is currently taken into operation on more and more lines inside and outside of Europe. Aided by the fact that the ETCS specifications are openly available, the number of suppliers is increasing. This, and the authors' specific experience with testing ETCS systems is the motivation to use ETCS as application example here (Stanley & IRSE, 2011). A particular emphasis will be put on conformity tests according to ETCS subset 076 (UNISIG, 2015).

The European Rail Traffic Management System (ERTMS) consists of the safe communication system GSM-Railway (GSM-R) and the European Train Control System (ETCS). ERTMS/ETCS should implement technical as well as operational interoperability on the trans-European railway network. This means that a train equipped with an ETCS on-board system should be able run on every ETCS line.

Technically, the ETCS on-board unit (OBU) consists of a central vital computer system, a spot transmission system from trackside transponders called EuroBalises, the wireless communication system EuroRadio which uses the GSM-R, one or two Driver-Machine-Interfaces (DMI), a Juridical Recording Unit (JRU) and a train interface unit (TIU). In the ETCS application level 1, EuroBalises or short Balises are used to transmit the permission to run, given by a conventional wayside signal, to the train. In the ETCS application levels 2 and 3, movement authorities are sent by the so-called radio block center (RBC)

Table 1. Communication-based train control systems using spot transmission

Acronym	Transmission	Country
AWS (Automatic Warning System)	Magnets	UK
ASFA (Anuncio de Señales y Frenado Automático – Display of signals and automatic braking)	Balises	Spain
EBICAB	Balises	Sweden, Norway, Bulgaria, Portugal
ETCS (European Train Control System) Level 1	Balises	Multiple
KVB (Contrôle de vitesse par balises – Speed control by Balises)	Balises	France
PZB (Punktförmige Zugbeeinflussung – Spot transmission train protection)	Switchable electromagnets	Germany, Austria
TBL (Transmission Balise-Locomotive)	Balises	Belgium
TPWS (Train Protection and Warning System)	Balises	UK, Australia
ZUB 121	Transponders	Switzerland
ZUB 132	Balises	Denmark

Table 2. Communication-based train control systems using continuous transmission

Acronym	Transmission	Country
ATB (Automatische treinbeïnvloeding – Automatic train protection)	Coded track circuits	Netherlands
BACC (Blocco automatico a correnti codificate – Automatic block with coded currents)	Coded track circuits	Italy
ETCS (European Train Control System) Level 2 and 3	Radio	Level 2: Multiple, Level 3: Sweden
EVM (Elektronikus Vonat Megàllitò Rendszer – Electronic train stop system)	Coded track circuits	Hungary
LZB (Linienförmige Zugbeeinflussung – continuous train protection)	Loop	Germany, Austria
PTC (Positive Train Control)	Radio	USA
TVM (Transmission voie-machine – Transmission track – machine)	Coded track circuits	France, Belgium

to the train and shown on-board by the DMI to the driver. ETCS trains can also run on lines equipped with existing national train control systems. For that, so-called specific transmission modules (STM) translate information coming from the existing system into ETCS-telegrams.

Six different companies have committed themselves to implement ETCS and to provide products to the railways in Europe. This leads to the need to prove that their products fulfill the following high-level requirements:

1. They have been implemented according to the European system requirement specification – the so-called conformity,
2. They interact technically in the correct way – the so-called technical interoperability,
3. They fulfill together the required operational functionality – the so-called operational interoperability,
4. They are doing what the railway needs to perform the operational tasks – the so-called operational serviceability, and
5. Finally they are doing all this under all conditions in a safe way.

The European Directive L51/1 (2012) requires all countries of Europe to implement ERTMS/ETCS as the railway management and train control system by setting the technical specification of interoperability (TSI) for command, control and signaling (TSI CCS - (EC, 2012)) into force. The technical requirements as well as the procedural definitions to bring ETCS into operation are given in the TSI CCS.

Testing and Tests

The standard EN 50126 (ISO, 2000) regulates how to handle the aspects of reliability, availability maintainability and safety (RAMS) in the development of railway applications. It is complemented by the EN 50128 (ISO 03 2012) addressing specifically the software of electronic control and protection systems. They specify the process to be used and put an emphasis on verification and validation activities.

Both sources of requirements – the TSI CCS as well as the standards EN 50126 and EN 50128– name testing as a technique by which properties of a system are demonstrated. Each test has a specific goal, which underlies the way the test is specified and the observations made during the test are evaluated. Some examples for different classes tests which differ in their goals are (c.f. the requirements on ETCS equipment above):

- Testing of *conformity* of products: Here, the goal is to show that a product fulfills exhaustively the corresponding system specification. The specification coverage is checked by tracing the test cases back to the system spec.
- Operational tests of the *track engineering*: The goal here is to show that the engineering of the track layout and the related elements of the train control system are suitable to fulfil the given operational objective. Such tests should cover lists of regular or disturbed operational tasks. They are normally performed in simulation environments and only rarely take the form of a set of precisely defined test cases.
- Operational tests of integrated trackside and on-board subsystems as known as *interoperability* tests: Here the goal is to show that products work together in the intended way. In comparison to the conformity tests which are more a kind of verification those tests are more validations. The test coverage is quite similar to the operational tests of the track engineering but is done in a test environment similar to the conformity test.

Further classes of tests exist, e.g. safety tests, security tests, load test etc., which are not in the focus of this chapter.

The Current ETCS Test Specification and Execution Process

Conformity Tests

The main requirements for ETCS equipment is given by its System Requirements Specification (SRS) (UNISIG, 2013). Before the products built according to the SRS may go into operation, their conformity with the SRS and their interoperability has to be proven by lab tests. The conformity tests were specified as test cases, parametrized and concatenated to test sequences in the so-called ETCS subset 076. This specification includes the Hardware-in-the-loop tests of the ERTMS/ETCS on-board equipment (UNISIG, 2015), the test plan (UNISIG, 2013), method of their creation (UNISIG, 2015) and the method of test (UNISIG, 2013). The specification of the reference lab architecture (UNISIG, 2015) completes the set of specifications.

For the better understanding of the following sections these steps will be discussed shortly in the following Figure 1.

The defined target of test sequences is to test each requirement of the SRS at least once (UNISIG, 2013). Firstly, to manage the amount of requirements of the SRS, 650 features have been identified. After this, the required positive and negative test cases have been created for each feature. Totally 1276 test cases have been generated. Equivalent test cases for different ETCS modes or levels have been merged to reach a first optimization and reduction of the number of test cases. This means that test cases which are applicable for different mode-level combinations are described as only one test case, if the feature is not dedicated to a specific mode-level combination. Important is the testing of the feature

Figure 1. Relation between SRS and Test Sequence

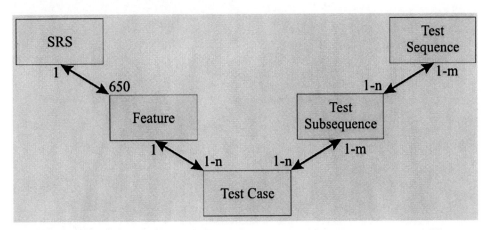

itself. Just for clarification, each ETCS mode is an operational state of the on-board equipment and the ETCS level is an overall degree of the usable functionality of ETCS. Due to the causal interaction of the test cases they are concatenated to test subsequences. Each test subsequence describes one mode or one level or one mode-level transition of the on-board equipment. For the execution of the tests the test subsequences are grouped into 62 test sequences, which all start at the powering on of the on-board equipment and end with the no power mode. The test subsequences are concatenated due to their start- and end-conditions to reach a consistent sequence of system states. The test sequences have been optimized to reduce the redundancy between them. Some subsequences do occur more than once, but only if that is needed to establish the preconditions of subsequences not otherwise covered. Up to the generation of test sequences, the specification is completely generic. At this stage the variables and parameters are filled with values, though some remain to be set dynamically during execution of the sequence. So the test sequences could be understood as operational test trips. The relation from the SRS up to the test sequence is shown in Figure 1.

Finally the fundamental structure of the test sequence should be clarified. Each test sequence simulates a test trip by stimulating the on-board equipment via the black-box-interfaces. In addition, the SRS-conformant reactions of the on-board equipment are defined in each test sequence. The reactions and the stimulating events are bound to the interface where they should be observed and evaluated or raised. Essentially the test sequences consist of the stimuli and the expected reactions of the on-board equipment. In the test sequence one stimulus or reaction is represented by a test step. Figure 2 shows the structure of a test sequence.

The test sequences contain up to several thousand test steps and their execution in real-time in the labs need up to several hours. A time rafting testing is not possible due to the fact that the real time behavior of the ETCS component is tested. As mentioned above the test sequences are implemented in the reference labs. Some of the test sequences have been executed successfully, but the stated problems of duration and unstable inputs on the user interface by the human being show that automation is needed. As soon as an input is missed or incorrect the complete test sequence must be repeated.

The tests defined by this method are documented in the ETCS subset 076 (UNISIG, 2015). These are used to proof the conformity of the constituent European Vital Computer (EVC) which is the core of the on-board unit.

Figure 2. Structure of a test sequence for conformity testing

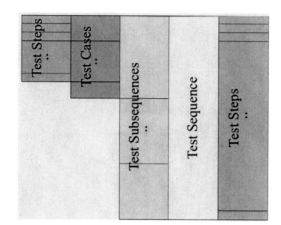

The complete equipment of a train is called an assembly. This contains all constituents as well as the specific engineering for the vehicle series. Basing on the conformity tests for constituents are interoperability tests for assemblies defined.

Interoperability and Operational Test Sequences

The conformity test sequences which have been discussed in the previous section fulfil the purpose to show that an application is realizing the specification sufficiently complete. They do not claim to be operationally reasonable. Thus, a railway undertaking tendering ERTMS/ETCS systems need to check whether these fulfil their operational requirements. These tests are a separate set of test sequences at the moment. They have to be defined by a similar methodical approach as show above, but they have to fulfil more requirements: The test sequences must represent the most typical or important scenarios of the operation of the railway. They need to show the fulfilment of the European requirements as well as the national add-ons.

Our approach is to use the test cases which have been defined for the conformity test for the operational test sequences as well. Some specific test steps and test cases are added to represent operational aspects which are not represented in the technical tests. The method for the generation of the test sequences is employed here, too. So the test cases consist out of technical and operational test steps. The test cases are concatenated to a realistic test sequence. This operational test sequence is formalized and filled with parameters according to the same rules as the conformity test sequences. Test sequences are looking quite similar to operational test sequences but describe specifically selected operational situations (Figure 3).

The main difference is that the definition of the test sequences is not optimized to fulfil the specification requirements using the shortest possible sequences. The optimization criteria are here to find as much relevant or important regular or disturbed operational scenarios to be tested.

The parameterization of the test sequences is done according to the operational environment. Average or standard parameters are typically used for this purpose. Extreme or rare parameters are to be avoided.

The main advantage of the described approach is that the basic database of the test cases as well as the testing environment in the lab which has been established for conformity tests can be used for operational tests or tests of interoperability, too.

Automated Testing

Figure 3. Structure of a test sequence for operational testing

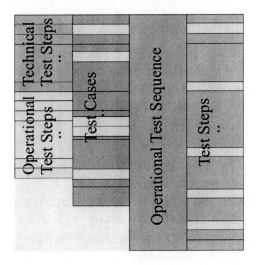

APPROACHES FOR TEST AUTOMATION

The Challenge of Test Effort, Cost and Duration

As already mentioned above, the aim is to reduce the effort and duration of the tests from specifying, execution up to the evaluation. Improving quality is another aim which is taken into account, too. Globally the approaches can be organized following the process of testing:

1. Automation of the test specification and test case generation
2. Automation of the test execution
3. Automation of the test analysis, evaluation and reporting

The three global steps are described more in detail in the following three sections.

Automation of Test Specification

Generally, in today's practice tests are generated manually with some software-support on an editorial level. The ETCS test specification has been developed over the last 14 years by a group of highly skilled engineers with a relatively low level of tool support.

The experience of test creation for several versions of ETCS shows the urgent need of special tools for editing test scenarios. For instance, the first approach to arrange test cases into executable sequences for the ETCS OBU, the Chinese postman algorithm has been used. This led to a stable test suite only after extensive work on parameter instantiation and calibration. Getting the timing, position and velocity parameters right proved to be rather difficult. Automatizing several of these steps seems possible, but requires dedicated procedures.

Advanced tools solving these problems should provide a beneficial interface between railway sector and test domain. A graphical user interface (GUI) should provide easy access to the test data for the railway expert. The logic in the background has to ensure formal correctness, adapt parametrizations and finally create a format which can directly read by a test bench. (Meyer zu Hörste, Jaschke, & Lemmer, 2003)

The tools which are in use today are providing only low level generation and syntax checking functionalities. Hence the quality of the test specification depends on the quality of the system specification. There is also need to apply formal methods to improve the quality of the entire specification, development and implementation process (Meyer zu Hörste, 2004).

Automation of Test Execution

The basic approach for the testing in a lab implements two kinds of simulated scenarios: The functional behavior is tested against the functional requirements without introducing disturbances. This kind of experiments can be used e. g. to test the conformity with the functional or technical system requirements. The second one is, that the behavior is tested under the real - i.e. disturbed - environmental influences against the appropriate safety and availability requirements. This kind of experiment can be used to test operational safety or availability. The stability and performance of operation concepts can be evaluated, too.

The relevant sources and typical modes of errors or disturbances have been studied carefully. Some important sources have been already identified: All modules or subsystems which perform a information transmission in a continuous or discrete way. Basically they have five types of errors:

1. Delay of a message
2. Change of the sequence of messages
3. Loss of a message
4. Change of message content
5. Sending of a wrong message

This behavior can be represented e. g. by fixed values calculated as a stochastic mean value or a stochastic value defined by a given distribution function. They can be independent from the environment or defined by location or time. These different behaviors will be represented by specific "Transmission Error Modules" which can be chosen related to the purpose of the experiment. Their implementation will start with the definition of generic modules, which fulfil the basic error modes. Two basic generic error-modules are radio communication errors and transponder errors. For the use of testing they will be parameterized, e. g. to represent the behavior of GSM-R or the EuroBalise. In the following steps of implementation these modules will be extended and refined (Meyer zu Hörste, Jaschke, & Lemmer, 2003).

Automation of Test Analysis, Evaluation and Reporting

To be able to make qualified assertions of standard conformance, several arguments have to be spelled out. On the one hand, the correctness and completeness, respectively sufficient coverage, of the test cases with respect to the specification has to be checked. This involves techniques and methods form the domain of model based testing. Currently, manually derived test suites are evaluated for their suitability. In future enhancements of the overall approach, also test case generation from the specification models is intended to be considered.

Adapter design and validation will have to cope with the common problems of crossing abstraction levels (namely atomicity and timing issues as well as value concretizations). For the internal-external adapter a monitoring concept which observes its operation dynamically is envisioned. The user interface of an interlocking system provides much information about internal states and thus qualifies as an adequate point of observation.

Reduction of Operational Field Tests

Operational field tests are done manually in general, so formalization as for lab tests required is not needed. But the definition of the operational sequence to be tested needs to be precise as well as the definition the reference results which is the basis for the evaluation.

An extension of the approach presented here can be used for the definition of the field test as well. All the information not needed for the tested are omitted. The test definition and the expected test results are given as a kind of check list. The advantage is here, that the basic database of the test cases can be used in a consistent way for both lab tests as well as field tests.

Solutions and Recommendations

While working for many years in the field of test automation the DLR has already developed a number of solutions. These are presented in this section.

The conformity tests take place today in one of the three independent Subset 076 test laboratories. One of them is located in Germany at the German Aerospace Centre (DLR) in Braunschweig. The two other ones are the labs of CEDEX in Spain and MULTITEL in Belgium.

The DLR Institute of Transportation Systems runs a rail laboratory for the research, testing and simulation of Train Traffic Control and Safety Systems, called RailSiTe® (Figure 4). The lab itself is a complex distributed system. It consists of more than ten computers, running up to 25 distinct software modules. Each module plays a part in a comprehensive simulation of the Railway system on train-side, trackside or air gap.

Automation of Test Specification with advanced methods:

This problem is the most challenging of the three which are described in the previous section. Ways to deal with it in practice are still under development. Some approaches are presented in the section on future research directions.

Automation of Test Execution in a fully automatic Test Bench

The first step for automation, which has been undertaken at the DLR was the automation of the test execution. Typically a test consists out of a number of Triggers to be sent to defined interfaces at defined moments. The resulting reactions of the system – at the same or other interfaces – are recorded and later compared with the expected reactions. The relation between the triggers among themselves as well as the reactions are defined by times e.g. "the drivers reactions must be recorded 5 seconds after the telegram has been received" or by distances as e.g. "400m after driving over the Balise the train has to stop". For the test execution the distance relations are converted into times by using the actual speed of the train.

The functional requirements for an On-Board reference test facility are defined in a specification standard published by the ERA (UNISIG, 2015). The test bench is structured in following way:

Figure 4. Structure of the test automation in the test bench

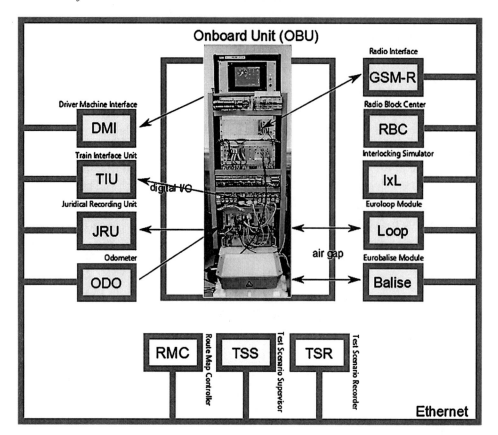

- The system under test (SuT) can be a real system or a simulated one. One of the simulated ones is the so-called Test Sequence Debugger, which is used to validate test sequences and test benches.
- All the real interfaces are connected to a tool which sends the stimuli and receives the reactions. Table 3 gives some more details about the interface tools.
- A central node controls all the interface tools.
- All triggers and reactions are time stamped and logged in database.

Figure 4 shows the general architecture of the test bench.

The central node called test bench controller tool has been realized in a way that all the different computers are remote controlled. This node knows how to start or even re-start a sequence and which software tools are needed. A list of test sequences can be pre-programed, too. By using this mechanism a 24/7 test execution can be realized.

The certification requirements define that the DMI has to be tested at the physical interface, so the inputs have to be done mechanically at the touchscreen or the soft-keys. Due to the same reason an optical recognition of the DMI events is required. To reach a high quality of the test and a high degree of automatic evaluation, the human operator providing manual input will be replaced by an industrial robot.

The robot will be developed for replacing the inputs of a driver, therefore enabling the test environment to be able to perform black box testing onboard units without human intervention. This will enable

Automated Testing

Table 3. Interface tools

Interface	Directionality	Implementation
Driver Machine Interface / DMI	Trigger and Reaction	Trigger: manual or by physical robot. Reaction: optical recorded by camera
Balise Transmission Module / BTM	Trigger and Reaction	Standardized Air gap
Loop Transmission Module / LTM	Trigger and Reaction	Standardized Air gap
Radio Transmission Module / RTM	Trigger and Reaction	Standardized Air gap
Specific Transmission Module / STM	Trigger and Reaction	Different e.g. Profibus
Train Interface Unit / TIU	Trigger and Reaction	Electrical contacts
Odometer Interface / ODO	Trigger only	Specific by supplier
Juridical Recording Unit / JRU	Reaction only	Standardized download tool interface

tests to be run 24hs, therefore leaving the time of ERTMS experts for the analysis of results and not on reproducing manually data entry in the tested equipment.

The robot consists of a number of separate parts:

- A rigid standing frame, which can be mounted physically to the system under test
- A two-dimensional positioning unit which positions the actuator at the right position in front of the DMI and obscures the visual area of the cameras as short as possible
- An actuator which is giving the physical inputs to the DMI with the correct pressure
- Power supply for the positioning unit and the actuator

One of the most demanding problems to be solved in the hardware-development of the robot is to define a frame and positioning unit which is sufficiently rigid to enable precise operation and on the other side light enough to be fast out of the visual area of the cameras. Figure5 shows the version 2.0 of the DLR DMI robot.

Based on the coordinates, actions and delays gathered by the DMI event recorder, a software kit will be developed in order to automate the control of the robot for each test run and DMI. This kit will allow converting the identified DMI data entry to robot control commands (actions) required to let it interact autonomously with the DMI. This interaction will also integrate a feedback response as soon as the action on the DMI is achieved. In other words, an action will be considered as a sequence of movements to be performed by the robot and will end with a feedback (acknowledge) sent by the robot module. During the movement, the stereo acquisition module will be used to detect the change on the DMI screen so as to inform the robot module whether the action is performed correctly or not.

Automation of Test Analysis, Evaluation and Reporting

The test specification requires a touch display or a softkey version as the interface between the driver and the machine. There is no option for a digital version of this interface, e.g. using the bus between the display and the core computer for stimulation. Thus a robot was installed to perform all necessary driver inputs during the test execution. Using a robot for inputs is a very good test for the grade of formaliza-

Figure 5. DMI robot

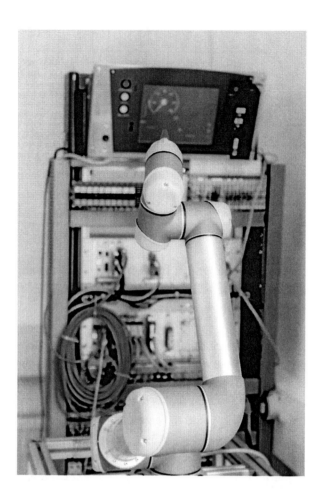

tion of the test description, but if something is not formal this will not work. This formalization is the mapping between the required inputs of the driver and a certain x,y and z coordinate on the touch screen display, depending on the mounting of the DMI. Accordingly all DMI input signals were categorized and formalized. Every category represents one input procedure on the DMI. The robot control software translates this procedure into a sequence of keystrokes, i.e. into concrete coordinates, and executes them. The architecture of the robot software is based on the layers for the event description.

A further configuration file keeps the geometric dimensions and the position of the display in relation to the robot. The coordinates needed for this configuration can be acquired by moving the robot manually to three points of the display. Every additional calculation is done automatically and the robot is ready for action. The formalization was done based on the test cases of Subset-076. I.e. the level of formalization was kept as abstract as possible. Thus the formalization can be reused for new scenarios which are based on the same test cases. Only a few events are depending on the specific context. Therefore only those has to be reviewed when they appear in new scenarios.

As an example, Figure 6 shows the Testscenario-Editor of the DLR. It uses "Drag and Drop" to add features to the scenario. All signals relevant for the test bench are added automatically. The root database is the subset-076 database of test cases; here it is used as a kind of definition of the ETCS language. The

Automated Testing

Figure 6. Screen-shot of the TS-Edit

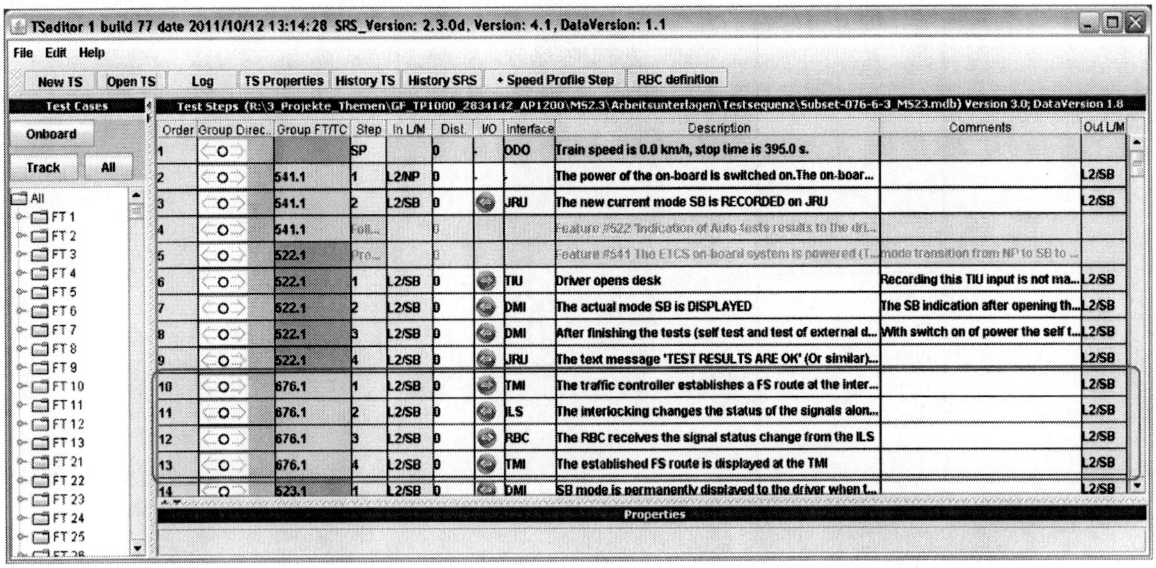

scenario section describes the single steps which have to be performed at a certain distance. Of course, the GUI still looks to be a little bit test benches affine, an interface based on a track layout, supporting a more intuitive way of scenario generation is part of an ongoing research project.

A specific feature of the editor is its support for multi-user editing. I.e. multiple users can create and edit test cases and test scenarios at the same time. This is particularly important if many partners are working in parallel on one set of scenarios. All versions and all changes are recorded to allow traceability of the complete process at any time.

On demand, scenarios can be exported to pdf (or other formats) to create a legal base. The editor is not limited to ETCS, the system under test is defined by the root database.

One specific challenge appeared in the process of the specification of ETCS tests in Europe: The teams specifying the tests are distributed over Europe and they need to evaluate whether their test sequences are correct. The DLR provided a worldwide unique tool for the test sequence evaluation: A web-based automatic test bench. The idea of this web-based tool is to have a fully automatic test bench at a remote location which can automatically execute newly generated test sequences. Using today's web-technologies this can be done by the automatic test bench and the Test Sequences Debugging (TSD) tool automatically, too. The only major difference is that a dedicated, secure storing area has to be implemented between the internet and the test bench (Figure 7).

The process of automatic evaluation works in the following way:

1. An editor of a test sequence copies the final test sequence in the standardized format in a dedicated web-space (in this case a DLR-TeamSite) in a defined folder
2. The following night, the test sequence is transferred into the automatic test bench
3. The test sequence is executed using the Test Sequence Debugging tool
4. The log-files are created in a defined format, which allows to track back the observations to the test sequence
5. The log-files are transferred back into the same folder on the TeamSite, where the TS is located

Figure 7. Structure of the test automation in the web test bench

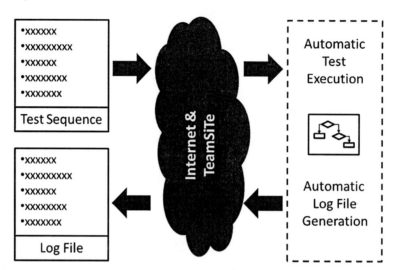

So at the next morning the editor can check the results which are created by his TS. The major advantage of this tool and procedure is that an editor of a test sequence does not need to spend working time in the laboratory for evaluating his sequence. The results are faster generation and evaluation of test sequences as well as reductions in cost.

FUTURE RESEARCH DIRECTIONS

The Benefits of Formalizing Specifications

Further improvements of testing depend on enhancements of the development process as a whole. As has become apparent in the previous section, if some step is to be automatized, all relevant input to the step has to be available in a form accessible to mechanical analysis. And if the derivation of tests shall be (partly) automated, first the specifications to be tested have to be formalized. That means, they must have a somewhat formal semantics, i.e., one must apply formal methods.

Albeit slowly, the use of formal methods in the development of rail equipment does increase.

Formalizing specifications has a benefit in itself, as ambiguities, omissions and inconsistencies are reduced when a specification is formulated in a rigorous or nearly rigorous notation. An example is the ongoing development of formalizing the ERTMS/ETCS SRS in the project openETCS (Project-Team, 2014).

Besides the effect of improving the quality, a formalized specification is a good basis for many further process enhancements. With respect to testing, formalized requirements permit at the very least a systematic derivation of test cases, and a sound coverage analysis. In favorable scenarios, test cases can even be generated automatically. This will be presented in more detail in this section. What is shown here relies on the formalization of specifications by executable models.

Automated Testing

Application Examples

One example where such models are used is the already mentioned project openETCS. It is concerned with establishing a tool chain for model-based development including verification, and applies the methods to develop software for the ETCS on-board unit. Another example, closer to practice, is the specification of interfaces of track-side equipment. What is different with track-side equipment compared to ETCS is the lesser degree of standardization: Interfaces and even functional architectures may differ, depending on the manufacturer. To improve compatibility, an approach currently employed by some infrastructure operators in the EULYNX initiative is to specify the interface behavior of equipment components. In both application examples, tests are generated from models and applied to implementations.

The examples for models used in this section are taken from the EULYNX application domain. Besides the introduction of model-based techniques, also another particularity of that application is interesting. The specification of interface behavior is done incrementally, i.e., only some of the interfaces of an interlocking system are specified (and shall be tested), while others remain to be considered some time in the future. The reason is that it is easier to specify the way an interlocking communicates with an RBC, or another interlocking, or a point machine controller, than to specify the functional behavior of an interlocking completely. And also for the manufacturers it is easier to implement a standard version of a specified *focus interface*, and not having to come up with a re-implementation of the full system. Though it is intended to have all interfaces standardized at some point in time, the incremental approach makes it possible to have practical improvements realized fast.

But there is a downside, too. When it comes to testing, the incremental specifications pose a severe problem. To drive the focus interface (and observe the correct interpretation of messages received over it), it is usually necessary to have access to (all the) other interfaces. Since not all of them are specified, the test cannot be specified completely, either. In the specifications, this incompleteness is escaped by introducing a *virtual* internal interface. This virtual interface subsumes everything which addresses the other, unspecified interfaces and also what goes on internally in the system. These virtual communications are not formally related to observables of the system. Somewhat descriptive names are used instead and left to be interpreted by the system implementers and testers. Practical testing has then to bridge the gap from the accessible interfaces to the virtual one from the specification. For that, the manufacturer has to provide the specifics of the system under test.

Despite this principal difficulty with an incremental specification style, model-based techniques are applicable. In fact, any other testing approach would face similar obstacles. And the EULYNX approach has already begun to be put into practice by the German Railways infrastructure branch.

The rest of this section presents a more detailed view of model-based development and the introduction of model-based testing techniques.

Model-Based Development and Testing

In general, there may be models of different kinds, capturing various aspects like timing, cost, power consumptions or other. For the use in testing, modeling the functionality is most helpful. If the model behavior is in a well-defined relation to the intended behavior of the system, the model can act as a reference to check the later implementation. By employing test generation procedures to such models, substantial parts of the necessary test derivation can be done by tools. The implementation is checked for conformity with the specification by applying test suites which cover the specification model (Weißleder & Schlingloff, 2011; Peleska, 2013).

If such an approach is taken, the overall effort is shifted from late phases like test case writing and error correction to the earlier phase of system specification. It should result in an overall reduction of cost and time, and should also increase the reliability of a complex combination of systems like ETCS.

Principles of Model-Based Testing: Modeling Domain

In present technical realizations of model-based testing (MBT), it is essential that the models are operational. "Operational"-in contrast to "descriptive"- means, that these models are executable (Weißleder & Schlingloff, 2011). They should be able to produce a behavior. This could be done by simulating them or by translating them to code. Thus, such models are not very different from programs. They are, however, not intended to replace the implementation or become part of it. Ideally, they should capture the essence of the system behavior and not its details. Things like message coding, computation efficiency, redundancy and hardware specifics should be left out. The models should be as abstract as possible while keeping a well-defined relation to the real behavior. I.e., the model should abstract from the concrete implementation.

In cases, it may be a challenging task to find a suitable abstract level. It must be both rich enough to enable a program-like representation of the system, and at the same time it must be simple enough so that the model is a specification that can be written, read and serve as a means of communication.

Often, specification models are defined in a graphical automaton-like format. UML/SysML state machines or activity diagrams are popular instances (Papyrus SysML editor, 2014). These have a finite, graphically represented control structure consisting of states/activities and transitions. Triggers, conditions and data operations are added in textual form and may employ a program-like notation.

Figure 8 gives an example of a part of a specification automaton. This is a SysML model of a component of an RBC which stores the state of a switch on which it is informed by messages from an interlocking to which the RBC is connected. These messages either refer directly to a particular field element ("Msg_S1" for the switch "S1"), or to a group of field elements ("Msg_G", where "S1" belong to the group "1"). Receiving a group message means that the interlocking has lost contact to the elements belonging to that group.

Test Generation

If the specification is given in the form of an executable model instead of a list of single requirements as in an ordinary specification, the notion of conformance has to be adapted. A list of requirements is satisfied, if each single entry in the list is covered. An implementation is conformant to a model, if the behavior resulting from the combination of the model constituents is the abstract image of the implementation behavior. This means, that a set of tests has to be constructed on the basis of the model which check this relation thoroughly. For a given testing problem, a precise meaning of "thoroughly" has to be fixed.

Widely-used criteria resemble coverage notions known from program testing. *Statement* or *branch cover* of a program translate, roughly, to *transition cover* of a state machine. It is the task of test generation, to identify executable paths of a model and produce test cases which drive the model so that, e.g., every transition of the model is fired by at least one test case. An implementation passes the tests, if it behaves like the model in all these cases.

Automated Testing

Figure 8. SysML model of an RBC component storing the state of a switch

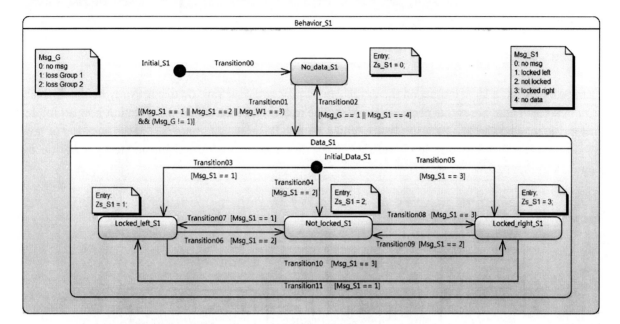

There are tools, even commercial ones, available which can analyze models and produce such test suites. This is a very difficult task. At its core, complex semantical and logical procedures are needed which compute stimuli to execute paths. In its practical application, it is also necessary to restrict the form of models to achieve a high success rate. Despite these restrictions and difficulties, even an incomplete test suite can be used. It is always possible to invest manual effort to complete a test suite.

Test Execution

Tests generated from an abstract, executable model can serve as a test specification, for verifying or validating a design artifact at a later stage. These test cases come with a complete description of the stimuli and success criteria. In simple cases, these cases can be directly applied to the test object. Most test generation tools offer support for test execution environments or are even integrated into one (e.g. the RT-Tester tool suite).

Depending on the kind of the abstraction relation between model and test object, these generated tests need not be directly applicable.

- **Interface:** Types or granularity of communications may differ. A solution, which is often applicable, inserts adapters. Complicated situations require to generate new data for the test cases.
- **Timing:** In particular in real-time systems, the timing of outputs of the test object may differ from that of the model. Permitting slack is a simple, yet sometimes too coarse remedy.
- **Value:** Numeric computations need not lead to the same result in model and test object. As with timing, one may weaken the acceptance criterion by allowing imprecision.

From these remarks it should be clear that the integration of MBT in a development should be planned with care.

Process Integration

Using models has a large impact on the development process, which we do discuss only in its relevance w.r.t. testing. There are two aspects to consider. The text above focused on the question how an implementation may be checked for compliance with a model. Besides that, there is also the question of how a model serving as a specification is verified. These usage scenarios for models and test generation are depicted in Figure 9.

Figure 9. Usage Scenarios for MBT

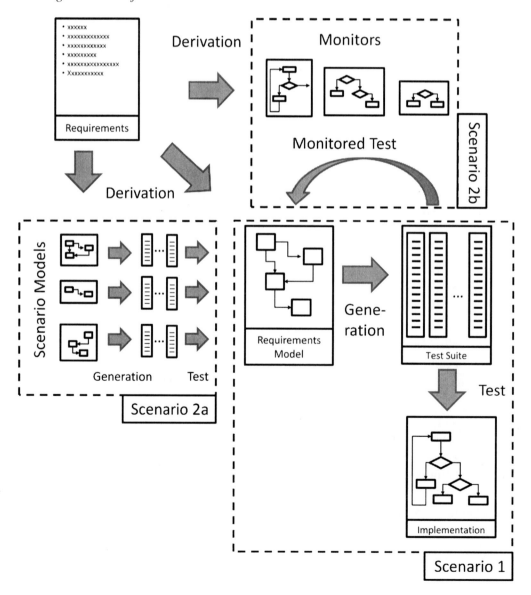

Automated Testing

The first aspect (Scenario 1), deriving tests to check an implementation for conformance with a model, constitutes the standard usage scenario of MBT. It fits smoothly into prevalent processes. The main tasks for integrating it are:

- Selecting a suitable technical definition of the test completeness criterion (e.g., transition cover) and demonstrating its adequacy
- Providing evidence that the test criterion is reached
- Making the test suite executable (see above)
- Defining and implementing change management for the tests. If the model has to be changed after the test suite has been generated from it, it might be difficult to make incremental changes to an automatically generated test suite.

The second aspect concerns the verification and tracing of the model in relation to higher-level specifications. The EN 50128 (ISO, 2012) says additionally that a requirement specification (for the software of a generic system, Section 7.2.2.4) has to be checked to be "complete, clear, precise, unequivocal, verifiable, testable, maintainable and feasible".

Not every model satisfies this. If it is executable, it is in some way consistent (*feasible, unequivocal*) and *complete*. For *compliance* and *tracing*, testing may be employed. One can test a model for required properties. Also for that task, automatic test generation can help. Deviating from the standard usage of such methods described above, there is a priori no model available against which one could test.

Either targeted test models are to be written which represent properties from the requirements in an operational form (Scenario 2a). Then, running tests generated from these targeted models can verify the respective property in the executable model, which plays the role of an implementation in that application of MBT. Figure 10 shows an example of a scenario model. It tests two transitions of the switch state controller model from *Figure 8*.[1] Or one generates a suite of tests which covers the model to be checked, and monitors for the satisfaction of requirement properties while executing the tests from the suite (Scenario 2b). This requires to derive monitors from the requirement specification.

Figure 10. Test scenario for switch controller model

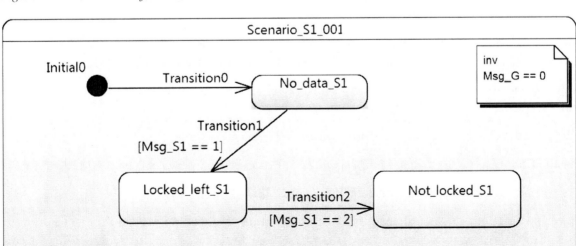

CONCLUSION

In current practice, highly automated, mechanically evaluated tests need strictly formal test cases which are unambiguously defined. If these preconditions are not met, the automation effort is high, so that it is often better to only aim for a partial solution. To get test cases of high quality, the choice of tools used in defining them is crucial: The tools must restrict the engineer to prevent unclear definitions being introduced. They must enable the domain expert to write test cases which can readily be translated for use in the test bench.

The DLR presented a number of solutions for evolution of test automation. This begins with an editor called TS-Edit for supporting the test specification and debugging. For the 24/7 automation of the test execution a specific robot and camera system has been set up to replace the human being as tester. A worldwide unique tool is used for the web-based validation of test sequences. This tool allows performing a test by sending the test sequence to a fully automated test bench via internet.

In the future, it is to be expected that model based techniques will enable considerable savings in the phases of defining, performing and evaluating tests. The use of such techniques, though, still has to be explored. And it is to be expected that specific adaptations are to be made to make them suitable for the railway sector.

The model-based testing will open new options for improving the testing. The generation of tests from formal models will reduce cost as well as improve the quality of the tests.

REFERENCES

EC. (2012, 01 25). 2012/88/EU. Commission decision of 25 January 2012 on the technical specification for interoperability relating to the control-command and signalling subsystems of the trans-European rail system. Brussels, Belgium: TSI CCS.

ISO. (2000). *EN 50126. Railway applications - The specification and demonstration of reliability, availability, maintainability and safety.* RAMS.

ISO. (03 2012). EN 50128. *Railway applications - Communication, signalling and processing systems - Software for railway control and protection systems.*

Meyer zu Hörste, M. (2004). *Methodische Analyse and generische Modellierung von Eisenbahnleit- und –sicherungssystemen [Dissertation].* VDI Fortschritt Berichte, Dusseldorf, Germany. (In German).

Meyer zu Hörste, M., Jaschke, K., & Lemmer, K. (2003). A test facility for ERTMS/ETCS conformity. In E. Schnieder, & G. Tarnai (Eds.), FORMS 2003.

Ning, B. (2010). *Advanced Train Control. Wessex.* WIT Press.

openETCS Project-Team. (2014, 12 12). *openETCS.* Retrieved from www.openetcs.org

Papyrus SysML editor. (2014). Retrieved http://eclipse.org/papyrus

Peleska, J. (2013). Industrial-Strength Model-Based Testing - State of the Art and Current Challenges. In A. K. Petrenko, & H. Schlingloff, (Eeds.), *Proceedings Eighth Workshop on Model-Based Testing, Rome, Italy, 17th March 2013. Electronic Proceedings in Theoretical Computer Science* (S. 3-28).

Schön, W., Larraufie, G., Moens, G., & Poré, J. (2013). *Railway Signalling and Automation* (Vol. 2). Paris: La Vie du Rail.

Schön, W., Larraufie, G., Moens, G., & Poré, J. (2013). *Railway Signalling and Automation* (Vol. 3). Paris: La Vie Du Rail.

Schön, W., Larraufie, G., Moens, G., & Poré, J. (2014). Railway Signalling and Automation (Vol. 1). Paris: La Vie du Rail.

Stanley, P., & IRSE (Eds.). (2011). *ETCS for Engineers.* Hamburg, Germany: Eurailpress / DVV Media Group.

UNISIG. (2013). ETCS Subset 026. ETCS System Requirement Specification (SRS) (Issue 3.3.0). (ERA, Ed.) Lille, Frankreich: ERA.

UNISIG. (2013). ETCS Subset 076-0. Test Plan (Issue 2.3.3). (ERA, Ed.) Lille, France: ERA.

UNISIG. (2013, 11 22). ETCS Subset 076-3. Methodology of Testing (Issue 2.3.11). (ERA, Ed.) Lille, France: ERA.

UNISIG. (2015). ETCS Subset 076-6-3. Test Sequences(Issue 3.0.0). (ERA, Ed.) Lille, Frankreich: ERA.

UNISIG. (2015). ETCS Subset 076-7. Test Sequence Validation and Evaluation / Scope of the Test. (ERA, Ed.) Lille, France: ERA.

UNISIG. (2015). ETCS Subset 094. Functional Requirements for an On-Board reference Test Facilitiy(Issue 3.0.0), 3.0.0. (ERA, Ed.) Lille, France: ERA.

Weißleder, S., & Schlingloff, H. (2011). Automatic Model-Based Test Generation from UML State Machines. In J. Zander, I. Schieferdecker, & P. J. Mosterman (Eds.), *Model-Based Testing for Embedded Systems* (pp. 77–109). Oxon, UK: CRC Press. doi:10.1201/b11321-5

KEY TERMS AND DEFINITIONS

Balise: A transponder specified by European standards which is placed between the rails and is used by train control systems to mark a position and to transmit train control information to the trains.

Conformity: A system has been implemented according to the European system requirement specification.

European Railway Agency (ERA): An Agency of the European Commission responsible among others for standardizing the European Train Control System ETCS. It is the so-called "system owner" of ETCS.

Interoperability: Subsystems provided by different manufacturers are working together in the intended operational manner.

Operational Interoperability: The two subsystems of a train control system (e.g. ETCS) fulfill together the operational functionality.

Safety: The property of a system to react in a way that no risks are created.

Serviceability: The train control system fulfil the operational need of the railway.

Technical Interoperability: The trackside and onboard subsystems of a train control ETCS interact technically in the correct way.

Validation: Proof of the operational usability.

Verification: Proof of the correctness and consistency between two development steps.

ENDNOTES

[1] The graphics in Figure 10 does not show the complete information of the scenario specification. The comment specifying "Msg_G == 0" as an invariant adds part of additional behavioral assumptions.

Chapter 13
Alertness Monitoring System for Vehicle Drivers using Physiological Signals

Anwesha Sengupta
Indian Institute of Technology Kharagpur, India

Anjith George
Indian Institute of Technology Kharagpur, India

Anirban Dasgupta
Indian Institute of Technology Kharagpur, India

Aritra Chaudhuri
Indian Institute of Technology Kharagpur, India

Bibek Kabi
Indian Institute of Technology Kharagpur, India

Aurobinda Routray
Indian Institute of Technology Kharagpur, India

ABSTRACT

The present chapter deals with the development of a robust real-time embedded system which can detect the level of drowsiness in automotive and locomotive drivers based on ocular images and speech signals of the driver. The system has been cross-validated using Electroencephalogram (EEG) as well as Psychomotor response tests. A ratio based on eyelid closure rates called PERcentage of eyelid CLOSure (PERCLOS) using Principal Component Analysis (PCA) and Support Vector Machine (SVM) is employed to determine the state of drowsiness. Besides, the voiced-to-unvoiced speech ratio has also been used. Source localization and synchronization of EEG signals have been employed for detection of various brain stages during various stages of fatigue and cross-validating the algorithms based in image and speech data. The synchronization has been represented in terms of a complex network and the parameters of the network have been used to trace the change in fatigue of sleep-deprived subjects. In addition, subjective feedback has also been obtained.

INTRODUCTION

Driver fatigue is one of the major causes of road and rail accidents across the world that often result in death or serious injury. Crashes caused by tired drivers are most likely to happen when driving happens in long and monotonous roads, or when the driving schedule interferes with the circadian rhythm of the driver, or when the driver has had poor quality or quantity of sleep. Alcohol or certain medications have also been pointed out as probable causes.

DOI: 10.4018/978-1-5225-0084-1.ch013

Railway accidents are also a potential threat to the safety and lives of passengers. Besides, disregard of signal by locomotive drivers at level crossing gates can endanger the lives of road users. Increase in rail traffic leads, congestion of railway routes necessitating frequent interaction of train drivers with the traffic controllers, increase of locomotive speeds etc. has emphasized the need for train drivers to stay alert at all times. Reports indicate that of all the accidents that pose a threat to railway safety and management, many can be attributed to human errors, including performance of the driver.

Lack of adequate sleep can lead to depreciation in mood (irritability, anxiety, lack of motivation), health (increased blood pressure and increased risk of heart attack) and performance (lack of concentration, drop in attention/ vigilance, increased reaction time). Hence sleepiness impairs the ability to execute attention-based activities such as driving.

Alertness level in human beings can be assessed using different measures such as Electroencephalogram (EEG)(M. Bundele & Banerjee, 2009; Cajochen, Zeitzer, Czeisler, & Dijk, 2000), blood samples(Penetar, McCann, & Thorne, 1993), ocular features(Arief, Purwanto, Pramadihanto, Sato, & Minato, 2009), speech signals(Dhupati, Kar, Rajaguru, & Routray, 2010)and skin conductance(M. Bundele & Banerjee, 2009). Methods based on EEG signals and blood samples have been reported to be most accurate for estimating the state of drowsiness (Cajochen et al., 2000; Matousek & Petersen, 1983). However, these methods are contact-based and have restricted feasibility of implementation in practical scenarios. There have been some studies (Hanowski & Bowman, 2008; Inchingolo & Spanio, 1985; Smith, Shah, & da Vitoria Lobo, 2003; Wierwille, Wreggit, Kirn, Ellsworth, & Fairbanks, 1984; Zhu & Ji, 2004) with regard to eyelid movements such as blink frequency, Average Eye-Closure Speed (AECS), PERCLOS, eye saccade etc. as quantitative measures of the alertness level of an individual. Of them, PERCLOS is reported to be the best and most robust measure for drowsiness detection (Bowman, Schaudt, & Hanowski, 2012; Hammoud, Witt, Dufour, Wilhelm, & Newman, 2008) whereas saccadic movements are reported to be the best indicator of the alertness level (Ueno & Uchikawa, 2004). PERCLOS is based on eye closure rates whereas Saccadic Rate (SR) is based on fast eye movements.

BACKGROUND

Fatigue and Its Hazards

With the development of technology and improvement of worldwide transportation systems, road accidents have become a huge cause for untoward loss of human lives. According to WHO statistics, every year across the world there are more than 120 million deaths due to traffic accidents, along with millions of people being injured or disabled. Global annual economic losses caused by traffic accidents have been estimated to be in the order of 518 billion dollars, of which the losses in developing countries account for nearly 100 billion dollars.

Fatigue in drivers has been widely reported as the cause behind 20-30% of road accidents across the world ("Driver Fatigue And Road Accidents," 1996; "Fatigue is a Major Cause in Truck Crashes," 1998; Saroj K.L. Lal & Craig, 2001). Fatigue has also been pointed out as a major contributing factor behind air crashes ("Airplane crashes are very often caused by fatigue I L.A. Fuel," n. d.) and locomotives. Since fatigue in drivers often prove fatal and can lead to loss or damage of lives and property, it is important that fatigue in drivers be detected early, so that appropriate countermeasures can be designed and employed. Fatigue in drivers is also one of the major causes of fatal accidents at air and sea. On an average, greater than 3000 train accidents occur annually in the United States, causing death, injury and damage to property (Jackson & Earl, 2006).

Accidents taking place in the railway system includes ignoring a signal (without involving substantial material damage) to a crash between trains. the cognitive loading on the train driver is caused by the mental processing of all information that the driver gets to see and hear, including taking note of signals, checking the speedometer for speed and stopping at the right stations following the timetable. Hence the attention of the driver is continually divided between driving the train (primary task) and taking note of information from the surroundings (secondary task) (People and Rail Systems: Human Factors at the Heart of the Railway, 2007).

A study of train accidents over a three-year period within an Australian public rail authority pointed out sustained attention as the most significant human factor contributing to all types of rail mishaps, including inattentiveness to railway signals (Edkins & Pollock, 1997). A locomotive driver's job necessitates psychological and cognitive skills such as the ability to stake awake, to remain attentive and to be able to cope with fatigue. Acute sleep deprivation and need for recovery has an adverse effect on a considerable number of train drivers (Zoer, Sluiter, & Frings-Dresen, 2014). Recent research involving heart-rate recordings from British railway drivers during normal duty runs has revealed the existence of fatigue in locomotive drivers in spite of the fact that driving modern locomotives does not involve the use of much physical energy (Grant, 1971). Such fatigue is likely to arise from monotony caused by prolonged inaction coupled with stress.

A study of a procedure for early detection of individual psychological deficits adversely influencing cognitive driving abilities in train drivers involving performance on attention and memory tests followed by extended psychological examination (De Valck, Elke; Smeekens, Ludo; Vantrappen, 2015) pointed out sleep disorders as one of the major causes of drop in cognitive abilities of train drivers. An attempt was made to investigate systems to detect and quantify the signs of fatigue due to induced sleep deprivation using non-invasive facial analytics combining color skin segmentation technique, connected component of binary image usage, and learning machine algorithm (Ibrahim, 2014). A method for detection of the early signs of fatigue in train drivers has been described in (Gulhane & Mohod, 2014) which is designed to send off an alarm to the control room as soon as symptoms of fatigue are detected in the driver. A real-time fatigue detection and alarm system for drivers has been proposed in (Abbood, Al-Nuaimy, Al-Ataby, Salem, & AlZubi, 2014) using physiological signs as such as pupil response, gaze patterns, steering reaction and EEG to build a behavioral model to determine the trend of growth of fatigue with time.

Fatigue has been classified as physical and mental fatigue; mental fatigue is believed to be psychological in nature and physical fatigue is taken to be synonymous to muscle fatigue. Mental fatigue is taken to be associated with unwillingness for any effort, reduced efficiency and alertness and impaired mental performance. Factors influencing mental fatigue include nutrition, physical health, environment, physical activity, and recuperation periods. Mental fatigue is more of a functional state, which grades in one direction into sleep, and in the other direction into a relaxed, restful condition, both of which are likely to diminish attention and alertness.

Muscular fatigue is the phenomenon of reduction in performance of a muscle after stress. Physical fatigue has been linked with decline in alertness, mental concentration and motivation, reduction in work output, weaker and slower muscular contractions, muscular tremor and localized pain. Hard physical work before driving can also increase the risks of experiencing fatigue during driving. Since fatigue affects attention and performance, it is important to consider 'vigilance' with which it overlaps. In general, vigilance refers to a general state of wakefulness marked by arousal or alertness, and the terms vigilance and arousal are considered identical in literature. Vigilance and task performance may be impaired by physical and mental fatigue, and other environmental factors such as noise, vibration, ambient temperature, frequency and environmental pollutants. Besides, feelings of sleepiness and fatigue may also be induced by sleep deprivation, and this phenomenon is known to be affected by circadian rhythms.

Fatigue is characterized by the lack of alertness and associated with drop in mental and physical performance and a reduced inclination to work (Grandjean, 1979). Driving is a complex cognitive activity, involving a number of simultaneous cognitive processes perception, psychomotor skills, reasoning abilities, auditory and visual processing, decision making and reaction to stimuli. Continuous and repeated execution of such skills may induce physical, mental and visual fatigue, depending upon the time scale (Macdonald, 1985). Activities such as changing of clutches and gear, moving the steering wheel, pulling of brakes etc., are likely to induce physical fatigue. Physical fatigue is characterized by declined muscle activities and efficiency in terms of movement, power and co-ordination, whereas visual fatigue is concerned with stress developed in the eyes. Since mental fatigue is characterized by a state of decreased cognitive performance (Matthews, Davies, Westerman, & Stammers, 2000), a long duration driving task is likely to induce mental fatigue as well.

Driving at night is likely to induce stress on the eyes (visual fatigue) due to the intermittent glare of light from the opposite direction. Visual fatigue is reflected in eye blink rate, PERCLOS and other ocular parameters. Sleep deprivation may increase visual fatigue, leading to an increase in blinking frequency and long duration blink (Caffier, Erdmann, & Ullsperger, 2003; Schleicher, Galley, Briest, & Galley, 2008). Fatigue in drivers is affected by sleep disorders, circadian disruption, sleep deprivation, irregular working hours, long and monotonous driving etc. Environmental stimuli and psychological factors like anxiety, mood, expectation etc. (Saroj K L Lal & Craig, 2001) are also known to bring about fatigue in drivers. Driving performance is also affected by circadian rhythm and is known to be the worst between 02.00am at night to 06.00am in the morning (Lenné, Triggs, & Redman, 1997). (Maycock, 1997) found that the sleep related accidents take a peak in the afternoon.

A boring view or a monotonous highway can lead to an illusory state called highway hypnosis (WILLIAMS, 1963) which slows down the reaction time of the driver. Subjective factors as age, gender, medication, stress, mood; personality etc. of the driver are also effective variables in fatigue generation. Fatigue in drivers may also be induced by extrinsic factors such as steering, brakes, tires etc., comfort inside the vehicle, music, cell phone, passengers and environmental factors such as road condition, traffic, weather, surroundings etc.

Measures of Fatigue

PERCLOS

PERCLOS is a drowsiness metric which was established by (Wierwille et al., 1998.). PERCLOS may be defined as the proportion of time the eyelids are at least 80% closed over the pupil. PERCLOS can be estimated from eye image sequences at sufficient frame rates (Wu, Sun, Xie, & Zhao, 2010). There are several steps involved before estimating the accurate PERCLOS value. The first step is face detection followed by eye detection and eye state classification into open and closed. Finally, the PERCLOS value is calculated as follows.

$$PERCLOS\ P = \frac{closed\ eyes\ count}{total\ eyes\ count} \qquad (1)$$

Since a higher value of *P* indicates higher drowsiness level and vice versa (Dinges, Mallis, Maislin, & Powell, 1998)it is essential that eye states are detected and classified accurately.

PERCLOS has often been reported in literature as a suitable indicator of drowsiness level. (Hong & Qin, 2007) developed an embedded system to detect driver's drowsiness by computing PERCLOS. The face was detected using Haar-like features and the eye was localized using horizontal and vertical variance projection. The eye states were classified by thresholding the eye region, followed by the application of Laplacian on the binary image and using a complexity function. An accuracy of about 85% was obtained using this method and the algorithm was tested for real-time applications. However, since the algorithm is dependent on color information and hence on illumination, its performance may be poor for on-board situations where problems may be caused by the presence of shadows. A video-based system of drowsy driver monitor named Copilot has been developed at the Robotics Institute in Carnegie Mellon University (Ayoob, Steinfeld, & Grace, 2003) estimate PERCLOS. However the issues such as illumination variation and head rotations have not been adequately addressed. PERCLOS has been compared with vertical eye movement frequencies to distinguish between wakeful and drowsy states of drivers in a driving simulator (Hayami, Matsunaga, Shidoji, & Matsuki, 2002). Eye state detection and PERCLOS computation have been carried out by a method of ellipse fitting using active illumination methods for locating the eyes (Bergasa & Nuevo, 2005). A real-time system for monitoring driver's attention level by tracking the eyes and computing PERCLOS has been described in (Ji & Yang, 2002). The present work tends to address the harmful effect of sustained exposure to active illumination techniques on the eyes by using passive illumination.

Speech and Alertness

Speech signal has been reported to be a significant indicator of reduced alertness in humans (Greeley & Friets, 2006; Whitmore & Fisher, 1996). Non-obtrusive nature of speech data and its sensor free application makes it advantageous for on-board assessment of alertness level of drivers. Moreover, speech is easier to record even under extreme environmental conditions and noisy surroundings like (Krajewski, Schnieder, Sommer, Batliner, & Schuller, 2012), high temperature, high humidity, crowded traffic, motor sound, horn sound etc. There are various frame-level-based acoustic features, which get affected by fatigue states (Krajewski, Batliner, & Golz, 2009; Krajewski, Wieland, & Batliner, 2008) such as pitch, autocorrelation function, duration of Voiced/Unvoiced (V/UV) segments, Mel Frequency Cepstrum Coefficients (MFCC), Linear Frequency Cepstrum Coefficients (LFCC) etc. The Voiced-to-Unvoiced Ratio (VUR) has been found to be an efficient and fast method for detecting the drowsiness level of an individual (Dhupati et al., 2010).

EEG and Fatigue

The link between changes in behavioral arousal and the EEG spectrum is strong enough for the EEG spectrum to be used as a direct indicator of arousal level. In general, EEG features in the frequency domain have been found to be more efficient and reliable than those in the time domain for prediction of the behavioral alertness level. Changes in EEG with vigilance have generally shown that distribution, amplitude and frequency of alpha waves in the EEG spectrum change with the onset of drowsiness. A change in the pattern of alpha wave distribution during driver fatigue has been reported (Saroj K L Lal & Craig, 2001) A positive relation between EEG power and cognitive performance in the alpha frequency range has been reported (Klimesch, 1999).

A considerable body of work has been carried out on EEG-based fatigue detection and various methods have been reported to find the changes in EEG signal characteristics during the onset of fatigue. Relative energy and ratio of relative energy of EEG waves belonging to various frequency bands (alpha, beta, beta/alpha ratio and (alpha+theta)/beta ratio) has often been used as an indicator of fatigue (Eoh, Chung, & Kim, 2005). The relative energy parameter (alpha+theta)/beta has been found to decrease with a decrease in alertness level (De Waard & Brookhuis, 1991).

The change in brain activity during a fatigue-inducing monotonous driving session has been investigated in (Jap, Lal, Fischer, & Bekiaris, 2007). (S. K. L. Lal, Craig, Boord, Kirkup, & Nguyen, 2003) describe an EEG based fatigue countermeasure algorithm and attempts to report its reliability. (Vuckovic, Radivojevic, Chen, & Popovic, 2002) present a method for classifying alert and drowsy states from full spectrum EEG recordings: cross spectral densities of full spectrum EEG are fed to an ANN with two discrete outputs: drowsy and alert. A method for automatic recognition of alertness level has been suggested in (Kiymik, Akin, & Subasi, 2004); Power Spectral Density obtained from Discrete Wavelet Transform of full spectrum EEG is fed to an ANN with three discrete outputs: alert, drowsy and sleep. A method for the automatic detection of onset of drowsiness during driving using Support Vector Machines to identify and classify EEG changes between alert and drowsy states has been proposed in (Yeo, Li, Shen, & Wilder-Smith, 2009).

Fluctuations in alertness level are accompanied by changes in the EEG power spectrum. The work in (Jung, Makeig, Stensmo, & Sejnowski, 1997) combines power spectrum estimation, principal component analysis and ANN to illustrate that an accurate estimation of the global level of alertness of an operator is possible using EEG recordings from two central scalp sites. Alertness detection procedures based on the spectral analysis of EEG signal have also been proposed (Álvarez Rueda, 2006; Jung et al., 1997). In (Makeig & Jung, 1995), minute-scale fluctuations in the normalized EEG log spectrum during drowsiness have been correlated with concurrent changes in level of performance for a sustained auditory detection task. An algorithm has been developed for automatic recognition of alertness level using full-spectrum EEG recording in (Kiymik et al., 2004). Time-frequency analysis of EEG signals and independent component analysis have been employed (Ruey Song Huang, Jung, Delorme, & Makeig, 2008) to analyze the tonic and phasic dynamics of EEG activities during a continuous compulsory tracking task. Relative spectral amplitudes in alpha and theta bands, as well as the mean frequency of the EEG spectrum, have been used to predict alertness level in an auditory response test (R S Huang, Tsai, & Kuo, 2001). The change in brain activity during a fatigue-inducing monotonous driving session has been investigated in (Gupta, Kar, Gupta, & Routray, 2010). Four frequency components were extracted by using FFT spectral analysis at the frontal, central, temporal and occipital brain sites.

Another significant method of analysis of EEG to detect fatigue levels includes use of wave entropy as the indicator of fatigue. Shannon Entropy, Renyi entropy, Tsallis entropy, Kullback–Leibler Entropy and Cross-Approximate Entropy have often been employed as indicators of fatigue (Papadelis et al., 2006, 2007). A method based on Shannon Entropy and Kullback-Leibler Entropy measures and alpha band relative energy for relative quantification of fatigue during driving has been proposed (Kar, Bhagat, & Routray, 2010).

A number of synchronization measures such as correlation (Chialvo, 2004), phase synchronization (Rosenblum, Pikovsky, & Kurths, 1996), synchronization likelihood (SL) (Montez, Linkenkaer-Hansen, van Dijk, & Stam, 2006; C. J. Stam & Van Dijk, 2002) etc. have been employed to find the connectivity between different cortical areas of the brain. SL measures both linear and non-linear interdependencies between two time series and is suitable for non-stationary signals like EEG (Montez et al., 2006; C. J. Stam & Van Dijk, 2002). (Ahmadlou & Adeli, 2012) have recently proposed the Visibility Graph Similarity (VGS), a generalized measure of synchronization that converts a time series into an undirected graph. Their work also showed that VGS gives more accurate measure of synchronization as compared to SL. The use of synchronization techniques for analyzing fatigue is a recent development (Kar, Bhagat, & Routray, 2010b; Kar & Routray, 2013; Sengupta, Routray, & Kar, 2013).

Synchronization between different brain areas can be formulated in terms of a complex network (C. J. Stam, 2005). Such networks are extensively used in different research areas as task classification, disease detection, sleep study etc. (C J Stam, Jones, Nolte, Breakspear, & Scheltens, 2007; C. J. Stam, 2004; Van Dongen & Dinges, 2000). Literature suggests the presence of a small-world structure in brain networks in terms of anatomical and functional connectivity (Strogatz, 2001). Various brain functions, including the cognitive, depend on the effective integration of brain networks and the functional interactions between different areas of the brain (Cornelis J Stam & Reijneveld, 2007).

Other Measures

Presently, Galvanic Skin Response (GSR) has been used in the railways for fatigue detection and as an indicator of loss of attention in human beings. The electrical conductance of the skin varies with its moisture level; since sweat glands are controlled by the sympathetic nervous system, skin conductance may be used as a indicator of psychological and physiological arousal of which fatigue or loss of attention is a special case.

Since skin conductance alone is not sufficient for detection of fatigue, a combination of skin conductance and pulse oximetry was used by Bundele et al. (M. M. Bundele, 2009) for detection of fatigue of vehicular drivers. An experiment was designed by Yoshihiro Shimomura et al. (Shimomura, Yoda, & Sugiura, 2008) to induce cognitive loading. Small changes were obtained in the frequency band 0.03-0.5 Hz after Fourier Transform of the GSR The (Mean + Standard Error) value of the power in various frequency bands was also found to increase with an increase in the complexity of the task. However, on-road driving is likely to be a more complex task than the one offered in the experiment and effect of fatigue on such drivers remains to be investigated. Besides, the changes on SCR was not investigated for all subjects.

MAIN FOCUS OF THE CHAPTER

This chapter describes a system for monitoring the drowsiness of automotive drivers with a special treatment for locomotive drivers. The system and algorithms have been experimentally validated using experiments designed to induce fatigue gradually through a set of tasks and activities continued for a long time along with sleep deprivation. The aim of the study is the development of a speech and image-based method for on-board alertness monitoring of drivers. Requisite ethical clearance was obtained prior to the experiment. Information was collected along various channels to find a correlation among changes in various kinds of physiological data during the course of the experiment. The experiment was conducted at Indian Institute of Technology, Kharagpur, India.

The present work employs PERCLOS and VUR for implementation, while synchronization has been utilized for cross-validation along with psychometric as well as subjective measures.

Experiment Design

The first step for designing a system for fatigue monitoring is the design of experiments. A popular method for inducing physical fatigue is sleep deprivation (Eoh et al., 2005). Twelve healthy male drivers were involved in the experiment. All the subjects were reported to have no sleep disorders. They were asked to refrain from any type of medicine and stimulus like alcohol, tea or coffee during the experiment. The experiment was conducted in two sessions with six subjects in each session.

The duration of the experiment was divided into 12 identical stages of three hours each. Each stage involved activities meant to induce mental, physical and visual fatigue in the subject. EEG data were recorded at the beginning of the experiment and at the final phase of each stage. The condition of the subject was monitored by a medical practitioner at the beginning of each stage. After the subject was declared fit, he was asked to perform the following predefined tasks:

- Physical exercise on a treadmill for 2-5 minutes to generate physical fatigue
- Simulated driving for about 30 minutes to generate physical, visual, and mental fatigue
- Auditory and visual tasks for 15 minutes to generate mental and visual fatigue
- A computerized game related to driving for about 20 minutes

A single stage of experiment lasted for about three hours. The subjects were allowed to read books or newspapers in the interval between two stages in order to keep them awake. Closed circuit cameras were used to monitor the subjects throughout the duration of the experiment. The experimental setup has been shown in Figure 1 and placement of subject in Figure 2.

The stages were continued for about 36 hours (12 stages) when most subjects complained of extreme fatigue.

Three sets of EEG were recorded in each stage, i.e.

- For 3 minutes during the computer game
- For 2 minutes after the game with open eyes and no activity condition
- For 2 minutes with closed eyes and no activity.

Nineteen scalp electrodes (Ag/AgCl, RMS, India) were used for EEG recording, in addition to reference and ground to collect the signals from locations Fp1, Fp2, F3, F4, F7, F8, Fz, T3, T4, T5, T6, C3, C4, Cz, P3, P4, Pz, O1, and O2 following the international 10–20 system. In this study, the 3 minute EEG recording during simulated computer driving game has been chosen for analysis.

In addition to the EEG recording, several physiological parameters were recorded during the various stages of the experiment.

Figure 1. Experimental setup for simulated driving

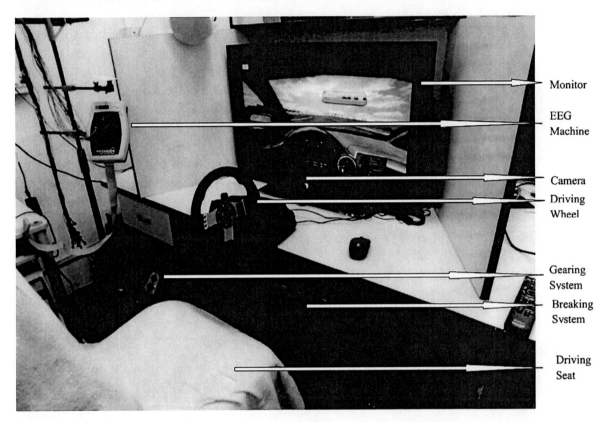

Figure 2. Placement of subject

1. Video Recording: Video images of each subject were recorded at all the stages using a digital color camera (model: JN-2019P) at 30 fps, synchronized with the EEG recordings.
2. Blood sample collection: Blood samples were collected from each subject at five instances during 36 hours (For example, blood samples were collected for subject 1 at the following instants: Day1: 08.30 hrs; 16.30 hrs; Day2:01.30 hrs; 10.30 hrs and 19.30 hrs).
3. Speech recording: Speech data were collected by vocal response of the sentence "Now the time is -------", at all stages of the experiment. This sentence was selected to find the duration of voiced and unvoiced segments.
4. Subjective assessment: Subjective feedback was obtained from the drivers in course of the experiment.
5. PVT Data: The subjects were asked to respond to four different auditory inputs (up, down, left and right) and movement of an object (up, down, left and right), generated randomly in a computer, and accordingly respond by pressing the four arrows in the keyboard. In both the tests, the response time (the time gap between command generation and response from the subjects) was computed.
6. ECG, Spirometry and Oximetry data were also recorded for correlation analysis

PERCLOS Estimation

The present work uses the following steps for PERCLOS estimation:

- Detection of face in real-time
- Selection of a Region Of Interest (ROI)
- Localization of the eye in the ROI (in daytime as well as with NIR lighting)
- Classification of the eye as open or closed. Here closed eye denotes at least 80% spatially closed over the pupil.
- Computation of PERCLOS value.

Detection of face and eyes accurately in various illumination conditions is a very challenging problem. The present study uses Near Infrared Red (NIR) illuminators for capturing the images at night and low light conditions. A 66.67% (two-third) overlapping time window of three minutes duration is used to compute the PERCLOS value for every minute.

REAL-TIME FACE DETECTION

Haar-like features (Viola & Jones, 2001) have been finally selected for face detection owing to its better performance in the particular context. Table 1 describes the framework of face and eye detection for estimating PERCLOS.

This method achieves higher detection rate, which is dependent on Scale Factor (SF), primarily because of the low resolution face detection (Figure 3) (Dasgupta, George, Happy, & Routray, 2013).

a_i's and b_i's for $i=1,2,3,4$ are obtained from the face detector based on Haar Classifier. ROI co-ordinates c_2 and c_3 are obtained as $c_2 = \dfrac{b_1 + b_2}{2}$ and $c_3 = \dfrac{b_3 + b_4}{2}$ respectively. The ROI co-ordinates d_i's and e_i's are obtained using the following:

Alertness Monitoring System for Vehicle Drivers using Physiological Signals

Table 1. Face and eye detection for estimating PERCLOS

The image obtained from the camera is scaled down by a factor of *k*. The original image is stored before it is scaled down.
- *Face is detected in the down-sampled image using the Haar classifier for face.*
- *The coordinates of the face are obtained from the Haar classifier in the down-sampled frame.*
- *An ROI is selected from the detected facial image, based on the fact that the eyes are located in the upper half of the face.*
- *ROI co-ordinates are extracted from the detected frame and remapped on to the original frame as shown in Figure 2.*
- *Eyes are then detected in the selected ROI.*

Figure 3. Scheme of eye detection

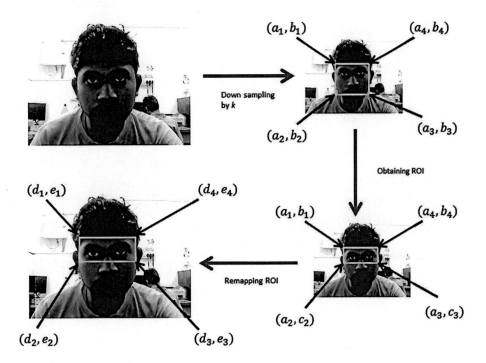

If $D = \begin{bmatrix} d_1 \\ d_2 \\ d_3 \\ d_4 \end{bmatrix}$, $E = \begin{bmatrix} e_1 \\ e_2 \\ e_3 \\ e_4 \end{bmatrix}$, $A = \begin{bmatrix} a_1 \\ a_2 \\ a_3 \\ a_4 \end{bmatrix}$, and $B = \begin{bmatrix} b_1 \\ c_2 \\ c_3 \\ b_4 \end{bmatrix}$ then

$$D = k \times A \tag{2}$$

$$E = k \times B \tag{3}$$

IN-PLANE ROTATED FACE DETECTION

Detection accuracy is found to be low for a moderate amount of in-plane rotation of the face of the subject. In (Dasgupta, George, Happy, Routray, & Shanker, 2013), an affine transformation based method has been used for the detection of in-plane rotated faces. In (2), the rotation matrix R is found from the angle of rotation θ of the image. If a point on the input image is $A = \begin{bmatrix} x \\ y \end{bmatrix}$ and the corresponding point on the affine transformed image is $B = \begin{bmatrix} x' \\ y' \end{bmatrix}$, then B can be related to A.

$$R = \begin{bmatrix} \cos\theta & -\sin\theta \\ \sin\theta & \cos\theta \end{bmatrix} \qquad (4)$$

$$A = RB \qquad (5)$$

There is inherent $\pm 15°$ rotation already provided in the Haar classifier, and hence the angles of rotation, θ are given values of $\pm 45°$ and $\pm 30°$ successively. After the face detection, the algorithm returns the ROI for eye detection in a de-rotated form (Table 2). Figure 4 shows tilted face detection using an affine transformation of the input image.

Real-Time Eye Detection

Once the face is localized and ROI has been remapped, the next step is to detect the eyes in the ROI. Subsequently the detected eye is classified as open or closed to obtain the PERCLOS value.

PCA Based Eye Detection

Haar-like features fail to provide a reliable accuracy in the context of eye detection as the Haar-like features present in the eye are less prominent as compared to face. Hence, PCA(Yang & Zhang, 2004) based eye detection has been adopted for day driving conditions owing to its superior real-time performance compared to other methods as reported in Zahran & Abbas (2009). The ROI is resized to a resolution of 200×70 using bi-cubic interpolation (Gonzalez, 2004). Block-wise search is carried out in each sub-window of size 50×40 pixels with a 10% overlap. The training is carried out using 460 eye images cropped from images taken from custom dataset, each of size 50×40.

Table 2. Algorithm: Detection of in-plane rotated faces

- The Haar classifier checks for the presence of frontal faces in the input image.
  - *If frontal face is found it follows the present scheme.*
  - *If frontal face is not found, a rotational matrix R is formed with θ=30°, to transform the image.*
  - *The Haar classifier searches for the face in the transformed image. If frontal face is found, it follows the scheme as given in* Table 1.
  - *If the frontal face is still not found, a rotational matrix R is formed with θ=−30° and θ=±45° successively to search for faces in the respective transformed frames.*
  - *If the Haar classifier is unable to detect the face for none of the set values of θ, the algorithm concludes that no face is found.*

Figure 4. In-plane rotated face detection

To examine the performance of eye detection, a test was conducted with 300 face images having 200 open and 100 closed eyes but different from those used in training. Figure 5 shows some detection results using PCA. Table 5 shows that the PCA technique is quite accurate for detecting the eye from the ROI.

Classification of Eye States

For accurate estimation of PERCLOS, the detected eye needs to be accurately classified into open or closed states. SVM is reported to be a robust classifier for a two-class problem (Bhowmick & Kumar,

Figure 5. Some detection results using PCA

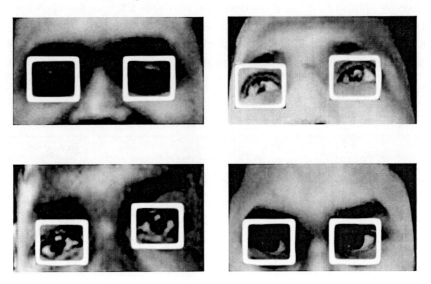

Table 5. Eye detection results using PCA

True Positives(T_p)	True Negatives (T_n)	False Positives (F_p)	False Negatives (F_n)	True Positive Rate (T_{pr})	False Positive Rate (F_{pr})
198	97	3	2	98.5%	2.02%

2009) and is used here for the purpose. In the present case, the eye state classification is a two-class classification problem where the class labels include open and closed eye classes.

The training of SVM has been carried out with 460 images (230 open and 230 closed eyes) taken from a database created using normal and NIR illumination. Some training images are shown in Figure 5. The testing is carried out using another 1700 images taken from the same source. Now the weight vectors, corresponding to each sample found in the eye detection phase, are fed to the SVM along with the ground truth. The training is carried out with different kernels and accuracy levels and the classification results are shown in Table 6.

The results reveal that the performance of the SVM is consistent under normal and NIR images. A third order polynomial kernel is found to provide the most accurate result and hence been implemented in the final algorithm.

Once the eyes are classified, the algorithm computes the PERCLOS value using the number of open and closed eye count over a sliding time window of 3 minutes duration. The real-time algorithm is deployed into a Single-Board Computer (SBC) having Intel Atom processor, with 1.6 GHz processing speed and 1 GB RAM. The overall processing speed is found to be 9.5 fps, which is quite appreciable for estimating PERCLOS.

Compensation in Illumination Variation

The detection accuracy is found to be low when the lighting conditions are extremely bright or dark, particularly with the occurrence of shadows. Compensation in illumination variation and effects of shadows is carried out using Bi-Histogram Equalization (BHE) (Dasgupta, George, Happy, & Routray, 2013). In case the face is not detected even after the geometric transformations, the algorithm considers the problem as an illumination variation issue and performs BHE as a preprocessing step. In this method, the input image is decomposed into two sub-images. One of the sub-images is the set of samples less than or equal to the mean image intensity whereas the other one is the set of samples greater than the mean image. The BHE is used to equalize the sub-images independently based on their respective

Table 6. Detection results with SVM

Kernel Function	True Positives(T_p)	True Negatives (T_n)	False Positives (F_p)	False Negatives (F_n)	True Positive Rate (T_{pr})	False Positive Rate (F_{pr})
Linear SVM	838	808	42	12	98.58%	4.94%
Quadratic	827	765	85	23	97.29%	10%
Cubic Polynomial	848	807	43	2	99.76%	5.32%

histograms under the constraint that the pixel intensities in the first subset are mapped onto the range from the minimum gray level to the input mean intensity while the pixel intensities in the second subset are mapped onto the range from the mean intensity to the maximum gray level. Hence, the resulting equalized sub-images are bounded by each other around the input mean. Hence the mean intensity is preserved which makes the method illumination invariant.

The input image X is subdivided into two sub-images X_L and X_U based on the mean intensity image X_m such that $X = X_L \cup X_U$ where

$$X_L = \{X(i,j) | X(i,j) \leq X_m, \forall X(i,j) \in X\} \quad (6)$$

$$X_U = \{X(i,j) | X(i,j) > X_m, \forall X(i,j) \in X\} \quad (7)$$

Now, histogram equalization is carried out on images X_l and X_U separately. A data set of 400 images is prepared of on-board images of a driver under extremely varying lighting conditions. The dataset is tested with the algorithm both before and after applying BHE. Using BHE, the detection accuracy was found to improve from 92% to 94%. However, the results reveal that there is scope of further improvement to compensate varying illumination levels and shadows. The pre-processing steps such as the geometric transformations and BHE, improves the accuracy but at the same time reduces the frame rate, owing to their computational burden.

EEG Analysis

The present work considers modelling of the brain states in various stages of sleep deprivation from EEG signals using distributed current dipole method and EEG synchronization.

EEG Preprocessing

The raw EEG signal is passed through a band pass filter with cutoff frequencies of 0.5Hz and 30Hz followed by normalization (conversion into a dataset with zero mean and unity standard deviation). This operation removes the power-line artifacts and ensures removal of any unwanted bias that might have been introduced during experimental recording. The EEG is then decomposed into various bands using the Discrete Wavelet Transform approach (Croft & Barry, 2000; Krishnaveni, Jayaraman, Anitha, & Ramadoss, 2006; Krishnaveni, Jayaraman, Aravind, Hariharasudhan, & Ramadoss, 2006).

Linear Distributed Current Dipole Approach - The LORETA Algorithm

The present work implements the Standardized LOw-REsolution brain electromagnetic Tomography Algorithm (sLORETA) (Pascual-Marqui, Michel, & Lehmann, 1994) to obtain the source conditions by constructing the solution of the inverse EEG problem. The problem description begins with a simulated surface grid of N_v voxels shaped like the human neo-cortex; where voxels may be defined as unit element of the 3D volume.

In the particular discretization employed, there are $N_v = 10014$ cortical gray-matter voxels. At each voxel, a local three-dimensional current vector $j(y,t) = (j_x(v,t), j_y(v,t), j_z(v,t))$ is assumed, where v is a voxel label; t denotes time.

The column vector $J(t)$ of all such current vectors (i.e., for all gray-matter voxels) may be denoted by

$$J(t) = (j(1,t), j(2,t), \ldots, j(N_v, t)) \tag{8}$$

The flow chart for Distributed Current Dipole Approach is shown in Figure 6.

The details of the sLORETA algorithm are beyond the scope of current work. In essence, the sLORETA algorithm finds out an estimation of the $J(t)$ vector by use of a Buckas-Gilbert Resolution matrix (Grech et al., 2008) constructed from the Lead Field Matrix (LFM). The algorithm has been implemented through Brainstorm ("Brainstorm | Real-time 3D graphics and virtual set solutions," n. d.).

Construction of Scouts

Difficulties present in source characterization through use of sLORETA are due to the large number of dipoles which again change with time. The storage requirement, for recording the evolution of the voxel contents with time, is exceedingly large. Therefore, scouts have been constructed below specific electrodes, namely CZ, C3, C4, FP1 and FP2 to obtain the source nature in these locations. These electrodes have been reported to be informative about the state of drowsiness in previous work (Saroj K.L. Lal et al., 2003)

In Figure 7, the five scouts are displayed in the following manner – in axial or top view of the grid, scout shown in blue color is the closest region to CZ electrode. The right sided sagittal view displays the

Figure 6. Flow chart for Distributed Source Localization

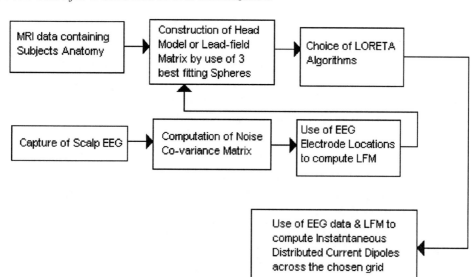

Figure 7. Scouts shown on the simulated cortex surface

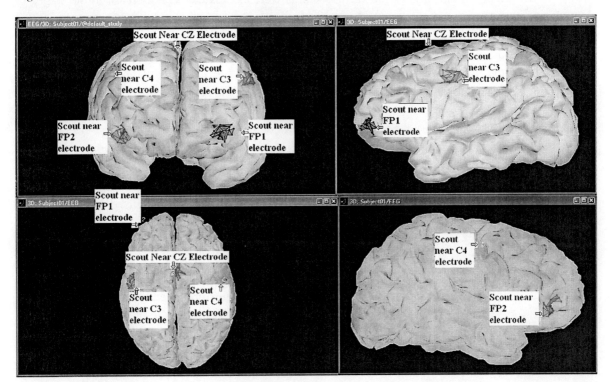

scout closest to C4 electrode in cyan. The left sided sagittal view, on the other hand, shows the region closest to C3 electrode in orange. The frontal coronal view shows the regions closest to electrodes FP1 & FP2, respectively, by red & green regions.

Calculation of the Renyi and Shannon Entropy of the Scouts Progression

Entropy is defined as a measure of information. The Shannon's entropy (*SE*) is a disorder quantifier (Shannon, 1948) and is a measure of flatness of energy spectrum in the wavelet domain. It is defined as

$$SE = -\sum_j p_j (\log p_j) \quad (9)$$

Another statistical measure closely related to SE is Renyi entropy (*RE*) (Rényi, 1961). The basic definition of *RE* is given by

$$RE = \frac{1}{1-q} \log \left[\sum_j p_j^q \right] \quad (10)$$

where p_j is the relative energy as described earlier and $q \in R$ is known as the entropic index. The parameter q confers generality to this information measure. The present study uses $q=2$ to calculate 2nd order entropy. Changes of the entropy for the various scouts were calculated, with increase in the stages of sleep deprivation and fatigue.

An overall increase in Renyi entropy values is observed for the scouts constructed below FP1 & FP2, whereas a decrease is observed in scouts below CZ, C3 & C4 (shown in Figure 8). The effect is also affected by the circadian rhythm of the subject. For all the subjects, standard deviation of the Renyi entropy values is found to be less than 0.05, leaving the nature of the curve similar (Chaudhuri, Routray, & Kar, 2012).

In this work, an attempt was taken to establish a non-chemical, non-invasive, and cost-effective way to observe the effects of sleep deprivation on the electrical activations occurring on human neo-cortex surface. The results clearly indicate that the estimated sources in frontal lobe vary with increasing randomness corresponding to increase of drowsiness. This may be attributed to the increased cognitive loading due to sleep deprivation and is found to be correlated with the increased synchronization observed in (Routray & Kar, 2012).

EEG Synchronization and Network Modeling

A Weighted Visibility Graph Similarity (WVGS) technique has been employed for studying fatigue and loss of alertness. This is a development on the earlier use of the Visibility Graph technique for EEG-based diagnosis of Alzheimer's Disease using a chaos-wavelet approach (Ahmadlou, Adeli, & Adeli, 2010). The Horizontal Visibility Graph Technique (G. Zhu, Li, & Wen, 2014)has also been employed for analyzing the synchronization between brain areas in an experiment involving sleep-deprived drivers.

Figure 8. The Variation of Renyi Entropy of the Scout Potentials through 11 stages of fatigue

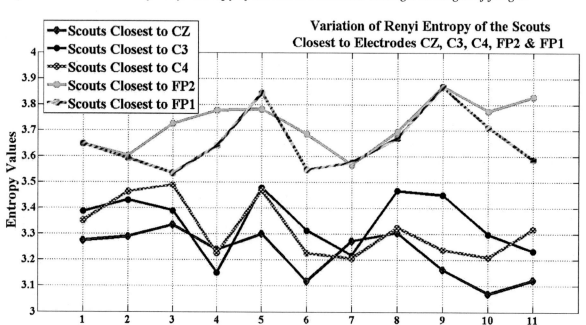

Weighted Visibility Graph Synchronization

$x_{k,i}$ represents M simultaneously recorded EEG signals of length N, where $k=1,2,3,\ldots,M$ and $i=1,2,3,\ldots,N$. From each time series a state vector $X_{k,i}$ is formed as

$$X_{k,i} = (x_{k,i}, x_{k,i+l}, x_{k,i+2l}, \ldots, x_{k,i+(m-1)l})$$
$$i = 1,2,3,\ldots, N-(m-1)l \tag{11}$$

where l is the lag and m is known as embedding dimension. These vectors form an m-dimensional lagged phase space of which $X_{k,i}$ is a point.

In other words, for each EEG record a matrix is constructed as

$$X_k = \begin{bmatrix} x_{k,1} & x_{k,1+l} & \cdots & x_{k,1+(m-1)l} \\ x_{k,2} & x_{k,2+l} & \cdots & x_{k,2+(m-1)l} \\ \cdots & \cdots & \cdots & \cdots \\ x_{k,N-(m-1)l} & x_{k,N-(m-1)l+l} & \cdots & x_{k,N} \end{bmatrix} \tag{12}$$

N_t identical instances are selected on each time series. For each time series k and each reference instance t, two windows w_1 and w_2 are defined as $w_1 \ll w_2 \ll N$.

The m dimensional distance vector $D_{i,j,m}$ between $X_{k,i}$ and $X_{k,j}$, $\frac{w_1}{2} \ll |i-j| \ll \frac{w_2}{2}$, is computed as

$$D_{i,j,m} = |X_{k,i} - X_{k,j}| \tag{13}$$

This forms the m-dimensional distance time series (DTS) that is required to find the visibility of the nodes at each time instant.

For each node, the visibility is observed along each dimension. Two nodes of the graph are connected along the dimension m if and only if:

$$d_{i,j+q,m} < d_{i,p,m} + \left(\frac{p-(q+j)}{p-q}\right)(d_{i,q,m} - d_{i,p,m}), \forall j \in Z^+ \tag{14}$$

where $j<(n-m)$ Visibility along different directions and the corresponding visibility graphs are shown in Figure 9.

The weighted visibility graph (WVG) (Figure 10) is obtained by calculating the mean visibility of each pair of nodes as

$$w_{i,j} = \frac{\sum \text{actual visibility in each dimension}}{m} \tag{15}$$

Figure 9. Visibility along different dimensions and corresponding visibility graphs

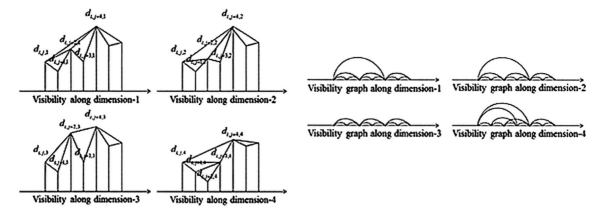

The degree of each node is calculated as a sum of weights connected to that node. This generates a degree sequence (DS) for each node at each instant of time t.

The synchronization $S^t_{k_1,k_2}$ between two signals $x_{(k_1,t)}$ and $x_{(k_2,t)}$ at instant t is measured as cross-correlation of the degree sequences of the pair of signals. The overall synchronization of the signals is measured as a mean of synchronization values obtained at different time instances.

$$S_{k_1,k_2} = \frac{\sum_t S^t_{k_1,k_2}}{N_t} \qquad (16)$$

Horizontal Visibility Graph Synchronization

For the Horizontal Visibility Graph (HVG) technique, (G. Zhu et al., 2014), the connection criterion between nodes in (11) is modified as,

Nodes i and j are connected if and only if $x_i > x_k$ and $x_j > x_k$ $\forall k \in (i,j)$ \qquad (17)

A time series and the corresponding horizontal visibility graph are shown in Figure 11.

Figure 10. Weighted visibility graph

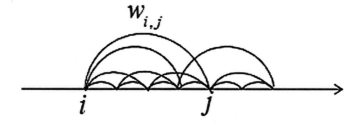

Alertness Monitoring System for Vehicle Drivers using Physiological Signals

Figure 11. Time series and corresponding Horizontal Visibility Graph

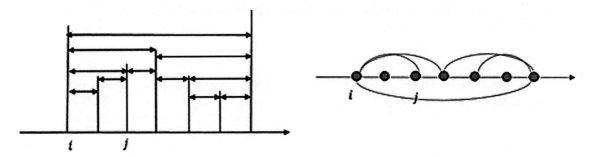

Construction of a Complex Network from Synchronization Values

For construction of the brain network, a threshold is applied to each element of the synchronization matrix. In the present case, the mean synchronization across all subjects and stages is taken as the threshold. If an entry of the synchronization matrix is greater than the threshold, it is replaced by 1, else 0 replaces it. A connection is assumed to exist between two nodes in the network if the corresponding entry in the adjacency matrix happens to be 1, and no connection exists otherwise.

Computation of Network Parameters

Brain networks may be quantified using a small number of network measures that are neuro-biologically significant and easy to compute (Sporns & Zwi, 2004, Bassett & Bullmore, 2006). Literature reports that the brain adheres to the twin principles of functional integration and functional segregation; functional segregation requires that cells with common functional properties be grouped together, and functional integration handles the interactions among specialized neuronal populations. A number of network parameters have been suggested in literature to help describe and define these activities(Bassett & Bullmore, 2006; Sporns & Zwi, 2004); for example, characteristic path length and density degree are taken to be the measures of integration between brain areas, whereas segregation between areas of the brain is indicated by the clustering coefficient (Rubinov & Sporns, 2010)

Functional integration measures indicate how fast the specialized information from various brain regions can be combined and are mostly based upon the concept of a path.

Characteristic path length (L) indicates the functional integration of the graph. The smaller the length, better is the integration and hence easier is the information transfer. For un-weighted networks, the characteristic path length is defined as (Rubinov & Sporns, 2010)

$$L = \frac{1}{n}\sum_{i \in N} L_i = \frac{1}{n} \frac{\sum_{j \in N, j \neq i} d_{ij}}{n-1} \qquad (18)$$

where N is the set of all nodes in the network and n is the number of nodes. L_i is the average distance between node i and all other nodes.

Functional segregation refers to specialized information processing that occurs within groups of brain regions (referred to as clusters or modules) that are densely interconnected.

Clustering Coefficient (C) measure of local structure of the network. For a node i, it is defined as the ratio of number of existing branches connected to i and the number of maximum possible branches between neighbors of node i. The mean clustering coefficient for the network reflects, on average, the prevalence of clustered connectivity around individual nodes. For an un-weighted network clustering coefficient is defined as (Rubinov & Sporns, 2010)

$$C = \frac{1}{n}\sum_{i \in N} C_i = \sum_{i \in N} \frac{2t_i}{k_i(k_i - 1)} \qquad (19)$$

Here C_i is the clustering coefficient of node i and $t_i = \frac{1}{2}\sum_{j,h \in N_e} a_{ij}a_{ih}a_{jh}$ is the geometric mean of triangles around node i. k_i is the degree of node i.

Small-World Nature of the Brain Network

The values of clustering coefficient is high for regular networks ($C \approx 3/4$) and low for random graphs (k/N), while those for path length are large for regular networks ($L \approx N/2k$) and low for random networks ($L \approx \ln(N)/\ln(k)$) (C. J. Stam, 2004). A small-world network has a high clustering coefficient and a small path length. That is to say, the clustering coefficient for a small-world network would be close to that of a regular network and the path length would be close to that for a random network (Watts & Strogatz, 1998). Such a network structure, which is neither regular nor random, has been suggested for networks in the brain (C J Stam et al., 2009).

Results

Adjacency matrices and for all subjects for 4 stages ((Stage 1-12.15pm, Stage 4-10.00pm, Stage 7-7.45am, Stage11-8.45pm) have been shown in Figure 12. The black squares correspond to 0 while the bright squares correspond to 1 in the adjacency matrix. It may be seen that the brightness increases with progression in stages, indicating an increase in synchronization. The mean of the synchronization values across all subjects and stages is considered as threshold for construction of adjacency matrices. The brain networks corresponding to the adjacency matrices have been shown in Figure 12. More significant synchronization has been observed in the parietal and occipital lobes. This may be explained by the fact that an important role is played by the parietal and occipital lobes in movement, cognition and visual information processing – the skills involved in execution of the experiment.

Variation of network parameters along the stages of the experiment for un-weighted networks has been shown in Figure 13. The characteristic path length (L) shows a decreasing trend over successive stages. This corresponds to the increase in synchronization values between electrodes in successive stages of the experiment. The connection strength between the electrodes increases with an increase in the synchronization and this facilitates the information transfer between different areas of the brain. The increasing trend in clustering coefficient (C) is indicative of tight coupling between the corresponding electrode regions that increases in successive stages with an increase in fatigue. However, some non-

Figure 12. Adjacency matrices and corresponding brain networks for 4 stages for 12 subjects for various frequency bands

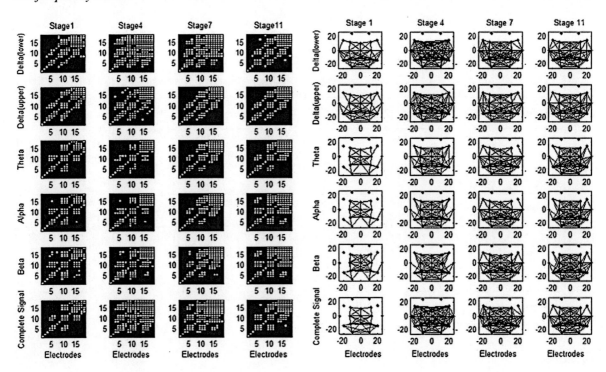

uniformity is also observed, since the effect of sleep deprivation and fatigue is also influenced by natural circadian rhythm of the subjects. The distortion in the trend may be attributed to the effect of variation of circadian rhythm upon the change in fatigue levels of the subjects.

Speech Signals and Alertness

In this work, the duration of voiced/unvoiced segments has been used as the frame-level-based feature to detect fatigue based on speech. This discrimination problem has been worked on extensively in (Ahmadi & Spanias, 1999). There are various algorithms for solving the classification problem (Alkulaibi, 1996; Lobo & Loizou, 2003; Shah & Iyer, 2004). In (Leonard Janer, L., e Bonet, J. J., & Lleida-Solano, 1996), wavelet based method is used for pitch and V/UV detection.. In this work wavelet transform has been used for voiced and unvoiced classification. The per wavelet band average energy distribution has been used as a feature for classification (Cai, Zhu, & Guo, 2007). The duration of voiced and unvoiced segments is calculated using the following steps:

1. Input speech is divided into frames each of window length of 25 ms
2. A 6 level discrete wavelet transform is applied
3. Per band energy is computed
4. Ratio of 62.5-1000 Hz bands energy to total energy is calculated
5. Classification of voiced or unvoiced segments based on the threshold

Figure 13. Variation of degree of connectivity, clustering coefficient and characteristic path length for all subjects and stages for various frequency bands for un-weighted networks

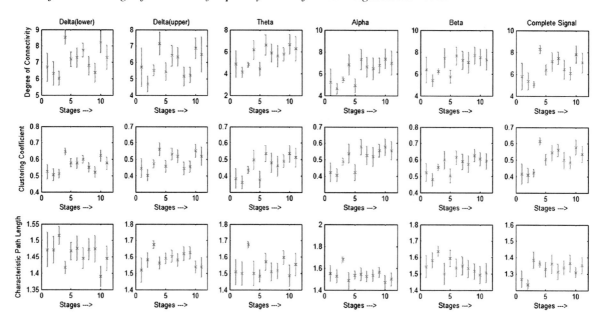

Results

A voiced segment has high energy in its fundamental and second harmonic frequency bands of the wavelet domain. An unvoiced segment has energy concentration in higher frequency bands. Since the fundamental frequency of voiced segments ranges from 60-500 Hz, the ratio between the energy of the bands between 62.5 Hz and 1000Hz to that of all bands is used as the parameter for voiced and unvoiced classification. Figure 14 shows the ratio of voiced and unvoiced segments for 11 stages. The ratio is observed to decrease with increase in fatigue.

Mel frequency cepstral coefficients (MFCC) features along with Gaussian mixture model (GMM) has also been used to segregate voiced and unvoiced parts of the speech (Table 7) (Quatieri, 2002).

After computing the MFCC features, GMM classifier is used to segregate voiced and unvoiced parts of speech. The log likelihood of each frame is computed using the GMMs and the frame belongs to the class which corresponds to the maximum value of log likelihood. The voiced-to-unvoiced ratio was found to decrease with a progression in stages (Figure 15).

Implementation for Automotive Drivers

The system consists of a microphone, a pair of speakers, processing board (SBC), power supply and an LCD. The block diagram is shown in Figure 16.

The system has been installed on a car. The microphone has been fixed on the dashboard while the speakers are kept at the bottom of the steering wheel. The power is drawn from the car battery at 12V DC. The driver is asked to speak for a duration of 20 seconds. Then the voice is analyzed for classification of the state of alertness of the drivers.

Figure 14. Ratio of voiced and unvoiced segments of 12 subjects for 11 stages

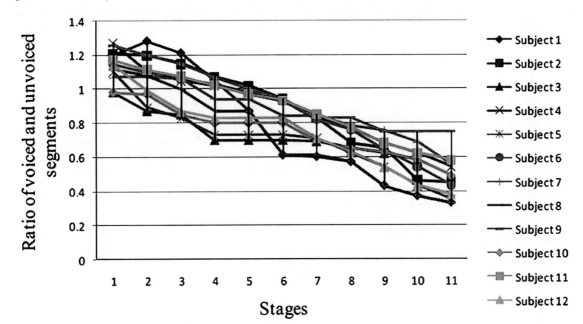

Figure 15. Mel-scale filter bank

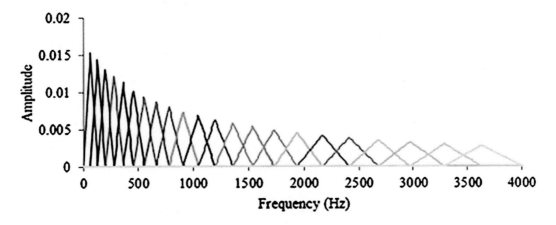

Proposed System Design for Locomotive Driving

The system design for analysis of fatigue in vehicle drivers can be extended to the case of locomotive drivers. The proposed system shall consist of a camera, an embedded processing unit, a near-infrared lighting system and voltage regulator. It will record the facial images continuously and the embedded platform will execute the real time video processing algorithms to track the eye and the pupil to estimate the PERCLOS. The values of PERCLOS will be used to assess the level of alertness. Both the day and night vision systems will be developed for the purpose. The image based algorithms will be validated by more accurate signals such as EEG signals and psychomotor tests. The proposed position of the camera is shown in Figure 17.

Table 7. Steps to calculate MFCC

- The discrete Short Time Fourier Transform (STFT), $X(n,w_k)$ the framed speech waveform $X(n)$ is computed A Hamming window $w[n]$ is used.

$$X(n, w_k) = \sum_{m=-\infty}^{\infty} x[m]w[n-m]e^{-jw_k m}$$

where $\omega_k = \dfrac{2\pi}{N} k$ and N is the length of the Discrete Fourier Transform (DFT).

- *The magnitude of $X(n,w_k)$ is then weighted by a series of filter frequency responses whose center frequencies and bandwidths roughly match those of the critical band filters. These filters follow the mel-scale so that the band edges and center frequencies of the filters are linear for low frequencies and logarithmically increase with increasing frequency. An example of mel-scale filter bank is shown in Figure 15.*
- *The energy of the STFT, weighted by each mel-scale filter frequency response, is calculated. The energy E_{mel} for each speech frame at time instant n and for the l-th mel-scale filter can be written as,*

$$E_{mel}(n,l) = \frac{1}{A_l} \sum_{k=L_l}^{U_l} |V_l(\omega_k) X(n,\omega_k)|^2$$

where the frequency response of the l-th mel-scale filter is denoted as $V_l(\omega)$, U_l and L_l denote the upper and lower frequency indices over which each filter is non zero and where

$$A_l = \sum_{k=L_l}^{U_l} |V_l(\omega_k)|^2$$

A_l *normalizes the filters according to their varying bandwidths so as to give equal energy for a flat input spectrum.*
- *The real cepstrum associated with $E_{mel}(n,l)$ (mel-cepstrum), is computed for the speech frame at time n by taking the discrete cosine transform of log compressed filter outputs. Hence,*

$$C_{mel}[n,m] = \frac{1}{R} \sum_{l=0}^{R-1} \log\{E_{mel}(n,l)\} \cos\left(\frac{2\pi}{R} lm\right)$$

where R represents number of filters. The higher order cepstra are numerically quite small and this results in a very wide range of variances when moving from low to high cepstral coefficients.
- *Therefore, the cepstral coefficients are re-scaled to have similar magnitudes. This is done by filtering the cepstral values C_{mel} to get filtered values using the following formula*

$$C'_{mel} = \left(1 + \frac{L}{2} \sin \frac{\pi n}{L}\right) C_{mel}$$

C'_{mel} is the final MFCC coefficient.

System Requirements

A flow diagram for the proposed setup for locomotive drivers is presented in Figure 18. The system consists of the following:

- Camera, an embedded processing unit, a near-infrared lighting system and voltage regulator
- Input supply: designed for taking a continuous load of 15/20 Watt under specified input and output conditions; typical operating voltage of the system is 12 V DC at a maximum of 1200 mA current drawn

Alertness Monitoring System for Vehicle Drivers using Physiological Signals

Figure 16. Block Diagram: system for automotive driving

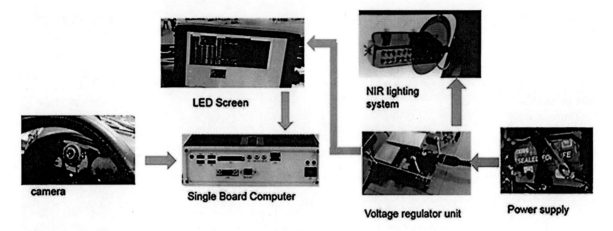

Figure 17. Camera placement on driver's cab on the locomotive

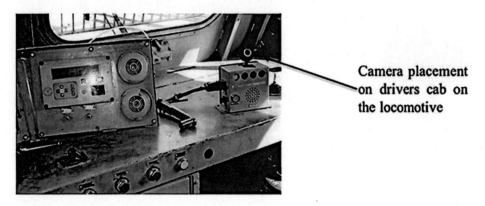

FUTURE RESEARCH DIRECTIONS

As it is evident from this chapter, that a system for detecting the drowsiness level has already been developed and tested. The same concept can be applied intelligently for implementing the system for locomotive driving. The major issues involved in such an implementation are:

- Noise in the driver cabin
- Huge amount of vibration
- Extremely high temperature
- Free movement of the train driver in his cabin

There is also scope of implementing additional channels of information for data fusion such as GSR to make the system more robust.

Figure 18. Proposed setup for locomotive drivers

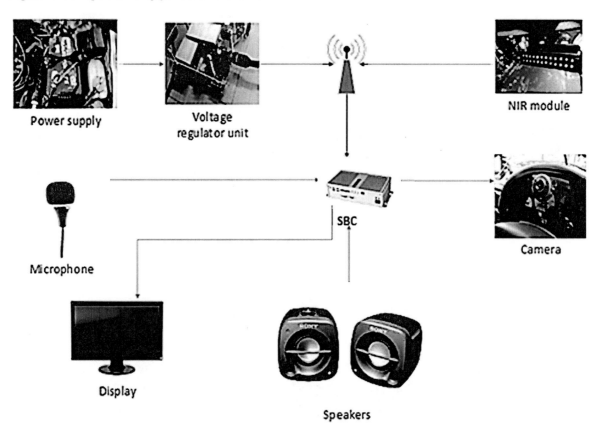

CONCLUSION

This chapter presents the development of a real-time embedded system to monitor the loss of attention of drivers. Image and speech based methods have been combined for the development of a robust real-time framework. EEG based methods have been implemented to validate the performance of the developed system. PERCLOS is used as an index to detect the evolution of fatigue states in image-based method. The ratio of voiced to unvoiced speech is used as the metric in the speech based implementation. The information from both modalities has been combined in the framework for robust estimation of driver alertness level. A sleep deprivation study has been conducted for 36 hours, where each of the 12 healthy subjects have been asked to perform various activities to induce fatigue in them. Physiological signals as well as non-contact measures based on image and voice are also captured during the experiment. High correlation has been observed among the physiological signals and non-contact measures in the fatigue level estimation. The developed system has been tested in laboratory as well as on-road conditions. The accuracy of the algorithm has been found to be well above the requirements. The feedback given to the driver can be useful in reducing the number of accidents. The main limitation of the proposed methodology is that, it detects the level of alertness in a diagnostic mode. Development of a predictive framework for human fatigue level can be attempted in future. The methodology established in this chapter can be extended to a multimodal alertness level prediction system for locomotive drivers, with the addition of prognostic models.

REFERENCES

Ahmadi, S., & Spanias, A. S. (1999). Cepstrum-based pitch detection using a new statistical V/UV classification algorithm. *IEEE Transactions on Speech and Audio Processing, 7*(3), 333–338. doi:10.1109/89.759042

Ahmadlou, M., & Adeli, H. (2012). Visibility graph similarity: A new measure of generalized synchronization in coupled dynamic systems. *Physica D. Nonlinear Phenomena, 241*(4), 326–332. doi:10.1016/j.physd.2011.09.008

Ahmadlou, M., Adeli, H., & Adeli, A. (2010). New diagnostic EEG markers of the Alzheimer's disease using visibility graph. *Journal of Neural Transmission (Vienna, Austria : 1996), 117*, 1099–1109. http://doi.org/<ALIGNMENT.qj></ALIGNMENT>10.1007/s00702-010-0450-3

Airplane crashes are very often caused by fatigue. (n. d.). *LAFuel.com*. Retrieved from http://www.lafuel.com/2013/04/-3&d=78

Alkulaibi, A. (1996). Fast HOS based simultaneous voiced/unvoiced detection and pitch estimation using 3-level binary speech signals. *Proceedings of the 8th IEEE signal processing workshop on Statistical Signal and array processing* (pp. 194-197). Retrieved from http://ieeexplore.ieee.org/xpls/abs_all.jsp?arnumber=534851

Álvarez Rueda, R. (2006, November 1). Assessing alertness from EEG power spectral bands. *Bibdigital.epn.edu*. Retrieved from http://bibdigital.epn.edu.ec/handle/15000/9872

Arief, Z., Purwanto, D., Pramadihanto, D., Sato, T., & Minato, K. (2009). Relation between eye movement and fatigue: Classification of morning and afternoon measurement based on Fuzzy rule. *Proceedings of the International Conference on Instrumentation, Communication, Information Technology, and Biomedical Engineering 2009, ICICI-BME 2009*. http://doi.org/ doi:10.1109/ICICI-BME.2009.5417286

Ayoob, E., Steinfeld, A., & Grace, R. (2003). Identification of an "appropriate" drowsy driver detection interface for commercial vehicle operations. *Proceedings of The Human Factors and Ergonomics society annual meeting, 47*(16), 1840-1844. Retrieved from http://pro.sagepub.com/content/47/16/1840.short

Bassett, D. S., & Bullmore, E. (2006). Small-world brain networks. *The Neuroscientist, 12*(6), 512–523. doi:10.1177/1073858406293182 PMID:17079517

Bergasa, L. M., & Nuevo, J. (2005). Real-time system for monitoring driver vigilance. *Proceedings of the IEEE International Symposium on Industrial Electronics* (Vol. III, pp. 1303–1308). http://doi.org/doi:<ALIGNMENT.qj></ALIGNMENT>10.1109/ISIE.2005.1529113

Bhowmick, B., & Kumar, K. S. C. (2009). Detection and classification of eye state in ir camera for driver drowsiness identification. *Proceedings of the 2009 IEEE International Conference on Signal and Image Processing ApplicationsConferenceICSIPA '09* (pp. 340–345). http://doi.org/ doi:10.1109/ICSIPA.2009.5478674

Bowman, D., Schaudt, W., & Hanowski, R. (2012). Advances in drowsy driver assistance systems through data fusion. In *Handbook of Intelligent Vehicles* (pp. 895-912). Retrieved from http://link.springer.com/10.1007/978-0-85729-085-4_34

Bundele, M., & Banerjee, R. (2009). Detection of fatigue of vehicular driver using skin conductance and oximetry pulse: a neural network approach. *Proceedings of the 11th International Conference on Information Integration and web-based applications & services* (pp. 739-744). Retrieved from http://dl.acm.org/citation.cfm?id=1806478

Caffier, P. P., Erdmann, U., & Ullsperger, P. (2003). Experimental evaluation of eye-blink parameters as a drowsiness measure. *European Journal of Applied Physiology, 89*(3), 319–325. doi:10.1007/s00421-003-0807-5 PMID:12736840

Cai, R. S., Zhu, Y. T., & Guo, Y. M. (2007). Wavelet-based multi-feature voiced/unvoiced speech classification algorithm. *Proceedings of theIET Conference on Wireless, Mobile and Sensor Networks 2007 (CCWMSN07)* (Vol. 2007, pp. 897–900). IEE. http://doi.org/ doi:10.1049/cp:20070294

Cajochen, C., Zeitzer, J. M., Czeisler, C. A., & Dijk, D. J. (2000). Dose-response relationship for light intensity and ocular and electroencephalographic correlates of human alertness. *Behavioural Brain Research, 115*(1), 75–83. doi:10.1016/S0166-4328(00)00236-9 PMID:10996410

Chaudhuri, A., Routray, A., & Kar, S. (2012). Effect of sleep deprivation on estimated distributed sources for Scalp EEG signals: A case study on human drivers. *Proceedings of the 2012 4th International Conference on Intelligent Human Computer Interaction (IHCI)* (pp. 1–6). IEEE. http://doi.org/ doi:10.1109/IHCI.2012.6481805

Chialvo, D. R. (2004). Critical brain networks. In A. Physica (Ed.), *Statistical Mechanics and its Applications* (Vol. 340, pp. 756–765).doi:10.1016/j.physa.2004.05.064

Croft, R. J., & Barry, R. J. (2000). Removal of ocular artifact from the EEG: A review. *Neurophysiologie Clinique, 30*(1), 5–19. doi:10.1016/S0987-7053(00)00055-1 PMID:10740792

Dasgupta, A., George, A., Happy, S. L., & Routray, A. (2013). A Vision-Based System for Monitoring the Loss of Attention in Automotive Drivers. *IEEE Transactions on Intelligent Transportation Systems, 15*(1), 1–14.doi:10.1109/TITS.2013.2271052

Dasgupta, A., George, A., Happy, S. L., Routray, A., & Shanker, T. (2013). An on-board vision based system for drowsiness detection in automotive drivers. *International Journal of Advances in Engineering Sciences and Applied Mathematics, 5*(2-3), 94–103. doi:10.1007/s12572-013-0086-2

De Waard, D., & Brookhuis, K. A. (1991). Assessing driver status: A demonstration experiment on the road. *Accident; Analysis and Prevention, 23*(4), 297–307. doi:10.1016/0001-4575(91)90007-R PMID:1883469

Dhupati, L. S., Kar, S., Rajaguru, A., & Routray, A. (2010). A novel drowsiness detection scheme based on speech analysis with validation using simultaneous EEG recordings. *Proceedings of the2010 IEEE International Conference on Automation Science and Engineering* (pp. 917–921). IEEE. http://doi.org/ doi:10.1109/COASE.2010.5584246

Dinges, D., Mallis, M., Maislin, G., & Powell, I. (1998). Evaluation of techniques for ocular measurement as an index of fatigue and the basis for alertness management. Retrieved from http://trid.trb.org/view.aspx?id=647942

Driver fatigue and road accidents. (n. d.). *ROSPA*. Retrieved from http://www.rospa.com/roadsafety/info/fatigue.pdf

Eoh, H. J., Chung, M. K., & Kim, S. H. (2005). Electroencephalographic study of drowsiness in simulated driving with sleep deprivation. *International Journal of Industrial Ergonomics, 35*(4), 307–320. doi:10.1016/j.ergon.2004.09.006

Fatigue is a Major Cause in Truck Crashes. (n. d.). *OPTAlert*. Retrieved from http://www.optalert.com/news/truck-crashes-fatigue

Gonzalez, R.C., Woods, R.E., & Eddins, S.L. (2004). Digital image processing using MATLAB. Retrieved from http://course.sdu.edu.cn/Download/a273bf58-2fc1-4bbf-92a3-fbaa8c54f534.pdf

Grandjean, E. (1979). Fatigue in industry. *British Journal of Industrial Medicine, 36*, 175–186. doi:10.2105/AJPH.12.3.212 PMID:40999

Grech, R., Cassar, T., Muscat, J., Camilleri, K. P., Fabri, S. G., Zervakis, M., & Vanrumste, B. et al. (2008). Review on solving the inverse problem in EEG source analysis. *Journal of Neuroengineering and Rehabilitation, 5*(1), 25. doi:10.1186/1743-0003-5-25 PMID:18990257

Greeley, H., & Friets, E. (2006). Detecting fatigue from voice using speech recognition. *Proceedings of the 2006 IEEE International Symposium on Signal Processing and Information technology.* Retrieved from http://ieeexplore.ieee.org/xpls/abs_all.jsp?arnumber=4042307

Gupta, S., Kar, S., Gupta, S., & Routray, A. (2010). Fatigue in human drivers: A study using ocular, psychometric, physiological signals. *Proceedings of the 2010 IEEE Students' Technology Symposium TechSym '10* (pp. 234–240). doi:<ALIGNMENT.qj></ALIGNMENT>10.1109/TECHSYM.2010.5469152

Hammoud, R., Witt, G., Dufour, R., Wilhelm, A., & Newman, T. (2008). On driver eye closure recognition for commercial vehicles. Retrieved from http://papers.sae.org/2008-01-2691/

Hanowski, R., & Bowman, D. (2008). PERCLOS+: Development of a robust field measure of driver drowsiness. *Proceedings of the 15th World Congress on Intelligent transport systems and ITS America's 2008 Annual meeting.* Retrieved from http://trid.trb.org/view.aspx?id=904975

Hayami, T., Matsunaga, K., Shidoji, K., & Matsuki, Y. (2002). Detecting drowsiness while driving by measuring eye movement - a pilot study. *Proceedings of the IEEE 5th International Conference on Intelligent Transportation Systems.* doi:<ALIGNMENT.qj></ALIGNMENT>10.1109/ITSC.2002.1041206

Hong, T., & Qin, H. (2007). Drivers drowsiness detection in embedded system. *Proceedings of the IEEE International conference on Vehicular Electronics and Safety ICVES '07.* Retrieved from http://ieeexplore.ieee.org/xpls/abs_all.jsp?arnumber=4456381

Huang, R. S., Jung, T. P., Delorme, A., & Makeig, S. (2008). Tonic and phasic electroencephalographic dynamics during continuous compensatory tracking. *NeuroImage, 39*(4), 1896–1909. doi:10.1016/j.neuroimage.2007.10.036 PMID:18083601

Huang, R. S., Tsai, L. L., & Kuo, C. J. (2001). Selection of valid and reliable EEG features for predicting auditory and visual alertness levels. *Proceedings of the National Science Council, Republic of China. Part B, Life Sciences, 25*, 17–25. PMID:11254168

Inchingolo, P., & Spanio, M. (1985). On the identification and analysis of saccadic eye movements-A quantitative study of the processing procedures. *IEEE Transactions on Biomedical Engineering*, BME-32(9), 683-695. Retrieved from http://ieeexplore.ieee.org/xpls/abs_all.jsp?arnumber=4122143

Jackson, C. A., & Earl, L. (2006). Prevalence of fatigue among commercial pilots. *Occupational Medicine (Oxford, England)*, 56(4), 263–268. doi:10.1093/occmed/kql021 PMID:16733255

Jap, B. T., Lal, S., Fischer, P., & Bekiaris, E. (2007). Using spectral analysis to extract frequency components from electroencephalography: Application for fatigue countermeasure in train drivers. *Proceedings of the 2nd International Conference on Wireless Broadband and Ultra Wideband Communications, AusWireless '07*.doi:<ALIGNMENT.qj></ALIGNMENT>10.1109/AUSWIRELESS.2007.83

Ji, Q., & Yang, X. (2002). Real-Time Eye, Gaze, and Face Pose Tracking for Monitoring Driver Vigilance. *Real-Time Imaging*, 8(5), 357–377. doi:10.1006/rtim.2002.0279

Jung, T. P., Makeig, S., Stensmo, M., & Sejnowski, T. J. (1997). Estimating alertness from the EEG power spectrum. *IEEE Transactions on Bio-Medical Engineering*, 44(1), 60–69. doi:10.1109/10.553713 PMID:9214784

Kar, S., Bhagat, M., & Routray, A. (2010). EEG signal analysis for the assessment and quantification of driver's fatigue. *Transportation Research Part F: Traffic Psychology and Behaviour*, 13(5), 297–306. doi:10.1016/j.trf.2010.06.006

Kar, S., & Routray, A. (2013). Effect of Sleep Deprivation on Functional Connectivity of EEG Channels. *IEEE Transactions on Systems, Man, and Cybernetics Systems*, 43(3), 666–672.doi:10.1109/TSMCA.2012.2207103

Kiymik, M. K., Akin, M., & Subasi, A. (2004). Automatic recognition of alertness level by using wavelet transform and artificial neural network. *Journal of Neuroscience Methods*, 139(2), 231–240. doi:10.1016/j.jneumeth.2004.04.027 PMID:15488236

Klimesch, W. (1999). EEG alpha and theta oscillations reflect cognitive and memory performance: A review and analysis. *Brain Research. Brain Research Reviews*, 29(2-3), 169–195. doi:10.1016/S0165-0173(98)00056-3 PMID:10209231

Krajewski, J., Batliner, A., & Golz, M. (2009). Acoustic sleepiness detection: Framework and validation of a speech-adapted pattern recognition approach. *Behavior Research Methods*, 41(3), 795–804. doi:10.3758/BRM.41.3.795 PMID:19587194

Krajewski, J., Schnieder, S., Sommer, D., Batliner, A., & Schuller, B. (2012). Applying multiple classifiers and non-linear dynamics features for detecting sleepiness from speech. *Neurocomputing*, 84, 65–75. doi:10.1016/j.neucom.2011.12.021

Krajewski, J., Wieland, R., & Batliner, A. (2008). An acoustic framework for detecting fatigue in speech based human-computer-interaction. In Computers Helping People with Special Needs, LNCS (Vol. 5105, pp. 54–61). doi:10.1007/978-3-540-70540-6_7

Krishnaveni, V., Jayaraman, S., Anitha, L., & Ramadoss, K. (2006). Removal of ocular artifacts from EEG using adaptive thresholding of wavelet coefficients. *Journal of Neural Engineering, 3*(4), 338–346. doi:10.1088/1741-2560/3/4/011 PMID:17124338

Krishnaveni, V., Jayaraman, S., Aravind, S., Hariharasudhan, V., & Ramadoss, K. (2006). Automatic identification and Removal of ocular artifacts from EEG using Wavelet transform. *Measurement Science Review, 6*, 45–57. Retrieved from http://www.freewebs.com/biomedical-eng/scprs/2.pdf

Lal, S. K. L., & Craig, A. (2001). A critical review of the psychophysiology of driver fatigue. *Biological Psychology, 55*(3), 173–194. doi:10.1016/S0301-0511(00)00085-5 PMID:11240213

Lal, S. K. L., Craig, A., Boord, P., Kirkup, L., & Nguyen, H. (2003). Development of an algorithm for an EEG-based driver fatigue countermeasure. *Journal of Safety Research, 34*(3), 321–328. doi:10.1016/S0022-4375(03)00027-6 PMID:12963079

Lenné, M. G., Triggs, T. J., & Redman, J. R. (1997). Time of day variations in driving performance. *Accident; Analysis and Prevention, 29*(4), 431–437. doi:10.1016/S0001-4575(97)00022-5 PMID:9248501

Leonard Janer, L., Bonet, J. J., & Lleida-Solano, E. (1996). Pitch detection and voiced/unvoiced decision algorithm based on wavelet transforms. *Proceedings of theFourth International Conference on Spoken Language Processing* (Vol. 2, p. 1209). doi:10.1109/ICSLP.1996.607825

Lobo, A., & Loizou, P. (2003). Voiced/unvoiced speech discrimination in noise using gabor atomic decomposition. *Acoustics, Speech, and Signal ....* Retrieved from http://ieeexplore.ieee.org/xpls/abs_all.jsp?arnumber=1198907

Macdonald, W. A. (1985). *Human Factors & Road Crashes - A Review of Their Relationship.*

Makeig, S., & Jung, T. P. (1995). Changes in alertness are a principal component of variance in the EEG spectrum. *Neuroreport, 7*(1), 213–216. doi:10.1097/00001756-199512000-00051 PMID:8742454

Matousek, M., & Petersen, I. (1983). A method for assessing alertness fluctuations from EEG spectra. *Electroencephalography and Clinical Neurophysiology, 55*(1), 108–113. doi:10.1016/0013-4694(83)90154-2 PMID:6185295

Matthews, G., Davies, D. R., Westerman, S. J., & Stammers, R. B. (2000). *Human performance: Cognition, stress, and individual differences.* East Sussex: Psychology Press.

Maycock, G. (1997). Sleepiness and driving: The experience of U.K. car drivers. *Accident; Analysis and Prevention, 29*(4), 453–462. doi:10.1016/S0001-4575(97)00024-9 PMID:9248503

Montez, T., Linkenkaer-Hansen, K., van Dijk, B. W., & Stam, C. J. (2006). Synchronization likelihood with explicit time-frequency priors. *NeuroImage, 33*(4), 1117–1125. doi:10.1016/j.neuroimage.2006.06.066 PMID:17023181

Papadelis, C., Chen, Z., Kourtidou-Papadeli, C., Bamidis, P. D., Chouvarda, I., Bekiaris, E., & Maglaveras, N. (2007). Monitoring sleepiness with on-board electrophysiological recordings for preventing sleep-deprived traffic accidents. *Clinical Neurophysiology, 118*(9), 1906–1922. doi:10.1016/j.clinph.2007.04.031 PMID:17652020

Papadelis, C., Kourtidou-Papadeli, C., Bamidis, P. D., Chouvarda, I., Koufogiannis, D., Bekiaris, E., & Maglaveras, N. (2006). Indicators of sleepiness in an ambulatory EEG study of night driving. *Proceedings of the Annual International Conference of the IEEE Engineering in Medicine and Biology Society* (Vol. 1, pp. 6201–6204). Doi:<ALIGNMENT.qj></ALIGNMENT>10.1109/IEMBS.2006.259614

Pascual-Marqui, R. D., Michel, C. M., & Lehmann, D. (1994). Low resolution electromagnetic tomography: A new method for localizing electrical activity in the brain. *International Journal of Psychophysiology*, *18*(1), 49–65. doi:10.1016/0167-8760(84)90014-X PMID:7876038

Penetar, D., McCann, U., & Thorne, D. (1993). Caffeine reversal of sleep deprivation effects on alertness and mood. *Psychopharmacology*, 112(2), 359-365. Retrieved from http://link.springer.com/article/10.1007/BF02244933

Quatieri, T. F. (2002). *Discrete-Time Speech Signal Processing: Principles and Practice*. Pearson Education. Retrieved from http://books.google.com/books?hl=en&lr=&id=UMR9ByupVy8C&pgis=1

Real-time 3D graphics and virtual set solutions. (n. d.). *Brainstorm*. Retrieved from http://neuroimage.usc.edu/brainstorm/

Rényi, A. (1961). On Measures of Entropy and Information. *Proceedings of the Fourth Berkeley Symposium on Mathematical Statistics and Probability* (*Vol. 1*, pp. 547–561). The Regents of the University of California.

Rosenblum, M., Pikovsky, A., & Kurths, J. (1996). Phase synchronization of chaotic oscillators. *Physical Review Letters*, *76*(11), 1804–1807. Retrieved from http://www.ncbi.nlm.nih.gov/pubmed/10060525 doi:10.1103/PhysRevLett.76.1804 PMID:10060525

Routray, A., & Kar, S. (2012). Classification of brain states using principal components analysis of cortical EEG synchronization and HMM. *Proceedings of the 2012 IEEE International Conference on Acoustics, Speech and Signal Processing (ICASSP)* (pp. 641–644). IEEE. http://doi.org/ doi:10.1109/ICASSP.2012.6287965

Rubinov, M., & Sporns, O. (2010). Complex network measures of brain connectivity: Uses and interpretations. *NeuroImage*, *52*(3), 1059–1069. doi:10.1016/j.neuroimage.2009.10.003 PMID:19819337

Schleicher, R., Galley, N., Briest, S., & Galley, L. (2008). Blinks and saccades as indicators of fatigue in sleepiness warnings: Looking tired? *Ergonomics*, *51*(7), 982–1010. doi:10.1080/00140130701817062 PMID:18568959

Sengupta, A., Routray, A., & Kar, S. (2013). Complex brain networks using Visibility Graph synchronization. *Proceedings of the2013 Annual IEEE India Conference (INDICON)* (pp. 1–4). IEEE. http://doi.org/ doi:10.1109/INDCON.2013.6726126

Shah, J., & Iyer, A. (2004). Robust voiced/unvoiced classification using novel features and gaussian mixture model. *Proceedings of the IEEE Conference on Acoustics, Speech and Signal Processing*.

Shannon, C. E. (1948). A Mathematical Theory of Communication. *ACM SIGMOBILE mobile computing and communications review*, *5*(1), 3–55.

Shimomura, Y., Yoda, T., & Sugiura, K. (2008). Use of frequency domain analysis of skin conductance for evaluation of mental workload. *The Journal of Physiology*. PMID:18832780

Smith, P., Shah, M., & da Vitoria Lobo, N. (2003). Determining driver visual attention with one camera. *IEEE Transactions on Intelligent Transportation Systems*, 4(4), 205–218. doi:10.1109/TITS.2003.821342

Sporns, O., & Zwi, J. D. (2004). The small world of the cerebral cortex. *Neuroinformatics*, 2(2), 145–162. doi:10.1385/NI:2:2:145 PMID:15319512

Stam, C. J. (2004). Functional connectivity patterns of human magnetoencephalographic recordings: A "small-world" network? *Neuroscience Letters*, 355(1-2), 25–28. doi:10.1016/j.neulet.2003.10.063 PMID:14729226

Stam, C. J. (2005). Nonlinear dynamical analysis of EEG and MEG: Review of an emerging field. *Clinical Neurophysiology*, 116(10), 2266–2301. doi:10.1016/j.clinph.2005.06.011 PMID:16115797

Stam, C. J., de Haan, W., Daffertshofer, A., Jones, B. F., Manshanden, I., van Cappellen van Walsum, A. M., & Scheltens, P. et al. (2009). Graph theoretical analysis of magnetoencephalographic functional connectivity in Alzheimer's disease. *Brain. Journal of Neurology*, 132, 213–224.doi:10.1093/brain/awn262 PMID:18952674

Stam, C. J., Jones, B. F., Nolte, G., Breakspear, M., & Scheltens, P. (2007). Small-world networks and functional connectivity in Alzheimer's disease. *Cerebral Cortex*, 17(1), 92–99. doi:10.1093/cercor/bhj127 PMID:16452642

Stam, C. J., & Reijneveld, J. C. (2007). Graph theoretical analysis of complex networks in the brain. *Nonlinear Biomedical Physics*, 1(1), 3. doi:10.1186/1753-4631-1-3 PMID:17908336

Stam, C. J., & Van Dijk, B. W. (2002). Synchronization likelihood: An unbiased measure of generalized synchronization in multivariate data sets. *Physica D. Nonlinear Phenomena*, 163(3-4), 236–251. doi:10.1016/S0167-2789(01)00386-4

Strogatz, S. H. (2001). Exploring complex networks. *Nature*, 410(6825), 268–276. doi:10.1038/35065725 PMID:11258382

Ueno, A., & Uchikawa, Y. (2004). Relation between human alertness, velocity wave profile of saccade, and performance of visual activities. *Proceedings of the Annual International Conference of the IEEE Engineering in Medicine and Biology Society* (Vol. 2, 933–935). Doi:<ALIGNMENT.qj></ALIGNMENT>10.1109/IEMBS.2004.1403313

Van Dongen, H. P. A., & Dinges, D. F. (2000). Circadian Rhythms in Fatigue, Alertness and Performance. In *Principles and Practice of Sleep Medicine* (pp. 391–399). Retrieved from http://www.nps.navy.mil/orfacpag/resumepages/projects/fatigue/dongen.pdf

Viola, P., & Jones, M. (2001). Rapid object detection using a boosted cascade of simple features. *Proceedings of the 2001 IEEE Computer Society Conference on Computer Vision and Pattern Recognition. CVPR '01* (Vol. 1, pp. I–511–I–518). http://doi.org/ doi:10.1109/CVPR.2001.990517

Vuckovic, A., Radivojevic, V., Chen, A. C. N., & Popovic, D. (2002). Automatic recognition of alertness and drowsiness from EEG by an artificial neural network. *Medical Engineering & Physics, 24*(5), 349–360. doi:10.1016/S1350-4533(02)00030-9 PMID:12052362

Watts, D. J., & Strogatz, S. H. (1998). Collective dynamics of "small-world" networks. *Nature, 393*(6684), 440–442. doi:10.1038/30918 PMID:9623998

Whitmore, J., & Fisher, S. (1996). Speech during sustained operations. *Speech Communication, 20*(1-2), 55–70. doi:10.1016/S0167-6393(96)00044-1

Wierwille, W. W., Wreggit, S. S., Kirn, C. L., Ellsworth, L. A., & Fairbanks, R. J. (n. d.). Research on vehicle-based driver status/performance monitoring; development, validation, and refinement of algorithms for detection of driver drowsiness (Final report). Retrieved from http://trid.trb.org/view.aspx?id=448128

Williams, G. W. (1963). Highway hypnosis: An hypothesis. *The International Journal of Clinical and Experimental Hypnosis, 11*(3), 143–151. doi:10.1080/00207146308409239 PMID:14050133

Wilson, J.R. (2007). *People and Rail Systems: Human Factors at the Heart of the Railway*. Ashgate Publishing, Ltd. Retrieved from https://books.google.com/books?hl=en&lr=&id=RLOrqGhSMOsC&pgis=1

Wu, Q., Sun, B., Xie, B., & Zhao, J. (2010). A PERCLOS-based driver fatigue recognition application for smart vehicle space. *Proceedings of the 3rd International Symposium on Information Processing ISIP '10* (pp. 437–441). doi:<ALIGNMENT.qj></ALIGNMENT>10.1109/ISIP.2010.116

Yang, J., & Zhang, D. (2004). Two-dimensional PCA: a new approach to appearance-based face representation and recognition. *IEEE Transactions on Pattern Analysis and Machine intelligence, 26*(1). Retrieved from http://ieeexplore.ieee.org/xpls/abs_all.jsp?arnumber=1261097

Yeo, M.V.M., Li, X., Shen, K., & Wilder-Smith, E.P.V. (2009). Can SVM be used for automatic EEG detection of drowsiness during car driving? *Safety Science, 47*(1), 115–124. doi:10.1016/j.ssci.2008.01.007

Zahran, E., & Abbas, A. (2009). High performance face recognition using PCA and ZM on fused LWIR and VISIBLE images on the wavelet domain. *Proceedings of the international conference on computer engineering & systems ICCES '09*. Retrieved from http://ieeexplore.ieee.org/xpls/abs_all.jsp?arnumber=5383223

Zhu, G., Li, Y., & Wen, P. P. (2014). Analysis and Classification of Sleep Stages Based on Difference Visibility Graphs from a Single Channel EEG Signal. *IEEE Journal of Biomedical and Health Informatics, 18*(6), 1813-1821. Doi:<ALIGNMENT.qj></ALIGNMENT>10.1109/JBHI.2014.2303991

Zhu, Z.Z.Z., & Ji, Q.J.Q. (2004). Real time and non-intrusive driver fatigue monitoring. *Proceedings of the 7th International IEEE Conference on Intelligent Transportation Systems (IEEE Cat. No.04TH8749)*. doi:<ALIGNMENT.qj></ALIGNMENT>10.1109/ITSC.2004.1398979

ADDITIONAL READING

Belyavin, A., & Wright, N. A. (1987). Changes in electrical activity of the brain with vigilance. *Electroencephalography and Clinical Neurophysiology, 66*(2), 137–144. doi:10.1016/0013-4694(87)90183-0 PMID:2431878

Bergasa, L. M., Nuevo, J., Sotelo, M. A., Barea, R., & Lopez, M. E. (2006). Real-time system for monitoring driver vigilance.. *IEEE Transactions on* Intelligent Transportation Systems, 7(1), 63–77.

Bhaduri, S., & Ghosh, D. (2014). Electroencephalographic Data Analysis with Visibility Graph Technique for Quantitative Assessment of Brain Dysfunction. *Clinical EEG and Neuroscience*. PMID:24781371

Dasgupta, A., George, A., Happy, S. L., & Routray, A. (2013). A Vision-Based System for Monitoring the Loss of Attention in Automotive Drivers.

Dasgupta, A., George, A., Happy, S. L., Routray, A., & Shanker, T. (2013). An on-board vision based system for drowsiness detection in automotive drivers. *International Journal of Advances in Engineering Sciences and Applied Mathematics, 5*(2-3), 94–103. doi:10.1007/s12572-013-0086-2

Dinges, D. F., & Grace, R. (1998). PERCLOS: A valid psychophysiological measure of alertness as assessed by psychomotor vigilance (Tech. Rep. MCRT-98-006). *Federal Highway Administration*, Office of motor carriers.

Dong, Y., Hu, Z., Uchimura, K., & Murayama, N. (2011). Driver inattention monitoring system for intelligent vehicles: A review. *IEEE Transactions on* Intelligent Transportation Systems, 12(2), 596–614.

Fioriti, V., Tofani, A., & Di Pietro, A. (2012). Discriminating Chaotic Time Series with Visibility Graph Eigenvalues. *Complex Systems, 21*(3).

Hanke, S., Oberleitner, A., Lurf, R., & König, G. (2013). *A Wearable Device for Realtime Assessment of Vigilance*. Biomedical Engineering/Biomedizinische Technik.

Heitmann, A., Guttkuhn, R., Aguirre, A., Trutschel, U., & Moore-Ede, M. (2001). Technologies for the monitoring and prevention of driver fatigue. *Proceedings of the First International Driving Symposium on Human Factors in Driver Assessment, Training and Vehicle Design* (pp. 81-86).

Jagannath, M., & Balasubramanian, V. (2014). Assessment of early onset of driver fatigue using multimodal fatigue measures in a static simulator. *Applied Ergonomics, 45*(4), 1140–1147. doi:10.1016/j.apergo.2014.02.001 PMID:24581559

Ji, Q., & Yang, X. (2002). Real-time eye, gaze, and face pose tracking for monitoring driver vigilance. *Real-Time Imaging, 8*(5), 357–377. doi:10.1006/rtim.2002.0279

Ji, Q., Zhu, Z., & Lan, P. (2004). Real-time nonintrusive monitoring and prediction of driver fatigue. *IEEE Transactions on* Vehicular Technology, 53(4), 1052–1068.

Mardi, Z., Ashtiani, S. N. M., & Mikaili, M. (2011). EEG-based Drowsiness Detection for Safe Driving Using Chaotic Features and Statistical Tests. *Journal of medical signals and sensors, 1*(2), 130.

Mehar, N., Zamir, S., Zulfiqar, A., Farouqui, S., Rehman, A., & Rashdi, M. A. (2013). Vigilance Estimation Using Brain Machine Interface. *World Applied Sciences Journal*, *27*(2), 148–154.

Oken, B. S., Salinsky, M. C., & Elsas, S. M. (2006). Vigilance, alertness, or sustained attention: Physiological basis and measurement. *Clinical Neurophysiology*, *117*(9), 1885–1901. doi:10.1016/j.clinph.2006.01.017 PMID:16581292

Pastor, G., Tejero, P., Choliz, M., & Roca, J. (2006). Rear-view mirror use, driver alertness and road type: An empirical study using EEG measures. *Transportation Research Part F: Traffic Psychology and Behaviour*, *9*(4), 286–297. doi:10.1016/j.trf.2006.01.007

Pei, Z., Zhenghe, S., & Yiming, Z. (2002). PERCLOS-based recognition algorithms of motor driver fatigue. *Zhongguo Nongye Daxue Xuebao*, *7*(2), 104–109.

Routray, A., & Kar, S. (2012, March). Classification of brain states using principal components analysis of cortical EEG synchronization and HMM. *Proceedings of the 2012 IEEE International Conference on Acoustics, Speech and Signal Processing (ICASSP)* (pp. 641-644). IEEE. doi:10.1109/ICASSP.2012.6287965

Sengupta, A., Routray, A., & Kar, S. (2013, December). Complex Brain Networks Using Visibility Graph Synchronization. *Proceedings of the 2013 Annual IEEE India Conference (INDICON)* (pp. 1-4). IEEE. doi:10.1109/INDCON.2013.6726126

Smith, P., Shah, M., & da Vitoria Lobo, N. (2000, September). Monitoring head/eye motion for driver alertness with one camera. *Proceedings of the International Conference on Pattern Recognition* (Vol. 4, pp. 4636-4636). IEEE Computer Society.

Ueno, H., Kaneda, M., & Tsukino, M. (1994, August). Development of drowsiness detection system. *Proceedings of the* Vehicle Navigation and Information Systems Conference (pp. 15–20). IEEE.

Wang, Q., Yang, J., Ren, M., & Zheng, Y. (2006, June). Driver fatigue detection: a survey. *Proceedings of the Sixth World Congress on Intelligent Control and Automation WCICA '06* (Vol. 2, pp. 8587-8591). IEEE.

Zhang, C., Wang, H., & Fu, R. (2014). Automated Detection of Driver Fatigue Based on Entropy and Complexity Measures.

Zhang, C., Yu, X., Yang, Y., & Xu, L. (2014). Phase Synchronization and Spectral Coherence Analysis of EEG Activity during Mental Fatigue. *Clinical EEG and Neuroscience*. PMID:24590874

KEY TERMS AND DEFINITIONS

Characteristic Path Length: Measure of functional integration in brain networks. The average shortest path length between all pairs of nodes in the network.

Clustering Coefficient: Measure of segregation in brain networks. The fraction of triangles around an individual mode in the network; equivalent to the fraction of the node's neighbors that are also neighbors of each other.

EEG: The electrical activity of the brain, recorded for a short period of time along the scalp. Measure of fluctuation of voltage resulting from ionic current flows within the neurons of the brain.

Entropy: Average amount of information contained in a sample drawn from a distribution or data stream. Measure of uncertainty of the source of information.

Fatigue: A physical and/or mental state of being tired and weak. May be caused by overwork, lack of sleep, anxiety, boredom, over/underactive thyroid glands, depression or certain medications.

PERCLOS: The percentage time where eyes are occluded at least by 80%.

Saccades: Saccadic movements are the ballistic movements of both eyes in the same direction which the subject is shifting the point of gaze. Eye saccadic movements are the one of the fastest movements human body can make.

Saccadic Ratio: The ratio of peak saccadic velocity to the saccadic duration.

Visibility Graph: A graph for a set of points in the Euclidean plane, where each node stands for the location of a point, and each edge represents a visible connection between them.

Section 4
Operation

Chapter 14
Railway Operations Models:
The OR Approach

Sundaravalli Narayanaswami
IIM Ahmedabad, India

ABSTRACT

This chapter is intended as an exposure to OR based methods, particularly the analytical approach to modelling railway operations. An overview of several planned operations in railway transportation is provided in an academic context. Some of the applications and the associated models are applied in realistic settings in the transportation industry, and also have demonstrated evidence of acceptance over a long number of years. Primary coverage is on transportation scheduling and the concise discussions are on planning phases, various operations that can be deterministically modeled and analysed, model development, few exercises and real-world stories, wherever appropriate. All sections are adequately provided with the list of references and an interested reader can benefit from a conceptual understanding to model development and to implement and deploy, under some prior knowledge on the basics and programming experience.

1. INTRODUCTION

A majority of railway operations modeling problems discuss transportation scheduling and to a lesser extent on other railway operations such as rolling stock and crew scheduling. Operations Research (OR) has been successfully employed in modeling and analysing a wide range of transportation problems for several decades. Many commercial solution vendors also have been applying OR techniques to produce superior and efficient software solutions for transportation problems. There are several reasons as to why OR is considered more appropriate for transportation problems. Transportation problems are generally large, highly complex, and there are multiple dependencies between the various sub-problems in transportation. In general, transportation majorly involves public systems and huge levels of frequent interactions with public support systems; which would mean multitude of issues and challenges. Specifically in transportation problems, a diverse set of stakeholders with huge conflicting agendas are involved, and different sections of societies are sought and output services benefit not necessarily those who paid

DOI: 10.4018/978-1-5225-0084-1.ch014

for it, but an entire society with little or no impact. Another aspect of transportation systems is that investments are extremely large, that it is impossible without huge funds, either on one's own capability or sources from multiple agencies. Public transportation systems are also responsible for building, maintenance and sharing of infrastructure with multiple stake-holders, that are sometimes chargeable and sometimes not. Railway transportation systems are quite distinct from other transportation modes, such as roadways, airways and waterways. Railways enjoy a dedicated right of ways of traffic movement, build, maintain their infrastructure and operate within their own facilities. Though the motive of transportation systems is not to make profits, considering the large project sizes and that most of the investments are drawn from external agencies, there is a pressure to recover loans and to self-sustain by generating the operational expenses and to generate surplus funds for future investments. This chapter is part method driven and part application driven. The next section covers an overview of Operations research, techniques and methods and few applications.

2. OPERATIONS RESEARCH: AN OVERVIEW

Operations Research is a scientific approach that helps decision makers and analysts to make informed decisions in order to improve the quality of operations in a cost effective manner. Most of the operations are concerned about allocating available resources to demands; the problems get more interesting when there are multiple conflicting demands, when the resources are insufficient or in huge excess in comparison to demands, when there are time restrictions, and when there are multiple conflicting demands. Broadly operations managers face two issues: one is to do decide between a set of choices, as which resource may be allocated to which demand and second is when, how long to allocate the resource to the demand (and / or the sequence in which resources are allocated to demands).

Forian et.al (1988) proposed the well-known topology of transportation models with two dimensions: procedures (activities to be performed, demand, generation) and perspectives (strategic, tactical and operational). Effective modeling using OR is to minimize the gaps by carefully matching procedures and perspectives. Specific aspects that are key determinants to right modeling are (i) decision making context, (ii) accuracy required, (iii) availability of suitable data, (iv) state of-the art in modeling, (v) available resources, (vi) data processing capabilities and requirements, and (vii) levels of training and skills. Modeling is a broader term that includes physical, conceptual, mathematical and other kinds of models. Physical models are generally not used in operational design or analysing in transportation; unless a break through archetype engineering technology is introduced in the system. To analyse large, complex systems such as railway transportation, it is necessary to understand the relationships of the components, the detailed operations and functionalities, and interactions between each other. A conceptual model helps in understanding such systems, define goals of each inter-related sub-systems in a view to analyse and evaluate the system on completion. A conceptual model is usually presented as a chart with systems components, or sequence of steps in an analysis, or flow of events. Inter-relations between various components, events that trigger any operations or decision making and all output possibilities are included.

The modeling process can be described using the figure below. Starting from an existing system or a blank space, an analyst makes assumptions that maps the real world system to sufficient details, yet simplified to compute and analyse. One of the assumptions of modeling in this chapter is that the system is deterministic. Brief discussions on in deterministic or stochastic variations shall be discussed later. Figure 1 is an illustration of the modeling process.

Figure 1. A modeling process as work flow

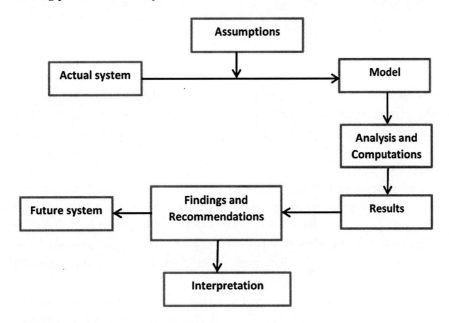

3. RAILWAY OPERATIONS PLANNING

While planning railway operations, one major consideration is the type of scheduling railway services. Broadly there are two types: (i) schedule based approach and (ii) tonnage based approach. In schedule based approach, train schedules are published as a timetable and it is customer effective. In tonnage based approach, trains are not dispatched, until sufficient tonnage has not arrived to load trains to their full capacity. In the Table 1, both approaches are compared.

A more often repeated classification of railway operations planning, as per the available literature are strategic planning, operational planning and realtime planning.

Table 1.

	Schedule Based Approach	Tonnage Based Approach
Number of trains operated in the domain	Fixed	Variable
Length of trains	Fixed	Variable
Operating costs	High	Low
Reliability of Services	High	Very less
Rolling stock utilization	High	Less
Deadheading of equipment / crew	Can be less	High
Frequency of services	High	Unpredictable
Transit / Halt time constraints	Rigid and hard	Soft
Traffic type	Generally for passenger	More applicable for freight

Train dispatching problem poses two questions:

1. When should a train be dispatched from a station?
2. When should a train arrive at station?

These two questions shall also respond to how long a train should halt at a station? And how much is the transit time of train from one station to another? The objective of train dispatching problem is to determine detailed train movements and timetables over a rail network under various (Sahin, 1999) operational constraints and restrictions in order to minimize both train deviations from the planned schedule and total delay as well. Depending on the planning phase, the train dispatching problem can be formulated at the strategic or operational or real-time stages. Several hard and soft constraints can be imposed on this problem. If trains travel on single-line tracks overtaking and cross over are permitted at specific locations (sidings or meet-points), which are conveniently located at regular intervals along the line. Delays or deviations from the planned schedule occur when trains traveling (Michaelis and Schobel, 2009) either in opposite directions or in the same direction meet, thus requiring one of the trains to be pulled over for the other to cross or overtake it.

Given a blocking plan, developing a train schedule is (Bussieck, 1998) perhaps the next-most important operational planning task faced by a railroad. The train scheduling problem is to determine train routes, their frequencies, and the days of operation of each train, aiming to minimize the cost of carrying blocks of cars from their origins to their destinations.

Once a railroad has identified a blocking plan, it must design a train schedule so that trains can efficiently carry blocks from their origins to their destinations with an aim to minimize the total costs involved (Keaton, 1989). The train schedule design problem, henceforth referred to as the train scheduling problem, determines:

1. How many trains to run;
2. The origin, destination, and route of each train;
3. The arrival and departure times of each train at each station that it stops;
4. The weekly operating schedule for each train; and
5. The assignment of blocks of cars to trains, so that the total cost of transportation (including distance, crew, locomotives, fuel, car hire, etc.) is the minimum possible.

These decisions need to be made on a weekly basis, as a train schedule repeats every week.

The train scheduling problem is a very large-scale network optimization problem with trillions of decision variables (Jespersen-Groth et. al., 2009). Its difficulty precluded the development of train scheduling algorithms that are much needed in practical deployment. Some attempts have been made in the past to solve this problem. Early research on train scheduling divide the train scheduling problem into two separate stages: the train design problem and the block routing problem, which are solved separately. The train design problem determines train routes and the block routing problem routes blocks over the trains formed. An iterative procedure solves each of the two stages in succession, using the solution from the other stage to guide the next iteration. More recent papers consider integrated train scheduling problems (Leuthi, 2009) that are solved by a variety of heuristic approaches. Some of the solutions are applicable in real-life, because they are not scalable for realistically large train scheduling problems, or they ignore the practical realities necessary to generate implementable solutions.

Railway Operations Models

The train scheduling problem comprises three entities:

1. The physical (railroad) network;
2. Trains that travel on the physical network; and
3. Blocks that travel on the trains.

As a train travels from its origin to its destination, it picks up blocks at various nodes it visits and may also drop off blocks at those nodes (Cordeau et. al., 1998). Typically, several blocks ride on a train at any segment. Likewise, a block may travel on several trains as it goes from its origin to its destination. Major decisions to be made in this problem are:

1. What is the number of trains to run?
2. What are the origin, destination, and route of each train?
3. What are the arrival and departure times at each station that a train visits?

The crew scheduling problem entails assigning crews to trains for each route on all zones. Crew scheduling has to comply with human resources regulations, and also crew costs should be minimal and train delays due to crew unavailability should be minimized. Considering the large size of the problem and the multiple inter dependencies, the transportation problem, as a whole is not tractable (Brucker, 2005). Hence it is divided into a set of sub-problems that are usually solved sequentially at various stages of planning. Strategic planning problems concern long-term decisions such as routing and network designs (Leander and Lukaszewicx, 2008). Strategic problems are aimed at maximizing service quality under resource restrictions. Tactical planning problems concern (Jovanovic and Harker, 1991) service decisions, such as service frequency, timetabling, and are solved seasonally with occasional updates. Operational planning problems involve executing the proposed tactical services at minimum cost. Some of the operational control problems are vehicle scheduling, maintenance scheduling, shunting of rolling stock (Budai et. al. 2009), rakes management, corridor (platform) occupancy and crew scheduling and rosters. At a further lower hierarchy are the real-time control operations, though not too evident in manually controlled heavy railways.

We present in this chapter few problems under each category, sometimes with a known solution approach and methodology to illustrate the problem sizes that are typically solved.

4. STRATEGIC PLANNING

Strategic level planning is concerned with transit routes design, network planning, infrastructural planning, human resources and rolling stock (capacity) estimation (Maroti, 2006) and planning, pricing of services and revenue management (Goossens, 2004). In short, strategic planning considers both capacity and demand over the entire network. A transportation network design problem shall decide the fixed routes, nodes and links to include in every fixed route, the service frequency in each route and finally estimate and plan the resource requirements to efficiently operate services on all routes. A typical scheduling problem can be formulated on a graph with nodes, links and subsequently routes. Let $G = (N, A)$ be a graph with a set of nodes, N, a set of links, A and a set of routes, R. In a railway network node is a station or a halt location, link is a track that connects nodes and route is a directed set of links that can typically form an origin – destination (O-D) pair (Caprara et. al., 2007) At the strategic planning level, demand is estimated over different periods of times, and accordingly service frequency is decided. Since transportation is a public service with limited scope of monetary profits, the number of services is decided to be higher than the maximum demand requirements.

As illustrated in Figure 2 above, as the demand frequency increases, utilization of the available capacity reaches a saturation level, without much respite on the delay in services. If service delay has to be kept low, utilization also has to be low to cater to frequent demand requests. In transportation parlance, schedule leeway has to be sufficiently large at strategic level planning so as to efficiently manage uncertain and emergency demands at the lower levels of planning hierarchy (Cacchiani, 2008). A network planning model is presented here.

It is assumed that the demand O-D matrix, between a set of nodes or zones for a particular time period is known or can be mapped using available data. The O-D matrix serves as the input at strategic planning of network planning operations. A most commonly used objective is to minimize the total cost or the total time taken to fulfill the demands. The generalized cost is found by applying different weights in the objective function to the different components of travel time such as travel time, initial waiting time, alight- time, boarding-time, transit time, and halt time separately. Some formulations also include the number of transfers as a component in the generalized cost (Albrecht, 2009). In addition to the objective, several other constraints are proposed in the problem; constraints are applied to:

1. Ensure that adequate access to specific nodes or zones that are of high demand;
2. Ensure minimum frequencies of service to specific nodes or links in the network; and
3. Any other design considerations such as infrastructure availability and right-of-way.

Initially, a set of routes is constructed and then a set of frequencies of operating traffic on these routes is determined based on predicted demands. This could be developed using heuristic procedures, so that all nodes and links are covered in the planned routes. Generally, traffic routes are grouped together based on certain criteria such as zones or operations and is called as a fleet (Dorfman and Medanic, 2004). When the routes, fleet and demands are known, frequencies are determined by minimizing total travel time, subject to fleet size constraints. This is given as

$$Min \sum_{i \in N} \sum_{j \in N} D_{ij} T_{ij}(f)$$

Figure 2. Capacity – Delay variations

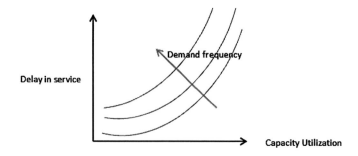

Railway Operations Models

subject to

$$\sum_{r \in R} \lceil RT_r f_r \rceil \leq totalfleetsize \tag{4.1}$$

where, D_{ij} is the demand between nodes i and j; T_{ij} is the travel time between nodes i and j, that also includes waiting times, as a function of frequencies of routes **f**. R is the set of routes, RT_r is the round trip time on route r and f_r is the frequency of route r.

The basic network planning problem can be addressed in different ways (Ingolitto, et. al., 2004):

1. Determining the links on the network to operate the routes and thereby fulfill demands;
2. Determining the optimal frequencies on each route;
3. Determining the minimum fleet size that can fulfill the demands on all routes; and
4. Determining the routes, when demand are known.

In further developments, objectives are combined; say to determine route and frequency together to fulfill all demands. Particularly in network planning of urban areas, routes are planned for integrating multi-modal transportation and to uniformly distribute traffic demands using a modal combination. Network planning is also performed to ensure last-mile connectivity. In this approach, also called as transit design problem, the objective is to maximize number of direct trips, for a given fleet and demand size. Then the number of transits, transit nodes, and transit time are determined, as required.

Recent studies examine network planning problem using a variety of approaches such as simulated annealing, Tabu search (Gorman, 1998), ant colony optimization (Ghoseiri and Morshedsolouk, 2006) genetic algorithms (Nachtigall and S. Voget, 1996) and (Salim and Cai, 1996), branch and bound, branch and cut, local search, and random search techniques (Huisman et. al., 2005) and mathematical solvers. Generally all the techniques begin by developing a set of skeleton routes followed by meta-heuristics to generate additional routes; the output is tested and evaluated through a network evaluation tool. Subsequent iterations are performed between the network analysis tool and the meta-heuristics to improve the solution quality.

5. TACTICAL PLANNING

Tactical planning is at the next lower level of hierarchy. Long term plans developed at the strategic levels are incorporated as tactical plans over a shorter period of time and with clearly defined operational goals (Jovanovic and Harker, 1991). At the tactical stages, demand sizes are better known, domain level issues are predictable, and therefore performance criteria can be defined more specifically. When a network plan is developed at the strategic level, timetabling is one of the tactical level problems that follow network planning (Michaelis and Schöbel, 2009). Other tactical stage problems are rolling stock planning, crew planning (Walker et. al., 2005), vehicle composition planning, corridor management, rake links management etc. We present here a basic model for timetabling. Timetabling is generating a schedule on each route, based on the desired service frequency. Routes, links and stations on each route, frequencies of services, are major inputs in tactical planning that are obtained from strategic plans (Chang and Chung,

2005). Timetabling generates schedules, ie., travel times of each train on each link, halt times of each train at each station, by adding slacks to each train's design running speed, so that service frequency and other strategic decisions (Liebchen and Stiller, 2009) are maintained. All timetables, generated for a reasonably long time, say a year, are made available to the public well in advance. Hence timetables are operator's public commitment to offer some sets of transport services.

Several variations of the problems using slack times, linking multiple modes, scheduling feeder networks have been researched. Usually timetables are cyclic, that the generated schedules repeat over a time period, say a week or 24 hours. Assumptions in generating timetables are constant arrival / departure / meet headways and a uniform demand over a time period. Some more complex issues of the timetable are related to integration, such as integration of multi-modes, integration of rakes allocation, integration of crew scheduling (Khan and Zhou, 2009) etc. Given R as the set of all trains, P is the set of all time-space paths, P_r is the set of all time-space paths of train r, P_s is the set of all time-space paths through station s, T is the set of time intervals indexed by \tau, c_p as the general cost of time-space path, r_p is the train which time-space path p belongs to, x_p is a 0-1 variable with $x_p = 1$, if train r_p is scheduled by time-space path p; 0 otherwise.

The objective is to minimize the general cost of all time-space paths

$$x_p + \sum_{p' \in P_{sip}} x_{p'} \leq 1, \forall s \in S, e \in E_{s,out}, p \in P_e \tag{5.1}$$

subject to:

1. Flow conservation constraints

$$\sum_{p \in P_r} x_p = 1, r \in R \tag{5.2}$$

2. Station capacity constraints

$$\sum_{r \in R} \sum_{p \in P_{si\tau}} x_p \leq num_s, \forall s \in S, \tau \in T \tag{5.3}$$

where num_s is the number of parallel tracks in station s

3. Minimum departure headway constraints

$$x_p + \sum_{p' \in P_{sip}} x_{p'} \leq 1, \forall s \in S, e \in E_{s,out}, p \in P_e \tag{5.4}$$

where $E_{s,out}$ is the set of section outflow from station s

Railway Operations Models

4. Minimum arrival headway constraints

$$x_p + \sum_{p=P_{s,p}} x_p \leq 1, \forall s \in S, p \in P_s \quad (5.5)$$

5. Safety constraints

$$\sum_{p \in P_{tw}} x_p = 0, \forall tw \in TW \quad (5.6)$$

where TW is the set of safety / maintenance windows.

6. Minimum halt time constraints

$$\sum_{p = P_s} x_p = 0, \forall s \in S \quad (5.7)$$

7. Decision variables

$$x_p = \{0,1\}, \forall p \in P \quad (5.8)$$

As observed the timetabling problem is a Mixed-integer large scale optimization problem and is known to be of NP-hard complexity. The most commonly used heuristic approach to solve these problems is branch-and bound or branch-and price algorithm.

6. OPERATIONAL PLANNING

Following the hierarchy of operations, when a timetable is generated by a railway operator for an available set of resources, and for a desired frequency of services, according to demands, operations are planned (Lindner and Zimmermann, 2005). Some of the operational plans are rakes link management, rolling stock planning, crew scheduling, roster planning, maintenance scheduling, and so on. The objective is to minimize total costs involved, with timetable as the primary input at the operational planning stages. Each of the planning problems at the operational control is supposed to give an insight into whether the timetable generated at the previous stage is feasible or not. For example, number of train units available is restricted by the available resource capacity. During train scheduling, it is understood for the desired frequency of services with the available resources, if it is feasible to develop a train schedule that can achieve the desired objective function or not (Burdett and Kozan 2008). Likewise, operational plan for all resources is developed and whenever infeasibility occurs, timetabling problem is revisited so that the complete operations can be feasibly planned.

The simplest form of vehicle scheduling problem can be modeled as below:

$$\min \sum_{(i,j) \in A} c_{ij} x_{ij} \tag{6.1}$$

$$\frac{d_j^{toLOC} - fromTime \leq D * BD2_j}{d_j^{toLOC} \geq toTime * conf_j}$$
$$toTime - a_j^{fromLOC} \leq D * AD1_j \quad \text{such that,}$$
$$AD1_j, BD2_j, conf_j \in \{0,1\}$$
$$\forall j = m+1, m+2...M$$

$$\sum_{j:(i,j) \in A} x_{ij} = 1, \forall i \in V \tag{6.2}$$

$$\sum_{i:(i,j) \in A} x_{ij} = 1, \forall j \in V \tag{6.3}$$

$$\sum_{(i,j) \in P} x_{ij} \leq |P| - 1, \forall P \in \pi \tag{6.4}$$

$$x_{ij} \in \{0,1\}, \forall (i,j) \in A \tag{6.5}$$

where A is the set of all arcs between all nodes in set V. P is the set of all paths, and x_{ij} is s path link and c_{ij} is the cost of traversing the path link x_{ij}. The objective of the model is to minimize the total costs. The first constraint ensures that each trip is covered only once and the second constraint ensures that only one train is involved in a trip. The third constraint is for flow conservation and the last constraint imposes binary restrictions. The model is known to be NP hard, but it has been successfully solved for a real-life problem that is reasonably large using a variety of relaxation techniques. The main challenge in the vehicle scheduling problem is exponential increase in the number of variables and constraints for very small increase in problem size. Several extensions of the basic model are possible by incorporating additional constraints to devise operational plans in rolling stock management, crew scheduling, corridor planning, maintenance scheduling (Heilporn et. al., 2008) and so on. Based on available resource capacity, feasibility analysis is done for all types of resources to finalize the operational level plans.

7. REAL TIME CONTROL

Disruptions to a planned schedule happen due to a wide variety of reasons, such as weather, accidents, technical faults, and so on (Wüest, et. al., 2009). Disruptions can lead to performance failure, where the pre-specified objective cannot be still maintained; or it could lead to conflicts, that are severe violations of safety constraints. An example of a violation of hard constraint is two trains occupying the same path length at the same time, which should be prevented by all means. A third consequence of disruption is a deadlock, where two trains that occupy two different path lengths mutually demand the other path length. When deadlocks happen, one of the two trains should backtrack so that the other train can make the demanded path length free for the first train. Deadlocks should be avoided, so that additional cost involved in backtracking can be prevented. Each of the three consequences of disruptions is handled in different resolution approaches (Mazzarello and Ottaviani, 2007). A rescheduling problem can be stated as, given static information such as timetable, railway infrastructure, train characteristics etc., and dynamic information such as disruptions, train positions at time t, a new conflict free schedule for the time window [t+a, t+b] has to be developed such that some performance index is minimized, where (a,b) is the conflict period.

A general form of rescheduling problem as a linear model is given below:

$$\min f(t,x) \tag{7.1}$$

$$\begin{aligned} &t_j \geq t_i + w_{ij}, \forall (i,j) \in F \\ &t_j \geq t_i + w_{ij} - M(1 - x_{ij,hk}), \forall (i,j)(h,k) \in F \\ &t_k \geq t_h + w_{hk} - M x_{ij,hk} \end{aligned} \tag{7.2}$$

$$\begin{aligned} &x_{ij,hk} = 1, if\ (i,j)\ is\ selected \\ &0, if\ (h,k)\ is\ selected \end{aligned} \tag{7.3}$$

where f(t,x) is some performance function in time and space, F is the set of nodes, A is the set of paths and w_{ij} is the time window of disruption. We present below a more elaborate extension of the model as a LP formulation. There are several other extensions available in existing literature.

One of the best known open problems in transportation scheduling is presented here, that the railroad industry has been trying to find a benchmark solution for decades. The problem is to scheduling the movement of trains optimally over a track in such a way that the trains do not collide, honor operational constraints and achieve maximum overall efficiency (or minimum overall delay). The problem is generally known as the *Dispatching* or *Meet-Pass Planning* problem, the tools used to address this problem in real-time are generally called *Movement Planners*, and the tools used to explore this problem on a planning basis are called *Line Capacity Models* or *Dispatching Models*. While many solutions exist for the planning mode of this problem, only after decades of work are a few solutions for the real-time variation now starting to enter operational use. The model presented here is based on Narayanaswami (2010) and is simple and small-sized in comparison to complexity of real-life problem instances. More recent heuristic models for rescheduling are discussed in Torquist. (2012) and Veelenturf et. al. (2012).

Problem Description

Given a conflict-free, feasible, bi-directional railway schedule between two terminals on a single track; trains run on different speeds on the track from one terminal (origin) to another (destination) terminal.

Feasibility of schedule refers to confirming to safety constraints, such as two trains cannot occupy a track segment simultaneously and there has to be a safe distance between trains that follow each other and that travel in opposite directions. Sidings (track segments of short lengths) are available at stations (scheduled halt locations) on the single track. Overtaking or cross-over of two trains can happen only at stations where sidings are available. Let this schedule be en-route and a disruption manifests as a track segment unavailability for a specifically known time-interval between two halt locations. Conversely, if a train is delayed at its origin terminal, the delay is also considered as track segment unavailability at that terminal. Impact of such disruption could be delay of one or more trains, conflict situations when one or more trains demanding to occupy a track segment simultaneously or a deadlock when two or more trains demand track segments mutually occupied by each other. The problem is to generate a conflict-free, feasible, complete reschedule of all trains in the operational domain with a quantifiable performance objective (Caimi et. al., 2009). The most common objective of this problem is to restore the original schedule at all stations and terminals, such that arrival and departure times for all trains at each station are close to the scheduled times. The problem setting is a linear layout of single track bi-directional railway traffic, as illustrated in Figure 3. Notation of data sets, parameters, decision variables and problem specific data is listed, followed by the objective function and constraints sets.

1.1 Notation

1.1.1 Sets and Indices

- T: Set of trains, i and $j \in T$, $T = \{1, 2, \ldots M\}$, and TU and TD are two partition sets of T.
- TU: Set of UP trains, $i \in TU$, $TU = \{1, 2, \ldots m\}$.
- TD: Set of DN trains, $j \in TD$, $TD = \{m + 1, m + 2, \ldots M\}$.
- S: Set of meet points $k \in S$, $S = \{1, 2, \ldots N\}$

1.1.2 Disruption data

- *fromLOC*: $fromLOC \in S$, the meet point where disruption starts.
- *toLOC*: $toLOC \in S$, the meet point where disruption ends, $toLOC = fromLOC + 1$, where *fromLOC* < *toLOC*.
- *fromTime*: Time at which disruption starts.
- *toTime*: Time at which disruption gets over.

Figure 3. Single track railway network with N stations and bi-directional traffic of M trains

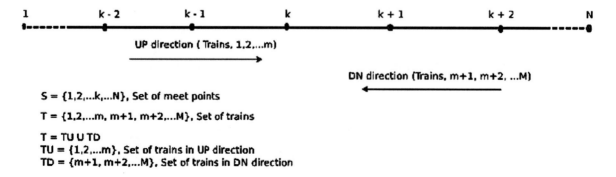

Railway Operations Models

1.1.3 Parameters

- a_i^k: Scheduled arrival time of train i at meet point k.
- d_i^k: Scheduled departure time of train i at meet point k.
- Tut_i^k: Minimum travel time of UP train i to reach meet point k from meet point $(k-1)$,

$\forall i = 1, 2, \ldots m$.

- Tdt_j^k: Minimum travel time of DN train j to reach meet point k from meet point $(k+1)$,

$\forall j = m+1, m+2, \ldots M$.

- Th_i^k: Minimum halt time of train i at meet point k.
- HA_{ij}^k: Minimum time difference between the arrivals of consecutive trains i and j of the same direction at meet point k.
- HD_{ij}^k: Minimum time difference between the departures of consecutive trains i and j of the same direction at meet point k.
- HM_{ij}^k: Minimum meet-time difference between the arrival of train i and the departure of train j at meet point k.
- w_i: Weight of train $i \in T$, in proportion to its operational importance.
- D: An arbitrarily large integer; included in such a manner that a constraint is binding if the 0-1 decision variable takes a value of 1 and becomes redundant, otherwise.
- $AU1_i$, $BU2_i$: 0-1 Conflict detection parameters for UP trains; they take up a value of 0 or 1 depending on the following conditions:

$$AU1_i = \begin{cases} 1 & if\ a_i^{toLOC} \geq fromTime \\ 0 & otherwise \end{cases} \quad BU2_i = \begin{cases} 1 & if\ d_i^{fromLOC} \leq toTime \\ 0 & otherwise \end{cases}$$

where a_i^k and d_i^k are defined for $k=toLOC$ and $k=fromLOC$ respectively

- $AD1_j$, $BD2_j$: 0-1 Conflict detection parameters for DN trains; they take up a value of 0 or 1 depending on the following conditions:

$$AD1_j = \begin{cases} 1 & if\ a_j^{fromLOC} \geq fromTime \\ 0 & otherwise \end{cases} \quad BD2_j = \begin{cases} 1 & if\ d_j^{toLOC} \leq toTime \\ 0 & otherwise \end{cases}$$

1.1.4 Decision Variables

- a_i^k: Actual (rescheduled) arrival time of train i at meet point k.
- d_i^k: Actual (rescheduled) departure time of train i from meet point k.
- Ax_{ij}^k: 0-1 Train flow variable, when both i and j are UP trains and equals 1 if train i arrives immediately before train j at meet point k and 0 otherwise.
- Dx_{ij}^k: 0-1 Train flow variable, when both i and j are UP trains and equals 1 if train i departs immediately before train j from meet point k and 0 otherwise.

- Ay_{ij}^k: 0-1 Train flow variable, when i is an UP train, j is a DN train and equals 1 if train i departs immediately before the arrival of train j at meet point k and 0 otherwise.
- Dy_{ij}^k: 0-1 Train flow variable, when i is an UP train, j is a DN train and equals 1 if train i arrives immediately before the departure of train j from meet point k and 0 otherwise.
- Az_{ij}^k: 0-1 Train flow variable, when both i and j are DN trains and equals 1 if train i arrives immediately before train j at meet point k and 0 otherwise.
- Dz_{ij}^k: 0-1 Train flow variable, when both i and j are DN trains and equals 1 if train i departs immediately before train j from meet point k and 0 otherwise.
- $conf_i$: 0-1 Conflict variable, and equals 1 if schedule of UP train i is conflicted because of disruption and 0 otherwise, where $i \in \{1 \ldots m\}$.

$$conf_i = (AU1_i \wedge BU2_i) \tag{7.4}$$

- $conf_j$: 0-1 Conflict variable, and equals 1 if schedule of DN train j is conflicted because of disruption and 0 otherwise, where $j \in \{m+1, \ldots M\}$.

$$conf_j = (AD1_j \wedge BD2_j) \tag{7.5}$$

1.2 Objective Function

The objective is to minimize the weighted sum of the difference between the actual arrival time at the destination and the scheduled arrival time at the destination for all trains.

$$Min \sum_{i \in TU} w_i * (\overline{a_i^N} - a_i^N) + \sum_{j \in TD} w_j * (\overline{a_j^1} - a_j^1) \tag{7.6}$$

subject to the following constraints:

1.3 Constraints

- *Minimum travel time constraints:*

The actual arrival time of a train at a station is determined by the minimum travel time of the train and the actual departure time of the train from the previous station in both directions.

$$\overline{d_i^{k-1}} + \overline{Tut_i^k} \leq \overline{a_i^k} \forall i = 1,2\ldots m \text{ and } \forall k = 2,3\ldots N \tag{7.7}$$

$$\overline{d_j^{k+1}} + \overline{Tdt_j^k} \leq \overline{a_j^k} \forall i = m+1, m+2\ldots M \text{ and } \forall k = 1,2\ldots N-1 \tag{7.8}$$

- *Minimum halt time constraints:*

A train has to halt at a station for a minimum halt duration.

Railway Operations Models

$$\overline{a_i^k} + \overline{Th_i^k} \leq \overline{d_i^k} \forall i = 1,2...M and \forall k = 2,3...N \tag{7.9}$$

- *Follow-up constraints for UP trains:*

There must be a minimum time difference called arrival headway and departure headways between two successive arrivals and departures respectively in the UP direction.

$$\overline{a_i^k} + HA_{ij}^k \leq \overline{a_j^k} + (1 - Ax_{ij}^k) * D, \forall i,j = 1,2...m, i \neq j and \forall k = 2,3...N \tag{7.10}$$

$$\overline{d_i^k} + HD_{ij}^k \leq \overline{d_j^k} + (1 - Dx_{ij}^k) * D, \forall i,j = 1,2...m, i \neq j and \forall k = 1,2...N-1 \tag{7.11}$$

$$\overline{d_j^k} + HD_{ij}^k \leq \overline{d_i^k} + (Dx_{ij}^k) * D, \forall i,j = 1,2...m, i \neq j and \forall k = 1,2...N-1 \tag{7.12}$$

$$Ax_{ij}^k, Dx_{ij}^k \in \{0,1\}, \forall i,j = 1,2...m and \forall k = 1,2...N-1 \tag{7.13}$$

- *Follow-up constraints for DN trains:*

There must be a minimum time difference called arrival headway departure headways between two successive arrivals and departures respectively in the DN direction.

$$\overline{a_i^k} + HA_{ij}^k \leq \overline{a_j^k} + (1 - Az_{ij}^k) * D, \forall i,j = m+1, m+2...M, i \neq j and \forall k = 1,2...N-1 \tag{7.14}$$

$$\overline{a_j^k} + HA_{ij}^k \leq \overline{a_i^{k-1}} + (Az_{ij}^k) * D, \forall i,j = m+1, m+2...M, i \neq j and \forall k = 1,2...N-1 \tag{7.15}$$

$$\overline{d_i^k} + HD_{ij}^k \leq \overline{d_j^k} + (1 - Dz_{ij}^k) * D, \forall i,j = m+1, m+2...M, i \neq j and \forall k = 2,3...N \tag{7.16}$$

$$\overline{d_j^k} + HD_{ij}^k \leq \overline{d_i^k} + (Dz_{ij}^k) * D, \forall i,j = m+1, m+2...M, i \neq j and \forall k = 2,3...N \tag{7.17}$$

$$Az_{ij}^k, Dz_{ij}^k \in \{0,1\}, \forall i,j = m+1, m+2...M and \forall k = 2,3...N \tag{7.18}$$

- *Meet Constraints:*

There must be a minimum time difference called Meet headway between two trains in opposite directions, when they meet at a station.

$$\overline{a_i^k} + HM_{ij}^k \leq \overline{d_j^k} + (1 - Dy_{ij}^k) * D \tag{7.19}$$

$$\overline{a_j^k} + HM_{ij}^k \leq \overline{d_i^k} + (Ay_{ij}^k) * D \tag{7.20}$$

$$\overline{d_i^k} + HM_{ij}^k \leq \overline{a_j^k} + (1 - Ay_{ij}^k) * D \tag{7.21}$$

$$\overline{d_j^k} + HM_{ij}^k \leq \overline{a_i^k} + (Dy_{ij}^k) * D \tag{7.22}$$

$$Ay_{ij}^k, Dy_{ij}^k \in \{0,1\} \tag{7.23}$$

$\forall i, j = m+1, m+2...M \text{ and } \forall k = 1,2...N-1$

- *UP Consistency Constraints:*

Expression 7 maintains that departure order at meet point $k - 1$ for all UP trains should follow the arrival orders of all UP trains at meet point k.

$$Dx_{ij}^{k-1} = Ax_{ij}^k \tag{7.24}$$

$\forall i = 1,2..m, j = 1,2...m, i \neq j \text{ and } \forall k = 2,3...N$

- *DOWN Consistency Constraints:*

Expression 8 maintains that departure order at meet point k for all DOWN trains should follow the arrival orders of all DOWN trains at meet point $k - 1$.

$$Dz_{ij}^k = Az_{ij}^{k-1} \tag{7.25}$$

$\forall i, j = m+1, m+2...M, i \neq j \text{ and } \forall k = 2,3...N$

- *Meet Consistency Constraints:*

Expression 9 maintains that the arrival-departure order at meet point k for a pair of UP and DOWN train should follow the departure-arrival order at an adjacent meet point for that pair of UP and DOWN train.

$$Ay_{ij}^k = Dy_{ij}^{k+1} \tag{7.26}$$

Railway Operations Models

$\forall i = 1, 2..m, j = m+1, m+2...M, i \neq j$ and $\forall k = 2, 3...N$

- *Rescheduling constraints:*

Expressions (10a) and (10b) denote that the actual arrival / departure time of any train at any meet point is at least the respective scheduled arrival /departure time of that train at that meet point.

$$\overline{a_i^k} \geq a_i^k \tag{7.27}$$

$\forall i = 1, 2...m$ and $k = 2, 3...N$

$$\overline{d_i^k} \geq d_i^k \tag{7.28}$$

$\forall i = 1, 2...m$ and $k = 2, 3...N$

- *Disruption Constraints in the UP direction:*

Expressions (11) identify trains that conflict in the UP direction and change their actual departure times from *fromLOC* to be atleast *toTime*. Departure times of all other UP trains are maintained.

$$a_i^{toLOC} - fromTime \leq D * AU1_i \tag{7.29}$$

$$toTime - d_i^{fromLOC} \leq D * BU2_i \tag{7.30}$$

$$conf_i = AU1_i \wedge BU2_i \tag{7.31}$$

$$AU1_i, BU2_i, conf_i \in \{0, 1\} \tag{7.32}$$

$$\overline{d_i^{fromLOC}} \geq toTime * conf_i \tag{7.33}$$

$\forall i = 1, 2..m$

- *Disruption Constraints in the DN direction:*

Expressions (12) identify trains that conflict in the DN direction and change their actual departure times from *toLOC* to be atleast *toTime*. Departure times of all other DN trains are maintained.

$$d_j^{toLOC} - fromTime \leq D * BD2_j \tag{7.34}$$

$$toTime - a_j^{fromLOC} \leq D * AD1_j \qquad (7.35)$$

$$conf_j = AD1_j \wedge BD2_j \qquad (7.36)$$

$$AD1_j, BD2_j, conf_j \in \{0,1\} \qquad (7.37)$$

$$\overline{d_j^{toLOC}} \geq toTime * conf_j \qquad (7.38)$$

$$\forall j = m+1, m+2...M$$

8. STRATEGIES OF PLANNING DESIGN, IMPLEMENTATION AND MANAGEMENT

The "largest and most aggregate" perspective of transportation management is the strategic management; this is a one-analysis, design and umbrella decision making applied to define of the supply policy for the organization; strategic planning is also done to resource acquisition and major investment allocation over long-time horizons. Analysis and decision making problems at this top level of the organization have system-wide long-lasting impacts and the planner has usually a very large degree of freedom since there are no other constraints other than resources and most organizations are willing to augment their available resources. These problems deal with network design and improvement, implementation of new service links, terminal capacity planning and location, fleet acquisition, etc. This strategic definition and analysis typically involves major capital investments, and it must take into consideration the state of the global transportation system, the possible variation of the transportation demand in relation with the demographic trends and the evolution of the land use pattern, etc. It must also consider the variability of the economic context (e.g., capital cost, inflation rate, cost of labor, and energy), the government context (e.g., transportation policy, regulation, support to some mode in particular), and the evolution of the financial status of the organization.

Tactical issues, with a "narrower" perspective, concern questions that are raised regarding planning for the allocation of material resources of the organization in order to improve its efficiency or productivity. These issues have medium-term planning horizons and deal mainly with the efficient use of existing, given resources rather than the acquisition of major new ones. When adopting such a tactical perspective, the organization addresses problems such as frequency choice for each service considered, broad vehicle routing, definition of tariffs, maintenance planning (vehicle fleet maintenance, road maintenance), etc. Of course, tactical analysis and decisions must also take into consideration the given financial capabilities of the organization and institutional constraints such as regulation with regard to authorized variations in service levels and fares.

Finally, when dealing with the "most disaggregated and narrowest" perspective- the operational one-the organization has to solve its very short-term problems and organize in detail its day-to-day activities taking as fixed or given what is not strictly part of the questions to be answered. The production of timetables and the detailed routing and scheduling of vehicles and crews are examples of such problems relevant

Railway Operations Models

to an operational perspective. To analyze and solve those problems, the planner/manager will consider most parts of the transportation system and environment as fixed or given, including all constraints such as the average demands relevant to the problem (for instance origin-destination demands), the available human resources, the union rules, the detailed maintenance requirements, etc. On the other hand, he or she will have to take into consideration various unfixed matters such as all possible geometric configurations of bus lines and the weekly and daily variations of the transportation demands; such matters are usually irrelevant when dealing with the tactical or strategic perspectives described above.

9. CONCLUSION

In this chapter, railway operations' planning is discussed in four broad hierarchical stages. Most of the current research publications address specific one or two problems in railway operations and propose a suitable solution approach for that problem. This chapter treats railway operations in a comprehensive manner, with know-how on developing a basic model at any stages of operations planning. The basic model is extensible to a more detailed and sophisticated one, by including additional constraints and objectives depending on problem specific data. Many articles are available in literature that deals with solution approaches and methodology to solve these models. Broadly, three different solution approaches are predominant both in research and practice. They are (i) computing exact solutions using mathematic solvers, (ii) computing approximate solutions using heuristic and meta-heuristic procedures and (iii) developing analytical solutions using simulation approaches. Many of the operations modeling problems fall under NP hard complexity class, which means that there is no known solution approach that can solve these problems in a polynomial time. As the problem sizes increase, computational time and power required to solve such problems increase exponentially. Therefore, it is extremely difficult to develop a complete operational plan for an entire railway transportation system. Hence, a variety and combination of tools, techniques and methodology are used both in research and practice to develop useful solutions that can be effectively utilized. Recent developments in this topic include intelligent approaches, using evolutionary algorithms. Some of the evolutionary models use genetic algorithms, fuzzy logic, simulated annealing, which attempt to generate hybrid solutions from an initial basic solution, thereby improving the objective function. Agents based methods are applied in distributed systems, where large scale systems that are physically distributed can be easily modeled and solved. Agents based methods also incorporate intelligent solution approaches and depict a host of features such as autonomy, scalability, inter-operability, extensibility which are very relevant in complex problems in transportation management. However, all these hybrid and advanced technologies deliver robust solutions, if and only if the underlying procedures are drawn from a theoretically well-grounded and relaxed mathematical model.

REFERENCES

Albrecht, T. (2009). The Influence of Anticipating Train Driving on the Dispatching Process in Railway Conflict Situations. *Networks and Spatial Economics*, *9*(1), 85–101. doi:10.1007/s11067-008-9089-0

Brucker, P., Heitmann, S., & Knust, S. (2005). *Scheduling Railway Traffic at a Construction site*. Springer Berlin Heidelberg. doi:10.1007/3-540-26686-0_15

Budai, G., Maróti, G., Dekker, R., Huisman, D., & Kroon, L. (2009). Rescheduling in Passenger Railways: The Rolling Stock Rebalancing Problem. *Journal of Scheduling*. doi:10.1007/ s10951-009-0133-9

Burdett, R., & Kozan, E. (2008). A sequencing approach for creating new train timetables. OR Spectrum. doi:10.1007/s00291-008-0143-6

Bussieck, M. (1998). Optimal Lines in Public Rail Transport [PhD thesis]. Technischen Universität Braunschweig, Braunschweig.

Cacchiani, V. (2008). Models and Algorithms for Combinatorial Optimization Problems Arising in Railway Applications [PhD thesis]. Universit di Bologna.

Cai, X., & Goh, C. J. (1994). A Fast Heuristic for the Train Scheduling Problem. *Computers & Operations Research*, *21*(5), 499–510. doi:10.1016/0305-0548(94)90099-X

Caimi, G., Fuchsberger, M., Burkolter, D., Herrmann, T., Wüst, R., & Roos, S. (2009). Conflict-free train scheduling in a compensation zone exploiting the speed profile. *Proceedings of the 3rd international seminar on railway operation modeling and analysis*.

Caprara, L., Kroon, G., Monaci, M., Peeters, M., & Toth, P. (2007). Passenger Railway Optimization. In C. Barnhart & G. Laporte (Eds.), *Handbooks in Operations Research and Management Science* (Vol. 14, pp. 129–187). Elsevier.

Chang, S. C., & Chung, Y. C. (2005). From Timetabling to train regulation-a new train operation model. *Information and Software Technology*, *47*(9), 575–585. doi:10.1016/j.infsof.2004.10.008

Cheng, Y. (1996). Optimal Train Traffic Rescheduling Simulation by a Knowledge-Based System Combined with Critical Path Method. *Simulation Practice and Theory*, *4*(6), 399–413. doi:10.1016/ S0928-4869(96)00034-1

Cheng, Y. (1998, November). Hybrid Simulation for Resolving Resource Conflicts in Train Traffic Rescheduling. *Computers in Industry*, *35*(3), 233–246. doi:10.1016/S0166-3615(97)00071-7

Cordeau, J. F., Toth, P., & Vigo, D. (1998, November). A Survey of Optimization models for Train Routing and Scheduling. *Transportation Science*, *32*(4), 380–404. doi:10.1287/trsc.32.4.380

Ghoseiri, K. & Morshedsolouk, F. (2006). ACS-TS: Train Scheduling using Ant Colony System. *Journal of Applied Mathematics and Decision Sciences*. doi:.10.1155/JAMDS/2006/95060

Goossens, J. W. (2004). Models and Algorithms for Railway Line Planning Problems [PhD thesis]. Universiteit Maastricht.

Gorman, M. (1998). An Application of Genetic and Tabu searches to the Freight Railroad Operating Plan Problem. *Annals of Operations Research*, *78*(19), 51–69. doi:10.1023/A:1018906301828

Heilporn, L., De Giovanni, L., & Labbé, M. (2008). De Giovanni, and M. Labbè. Optimization Models for the Single Delay Management Problem in Public Transportation. *European Journal of Operational Research*, *189*(3), 762–774. doi:10.1016/j.ejor.2006.10.065

Hüisman, D., Kroon, L. G., Lentink, R. M., & Vromans, M. C. J. M. (2005). *Operations Research in Passenger Railway Transportation. Technical report, Erasmus Research Institute of Management (ERIM).* Rotterdam School of Management.

Jespersen-Groth, J., Pottho, D., Clausen, J., Huisman, D., Kroon, L. G., Maroti, G., & Nielsen, M. N. (2009). Disruption Management in Passenger Railway Transportation. Robust and Online Large-Scale Optimization. Springer Berlin / Heidelberg.

Jovanovic, D., & Harker, P. (1991). Tactical Scheduling of Rail Operations: SCAN I system. *Transportation Science*, 25(1), 46–64. doi:10.1287/trsc.25.1.46

Keaton, M. H. (1989). Designing optimal railroad operating plans: Lagrangian relaxation and Heuristic approaches. *Transportation research Part B: Methodological*, 23 B(6), 415-431.

Khan, M. B., & Zhou, X. (2009). Stochastic Optimization Model and Solution Algorithm for Robust Double-Track Train-Timetabling Problem. *IEEE Transactions on Intelligent Transportation Systems*, 10(3). doi:10.1109/TITS.2009.2030588

Leander, P. & Lukaszewicz, P. (2008). EETROP: Energy Efficient Train Operation-State of the art of train ECO-operation (Technical Report 20).

Liebchen & Stiller. S. (2009) Delay resistant timetabling. *Public transport*, 1(1):55-72. doi:10.1007/s12469-008-0004-3

Lindner, T. & Zimmermann, U. (2005). Cost Optimal Periodic Train Scheduling. *Mathematical Methods of Operations Research*, 62(2), 281-295.

Luèthi, M. (2009). Improving the Efficiency of Heavily Used Railway Networks through Integrated Real-Time Rescheduling [PhD thesis]. Swiss Federal Institute Of Technology, ETH Zurich.

Maròti. (2006). Operations Research Models for Railway Rolling Stock Planning [PhD thesis]. Technische Universiteit, Eindhoven, Amsterdam.

Mazzarello, M., & Ottaviani, E. (2007). A Traffic Management System for Real-time Traffic Optimisation in Railways. *Transportation Research Part B: Methodological*, 41(2), 246–274. doi:10.1016/j.trb.2006.02.005

Michaelis, M., & Schöbel, A. (2009). Integrating line planning, timetabling, and vehicle scheduling: a customer-oriented heuristic. *Public Transport*, 1(3). doi:10.1007/s12469-009-0014-9

Müller-Hannemann, M., Schulz, F., Wagner, D., & Zaroliagis, C. D. (2004). Timetable Information: Models and Algorithms. In ATMOS (pp. 67-90).

Nachtigall, K., & Voget, S. (1996). A Genetic Algorithm Approach to Periodic Railway Synchronization. *Computers & Operations Research*, 23(5), 453–463. doi:10.1016/0305-0548(95)00032-1

Sahin, I. (1999). Railway Traffic Control and Train Scheduling based on Inter-train Conflict Management. *Transportation Research Part B: Methodological*, 33(7), 511–534. doi:10.1016/S0191-2615(99)00004-1

Salim, V., & Cai, X. (1997). A Genetic Algorithm for Railway Scheduling with Environmental Considerations. *Environmental Modelling & Software*, 12(4), 301–310. doi:10.1016/S1364-8152(97)00026-1

Sundaravalli Narayanaswami. (2010). *Dynamic and Realtime Rescheduling Models: An Empirical Analysis from Railway Transportation*. Saarbuchen, Germany: Lambart Academic Publishers.

Törnquist, J. (2012). Design of an effective algorithm for fast response to the rescheduling of railway traffic during disturbances. *Transportation Research Part C, Emerging Technologies*, *20*(1), 62–78. doi:10.1016/j.trc.2010.12.004

Veelenturf, L. P., Nielsen, L. K., Maroti, G., & Kroon, L. G. (2011). Passenger oriented disruption management by adapting stopping patterns and rolling stock schedules. *Proceedings of the 4th International Seminar on Railway Operations Modelling and Analysis RailRome '11*.

Walker, G., Snowdon, J. N., & Ryan, D. M. (2005). Simultaneous Disruption Recovery of a Train Timetable and Crew Roster in Real Time. *Computers & Operations Research*, *32*(8), 2077–2094. doi:10.1016/j.cor.2004.02.001

Wüest, R., Laube, F., Roos, S., & Caimi, G. (2008). Sustainable Global Service Intention as objective for Controlling Railway Network Operations in Real Time. *Proceedings of the 8th World Congress of Railway Research (WCRR)*, Seoul, Korea.

Yoko, T., & Norio, T. (2005). Robustness Indices for Train Rescheduling. *Proceedings of the 1st International Seminar on Railway Operations Modelling and Analysis*, Delft, The Netherlands.

Chapter 15
A General Simulation Modelling Framework for Train Timetabling Problem

Özgür Yalçınkaya
Dokuz Eylül University, Turkey

ABSTRACT

One of the most important problems encountered and needed to be solved in railway systems is train timetabling (scheduling) problem. This is the problem of determining a feasible timetable for sets of trains which does not violate track capacities and additionally satisfies some operational constraints of the railway system. In this chapter, a feasible timetable generator framework for stochastic simulation modelling is introduced. The objective is to obtain a feasible train timetable for all trains in the railway system, which includes train arrival and departure times at all visited stations and calculated average train travel time. Although this chapter focuses on train timetabling (scheduling) problem, the developed general framework can also be used for train dispatching (rescheduling) problem if the model can be fed by the real-time data. Since, the developed simulation model includes stochastic events, and it can easily cope with the disturbances that occur in the railway systems, it can be used for dispatching.

INTRODUCTION

Management of railway systems is an important issue of transport systems. One of the important problems in management of railway systems is the train timetabling (scheduling) problem. This is the problem of determining a timetable for sets of trains that does not violate track capacities and satisfy some operational constraints of the railway system. Several variations of the problem can be considered, mainly depending on the objective function to be optimized, decision variables, constraints and complexity of the studied railway network. Several names have been given to the problem widely using three-word phrases:

DOI: 10.4018/978-1-5225-0084-1.ch015

beginning with *Train / Railway*
going on with *Timetabling / Scheduling / Dispatching / Rescheduling / Planning / Pathing*
and ending with *Problem*
words with a few exceptions.

A general most common train timetabling problem in the literature considers a single track linking two major stations with a number of intermediate stations in between (Caprara et al., 2002). It is assumed that $S = \{1, ..., s\}$ represents the set of stations, numbered according to the order in which they appear along the rail line. In particular, 1 and s denote the initial and final stations, respectively. Analogously, it is assumed that $T = \{1, ..., t\}$ denotes the set of trains which are candidate to be run in a given time horizon. For each train $j \in T$, a starting station f_j and an ending station l_j ($l_j > f_j$) are given. Let $S^j = \{f_j, ..., l_j\} \subseteq S$ be the ordered set of stations visited by train j. A timetable defines, for each train $j \in T$, the arrival and departure times for the stations $f_j, f_j+1, ... l_j-1, l_j$. The running time of train j in the timetable is the time elapsed between origin station and destination station of the train (Caprara et al., 2002). This general problem can be more sophisticated by adding some real life behaviour of railway systems or relaxing some assumptions made related with the railway system under consideration.

The problem has been studied by researchers and so far many efforts have been spent on it. In early years, due to the limitations of computers' abilities and the complexity of the problem, the problem was relaxed by unrealistic assumptions and generally deterministic models were studied. Depending on the increasing computer capabilities more realistic models were developed. Although simulation for modelling has been used in some articles, none of them includes a comprehensive framework. This has been the main motivation for the authors to develop a feasible timetable generator simulation modelling framework.

In this chapter, a feasible timetable generator framework for stochastic simulation modelling is developed for obtaining a feasible train timetable for all trains in a railway system. This framework includes train arrival and departure times for all stations visited by each train and calculated average train travel time. A general stochastic simulation modelling framework is developed and explained step by step in order to guide to researchers who aim to develop a simulation model of railway transportation systems. By using this framework all the railway systems can be modelled with only problem and infrastructure specific modifications and feasible solutions are easily obtained. In order to avoid a deadlock, a general *blockage preventive algorithm* is also developed and embedded into the simulation model.

In next section the literature on the problem is given. After that, the simulation modelling framework is demonstrated in detail on a hypothetic problem. Next, the obtained results are discussed. Concluding remarks and future work directions are exhibited in the last parts of the chapter.

BACKGROUND

The studies on the train timetabling problem aim at achieving a train timetable with arrival and departure times of all trains at the visited stations in the system. These studies generally begin with a planned *infeasible* initial (draft) timetable with many conflicts needed to be solved. After these conflicts are solved a *feasible* train timetable is composed, and the train operating authority runs the trains according to the feasible timetable.

In review papers: Assad (1980), Cordeau et al. (1998), Newman et al. (2002) and Caprara et al. (2007) the train timetabling problem was considered with some railway optimization problems. Cacchiani and Toth (2012) surveyed the main studies dealing with the train timetabling problem in its nominal and robust versions.

In papers that focused on the train timetabling problem mathematical models are: Frank (1966), Szpigel (1973), Mees (1991), Jovanovic and Harker (1991), and Odijk (1996). On the other hand Higgins et al. (1997), Brännlund et al. (1998), Tormos et al. (2008) and Liu and Kozan (2009) used heuristics and metaheuristics as solution approaches.

Caprara et al. (2001, 2002) concentrated on train timetabling problem relevant to a single, one way track linking two major stations with a number of intermediate stations between them. A graph theoretic formulation was proposed for the problem using a directed multigraph in which nodes correspond to departures or arrivals at a certain station at a given time instant. The objective was to maximize sum of the profits of the scheduled trains. Caprara et al. (2006) extended the problem considered by Caprara et al. (2002), by taking into account additional real world constraints. On the other hand, Cacchiani et al. (2008) proposed heuristic and exact algorithms for the periodic and non-periodic train timetabling problem on a corridor to maximize the sum of the profits of the scheduled trains. The heuristic and the exact algorithms were based on the solution of the relaxation of an integer linear programming formulation in which each variable corresponds to a full timetable for a train. This approach was in contrast with previous approaches proposed by Caprara et al. (2001, 2002, 2006) so that these authors had considered the same problem and used integer linear programming formulations in which each variable was associated with a departure and/or arrival of a train at a specific station in a specific time instant.

Zhou and Zhong (2007) focused on single track and proposed a generalized resource constrained project scheduling formulation for train timetabling problem. The developed algorithm chronologically added precedence relation constraints between conflicting trains to eliminate conflicts, and the resulting sub-problems were solved by the longest path algorithm to determine the earliest start times for each train in different segments. Castillo et al. (2009) used an optimization method to solve train timetabling problem for a single tracked bidirectional line, similar to the one presented by Zhou and Zhong (2007) but more complex, and discuss the problem of sensitivity analysis. A three stage method is proposed to deal with the problem and a sequential combination of objective functions is used for solution.

In recent years, Zhou and Zhong (2005), Liebchen (2008), and Lee and Chen (2009) have spent efforts to optimize multi objective train scheduling problems.

In a few papers a simulation model was developed for train timetabling problem. Wong and Rosser (1978) are the first authors who developed a simulation model for train scheduling problem. The output of the simulation model comprised a pictorial representation of the pattern of train movements as well as detailed statistics for each train. The problem was to determine where a crossing or overtaking should be allowed to occur, and the objective was to minimize the sum of weighted costs of delaying trains at passing loops where the weights chosen reflected the importance of each type of train. To improve the system performance, train starting times were varied, and one train at a time heuristic iterative procedure was used for improvements. Petersen and Taylor (1982) presented a state space description for the problem of moving trains over a line, and an algebraic description of the relationships that must hold for feasibility and safety considerations was given. The line blockage problem at high traffic intensities was discussed under conditions that ensure the blockage not to occur. The objective of the study was to minimize the terminating times of the trains. Geske (2006) focused on the railway scheduling problem and developed a constraint based deterministic simulation model with the objective of reducing the lateness of trains. Selecting alternative paths in stations was an optimization task to reduce lateness and to find a conflict free solution. The results of the proposed sequentially train scheduling heuristic was compared with those of a genetic algorithm.

Two books gave comprehensive knowledge on the problem. Pachl (2009) provided basic knowledge in the science of railway operation in a close connection to signalling principles and traffic control technologies. Hansen and Pachl (2014) described the methods of railway timetabling, operations analysis and modelling, simulation and traffic management in order to stimulate their broader application in practice.

Above, a brief review of researches related to the problem has been presented. While analytical results were obtained by exact algorithms, simulation models and meta-heuristics with approximate outcomes were also employed. The metaheuristics were employed by the researchers in the relevant area after 1990s, multi objectives were optimized after 2000s. On the other hand, in a few papers simulation models were constructed for the scheduling problem. Simulation models are flexible and solve real problems without making many restricting assumptions as in most analytical models. Although simulation for modelling has been used in some articles, none of them included a comprehensive framework. This has been the main motivation for the authors to develop a feasible timetable generator simulation modelling framework.

A FEASIBLE TIMETABLE GENERATOR SIMULATION MODELLING FRAMEWORK

In this section, a feasible timetable generator simulation modelling framework for the train timetabling problem is given. The objective is to obtain a feasible train timetable for all trains in the system. The feasible train timetable includes train arrival and departure times at all visited stations and additionally calculated average train travel time. This section involves two subsections. In the first subsection, a hypothetic problem is introduced. In the second subsection, the simulation modelling framework is developed and applied on the hypothetic problem.

A Hypothetic Train Timetabling Problem

The proposed simulation modelling framework is implemented on a hypothetic problem (Yalçınkaya, 2010; Yalçınkaya & Bayhan 2012) that is similar with the common problem studied in the literature. The infrastructure in the studied problem has a line structure inspired by a real railway line system, and the system has a planned initial timetable with arrival and departure times of trains only at two end stations of the infrastructure.

Railway Infrastructure

The railway, which is inspired by a real line, is a bidirectional single track corridor as analogous to many studied lines in the literature and in real railway systems. The line-station diagram of the bidirectional single track corridor and the infrastructure of stations are shown in Figure 1. There are 10 real stations on the corridor that are labelled as S_i ($i = 1, 2, ..., 10$) from the east to the west. The corridor has two terminuses, TS_1 and TS_{10}, which indicate the beginning and the finishing points of it.

Figure 1. Line-station diagram of the bidirectional single track corridor

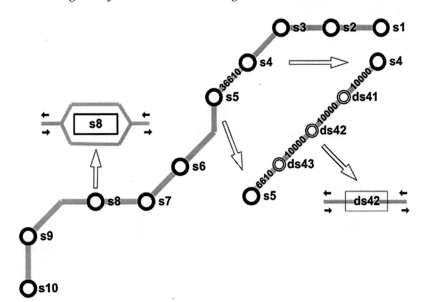

Track lengths between the real stations are given in Table 1, as it is seen, the total track length from TS_1 to TS_{10} is 286270 meters. Since all the real stations have 200 meters platform, the whole length of the corridor is 288270 meters.

Table 1. Track lengths between the real stations

To From	TS_1 (*East*)	S_1	S_2	S_3	S_4	S_5	S_6	S_7	S_8	S_9	S_{10}	TS_{10} (*West*)
TS_1 (*East*)	0	500	28070	60170	88400	125210	170060	197060	214460	243560	285770	286270
S_1	500	0	27570	59670	87900	124710	169560	196560	213960	243060	285270	285770
S_2	28070	27570	0	32100	60330	97140	141990	168990	186390	215490	257700	258200
S_3	60170	59670	32100	0	28230	65040	109890	136890	154290	183390	225600	226100
S_4	88400	87900	60330	28230	0	36810	81660	108660	126060	155160	197370	197870
S_5	125210	124710	97140	65040	36810	0	44850	71850	89250	118350	160560	161060
S_6	170060	169560	141990	109890	81660	44850	0	27000	44400	73500	115710	116210
S_7	197060	196560	168990	136890	108660	71850	27000	0	17400	46500	88710	89210
S_8	214460	213960	186390	154290	126060	89250	44400	17400	0	29100	71310	71810
S_9	243560	243060	215490	183390	155160	118350	73500	46500	29100	0	42210	42710
S_{10}	285770	285270	257700	225600	197370	160560	115710	88710	71310	42210	0	500
TS_{10} (*West*)	286270	285770	258200	226100	197870	161060	116210	89210	71810	42710	500	0

Planned Initial Train Timetable

The train arrival and departure times at two end stations are given in Table 2, where WB_i ($i = 1, 2, ..., 10$) indicates a *westbound* train that begins its trip from the first real station on the *east* of the corridor and plans to finish at the first real station on the *west* of the corridor. On the other hand, EB_i ($i = 1, 2, ..., 10$) indicates an *eastbound* train which has an opposite direction to WB trains.

A Feasible Timetable Generator Simulation Model

The simulation model is developed by using ARENA discrete event simulation software (Kelton et al., 2009) in a modular manner. In ARENA, the user builds an experiment model by placing boxes of different shapes that represent processes or logic. Connector lines are used to join these boxes together and specify the flow of entities. While boxes have specific actions relative to entities, flow, and timing, the precise representation of each boxes and entity relative to real-life objects is subject to the modeller. Statistical data can be recorded and outputted as reports.

First, the railway infrastructure including links, intersections and stations is modelled, and then track failures and repairs are included. After that, train movement logic on the infrastructure is modelled. Author use a rule for track allocation to candidate trains that are the trains waiting at neighbour stations of the track to use it. Next, fixed train speeds are relaxed, and additional unplanned delays at the stations are inserted. Then, the number of trains in the system is increased and randomness is added to the planned initial train timetable. As a last step, animation of the system is developed.

Assumptions are made during the modelling phase of the simulation model. It must be noted that many of assumptions are the fundamental assumptions made by the existing studies in the literature. Some of the assumptions made for the simulation model are:

- The unit for length and time is meter and second, respectively.
- It takes 32 seconds for trains to reach the real stations S_1 and S_{10} from the park area, then the trains wait 568 seconds at these stations, i.e., they spend totally 600 seconds (10 minutes) as a dwell

Table 2. Planned initial train timetable

Station	Train	Arrival Time	Departure Time	Station	Train	Arrival Time	Departure Time
S_1	WB_1	00:00	00:10	S_{10}	EB_1	00:00	00:10
	WB_2	02:00	02:10		EB_2	02:00	02:10
	WB_3	04:00	04:10		EB_3	04:00	04:10
	WB_4	06:00	06:10		EB_4	06:00	06:10
	WB_5	08:00	08:10		EB_5	08:00	08:10
	WB_6	10:00	10:10		EB_6	10:00	10:10
	WB_7	12:00	12:10		EB_7	12:00	12:10
	WB_8	14:00	14:10		EB_8	14:00	14:10
	WB_9	16:00	16:10		EB_9	16:00	16:10
	WB_{10}	18:00	18:10		EB_{10}	18:00	18:10

A General Simulation Modelling Framework for Train Timetabling Problem

time. First trips are planned to begin at 00:10:00 o'clock. But due to additional unplanned delays at the stations lateness may occur.

- Time spent for reaching to a terminus (TS_1 or TS_{10}) from the park area is negligible.
- The WB trains' departure station is S_1 and destination is S_{10}, and the EB trains' departure station is S_{10} and destination is S_1.
- There will be 20 trains running in a day, 10 of them are the WB and the other 10 are the EB trains.
- All the trains are the same type.
- Passengers are ignored at this level of the model.
- There is time headway (40 seconds) between two consecutive trains at a station, which have the same trip direction, in order to have a safe trip.
- More than one train that have the same direction can use the same track with distance headway (1000 metres) between them.
- The train lengths are 50 meters.
- Earliness and lateness time in the planned initial train timetable, due to some uncontrollable events that may occur outside of the corridor, is uniformly distributed between -900 and +900 seconds.

Railway Infrastructure Modelling

The detailed line-station diagram of the corridor is denoted in Figures 2-4. The letter "E" indicates the east and the letter "W" indicates the west directions.

The railway infrastructure is a union of intersections and links, and modelled via the *Networks Element* of the ARENA.

Figure 2. Detailed line-station diagram of the corridor from S_1 to S_3

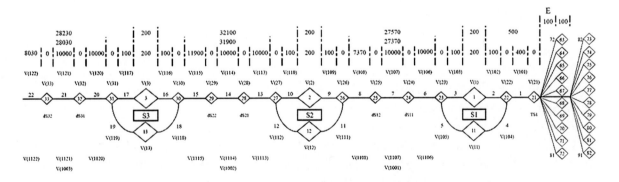

Figure 3. Detailed line-station diagram of the corridor from S_4 to S_7

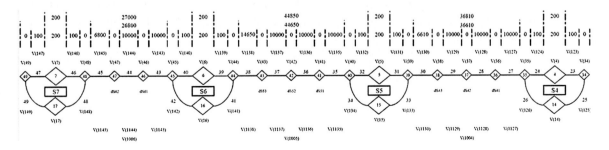

Figure 4. Detailed line-station diagram of the corridor from S_8 to S_{10}

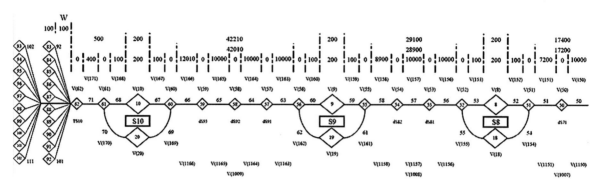

The links are track parts on which train traverses during its trip from a station to another neighbour station. The links are numbered from 1 to 111, and shown as lines in Figures 2-4, and are modelled via the *Links Element* of the ARENA.

The intersections, the connection points of the links, are numbered from 1 to 102 and shown in lozenge shape in Figures 2-4. The intersections are modelled via the *Intersections Element* of the ARENA. There are three kinds of intersections in the model:

- The first, big lozenge shapes denote the intersections related to the real stations, and have lengths in meter. For instance, the big lozenge shape numbered as 2 is related to the first part of S_2 and connects link 9 with link 10.
- The second, small lozenge shapes denote the intersections that are only used for connecting the links and dummy stations that are located on the tracks between the real stations to keep a train wait during the repairing of a track failure, and have no length. For instance, the small lozenge shape numbered as 22 connects link 1 with link 2 and link 4.
- The third, intersections numbered from 63 to 102 are related to the park areas where the empty trains can park.

The stations are locations where a train can stop for boarding and alighting events, for parking or for waiting until a failure is accomplished. The real stations, S_i ($i = 1, 2, ..., 10$), are interrelated with two intersections. The dummy stations, dS_{ij} ($j = 1$ for $i = 7$; $j = 1, 2$ for $i = 1, 2, 3, 6, 8$; $j = 1, 2, 3$ for $i = 4, 5, 9$), are located on the tracks between the real stations to keep a train wait during the repairing of a failure, if the failure occurs while a train is traversing between the real stations. The stations are modelled via the *Stations Element* of the ARENA.

The links and the intersections related to the corridor are shown in Table 3, and the stations related to the corridor are depicted in Table 4. In *Description* column of the Table 3, a brief explanation is given to define the corresponding link and intersection. In this column, while describing an intersection the dummy station related with the described intersection is given.

In the simulation model the links, the intersections and track failures are controlled via variables. For instance, link 2 is controlled by a variable that has number 102 in the *Variables Element* of the simulation model and demonstrated by V(102) in Figure 2. The *Variables Element* of the ARENA is

Table 3. Links and intersections related to the corridor

	Name	Description
Link number		
i = 1, ..., 111	Lnk(i)	Link (*i*) in the railway network that is the track part trains traverse on
Intersection number		
i = 1,..., 10	Int(i)_S(i)_1	Intersection (*i*) interrelated with first part of station (*i*)
i = 11,..., 20	Int(i)_S(i-10)_2	Intersection (*i*) interrelated with second part of station (*i* -10)
21	Int21_TS1	Intersection interrelated with TS1
i = 22, 23, 26, 27, 30, 31, 34, 35, 39, 40, 44, 45, 48, 49, 51, 52, 55, 56, 60, 61	Int(i)	Intersection (*i*) in the railway network that is connecting neighbour links
24	Int24_dS11	dS11
25	Int25_dS12	dS12
28	Int28_dS21	dS21
29	Int29_dS22	dS22
32	Int32_dS31	dS31
33	Int33_dS32	dS32
36	Int36_dS41	dS41
37	Int37_dS42	dS42
38	Int38_dS43	dS43
41	Int41_dS51	dS51
42	Int42_dS52	dS52
43	Int43_dS53	dS53
46	Int46_dS61	dS61
47	Int47_dS62	dS62
50	Int50_dS71	dS71
53	Int53_dS81	dS81
54	Int54_dS82	dS82
57	Int57_dS91	dS91
58	Int58_dS92	dS92
59	Int59_dS93	dS93
62	Int62_TS10	Intersection interrelated with TS10
i = 63,..., 102	Int(i)_park	Intersection (*i*) interrelated with park area

used for defining the variables. The variables related to the corridor model are given in Table 5. In this table, *Description* column defines the dummy station related with the intersection that is controlled by the corresponding variable.

Table 4. Stations related to the corridor

Station number	Name	Description
$i = 1,..., 10$	S(i)_1	First part of real station (i)
$i = 11,..., 20$	S(i-10)_2	Second part of real station (i -10)
21	TS1	Terminus TS1
24	dS11	Dummy station
25	dS12	Dummy station
28	dS21	Dummy station
29	dS22	Dummy station
32	dS31	Dummy station
33	dS32	Dummy station
36	dS41	Dummy station
37	dS42	Dummy station
38	dS43	Dummy station
41	dS51	Dummy station
42	dS52	Dummy station
43	dS53	Dummy station
46	dS61	Dummy station
47	dS62	Dummy station
50	dS71	Dummy station
53	dS81	Dummy station
54	dS82	Dummy station
57	dS91	Dummy station
58	dS92	Dummy station
59	dS93	Dummy station
62	TS10	Terminus TS10
$i = 63,..., 102$	Sta(i)_park	Station (i) interrelated with park area

Assumptions related to the railway infrastructure part of the simulation model are:

- The railway system is a single track line, a corridor.
- The traffic on tracks is bidirectional.
- All the real stations have 200 meters platforms for boarding and alighting events.
- There are 10 real stations and 20 dummy stations on the corridor, that is, the corridor is 288270 meter long.
- The terminuses (TS_1 and TS_{10}) have infinite train capacity.
- The terminus TS_1 is located on the east point, and the terminus TS_{10} is located on the west point of the corridor.

A General Simulation Modelling Framework for Train Timetabling Problem

Table 5. Variables related to the corridor

Variable Number	Name	Description
i = 1,..., 10	V(i)_S(i)_1_Int(i)	Variable (*i*) controls intersection (*i*) interrelated with first part of station (*i*)
i = 11,..., 20	V(i)_S(i-10)_2_Int(i)	Variable (*i*) controls intersection (*i*) interrelated with second part of station (*i* -10)
21	V21_TS1_Int21	Variable controls intersection 21 interrelated with TS1
i = 22, 23, 26, 27, 30, 31, 34, 35, 39, 40, 44, 45, 48, 49, 51, 52, 55, 56, 60, 61	Var(i)_Int(i)	Variable (*i*) controls intersection (*i*)
24	V24_dS11_Int24	dS11
25	V25_dS12_Int25	dS12
28	V28_dS21_Int28	dS21
29	V29_dS22_Int29	dS22
32	V32_dS31_Int32	dS31
33	V33_dS32_Int33	dS32
36	V36_dS41_Int36	dS41
37	V37_dS42_Int37	dS42
38	V38_dS43_Int38	dS43
41	V41_dS51_Int41	dS51
42	V42_dS52_Int42	dS52
43	V43_dS53_Int43	dS53
46	V46_dS61_Int46	dS61
47	V47_dS62_Int47	dS62
50	V50_dS71_Int50	dS71
53	V53_dS81_Int53	dS81
54	V54_dS82_Int54	dS82
57	V57_dS91_Int57	dS91
58	V58_dS92_Int58	dS92
59	V59_dS93_Int59	dS93
62	V62_TS10_Int42	Variable controls intersection 62 interrelated with TS10
i = 101,..., 171	V(i)_Lnk(i-100)	Variable (*i*) controls link (*i* -100)

- Every middle real station has capacity of two trains, that is, there will be at most two trains at a real station at the same time.
- Every dummy station has capacity of one train, that is, there will be at most one train at a dummy station at a specific time.
- There are 100 meters length park areas near the terminuses. These park areas are used by a train, which finished its trip, while leaving the corridor, or waiting to enter the corridor.
- Distance between these park areas and terminuses are 100 meters.

Track Failure Modelling

Track failure is an event that prevents a train to occupy the impaired track for a trip. The train can use the track after it is repaired. Distributions for failure times and repair times for the failed tracks are depicted in Table 6. In the first three columns, variable number, location and length of the tracks are given. These tracks are ranked according to their lengths. The shortest one has rank 1 and lies between S_7 and S_8, and is selected to be the *base track*. The longest one has rank 9 and lies between S_5 and S_6. The ratios are obtained by dividing the lengths of the tracks to the length of the *base track*.

We divided the long tracks into smaller parts (short tracks) and located the dummy stations between these short tracks as depicted in Figures 2-4. After obtaining failures for tracks according to the distributions shown in Table 6, these failures are transferred to the short tracks with the probabilities exhibited in Table 7.

In the simulation model, the track failures are controlled via variables. If a failure occurs in a track part, trains are prevented to use this part until it is repaired. The variables related to the track failure model are exhibited in Table 8.

The line-station diagram of the track between S_5 and S_6 is given in Figure 5, and the SIMAN View (Kelton et al., 2009) of the failure model logic related to the track between S_5 and S_6 is given in Computer Code 1.

Computer Code 1:

```
1$ DELAY: Expo(33283):2$;
2$ ASSIGN: V1005_fS5=1:3$;
3$ BRANCH, 1:
 With,0.25:4$;
 With,0.25:5$;
 With,0.25:6$;
 With,0.25:7$;
4$ ASSIGN: V1135_fLnk35=1:8$;
```

Table 6. Failure times and repair times distributions

Variable Number	Location	Length (Meter)	Rank	Ratio	1/Ratio	Failure Time Distribution	Repair Time Distribution
1001	S1-S2	27370	3	1.591	0.628	Expo (54296)	Expo (2864)
1002	S2-S3	31900	6	1.855	0.539	Expo (46586)	Expo (3338)
1003	S3-S4	28030	4	1.630	0.614	Expo (53017)	Expo (2933)
1004	S4-S5	36610	7	2.128	0.470	Expo (40592)	Expo (3831)
1005	S5-S6	44650	9	2.596	0.385	Expo (33283)	Expo (4673)
1006	S6-S7	26800	2	1.558	0.642	Expo (55451)	Expo (2805)
1007	S7-S8	17200	1	1.000	1.000	Expo (86400)	Expo (1800)
1008	S8-S9	28900	5	1.680	0.595	Expo (51421)	Expo (3024)
1009	S9-S10	42010	8	2.442	0.409	Expo (35374)	Expo (4396)

```
5$ ASSIGN: V1136_fLnk36=1:8$;
6$ ASSIGN: V1137_fLnk37=1:8$;
7$ ASSIGN: V1138_fLnk38=1:8$;
8$ DELAY: Expo(4673):9$;
9$ ASSIGN: V1005_fS5=0:
 V1135_fLnk35=0:
```

Table 7. Failure probabilities of the short tracks

Variable Number of Long Track	Location of Long Track	Length of Long Track (Meter)	Link Number	Variable Number of Short Track	Length of Short Track (Meter)	Probability (%)	Location of Short Track	
1001	S1-S2	27370	6	1106	10000	33	S1	-dS11
			7	1107	10000	33	dS11	-dS12
			8	1108	7370	34	dS12	-S2
1002	S2-S3	31900	13	1113	10000	33	S2	-dS21
			14	1114	10000	33	dS21	-dS22
			15	1115	11900	34	dS22	-S3
1003	S3-S4	28030	20	1120	10000	33	S3	-dS31
			21	1121	10000	33	dS31	-dS32
			22	1122	8030	34	dS32	-S4
1004	S4-S5	36610	27	1127	10000	25	S4	-dS41
			28	1128	10000	25	dS41	-dS42
			29	1129	10000	25	dS42	-dS43
			30	1130	6610	25	dS43	-S5
1005	S5-S6	44650	35	1135	10000	25	S5	-dS51
			36	1136	10000	25	dS51	-dS52
			37	1137	10000	25	dS52	-dS53
			38	1138	14650	25	dS53	-S6
1006	S6-S7	26800	43	1143	10000	33	S6	-dS61
			44	1144	10000	33	dS61	-dS62
			45	1145	6800	34	dS62	-S7
1007	S7-S8	17200	50	1150	10000	50	S7	-dS71
			51	1151	7200	50	dS71	-S8
1008	S8-S9	28900	56	1156	10000	33	S8	-dS81
			57	1157	10000	33	dS81	-dS82
			58	1158	8900	34	dS82	-S9
1009	S9-S10	42010	63	1163	10000	25	S9	-dS91
			64	1164	10000	25	dS91	-dS92
			65	1165	10000	25	dS92	-dS93
			66	1166	12010	25	dS93	-S10

Table 8. Variables related to the track failure model

Variable Number	Name	Description
i = 1001, ..., 1009	V(i)_fS(i-1000)	Variable (*i*) controls track failure between S(*i*) and S(*i* +1), default value is 0, takes value 1 when there is a failure, and after repair again takes value 0
i = 1106, 1107, 1108, 1113, 1114, 1115, 1120, 1121, 1122, 1127, 1128, 1129, 1130, 1135, 1136, 1137, 1138, 1143, 1144, 1145, 1150, 1151, 1156, 1157, 1158, 1163, 1164, 1165, 1166	V(i)_fLnk(i-1100)	Variable (*i*) controls track failure interrelated with link (*i* -1100), default value is 0, takes value 1 when there is a failure, and after repair again takes value 0

Figure 5. The line-station diagram of the track between S_5 and S_6

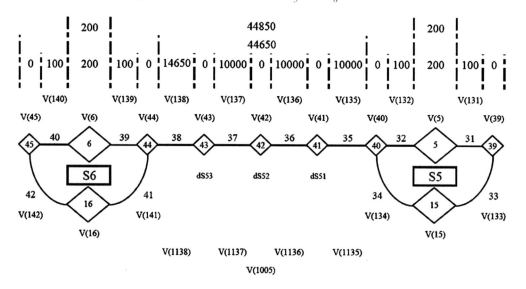

```
V1136_fLnk36=0:
V1137_fLnk37=0:
V1138_fLnk38=0:10$;
10$ BRANCH, 1:
If,(#ofWestboundTrains+#ofEastboundTrains)==0:11$;
Else:1$;
11$ DISPOSE:
```

The event diagrams for *track failure event* and *track repair event* are exhibited in Figures 6 and 7 respectively.

Figure 6. Flowchart for track failure event

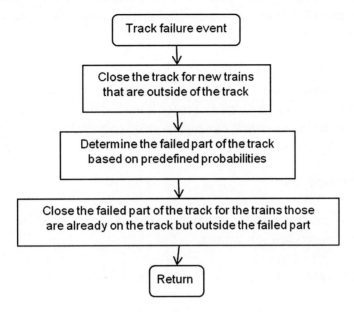

Figure 7. Flowchart for track repair event

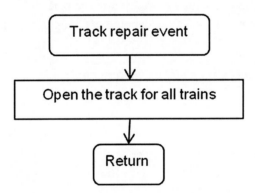

Assumptions related to the track failure part of the simulation model are:

- There will be track failures that will stop traffic on the related track. The failure and repair times distributions are given in Table 6.
- The failure time of the *base track* is distributed exponentially with a mean of 86400 seconds (24 hours), that is, it is expected to observe one failure for the *base track* in a day.
- Failure times for other tracks are also exponentially distributed with a mean of 86400 seconds (24 hours) times 1/ratio value. That is, failure times of other tracks have expected values inverse ratio to their lengths, namely the longer the track part the more frequently the track failure occurrence. For instance, it is assumed that failure time of the longest track is distributed exponentially with a mean of 33283 seconds (9.25 hours), that is, it is expected to observe more than two (24 / 9.25 = 2.6) failures in a day for the longest track.

- The repair time distribution is exponential with 1800 seconds (0.5 hours) mean for the *base track*.
- Repair times for other tracks are also exponentially distributed with a mean of 1800 seconds (0.5 hours) times related ratio value.
- After failures for the tracks were created according to the probability distributions shown in Table 6, these failures are transferred to the links due to the probabilities given in Table 7.
- If a track failure happens while a train is traversing on this track and if the next station is a dummy one, train goes to next dummy station and a check is made if the failure is on the train's destination direction or not. If the failure is on its destination side, the train waits until failure is repaired; else the train goes on its trip.

Train Movement Modelling

The train movement logic on the infrastructure is explained on some parts of the corridor via computer codes and event diagrams in below subsections.

Train Movement Logic from Park Area to a Real Station via a Terminus

Assume that a train is moving from the park area located at the east to reach S_1 via TS_1. In Figure 8, the line-station diagram of TS_1 and its neighbourhood is given.

The SIMAN Views of the train movement logic from the park area to TS_1 is depicted in Computer Code 2, and the train movement logic at TS_1 is denoted in Computer Code 3.

Figure 8. Line-station diagram of TS_1 and its neighborhood

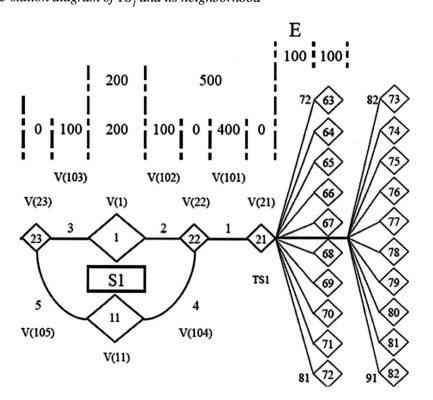

Computer Code 2:

```
1$        CREATE,        1:2$;
2$        BRANCH,        1:
              If,(ed(211)==1):3$;
              If,(ed(212)==1):4$;
              Else:5$;
3$        ASSIGN:        m=21:
                         FromStation=63:
                         ToStation=21:
                         Destination=1:
                         v(21)=v(21)+1:
                         v(101)=v(101)+1:
                         v(22)=v(22)+1:
                         v(102)=v(102)+1:
                         v(1)=v(1)+1:
                         v(103)=v(103)+1:
                         #ofWestboundTrains=#ofWestboundTrains-1:6$;
4$        ASSIGN:        m=21:
                         FromStation=63:
                         ToStation=21:
                         Destination=11:
                         v(21)=v(21)+1:
                         v(101)=v(101)+1:
                         v(22)=v(22)+1:
                         v(104)=v(104)+1:
                         v(11)=v(11)+1:
                         v(105)=v(105)+1:
                         #ofWestboundTrains=#ofWestboundTrains-1:6$;
5$        SCAN:          (ed(211)==1).or.(ed(212)==1):2$;
6$        DUPLICATE:     7$;
                         1:9$;
7$        REQUEST,       1:TrainFleet(sds,Train#):8$;
8$        TRANSPORT:     TrainFleet,TS1;
9$        BRANCH,        1:
              If,#ofWestboundTrains==0:11$;
              Else:10$;
10$       DELAY:         7200+unif(-900,900):2$;
11$       DISPOSE:
EXPRESSIONS:
211,FromParkViaTS1ToS1p1,
((ed(21101)==1).and.(ed(21102)==1)):
21101,FromParkViaTS1ToS1p1_1,
(v(21)==0).and.(v(101)==0).and.(v(22)==0).and.(v(102)==0).and.(v(1)==0).and.
```

(v(103)==0).and.((v(11)<=0).or.(v(2)>=0).or.(v(12)>=0)).and.((v(11)<=0).
or.((v(2)==0).or.(v(12)==0)).or.(v(3)>=0).or.(v(13)>=0)).and.((v(11)<=0).
or.((v(2)==0).or.(v(12)==0).or.(v(3)==0).or.(v(13)==0)).or.(v(4)>=0).
or.(v(14)>=0)).and.((v(11)<=0).or.((v(2)==0).or.(v(12)==0).or.(v(3)==0).
or.(v(13)==0).or.(v(4)==0).or.(v(14)==0)).or.(v(5)>=0).or.(v(15)>=0)).
and.((v(11)<=0).or.((v(2)==0).or.(v(12)==0).or.(v(3)==0).or.(v(13)==0).
or.(v(4)==0).or.(v(14)==0).or.(v(5)==0).or.(v(15)==0)).or.(v(6)>=0).
or.(v(16)>=0)).and.((v(11)<=0).or.((v(2)==0).or.(v(12)==0).or.(v(3)==0).
or.(v(13)==0).or.(v(4)==0).or.(v(14)==0).or.(v(5)==0).or.(v(15)==0).
or.(v(6)==0).or.(v(16)==0)).or.(v(7)>=0).or.(v(17)>=0)).and.((v(11)<=0).
or.((v(2)==0).or.(v(12)==0).or.(v(3)==0).or.(v(13)==0).or.(v(4)==0).
or.(v(14)==0).or.(v(5)==0).or.(v(15)==0).or.(v(6)==0).or.(v(16)==0).
or.(v(7)==0).or.(v(17)==0)).or.(v(8)>=0).or.(v(18)>=0)):
21102,FromParkViaTS1ToS1p1_2,
((v(11)<=0).or.((v(2)==0).or.(v(12)==0).or.(v(3)==0).or.(v(13)==0).
or.(v(4)==0).or.(v(14)==0).or.(v(5)==0).or.(v(15)==0).or.(v(6)==0).
or.(v(16)==0).or.(v(7)==0).or.(v(17)==0).or.(v(8)==0).or.(v(18)==0)).
or.(v(9)>=0).or.(v(19)>=0)).and.((v(11)<=0).or.((v(2)==0).or.(v(12)==0).
or.(v(3)==0).or.(v(13)==0).or.(v(4)==0).or.(v(14)==0).or.(v(5)==0).
or.(v(15)==0).or.(v(6)==0).or.(v(16)==0).or.(v(7)==0).or.(v(17)==0).
or.(v(8)==0).or.(v(18)==0).or.(v(9)==0).or.(v(19)==0)).or.(v(10)>=0).
or.(v(20)>=0)):
212,FromParkViaTS1ToS1p2,
((ed(21201)==1).and.(ed(21202)==1)):
21201,FromParkViaTS1ToS1p2_1,
(v(21)==0).and.(v(101)==0).and.(v(22)==0).and.(v(104)==0).and.(v(11)==0).
and.(v(105)==0).and.((v(1)<=0).or.(v(2)>=0).or.(v(12)>=0)).and.((v(1)<=0).
or.((v(2)==0).or.(v(12)==0)).or.(v(3)>=0).or.(v(13)>=0)).and.((v(1)<=0).
or.((v(2)==0).or.(v(12)==0).or.(v(3)==0).or.(v(13)==0)).or.(v(4)>=0).
or.(v(14)>=0)).and.((v(1)<=0).or.((v(2)==0).or.(v(12)==0).or.(v(3)==0).
or.(v(13)==0).or.(v(4)==0).or.(v(14)==0)).or.(v(5)>=0).or.(v(15)>=0)).and.
((v(1)<=0).or.((v(2)==0).or.(v(12)==0).or.(v(3)==0).or.(v(13)==0).or.(v(4)==0).
or.(v(14)==0).or.(v(5)==0).or.(v(15)==0)).or.(v(6)>=0).or.(v(16)>=0)).and.
((v(1)<=0).or.((v(2)==0).or.(v(12)==0).or.(v(3)==0).or.(v(13)==0).or.(v(4)==0).
or.(v(14)==0).or.(v(5)==0).or.(v(15)==0).or.(v(6)==0).or.(v(16)==0)).
or.(v(7)>=0).or.(v(17)>=0)).and.((v(1)<=0).or.((v(2)==0).or.(v(12)==0).
or.(v(3)==0).or.(v(13)==0).or.(v(4)==0).or.(v(14)==0).or.(v(5)==0).
or.(v(15)==0).or.(v(6)==0).or.(v(16)==0).or.(v(7)==0).or.(v(17)==0)).
or.(v(8)>=0).or.(v(18)>=0)):
21202,FromParkViaTS1ToS1p2_2,
((v(1)<=0).or.((v(2)==0).or.(v(12)==0).or.(v(3)==0).or.(v(13)==0).or.(v(4)==0).
or.(v(14)==0).or.(v(5)==0).or.(v(15)==0).or.(v(6)==0).or.(v(16)==0).
or.(v(7)==0).or.(v(17)==0).or.(v(8)==0).or.(v(18)==0)).or.(v(9)>=0).
or.(v(19)>=0)).and.((v(1)<=0).or.((v(2)==0).or.(v(12)==0).or.(v(3)==0).

or.(v(13)==0).or.(v(4)==0).or.(v(14)==0).or.(v(5)==0).or.(v(15)==0).
or.(v(6)==0).or.(v(16)==0).or.(v(7)==0).or.(v(17)==0).or.(v(8)==0).
or.(v(18)==0).or.(v(9)==0).or.(v(19)==0)).or.(v(10)>=0).or.(v(20)>=0)):

Computer Code 3:

```
1$       STATION,        TS1;2$;
2$       BRANCH,         1:
             If,FromStation==1:3$;
             If,FromStation==11:3$;
             Else:7$;
3$       DUPLICATE:      4$;
                         1:6$;
4$       ASSIGN:    TravelTimeOfTrain(Train#,1)=tnow-TimeIn:12$;
5$       DELAY:          5:6$;
6$       ASSIGN:         v(22)=v(22)+1:
                         v(101)=v(101)+1:
                         v(21)=v(21)+1:23$;
7$       BRANCH,         1:
             If,Destination==1:8$;
             If,Destination==11:10$;
8$       ASSIGN:         FromStation=21:
                         ToStation=1:
                         TimeIn=tnow:9$;
9$       TRANSPORT:      TrainFleet,S1_1;
10$      ASSIGN:         FromStation=21:
                         ToStation=11:
                         TimeIn=tnow:$11;
11$      TRANSPORT:      TrainFleet,S1_2;
12$      BRANCH,         1:
             If,(nx(73)==0).and.(ndx(82)==0):13$;
             If,(nx(74)==0).and.(ndx(83)==0):14$;
             If,(nx(75)==0).and.(ndx(84)==0):15$;
             If,(nx(76)==0).and.(ndx(85)==0):16$;
             If,(nx(77)==0).and.(ndx(86)==0):17$;
             If,(nx(78)==0).and.(ndx(87)==0):18$;
             If,(nx(79)==0).and.(ndx(88)==0):19$;
             If,(nx(80)==0).and.(ndx(89)==0):20$;
             If,(nx(81)==0).and.(ndx(90)==0):21$;
             If,(nx(82)==0).and.(ndx(91)==0):22$;
13$      TRANSPORT:      TrainFleet,Sta73_park;
14$      TRANSPORT:      TrainFleet,Sta74_park;
15$      TRANSPORT:      TrainFleet,Sta75_park;
16$      TRANSPORT:      TrainFleet,Sta76_park;
```

```
17$     TRANSPORT:     TrainFleet,Sta77_park;
18$     TRANSPORT:     TrainFleet,Sta78_park;
19$     TRANSPORT:     TrainFleet,Sta79_park;
20$     TRANSPORT:     TrainFleet,Sta80_park;
21$     TRANSPORT:     TrainFleet,Sta81_park;
22$     TRANSPORT:     TrainFleet,Sta82_park;
23$     DISPOSE:
```

The event diagrams for train movement from the park area event and train arrival to the terminus event are given in Figures 9 and 10 respectively.

Figure 9. Flowchart for train movement from the park area event

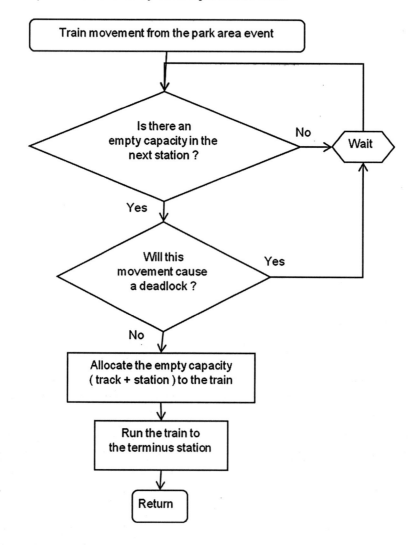

Figure 10. Flowchart for train arrival to the terminus event

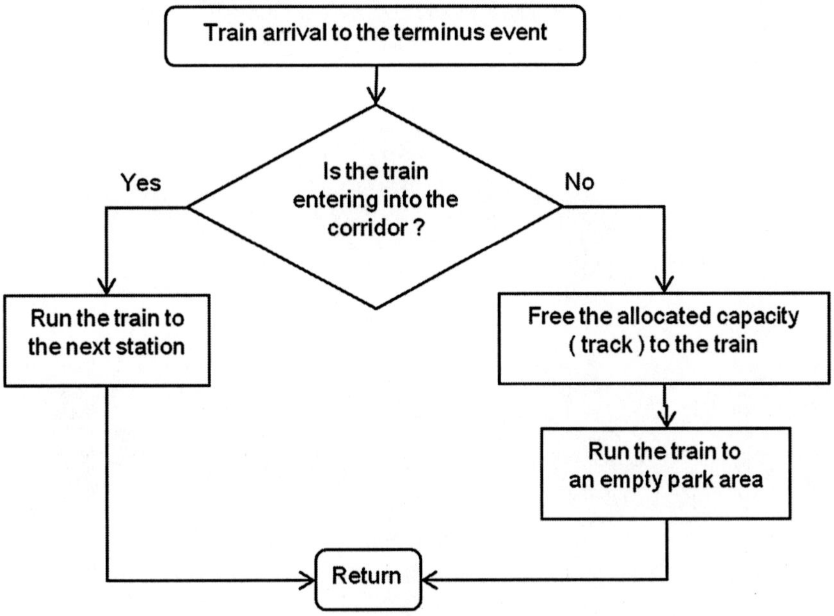

Train Movement Logic at a Real Station

Assume that a train is just arrived to the first part of the real station S_5 that is interrelated with the intersection 5 in the simulation model. The line-station diagram of S_5 and its neighbourhood is given in Figure 3. The SIMAN View of the train movement logic at the first part of S_5 is shown in Computer Code 4.

Computer Code 4:

```
1$      STATION,        S5_1:2$;
2$      BRANCH,         1:
                If,FromStation==4:3$;
                If,FromStation==14:3$;
                If,FromStation==6:5$;
                If,FromStation==16:5$;
3$      ASSIGN:         TrainArrivalTime(Train#,14)=tnow:
                        v(35)=v(35)-1:
                        v(127)=v(127)-1:
                        v(36)=v(36)-1:
                        v(128)=v(128)-1:
                        v(37)=v(37)-1:
                        v(129)=v(129)-1:
                        v(38)=v(38)-1:
                        v(130)=v(130)-1:
                        v(39)=v(39)-1:
```

		FromStation=5:4$;
4$	DELAY:	600+expo(90):7$;
5$	ASSIGN:	TrainArrivalTime(Train#,14)=tnow:
		v(44)=v(44)+1:
		v(138)=v(138)+1:
		v(43)=v(43)+1:
		v(137)=v(137)+1:
		v(42)=v(42)+1:
		v(136)=v(136)+1:
		v(41)=v(41)+1:
		v(135)=v(135)+1:
		v(40)=v(40)+1:
		FromStation=5:6$;
6$	DELAY:	600+expo(90):29$;
7$	ASSIGN:	v(2000)=v(2000)+1:
		TrainOrderInQ=v(2000):8$;
8$	QUEUE,	Q5:9$;
9$	SCAN:	(v(1005)==0):10$;
10$	SEIZE,	R5,1:11$;
11$	DELAY:	0.000001:12$;
12$	BRANCH,	1:
		If,FromStation==5:13$;
		If,FromStation==15:16$:
		If,FromStation==6:21$;
		If,FromStation==16:24$;
13$	BRANCH,	1:
		If,(ed(51)==1):14$;
		If,(ed(52)==1):15$;
		Else:19$;
14$	RELEASE:	R5,1:51$;
15$	RELEASE:	R5,1:52$;
16$	BRANCH,	1:
		If,(ed(51)==1):17$;
		If,(ed(52)==1):18$;
		Else:19$;
17$	RELEASE:	R5,1:*related with 2.part of S5*;
18$	RELEASE:	R5,1:*related with 2.part of S5*;
19$	RELEASE:	R5,1:20$;
20$	SCAN:	(ed(51)==1).or.(ed(52)==1):8$;
21$	BRANCH,	1:
		If,(ed(63)==1):22$;
		If,(ed(64)==1):23$;
		Else:27$;
22$	RELEASE:	R5,1:*related with 1.part of S6*;

A General Simulation Modelling Framework for Train Timetabling Problem

```
23$     RELEASE:        R5,1:related with 1.part of S6;
24$     BRANCH,         1:
            If,(ed(63)==1):25$;
            If,(ed(64)==1):26$;
            Else:27$;
25$     RELEASE:        R5,1:related with 2.part of S6;
26$     RELEASE:        R5,1:related with 2.part of S6;
27$     RELEASE:        R5,1:28$;
28$     SCAN:           (ed(63)==1).or.(ed(64)==1):8$;
29$     ASSIGN:         v(2000)=v(2000)+1:
                        TrainOrderInQ=v(2000):30$;
30$     QUEUE,          Q4:31$;
31$     SCAN:           (v(1004)==0):32$;
32$     SEIZE,          R4,1:33$;
33$     DELAY:          0.000001:34$;
34$     BRANCH,         1:
            If,FromStation==4:35$;
            If,FromStation==14:38$;
            If,FromStation==5:43$;
            If,FromStation==15:46$;
35$     BRANCH,         1:
            If,(ed(41)==1):36$;
            If,(ed(42)==1):37$;
            Else:41$;
36$     RELEASE:        R4,1:related with 1.part of S4;
37$     RELEASE:        R4,1:related with 1.part of S4;
38$     BRANCH,         1:
            If,(ed(41)==1):39$;
            If,(ed(42)==1):40$;
            Else:41$;
39$     RELEASE:        R4,1:related with 2.part of S4;
40$     RELEASE:        R4,1:related with 2.part of S4;
41$     RELEASE:        R4,1:42$;
42$     SCAN:           (ed(41)==1).or.(ed(42)==1):30$;
43$     BRANCH,         1:
            If,(ed(53)==1):44$;
            If,(ed(54)==1):45$;
            Else:49$;
44$     RELEASE:        R4,1:57$;
45$     RELEASE:        R4,1:58$;
46$     BRANCH,         1:
            If,(ed(53)==1):47$;
            If,(ed(54)==1):48$;
            Else:49$;
```

47$	RELEASE:	R4,1:*related with 2.part of S5;*
48$	RELEASE:	R4,1:*related with 2.part of S5;*
49$	RELEASE:	R4,1:50$;
50$	SCAN:	(ed(53)==1).or.(ed(54)==1):30$;
51$	ASSIGN:	ToStation=6:
		v(40)=v(40)+1:
		v(135)=v(135)+1:
		v(41)=v(41)+1:
		v(136)=v(136)+1:
		v(42)=v(42)+1:
		v(137)=v(137)+1:
		v(43)=v(43)+1:
		v(138)=v(138)+1:
		v(44)=v(44)+1:
		v(139)=v(139)+1:
		v(6)=v(6)+1:
		v(140)=v(140)+1:
		TrainDepartureTime(Train#,14)=tnow:53$;
52$	ASSIGN:	ToStation=16:
		v(40)=v(40)+1:
		v(135)=v(135)+1:
		v(41)=v(41)+1:
		v(136)=v(136)+1:
		v(42)=v(42)+1:
		v(137)=v(137)+1:
		v(43)=v(43)+1:
		v(138)=v(138)+1:
		v(44)=v(44)+1:
		v(141)=v(141)+1:
		v(16)=v(16)+1:
		v(142)=v(142)+1:
		TrainDepartureTime(Train#,14)=tnow:53$;
53$	DUPLICATE:	54$;
		1:55$;
54$	TRANSPORT:	TrainFleet,dS51,unif(25.00,30.56);
55$	DELAY:	10:56$;
56$	ASSIGN:	v(131)=v(131)-1:
		v(5)=v(5)-1:
		v(132)=v(132)-1:63$;
57$	ASSIGN:	ToStation=4:
		v(39)=v(39)-1:
		v(130)=v(130)-1:
		v(38)=v(38)-1:
		v(129)=v(129)-1:

A General Simulation Modelling Framework for Train Timetabling Problem

```
                              v(37)=v(37)-1:
                              v(128)=v(128)-1:
                              v(36)=v(36)-1:
                              v(127)=v(127)-1:
                              v(35)=v(35)-1:
                              v(124)=v(124)-1:
                              v(4)=v(4)-1:
                              v(123)=v(123)-1:
                              TrainDepartureTime(Train#,14)=tnow:59$;
58$         ASSIGN:           ToStation=14:
                              v(39)=v(39)-1:
                              v(130)=v(130)-1:
                              v(38)=v(38)-1:
                              v(129)=v(129)-1:
                              v(37)=v(37)-1:
                              v(128)=v(128)-1:
                              v(36)=v(36)-1:
                              v(127)=v(127)-1:
                              v(35)=v(35)-1:
                              v(126)=v(126)-1:
                              v(14)=v(14)-1:
                              v(125)=v(125)-1:
                              TrainDepartureTime(Train#,14)=tnow:59$;
59$         DUPLICATE:        60$;
                              1:61$;
60$         TRANSPORT:        TrainFleet,dS43,unif(25.00,30.56);
61$         DELAY:            10:62$;
62$         ASSIGN:           v(132)=v(132)+1:
                              v(5)=v(5)+1:
                              v(131)=v(131)+1:63$;
63$         DISPOSE:
EXPRESSIONS:
41,FromS4p1OrS4p2ToS5p1,
(v(35)>=0).and.(v(127)>=0).and.(v(36)>=0).and.(v(128)>=0).and.(v(37)>=0).and.
(v(129)>=0).and.(v(38)>=0).and.(v(130)>=0).and.(v(39)>=0).and.(v(131)==0).and.
(v(5)==0).and.(v(132)==0).and.(v(1004)==0).and.((v(15)<=0).or.(v(6)>=0).or.
(v(16)>=0)).and.((v(15)<=0).or.((v(6)==0).or.(v(16)==0)).or.(v(7)>=0).or.
(v(17)>=0)).and.((v(15)<=0).or.((v(6)==0).or.(v(16)==0).or.(v(7)==0).or.
(v(17)==0)).or.(v(8)>=0).or.(v(18)>=0)).and.((v(15)<=0).or.((v(6)==0).or.
(v(16)==0).or.(v(7)==0).or.(v(17)==0).or.(v(8)==0).or.(v(18)==0)).or.(v(9)>=0).
or.(v(19)>=0)).and.((v(15)<=0).or.((v(6)==0).or.(v(16)==0).or.(v(7)==0).or.
(v(17)==0).or.(v(8)==0).or.(v(18)==0).or.(v(9)==0).or.(v(19)==0)).or.
(v(109)>=0).or.(v(20)>=0)):
42,FromS4p1OrS4p2ToS5p2,
```

(v(35)>=0).and.(v(127)>=0).and.(v(36)>=0).and.(v(128)>=0).and.(v(37)>=0).and.
(v(129)>=0).and.(v(38)>=0).and.(v(130)>=0).and.(v(39)>=0).and.(v(133)==0).and.
(v(15)==0).and.(v(134)==0).and.(v(1004)==0).and.((v(5)<=0).or.(v(6)>=0).or.
(v(16)>=0)).and.((v(5)<=0).or.((v(6)==0).or.(v(16)==0)).or.(v(7)>=0).or.
(v(17)>=0)).and.((v(5)<=0).or.((v(6)==0).or.(v(16)==0).or.(v(7)==0).or.
(v(17)==0)).or.(v(8)>=0).or.(v(18)>=0)).and.((v(5)<=0).or.((v(6)==0).or.
(v(16)==0).or.(v(7)==0).or.(v(17)==0).or.(v(8)==0).or.(v(18)==0))).or.(v(9)>=0).
or.(v(19)>=0)).and.((v(5)<=0).or.((v(6)==0).or.(v(16)==0).or.(v(7)==0).or.
(v(17)==0).or.(v(8)==0).or.(v(18)==0).or.(v(9)==0).or.(v(19)==0)).or.
(v(109)>=0).or.(v(20)>=0)):
51,FromS5p1OrS5p2ToS6p1,
(v(40)>=0).and.(v(135)>=0).and.(v(41)>=0).and.(v(136)>=0).and.(v(42)>=0).and.
(v(137)>=0).and.(v(43)>=0).and.(v(138)>=0).and.(v(44)>=0).and.(v(139)==0).and.
(v(6)==0).and.(v(140)==0).and.(v(1005)==0).and.((v(16)<=0).or.(v(7)>=0).or.
(v(17)>=0)).and.((v(16)<=0).or.((v(7)==0).or.(v(17)==0)).or.(v(8)>=0).or.
(v(18)>=0)).and.((v(16)<=0).or.((v(7)==0).or.(v(17)==0).or.(v(8)==0).or.
(v(18)==0)).or.(v(9)>=0).or.(v(19)>=0)).and.((v(16)<=0).or.((v(7)==0).or.
(v(17)==0).or.(v(8)==0).or.(v(18)==0).or.(v(9)==0).or.(v(19)==0)).or.
(v(10)>=0).or.(v(20)>=0)):
52,FromS5p1OrS5p2ToS6p2,
(v(40)>=0).and.(v(135)>=0).and.(v(41)>=0).and.(v(136)>=0).and.(v(42)>=0).and.
(v(137)>=0).and.(v(43)>=0).and.(v(138)>=0).and.(v(44)>=0).and.(v(141)==0).and.
(v(16)==0).and.(v(142)==0).and.(v(1005)==0).and.((v(6)<=0).or.(v(7)>=0).or.
(v(17)>=0)).and.((v(6)<=0).or.((v(7)==0).or.(v(17)==0)).or.(v(8)>=0).or.
(v(18)>=0)).and.((v(6)<=0).or.((v(7)==0).or.(v(17)==0).or.(v(8)==0).or.
(v(18)==0)).or.(v(9)>=0).or.(v(19)>=0)).and.((v(6)<=0).or.((v(7)==0).or.
(v(17)==0).or.(v(8)==0).or.(v(18)==0).or.(v(9)==0).or.(v(19)==0)).or.
(v(10)>=0).or.(v(20)>=0)):
53,FromS5p1OrS5p2ToS4p1,
(v(39)<=0).and.(v(130)<=0).and.(v(38)<=0).and.(v(129)<=0).and.(v(37)<=0).and.
(v(128)<=0).and.(v(36)<=0).and.(v(127)<=0).and.(v(35)<=0).and.(v(124)==0).and.
(v(4)==0).and.(v(123)==0).and.(v(1004)==0).and.((v(14)>=0).or.(v(3)<=0).or.
(v(13)<=0)).and.((v(14)>=0).or.((v(3)==0).or.(v(13)==0)).or.(v(2)<=0).or.
(v(12)<=0)).and.((v(14)>=0).or.((v(3)==0).or.(v(13)==0).or.(v(2)==0).or.
(v(12)==0)).or.(v(1)<=0).or.(v(11)<=0)):
54,FromS5p1OrS5p2ToS4p2,
(v(39)<=0).and.(v(130)<=0).and.(v(38)<=0).and.(v(129)<=0).and.(v(37)<=0).and.
(v(128)<=0).and.(v(36)<=0).and.(v(127)<=0).and.(v(35)<=0).and.(v(126)==0).and.
(v(14)==0).and.(v(125)==0).and.(v(1004)==0).and.((v(4)>=0).or.(v(3)<=0).or.
(v(13)<=0)).and.((v(4)>=0).or.((v(3)==0).or.(v(13)==0)).or.(v(2)<=0).or.
(v(12)<=0)).and.((v(4)>=0).or.((v(3)==0).or.(v(13)==0).or.(v(2)==0).or.
(v(12)==0)).or.(v(1)<=0).or.(v(11)<=0)):
63,FromS6p1OrS6p2ToS5p1,
(v(44)<=0).and.(v(138)<=0).and.(v(43)<=0).and.(v(137)<=0).and.(v(42)<=0).and.
(v(136)<=0).and.(v(41)<=0).and.(v(135)<=0).and.(v(40)<=0).and.(v(132)==0).and.

A General Simulation Modelling Framework for Train Timetabling Problem

```
(v(5)==0).and.(v(131)==0).and.(v(1005)==0).and.((v(15)>=0).or.(v(4)<=0).or.
(v(14)<=0)).and.((v(15)>=0).or.((v(4)==0).or.(v(14)==0)).or.(v(3)<=0).or.
(v(13)<=0)).and.((v(15)>=0).or.((v(4)==0).or.(v(14)==0)).or.(v(3)==0).or.
(v(13)==0)).or.(v(2)<=0).or.(v(12)<=0)).and.((v(15)>=0).or.((v(4)==0).or.
(v(14)==0).or.(v(3)==0).or.(v(13)==0).or.(v(2)==0).or.(v(12)==0)).or.(v(1)<=0).
or.(v(11)<=0)):
64,FromS6p1OrS6p2ToS5p2,
(v(44)<=0).and.(v(138)<=0).and.(v(43)<=0).and.(v(137)<=0).and.(v(42)<=0).and.
(v(136)<=0).and.(v(41)<=0).and.(v(135)<=0).and.(v(40)<=0).and.(v(134)==0).and.
(v(15)==0).and.(v(133)==0).and.(v(1005)==0).and.((v(5)>=0).or.(v(4)<=0).or.
(v(14)<=0)).and.((v(5)>=0).or.((v(4)==0).or.(v(14)==0)).or.(v(3)<=0).or.
(v(13)<=0)).and.((v(5)>=0).or.((v(4)==0).or.(v(14)==0)).or.(v(3)==0).or.
(v(13)==0)).or.(v(2)<=0).or.(v(12)<=0)).and.((v(5)>=0).or.((v(4)==0).or.
(v(14)==0).or.(v(3)==0).or.(v(13)==0).or.(v(2)==0).or.(v(12)==0)).or.(v(1)<=0).
or.(v(11)<=0)):
```

The event diagrams for train arrival to the real station event and train departure from the real station event are exhibited in Figure 11 and 12 respectively.

Train Movement Logic at a Dummy Station

Assume that a train is just arrived to the dummy station dS_{51}. The line-station diagram of dS_{51} and its neighbourhood is depicted in Figure 3. The SIMAN View of the train movement logic at the dummy station is shown in Computer Code 5.

Computer Code 5:

```
0$       STATION,       dS51:1$;
1$       ASSIGN:        TrainArrivalTime(Train#,15)=tnow:2$;
2$       BRANCH,        1:
             If,ToStation==5:3$;
             If,ToStation==15:3$;
             If,ToStation==6:4$;
             If,ToStation==16:4$;
3$       SCAN:          v(1135)==0:5$;
4$       SCAN:          v(1136)==0:5$;
5$       ASSIGN:        TrainDepartureTime(Train#,15)=tnow:6$;
6$       BRANCH,        1:
             If,ToStation==5:7$;
             If,ToStation==15:8$;
             If,ToStation==6:9$;
             If,ToStation==16:9$;
7$       TRANSPORT:     TrainFleet,S5_1,unif(23.6,26.4);
8$       TRANSPORT:     TrainFleet,S5_2,unif(23.6,26.4);
9$       TRANSPORT:     TrainFleet,dS52,unif(23.6,26.4);
```

Figure 11. Flowchart for train arrival to the real station event

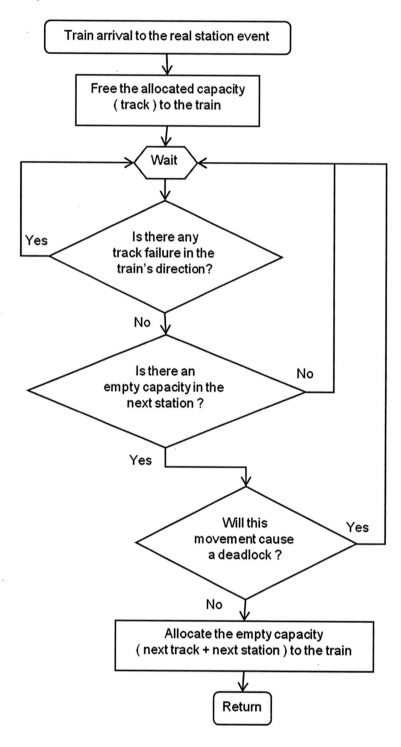

A General Simulation Modelling Framework for Train Timetabling Problem

Figure 12. Flowchart for train departure from a real station event

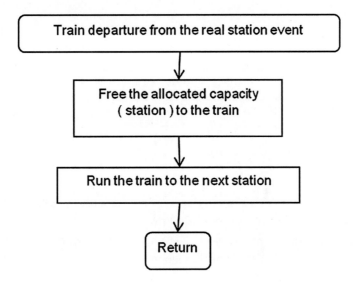

The event diagrams for train arrival to the dummy station event and train departure from the dummy station event are exhibited in Figures 13 and 14 respectively.

Assumptions related to the train movement part of the simulation model are as follows:

- Each train stops at real stations except terminuses.
- A train stops at a dummy station if there is a failure in a track placed in front of that train.
- Trains' speeds are uniformly distributed over an interval (90 kilometres/hour, 110 kilometres/hour).
- Trains that have reverse directions can meet and cross each other only at the real stations.

Figure 13. Flowchart for train arrival to the dummy station event

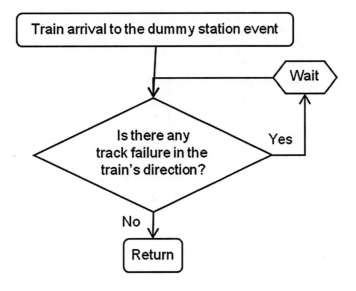

Figure 14. Flowchart for train departure from the dummy station event

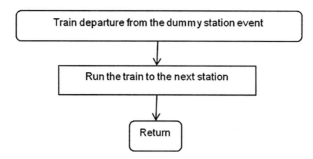

- Dwell times for each station are 600 seconds (10 minutes). That is each train will stop at least 600 seconds at the all stations for boarding and alighting events.
- To represent unplanned delays at a station, delay time is defined. It is assumed that delay time is exponentially distributed with a mean of 90 seconds. Delay time is added to the dwell times. Due to this unplanned delay, overtaking is possible.
- Trains that have same directions can meet and overtake each other only at the real stations.
- Track occupying decision is taken at the real stations based on the answers given the following questions:
  - Are the links and intersections suitable?
  - Does a track failure exist?
  - Does this decision cause a deadlock?
- First come first served (FCFS) dispatching rule is used to select one train among candidate trains. Candidate trains are the trains waiting at neighbour real stations of the long track that want to use the same long track and have finished waiting for dwell time and additional unplanned delay time. Namely they are the trains ready to travel and deserve to begin checking the conditions. If the all conditions to move are suitable for a candidate train, which arrived first to one of the neighbour real station of the long track it will begin to trip, else the same check is made for another train arrived second. Checking goes on until a suitable train is found.

Blockage Preventive Algorithm

A common potential deadlock is exhibited in Figure 15 where there are four trains: two trains are the WB trains and located at $S(i)$ and the other trains are the EB trains and located at $S(i+1)$. As can be seen in this figure, the system has a deadlock. Deadlock situation goes on until one of those trains reverses its direction.

Another example of a deadlock is shown in Figure 16. There are two WB trains at $S(i)$, two EB trains at $S(i+2)$ and two trains (not important to be a WB or an EB train) at $S(i+1)$.

Since we avoid reversing the direction of trains, in order to obtain a feasible train timetable we must take course of action to prevent a deadlock. In a recent study, Pachl (2011) reviewed the existing strategies proposed to prevent the deadlock problem in railway systems. The author concluded that the proposed models are based on the game board philosophy and are not suitable to be used in control logic of real simulation systems. Pachl (2011) proposed a rule-based deadlock avoidance method for simulation systems which follows the idea that a specified number of track sections ahead of a train must be reserved

Figure 15. An example of a deadlock

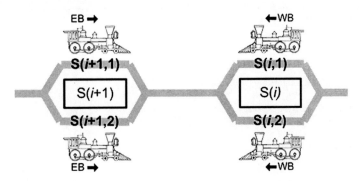

Figure 16. Deadlock between six trains

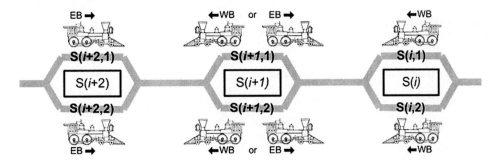

before this train is allowed to enter a track section. The number of track sections to be reserved depends on a set of logical rules.

Distinctively, in this study the long track sections are controlled in order to prevent a conflict which occurs if the long track section is used at the same time by the two trains in opposite directions. On the other hand, in order to avoid a deadlock that prevents the movement of the trains, a *blockage preventive algorithm* is developed that does not control the track sections but controls the real stations. The algorithm given in Table 9 follows the idea that the whole real stations in the direction of the train are checked before permitting the train to departure from its current real station.

The algorithm first checks the next real station in front of the train while the train is in a real station. For this train, if there is an empty capacity in the next real station, the algorithm checks whether it is the last real station or not. If the next real station is the last real station, the checking finishes and the train goes on its trip, otherwise the checking goes on for the real station behind the next real station and checking lasts until the last real station is checked. The empty capacities of the checked stations and also the directions of the trains that occupy the checked stations are important. The algorithm checks the whole real stations in front of the train; it never permits a deadlock and guaranties to obtain a feasible train timetable. The aim here is to prevent any train departure in advance that will cause a deadlock in the future.

Table 9. Blockage preventive algorithm

Assume that a WB train is at $S(i, j)$ real station and aimed to travel to $S(i+1, 1)$ real station where $i = 1, ..., 9$ denotes the station number and $j = 1, 2$ denotes the platform number.
Step = 1
$S(i+1, 1)$ **empty ?**
Yes: Go next.
No: Prohibit the train to travel, STOP the algorithm for that train.
$i+1 = 10$ **?**
Yes: Permit the train to travel, STOP checking blockage.
No: Go next.
Step = 2
$S(i+1, 2)$ **empty *or* allocated to an EB train ?**
or $S(i+2, j)$ **empty *or* allocated to a WB train ?**
Yes: Go next.
No: Prohibit the train to travel, STOP the algorithm for that train.
$i+2 = 10$ **?**
Yes: Permit the train to travel, STOP checking blockage.
No: Go next.
Step = 3
$S(i+1, 2)$ **empty *or* allocated to an EB train ?**
or **{ $S(i+2, j)$ empty? }**
or $S(i+3, j)$ **empty *or* allocated to a WB train ?**
Yes: Go next.
No: Prohibit the train to travel, STOP the algorithm for that train.
$i+3 = 10$ **?**
Yes: Permit the train to travel, STOP checking blockage.
No: Go next.
From Step = 4 to Step = 8
$S(i+1, 2)$ **empty *or* allocated to an EB train ?**
or **{ $S(i+k, j)$ empty?; $k = 2, ..., $ (Step-1)}**
or $S(i+\text{Step}, j)$ **empty *or* allocated to a WB train ?**
Yes: Go next.
No: Prohibit the train to travel, STOP the algorithm for that train.
$i+\text{Step} = 10$ **?**
Yes: Permit the train to travel, STOP checking blockage.
No: Go next.
Step = 9
$S(i+1, 2)$ **empty *or* allocated to an EB train ?**
or **{ $S(i+k, j)$ empty?; $k = 2, ..., $ (Step-1)}**
or $S(i+\text{Step}, j)$ **empty *or* allocated to a WB train ?**
Yes: Permit the train to travel, STOP the algorithm for that train.
No: Prohibit the train to travel, STOP the algorithm for that train.

Verification of the Simulation Model

The simulation model is verified by developing the model in a modular manner, using interactive debuggers, substituting constants for random variables, manually checking the results and animating the system. In order to develop the simulation model in a modular manner as explained in detail in above sections a step by step approach that gives ability to systematically model a complex system was used.

First the railway network that includes the links, the intersections and the stations was modelled. Then, the track failures and repairs were added, and the train movement logic on the railway corridor was modelled. Next, we added a rule for track allocation to the candidate trains. The fixed train speeds were relaxed, and then additional unplanned delays at the stations were modelled. The number of trains in the system was increased and randomness was added to the planned initial train timetable. As a last

A General Simulation Modelling Framework for Train Timetabling Problem

step, animation of the system was developed. The animation part of the simulation model was built by using the *Animate* tool of the ARENA, to see if model is working as intended, and to understand the system clearer.

SOLUTIONS AND RECOMMENDATIONS

In this section some results of the simulation model are discussed. As a first step, the authors begin with a planned initial train timetable. This timetable is infeasible since it includes conflicts. Secondly, they focus on a feasible, conflict free initial train timetable which is obtained by the deterministic simulation model. Lastly, a feasible train timetable that is obtained by the stochastic simulation model is given in detail.

Infeasible Planned Initial Train Timetable

The trains are scheduled on a bidirectional single track corridor regarding the initial train timetable, where all the inputs are deterministic, given in Table 2. The corridor has 10 real and 20 dummy, totally 30 stations. There is no randomness in the planned train arrival and departure times, failures and repairs are excluded, train speeds are fixed at 100 kilometres/hour, and no additional delay is added to the dwell times.

The discussion is begun on an empty infrastructure, on which only WB_1 train is running. The simulation model is run for this scenario and the train-station diagram exhibited in Figure 17 is obtained. The train travel time for WB_1 is calculated as 16350 seconds.

The next scenario is related to EB_1 train that is running on the empty infrastructure. The train-station diagram exhibited in Figure 18 is obtained by the deterministic simulation model. The calculated train travel time is the same as the previous one, 16350 seconds.

After that, the train-station diagram for the planned initial train timetable given in Table 2 is manually obtained. This diagram is based on WB_1 and EB_1 timetables, and depicted in Figure 19.

Figure 17. Train-station diagram for WB_1

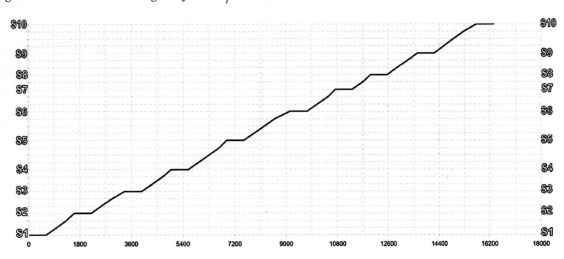

Figure 18. Train-station diagram for EB$_1$

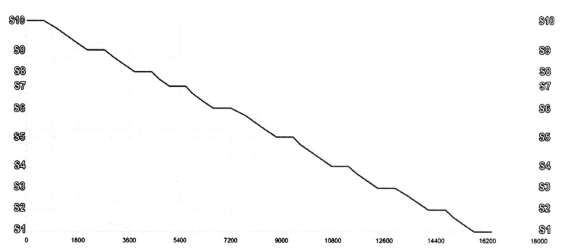

Figure 19. Infeasible train-station diagram for the planned initial train timetable

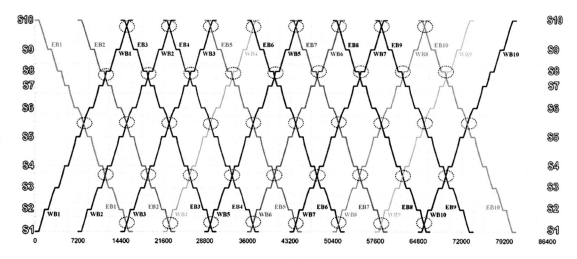

The timetable in Figure 19 is not conflict free, it is infeasible. The conflict locations are indicated by dotted line circles, and it is calculated that there are 44 conflicts to be solved in such a deterministic system. To display the problem clearer the conflicts between WB$_3$ and some EB trains are displayed in Figure 20.

As it is seen in this figure, WB$_3$ will have the first conflict with EB$_1$ train between real stations S$_1$ and S$_2$. The other conflicts will be with EB$_2$ between S$_3$ and S$_4$, with EB$_3$ between the S$_5$ and S$_6$, with EB$_4$ between S$_7$ and S$_8$ and with EB$_5$ between S$_9$ and S$_{10}$, respectively.

The infeasible planned initial train timetable is given in Table 10.

Figure 20. Infeasible train-station diagram for WB₃

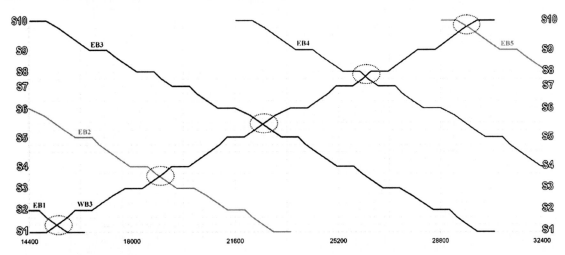

Figure 21. Feasible train-station diagram for the planned initial train timetable

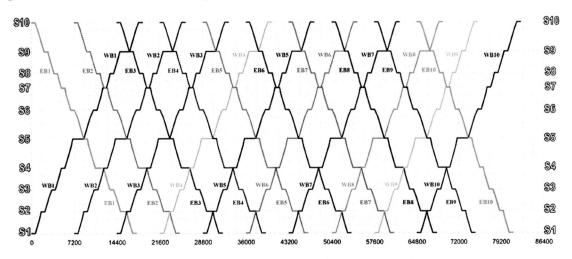

Feasible Planned Initial Train Timetable

The developed simulation model framework, which is explained in detail in above sections, has the ability to solve the conflicts and to create a conflict free feasible train timetable. The link controls via the variables in the framework prevents the usage of the same long track part at the same time by the trains that have an opposite directions (the cause of the conflicts).

In this scenario all the inputs are taken to be deterministic. That is, randomness in the planned train arrival and departure times is not allowed. Train speeds are fixed at 100 kilometres/hour. The simulation model is run for these deterministic input values. The feasible train-station diagram obtained by the deterministic simulation model is given in Figure 21.

Figure 22 shows the conflict free feasible train station diagram for WB_3. It is seen that EB_1 train waits at S_2 and permits WB_3 to travel from S_1 to S_2, and EB_2 waits WB_3 at S_4 for emptying the track part

Table 10. Infeasible planned initial train timetable

Train#	WB/EB	S1		S2		S3		S4		S5		S6		S7		S8		S9		S10		Travel Time
		ArT	DpT	ArT	DpT	ArT	DpT	ArT	DpT	ArT	DpT	ArT	DpT	ArT	DpT	ArT	DpT	ArT	DpT	ArT	DpT	
1	WB1	32	600	1600	2200	3363	3963	4986	5586	6919	7519	9140	9740	10719	11319	11952	12552	13607	14207	15734	16334	04:32:30
2	EB1	15734	16334	14134	14734	12371	12971	10748	11348	8816	9416	6595	7195	5016	5616	3782	4382	2127	2727	32	600	04:32:30
3	WB2	7232	7801	8800	9400	10563	11163	12186	12786	14119	14719	16340	16940	17919	18519	19152	19752	20807	21407	22934	23534	04:32:30
4	EB2	22934	23534	21334	21934	19571	20171	17948	18548	16016	16616	13795	14395	12216	12816	10982	11582	9327	9927	7232	7801	04:32:30
5	WB3	14432	15001	16000	16600	17763	18363	19386	19986	21319	21919	23540	24140	25119	25719	26352	26952	28007	28607	30134	30734	04:32:30
6	EB3	30134	30734	28534	29134	26771	27371	25148	25748	23216	23816	20995	21595	19416	20016	18182	18782	16527	17127	14432	15001	04:32:30
7	WB4	21632	22201	23200	23800	24963	25563	26586	27186	28519	29119	30740	31340	32319	32919	33552	34152	35207	35807	37334	37934	04:32:30
8	EB4	37334	37934	35734	36334	33971	34571	32348	32948	30416	31016	28195	28795	26616	27216	25382	25982	23727	24327	21632	22201	04:32:30
9	WB5	28832	29401	30400	31000	32163	32763	33786	34386	35719	36319	37940	38540	39519	40119	40752	41352	42407	43007	44534	45134	04:32:30
10	EB5	44534	45134	42934	43534	41171	41771	39548	40148	37616	38216	35395	35995	33816	34416	32582	33182	30927	31527	28832	29401	04:32:30
11	WB6	36032	36601	37600	38200	39363	39963	40986	41586	42919	43519	45140	45740	46719	47319	47952	48552	49607	50207	51734	52334	04:32:30
12	EB6	51734	52334	50134	50734	48371	48971	46748	47348	44816	45416	42595	43195	41016	41616	39782	40382	38127	38727	36032	36601	04:32:30
13	WB7	43232	43801	44800	45400	46563	47163	48186	48786	50119	50719	52340	52940	53919	54519	55152	55752	56807	57407	58934	59534	04:32:30
14	EB7	58934	59534	57334	57934	55571	56171	53948	54548	52016	52616	49795	50395	48216	48816	46982	47582	45327	45927	43232	43801	04:32:30
15	WB8	50432	51001	52000	52600	53763	54363	55386	55986	57319	57919	59540	60140	61119	61719	62352	62952	64007	64607	66134	66734	04:32:30
16	EB8	66134	66734	64534	65134	62771	63371	61148	61748	59216	59816	56995	57595	55416	56016	54182	54782	52527	53127	50432	51001	04:32:30
17	WB9	57632	58201	59200	59800	60963	61563	62586	63186	64519	65119	66740	67340	68319	68919	69552	70152	71207	71807	73334	73934	04:32:30
18	EB9	73334	73934	71734	72334	69971	70571	68348	68948	66416	67016	64195	64795	62616	63216	61382	61982	59727	60327	57632	58201	04:32:30
19	WB10	64832	65401	66400	67000	68163	68763	69786	70386	71719	72319	73940	74540	75519	76119	76752	77352	78407	79007	80534	81134	04:32:30
20	EB10	80534	81134	78934	79534	77171	77771	75548	76148	73616	74216	71395	71995	69816	70416	68582	69182	66927	67527	64832	65401	04:32:30
																		Average train travel time				16350

Figure 22. Feasible train-station diagram for WB₃

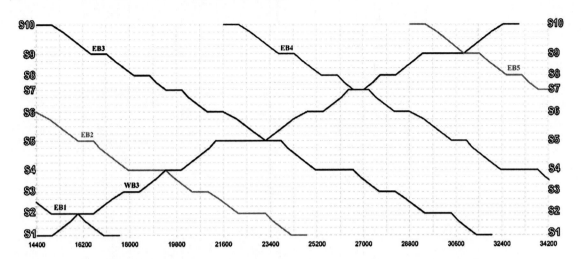

between S_4 and S_3. On the other hand WB_3 waits at S_5 to permit EB_3 to travel from S_6 to S_5. Since WB_3 spends additional time while waiting EB_3, WB_3 meets EB_4 at S_7 and passes without spending additional time. Next, WB_3 waits EB_5 at S_9 and then finishes its trip.

The feasible train timetable calculated by the deterministic simulation model is given in Table 11. The calculated average train travel time for this scenario is 17956 seconds.

In order to observe the change in the computer running time versus the number of trains in the deterministic simulation model, the model is run for different number of trains, the result are shown in Figure 23. As it is seen in Figure 23(a) the computer running time is increasing nonlinearly when the number of trains in the system is increased. On the other hand, Figure 23(b) shows the increments in computer running time versus the added two trains. It is seen that the amount of increment caused by addition of two trains into the system depends on the number of current trains in the system. To make it clearer, it is assumed that there are four trains in the system. In this case, the computer running time is 9.0 seconds. When two new trains are added in the system (i.e. there will be six trains in the system), the computer running time will be 15.0 seconds (i.e. the increase in computer running time will be 6.0 seconds). On the other hand the increment will be 17.4 seconds if two trains are added into the system which has 18 trains.

Feasible Train Timetable

In this section a feasible train timetable obtained by the stochastic simulation model is focused. The train-station diagram of the feasible solution of which the calculated average train timetable is 24218 seconds is given in Figure 24, and the related train timetable is depicted in Table 12.

Disturbance Management

In order to see what happens in the system after a failure has occurred, how the disturbance managed, Figure 25 should be examined. In this figure, while the simulation model is running through 39600-

Table 11. Feasible planned initial train timetable

Train#	WB/EB	S1		S2		S3		S4		S5		S6		S7		S8		S9		S10		Travel time
		ArT	DpT	ArT	DpT	ArT	DpT	ArT	DpT	ArT	DpT	ArT	DpT	ArT	DpT	ArT	DpT	ArT	DpT	ArT	DpT	
1	WB1	32	600	1600	2200	3363	3963	4986	5586	6919	8816	10438	11038	12017	12617	13250	13850	14905	16527	18053	18653	18670	05:11:10
2	EB1	17000	17600	14972	16000	13209	13809	10748	12186	8816	9416	6595	7195	5016	5616	3782	4382	2127	2727	32	600	17616	04:53:36
3	WB2	7232	7801	8800	9400	10563	11163	12186	12786	14118	16016	17638	18238	19217	19817	20450	21050	22105	23727	25254	25854	18670	05:11:10
4	EB2	24200	24800	22172	23200	20409	21009	17948	19386	16016	16616	13794	14394	12215	12815	10982	11582	9327	9927	7232	7801	17616	04:53:36
5	WB3	14431	15000	16000	16600	17763	18363	19386	19986	21318	23216	24838	25438	26417	27017	27650	28250	29305	30927	32454	33054	18670	05:11:10
6	EB3	31400	32000	29372	30400	27609	28209	25148	26586	23216	23816	20994	21594	19415	20015	18182	18782	16527	17127	14431	15000	17616	04:53:36
7	WB4	21631	22200	23200	23800	24963	25563	26586	27186	28518	30416	32038	32638	33617	34217	34850	35450	36505	38127	39654	40254	18670	05:11:10
8	EB4	38600	39200	36572	37600	34809	35409	32348	33786	30416	31016	28194	28794	26615	27215	25382	25982	23727	24327	21631	22200	17616	04:53:36
9	WB5	28831	29400	30400	31000	32163	32763	33786	34386	35718	37616	39238	39838	40817	41417	42050	42650	43705	45327	46854	47454	18670	05:11:10
10	EB5	45800	46400	43772	44800	42009	42609	39548	40986	37616	38216	35394	35994	33815	34415	32582	33182	30927	31527	28831	29400	17616	04:53:36
11	WB6	36031	36600	37600	38200	39363	39963	40986	41586	42918	44816	46438	47038	48017	48617	49250	49850	50905	52527	54054	54654	18670	05:11:10
12	EB6	53000	53600	50972	52000	49210	49810	46748	48186	44816	45416	42594	43194	41015	41615	39782	40382	38127	38727	36031	36600	17616	04:53:36
13	WB7	43231	43800	44800	45400	46563	47163	48186	48786	50118	52016	53638	54238	55217	55817	56450	57050	58105	59727	61254	61854	18670	05:11:10
14	EB7	60200	60800	58172	59200	56410	57010	53948	55386	52016	52616	49794	50394	48215	48815	46982	47582	45327	45927	43231	43800	17616	04:53:36
15	WB8	50431	51000	52000	52600	53763	54363	55386	55986	57318	59216	60838	61438	62417	63017	63650	64250	65305	66927	68454	69054	18670	05:11:10
16	EB8	67400	68000	65372	66400	63610	64210	61148	62586	59216	59816	56994	57594	55415	56015	54182	54782	52527	53127	50431	51000	17616	04:53:36
17	WB9	57631	58200	59200	59800	60963	61563	62586	63186	64518	66416	68038	68638	69617	70217	70850	71450	72505	73105	74632	75232	17648	04:54:08
18	EB9	74172	74772	72572	73172	70810	71410	68348	69786	66416	67016	64194	64794	62615	63215	61382	61982	59727	60327	57631	58200	17188	04:46:28
19	WB10	64832	65401	66400	67000	68163	68763	69786	70386	71718	73616	75238	75838	76817	77417	78051	78651	79705	80305	81832	82432	17648	04:54:08
20	EB10	80534	81134	78935	79535	77172	77772	75548	76148	73616	74216	71395	71995	69815	70415	68582	69182	66927	67527	64832	65401	16350	04:32:30
																				Average train travel time		17956	04:59:17

372

A General Simulation Modelling Framework for Train Timetabling Problem

Figure 23. Computer running time for 20 replications versus the number of trains in the system

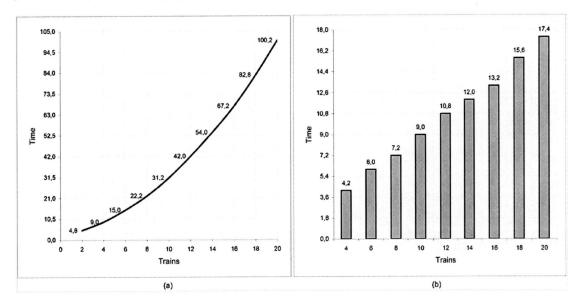

Figure 24. Feasible train-station diagram

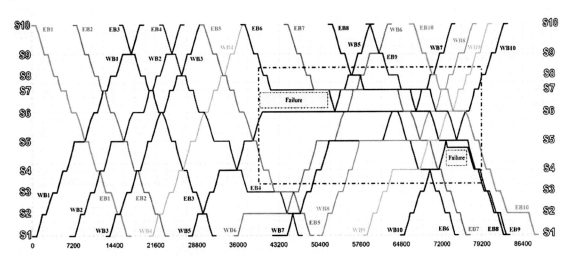

79200 seconds a part of the system between the stations dS_{31} and dS_{81} is displayed. As can be seen in the dotted line circle denoted by 1, a failure occurred after EB_8 train begins its trip from S_5 to S_4. Therefore, EB_8 waits at the dummy station dS_{43} during the repair, and then EB_8 and EB_9 trains traverse on the track part between S_5 and S_4. More than one train that have the same direction can use the same track with headway between them.

While the simulation model is running through 39600-70200 seconds, a part of the system between dS_{43} and dS_{71} is displayed in Figure 26, which is the dotted line rectangle denoted by 2 in Figure 25.

Figure 25. Feasible train-station diagram for dS_{31}-dS_{81} part from 39600 to 79200 seconds

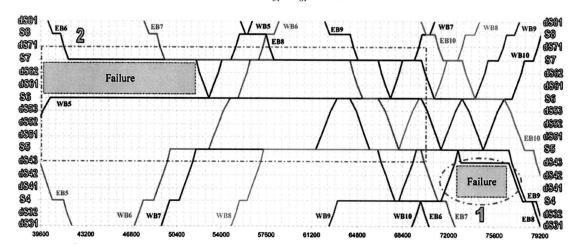

Figure 26. Feasible train-station diagram for dS_{43}-dS_{71} part from 39600 to 70200 seconds

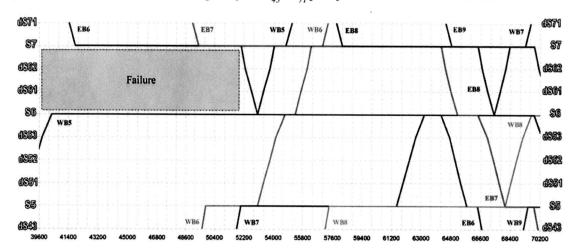

There is a track failure between S_6 and S_7 that prevents trains to be run. After the track is repaired the trains can travel. But at that time there are more than one candidate trains waiting for using the repaired track part. This occurred disturbance should be managed in order to have a feasible timetable.

To make clearer how the disturbance is managed, the occurred important events from 39600 to 70200 seconds on the track part between dS_{43} and dS_{71} are explained. The positions of trains p_i ($i = 1, ..., 39$) are exhibited step by step in Figures 27-32. The trains that have waited for the track repair are shown on the left side of the figures.

- In the first position (p_1) the time is just after 39600 seconds, all the three stations (S_5, S_6 and S_7) are empty and WB_5 is travelling between S_5 and S_6. Additionally, there is a track failure event between S_6 and S_7, and the track is prevented to be used.
- In second position (p_2), WB_5 has reached to S_6 and the failure is still going on.

A General Simulation Modelling Framework for Train Timetabling Problem

Table 12. Feasible train timetable

Train#	WB/EB	S1 ArT	S1 DpT	S2 ArT	S2 DpT	S3 ArT	S3 DpT	S4 ArT	S4 DpT	S5 ArT	S5 DpT	S6 ArT	S6 DpT	S7 ArT	S7 DpT	S8 ArT	S8 DpT	S9 ArT	S9 DpT	S10 ArT	S10 DpT	Travel time
1	WB1	32	710	1718	2370	3610	4334	5369	6058	7405	9305	10892	11500	12493	13179	13809	14801	15907	17423	18945	19620	19637	05:27:17
2	EB1	16591	17269	14957	15573	12943	13853	11303	11937	9305	9951	6932	7740	5097	5980	3829	4447	2159	2797	32	650	17285	04:48:05
3	WB2	6548	7504	8509	9188	10290	10922	11937	12622	13893	16067	17704	19598	20541	21152	21829	22458	23558	24545	26036	26679	20180	05:36:20
4	EB2	23290	23924	21554	22270	19653	20448	17916	18623	16067	16670	13801	14437	12227	12854	11029	11639	9325	9968	7283	7857	16690	04:38:10
5	WB3	13020	13591	14638	15388	16611	17340	18326	18976	20294	20895	22540	23171	24136	24749	25406	26202	27217	27838	29383	30010	17038	04:43:58
6	EB3	31527	32135	29896	30517	28123	28777	26433	27125	24447	25093	22209	22849	20483	21226	19079	19838	17423	18049	15379	15958	16803	04:40:03
7	WB4	20482	21132	22091	22868	24065	24696	25727	26433	27761	28604	30172	30819	31800	32469	33108	33738	34753	35360	36890	37550	17115	04:45:15
8	EB4	46640	47324	44649	45661	37693	43468	35829	36631	31849	34501	29219	30172	27521	28253	26202	26909	24545	25155	22421	23018	24950	06:55:50
9	WB5	27385	28019	29001	29896	31104	31747	32767	35829	37166	38678	40369	53030	54044	54747	55439	56092	57165	57770	59293	59933	32596	09:03:16
10	EB5	48714	49327	46972	47699	42436	45825	40544	41435	38678	39280	36323	37067	34621	35323	33398	34008	31670	32329	29492	30072	19883	05:31:23
11	WB6	35341	35989	36985	44649	45825	46653	47731	48549	49867	53040	54700	55333	56306	56976	57567	58193	59230	60861	62387	63080	27786	07:43:06
12	EB6	74382	75006	72588	73402	70830	71441	67792	69818	65769	66416	53030	64195	41820	52014	40556	41181	38850	39521	36611	37316	38443	10:40:43
13	WB7	42111	44659	45661	46972	48174	48822	49834	50604	52001	61513	63183	67413	68363	69297	69884	70503	71577	72320	73816	74426	32362	08:59:22
14	EB7	76396	77102	74750	75359	72906	73599	71171	71812	68094	69828	65202	66426	49420	64205	48164	48787	46450	47068	44232	44874	32917	09:08:37
15	WB8	49975	50809	51791	52485	53650	54414	55464	56081	57392	68094	69661	70854	71854	72687	73278	74006	75089	76121	77664	78281	28354	07:52:34
16	EB8	83023	83697	81360	82003	79611	80222	77844	78578	71400	72740	67413	69838	58193	66436	55932	57567	54105	54822	51945	52603	31800	08:50:00
17	WB9	58059	58681	59752	60392	61551	62183	63237	68104	69470	71400	73043	74060	75066	75882	76526	77146	78229	78912	80349	81026	23015	06:23:35
18	EB9	83488	84113	81735	82464	79913	80555	78091	78897	74716	76741	70854	73043	64837	69848	62580	64215	60861	61498	58604	59293	25557	07:05:57
19	WB10	64678	65342	66358	66960	68129	68751	69818	71410	72740	74716	76355	76967	77940	78636	79243	79855	80986	81638	83094	83775	19144	05:19:04
20	EB10	88473	89102	83680	87468	81722	82446	80019	80686	77981	78685	74060	76355	71558	73053	70199	70907	68497	69114	66341	67052	22808	06:20:08
																			Average train travel time			24218	06:43:39

375

- Next, EB_6 has reached to S_7 (p_3), and then the other part of S_7 is occupied by EB_7 (p_4), the trains are waiting because the failure is still going on.
- Later, WB_6 occupies S_5 (p_5), and then the other part of the S_5 is occupied by WB_7 (p_6). Although there is an empty capacity at S_6, both WB_6 and WB_7 stop at S_5, since a movement from S_5 to S_6 will cause a blockage. The failure is still going on, and the *five* trains are waiting for the track repair event.
- Then, it is seen that the failure track has been repaired and has opened for the candidate trains (p_7). Although the first train in the candidate trains queue is WB_5, its move will cause a blockage. Thus, another candidate train EB_6 moves and uses the repaired track part.
- The later changes in positions (from p_8 to p_{12}) of these five trains, till one of them leave the track part can be seen in Figure 28.
- The first of the five trains which leave the track part is WB_5. After WB_5 has left there remain four trains (p_{13}).
- Then WB_6 has left, and there remain three trains which have waited for the track repair (p_{16}).
- After that two new trains entered to the track part. WB_8 entered from S_5 (p_{17}) and EB_8 entered from S_7 (p_{18}).
- The changes in positions (from p_{19} to p_{22}) of the trains, till a new train entrance can be followed from Figures 29 and 30.
- Afterward, EB_9 entered from S_7 (p_{23}).
- The changes in positions (p_{24} and p_{25}) of the trains, before one of them leaving can be seen in Figure 30.
- Then, EB_6 has left from S_5 (p_{26}), and there remain two trains which have waited for the track repair event.
- The changes in positions (from p_{27} to p_{33}) of the trains, till one of them leave the track part can be followed from Figures 30 and 31.
- Next, WB_7 has left from S_7 (p_{34}), now there remain only one train that has waited for the track repair event.
- After that, WB_9 entered from S_5 (p_{35}).
- The changes in positions (p_{36}) of the trains, before one of them leaving the track part can be seen in Figure 32.
- Then, EB_7 that is the last train entered this track part before its repair has left from S_5 (p_{37}), and there remain no train which has waited for the track repair event.
- The changes in positions (p_{38} and p_{39}) of the trains, till the simulation time reaches 70200 can be followed from Figure 32. Now there are four trains which have entered to this part after the repair of the part.

FUTURE RESEARCH DIRECTIONS

Future work directions can be as follows:

- The problem has been dealt with from the service provider (train operating authority) point of view, but there are also the service users (passengers or freight transporting companies) in the railway system. The simulation modelling framework can be extended by including the service users.

A General Simulation Modelling Framework for Train Timetabling Problem

Figure 27. Train positions from p_1 to p_7 on the part dS_{43}-dS_{71} from 39600 to 70200 seconds

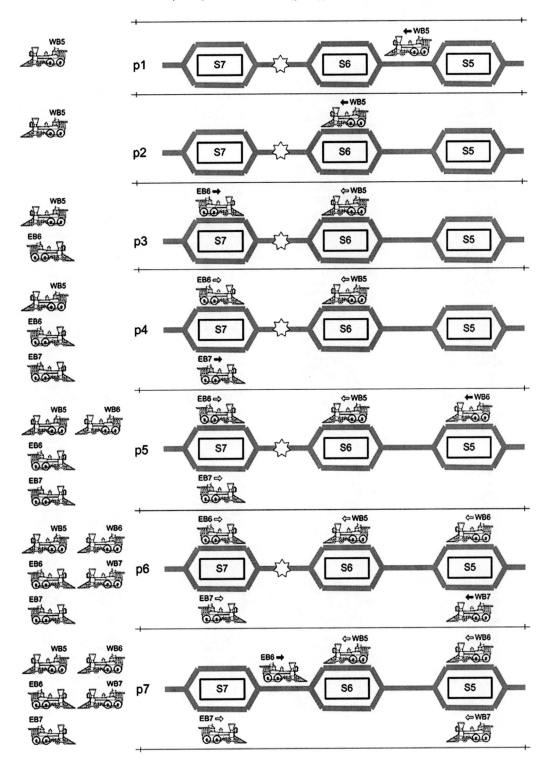

Figure 28. Train positions from p_8 to p_{14} on the part dS_{43}-dS_{71} from 39600 to 70200 seconds

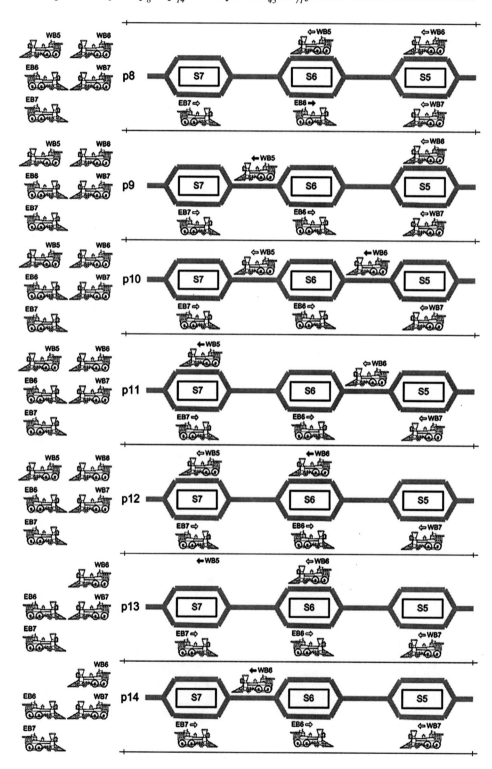

A General Simulation Modelling Framework for Train Timetabling Problem

Figure 29. Train positions from p_{15} to p_{21} on the part dS_{43}-dS_{71} from 39600 to 70200 seconds

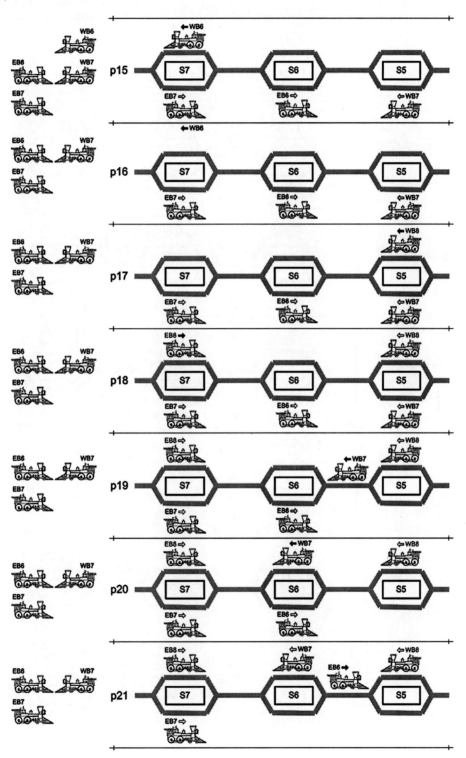

379

Figure 30. Train positions from p_{22} to p_{28} on the part dS_{43}-dS_{71} from 39600 to 70200 seconds

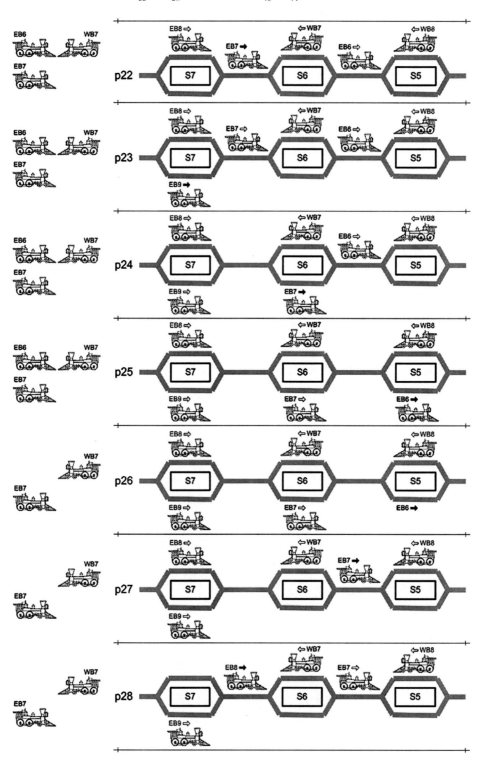

A General Simulation Modelling Framework for Train Timetabling Problem

Figure 31. Train positions from p_{29} to p_{35} on the part dS_{43}-dS_{71} from 39600 to 70200 seconds

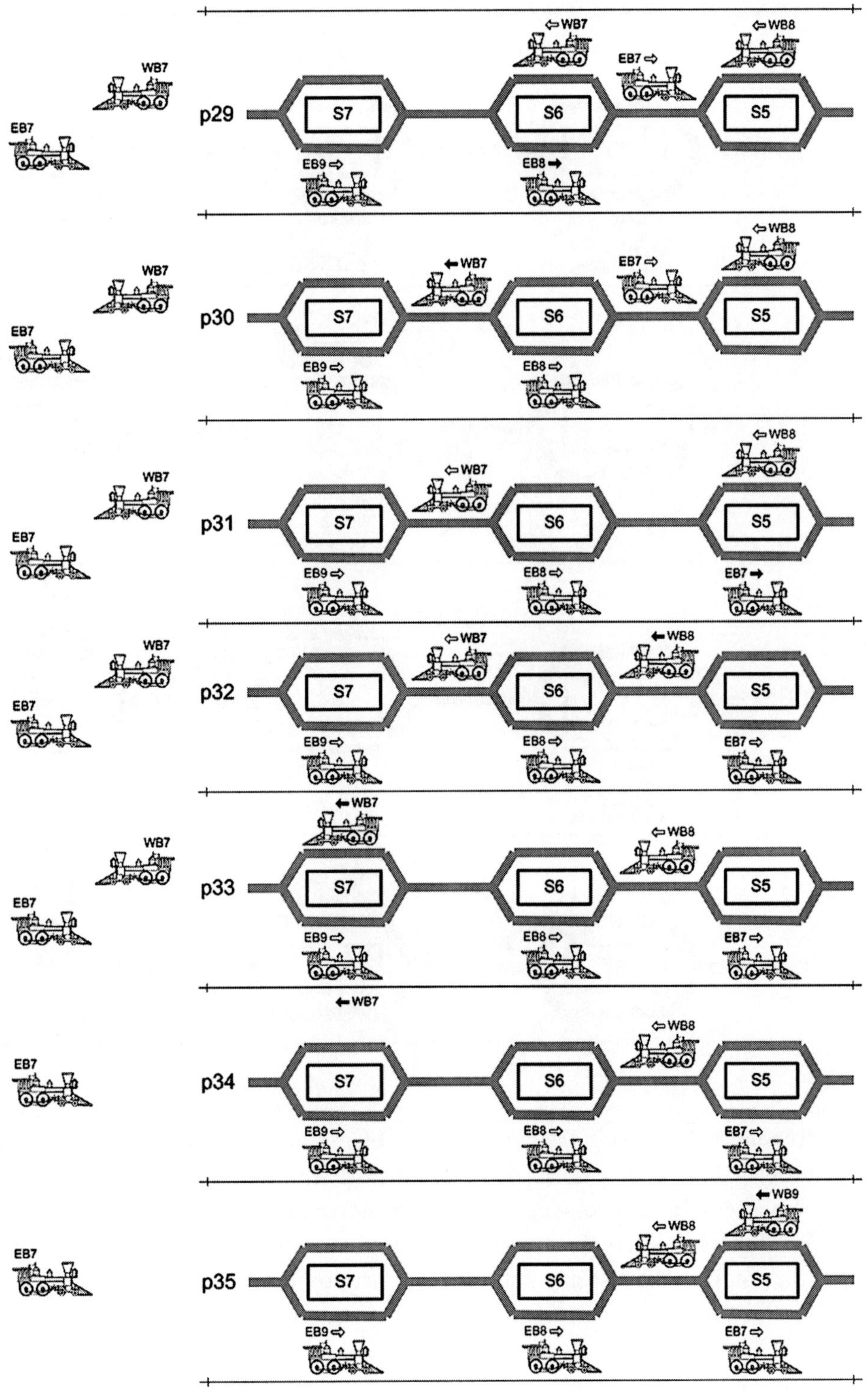

Figure 32. Train positions from p_{36} to p_{39} on the part dS_{43}-dS_{71} from 39600 to 70200 seconds

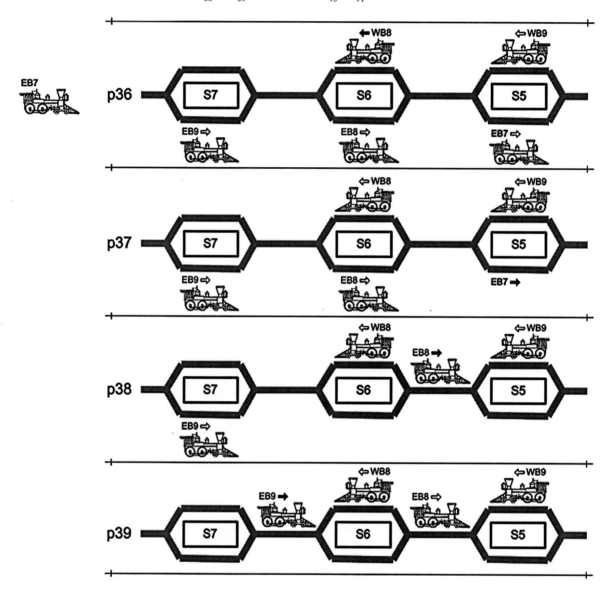

- The infrastructure considered in the hypothetic problem is a bidirectional single track corridor. It can be extended to have double track parts that can also be modelled as one way. Additionally, the corridor can be extended to a network by adding other capillary lines which may be interurban or urban. Some of these capillary lines may cut the corridor, some may use only a part of tracks and stations of the corridor, and also any station of the corridor may be a last station of a capillary line. The developed framework can be also used to model such network.
- The simulation modelling framework can be extended by including different train types, which have different priorities and different speed limits and visit different stations.

- If long multi platforms in where more than one train can be accommodated are regarded, the problem of sequencing of trains on the multi platforms arises and new decision variables are needed to solve the problem.

CONCLUSION

The train timetabling (scheduling) problem is the problem of determining a timetable for a set of trains which satisfies some operational constraints without violating track capacities. The objective is to prepare a feasible train timetable that includes arrival and departure times of all trains at the visited stations.

Simulation gives a chance to researchers to model complex problems that have stochastic nature. Although simulation modelling has been used in a few articles those focused on the problem, none of them includes a comprehensive framework. In this study, a comprehensive feasible timetable generator simulation modelling framework for the train timetabling (scheduling) problem is developed. The simulation model is developed to cope with the disturbances, therefore stochastic events were allowed in the simulation model. Managing disturbances is also the interest of dispatching (rescheduling). Therefore, the simulation framework can also be used for the train dispatching (rescheduling) problem if it can be feed by the real time data.

By using the presented framework, all the railway transportation systems can be modelled with only problem and infrastructure specific modifications and feasible solutions can easily be attained. In order to avoid a deadlock, a general *blockage preventive algorithm* is developed. Also this algorithm can be embedded in the simulation model and can be easily adapted to infrastructure specific modifications.

ACKNOWLEDGMENT

This chapter is supported by *The Scientific and Technological Research Council of Turkey* (TÜBİTAK).

The author would like to express his deepest thanks and gratitude to Prof. Dr. Günhan Miraç Bayhan who is the supervisor of him during his Bachelor of Science, Master of Science and Doctor of Philosophy degrees and a retired faculty of *Industrial Engineering Department of Dokuz Eylül University (Turkey)* for her guidance, support, encouragement and valuable advice throughout the progress of academic studies of the author.

The author would like to express his sincere thanks to Prof. Dr.-Ing Thomas Siefer who is the managing director of *Institute of Transport, Railway Construction and Operation*, and Prof. Dr.-Ing. Jörn Pachl who is the managing director of *Institute of Railway Systems Engineering and Traffic Safety* for their kindness, hospitality and supports during Postdoctoral Academic Researches of the author at *Technical University of Braunschweig (Germany)*.

REFERENCES

Assad, A.A. (1980). Models for rail transportation. *Transportation Research A*, 14 (A), 205-220.

Brännlund, U., Lindberg, P. O., Nou, A., & Nilsson, J.-E. (1998). Railway timetabling using lagrangian relaxation. *Transportation Science, 32*(4), 358–369. doi:10.1287/trsc.32.4.358

Cacchiani, V., Caprara, A., & Toth, P. (2008). A column generation approach to train timetabling on a corridor. *4OR: A Quarterly Journal of Operations Research, 6*(2), 125-142.

Cacchiani, V., & Toth, P. (2012). Nominal and robust train timetabling problems. *European Journal of Operational Research, 219*(3), 727–737. doi:10.1016/j.ejor.2011.11.003

Caprara, A., Fischetti, M., Guida, P. L., Monaci, M., Sacco, G., & Toth, P. (2001). Solution of real-world train timetabling problems. *Proceedings of the 34th Hawaii International Conference on System Sciences* (pp. 1-10).

Caprara, A., Fischetti, M., & Toth, P. (2002). Modeling and solving the train timetabling problem. *Operations Research, 50*(5), 851–861. doi:10.1287/opre.50.5.851.362

Caprara, A., Kroon, L., Monaci, M., Peeters, M., & Toth, P. (2007). Passenger railway optimization. *Handbooks in Operations Research and Management Science, 14*, 129–187. doi:10.1016/S0927-0507(06)14003-7

Caprara, A., Monaci, M., Toth, P., & Guida, P. L. (2006). A lagrangian heuristic algorithm for a real-world train timetabling problem. *Discrete Applied Mathematics, 154*(5), 738–753. doi:10.1016/j.dam.2005.05.026

Castillo, E., Gallego, I., Ureña, J. M., & Coronado, J. M. (2009). Timetabling optimization of a single railway track line with sensitivity analysis. *Top (Madrid), 17*(2), 256–287. doi:10.1007/s11750-008-0057-0

Cordeau, J.-F., Toth, P., & Vigo, D. (1998). A survey of optimization models for train routing and scheduling. *Transportation Science, 32*(4), 380–404. doi:10.1287/trsc.32.4.380

Frank, O. (1966). Two-way traffic on a single line of railway. *Operations Research, 14*(5), 801–811. doi:10.1287/opre.14.5.801

Geske, U. (2006). Railway scheduling with declarative constraint programming. InDeclarative Programming for Knowledge Management, *LNCS* (Vol. 4369, pp. 117–134).

Hansen, I. A., & Pachl, J. (2014). *Railway Timetabling & Operations. Analysis - Modelling - Optimisation - Simulation - Performance Evaluation*. Germany: Eurailpress.

Higgins, A., Kozan, E., & Ferreira, L. (1997). Heuristic techniques for single line train scheduling. *Journal of Heuristics, 3*(1), 43–62. doi:10.1023/A:1009672832658

Jovanovic, D., & Harker, P. T. (1991). Tactical scheduling of rail operations: The SCAN I system. *Transportation Science, 25*(1), 46–64. doi:10.1287/trsc.25.1.46

Kelton, W., Sadowski, R., & Swets, N. (2009). *Simulation with Arena*. USA: McGraw-Hill Higher Education.

Lee, Y., & Chen, C.-Y. (2009). A heuristic for the train pathing and timetabling problem. *Transportation Research Part B: Methodological, 43*(8-9), 837–851. doi:10.1016/j.trb.2009.01.009

Liebchen, C. (2008). The first optimized railway timetable in practice. *Transportation Science, 42*(4), 420–435. doi:10.1287/trsc.1080.0240

Liu, S. Q., & Kozan, E. (2009). Scheduling trains as a blocking parallel-machine job shop scheduling problem. *Computers & Operations Research, 36*(10), 2840–2852. doi:10.1016/j.cor.2008.12.012

Mees, A. I. (1991). Railway scheduling by network optimization. *Mathematical and Computer Modelling, 15*(1), 33–42. doi:10.1016/0895-7177(91)90014-X

Newman, A. M., Nozick, L. K., & Yano, C. A. (2002). Optimization in the rail industry. In Handbook of Applied Optimization (pp. 704-719).

Odijk, M. A. (1996). A constraint generation algorithm for construction of periodic railway timetables. *Transportation Research Part B: Methodological, 30*(6), 455–464. doi:10.1016/0191-2615(96)00005-7

Pachl, J. (2009). *Railway Operation and Control*. USA: VTD Rail Publishing.

Pachl, J. (2011). Deadlock avoidance in railroad operations simulations. In *Transportation Research Board 90th Annual Meeting* (Paper No: 11-0175). Washington DC, USA.

Petersen, E. R., & Taylor, A. J. (1982). A structured model for rail line simulation and optimization. *Transportation Science, 16*(2), 192–206. doi:10.1287/trsc.16.2.192

Szpigel, B. (1973). Optimal train scheduling on a single track railway. *Operations Research, 72*, 343–352.

Tormos, P., Lova, A., Barber, F., Ingolotti, L., Abril, M., & Salido, M. A. (2008). A genetic algorithm for railway scheduling problems. *Studies in Computational Intelligence, 128*, 255–276. doi:10.1007/978-3-540-78985-7_10

Wong, O., & Rosser, M. (1978). Improving system performance for a single line railway with passing loops. *New Zealand Operational Research, 6*, 137–155.

Yalçınkaya, Ö. (2010). *A feasible timetable generator simulation modelling framework and simulation integrated genetic and hybrid genetic algorithms for train scheduling problem* [Unpublished doctoral dissertation]. Dokuz Eylül University, İzmir, Turkey.

Yalçınkaya, Ö., & Bayhan, G. M. (2012). A feasible timetable generator simulation modelling framework for train scheduling problem. *Simulation Modelling Practice and Theory, 20*(1), 124–141. doi:10.1016/j.simpat.2011.09.005

Zhou, X., & Zhong, M. (2005). Bicriteria train scheduling for high-speed passenger railroad planning applications. *European Journal of Operational Research, 167*(3), 752–771. doi:10.1016/j.ejor.2004.07.019

Zhou, X., & Zhong, M. (2007). Single-track train timetabling with guaranteed optimality: Branch-and-bound algorithms with enhanced lower bounds. *Transportation Research Part B: Methodological, 43*(3), 320–341. doi:10.1016/j.trb.2006.05.003

KEY TERMS AND DEFINITIONS

Deadlock: A traffic situation involving a set of trains each claiming a resource which is not available, leading to the impossibility of further movement by any train in the set.

Dwell Time: The elapsed time that a train stops in a station for boarding and alighting events.

Feasible Solution: An assignment of values to the decision variables such that none of the constraints is violated.

Headway: The space or time interval between two successive trains.

Modelling of Railway Infrastructure: The transition of real world railway infrastructure into a computerized infrastructure model.

Simulation Model: A representation of a system that can be used to mimic the processes of the system under varying circumstances. It is usually operated subject to stochastic disturbances.

Schedule: The designation of train description, day, route, speed, arrival and departure times of a train.

Timetable: A document that contains the schedules of all trains of a line.

Chapter 16
Intelligent Transportation Systems:
The State of the Art in Railways

Sundaravalli Narayanaswami
Indian Institute of Management, India

ABSTRACT

"Intelligent Transportation systems" is what everyone wants to know about, and about which very little is available as know-how. ITS technologies and monitoring systems are quite popular and reasonably well deployed in developed countries, particularly the roadways and airways. ITS holds a greater promise than ever before, as both availability of niche technologies and demand for more efficient transportation systems have increased multi-fold in recent years. Of late, there are huge railway projects all over the world that spans through several techniques, such as light / heavy rails, monorails etc. Apart from the social benefits that can be envisaged, these projects are genuine examples of public-private partnerships along with global business operations. Many of these projects demonstrate a classy trend of moving towards automation of operations of very large scales. Few agent architectures are discussed in brief in this chapter.

ITS: AN INTRODUCTION

It has become increasingly clear that the issue of efficiently moving people and goods is far more complex than previously imagined. The problem is not just confined to surface transportation, that is, vehicles and roadways; it affects trains, passenger planes, air cargo, ferries, ships, pipelines and all available and currently utilized transportation modes. This realization had helped foster the broader notion of intelligent transportation system (ITS). Clearly, to synthesize a genuine and realistic ITS system, one must adopt a holistic approach that takes into consideration complex and asynchronous interdependencies between the many transportation modes and guided by a fundamental goal, namely, minimize the transit time for all travelers and merchandise in transit, subject to fair distribution of the available resources (Chowdhary and Sadek, 2003). In the not-too-distant future, one might envisage a routine space travel to

DOI: 10.4018/978-1-5225-0084-1.ch016

the moon and other artificial satellites and planets in our solar system, which need to be accommodated within the ITS system. It is hoped that ITS caters to our futuristic demands (Konstadinos G. Goulias, 2007); hence the unique attributes of space travel and the likes should be taken into consideration today, while planning out the fundamentals of the ITS architecture for a seamless future integration. The motivation and rationality behind ITS, presumably rests on two key scientific and engineering advances. The first is from computational advances - the increased availability of computing power in the form of powerful desktop workstations and mobile laptops, palmtops, and handheld personal digital assistants (PDAs). The second is from communication advances - the increasing availability of networking, both wire-line and wireless control. However, there are more to ITS developments than mere availability of computing power and communications network. For instance, US rail industry reports that locomotives sit idle as much as 40% due to bottlenecks in the rail corridors due to poor information, coordination, and control. With huge average fuel expenditures of many railroad operators (Ganti et al., 2010), the extent of fuel wastage is staggering with idle running locomotive engines. By any standards, such inefficiency is not sustainable. Additionally to hardware technologies, the key to successful solutions to complex transportation problems lies in comprehensive understanding of the control and coordination algorithms. While abstract, these algorithms serve to unify the computing and networking resources in a synergistic, nontangible manner.

ITS is an evolving scientific and engineering discipline. The primary goals of ITS are:

1. To minimize the travel time of all travelers,
2. To provide safe, secure and reliable merchandising for service providers and end users,
3. To enable optimal utilization of available resources, especially under the scenario of increasing travel speeds and large number of travelers, and
4. To offer a precise and timely information to travelers.

To achieve this goal, it is absolutely essential for a seamless integration of the different transportation modes, including vehicular traffic, trains, cargo air transport, passenger air transport, marine ferries, and others through asynchronous distributed control and coordination algorithms, subject to societal norms, policies, and guidelines. Such an integration will enable a traveler to:

1. Gain access to accurate information of any transportation mode from any location in the system at all times,
2. Deduce the most efficient route or reroute across all different transportation modes using personalized decision support systems, and
3. Book or cancel reservations, dynamically on any transportation system.

Therefore, ITS encompasses:

1. All subareas of transportation management covering both surface and non-surface transport management,
2. Traffic signaling,
3. Travel management, such as multimodal traveler information,
4. Public transportation and transit management,
5. Safety management, which subsumes incidents, railroad grade crossings,

6. Emergency services,
7. Advanced vehicle control, and
8. Fare payment and toll collection.

RECENT DEVELOPMENTS OF ITS: AN OVERVIEW

There are two distinct perspectives to understand ITS developments. The first perspective is through a review of the scientific and academic literature. Current ITS research is focused in the areas of system planning, vehicular traffic modeling, system evaluation, vehicle tracking, autonomous driving and GPS-based guidance (Guha et al., 2010), signal control, braking, lane detection and steering control, intelligent cruise control, disseminating road work information to drivers, noise pollution (Rana et al., 2010), platooning, simulators for evaluating ITS systems, and training ITS personnel. Several developments are worth reporting. Copper cables are increasingly being replaced with wireless and optical fibers to improve communications in rapid transit systems. Road signaling systems as traffic lights are constantly being recomputed and adjusted to enhance traffic flow and pack vehicles more closely on the roadway (Bhoraskar et al., 2012). High speed cameras are used for regulating road traffic under autonomous maneuvering studies. Perhaps the most important accomplishment in ITS to date and across all transportation modes has been in providing consumers with relevant information, under the rubric of information technology. The availability of arrival and departure information of flights and trains through the Internet from homes, businesses, and on mobile phones (Eriksson et al., 2008), worldwide; automatic notifications of delays and cancellations; facility to reserve seats and call taxicabs through SMS on mobile phones (Anderson et al., 2009); and the ability to get street directions and maps from a starting point to a destination station on the Internet represents remarkable privileges.

The second perspective is regarding the critical examination of the products and services that have been introduced into the market by the ITS industry in the recent decades. For example, in many cities, there are toll-free phone numbers and public web sites where one can obtain Urban Mobility Report with up-to-date information on transportation congestion. These services are immensely valuable but they lack in three fronts, especially with regard to developing countries. First, the information is limited to the major roadways or railways, while that for the many more secondary railroads is either missing or outdated. Second, the time necessary to broadcast complete information on all of the surface transport systems is usually so long that its usefulness is severely limiting. Third, the information is not dynamic in that it may change substantially due to accidents or incidents while in transit. Today, a number of trucks, airport shuttle service vans, and luxury automobiles are equipped with GPS-based navigation aids (Guha et al., 2010). The principal difficulty again is the lack of accurate and dynamic information regarding congestion and road closures in a timely manner. A number of highway authorities have turned to AM radio to broadcast ramp closures, lane constrictions due to construction, and other relevant highway conditions. This effort, while of great value, has encountered problems where drivers are unable to listen to the broadcast in a timely manner due to electromagnetic interference (Williams, 2008). Among many transportation communities, there is a strong drive to mount cameras along highways and freeways, feed the signals back to a centralized traffic control center, and monitor congestion. While installation and maintenance costs are high, the cameras can be immensely valuable in zooming onto accident sites and guiding en route police and paramedics or reading license plates of vehicles under suspicion. However, relative to the issue of congestion monitoring and control, other devices including optical fibers embedded underneath the pavements are far superior in that they are relatively inexpensive and highly reliable (Pattara-Atikom, 2007). There is a growing private sector travel speed data market. Land use (or congestion) pricing using electronic toll collection systems are in vogue in many developed nations and are considered as a major ITS achievement. Delays are reduced significantly as the need to halt for cash transactions is eliminated. The concept of toll collection, electronic or otherwise, as currently deployed, is inefficient in that it slows down traffic unnecessarily, is unsafe in that it fosters a dangerous driving environment, and is environmentally unsound. Research in railroad algorithms had revealed that the idle waiting times of trains under centralized scheduling are much higher than that under distributed scheduling. The result is inevitable congestion and delay.

DESIGN ISSUES AND CHALLENGES

History teaches us that the single greatest obstacle to our progress is our own inability to envision the future and our tacit refusal to believe in the promises. The first challenge is that there is a strong inherent barrier to envision how ITS can bring to fruition revolutionary changes that will yield quantum improvement to our quality of life and exceptional benefits to society today and well into the future. Where belief is lacking, there is little motivation to expend the enormous effort that is required for success. As a result, even the greatest idea will fizzle. Ironically, however, every great concept and principle, without exception, in any field including science, mathematics, and engineering has begun its journey in the world of intuition and imagination. It is only through great effort that these concepts had emerged to become the foundation and pillars of society.

The second challenge may be characterized as a self-fulfilling prophecy. Leading transportation personnel have come to believe that the real-world problems are so complex that they can never be solved by ITS. Many transportation officials believe that it is not possible to obtain realistic traffic demands for any ITS project; thus, the knowledge of the necessary design decisions and options is elusive and, therefore, the pursuit is pointless. Consider the following real anecdote. In 1995–1996, the section of Interstate 10 (I-10) from the Phoenix airport to Queen Creek in Arizona had only two lanes each way and was often severely clogged during rush hours. Arizona was experiencing phenomenal growth and new communities were popping up everywhere along the section of I-10. At a great expense and inconvenience to all, the Arizona Department of Transportation embarked on an ambitious project to widen I-10 to three lanes each way. After approximately one year of work, the freeway was finally reopened with all six lanes. At first, it was an amazing improvement. Congestion had virtually disappeared and commute along I-10 became a breeze. Within six months, congestion had returned with a vengeance and the situation was back to square one. The population explosion in barely a year and a half had brought the six-lane I-10 to a grinding halt. Thus, the frustrations of the transportation officials are clearly understandable.

The third challenge relates to technology requirements for development as well as training purposes. It was assessed in the earlier decades that sophisticated simulators were required to study and assess ITS projects. Some of the most sophisticated simulators that were available commercially did not qualify for real-life applications. It was conclusive that a thorough simulation study of any project requires a team of individuals with genuine knowledge and expertise to develop a simulator from scratch, tailored to the needs of the project. No "off-the-shelf" simulator can be expected to offer the level of flexibility necessary for an ITS project. By definition, the full dimension of a complex problem is never known a priori, implying that it is very difficult, if not impossible, for an "off-the-shelf" simulator with a limited

Intelligent Transportation Systems

set of options to anticipate all of the sophisticated needs of a given problem. In order to pioneer in ITS research and high-tech industry, a serious and extraordinary initiative is imperative to (1) bring behavior modeling and asynchronous, distributed simulation into mainstream research and education, and (2) cultivate real talent in this field through education and training.

The fourth challenge is the absence of proper nurturing of ITS research and development and transportation personnel. Enormous sums of money and effort have been poured into hardware, such as high speed cameras and signaling systems, for the purpose of monitoring congestions and detecting accidents (Hoh et al., 2008). While cameras by themselves do not represent true ITS advancement, their obsessive nationwide deployment definitely serves to distract us from the real and necessary improvements. In this era of extreme competition for high-tech industrial dominance, much can be achieved by ITS engineer-scientists to diligently pursue the real and formidable challenges in ITS.

DESIGN CHALLENGES

All petrol and diesel engines in locomotives, automobiles and aircrafts operate on technology that was discovered hundreds of years ago. Current developments are insignificant increments. New engine designs with higher efficiencies are the need for today's ITS.

Transport infrastructure, including bridges are still stagnant, stationary and static. Automated computer programs such as finite element analysis, usage of strain gauges to collect loading data can help infrastructure to react dynamically to changing environmental parameters. Self-protection can thus be imparted to a fair extent.

Electric automobiles, especially hybrid cars that operate both on gasoline and electricity are gaining popularity. While electric operated vehicles eliminate emissions (US Department of Transportation, 2010) during running and idling operational phases, plug-in hybrid electric vehicles (PHEV) reduce greenhouse gas (GHG) emissions considerably, when charged overnight. Issues yet to be resolved are variability in electricity consumption, fossil fuel adaption and difficulty in directly connecting fuel and electricity consumption. In essence, these vehicles redistribute the pollution, away from urban areas and into rural areas where the power-generating plants may be located. The net pollution remains unchanged. Of course, if the world's electrical energy could be derived entirely from nuclear and hydroelectric sources, the emission of carbon dioxide, carbon monoxide, sulfur dioxide, and other poisonous gases from burning fossil fuels would be eliminated (Jurdak et al., 2010). Thus, the challenge to design vehicles that can lower the total emissions continues to remain at the forefront.

A large percentage of vehicular accidents that occur at night are attributed to human inability to precisely gauge depth - the distance of an object directly ahead of us, either receding away or approaching toward us (Koukoumidis et al., 2011). The problem shall get worse, implying an enormous challenge to study and mitigate the problem unless sophisticated depth gauging systems are invented.

Maintenance of traffic infrastructure is another major issue. Traffic infrastructure, being capital intensive has been built doyens earlier and maintenance is reduced to superficial repairing of broken structure in a patch and pray mode.

The most popular paradox of ITS is "To speed or not to". Theory at one extreme says that speed kills and at the other end, is an equally scientific fact that higher speeds imply shorter travel times, which in turn is a competitive edge for business and also greatly increased quality of life. The ITS challenge and an opportunity is to create new thinking and extraordinary design principles for manufacturing and operating high speed safe and reliable vehicles.

Improved braking significantly lessens the impact of many accidents involving automobiles, trucks, trains, and even planes. A formidable challenge is to develop a radically new braking principle to reduce braking distance and stopping time significantly. If operational safety is also incorporated, the system can preempt many accidents.

Vehicular accidents throughout the world range from fender benders to minor injuries, long-term medical incapacitation, and, sometimes, death. The challenge for ITS is to understand the cause and nature of accidents at a much deeper level and yield revolutionary new techniques to mitigate or preempt them.

Over two-thirds of the world is covered by water, boats, aquatic vessels, and ships have been around forever. High speed water based transport should be envisaged. The benefits would be so immense that there may arise a fundamental shift in our mindset. Access to virtually every corner of the earth, in reasonable time, would become an everyday reality. Given that vast areas of the earth are covered by oceans, most of the population centers located by the water, and that most of the world's waters are interconnected, transportation routes and even our living environments are likely to incur a qualitative change.

Another popular ITS idea is personalized rapid transit (PRT), that first surfaced in the 1970s to 1990s. Under PRT, upon arrival at a city, a traveler climbs onto a two- or three-seater private automated car at designated stations at the airport and then punches in his or her desired destination from a series of choices around the city including hotels, convention center, shopping malls and eateries, museums, government complexes, and major downtown office buildings. The car glides on an above-ground guideway and automatically and safely switches within the loop guide way to reach the destination station (Dailey, 2001). Following the traveler's disembarkation, the car may either be immediately reutilized or later. PRT promises to greatly reduce congestion in highly populated urban centers. The challenge has been the lack of practical PRT architectures and the absence of a scientific approach to systematically analyze the needs to build PRT networks in a given environment.

Unlike in the past, today's vehicles are significantly more sophisticated, stemming from the use of fast computers and complex software. When an accident occurs, despite the fast computer controls, issue is extremely complex and is beyond the realm of logical flow of causal analysis (Yin et al., 2007). The challenge is to develop high tech vehicular and communication technology that can also impart safety.

Another formidable ITS challenge is preparedness for natural hazards. Much rapidity is required to restore infrastructure that are damaged by natural calamities; however robustness is desired for systems to operate in self-protective manner.

By their very nature, transportation systems span nontrivial physical distances. Goods and people are transported from one physical location to another, through routes that pass through one or more interchange points, also known as hubs in the literature. Hubs also facilitate sharing of resources, refueling, preventive maintenance, and so on. Therefore, the constituent entities of transportation systems are geographically dispersed. Prior to the discovery of electromagnetic communication, information on the transport of goods and people was propagated along with the transported material itself. Hence the transit speeds of the good and people and the information about their transport were similar.

The first major revolution was electromagnetic communication, which enabled the propagation of information about the movement of goods and people significantly faster than the actual transport of the material at limited speeds. Conceptually, a modern transportation network has two principal components: an information network that transports pure electromagnetic energy and a material transport network

Intelligent Transportation Systems

that carries goods and people. ITS can help make transportation network more efficient, by properly utilizing the information network.

The second major revolution is computing engines in transportation systems. Fast and precise computers are exploited to efficiently control and coordinate the transport of goods and people across the system. Excellent examples of computer usage in transportation in use today include the centralized control for railways, centralized air traffic controllers, and traffic management centers for automobiles. The coordination and control functions in such systems under the centralized paradigm are cost effective, and real-time. The inherent drawback of such systems is slow decision making, which means inaccuracy and imprecision. Problems are more pronounced with higher number of vehicles and increase in demand (Balan et al., 2011). Recently, few organizations are adopting a distributed control for dynamic decision making, which is effective and economical.

The third revolution, which is most exciting as well as most complex, is the integration of computing and communication systems. The genesis of a full- fledged ITS is achieved by the transformation of the centralized paradigm to the asynchronous, distributed paradigm that will integrate fast computers and high-performance computer networks through novel computer algorithms.

REAL WORLD ITS: A PRAGMATIC APPROACH FROM PRINCIPLES

There are certain major characteristics of ITS, which are becoming increasingly important. They are listed and discussed in brief below:

Real-time Information Interchange: The ITS architecture's greatest expectation is to provide logically relevant and accurate information in a timely manner to operators and needy travelers in a user friendly format that they can utilize to determine alternatives and re-compute their plans (Lin et al., 1999).

Automated computation: Unlike in many transportation systems across the world where the decision making and the computation of the arrival times are still estimated manually, future system architectures must employ automated decision-making computer systems to yield accurate information and achieve precise control and coordination.

Information Accuracy: Information must be accurate, timely, relevant, and consistent. Otherwise, a ITS architecture might face a premature extinction.

Operator independence: Except under true emergency conditions, the ITS architecture should neither attempt to control nor dictate a driver's behavior for two reasons (Mohan et al., 2008). First, no centralized authority can reliably and always know with certainty the goals, objectives, and thinking of every driver and, therefore, any attempt to control him or her will invariably be based on erroneous assumptions. Second, in the long run, drivers will resent the intrusion on their freedom to make their own decisions and will ultimately abandon ITS. The key principle is to provide drivers with as much relevant information as possible and equip them with appropriate networking and computing resources so that they can determine and freely choose the right course of action under the given circumstances.

Demand for flexibility and freedom of choice: The frequent lack of flexibility, the absence of personalized services, and the availability of mostly inaccurate estimates are increasingly being rejected by the customers of transportation systems (Thies and Davis, 2011). There is strong unwillingness to accept the traditional excuses of limited computational ability and network bandwidth. There is an increasing demand for flexibility, freedom of choice, and personalized service, and the trend is likely to continue into the future.

Accurate and up-to-date information: In the traditional approach, data are first collected at a centralized unit, processed, and the resulting information is disseminated to the geographically dispersed customers. However in distributed, dynamic systems, this delay implies that the information received by the customer has incurred latency and is, in essence, inaccurate and imprecise. The degree of the error due to latency is a function of the length of the delay, relative to the dynamic nature of the system, and the resolution of accuracy (Thiagarajan et al., 2009). Thus, latency is fundamental to every transportation system, and future system architectures must focus on distributed schemes that aim at eliminating all sources of latency, where possible, and realizing efficient, accurate, and timely decisions.

Focus on consumer of services: The ITS architecture must be fundamentally centered on each individual driver or traveler, subject to safety and fair resource availability for all. In addition to gross metrics such as resource utilization, average travel time, congestion reduction, and energy saving, that are important to the planners and operators (Singh et al., 2012), customer satisfaction is equally important.

Transportation network characteristics: In ITS, information is also transported along with people and freight. While carried along with the goods and people, information transmission facilitates dynamic, travel-related computations and decision making. Also, since such computations along with communications are permitted while in motion using wireless or infrared techniques, centralized information gathering and decision-making may be eliminated. Decreasing physical size and cost of hardware, increasing capability, and lower power consumption in computer designs encourage practicality in transportation networks.

Algorithms for control, coordination, and resource management: Since the constituent units and the resources of any transportation system are geographically dispersed, it is logical for future system architectures to exploit asynchronous, autonomous and distributed algorithms. By design, an asynchronous, distributed, and autonomous algorithm for a transportation system must necessarily reflect the highest, meta-level purpose or intent of the system. Parallel and local computations must be maximized while minimizing the communications between the entities, thereby implying high throughput, robustness, performance scalability and stability.

Simulation approach: Today's transportation systems are characterized by an increasing size and complexity, which implies a large number of variables and parameters, a wide variation in their values, a great diversity in the behaviors, and with restrictive results of analytical efforts. ITS systems are likely to be complex, implying that modeling and large-scale asynchronous distributed simulation may be the most logical and the only mechanism to study them objectively. Key benefits of modeling and simulation are many (Zhou et al., 2009). First, they enable one to detect design errors, prior to developing a prototype, in a cost-effective manner. Second, simulation of system operations may identify potential problems, including rare and otherwise elusive ones, during operations. Third, analysis of simulation results may yield performance estimates of the target system architecture and potential for growth.

Continual error detection and correction: By their very nature, complex systems are prone to design errors that manifest irregularly during operations and elude detection, but are severely damaging. To address this weakness, ITS architectures must incorporate automatic error detection and correction mechanisms in a continual manner, so that errors do not propagate in the system.

Inconsistencies verification: Any ITS architecture must be based on sound logical principles, and where established norms pose inconsistencies, they must be carefully analyzed. For example, virtually everywhere in United States road traffic, high occupancy vehicle (HOV) lanes, where present, are located at the extreme left. Since pavements to the left are designated fast lanes, the immediate connotation is that the HOV is a fast lane. This is reinforced by the fact that, often, fast non-HOV cars use the HOV lane to pass slower cars in the left lane. Furthermore, to get to this lane, an HOV entering the highway from a ramp typically located on the right must maneuver through fast-moving cars in the left lanes, if at all possible on a congested day, leading one to infer an underlying assumption that HOV drivers intend to drive fast. The assumption may be seriously wrong since many of the HOV vehicles may represent families traveling together who would prefer to drive at the speed limit and avoid accidents. To justify such driving behavior is very difficult, and for ITS to make a genuine difference, these problems must be addressed scientifically and objectively. Similar analysis should be adopted in other transportation modes also.

ITS MODELLING TECHNIQUES

Traditional real-world transportation systems are best understood by developing analytical models that attempt to replicate the system behavior through exact equations and as many system constraints as possible to capture limitations in the system. The models are then solved using mathematical techniques or heuristic or meta-heuristic procedures. This approach has been adequately successful for long and may continue to be effective in many disciplines. However, in all over the World, demands in real systems are becoming incredibly large and complex and resources are becoming miniscule. Therefore, analytical efforts are becoming futile and restrictive; given the increasing size and complexity of modern transportation systems lead to large number of variables and parameters, wide variation in their values, and greater diversity in the traffic behaviors. With advent of technology, it is anticipated that ITS based transportation in future is expected to be far more complex, implying a need for more advanced, sophisticated modeling as well as analytical tools.

Modeling refers to the representation of a system in a computational notation and not the physical replica. The fundamental goal is to represent an operational replica of the transportation systems architecture including all of its constituent components, as accurately and correctly as possible in a host computer. As a preface to intelligent transportation modeling, it must be emphasized that a complete theoretical understanding of mobility patterns and characteristics is near impossible from the realms of traffic science. Despite of this, there have been several tools and techniques to construct and analyze a framework of useful theories from traffic observations and experimentation. Such theories can be classified neither as deductive (i.e., derived from well-established theories), nor as inductive (black box). They are rather intermediate; basic mathematical structures are employed, and specific flow properties are incorporated as system constraints, as determined from empirical or experimental data. Traffic flow abides by the physical conservation of vehicle's equation, which again is incorporated as a constraint; therefore the main challenge in traffic flow research is to identify the complexity of problem that can be handled in a reasonable time using the available computational resources and accordingly propose an objective to best achieve the system goals. Understanding traffic flow and feasibility characteristics such as headway distributions, relation between occupancies, acceleration, deceleration speeds, capacity distributions and knowledge of the analytical tools such as queuing schemes, scheduling approaches, simulation techniques to predict the traffic and transportation dynamics under given demand, supply and control conditions, is an essential requirement for the planning, design and operation of a transportation system. The next challenge, at a slightly higher level relates to the motivation behind Intelligent transportation systems. The real challenge in a transportation system arises due to unexpected events. An ITS is expected to maintain the system robust and fail-proof in spite of stochastic disturbances. Transportation researchers have to decide levels of trade-offs between complexity of models and forecastability. For the analysis of a simple arterial, on-ramp, or an intersection area, as well as for studying traffic flow operations in urban or long distance networks, and also with people or freight movements, being able to predict the stochastic traffic elements is an essential factor in the analysis of the system.

ITS MODELS

Research studies classify traffic systems based on their microscopic and macroscopic characteristics. Microscopic variables such as (arrival, departure, and meet) headways, vehicle design and operational speeds and acceleration and deceleration patterns pertain to the individual driver–vehicle unit in relation to the other vehicle movements and instantaneous parameters at that moment. Microscopic models describe the individual vehicle behaviors with the given vehicle design and operational features in relation to the infrastructure and other system elements in the flow (Papageorgiou et al., 2006). On the contrary, macroscopic characteristics pertain to traffic flow properties in the system as a whole, say, at a cross-section at a particular time instant. Examples of macroscopic characteristics are average and maximum flow, time-mean speed, density, and space-mean speed. Macroscopic models describe traffic flow given the infrastructural capacities and average demands and usage patterns.

MICROSCOPIC AND MACROSCOPIC TRAFFIC FLOW MODELS

In general, microscopic models incorporate two tasks concerned with individual vehicle units: one is about longitudinal or forward movement (acceleration, speed, halt time, and distance to cover, halt and destination locations) and the other is about latitudinal or lateral movements (track changing, follow-up or overtake or meeting decisions). In microscopic modeling, longitudinal movements are determined by design values, but are reduced to operational values by system parameters. For example, typical design headways between two consecutive trains in metros are 90 seconds, but trains operate in 3 to 5 minutes interval, depending on traffic demands at a particular time. Lateral movements are either planned or reactive in response to other traffic movements. Macroscopic models are concerned with number of vehicle units and frequency of operations which are indicators of feasible transportation capacity and density of operations, respectively. To a large extent, macroscopic system parameters depend on trade-offs between periodic traffic demands and commercial viability. The target-speed attainable by a vehicle is called free speed. In real life, the free speed varies across vehicles, driver behavior and also based on other criteria, such as traffic density and weather. Generally in deterministic mathematical modeling, it is assumed that the free speeds are constants. In stochastic modeling of transportation, many traffic parameters are considered as random variables.

ITS is concerned about monitoring of systems, data collection and information interchange, and intervention of systems, all in real-time. By real-time, it is essentially meant that the maximum response time should be within a specified value, which is verified and validated over a large number of test cases. Particularly in railway transportation, the issues are more complex because there are multiple interdependencies between several sub-systems, which are complex on their own. In such large and complex systems, manual operations are also extremely hard and take lots of time and efforts. Increasing the multi-interdependencies in railway transportation systems has resulted in a hugely complex system

Intelligent Transportation Systems

which cannot be humanly managed and controlled. Dynamic and automated signaling systems have resulted in various versions of - Automatic Train Control (ATC), Automatic Train Operation (ATO), Automatic Train Supervision (ATS) and Automatic Train Protection (ATP) systems. And these are the most recent and popular paradigms of automation being used in current railway industries (Narayanaswami & Mohan, 2013); in order to have a complete interoperable and automated railway system, more rigorous research is needed.

Different technologies have been employed for automation of railway operations. One of the most popular ITS technology used in airway operational management is Agents based distributed systems. Agents systems are autonomous, scalable, inter-operable and extensible. Moreover, because of its inherent characteristics, agent technology can be used to design distributed systems. Agent technology is rapidly emerging as a powerful computing paradigm to cope with the complexity in dynamic distributed systems, such as traffic control and management systems, of which multi-agent systems and mobile agents are immensely used in realistic applications. Over the last decade, agent technology has shown great potential for solving problems in large scale distributed systems. One reason for the increasing number of applications using multi-agent technology in transportation operations is that the inherent distribution allows for a natural decomposition of the transport system into multiple sub-systems, each implemented through agents that collaborate, coordinate and compete (negotiate) with each other to achieve a desired global objective.

Multi-agent systems can significantly enhance the design and analysis of problem domains of the following characteristics:

1. A geographically or physically distributed problem domain
2. A dynamic environment for all sub-systems
3. Frequent or continuous interactions between sub-systems

The challenges in such problem domains are the system should be monitored in real-time, data collection, inference and intervening the system should all happen dynamically and in real-time. Such an application in transportation is termed as an ITS. In addition, traffic and transportation systems are extremely dynamic, with both locational and time changes. Agent technology is capable of modeling such dynamic variations. Hence the two characteristics of transportation that make agent technology a more appropriate technology are geographical distribution and dynamic changes. Agent technology is also capable of decomposing a large system into multitudes of sub-systems that can interact with each other. Efficient interoperability helps these distributed multi-agent sub-systems to coordinate and optimize the overall objective.

There are several agent based approaches in monitoring and controlling transportation operations. A set of autonomous agent sub-systems is linked to each other in a distributed application. Agent sub-systems coordinate their activities and share the system's resources, information, etc. Using the shared resources, transportation systems are continuously monitored, data is collected, and reactive solutions are developed and communicated to interfere with transportation systems. An agent is an autonomous computer program that accomplishes assigned tasks on behalf of an agent system. A multi-agent system is a system that consists of multiple interacting intelligent agents. In general, multi-agent technology is used for modeling distributed system in order to solve the problems that are difficult for an individual agent or monolithic system to solve. A multi-agent system could be deployed with both mobile and stationary agents. A stationary agent resides on a local server and communicates with distributed entities of the transportation system by sending and receiving messages. A mobile agent does not send or receive signals to other entities, but it gets created dynamically, moves to where the entities are located, autonomously installs itself on the entity, performs necessary operations, uninstalls itself and leaves the entity. Essentially a distributed multi-agent system can be implemented using either stationary or mobile or both types of agents. Also such agents can operate in a co-operative or a competitive manner. Co-operative agents share common goals, so they co-ordinate with each other to attain a common goal. Competitive agents have contradicting goals and their behavior varies according to applications. The common behaviors are negotiation, fair competition or manipulative competition.

Agent technology has significantly contributed to the advancement of traffic management in all modes of transportation, particularly in airways. However, the design, implementation, and application of agent-based approaches in railway transportation are still immature, and there is a severe lacunae in the available literature. Particularly certain characteristics of agent technology deserve a good deliberation; they are autonomy, ability to handle uncertainty, interoperability, scalability and extensibility. Railway transportation systems are large and widely distributed and so it is physically difficult to monitor the systems on a continuous basis. Autonomous agents eliminate the monotony of monitoring tasks otherwise performed manually. Agents also collect data, and they can be designed to collect data on a continuous basis or at a pre-decided periodicity. The collected data is processed by the agents themselves; several off-the shelf or bespoke algorithms are implemented for data processing, as required in the applications. When inferences are made, agents communicate decisions, implement those decisions and uninstall unwanted instances of the software. The autonomy and mobility characteristics of agents make them highly capable of handling uncertainties. As the agents monitor systems continuously, exceptions or incidents are immediately detected, inferences are made and remedial actions are immediately implemented.

Moreover, multi-agent systems handle incidents much differently from conventional programming paradigms. When multiple incidents occur simultaneously, agents systems instantiate instances of themselves. Each agent instance addresses an incident each simultaneously, thereby resolving all incidents rapidly. This feature contributes to the scalability of agent systems. In a large domain with dense transportation network, there are more number of entities to be monitored and controlled. MAS create instances to accomplish those tasks, thereby the system as a whole is scalable.

Also agent systems are implemented as a class, so that systmes are portable onto different platforms. This is particularly important in transportation systems as transport operators are likely to use different operating systems at different locations. When agents systems have to be executed on them, there has to be compatibility. Since agents are implemented as class files, irrespective of the computational environments, agent instances can get installed on the distributed systems and execute their tasks. Agents are also extensible, that incorporating additional features essentially means to enhancing additional behaviors. Each agent behavior amounts to accomplishing certain sets of tasks by the agent systems. When agents are required to perform additional tasks, additional behaviors are incorporated in the agent systems.

Multi-agent systems offer significant advantages such as autonomy, inter-operability, scalability and extensibility, which conventional programming paradigms are unable to provide. However, MASs have certain limitations in their ability to deal with uncertainty in dynamic environments. To overcome this weakness, it is proposed to integrate mobile communications technology with multi-agent systems to increase the flexibility and adaptability of large scale traffic management systems. Agent mobility provides additional advantages to address unexpected challenges in traffic control and management. Mobility enables quick incident diagnosis, dynamic system configuration, and deployment of new algorithms for conflicts resolution in a dynamic manner,

Many existing agent-based traffic management systems lack in considering interoperability between agent systems at the various platforms distributed over a network. Interoperability is a critical necessity in making decisions based on information across systems, to tackle interoperability issue, IEEE FIPA (Foundation for Intelligent Physical Agents), which is a consortium of companies, government agencies, and schools, has been employed to produce software standards for heterogeneous, interacting agents based systems. FIPA standards guarantee the interoperability between agents by coordinating different aspects of systems, such as system architecture, agent communication, agent management, and agent messaging. Finally less is being utilized on the capabilities and scalability design of multi-agents systems; as most agent systems are operated at the experimental levels and are not deployed on realistic applications, and incidents are generally created for empirical purposes. In a realistic setting, problems are more complex and large, that the resolution process has to scale.

Another microscopic modeling of transportation is through simulation techniques. Simulation refers to the prototype execution of the model of the target system design on the host computer, under given input stimuli, and the collection and analysis of the simulation results. The benefits of modeling using simulation are many. First, they enable one to detect design errors, prior to developing a prototype, in a cost-effective manner. Second, simulation of system operations may identify potential problems, including rare and otherwise elusive ones, during operations. Third, analysis of simulation results may yield performance estimates of the target system architecture. Unlike in the past, the increased speed and precision of today's computers promises the development of high-fidelity models of transportation systems, ones that yield reasonably accurate results, quickly. This, in turn, simulation models would permit system architects to study the performance impact of a wide variation of the key parameters, quickly and, in a few cases, even in real time or faster than real time. Thus, a qualitative improvement in system design may be achieved. In many cases, unexpected variations in external stress may be simulated quickly to yield appropriate system parameter values, which are then adopted into the system to enable it to counteract the external stress successfully. In simple terms, simulation permits posing "what if" questions to the system, observing and interpreting the results. This allows designers and operators to conceive hypothetical, yet realistic, scenarios that may arise in the future; develop solutions and strategies; and test them off-line, i.e., non-operational mode, in a cost-effective and safe environment. Last, the design of new performance metrics, if facilitated can help gain a better understanding of the nature of the system behavior.

CHARACTAERISTICS OF ITS MODELS

Traffic flow and microsimulation models designed to characterize the complex transportation systems have become an essential tool in traffic flow analysis and experimentation (Taplin, 1999). Such tools are employed in a number of applications, such as:

1. Evaluation of dynamic traffic management,
2. Design and testing of new transportation facilities, and
3. Operational flow models such as traffic state estimation and prediction, model-based traffic control, analysis and optimization, and dynamic traffic assignment.

Types and requirements of an application mainly determine which model is more suitable. Complexity of the model, thereby the computational resources and time for completion depend mainly on the required accuracy levels, available data and the network types.

Among microscopic models, simulation models are more time consuming than agent based models. This is due to the fact that simulation requires multiple runs to get statistically valid results, particularly in a class called as stochastic micro-simulation models. However, their features and the applications are also different. Simulation models are generally used in planning stages, sometimes at the macroscopic level also. Agent based models are applicable at real-time planning and control. There are also hybrid ITS models, which are known agent based simulations.

Finally, calibration, validation, verification and testing of microscopic models and their solutions are generally laborious. It is mainly because microscopic models operate in realistic settings which are much more complex than the empirical settings, where they are developed and tested. And it is very difficult to simulate all possible problem scenarios and also to create real-life sized problem scenarios.

Macroscopic models are developed for large scale, network-wide applications, where macroscopic characteristics of the flow are of prime interest (Hueper et al., 2006). Macroscopic models are too coarse to correctly describe microscopic details such as impacts caused by changes in traffic conditions. Macroscopic models are assumed to describe macroscopic characteristics of traffic flow more accurately. Calibration of macroscopic models is relatively simple (compared to microscopic models) using loop detector data. Mostly, speed-density relations derived from observations are required. Macroscopic models can also be applied for congestion propagation reliably, without the need for in-depth model calibration. Macroscopic models are used for planning at the strategic levels.

Mobile-C (Bo-Chen, 2009) is a IEEE-FIPA compliant application for distributed traffic detection and transportation management, integrating mobile agent technology with multi-agent. Mobile-C supports both stationary agents and mobile agents. Mobile agents enhance the ability of a traffic management system to deal with the uncertainty in a dynamic environment. It is designed to be compliant with IEEE FIPA standards both at agent-level and agent platform level. Compliance with IEEE FIPA standards ensures interoperability; integrates multiple detection systems, which enables traffic operators to have a comprehensive view of a traffic system. The open architecture of Mobile-C allows new detection systems to be easily added by wrapping them into sub-agent systems. Using mobile agents for dynamic algorithm and operation deployment has been simulated through a laser-based vehicle detection system, laser detector agency, and a higher level agency. The simulation results showed that mobile agents provide an effective way for dynamic software component deployment.

DARYN (RV Iyer et al., 1995) is a distributed decision-making algorithm for railway networks in scheduling "point-to-point" trains using centralized decision making. Several approaches in distributed decision making suffer from scalability, as the execution time and the memory requirements increase nonlinearly as the system grows in size. In DARYN, the overall decision process is analyzed and distributed onto every natural entity of the system using a simple cost function. The decision process for every train is executed by an on-board processor that negotiates, dynamically and progressively, for temporary ownership of the tracks with the respective station controlling the tracks, through explicit processor to processor communication primitives. This processor then computes its own route utilizing the results of its negotiation, its knowledge of the track layout of the entire system, and its evaluation of the cost function. Every station's decision process is also executed by a dedicated processor that, in addition, maintains absolute control over a given set of tracks and participates in the negotiation with the trains.

RYNSORD (Lee and Ghosh, 1998) is another novel Decentralized algorithm with SOft Reservation for efficient scheduling and congestion mitigation in RailwaY Networks (RYNSORD), wherein every train utilizes look-ahead to dynamically reschedule its route. Thus, a train, which is currently at position X and headed for the ultimate destination Y, first requests and reserves N tracks, i.e., the look-ahead, ahead of its current position for use at appropriate future times. The train moves through the N tracks and, upon completion, it again requests and reserves N subsequent tracks ahead of itself. The process continues until the train reaches its destination. This approach is considered soft, ie., less abrupt and more flexible to negotiations between train and stations. RYNSORD is modeled and implemented on a network of SUN Sparc workstations, configured as a loosely coupled parallel processor. Additionally there are several plug-in architectures of Multi-agent systems available through open source platforms and proprietary channels. Some of them are Zeus, DECAF, JADE. Several transportation applications have been developed using these architectures.

FUTURE OF ITS

Intelligent transportation system (ITS) promises a very exciting future, in two key fronts. First, several key concepts underlying the intelligent transportation system are extensible beyond vehicular traffic and trains to other modes of transportation. Many issues involving cargo air transport, passenger air transport, marine ferries, and personalized rapid transit (PRT) system and even space travel and Mars exploration can be modeled using agent features. The need for intelligent transportation becomes extremely vital under three scenarios—increased travel speeds, a significant increase in the number of travelers, and increased demand for precise and timely information by travelers. Certainly advancement in the frontiers of science and engineering technology has led into both theoretical and technological innovations in intelligent transportation systems. Multi-modal integration in Intelligent transportation systems is considered more beneficial to the transportation sector. Modal integration offers (i) improved access to accurate and timely status information of any transportation mode, anywhere in the world, from any point in the system, and (ii) reliable dynamic routing over multiple modes. Utilizing intelligent, personalized decision aids, the traveler may process the available information to compute the most efficient route or reroute across all different transportation modes, including air, railways, automobiles, ferries, and so on. The most frequent causes for re-planning include changes in the traveler's intention and needs and unscheduled delays in a currently reserved transportation system.

Second, ITS systems are complex, very expensive, and, once deployed, it is logical to expect them to remain in service for a reasonably long period. Therefore, it is absolutely essential to develop a sound and comprehensive understanding since such systems must be amenable to enhancements as the needs evolve with time. Behavior modeling and asynchronous distributed simulation are useful tools to understand system architecture and design trade-offs. Asynchronous distributed simulation considerably reduces computational time, which is otherwise not feasible with current uniprocessor simulators. Intelligent transportation systems offer a wide scope for future research and practical deployment.

REFERENCES

Anderson, R. E., Poon, A., Lustig, C., Brunette, W., Borriello, G., & Kolko, B. E. Building a transportation information system using only GPS and basic SMS infrastructure. Proceedings of ICTD'09. doi:10.1109/ICTD.2009.5426678

Balan, R., Khoa, N., & Lingxiao, J. (2011, June). Real-time trip information service for a large taxi fleet. *Proceedings of Mobisys, 11*. doi:10.1109/ITST.2007.4295824

Bhoraskar, R., Vankadhara, N., Raman, B., & Kulkarni, P. (2012, January). Wolverine: Traffic and road condition estimation using smartphone sensors. Proceedings of WISARD.

Chen, B., Cheng, H. H., & Palen, J. (2009, February). Integrating mobile agent technology with multi-agent systems for distributed traffic detection and management systems. *Transportation Research Part C, Emerging Technologies, 17*(1), 1–10. doi:10.1016/j.trc.2008.04.003

Chowdhary, M. A., & Sadek, A. (2008). *Fundamentals of Intelligent Transportation systems planning*. US: Artech House Inc.

Dailey, D. J., Maclean, S. D., Cathey, F. W., & Wall, Z. R. (2001). Transit vehicle arrival prediction: An algorithm and a large scale implementation. *Transportation Research Record*.

Eriksson, J., Girod, L., Hull, B., Newton, R., Madden, S., & Balakrishnan, H. (2008). The Pothole Patrol: Using a Mobile Sensor Network for Road Surface Monitoring. *Proceedings of the sixth annual international conference on mobile systems, applications and services MobiSys'08*.

Ganti, R., Pham, N., Ahmadi, H., Nangia, S., & Abdelzaher, T. GreenGPS: a participatory sensing fuel-efficient maps application. Proceedings of the sixth annual international conference on mobile systems, applications and services MobiSys'10. doi:10.1145/1814433.1814450

Guha, S., Plarre, K., Lissner, D., Mitra, S., Krishna, B., Dutta, P., & Kumar, S. (2010, November). Autowitness: locating and tracking stolen property while tolerating GPS and radio outages. Proceedings of the 8th ACM conference on embedded networked sensor systems (pp. 29-42). ACM. doi:10.1145/1869983.1869988

Hoh, B., Gruteser, M., Herring, R., Ban, J., Work, D., Herrera, J.,..., Jacobson, Q. (2008). Virtual trip lines for distributed privacy-preserving traffic monitoring. Proceedings of the 6th international conference on mobile systems, applications and services Mobisys '08.

Hueper, J., Dervisoglu, G., Muralidharan, A., Gomes, G., Horowitz, R., & Varaiya, P. (2009). Macroscopic Modeling and Simulation of Freeway Traffic Flow. *Proceedings of the 12th IFAC Symposium on Control in Transportation Systems* (pp. 112-116).

Iyer, R.V.; Ghosh, S. (1995, February). DARYN-a distributed decision-making algorithm for railway networks: modeling and simulation. *IEEE Transactions on Vehicular Technology, 44*(1), 180-191.

Jurdak, R., Corke, P., Dharman, D., & Salagnac, G. Adaptive GPS duty cycling and radio ranging for energy-efficient localization. *Proceedings of the 8th ACM conference on embedded networked sensor systems* (pp. 57-70). ACM. doi:10.1145/1869983.1869990

Konstadinos, G.G. (2007, June). *Travel Behavior and Demand Analysis and Prediction* (Working Paper). Department of Geography, University of California Santa Barbara.

Koukoumidis, E., Peh, L., & Martonosi, M. (2011, June). Signalguru: Leveraging mobile phones for collaborative traffic signal schedule advisory. Proceedings of Mobisys.

Lee, T., & Ghosh, S. (1998, November). RYNSORD: A novel, decentralized algorithm for railway networks with "soft reservation." *IEEE Transactions on Vehicular Technology, 47*(4), 201–222. doi:10.1109/25.728526

Lin, W., & Zeng, J. (1999). Experimental study of real-time bus arrival time prediction with GPS data. *Transportation Research Record*, 1666, 101–109.

Mohan, P., Padmanabhan, V., & Ramjee, R. (2008). Nericell: Rich monitoring of road and traffic conditions using mobile smartphones. In *SenSys*. ACM. doi:10.1145/1460412.1460444

Narayanaswami, S., & Mohan, S. (2013). The roles of ICT in driverless, automated railway operations. *International Journal of Logistics System and Management, 14*(4), 490–503. doi:10.1504/IJLSM.2013.052749

Papageorgiou, G., Damianou, P., & Pitsilides, A. (2006, July 6-7). A Microscopic Traffic Simulation Model for Transportation Planning in Cyprus, Ayia Napa, Cyprus.

Pattara-Atikom, W., & Peachavanish, R. Estimating road traffic congestion from cell dwell time using neural network. *Proceedings of the 7th International Conference on ITS ITST '07*.

Rana, R., Chou, C., Kanhere, S., Bulusu, N., & Hu, W. (2010, April). Earphone: an end-to-end participatory urban noise mapping system. Proceedings of IPSN.

Thies W., Ratan, A.L., & Davis, J. (2011). Paid crowdsourcing as a vehicle for global development. *Proceedings of theCHI Workshop on Crowd-sourcing and Human Computation*.

Singh, A., Singh, P., Yadav, K., Naik, V., & Chandra, U. (2012, July). Low energy and sufficiently accurate localization for non-smartphones. Proceedings of MDM.

Taplin, J. H. E. (1999). Simulation Models of Traffic Flow. *Proceedings of the 34th Annual Conference "OR in the New Millennium" ORSNZ '99*, Hamilton, New Zealand (pp. 175-184). Operational Research Society of New Zealand.

Thiagarajan, A., Ravindranath, L., Balakrishnan, H., Madden, S., & Girod, L. (2011). Accurate, low-energy trajectory mapping for mobile devices. Proceedings of NSDI.

Thiagarajan, A., Ravindranath, L., LaCurts, K., Madden, S., Balakrishnan, H., Toledo, S., & Eriksson, J. (2009, November). Vtrack: Accurate, energy-aware road traffic delay estimation using mobile phones. Proceedings of Sensys.

Transportation's Role in Reducing US Greenhouse Gas Emissions Volume 1: Synthesis Report. (2010, April). *US Department of Transportation*. Retrieved from http://ntl.bts.gov/lib/32000/32700/32779/DOT_Climate_Change_Report_-_April_2010_-_Volume_1_and_2.pdf

Williams, B. (2008). Intelligent transportation systems standards. London: Artech House.

Yin, X., Han, J., & Yu, P. S. Truth discovery with multiple conflicting information providers on the web. Proceedings of SIGKDD'07. doi:10.1145/1281192.1281309

Zhou, Q., Prasanna, V. K., Wang, Y., & Chang, H. (2009, September). A Semantic Framework for Integrated Modeling and Simulation of Transportation Systems. *Proceedings of the IFAC Symposium on Control in Transportation Systems (CTS '09)*.

Chapter 17
Integrated Traffic Management using Data from Traffic, Asset Conditions, Energy and Emissions

Thomas Böhm
German Aerospace Center (DLR), Germany

Christoph Lackhove
German Aerospace Center (DLR), Germany

Michael Meyer zu Hörste
German Aerospace Center (DLR), Germany

ABSTRACT

The traffic management is the core of the railway operations control technology. It receives the timetable information as a target definition and advises the command control and signaling systems to execute the rail traffic. Hence the traffic management system (TMS) has to take into account many sources of requests towards the traffic operation e.g. coming from the maintenance planning or the power supply system and to optimize the operation with respect to many criteria as e.g. punctuality, energy consumption, capacity and infrastructure wear. This chapter shows the sources of information for the TMS as well the resulting criteria. The final approach to configure a specific has to be done with respect to a specific application.

INTRODUCTION

Railways face several new and growing challenges. On the one hand, railway has to play an essential part in meeting the future mobility demands. This is especially true, when it comes to sustainable and environmentally friendly solutions. Therefore, the offered connections and services have to meet the demand at every level. Growth has to take place and the modal split has to develop in favor of railways. On the other hand, in many countries such as Germany, there are very limited possibilities to expand

DOI: 10.4018/978-1-5225-0084-1.ch017

the infrastructure accordingly. In order to fulfil those targets anyway, the existing infrastructure has to be utilized in the best possible way.

The traffic management in public transport has the task to ensure operation according to the timetable. This is at least the classical understanding. Today the traffic management widens. A suitable Traffic Management System (TMS) is crucial for the successful introduction of new technologies in order to meet future demands. One example is higher automated rail traffic using an ATO (Automatic Train Operation) and a moving block operation supported by a moving block interlocking and trainborne integrity supervision. To become effective all these technologies need a suitable TMS. As long as the TMS schedules the traffic in a classical operation, no improvement of performance or capacity can be achieved. The TMS is the link between the traffic planning in the different stages from network structure to the timetable adaptation on the one hand and the command, control and signaling systems on the other hand. Today the traffic management needs to incorporate more and more information and has to be enabled to proactively adapt in this system of increasing complexity. It is also a main source for passenger information and guidance.

The goals of the future TMS are utilize the track capacity, enable economical and sustainable operations and to ultimately ensure to meet the travelers expectations. In the following, a structure for such a TMS is proposed and key elements are identified. Also, the additional functionality is presented and future research areas are pointed out.

BACKGROUND: PROPOSED STRUCTURE AND ELEMENTS OF A (FUTURE) TMS

Modern TMS and even more future TMS are receiving or requesting more and more data from different sources. Age, quality and content of those data will be spread more than in past. The Figure 1 shows some of the sources and receivers of information of the TMS in the real-time environment. Naturally the TMS receives the timetable and network structure from the planning and collects data for the later analysis and controlling. The TMS is the core of the integrated mobility management in middle of Figure 1. It receives and sends information to individual customer interfaces as smart phones as well as public ones as the information pages in the internet. On the other side of the middle column it exchanges information with the control, command and signaling systems which are the interlocking and train control systems. Those systems can be trackside or onboard of the trains, too.

To make the optimal decision in the case of deviations from the original timetable the TMS has to take into consideration information from many other sources. Four examples are given in Figure 1. The timetables or – if available – the real-time information from other transportation systems have to be used to ensure the best service for the traveler. The maintenance management can deliver information about the status of the infrastructure and vehicles. It will be able in the future to give a prognosis, too. The power and energy management can support the TMS to decide in an energy-efficient way. Finally the interface for collaborative decision making can help to bridge the gap to other decision processes. Therefore the traffic management of public transport systems has to ensure that the transport system delivers the best quality of service to the customer. It has to take into account influences from many sources:

- Customer demand
- Current traffic status of the transport system by the command, control and signaling system

Figure 1. Real-time data exchange of the TMS

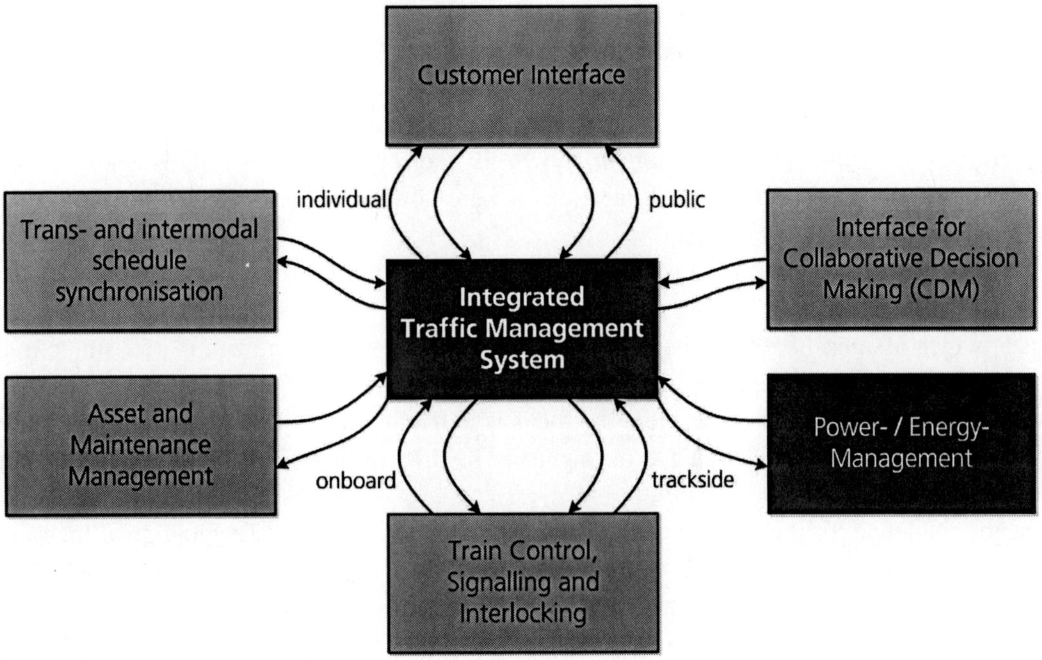

- Asset conditions and predictions
- Power/energy management
- Maintenance management
- Minimized power consumption
- Local reduction of noise and other emissions

In the following sections some of these data exchange partners should be discussed more in detail.

The core of the TMS has to take a multi-parameter, multi-target optimization. The output is finally the recommendation for trains how to run, which is typically a "delta" to the planned schedule. Generally the principles of traffic management can be applied without regular schedule, too. The traffic management has to take into account many different inputs:

- Static data as e.g. schedules, permitted speed profiles and noise sensitive areas
- Dynamic data as e.g. demand, train positions and speeds, prediction of train movements and arrival times, track conditions and weather

This can lead to conflicts of objectives as for example in the case of a small delay of some minutes a speed increase and dynamic driving can help to reduce the delay, while noise and emission reduction as well as minimized energy demand can be achieved by coasting.

The state of the art for TMS is to optimize with respect to the own timetable and real-time data. It exchanges timetable and real-time data with other scheduled modes of transport. Architectures, databases and formats for those applications are in regular use. Different approaches have been discussed already and are partially in use already (Hansen & Pachl, 2014). As examples of scientific approaches may be mentioned here the fuzzy-Petrinet approach by (Fay & Schnieder, 1999) and the genetic algorithm approach by (Wegele, 2008).

The energy consumption of a train depending on the driving style can be calculated or – to be more precise – estimated. The resulting demanding task for the TMS in the case of reactions to a disturbance is to take into account the increasing or decreasing energy consumption. The use of very short-term energy storages as double-layer capacitors or flywheels in electrical powered trains leads to new optimization algorithms. One of the central questions is here how long the timeframe for the operational prediction has to be.

The basic approach to optimize a train run on a given route has been analyzed by (Howlett & Pudney, 1995). The resulting optimization algorithm has been extended by (Schnieder, 2001) to a multi-tier approach, which assumes that there is on the first and lowest tier an optimization of a single train run by using information about the track and the signaling in front of the train. The next tier has the focus of several trains running on one line. Here the focus is in having the highest possible number of trains running over the line without stopping at the signals. The third has the focus on an area of the network or an entire network.

A major challenge for the railway of the future is reduction of noise. Technological measures are available but they are expensive as e.g. noise protection walls or need long time to be introduced as e.g. new composite brake shoes. Consequently are operational measures very important to reach additional improvements. Those will be desperately needed as the demand for rail freight traffic is expected to grow in the European transit countries (i.e. Netherlands, Germany, Belgium, Switzerland, and Austria) and further adverse noise effects from freight traffic on residents is unlikely to be tolerated. Even more, the replacement of gray iron brake shoes with composite brake shoes for the entire rolling stock will half noise emissions but the hen reached level still will not be sufficient to what e.g. the WHO advises (WHO, 2011).

The integrated traffic management takes into account the current condition as well as the wear and tear prediction caused by the operational use. Both processes as shown in Figure 2 for the TMS and Figure 3 for the asset and maintenance management are typically not coupled – at least not in real-time. But in the case that an element comes close to the need of maintenance an exchange between TMS and maintenance planning can help to reach an optimum for both.

Figure 2. Control cycle of the TMS

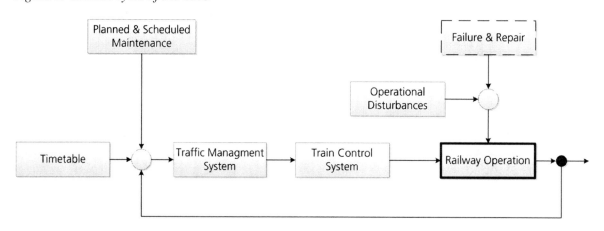

Figure 3. Control cycle of the maintenance and asset management

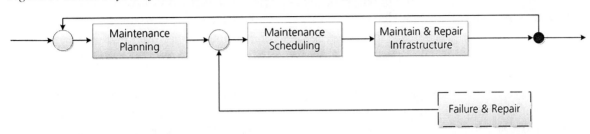

This should be explained by an example: If a switch is near to the limit of maintenance, the number of train runs over it can be limited. So the given switch could be assumed to stand either 10 long and heavy freight trains or 50 light and medium fast regional passenger trains or 25 high speed trains. So the Traffic Management has the task to decide which train in which order can pass the switch.

CORE FUNCTIONS OF A FUTURE TMS

The main tasks of the TMS are to support the dispatcher to make the optimal decisions and to inform the traveler about the ongoing and possible changes of the journey. It is the goal of the TMS to maintain a travel experience without stress even if disturbances take place. Therefore, the current situation ("Now Cast") and upcoming scenarios, also based on decisions to be made ("Forecast") have to be very well known as a core function.

The Now Cast includes the position and status of each train as well as the status of the infrastructure. It is important that the position of the trains is not simply approximated but measured e.g. with a GNSS. Furthermore, all relevant passenger information like destination, planned transfers and connections, and traveler preferences are input to decision making to identify the optimal service for the travelers. The Forecast predicts arrival and departure times at the stations. As the time of arrival is the key element for any further decisions, it is evident that a qualitative or quantitative statement concerning the probability of the prediction has to be given. In case of any predicted missing connections, the travel is adjusted and the customer is informed. Whether a connection is held or not depends also on the predicted departure time of the connecting trains. On the one hand, they depend on the operation schedule for staff and vehicles, operational constraints like occupied line sections ahead, or the technical status of vehicles and infrastructure. On the other hand, the departure times are subject to decision of the TMS.

Any decisions of the TMS, e.g. overtaking of freight trains or letting a passenger train wait for a connection, are integrated in the prediction of the operational status in the network. New computational capabilities will soon enable autonomous decision making. It is also important that such a system is accepted by the users, i.e. the dispatchers. The acceptance will merely be reached if the behavior of the TMS seems as a Black Box to the dispatchers. Instead, they shall be given support. First of all, upcoming conflicts can be shown to the personal early. Conflicts can be solved before they appear. Secondly, the outcome of different options, e.g. minimizing delay, energy demand, or noise emission, can be evaluated. Therefore, the quality of the decisions can be improved.

As the predicted arrival and departure times are shared among the whole travel chain, other transport services are also in the position to adapt. Thus, not only the arrival time at the final destination within the railway system will be known. It will then be possible to predict and alter the travel chain door-to-door. In the long run, this mechanism will enable an overall optimization of the transport system as a whole.

To improve the quality of the decisions based on the prediction, information other than status of vehicles and infrastructure has to be taken into account. Those are on the one hand the expected traveler demand, e.g. influenced by external events. This enables the TMS to predict the number of passengers and the transfer margins at connections. With a high emergence of travelers at a station e.g. due to a concert or sports event, changing times between trains might be significantly longer than usual. On the other hand, information about connected traffic systems (delays, traffic jams) becomes crucial. In case the demand for a connected transport mode exceeds its capacities, the future system wide optimization will share the load. Thus, sudden peaks in demand may occur because a part of the demand of an overloaded transport mode might be switched to rail and vice versa. In order to be able to adjust the operations in a proactive way, it is mandatory to share information about demand and capacity utilization freely. Standardized data interfaces will be needed to enable this new level of traffic management.

The demand for a line and the number of actual or predicted travelers also deliver input to optimize maintenance strategies (see Figure 4). Negative impact of unplanned maintenance work on the level of service can be minimized. Figure 5 shows the data flow of such a new, integrated traffic management system.

Another key element will be collaborative decision making at intermodal platforms like train stations or airports. Here, many different interest groups meet as partners as well as competitors. E.g. decisions have to be made regarding connecting trains, busses or planes in case of delays. The decision making process has to fulfil three requirements. First of all, decisions have to be made in time in order to be able to manage the situation anyhow. Secondly, the decisions have to meet the needs of the travelers. Last but not least, discrimination of one or several parties must not take place in order to ensure an enduring

Figure 4. Integrated Control Cycle of the TMS and Maintenance

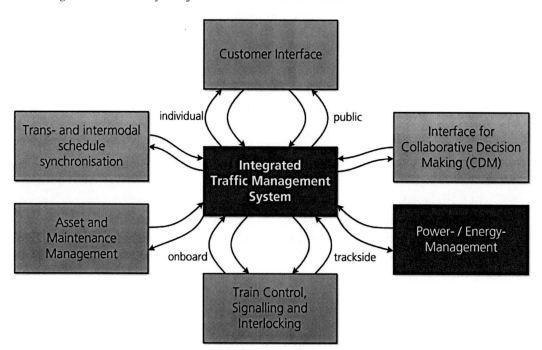

Figure 5. Data flow of new TMS

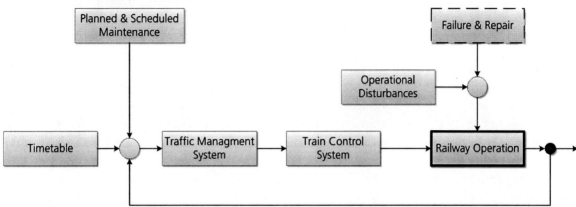

cooperation between the parties. To meet these requirements, a very transparent and standardized process has to be established using key performance indicators.

It becomes clear that the tasks of the scheduler become more and more complex. Through the integration of the railways in intermodal travel chains and optimization of the entire traffic system, the demand becomes highly dynamic. At the same time, the required level of service grows. On the other hand, new technologies have broadened the possibilities to support dispatchers. A key factor is to gain a better understanding of the user. This potential has to be used in order to design systems and human-machine interfaces in a way that fits their users perfectly. Safe and smooth rail transportation shall be reached.

Optimization of Multi-Train Energy-Efficiency

The project Next Generation Train (NGT) of the German Aerospace Centre (DLR) has shown that there are several tiers of energy-optimization relevant for an integrated optimization of railways (Scheier, Schumann, Meyer zu Hörste, Dittus, & Winter, 2014). The main results are shown in Table 1.

The final purpose of the work here is to develop a concept which extends the single train/single line approach by (Howlett & Pudney, 1995) to a global optimization of all trains in a network segment. This again feeds back to the on-line dispatching as well as the development of energy-optimized timetable design (Hansen & Pachl, 2014).

Table 1. Tiers of train driver assistance systems

	Focus	**Purpose**
Tier 0	Inside the train: Propulsion, Heating, Climatisation, Ventilation, Doors, etc.	• Balance of the energy consumption and reducing peaks. • Use on on-board storage for collecting recuperated energy for acceleration and/or coasting (in electrified trains) • Hybridization concepts.
Tier 1	Single train on a single line	• Taking into account static information: Track data like speed profiles, curves, gradients, block length etc. • Taking into account dynamic information: Signal aspects.
Tier 2	Multiple trains on a single line	Taking into account dynamic information: Train positions and speeds.
Tier 3	Multiple trains on a network segment	Taking into account network information: Static and dynamic conflicts.

Operational Noise Reduction

Operational measures to reduce noise emissions mean to avoid braking or accelerating in certain areas as well speed restrictions. Noise emissions of a train can be reduced easily by reducing speed. The rolling noise reduces by approximately 10dB if the velocity is halved. If such a strategy is applied to a line, its performance will drop massively, especially if it is applied only to freight trains as the speed margin between freight and passenger trains increases. So, this is not a feasible way to solve the noise problem of rail freight traffic. Thus, a more sophisticated approach has to be chosen to gain the potentials of operational noise reduction (Mönsters, Linder, & Lackhove, 2013). An example of different speed profiles are shown in Figure 6.

Much more promising is an approach which uses similar methods as energy efficient driving. To guarantee a certain level of service and stability of the schedule, time margins are included in each schedule. As long as the operations are on time, these time margins can be applied to areas along the line which are very noise sensitive. This allows for a reduced speed and therefore reduced emissions while ensuring keeping the operations on time. In order not to influence other trains in nodes, not only the arrival time at the platform has to be considered but the scheduled latest and earliest passing of each node.

This approach can be applied on regular passenger trains. It is not sufficient for delayed operations and most freight trains. To utilize the approach also for these cases, trains running ahead and behind the respective train have to be examined. To yield the optimal movement for the respective train, the latest and the earliest point in time to reach the block signal are calculated. The constraints are that neither the following train should be forced to slow down nor the respective train should reach the block signal before the block ahead is cleared. In case the calculation yields a margin between the times to which the block ahead is cleared and the respective block will be occupied by the following train, this time margin

Figure 6. Example of speed profiles (standard, reduced noise emission, reduced noise emission and energy demand)

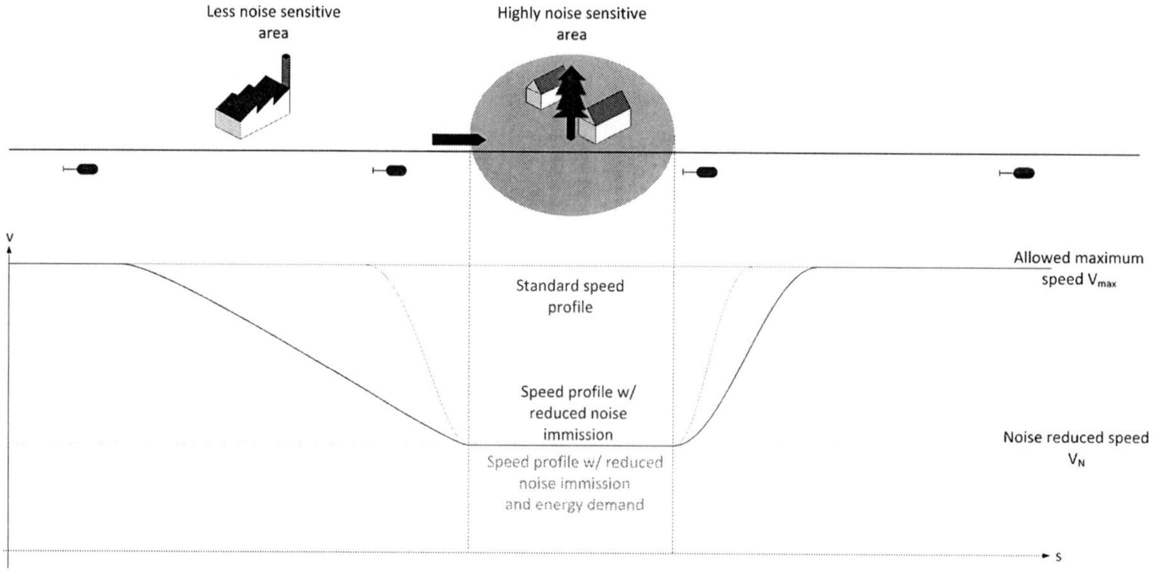

Integrated Traffic Management using Data from Traffic, Asset Conditions, Energy and Emissions

can be used to reduce the speed in noise sensitive areas. See Figure 7 for a schematic representation. Hence the line performance is not diminished, yet noise emission in sensitive areas can be minimized as shown in Figure 8.

Also, this approach suits very well to traffic disturbances as additional time margins can be used behind trains causing the disturbance. Furthermore, complete stops of trains are systematically minimized. This avoids cumulating delays stemming from completely stopped trains with limited ability to accelerate. The operations become more stable towards disturbances. This enhances the level of service. Furthermore, breaking and accelerating during the railway operation are potentially louder and more disturbing for residents than e.g. coasting or cruising trains. This is especially true if older freight trains come to a complete stop. Thus, a system which prevents complete stops and minimizes unnecessary acceleration and breaking adds substantially to the reduction of railway noise.

Technically, the TMS predicts train movements and calculates optimized target points for each train in the node, network or line. These target points contain the time at which the block signal ahead has

Figure 7. Scheme of calculation of time margins during train operations

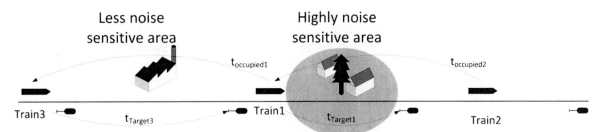

Figure 8. Schemes of track occupations (Standard, utilized time margins for reduced noise emissions)

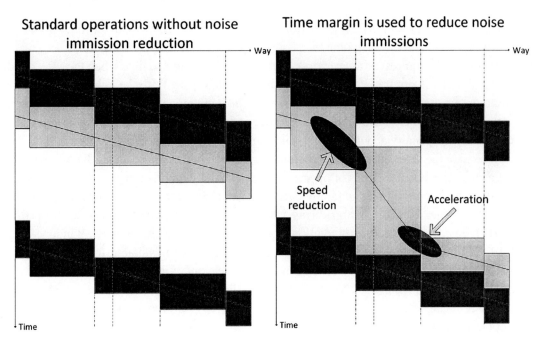

to be passed in order to minimize noise and energy demand under the constraint of keeping the operations stable. It is then sent to the respective train, e.g. using the open standard EETROP. On board of each train, these target points are used to compute an optimal speed profile. This can either be done by a driver assistance system or an automatic train operation system.

As mentioned above, the goals and principles of the described approach for operational noise reduction are close to those of energy efficient driving. Both have in common that unnecessary stops are avoided. Accelerating and breaking is reduced to obtain a smooth speed profile. Nevertheless, the resulting speed profiles will likely differ. Profiles optimized towards a minimal energy demand will spread time margins across the whole line whereas low noise emission speed profiles will use available time margins in noise sensitive areas. In the future, it should be examined how far a low noise speed profile is also energy efficient and vice versa. The goal must be to integrate these goals to gain a universal strategy on how to generate optimal speed profiles.

Asset Condition Monitoring and Prediction

The integrated traffic management takes into account the current condition as well as the wear and tear prediction caused by the operational use. While the asset management takes into account information provided by the TMS. This is supposed to be an on demand data exchange rather than a permanent real time interaction. Because neither the TMS nor the asset management need each other's data all the time. The interaction will serve both sides in the following ways:

- The TMS uses asset condition information to optimize travel time, train types and routes in case of condition related speed reductions at the track, scheduled maintenance activities or failures, whether those are unexpected or predicted.
- The asset management and especially the maintenance planning uses traffic information to schedule maintenance activities in order to keep traffic disturbances at a minimum.

The asset management uses traffic information (past and future) to improve asset condition diagnosis and prediction models in order to become more precise and more reliable (cf. Figure 9).

Therefore, the TMS requires the asset (type), its location, the estimated time when the asset condition will negatively affect a fluent traffic and the necessary time to repair. With the help of this information the TMS defines one or more optimal time windows in which maintenance activities result in minimum disturbances. Note, that other or even multiple optimization aims are also possible, e.g. cumulated travel time of all passengers, penalty and delay fees, noise emissions.

On the other side the asset management requires a track reference, the past and the estimated traffic (number of axis, cumulated axis load) or operational (e.g. switch repositioning) load. This is taken into account when models of condition and failure diagnosis and prediction are developed, like described in (Beck, Jäger, & Lemmer, 2007), (Chattopadhyay & Kumar, 2009), (Gutsche & Böhm, 2011) or (Böhm & Gutsche, 2011). Besides the condition prediction modelling, the optimal maintenance windows provided by the TMS are used to actually schedule maintenance in regards to available maintenance resource like machines and staff, prioritisation between multiple assets and combination of multiple activities. Of cause, the scheduling results need to be send back to the TMS. Additionally, the future traffic load enables the asset management to develop a reinvestment strategy tailored according to traffic demands and asset condition development.

Figure 9. Example how TMS information can improve the precision of asset condition prediction

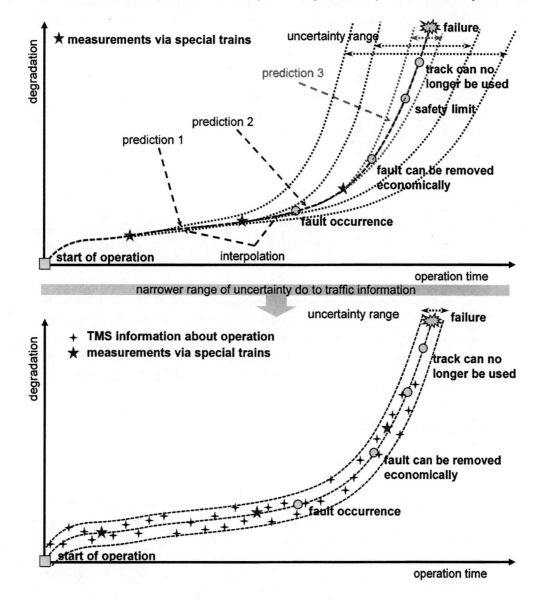

The development of a failure prediction model for the switch engine (point machine) will serve as one example for the interaction between the TMS and asset management. The failure prediction is primarily based on electric current measurements from the switch engine. Certain patterns are derived from these measurements indicating faults and upcoming failures (see (Böhm & Gutsche, 2011) for more information). But over time the patterns tend to change, while the model cannot anticipate this development. Hence, the prediction becomes more and more imprecise. The model can be adapted manually, but this may become a tedious task if hundreds or thousands of switches have individual prediction models. Since one of the reasons for the changing patterns is the continuous degeneration due to wear from traffic loads a much better way would be to integrate this into the model. Therefore, the cumulated axis load and the number of switch repositioning need to be available for each switch and the complete time interval, of which the current measurements are used. The data is analysed together and a combined, self-adapting prediction model is developed, e.g. similar to the integration of weather data in (Böhm, 2012).

The track condition, in particular flat spots (Figure 10), will serve as another example. The spots are the result of wheel spin from starting trains. They can be detected easily with sensors equipped on measurement trains or regular in service trains. They develop dynamically and provide a safety risk if they reach a certain size. When not removed via grinding the spots lead to cracks in the rail. Therefore, infrastructure managers defined different thresholds each inducing different actions. For example, at first the speed in the track section with the spots needs to be reduced. If further unrepaired the track use has to be prohibited. To avoid any such capacity reductions the maintenance management will provide the estimated time when the first threshold of capacity reduction is reach together with the track segment and the time necessary for grinding. Ideally the asset and maintenance management has already identified other activities in that track or in connecting track segments, so that activates can be combined and the track is blocked only once. TMS identifies several time slots for grinding which have a minimal affect according to the parameters used. With the time slots the maintenance management optimizes its activity schedule and reports the actual trajectory for the grinding. The TMS uses this to inform passengers and user if necessary.

The described interaction can vary in frequency. If both sides use intelligent decision algorithms it may be a rather short time interaction and process. In a fully automated holistic system this may even make ad-hoc traffic management possible. In a close interaction the TMS can decide whether to send a few heavy trains over the degraded track segments or a much higher number of short and very light trains - or anything in between that serves the customer, the safety and the capacity needs.

FUTURE RESEARCH DIRECTIONS

Integrated Noise and Energy Optimization

Taking into account the work described in the field of noise reduction and the optimization of the energy-consumption there is a need to develop only ONE integrated advice for the train driver to drive the train. Hence there is a need to integrate the relevant indicator for noise and for the energy consumption. The current work indicates that this system of indicators needs to be adjusted individually for different

Figure 10. Example of flat spots on the rail

railways. The basis for this work is developed in the energy-optimization of the timetable design as well the on-line dispatching in an entire network segment (Scheier, Schumann, Meyer zu Hörste, Dittus, & Winter, 2014).

Dispatcher Assistance and Human Factors

Many scientific studies show that it will be difficult – if not impossible to replace the human traffic management personnel like dispatchers completely by automation. Automation is well suited to handle normal situations and typical disturbances. The human operators are – if well experienced – the better choice for untypical and extreme disturbances. But to be able to act under those conditions they need to be aware about the situation and have the experience ready to be used. Hence they need to work in the before mentioned typical disturbances as well. Consequently, one field of future work will be the development of ergonomic and usable assistance for dispatchers taking into account the human factor. The work in this direction started already (Naumann, Grippenkoven, Giesemann, Stein, & Dietsch, 2013).

Robust and Resilient Management of Railway Operation

An extended view on the traffic management has been developed by the Air Traffic Management (ATM). In case of a disturbance which does not exceed certain limits the ATM has to be robust, which means bringing the operation back to the planned status. Newly the system has to be resilient, which means it has to bring the operation into a new stable status, which is as far as possible fulfilling the operational aim (Gluchshenko & Förster, 2013). Railway traffic management systems as discussed in this chapter should be robust and resilient in the future. Research in this direction just started.

CONCLUSION AND PERSPECTIVE

To reach an integrated mobility management a TMS is highly relevant. It needs to take into account many sources of information and to optimize under conflicts of objectives. New approaches for data analysis and decision taking are currently under scientific development.

External influences like reduced thresholds for noise and emission as well as continuously increasing energy cost are leading to new conditions for the railway operation. Besides network capacity and punctuality are new indicators like noise and energy consumption relevant for the operation. Hence new concepts and algorithms are needed to optimize the integrated traffic management systems of the future.

Asset and maintenance management is already well known for the railways. But the processes and management are not or only loosely coupled in real-time with the current traffic management. Future integrated TMS need interact more closely with the asset and maintenance planning. For this purpose is an improved condition monitoring as well as prognosis of the condition required.

The improvement of the ergonomics and usability of the assistance for traffic management personnel taking into account the human factor is one of the major works for the next future.

REFERENCES

Beck, K., Jäger, B., & Lemmer, K. (2007). *Optimisation of point life cycle costs through load-dependent maintenance* (M. C. Forde, Ed.). Edinburg, UK: Engineering Technics Press Edinburgh.

Böhm, T. (2012). *Accuracy Improvement of Condition Diagnosis of Railway Switches via External Data Integration* (C. Boller, Ed.). Germany.

Böhm, T., & Gutsche, K. (2011). *Diagnosis and Prediction for a Successful Management of Railway Infrastructure* (M. Singh, J. P. Liyanage, & R. B. Rao, Eds.). Kolkata, India: Publishing Services PL.

Chattopadhyay, G., & Kumar, S. (2009). Parameter Estimation for Rail Degradation Model. *International Journal of Performability Engineering*, 5(2), 119–130.

Fay, A., & Schnieder, E. (1999). Knowledge-based decision support system for real-time train traffic control. In N. Wilson (Ed.), Computer-Aided Transit Scheduling (pp. 347-370). Berlin: Springer. doi:10.1007/978-3-642-85970-0_17

Gluchschenko, O., & Förster, P. (2013). Performance based approach to investigate resilience and robustness of an ATM System. *ATM Seminar*. Chicago, IL, USA.

Gutsche, K., & Böhm, T. (2011). e-Maintenance of Railway Assets Based on a Reliable Condition Prediction. *International Journal of Performability Engineering*, 7(6), 573–582.

Hansen, I., & Pachl, J. (2014). Railway Timetabling & Operations: Analysis, Modelling, Optimisation, Simulation, Performance Evaluation (2nd ed.). Hamburg: Eurailpress/DVV Media Group.

Howlett, P. G., & Pudney, P. (1995). *Energy-Efficient Train Control*. Berlin, Heidelberg, New York: Springer. doi:10.1007/978-1-4471-3084-0

Mönsters, M., Linder, C., & Lackhove, C. (2013). Betriebliche Ansätze zur Minderung von Schienenverkehrslärm: Untersuchung der Auswirkungen einer Verringerung der Fahrgeschwindigkeit von Güterzügen auf die Leistungsfähigkeit verschiedener Streckenstandards. *Der Eisenbahningenieur (EI)* (pp. 16-19).

Naumann, A., Grippenkoven, J., Giesemann, S., Stein, J., & Dietsch, S. (2013). Rail Human Factors: Human-centred design for railway systems. *Proceedings of the 12th IFAC Symposium on Analysis, Design and Evaluation of Human-Machine Systems,* Las Vegas, NV, USA. IFAC.

Scheier, B., Schumann, T., Meyer zu Hörste, M., Dittus, H., & Winter, J. (2014). Wissenschaftliche Ansätze für einen energieoptimierten Eisenbahnbetrieb (Scientific Approaches for an energy-efficient railway operation). In Eisenbahn Ingenieur Kalender [Railway Engineer Calendar] (pp. 265-278). Hamburg: Media Group | Eurailpress (In German).

Schnieder, E. (2001). Section 6.43.36.5: Train and Railway Operations Control. In *UNESCO, Encyclopedia of Life Support Systems*. Paris: UNESCO.

Wegele, S. (2008). Automatic dispatching of train operations using a hybrid optimisation method. *Proceedings of the 8th World Conrgress on Rail Research (WCRR)(CD-Rom)*. Seoul, South Korea: Korea Railroad Corporation; Korea Rail Network Authority; Korea Railroad Research Institute.

WHO. (2011). Burden of disease from environmental noise - Quantification of healthy life years lost in Europe. New York: World Health Organisation (WHO).

KEY TERMS AND DEFINITIONS

Asset and Maintenance Management System (AMM): The systems, which are collecting the status of the assets and planning the maintenance activities.

Condition-Based Maintenance: Performing the maintenance works according to the condition or status of the asset and not (only) according to a pre-planned schedule.

Command-Control and Signaling System (CCS): The systems, which are ensuring the safe operation of the railways as e.g. the train control system or the interlocking.

Disturbance: Any external or internal event or influence, which leads to difference between the timetable and the actual train run.

Energy-Efficiency: Train driving style which leads to the minimal energy consumption.

Prediction: Estimating the remaining time until the maintenance of an asset is required.

Traffic Management System (TMS): A system, which advises the command-control and signaling system and receives for this purpose information about the current traffic situation and future demand and disturbances from other systems.

Chapter 18
Disruption Management in Urban Rail Transit System:
A Simulation Based Optimization Approach

Erfan Hassannayebi
Tarbiat Modares University, Iran

Arman Sajedinejad
Research Institute for Information Science and Technology (IRANDOC), Iran

Soheil Mardani
Tarbiat Modares University, Iran

ABSTRACT

The process of disruption management in rail transit systems faces challenging issues such as the unpredictable occurrence time, the consequences and the uncertain duration of disturbance or recovery time. The objective of this chapter is to adopt a discrete-event object-oriented simulation system, which applies the optimization algorithms in order to compensate the system performance after disruption. A line blockage disruption is investigated. The uncertainty associated with blockage recovery time is considered with several probabilistic scenarios. The disruption management model presented here combines short-turning and station-skipping control strategies with the objective to decrease the average passengers' waiting time. A variable neighborhood search (VNS) algorithm is proposed to minimize the average waiting time. The computational experiments on real instances derived from Tehran Metropolitan Railway are applied in the proposed model and the advantages of the implementing the optimized single and combined short-turning and stop-skipping strategies are listed.

INTRODUCTION

The optimization of railway traffic in public transportation systems is facing a growing attention both between practitioners and scholars. Dealing with such real practical problems necessitates the adoption of efficient modeling framework that can be used in different normal and disruption traffic situations.

DOI: 10.4018/978-1-5225-0084-1.ch018

Disruption Management in Urban Rail Transit System

In this regard, Disruption Management (DM) is defined as the process of selecting an appropriate set of strategies and corrective actions after the occurrence of a disorder or deviation from initial plan. It is a critical task in rail systems particularly within congested metropolitan areas. The objective of disruption management is to return the disrupted status to its planned operation status while minimizing all the negative impacts caused by disruptions and recovery costs (Yu & Qi, 2004). The DM concept has a great and growing body of literature on supply chain management and air transportation system. An increasing number of published scientific papers of disruption management in the railway transportation context indicate the importance of effective handling of disruptions. An urban metro system is characterized by its relatively short headway and non-periodic schedules. Major causes of disruptions in metro systems are: peak-hour congestion, unexpected demand increase, train malfunction, signaling failure and line blockage. The unexpected long waiting time, overload and delay severely impact the reliability of transit service. Disruption management has turned into a challenging problem in the operations and planning of urban rail transit systems due to the stochastic variations in passenger demand. In metro systems, a number of strategies are available to manage disturbances: temporary holding of trains, deadheading, short-turning, cancellation of a service, termination of a train, skipping stations, diverting trains, change the order of trains, using a stand-by or back-up train set, and exchanging stop patterns. Carrel, Mishalani, Wilson, Attanucci, and Rahbee (2010) proposed a comprehensive framework for investigating the decision factors and major considerations in service control on high-frequency metro lines. Pender, Currie, Delbosc, and Shiwakoti (2013) presented an international survey of practices in rail disruption management and categorized all responses of passenger rail transit organizations to unexpected disruptions according to the important disruption characteristics: the source, occurrence time, duration, and location.

The operational and passenger costs can be decreased by the means of scheduling strategies customized to the demand profile. The operational efficiency of the public transit services can be improved through demand pattern information over the route. Stop-skipping services are important operation control strategies which are capable of reducing the operating costs and consequently passenger traveling time. It is a well-known service pattern applied in both rail and bus transit systems during the both normal and disrupted situations. Short-turn services are another type of operation control strategies applied where there is a low passenger demand across the part of the route. The application of short-turning services is a tactical control policy that is beneficial in the case that high demand zones need to be supplied. Redirecting the flow of trains in the subway lines is an infrequent control strategy because of its difficulties and complexities. Moreover, short-turn services can help reduce the overcrowding after disruption.

Operational flexibility is an important issue that ensures that there is sufficient capacity to accommodate all the passengers after disruptive events. Immediate response to sudden changes in demand is needed by optimizing the stop-skipping services and by altering the service to redirect trains to congested areas. It should be noted that, usually a combination of transit operating strategies contribute to keep passenger loads and the passenger waiting times in balance (Chen & Niu, 2009).

Although sufficient literature exists on the optimization of control strategies in public transportation networks, not enough studies are conducted on optimization of train schedules under demand and operation uncertainties. The effectiveness of the control strategies relies upon a holistic approach of the whole system. The real-time process of testing and evaluating the disrupted situations is extremely difficult for human dispatchers even with the assistance of advanced information systems such as automatic vehicle location (AVL). Also, to make efficient decisions, computer based decision support systems are needed to help dispatchers to optimize the real-time control decisions. The objectives of this chapter are the simulation based analysis of primary delay and the secondary demand disruption caused by line blockade, together with developing an efficient solution method to optimize the single and combined disruption recovery strategies to be applied in the disturbed situations effectively in the metro systems through a discrete event simulation model of passenger flows and train services under stochastic disturbances. The running time of trains and dwell times are stochastic and variable over the period of study. The research question is how to optimize disruption management strategies in order to improve the effectiveness and robustness of the train timetable under demand and disruption uncertainties. Here, the minimization of average waiting time of passengers is of concern.

The main contribution of this chapter consists of the development of a new integrated simulation-based optimization approach for the rail disruption management problem. For this purpose, a variable neighborhood search algorithm is proposed to solve the combinatorial optimization problem efficiently.

The remainder of this chapter is organized as follows: the next section provides a comprehensive review of related works and the main contributions. A general description of the problem and assumptions are presented in Section 3. The components of the proposed discrete-event and object-oriented simulation system are described in Section 4. The solution methodology is explained in Section 5. The computational results accompanied with a discussion of the real instance of Tehran metropolitan network are given in Section 5 and the article ends with the conclusions and further research suggested in Section 6 and 7 respectively.

BACKGROUND

The operational control strategies are adopted in order to maintain headway regularity and minimize the undesirable direct and indirect effects of service disturbance on passengers. Table 1 presents a taxonomy of the literature on the demand-oriented real-time train timetabling and control problems. Ghoneim and Wirasinghe (1986) investigated the zone scheduling problem in urban rail. In their scenario the entire line is separated into a number of zones, defined based on the demand pattern. The trains must stop at all stations within each zone and then use stop-skipping service to reach to the terminus. The main advantages of this strategy are: reducing of the required number of trains and drivers, and passenger trip times. The decision variables are the fleet size, the number of zones and their configuration and the train headway at each zone. Ghoneim and Wirasinghe (1986) presented a dynamic programming approach for zone scheduling in order to minimize the total transportation cost.

Due to the stochastic variability in passenger demand and uncertainty in the train operations, the headways will unavoidably become irregular. The vehicle holding is one of the most frequently adopted real-time control strategies in public transportation systems. Here, a vehicle which is prepared to leave a station may be held for a period of time with the purpose of adjusting the headways by decreasing the passengers' waiting time. Puong (2001) focused on the development of integrating real-time deterministic train holding and short-turning models for rail transit systems with the objective to minimize both the passenger in-platform and in-train waiting times. Eberlein, Wilson, and Bernstein (2001) presented an extended formulation of holding problem with real-time information. The holding problem is formulated as a deterministic quadratic programming model in a rolling horizon setting. It is deduced that the holding solution is mostly independent of the passenger demand pattern across stations.

Skip and stop operation mode is a well-known method in increasing the commercial speed in public transportation system (Vuchic, 1973). Y. Li and Gendreau (1991) developed a binary stochastic program-

Disruption Management in Urban Rail Transit System

Table 1. A classification of related studies disruption management

Reference	Modeling Approach	Solution Approach	Objective Function (s)	Capacity Constraints Train	Capacity Constraints Depot	Control Strategies
(Eberlein et al., 1998)	Mixed integer non-linear programming model (MINLP)	Exact methods	Total passenger cost	√	-	Deadheading and Stop-skipping services
(Paolucci & Pesenti, 1999)	Object oriented DES	-	Deviations from the nominal timetable	-	-	Stop-skipping services
(O'Dell & Wilson, 1999)	Mixed integer programming model (MIP)	GAMS optimization software	Passenger waiting time	√	-	Holding and short-turning services
(Shen & Wilson, 2001)	Mixed integer programming model (MIP)	CPLEX Solver	Passenger waiting time	√	-	Holding, expressing and short-turning services
Vázquez-Abad and Zubieta (2005)	Discrete event simulation system (DES)	Gradient-Based Methods	Train operational costs and the passengers expected waiting time	-	-	Headway adjustment
Koutsopoulos and Wang (2007)	Discrete event simulation system (DES)	-	Average passenger waiting time and travel times	√	-	Expressing services and Holding
Flamini and Pacciarelli (2008)	Job shop	Heuristic	Regularity and punctuality	-	√	Rerouting and rescheduling
Mannino and Mascis (2009)	Blocking, no-wait job-shop	Branch and Bound (B&B)	Weighted sum of punctuality and regularity costs	-	√	Rerouting and rescheduling
(Lin & Sheu, 2010)	Non-linear stochastic optimization	Dual heuristic dynamic programming	Maintaining schedule/headway adherence	-	-	Running time and dwell time adjustment
Grube et al. (2011)	Object-oriented and event-driven dynamic simulation	Control Algorithms Programmed in MATLAB	Passengers' waiting Time	√	-	Dispatching time and Train speed
(Canca, Barrena, Zarzo, Ortega, & Algaba, 2012)	Mixed integer programming model (MIP)	CPLEX Solver	Minimizing the demand congestion and maintaining a level of quality of service	√	-	Inserting special train services, deadheading and short-turning
Canca, Barrena, Laporte, and Ortega (2014)	Mixed integer non-linear programming model (MINLP)	GAMS Solver	Passenger waiting time	√	-	Short turning services
Hassannayebi, Sajedinejad, and Mardani (2014)	Discrete event simulation system (DES)	Genetic Algorithm (GA)	Expected waiting time per passenger	√	√	Headway adjustment
(Jamili, Ghannadpour, and Ghorshinezhad (2014))	fuzzy Mixed integer programming model (MIP)	LINGO Solver	Multi-objectives (minimize the number of stops at stations, maximize the headway and minimize the differences among consecutive stop-skip patterns)	-	-	Stop-skipping services
This study	Discrete event simulation system (DES)	Variable neighborhood search (VNS) algorithm	Average waiting time of passengers	√	√	Short turning and stop skipping services

ming model to solve the real-time stop-skipping control problem on a transit bus route. The presented optimization model aims to minimize of the schedule deviation and the unsupplied passenger demand.

Deadheading and expressing are the special cases of stop-skipping control strategies. Deadhead run means that a train departs empty from the dispatching terminal through a number of stations with the purpose of saving running time and consequently reduces preceding headways at later stations. The problem here is to decide at any given time as to: which train should be deadheaded and how many stations should be skipped (Eberlein, Wilson, Barnhart, & Bernstein, 1998). Expressing a train means that is can skip stations similar to deadheading strategy. The skipped-stops are mostly at stations with negligible passenger loads and limited connecting transfer lines. The important difference between expressing and deadheading strategies is that expressing can begin at any station, not merely the terminal, and the train usually does not run empty (Eberlein, Wilson, & Bernstein, 1999). Thus, passengers can get on an express train but not a deadhead one. The real-time expressing problem is to determine, which vehicle should be expressed, how many stops should be skipped, and at which stop the expressing service should begin. Eberlein et al. (1998) studied the real-time deadheading problem in transit operations control and developed a nonlinear integer programming formulation. They present a general formulation and optimally solved a special case of this general formulation where vehicle dwell times, speeds, and passenger arrival rates remain constant. Eberlein et al. (1999) presented mathematical models and a heuristic algorithm to optimize the combination of real-time deadheading, expressing, and train holding strategies. They conclude that a coordinated integration of disruption recovery approaches can result in the most effective and robust control strategies.

Short-turning services are used here in order to attain higher frequencies in a part of the lines with higher demand. Short-turning includes selecting a subset of the vehicles to perform short cycles on routes with high demand (Cortés, Jara-Díaz, & Tirachini, 2011). Furth (1987) investigated the optimization of short turn services that aims to achieve load balance and minimize total cost. Deckoff (1990) developed a model that integrates deadheading and short-turning services in order to determine the zone configurations, the frequencies inside and outside the high demand zone and the vehicle capacities. They concluded that the results of combined strategy contribute to cases where demand fluctuations are observed within and between the zones. A service planning model is presented by Delle Site and Filippi (1998) for bus routes where the vehicle size is a variable. The service patterns include short-turn strategies. The objectives are minimization of the average waiting times and operation costs. Cao, Yuan, and Li (2014) proposed a multi-objective optimization model in order to minimize the waiting and traveling times of all passengers and the train's journey time.

The successful application and implementation of the integrated control strategies in public transportation system are reported in the literature: O'Dell and Wilson (1999) employed real-time holding and short-turning strategies for rail transit system during disruptions. They presented deterministic mixed integer programming formulations for minimizing the passenger waiting time and four scenarios of line blockage are considered based on delays of 10 and 20 minutes at two incident locations. Shen (2000) studied the real-time disruption management problem in rail transit system by considering the combination of holding, expressing and short- turning strategies. The objective function is minimizing the sum of the waiting time and weighted in-vehicle delay. Disruption duration is considered as a constant but the robustness of the control strategies are tested through a sensitivity analysis. Shen and Wilson (2001) presented a mixed integer programming formulation for optimal real-time disruption control including holding, expressing and short-turning strategies. A sensitivity analysis is conducted to examine the effect of the deterministic disruption duration on the performance of control strategies. The results indicate that the solution obtained from holding and expressing strategies are fairly robust, nevertheless the efficiency of short-turning strategy is rather sensitive to the accuracy of the disruption duration estimation.

Chen and Niu (2009) presented a transit scheduling model which includes short-turn services along the bus route. The decision variables in short-turn service are the location of the turn-back point and the route schedule which can minimize the total cost of operators and passengers. The problem is formulated as a nonlinear integer programming and a genetic algorithm is developed to solve a real-world case. Ulusoy, Chien, and Wei (2010) presented a mathematical model to optimize short-turn and express services where demand pattern is heterogeneous. They inferred that the optimized combined service strategies decrease the total cost. A short turning model is proposed by Tirachini, Cortés, and Jara-Díaz (2011) in a bus line. The optimization model incorporates the demand information to increase the service frequency subject to the demand pattern. Cortés et al. (2011) developed a combined short turn and deadhead for a single transit line.

In a study by Ding and Chien (2001), a real-time headway control model is presented to maintain the desired headways through minimizing the total headway variance over all stations. The proposed real-time control model is evaluated on a light rail transit (LRT) route in the city of Newark, New Jersey by simulation system. The simulation results prove that the expected passenger waiting time can be considerably reduced by using the control model.

Flamini and Pacciarelli (2008) addressed a real-time routing and sequencing problem of a metro terminal with the objective of optimizing punctuality or tardiness/earliness and regularity with respect to deviation from planned headway. The problem is formulated as a bi-criteria blocking job shop scheduling problem through the alternative graph model by Mascis and Pacciarelli (2002) subject to special constraints. A successful application of optimization methods for real-time traffic control in metro stations is presented by Mannino and Mascis (2009), who developed a real-time routing and scheduling model to operate trains by maximizing the weighted sum of punctuality and regularity costs.

Tanaka, Kumazawa, and Koseki (2009) presented a train rescheduling model under disruption for inter-city railways to minimize passenger inconvenience. The proposed method takes into account the explicit modeling of passenger flow dynamics. Sánchez, Ortega-Mier, and Arranz (2011) developed a discrete-event simulation model in order to evaluate incidents in the railway systems. The simulator has been employed to measure different control policies when short incidents happen. Grube, Núñez, and Cipriano (2011) presented an object-oriented and event-driven dynamic simulator for metro network systems, with the purpose of evaluating real-time control strategies. The control variables are the train speed, the temporary holding time, and the departure time from terminals. Simulation results indicate that using a time-dependent holding strategy improves the passengers' waiting time.

Carbone, Papa, and Sacco (2012) presented an optimization approach for delay recovery in urban metro systems in order to improve the recoverability of the delays. A mathematical programming formulation is proposed to find the train's optimal kinematic variables which include acceleration, deceleration and speed, and the stop times. The problem is solved through pattern search method as a gradient based optimization technique. Canca et al. (2012) introduced optimal train reallocation strategies and scheduling shuttle trains under service disruptions on a double direction line with the purpose of minimizing the demand congestion and maintaining a level of service quality. They proposed a mixed integer programming model to determine the optimal specifications of the short-turning and deadheading strategies with the capability of generating both periodic and non-periodic train timetables.

Canca et al. (2012) studied the scheduling of additional train services; namely shuttles, with the objective of distributing transportation supply efficiently. To achieve this objective, an optimization model is proposed by incorporating deadheading and short-turning strategies, attaining lower headway at crowded locations with minimum changes imposed to the planned services.

Cadarso, Marín, and Maróti (2013) investigated the disruption recovery problem of rapid transit rail systems. The model deals with disruptions that cause completely or partially blockage of tracks between two consecutive stations for a certain time period. The main contribution is developing an integrated optimization model for the train timetabling and rolling stock scheduling problems with respect to passenger demand behavior in the case of disruption. Computational efforts are reported on real instances of the Spanish rail operator RENFE. Recently Canca et al. (2014) proposed a short-turning policy for the management of demand disruptions in rapid transit systems with the objective of decreasing the passenger waiting time while maintaining certain level of service. In this short-turning policy, a number of vehicles operate with short cycles in order to decrease the headway in the partial route of the lines. The decision variables include the location of turn-back points, departure and arrival times and short-turning offsets.

There is a few numbers of studies that have considered a combination of short-turning and stop-skipping strategies in a dynamic and stochastic setting for rapid rail transit systems. The explicit effects of line blockage and its unknown duration on the passenger flows are rarely investigated by the existing works. Most of the current studies in the field of railway disruption management ignore the dynamicity and stochastic nature of disturbances. The simulation models are capable to handle the interaction between trains and passengers. This study addresses the simulation-based analysis of single and combined control strategies via proposing optimization methods subject to the train and the related infrastructure capacity constraints. A variable neighborhood search algorithm is adopted to solve the disruption management decisions.

Based on the above mentioned arguments as the motivation, the objectives sought here are:

- To formulate the dynamics of single and combined stop-skipping and short-turning control policies through a discrete event simulation model.
- To optimize the system performance in view of passengers in highly uncertain situation
- To conduct a simulation analysis in order to compare the performance of different single and combined control strategies.
- Evaluate the optimal timetable generated by variable neighborhood search algorithm through simulation analysis.

PROBLEM STATEMENT

In this section, the network structure and formulation for the optimizing control strategies under stochastic passenger demands are discussed. The considered rail transit network includes a single one-way loop (Figure 1). In order to maintain the operational feasibility of the train timetable, several requirements should be meet: safety rules (e.g. headway time), service levels (e.g. waiting time), and the capacity constraints (station and terminal parking capacity and the maximum train seat capacity). The safety rules require that any two consecutive train services be separated by a minimum headway time (h_{min}). Intermediate stations can accept only one train at a time and no overtaking can occur in the network. The shunting capacity (*SC*) of each terminal is limited, here it equals 2. The fleet is assumed to be homogeneous with finite capacity (C).

Figure 1. The rail network structure

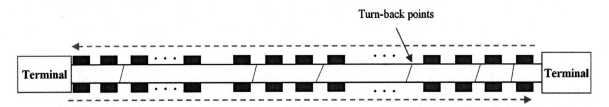

Train movements across the railway network are managed by operation control center. The system consists of remote control hardware, decision-support programs, and human dispatchers. In the rail network, the turn-back points are located to facilitate the short-turning operations at predefined stations. It is assumed that the train stop at every station is technically possible. In general, trains may have non-stop service at some stations with low traffic demand. Typically during the disruption and under some circumstances, trains skip a few stations in order to increase passenger satisfaction and to supply demand in a more efficient manner. The short turn services are a part of partial service operation applicable during the disruption. In this study, both of these operational recovery policies are incorporated in the simulation system to support the rail planners' decisions. The assumptions of the train operation control model are provided in the following sub-sections.

Passenger Dynamics

At the aggregate level, passenger demand can be characterized by passenger arrival rates and alighting ratios for each station. The passenger arrival process is regarded as being a non-stationary period-dependent Poisson distribution. Therefore, a piece-wise linear approximation of the dynamic arrival rate is applied here. For future reference, let λ^t and ρ^t denote the arrival rates and alighting ratio at period $t \in T$, respectively. The mean arrival rates vary over the stations in a dynamic manner.

Stochastic Dwell Time Function

The dwell function proposed by Shen and Wilson (2001) is applied here to approximate the station stop time. Here, dwell time is a piece-wise linear function of the number of passengers alighting (A) and boarding (B) and the passenger crowding situations. Under non-crowded flow conditions, the number of passenger alighting and boarding may not have a significant effect on the dwell time. The presented dwell function consists of a minimum dwell time for non-crowded condition (D_o). The dwell time can be extended with respect to the alighting and boarding time of passengers in crowded condition; therefore their coefficients under crowded condition symbolized as D_2 and D_3 are applied. The term D_l denotes a constant for crowded condition.

$$\text{Dwell time} = \max\{D_o, D_l + D_2 A + D_3 B\} \tag{1}$$

Running Time Function

The stochastic running times are considered in the implementation. The researchers consider distances between any consecutive stations k and $k+1$ (D_k). Let μ and σ^2 be the mean and standard deviation of the running time distribution. Here it is assumed that train running time between any successive stations follow a normal distribution with average $\mu=D_k/V_a$ and variance σ^2 (sec^2) where the V_a and V_m are defined as average and maximum speed of trains respectively. To ensure that all trains cannot exceed their maximum speed, the minimum running time (D_k/V_m) is used in Equation (2). A similar running time function is proposed by Nie and Hansen (2005).

$$Run_k = \max\{\frac{D_k}{V_m}, N \sim (\frac{D_k}{V_a}, \sigma^2)\} \tag{2}$$

Stop-Skip and Short Turn Operations

Congestions occur if the train supply is not sufficient to meet the passenger demand. In order to avoid secondary disruptions as a result of congestion, the station-skipping and short-turn operations are employed. The stop-skipping strategy is commonly adopted in rail transit lines, which allow trains to skip certain low demand stations and reduce the traveling time (Wang, De Schutter, van den Boom, Ning, & Tang, 2013). In this strategy, each train passes through only a subset of the stations along the route. When a train i skips station k, the train running time on the neighborhood block sections decreases by α seconds as presented in Equation (3).

$$x_{ik} = 1 \rightarrow Run_{k-1} = Run_{k-1} - \alpha \text{ and } Run_k = Run_k - \alpha \tag{3}$$

The options of the short-turn services differs seriously depending on the disruption time, duration, location and the current location of the trains. In this regard, two kinds of short-turning services are usually applied: short-turning in front of, or after, the location of the line blockage. This study considers both of them into the modeling framework and optimization method.

Assumptions

The decision support system that controls the stop-skip and short turn operations is available for both inbound and outbound routes. It is assumed that the disruption occurs at peak hours and trains run with minimum headway, where the transit route experiences a service disruption of an uncertain duration. Upon this occurrence, the skip-stop and short turning decisions are made based on the status of the trains.

During the normal situation, trains operate on the route without any skipping or short-turning in any station until they reach terminals. However, when a severe disruption occurs, the rail management employs control strategies such as station-skipping and short turning for transit operations during disruptions.

The following assumptions are made for the stop-skipping decisions:

1. Before the trains leave the terminals, the rail management provides information of stop-skipping decisions for passengers.
2. Train capacity is considered explicitly to explore the congestion and oversaturation effects in the transit line.
3. There is a minimum number of train services that must stop at all station.

SIMULATION MODEL

In the absence of tractable mathematical models for solving the complex train timetabling problem, the simulation based optimization is a flexible and powerful alternative to tackle the challenges. The discrete-event simulation (DES) modeling is a well-known approach with a wide-range of applications in traffic management of rail system (see details by Paolucci and Pesenti (1999), Suhl, Mellouli, Biederbick, and Goecke (2001), I. Hansen and Pachl (2008), Jia, Mao, Liu, Chen, and Ding (2008), Sajedinejad, Mardani, Hasannayebi, Mir Mohammadi K, and Kabirian (2011), Motraghi and Marinov (2012) and Hasannayebi, Sajedinejad, Mardani, and Mohammadi (2012). Based on the operation characteristics of urban transit system, an object-oriented and discrete-event simulation model is developed in a commercial discrete-event simulation software; Enterprise Dynamics (Hullinger, 1999). The object-oriented architecture of Enterprise Dynamics provides users the enhanced capability and flexibility of modeling complex systems. This developed discrete-event simulation system is composed of rail resources and moving trains. Block sections and platforms are single-capacity objects. For more detail of the developed discrete-event simulation model see Hassannayebi et al. (2014).

The main functionality of an object-oriented discrete-event simulation system is its capability of defining the behavior of objects. This modeling concept also makes model building and examination easier and more effective. In Enterprise Dynamics simulation system, an event list is defined for each object to define the behavior. This event list defines how an object must react to different events during a simulation run. Two practical event handlers are On-Entered and On-Exited. The On-Entered event handler will be executed for a station object when a train enters into it. The behavior of the main objects and the event handlers are summarized in Table 2.

The developed simulation system comprises of the process of generating the passenger demand and controlling the train operational services. Attributes and global variables are defined with the purpose of handling these discrete event processes. The attributes of trains include the arrival and departure times and the current passenger load at time $t(load(t))$. Station related attributes include the number of waiting passengers at time t on the platforms ($q(t)$) and the stop-skip and short turn decision variables.

Main events, which are defined in the simulation system includes passenger flow and train operation. The events that concern the train operation are the arrival, departure and short turn events. The discrete events of the passenger flow include the arrival of passengers to a queue, the boarding event of passengers and alighting event. The boarding and alighting of passengers happens as discrete events. The flow diagrams of the passenger arrival, boarding and alighting are shown in Figure 2 and Figure 3. The flow diagram of periodic train operations is depicted in Figure 4. When new passengers arrive at the stops, they put in a station queue and wait until a train arrives. All passengers board the train according to the first in first serve (FCFS) sequence. The exceed number of passenger must wait until next train arrival according to the available capacity of train and the alighting fraction (Figure 5). When a person board the train, the load attribute will be increases and concurrently the queue length attribute decreases by one.

Table 2. The behavior of the main objects and the event handlers in proposed metro simulation system

Entity (Object) \ Event	On-Entered and On-Entering Event Handler (Triggers on Entry of Objects)	On-Exited and On-Exiting Event Handler (Triggers on Exit of Objects)
Passengers	– Update the queue length attribute of the station	– Update the load attribute of the train
Trains	– Update the occupancy attribute of the station – Check the correspond decision variable for the stop and skip service	– Update the occupancy attribute of the station – Check the correspond decision variable for the short turn service and redirect the train
Station (platform)	– Calculating the dwell time of the trains according to the Equation (1) – Creating the exit event for entered train – Updating the load attribute of the trains according to the alighting ratio of the current period ($\rho^{(p)}\rho^{(t)}$) – Count the number of passed train	– Updating the load attribute of the trains ($b_i(t)$) according to the dwell time – Updating the station occupancy attribute
Block section	– Calculating the running time of the trains according to the Equation (2) and (3) – Updating the block section occupancy attribute – Creating exit events for entered train after the dwell time	– Updating track occupancy attribute
Headway controller	– Creating dispatch event according to the minimum safety headway (h_{min})	– Calculating the actual departure time of the trains and the headways

Figure 2. Flow diagram of the passenger arrival and alighting events

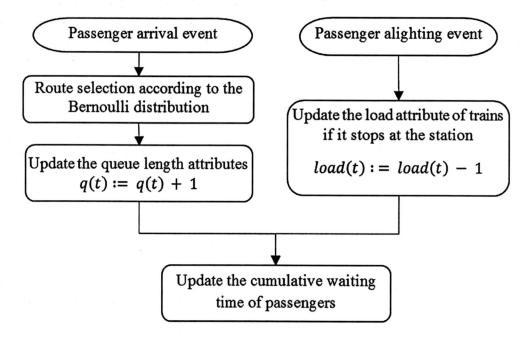

OPTIMIZATION APPROACH

Simulation-based optimization approach is an emerging research field which incorporates optimization techniques into simulation modeling. Given the stochastic, discrete and dynamic nature of the optimization problem, a variable neighborhood search method is proposed to solve the practically-sized instances

Figure 3. Flow diagram of the passenger boarding events

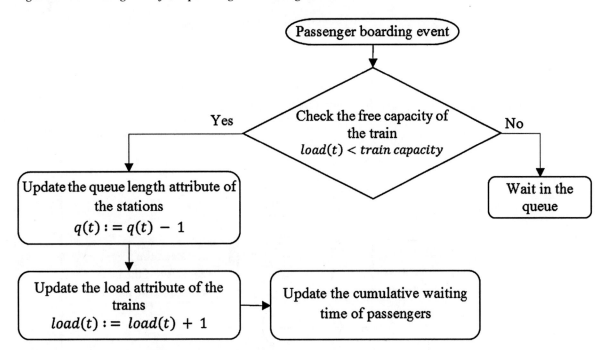

of the problem. Each solution is represented by four binary matrixes including the zero-one decision variables each with its own structure as shown in Figure 6.

The first two matrixes represent the short-turn decisions for inbound and outbound routes. The next two matrixes hold the information of stop-skipping services for each route. Let K denotes the set of all station and K_S represents the subset of all station ($K_S \subseteq K$) that skipping service is allowed for trains. The decision variables are stop-skip services for every train on each station. Let x_{ik} represents the stop-skip decision variables. If i^{th} train that skips station k then variable x_{ik} is one and otherwise is zero. Normally, stations with less demand have more probability to be skipped by trains. In this regard, the proposed variable neighborhood search algorithm utilizes a probabilistic way to decide skip and stop services. Therefore, the probability of that a station visited by a train is calculated by (4).

$$P(x_{ik} = 0) = \frac{\lambda_k^{(t)}}{\sum_{k' \in K_s} \lambda_{k'}^{(t)}} \quad t \in T \tag{4}$$

It is also assumed that no more than two consecutive trains can skip a station. In other words, we have the following equation as a constraint in the simulation model:

$$\sum_{i=1}^{j+2} x_{ik} \leq 2 \quad \forall k, j \tag{5}$$

Figure 4. Flow diagram of the train arrival and departure events in stations

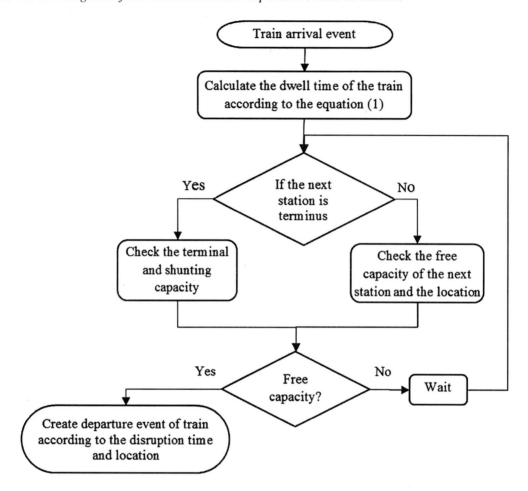

The Solution Method

It is obvious that achieving proper solutions in the optimization algorithms is not independent of the initial solution, so that two optimization stages are applied in this problem (Figure 7):

- Where a variety of initial solutions are generated and for each solution the objective/function is calculated from the simulated model
- Where an algorithm is designed for optimization based on variable neighborhood search.

The Initial Solution Generation

Here in the first phase of optimization, Q_{init} initial solutions are produced randomly (Figure 7). The value of Q_{init} is determined with respect to the size of solution space. In the massive solution space, any bigger Q_{init} can lead algorithm to achieve a stronger base for the second stage of simulation. On the other hand, very large value of Q_{init} can assign a portion of optimization effort to itself. Let M_1, M_2, M_3 and M_4 denoted the number of inbound and outbound stations in which short-turning and skipping are

Figure 5. Flow diagram of the train arrival and departure events in block sections

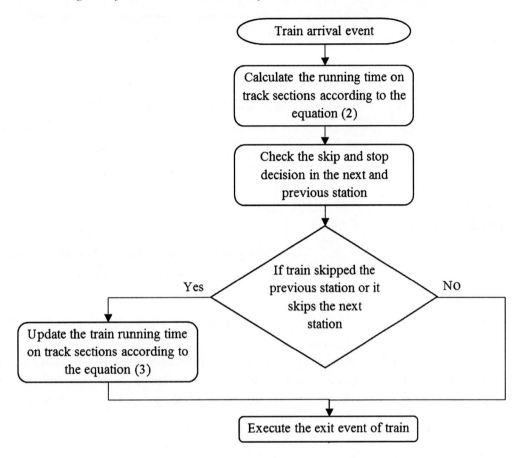

Figure 6. Solution encoding

Station number:	1	2	3	...	k	...	m-1	m
Train 1	1	0	1	...	1	...	0	0
Train 2	0	0	0	...	0	...	1	0
Train 3	1	0	1	...	1	...	0	1
⋮	⋮				⋮		⋮	
Train n-1	1	0	1	...	1	...	1	0
Train n	1	0	0	...	1	...	0	1

Figure 7. The proposed two-level optimization algorithm

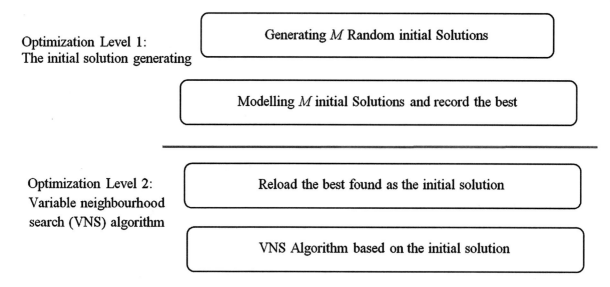

allowed. By assuming a maximum of *n* train service over each station, the solution space would have $2^{(M_1+M_2+M_3+M_4)n}$ different potential solutions.

Each solution is simulated R_{ep} number of replications and the average output is considered as the objective function value. The R_{ep} value corresponds to the objective function obtained from simulation with a correlation above 95%. The best found solution is denoted by f_{best}. The pseudo code of initial solution generation is illustrated in Table 3.

Variable Neighborhood Search (VNS) Algorithm

Local search optimization algorithms perform a series of local moves in an initial solution which decrease each time the cost function until a local optima solution is found. Explicitly, at every iteration of a local search algorithm a better-quality solution *x'* in the neighborhood *N(x)* of the solution *x* is found, until no more improvements are obtained. An effective way of avoiding to be trapped in a local optimum is to change the neighborhood structure in systematic way. A famous type of this approach is Variable Neigh-

Table 3. Pseudo code of initial solution generation

Random Search (Q_{init}, x_o)
begin
x_1 := Random (0, 1) ;
k := 1 ;
f_{best} := Inf
while k <= Q_{init} do
$f(x_k)$:= Simulate (x_k) ;
if $f(x_k) < f_{best}$ then x_{best} := x_k ;
k := k + 1 ;
x_k := Random ;
return (x_{best}) ;
end.

borhood Search (VNS) which is introduced by Mladenović and Hansen (1997) and rapidly developed both in its methods and its applications (P. Hansen, Mladenović, & Pérez, 2010).

In opposition to local search algorithms, VNS methods search gradually distant neighborhoods of the current solution. VNS has different variants and extensions. In this study we apply the basic VNS that is in fact a descent search method called variable neighborhood descent (VND) method. Let N_l denotes a finite set of specified neighborhood structures ($l=1,2,\ldots,l_{max}$). Let $N_l(x)$ represents the set of solutions in the lth neighborhood of x. In variable neighborhood descent method an initial solution x is given and the change of neighborhoods is made in a deterministic way. The general pseudo code of variable neighborhood descent method and its steps are presented in Table 4 (P. Hansen & Mladenović, 2001).

The steps of the proposed variable neighborhood search method are presented in Table 4. The search algorithm starts with the best generating random solutions in previous phase. As seen in Table 4, in the mentioned algorithm searching to find a desired solution begins from one found best solution. This point is selected among Q_{init} solutions which in the previous algorithm phase were produced in the random manner but within the required regulations and limitations necessary for a good solution. The best mentioned random solution is the initial solution at phase II of optimization. With respect to the high number of neighborhoods of each solution, searching among all of them to find the best solution is very time consuming. Here a few types of neighborhoods are defined:

- Neighborhoods with one distance: differs with one constituent element regarding good solutions, i.e only one specific train crosses a station while in the initial solution this train has made a stop in that station. The rest of stop and turns patterns among the solutions are the same.
- Neighborhoods with the D_{Radios} distance from the best found solution defined with D_{Radios} amount of different (zero or one) elements. Accordingly the number of D_{Radios} neighborhoods in every solution are many, e.g. first neighborhood size is $(M_1+M_2+M_3+M_4)n$ and the D_{Radios} neighborhood size equals to

$$\binom{(M_1+M_2+M_3+M_4)n}{D_{Radios}}.$$

Table 4. Pseudo code of variable neighborhood descent (VND) method

VND (k_{max}, x_o)
begin
$x:=x_o$;
$k:=1$; $D_{Radios}:=1$; $Rmax:= D_{Radios} * \alpha$;
while " Termination Condition " **do**
while $k <= R_{max}$ **do**
$x':=$ **Local Search** (x, D_{Radios});
if $f(x') < f(x)$ **then** $x:=x'$; $k:=1$; $D_{Radios}:=1$; $R_{max}:= D_{Radios} * \alpha$;
else $k:=k+1$;
$D_{Radios}:= \max(D_{max}, D_{Radios}+1)$; $R_{max}:= R_{max}:= D_{Radios}*\alpha$ **go to** 50
return (x_{best}) ;
end.

To search for a better and more effective solution space, only the number of random R_{max} from each best neighborhood solution is desirable. If no best solution is found in the first random R_{max} neighborhood the algorithm would search the best solutions in the second, third and so on. This type of optimization can stop the algorithm from being blocked in the local optimum. The maximum researchable neighborhoods are expressed by D_{max} parameter, indicating that the algorithm searches up to D_{max} neighborhoods and continuous its search in the same space. Now, since the farther neighbors contain more solutions, the R_{max} could depend on D_{Radios} volume and with an increase in the number of the D_{Radios} the numbers of random solutions in searching become more exploratory. The simplest assumption for R_{max} value can be expressed as follows:

$$R_{max} = \alpha . D_{Radios} \qquad (6)$$

where α is a value bigger than 1.

A termination condition could be varying from an algorithm to another. In case of disruption management, there is a need to reschedule the plan of the trains in the minimum possible time. Regards this, the condition of terminating the optimization can be defined based on the running time of the algorithm or also on the pre-defined finite number of iterations. It is obvious that assigning more time to optimization algorithm lead us to the better solution but necessity of achieving a good solution in a reasonable time is more important in the real world problem. A schematic representation of VNS search method is illustrated in (Figure 8). As observed in the (Figure 8), the second good solution is extracted locally from first optimum solution and it can find the better local optimum solution in a smaller neighborhood. http://nl.mathworks.com

We compared the performance of the proposed heuristic search method with the state of the art simulation-optimization methods embedded in OptQuest. OptQuest is general purpose optimizer, which is a registered solution of OptTek Systems, Inc. OptQuest has been applied in different real world applications of simulation based optimization (see for example Quaglietta (2013) and S. Li and Wang (2013)). It optimizes the system performance using a black box approach. In order to find the near optimal solution, OptQuest combines the meta-heuristics algorithms including Scatter Search, Genetic Algorithm, and Tabu Search (Kleijnen & Wan, 2007). The OptQuest optimization framework also incorporates neural network learning algorithms to improve its functionality. OptQuest has been integrated with different discrete event simulation soft-wares (Laguna & Marti, 2002). For a review of simulation embedded optimization packages see Möller (2014). In this study, E.D. OptQuest is used. E.D. OptQuest takes the advantages of the decision-support features of the Enterprise Dynamics simulation software with the use of global optimization algorithms.

REAL CASE

The methodology proposed in this chapter is applied in the real data of the underground rail systems of Tehran, IRAN. Line 1 of Tehran metro network is suggested as the real case. Its length is 39 km with 29 stations. Distances between stations are presented in Table 6. General specification of this real case is presented in Table 7. The total number of trips in Tehran metropolitan network during the 2010-2011 is illustrated in Figure 9.

Figure 8. A schematic representation of VNS search method (source of picture: http://nl.mathworks.com

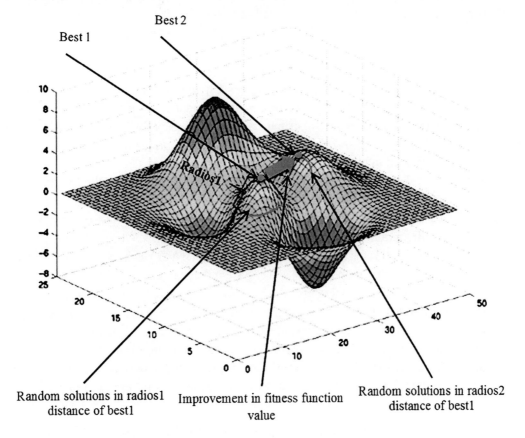

Figure 9. Number of trips in Tehran metropolitan network

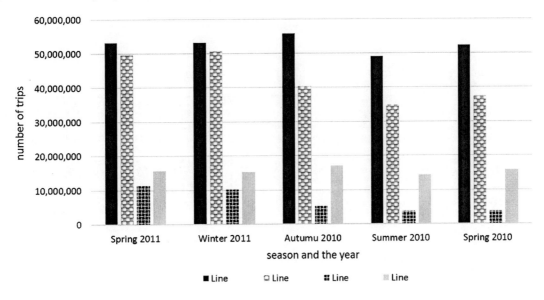

Dwell time approximation for non-crowded condition (D_o) in each station is 15 seconds. Demand profiles of stations with high demand are illustrated in Figure 10. The list of stations which short turn operation is allowed is {7, 12, 18, and 24}. The allowed stations to be skipped by trains are presented in Table 5.

The passenger arrival process is considered a Poisson stochastic process with period-dependent arrival rates. The study period is divided into equal intervals (θ=1 hour), and the mean arrival rate is constant during each interval. Let γ_k denotes the fraction of inbound demand at station k. Each passenger choices the inbound path according to a Bernoulli distribution proportioned by the inbound demand. According to these stochastic behaviors of passenger, Figure 12 shows these probabilities. Therefore, the arrival rate at each station is divided into inbound and outbound demands.

The classification of disorders is based on the duration of the disruption. In the first category, a minor disruption occurs and the resolution will take a few minutes to an hour. The disorders where their recovery duration casts between a few hours to a few days are in the next category. In this study, the focus is on the minor disruptions while the recovery time is uncertain.

Tehran metro system, applies the automatic control protection (ATP) technology to control the movement of trains. The automatic train operation (ATO) system that can manage the trains automatically is not implemented yet. It is an operational safety system adopted to automate train operations. Tehran metro network has issued different daily disruption reports. A study of recorded disruptions of passenger traffic and the accidents in the city of Tehran suggest the potential benefits of using decision support systems.

The control room is a command center where the deviation of train is controlled with a centralized system, a part of Tehran Traffic Control Centre. Tehran Metro and suburban traffic control center is equipped with a computer network, software and hardware which control the movements of the trains.

Table 5. The allowed stations to be skipped by trains (line 1 of Tehran Metro network)

Inbound Route	8	20	22	26
Outbound Route	3	8	14	26

Table 6. Distance between stations in line 1 of Tehran Metro network

Station Number	1	2	3	4	5	6	7	8	9	10	11	12	13	14	15
Distance (m)	1026	879	1337	1312	1335	1096	934	899	1013	726	936	967	621	889	
Station Number	15	16	17	18	19	20	21	22	23	24	25	26	27	28	29
Distance (m)	1194	864	533	609	1168	769	1137	1250	2199	1691	4706	2402	1075	1 901	

Table 7. Operational data and specification of Line 1 (Tehran Metro network)

Parameters	h_{min}(minutes)	V_a(km/h)	V_m(km/h)	σ^2(sec^2)	α (sec)	C	SC	D_0	D_1	D_2	D_3
Value	4	50	80	30	15	2000	2	15	5.5	0.08	0.07

Disruption Management in Urban Rail Transit System

Figure 10. Demand profile for a sub-set of stations of the line 1 (Tehran Metro network)

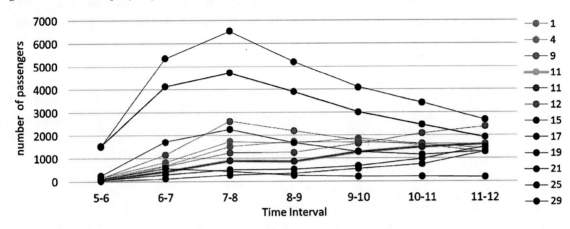

Figure 11. Demand pattern over stations of the transit routes (Tehran Metro network)

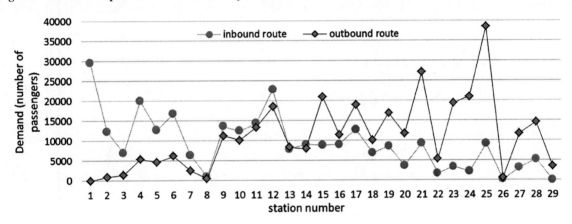

Figure 12. The probability of choosing inbound path for the passengers (Tehran Metro network)

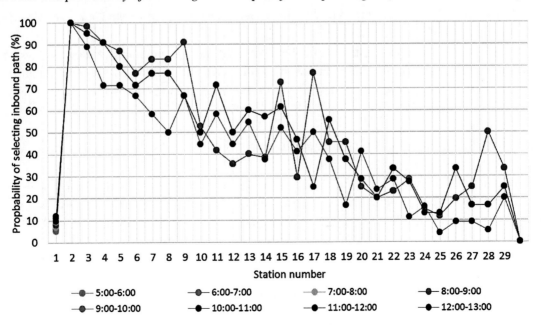

The software allows the user to specify the path for the trains. In the event of disruptions, the command centers, and traffic management center are responsible for the situation. Currently, manual recovery actions are performed during the disruption by the experts. Tehran metro network claims that the disturbances occur repeatedly in the network. The research survey in 2009 is conducted on the correlation between the incidence factors such as journey time, number of passengers transported and intervals between train services. The results of survey indicate that in 62% of cases, human factors, equipment failure had direct effects on the occurrence of disruptions and in 38% of cases the passengers are the source of disturbance due to lack of public awareness and information observable by the passengers. Due to passenger crowding, after the incident, the delays are propagated in the network. A major disruption occurred on 15 April 2012 because of a heavy rainfall in Tehran, which caused a flash flood in the Tehran metro tunnels (Taghizadeh, Soleimani, & Ardalan, 2013). After this incident of such scale, the rail planners seek to develop disruption management system in an online setting. It is believed that the Tehran metro needs a decision support system in case of emergencies that can be useful in case of daily disruption. The stop and skip service is one of the most common control strategies applied in Tehran metro network. The rail infrastructure promotes the short turn services in order to increase the frequency of the service permanently.

Case Data and Specifications

The major cause of disruptions in Tehran Metro network is damage to the line that causes the blockage. It is assumed that the specifications of the disruption are given. This disruption is a temporary blockage of one direction between station 13 and 14 in the first peak period. It is assumed that the line blockage occurs at 8:30 AM. The recovery time follows a discrete probability distribution. At the worse-case the recovery time is about 60 minutes and most likely it takes 30 minutes with probability of 0.3 and 0.5, respectively. At the best case, the recovery time is 10 minutes with probability of 0.2. Therefore, the expected duration of the line blockage is 30 minutes.

In order to efficiently optimize the control strategies over the inbound and outbound routes in the peak hours, the passenger demand must be analyzed in time and space. The demand profile during the first peak in a sub-set of stations of the line *1* (Tehran Metro network) is illustrated in Figure 10. Stations with higher demands are 1, 12, 21 and 25.

Figure 11 shows the demand patterns for both the inbound and outbound routes. No systematic profile can be observed from the beginning of the route to the end and the passenger demand fluctuates significantly in the stations of the route. For both the inbound and outbound routes, there is a high demand in downtown areas.

Measuring the Disruption Consequences

In this section, we briefly provide a simulation-based analysis of disruption consequences. The disruption occurs at the 12600 sec after beginning of the simulation run. As mentioned in the previous sections, the recovery time is uncertain with discrete probability distribution. A sensitivity analysis is used to examine the effect of the disruption duration The impact of recovery time scenarios on the average waiting time of passengers near the disruption location (station 13) is significant when no control strategy is employed (Figure 13). The simulation experiments validates the fact that the system need longer time to reach to normal situation after a sever disturbance. The average waiting times of passengers all over the route are affected by the duration of the disturbance (Table 8). In the first scenario, the average passenger waiting time increased by 1.02% compared with the normal scenario. For the second scenario the disruption takes 30 minutes and the average passenger waiting time amplified by 8.20% compared with the normal situation. In the third case which imposes the longer duration of disturbance, the passengers experience waiting time more than 58.04% of waiting time during the normal operation. As one can see, the variance of waiting time is increased considerably by longer duration of disturbance. Therefore, in the case of longer disruptions the system experiences an unstable condition. In order to achieve even headways, a new robust operational plan is needed. Furthermore, spreading an extended delay makes the headways uneven, which results in congestion at stations ahead of the blockage (like station 13). In this case and due to train capacity limit, the short-turning service provides effective delay control strategy.

Experiment Setting

The period of warm-up is considered before the start time of the disruption. The replication length is 6 hours and the simulation model will be executed for 10 replications (Table 9).

Figure 13. The impact of recovery time scenarios on the average waiting time of passengers (station 13)

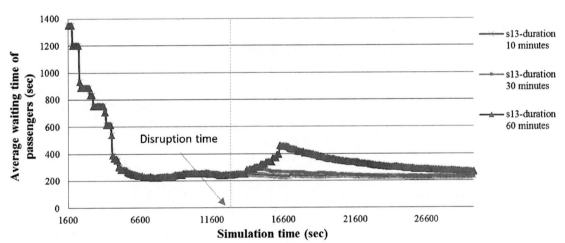

Table 8. Statistical analysis of the disruption duration on the waiting time of passengers

Disruption Duration	Average Passenger Waiting Time (sec/passenger)	Standard Deviation (sec/passenger)	Lower Bound (95%) (sec/passenger)	Upper Bound (95%) (sec/passenger)	Minimum (sec/passenger)	Maximum (sec/passenger)
0 minutes	215.82	8.29	214.20	217.45	199.85	242.69
10 minutes	218.03	8.77	216.31	219.75	199.76	247.16
30 minutes	233.52	7.22	232.1	234.93	214.67	249.99
60 minutes	341.08	18.33	337.49	344.68	311.56	389.05

Table 9. Experimental setting of simulation analysis and VNS algorithm

Parameter	Value
Q_{init}	50
R_{ep}	10
D_{max}	10
α	10
Warm-up period	3.5 hours
Replication length	6 hours
Termination Condition	600 iteration

SOLUTIONS AND RECOMMENDATIONS

The comparison of control strategies are provided in this section. To investigate the impact of single control strategies and combined ones, we test the following single and combined disruption handling scenarios:

- No control strategies (N)
- Only station-skipping strategy (SS)
- Only short-turning strategy (ST)
- Combined station-skipping and short-turning strategies (C)

The effect of the uncertainty in disruption duration is reported in Veelenturf et al. (2012) as a challenging issue in train rescheduling. Therefore, in order to compare the performance of the control strategies, the optimization algorithm is performed on two scenarios. The first scenario, the disruption duration is its expected value and in the second scenario disruption, duration is uncertain with known probability distribution.

Certain Recovery Time

The optimization algorithm runs for the case when the disruption duration is certain. The stochastic variability of the train running time and the passenger demand are still hold in the simulation model. The skip and stop and short-turning decisions are adapted based on the expected duration of the disruption. The primary measure in evaluating the quality of a timetable is the expected waiting time of passengers. The result of simulation experiment on the best solution obtained from VNS algorithm is presented in Table 10. The best solutions are simulated for 100 replications. The statistical report includes the average passenger waiting time (sec/passenger) and the 95% confidence interval for different optimized control strategies. According to the simulation results after 3000 iterations of the VNS algorithm, the combined station-skipping and short-turn control strategies provides about 5.18% reduction in average waiting time of passengers over no control scenario. The above findings suggest that the optimized *SS* and *ST* single strategies decrease the average waiting time by 19.5% and 20.19% respectively. The result indicates that under certain recovery time, the single strategies (*SS* and *ST*) outperformed the optimized combined strategies (*C*) in term of average waiting time of passengers. The stop-skip policy generates

Table 10. Statistical analysis of the best solution obtained from different control strategies (deterministic recovery time)

Strategy	Average Passenger Waiting Time (sec/passenger)	Standard Deviation (sec/passenger)	Lower Bound (95%) (sec/passenger)	Upper Bound (95%) (sec/passenger)	Minimum (sec/passenger)	Maximum (sec/passenger)
No control strategies (N)	305.59	43.46	297.07	314.12	232.80	541.01
Only station-skipping strategy (SS)	246.01	9.63	244.12	247.9	225.58	269.81
Only short-turning strategy (ST)	243.89	21.74	239.63	248.16	201.51	315.24
Combined station-skipping and short-turning strategies (C)	289.77	65.52	276.93	302.62	213.97	779.12

solution with lower variance in average waiting time and it could find a more robust solution compared with other control strategies.

The comparison of control strategies in term of algorithm convergence indicates that the stop-skipping strategy could reach to good solution faster than other control strategies although the short-turn policy reduced the average waiting time more than the other control strategies (Figure 14). The outcomes indicate that the combined control strategies dominated by the optimized single control strategies. It is because that the optimization of the combined control strategy needs more computational time and effort. In the case of certain recovery time, it is expected that the optimized short-turn policy can reduce the average waiting time efficiently. Comparing the results of different single and combined control strategies, the short turning strategy provides modest benefit beyond the stop-skip and the combined strategies where the recovery time is deterministic. The standard deviation of stop-skip control policy is also the lowest compared with the other control strategies.

Figure 14. Comparison of control strategies in term of solution convergence (certain recovery time)

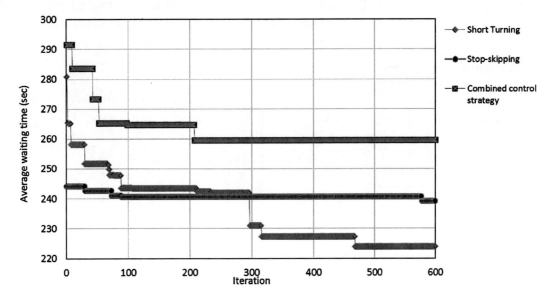

Uncertain Recovery Time

In the above simulation-based analysis the uncertainty in recovery time was ignored. In real situation, however, the recovery time is usually uncertain. We propose a framework for dealing with the uncertainty about the duration of the disruption. In performing the simulation experiments, different types of randomness are introduced in the model. The passenger arrival rates and alighting fractions and the train travel and dwell times are the stochastic elements of the simulation model. The random recovery time is also considered with the purpose of enhance the practical aspects of the model. Here, a simulation approach is proposed that explicitly deals with the uncertain duration of the disruption. In this simulation analysis, the same performance measure (average waiting time) is being undertaken to resolve the disruption under probabilistic duration of the disruption. With all these uncertainty in place, the skip and stop and the short turning decisions are made based on the average output of the simulation model. The examination of how sensitive the control policies are in the simulation model is of concern. Statistics on average passenger waiting time (sec/passenger) for different control strategies are summarized in Table 11. Among the single control strategies, the performance of the station skipping and short-turning strategies, is relatively similar in terms of average waiting time of passengers, while short-turning does slightly better compared with the best solution obtained by the optimized station skipping control policy. The authors claim that the combined control strategy (*C*) is less effective than the single optimized control strategies due to the complexity of the demand pattern and the uncertainty of recovery time. On the other hand, the time limitation forces the decision maker to choose between alternative solutions in a short time. Therefore, a fast solution method is more practical for the real-time management of the disorders. The simulation results show that among three control strategies, the combined one (*C*) provides highest variance in passenger waiting time and the stop-skipping strategy have more capability of controlling the delays at the beginning of the disturbance.

According to the simulation results after 3000 iterations of the VNS algorithm, the combined station-skipping and short-turn control strategies provides about 4.33% reduction in average waiting time of passengers over no control scenario. The *SS* and *ST* single strategies decrease the average waiting time by 21.06% and 17.54% respectively. The result indicates that under uncertain recovery time the single strategies (*SS* and *ST*) outperformed the optimized combined strategies (*C*) in term of average waiting time of passengers. Figure 15 shows the comparison of control strategies in term of algorithm conver-

Table 11. Statistical analysis of the best solution obtained from different control strategies (uncertain recovery time)

Strategy	Average Passenger Waiting Time (sec/passenger)	Standard Deviation (sec/passenger)	Lower Bound (95%) (sec/passenger)	Upper Bound (95%) (sec/passenger)	Minimum (sec/passenger)	Maximum (sec/passenger)
No control strategies (*N*)	342.00	87.13	324.91	359.09	227.29	731.65
Only station-skipping strategy (*SS*)	269.99	46.42	260.89	279.09	217.22	390.10
Only short-turning strategy (*ST*)	282.03	44.24	273.35	290.71	218.27	407.42
Combined station-skipping and short-turning strategies (*C*)	327.19	76.54	312.18	342.2	224.14	603.93

Figure 15. Comparison of control strategies in term of solution convergence (uncertain recovery time)

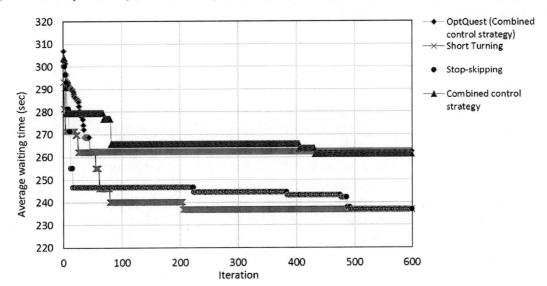

gence. The stop-skipping strategy could reach to good solution faster than other control strategies and OptQuest. Moreover, both single control strategies (*SS* and *ST*) could find the better solution compared with combined control strategies and OptQuest. According to the obtained results, the effectiveness of short-turning and stop skipping control strategies are sensitive to the uncertainty of the disruption duration.

FUTURE RESEARCH DIRECTIONS

The future study should address the incorporation of other control strategies such as expressing, holding and deadheading in combination and providing more detail comparisons of disruption recovery methods. A further study would be to investigate the disruption management as a bi-criteria optimization problem where both the operational and passenger costs are minimized. Furthermore, the variable neighborhood search optimization method can be combined with powerful meta-heuristic algorithms in order to find near-optimal solutions. A deep understanding of the impact of randomness on the models and algorithms is an interesting research direction. Relatively a few number of studies have considered robust planning approach in urban transit system with capacity constraints. To remedy this shortfall, a robust timetable or operation plan can be produced to absorb minor disturbances. The design of such a timetable that guarantees expected passenger traveling time is an important future research field.

CONCLUSION

Railway disruption management refers to the scientific efforts that consider the impact of disruptions in modeling and seeks to optimize control strategies in order to guarantee the best possible service for the passengers. The distinctive features of this problem are the inherent dynamic and stochastic nature of passenger demands and train operations. A few scientific works exist where the combination of short-turning and stop-skipping strategies in a dynamic and stochastic nature are addressed.

In this chapter, a simulation-optimization approach is presented for managing disruption, and to distribute transportation supply in an efficient manner in the metro lines. For this purpose, single and combined short-turning and stop-skipping control strategies are optimized through a discrete event simulation system which adopts a variable neighborhood search method. It was assumed that a part of the inbound route was blocked temporary and the train flow is subjected to the uncertainty in disruption duration. In order to solve the model, a solution framework is presented to obtain alternative solutions in reasonable time that are helpful in management of disorders. This is a great advantage in comparison to the traditional manual re-scheduling process in rail planning. The proposed disruption recovery model was tested on realistic instances of Tehran metro. A quantitative performance analysis was presented for comparing the proposed solution method with the state of the art simulation-based optimization algorithms embedded in OptQuest package. The results indicate that the single control strategies (short-turning with station-skipping) outperformed the combined control strategy with respect to time limitation. Furthermore, outcomes from the model application suggested that single optimized control strategies which minimize the average waiting time are adequately effective to be implemented by rail transit practitioners. The proposed VNS algorithm could efficiently handle the passenger waiting time through deciding the short-turn and stop-skip variables where uncertainty in disruption duration exists.

REFERENCES

Cadarso, L., Marín, Á., & Maróti, G. (2013). Recovery of disruptions in rapid transit networks. *Transportation Research Part E, Logistics and Transportation Review*, *53*, 15–33. doi:10.1016/j.tre.2013.01.013

Canca, D., Barrena, E., Laporte, G., & Ortega, F. A. (2014). A short-turning policy for the management of demand disruptions in rapid transit systems. *Annals of Operations Research*, 2014, 1–22.

Canca, D., Barrena, E., Zarzo, A., Ortega, F., & Algaba, E. (2012). Optimal train reallocation strategies under service disruptions. *Procedia: Social and Behavioral Sciences*, *54*, 402–413. doi:10.1016/j.sbspro.2012.09.759

Cao, Z., Yuan, Z., & Li, D. (2014). Estimation method for a skip-stop operation strategy for urban rail transit in China. *Journal of Modern Transportation*, *22*(3), 174–182. doi:10.1007/s40534-014-0059-6

Carbone, A., Papa, F., & Sacco, N. (2012). An Optimization Approach for Delay Recovery in Urban Metro Transportation Systems.

Carrel, A., Mishalani, R. G., Wilson, N. H., Attanucci, J. P., & Rahbee, A. B. (2010). Decision Factors in Service Control on High-Frequency Metro Line. *Transportation Research Record: Journal of the Transportation Research Board*, *2146*(1), 52–59. doi:10.3141/2146-07

Chen, M., & Niu, H. (2009). Modeling Transit Scheduling Problem with Short-Turn Strategy For A Congested Public Bus Line. *Paper presented at the Logistics@ sThe Emerging Frontiers of Transportation and Development in China*. doi:10.1061/40996(330)631

Cortés, C. E., Jara-Díaz, S., & Tirachini, A. (2011). Integrating short turning and deadheading in the optimization of transit services. *Transportation Research Part A, Policy and Practice*, *45*(5), 419–434. doi:10.1016/j.tra.2011.02.002

Deckoff, A. A. (1990). *The short-turn as a real time transit operating strategy*. Massachusetts Institute of Technology.

Delle Site, P., & Filippi, F. (1998). Service optimization for bus corridors with short-turn strategies and variable vehicle size. *Transportation Research Part A, Policy and Practice, 32*(1), 19–38. doi:10.1016/S0965-8564(97)00016-5

Ding, Y., & Chien, S. I. (2001). Improving transit service quality and headway regularity with real-time control. *Transportation Research Record: Journal of the Transportation Research Board, 1760*(1), 161–170. doi:10.3141/1760-21

Eberlein, X. J., Wilson, N. H., Barnhart, C., & Bernstein, D. (1998). The real-time deadheading problem in transit operations control. *Transportation Research Part B: Methodological, 32*(2), 77–100. doi:10.1016/S0191-2615(97)00013-1

Eberlein, X. J., Wilson, N. H., & Bernstein, D. (1999). *Modeling real-time control strategies in public transit operations Computer-aided transit scheduling* (pp. 325–346). Springer.

Eberlein, X. J., Wilson, N. H., & Bernstein, D. (2001). The Holding Problem with Real–Time Information Available. *Transportation Science, 35*(1), 1–18. doi:10.1287/trsc.35.1.1.10143

Flamini, M., & Pacciarelli, D. (2008). Real time management of a metro rail terminus. *European Journal of Operational Research, 189*(3), 746–761. doi:10.1016/j.ejor.2006.09.098

Furth, P. (1987). Short turning on transit routes. *Transportation Research Record, 1987*, 1108.

Ghoneim, N., & Wirasinghe, S. (1986). Optimum zone structure during peak periods for existing urban rail lines. *Transportation Research Part B: Methodological, 20*(1), 7–18. doi:10.1016/0191-2615(86)90032-9

Grube, P., Núñez, F., & Cipriano, A. (2011). An event-driven simulator for multi-line metro systems and its application to Santiago de Chile metropolitan rail network. *Simulation Modelling Practice and Theory, 19*(1), 393–405. doi:10.1016/j.simpat.2010.07.012

Hansen, I., & Pachl, J. (2008). *Railway timetable and traffic: analysis, modelling, simulation*. Hamburg, Germany: Eurailpress.

Hansen, P., & Mladenović, N. (2001). Variable neighborhood search: Principles and applications. *European Journal of Operational Research, 130*(3), 449–467. doi:10.1016/S0377-2217(00)00100-4

Hansen, P., Mladenović, N., & Pérez, J. A. M. (2010). Variable neighbourhood search: Methods and applications. *Annals of Operations Research, 175*(1), 367–407. doi:10.1007/s10479-009-0657-6

Hasannayebi, E., Sajedinejad, A., Mardani, S., & Mohammadi, K. (2012). *An integrated simulation model and evolutionary algorithm for train timetabling problem with considering train stops for praying*. Paper presented at the Simulation Conference (WSC). *Proceedings of the, 2012*(Winter).

Hassannayebi, E., Sajedinejad, A., & Mardani, S. (2014). Urban rail transit planning using a two-stage simulation-based optimization approach. *Simulation Modelling Practice and Theory, 49*, 151–166. doi:10.1016/j.simpat.2014.09.004

Hullinger, D. R. (1999). Taylor enterprise dynamics. *Paper presented at theSimulation Conference.*

Jamili, A., Ghannadpour, S., & Ghorshinezhad, M. (2014). The optimization of train timetable stop-skipping patterns in urban railway operations.

Jia, W., Mao, B., Liu, H., Chen, S., & Ding, Y. (2008). Service robustness analysis of trains by a simulation method. *Paper presented at the Traffic and Transportation Studies.* doi:10.1061/40995(322)70

Kleijnen, J. P., & Wan, J. (2007). Optimization of simulated systems: OptQuest and alternatives. *Simulation Modelling Practice and Theory*, *15*(3), 354–362. doi:10.1016/j.simpat.2006.11.001

Koutsopoulos, H. N., & Wang, Z. (2007). Simulation of urban rail operations: Application framework. *Transportation Research Record: Journal of the Transportation Research Board*, *2006*(1), 84–91. doi:10.3141/2006-10

Laguna, M., & Marti, R. (2002). *The OptQuest callable library Optimization software class libraries* (pp. 193–218). Springer.

Li, S., & Wang, C. (2013). Evaluation of Postponement Strategies in Benetton using OptQuest Simulation. *International Journal of Advancements in Computing Technology*, *5*(7).

Li, Y., & Gendreau, M. (1991). Real time scheduling on a transit bus route: a 0-1 stochastic programming model.

Lin, W., & Sheu, J. (2010). Automatic train regulation for metro lines using dual heuristic dynamic programming. *Proceedings of the Institution of Mechanical Engineers. Part F, Journal of Rail and Rapid Transit*, *224*(1), 15–23. doi:10.1243/09544097JRRT283

Mannino, C., & Mascis, A. (2009). Optimal real-time traffic control in metro stations. *Operations Research*, *57*(4), 1026–1039. doi:10.1287/opre.1080.0642

Mascis, A., & Pacciarelli, D. (2002). Job-shop scheduling with blocking and no-wait constraints. *European Journal of Operational Research*, *143*(3), 498–517. doi:10.1016/S0377-2217(01)00338-1

Mladenović, N., & Hansen, P. (1997). Variable neighborhood search. *Computers & Operations Research*, *24*(11), 1097–1100. doi:10.1016/S0305-0548(97)00031-2

Möller, D. P. (2014). Simulation Tools in Transportation *Introduction to Transportation Analysis. Modelling and Simulation (Anaheim)*, *2014*, 195–228.

Motraghi, A., & Marinov, M. V. (2012). Analysis of urban freight by rail using event based simulation. *Simulation Modelling Practice and Theory*, *25*, 73–89. doi:10.1016/j.simpat.2012.02.009

Nie, L., & Hansen, I. A. (2005). System analysis of train operations and track occupancy at railway stations. *European Journal of Transport and Infrastructure Research*, *5*(1), 31–54.

O'Dell, S. W., & Wilson, N. H. (1999). *Optimal real-time control strategies for rail transit operations during disruptions Computer-aided transit scheduling* (pp. 299–323). Springer.

Paolucci, M., & Pesenti, R. (1999). An object-oriented approach to discrete-event simulation applied to underground railway systems. *Simulation*, *72*(6), 372–383. doi:10.1177/003754979907200601

Pender, B., Currie, G., Delbosc, A., & Shiwakoti, N. (2013). Disruption Recovery in Passenger Railways. *Transportation Research Record: Journal of the Transportation Research Board, 2353*(1), 22–32. doi:10.3141/2353-03

Puong, A. (2001). *A Real-Time Train Holding Model for Rail Transit Systems*. Massachusetts Institute of Technology, Department of Civil and Environmental Engineering.

Quaglietta, E. (2013). A simulation-based approach for the optimal design of signaling block layout in railway networks. *Simulation Modelling Practice and Theory*.

Sajedinejad, A., Mardani, S., Hasannayebi, E., & Mir Mohammadi, K., S., & Kabirian, A. (2011). SIMARAIL: simulation based optimization software for scheduling railway network. *Paper presented at the Simulation Conference (WSC)*. doi:10.1109/WSC.2011.6148066

Sánchez, Á. G., Ortega-Mier, M., & Arranz, R. (2011). Discrete-Event Simulation Models for Assessing Incidents in Railway Systems. *International Journal of Information Systems and Supply Chain Management, 4*(2), 1–14. doi:10.4018/jisscm.2011040101

Shen, S. (2000). *Integrated real-time disruption recovery strategies: a model for rail transit systems*. Massachusetts Institute of Technology.

Shen, S., & Wilson, N. H. (2001). *An optimal integrated real-time disruption control model for rail transit systems Computer-aided scheduling of public transport* (pp. 335–363). Springer. doi:10.1007/978-3-642-56423-9_19

Suhl, L., Mellouli, T., Biederbick, C., & Goecke, J. (2001). *Managing and preventing delays in railway traffic by simulation and optimization Mathematical methods on optimization in transportation systems* (pp. 3–16). Springer. doi:10.1007/978-1-4757-3357-0_1

Taghizadeh, A. O., Soleimani, S. V., & Ardalan, A. (2013). Lessons from a flash flood in Tehran subway, Iran. *PLoS Currents, 2013*, 5.

Tanaka, S., Kumazawa, K., & Koseki, T. (2009). *Passenger flow analysis for train rescheduling and its evaluation.Paper presented at theInternational Symposium on Speed-up, Safety and Service Technology for Railway and Maglev Systems.*

Tirachini, A., Cortés, C. E., & Jara-Díaz, S. R. (2011). Optimal design and benefits of a short turning strategy for a bus corridor. *Transportation, 38*(1), 169–189. doi:10.1007/s11116-010-9287-8

Ulusoy, Y. Y., Chien, S. I.-J., & Wei, C.-H. (2010). Optimal all-stop, short-turn, and express transit services under heterogeneous demand. *Transportation Research Record: Journal of the Transportation Research Board, 2197*(1), 8–18. doi:10.3141/2197-02

Vázquez-Abad, F. J., & Zubieta, L. (2005). Ghost simulation model for the optimization of an urban subway system. *Discrete Event Dynamic Systems, 15*(3), 207–235. doi:10.1007/s10626-005-2865-9

Veelenturf, L. P., Potthoff, D., Huisman, D., Kroon, L. G., Maróti, G., & Wagelmans, A. P. (2012). *A recoverable robust solution approach for real-time railway crew rescheduling: Technical report*. Erasmus University Rotterdam.

Vuchic, V. R. (1973). Skip-stop operation as a method for transit speed increase. *Traffic Quarterly, 27*(2), 307–327.

Wang, Y., De Schutter, B., van den Boom, T., Ning, B., & Tang, T. (2013). Real-time scheduling for single lines in urban rail transit systems. *Paper presented at the 2013 IEEE International Conference on Intelligent Rail Transportation (ICIRT)*. doi:10.1109/ICIRT.2013.6696258

Yu, G., & Qi, X. (2004). *Disruption management*. World Scientific. doi:10.1142/5632

KEY TERMS AND DEFINITIONS

Block Section: It is referred to conventional fixed-block signaling system which divides the railway infrastructure into a number of block sections.

Discrete Event System: It is referred to the system where states are updated at discrete time points (events).

Disruption Management: It refers to the decision support methodologies that can be used in order to return the disrupted status into its planned operation.

Minimum Headway Time: The minimal safety time difference between two successive trains.

Recovery Task: An intervention in real-time planning process is called a recovery task or transposition.

Short Turn: It is referred to shortening of the route of a train before it reaches to the terminal.

Stop Skipping: A control strategy that determines a subset of stations to be visited by a train.

Train Control System: Comprises of train protection systems that automatically authorize the movement of trains safely.

Section 5
Engineering

Chapter 19
Steady State Modeling of Electric Railway Power Supply Systems for Planning and Operation Purposes

Pablo Arboleya
University of Oviedo, Spain

ABSTRACT

This chapter describes the different electric railway power supply systems and their main characteristics from the point of view of the power flow modeling and simulation. It considers the DC traction systems and also the AC ones, explaining the different elements embedded in the network and proposing steady state electrical models of each element. The basic methods for modeling the train behavior in terms of demanded/regenerated power are also detailed. Finally, the procedures to simulate the trains motion into the electrical network and how their motion and power demand affect to the electrical variables will be unraveled, explaining how to merge all the models in a system of equations.

INTRODUCTION

An electric locomotive can be defined as locomotive powered by electricity. This electricity can be obtained from a railway power supply system (RPSS) or from an on-board accumulator like a battery, fuel cell or ultracapacitor. Robert Davidson built the first electric locomotive in 1837 and it used the second option, galvanic type batteries powered it. However, the first electric locomotive using a RPSS date from 1879.Werner von Siemens built it and, it was fed by a 150V DC RPSS connected to the train through a contact roller (Day & McNeil, 1966). Most of the RPSS are based in overhead lines or third rail systems; in this case, the third rail was the chosen option. The traction system employed by Siemens locomotive was based in a series-wound motor of 2.2kW.

DOI: 10.4018/978-1-5225-0084-1.ch019

Steady State Modeling of Electric Railway Power Supply Systems for Planning and Operation Purposes

The fact that electric locomotives were much cleaner than steam ones launched the development of this kind of systems for solving the ventilation problems in tunnels applications. In 1895 a four-mile stretch with multiple tunnels of the Baltimore & Ohio Rail Road was electrified using an overhead distribution system of 675V DC and an electric locomotive built by General Electric with a power output of 4270kW. In 1898, the first electric AC locomotive was inaugurated, it was supplied by a 3000V and 15Hz three-phase RPSS.

The electrification of the railway tracks increased rapidly and by the year 2006, the 25% of all the railway power supply systems were electrified, a total of 240000km and approximately the 50% of all rail transport (Frey, 2012). In present days, countries like India are electrifying lines at a pace of 2000km/year. The drastic increase of the electrified railways was driven basically by the lower running and maintenance cost of electrical locomotives compared with diesel ones and the higher power/weight ratio of the electric locomotives, besides the reduction of noise, pollution and energetic dependence.

To date, many different railway power supply systems with different voltages, frequencies and constructive characteristics have been developed. In Table 1, the main electrification systems are represented.

Both AC and DC systems are quite extended all around the world. For planning and operating these kinds of systems, it is necessary the development of accurate models for obtaining voltage profiles, current and power flows through the lines, trains, substations and the rest of the devices embedded in the RPSS (Falvo, Lamedica, Bartoni, & Maranzano, 2011). For this purpose the power flow analyses has been the most used tool in DC systems (Arboleya, Diaz, & Coto, 2012) and also in AC systems (Pilo, Rouco, Fernandez, & Abrahamsson, 2012). It is crucial then, to know exactly how to model the topology and the devices installed in the systems as well as the trains. This chapter will be focused in the electrical network analysis. A brief idea of how the train behavior could be modeled and implemented presenting simplified models of the traction, braking systems, but a detailed mechanical model of the train will not be explained.

The present chapter is structured as follows, in the next section, the basic feeding circuits will be presented, and in this case no distinction will be made between AC and DC because the same topology could be applied to different systems. Then the AC railway power supply systems will be described, especially those based in 25kV, 50Hz and 2x25kV, 50Hz explaining the most typical configurations, feeding systems and embedded devices. Next the DC traction systems will be described. After that, a brief introduction to train simulation and modeling will be done, and some methods for positioning moving loads in electrical networks for power flow purposes will be explained emphasizing one of them based

Table 1. Rated, highest and lowest voltages in most common railway power supply systems

	Rated Voltage	Lowest Non-Permanent Voltage	Lowest Permanent Voltage	Highest Permanent Voltage	Highest Non-Permanent Voltage
DC	750	500V	500V	900V	1kV
	1500	1000V	1000V	1800V	1950V
	3000	2000V	2000V	3600V	4000V
AC	15kV, 16.7Hz	11kV	12kV	17.25kV	18kV
	25kV, 50Hz	17.5kV	19kV	27.5kV	29kV
	2x25kV, 50Hz	42kV	45kV	55kV	58kV

in graph theory. In the next section, the electrical models of the railway power supply systems elements like electrical lines, power transformers, autotransformers and rectifiers and converters will be presented. Finally the models of the main DC and AC systems will be described.

GENERAL FEEDING SCHEMES

Electric traction power supply systems are divided in electrical sections; one or more substations or a dedicated feeder supplies each electrical section. The main characteristic of the electrical sections is that they can be easily isolated from the rest of the system in case of failure or permanent short circuit. In such cases all the traction system except the isolated section can be normally operated until the problem is solved. The different electrical sections use to be isolated from other sections by means of neutral sections; these neutral sections prevent successive substation supplied electrical sections with different voltages being connected. However, for operational purposes, two different sections can be connected through track sectioning stations by means of section insulators or insulated overlapping sections. In case of more than one track, they can be connected through track paralleling stations. Sometimes, track sectioning stations and track paralleling stations are in the same cabin, in such cases they are called track sectioning and paralleling points.

In the Figure 1, four different feeding schemes are represented, for this example a double track line has been considered and for the sake of simplicity it has been only represented the contact line schemes without the returning circuit. These schemes are the most representative, and any other scheme could be obtained as a combination of the represented ones. Hill (1994) and Kiessling et al. (2009) presented a detailed analysis of all possible configurations.

In Figure 1 a), the single-end feed is represented. In this specific case, all sections are electrically independent because between successive sections the contact line is isolated. Each substation has two different incoming supplies but due to the normally open isolator each incoming supply feeds just one section. In each substation, the normally closed isolators produce a cross coupling and both tracks are connected to the same electrical point. In case of failure in one of the tracks, the fault could be easily isolated because there is one circuit breaker per track and section.

Figure 1 b) represents a double-end feed with longitudinal coupling. In this case, all sections are electrically connected. The two tracks of sections 1 and 2 are cross-coupled in substation 1 and the two tracks of sections 3 and 4 are cross-coupled in substation 2. In addition, there is a sectioning point allowing the longitudinal coupling of both tracks of sections 2 and 3 independently. Due to the longitudinal coupling, the voltage drop in this kind of scheme is lower than the previous one, because both substations feed the trains simultaneously. It must be pointed out that the power flow in the sections is not unidirectional like the single end feed scheme. This scheme cannot be applied when two adjacent substations are supplied by voltages with different magnitude or angle. For power flow analysis, these kind of double-feed systems are more complex because unlike the single-end feed systems, they could create meshed networks with quite complex topologies.

Another double-end feed system is represented in Figure 1 c), in this example, sections 2 and 3 are electrically isolated due to the normally open isolator, however a longitudinal coupling could be an option switching the mentioned isolators. Both tracks of the same sections are cross-coupled at substation level and also at paralleling point level. In this scheme, the two incoming supplies of each substation feeds simultaneously the two sections connected to the same substation. A similar feeding system is

Figure 1. Representation of most common feeding schemes in railway power supply systems

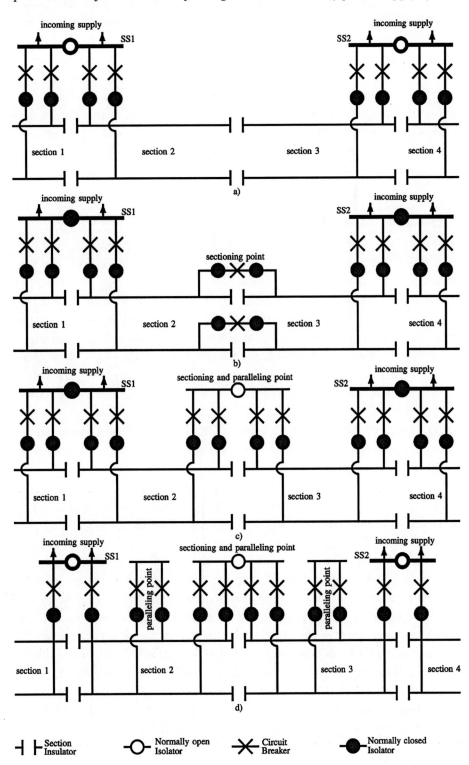

represented in Figure 1 d), it is a double-feed system with a deactivated longitudinal coupling, but in this case there are several cross coupling along the tracks. This configuration reduces also the voltage drop in the contact lines because the train in one track can be fed by the contact lines of both tracks. Other substantial difference can be found at substation level. In this example, there is just on circuit breaker per section meaning that the cross coupling can not be deactivated at substation level, if a fault occurs in one of the tracks, both of them must be disconnected.

The possibility of adding reinforcing feeder lines in parallel with the contact lines to reduce the voltage drop must be also commented. These reinforcing feeders are connected to the same substations and coupled with the contact line at specific intervals. The reinforcement can be done through all the length of the section or just near the substations.

AC TRACTION SYSTEMS

The most common AC feeding system is the one based in 25kV and 50Hz, except in countries like Germany or Sweden where the most used system is based on 15kV and 16.7Hz. The two traction networks are similar but the way in which these systems are fed is completely different. While the 25kV, 50Hz systems are connected to the transport or the distribution network using power transformers, the 15kV 16.7Hz systems are fed through their own single-phase transmission system connected to dedicated power plants or to the transmission system through frequency converter stations. These converter stations can be central converter stations or decentralized converter stations depending on their sizes. Central converter stations supply energy to several substations while decentralized converter stations supply energy directly to the traction system. The conversion from 50Hz to 16,7Hz can be done by means of rotating converter stations or static converter stations. The former use a three-phase motor connected to the 50Hz system coupled mechanically with a synchronous single-phase machine that produces power at 16.7Hz, the latter use power converters.

Depending on the selected railway power supply system, different feeding schemes can be used. For instance, using a 15kV, 16.7Hz system, where all sections are fed with the same voltage, the traction network could be operated with the longitudinal coupling between sections. The longitudinal coupling could be used also in 25kV, 50Hz systems, when all the substation transformers are fed using the same voltage angle. However, in some cases, for balancing the voltages in the distribution or transmission network feeding the AC system, different substations are fed with different phases. In such cases the use of single-end schemes in the traction network without longitudinal coupling is necessary in order to avoid short circuit currents caused by different voltages in adjacent sections. In the work of Kneschke (1985), a comprehensive analysis of special transformers applied for voltage unbalance compensation is carried out considering among others, Scott-connected, modified Woodbridge-connected, LeBlanc-connected, Wye-Delta symmetrical-connected and Wye-Wye asymmetrical-connected power transformers. In this chapter conventional power transformers will be employed, however the guidelines supplied for developing the models using conventional power transformers can be applied to other more complex devices.

In the Figure 2, four typical configurations are represented, all of them are supposed to be 25kV and 50Hz systems, and the train consumes 400A in all cases. In this paragraph the different feeding and returning paths of the current consumed by the train will be explained. The simplest configuration is the one depicted in Figure 2 a), it is known as direct connection and it uses a conventional single-phase power transformer. The primary winding connected to two of the transmission or distribution system phases, and the secondary winding connected directly to the contact line and the rails. The main advantage of this solution is its simplicity and its low cost. However, it presents serious drawbacks. The rails are the main returning path (in this case 300A returns through the rails). The current through the rails produce a big amount of electrical losses due to their high impedance. The appearance of dangerous touch potentials in the rails is another serious disadvantage. In addition, the rails are not isolated from ground, so there is a non-negligible amount of stray current (100 A), generating interferences with telecommunication systems and corroding nearby metallic elements. The scheme represented in Figure 2 b) partially solves this problem. In this case, booster transformers are installed between the different isolated sections of contact lines and rails. The booster transformers are placed at specific intervals and they are transformers with unitary turn ratio forcing the return current to circulate through the rails and eliminating the stray currents. In this situation, in the point where the train is connected with the rails, part of the current (300A) returns through the rail and the other part (100A) is stray current. However, the booster transformer forces the stray current to return to the rails and circulate through its secondary winding. This solution is more expensive, it solves the problem of the interferences but the losses and touch potentials in the rails are still high. The use of a return conductor as it is represented in Figure 2 c). The return conductor reduces the return impedance and the stray currents because most of the current will return through it. The installation of the return conductor can be combined with the installation of

Figure 2. AC 25kV feeding systems

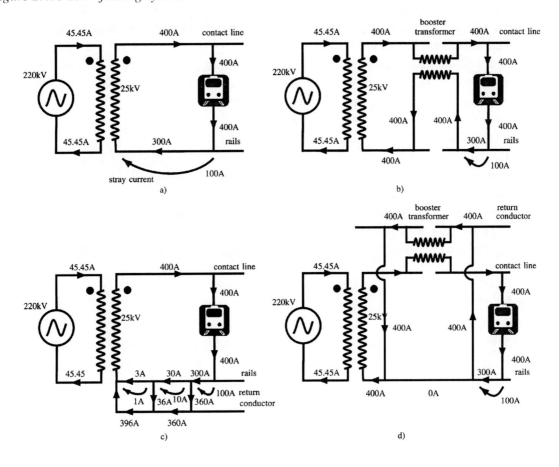

booster transformers as it is shown in Figure 2d), in such case, the booster transformers force the return current to flow through the return conductor avoiding the existence of stray currents and eliminating the telecommunication interference problems. The dangerous touch potentials are also reduced because the returning path impedance is also reduced due to the return conductor.

The 2x25kV and 50Hz bi-level scheme represented in Figure 3. In this configuration the secondary windings of the power transformers placed at the substations are equipped with a central tap connected to the rails, the other two poles are connected to the contact line and the so called negative feeder. The voltage between the contact line and the rails is 25kV and the voltage between the negative feeder and the rails is -25kV (Raygani, Tahavorgar, Fazel, & Moaveni, 2012). Even when the voltage between the contact line and the negative feeder is 50kV, the trains are fed at 25kV, so the same train could be used in a 25kV standard system or in a 2x25kV system. The most distinctive feature of this scheme is the use of autotransformers with a unity turn ratio, connected to overhead conductor, the rails and the negative feeder at specific intervals. The connection of the autotransformers divide the electrical section fed by one substation in different cells. In Figure 3 just two cells are represented, the first one between the substation and the first autotransformer and the second one between the two autotransformers. In this specific case of study the train is located in the second cell. Like in the previous examples, the train is consuming a current $I = 400A$, part of this current aI is supplied by the first autotransformer and the other part $(1-a)I$ is supplied by the second one. The value of a, depends on the location of the train in the cell. In this case a value of $a = 0.625$ has been assumed, so the first autotransformer supply 250A and the second one 150A. The return current flows through the rails from the train to the nearest autotransformers, then it is split in two, and the current flowing through the rails in the first cell is zero. This is one of the advantages of this scheme; the current through the rails is reduced, so the rails potentials and the losses are reduced too. The bi-level scheme reduces also the stray currents generating a lower level of interference with the telecommunication systems. In addition, it must be remarked that the current flowing from the second cell (the one with the train) to the power transformer through the contact line and the negative feeder is half of the current consumed by the train (in this case 200A). The feeding and return system in empty cells is a 50kV system instead a 25kV one, so the transport capacity of the line is increased and the distance between adjacent substations can be increased too. Summarizing, the current flow in this scheme can be described with five different loops, two of them in the first cell (the empty one) with a current of $I/2$. The other three current loops in the cell with the train (the second one), two of them are $(1-a)I$ current loops, and the third one an aI current loop.

Bi-level systems are not restricted to 25kV systems, they can be extended to other AC traction networks like 15kV networks obtaining a 2x15kV network. Bi-voltage systems can be also employed in DC traction systems using the same scheme, but replacing the autotransformers by DC/DC controlled converters.

DC TRACTION SYSTEMS

DC traction systems are connected to the AC distribution system through a power transformer and AC/DC rectifier equipment. Two common configurations are the six pulses pulses and twelve pulses non-controlled rectifiers. In Figure 4, both configurations are represented. In the work of Tzeng, Chen and Wu (1995, 1998), and Pozzobon (1998), a detailed analysis of these topologies is carried out. There exist also more complex AC/DC conversion topologies based on the use of other transformers configurations for rectifiers with more than 6 pulses and controlled semiconductor devices like thyristors or even IGBT

Figure 3. AC 2x25kV feeding systems

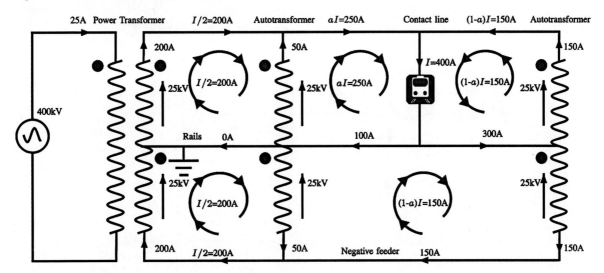

based bridges. The use of controlled devices is necessary when the substations must work in reversible way injecting power from the DC traction systems into the AC network. This phenomenon occurs when trains with regenerative braking circulate through the DC traction system. In such case, if the regenerated power of one braking train is not consumed by another near unit in traction mode, the contact line voltage will increase. When the voltage at a substation is higher than a given threshold, the controlled rectifier is activated injecting power from the DC system into the AC one. In such cases the most usual solution is the installation of the controlled rectifier in parallel to the non controlled one, so the non controlled is used for fed the DC system from the AC one, and the controlled rectifier is used for reversible power flows. This configuration is studied in (Tzeng, Wu & Chen, 1998) using a thyristor-controlled rectifier in parallel with a non-controlled diode bridge.

In DC traction networks, is possible to operate the system using the single-end feeding scheme, but in most of the cases the operation in double-end feeding mode is also feasible. This feature increases the complexity of the power flow analysis due to the meshes that appear in the electrical circuit. From the point of view of the mathematical problem, the radial configuration derived from the single-end feeding scheme is much more simple even when the voltage drops are higher using this configuration.

The grounding system in DC traction networks can be classified into three basic types. The first one uses a solid grounded connection of the rectifier negative pole. This system avoids the appearance of high voltage potentials in the rails, however a non-negligible percentage of the returning current is stray current generating interferences in the telecommunication systems and corrosion in the nearby metallic pipelines. One solution to overcome this drawback is the use of isolated negative feeder. However, in this case, dangerous potentials can arise in the rails. The third group of grounding systems uses a system that can connect or disconnect from the ground the rectifier negative pole. Using this system, under normal operation, when the potential in the rail is reduced the rectifier negative pole is isolated and the stray currents are negligible. When a dangerous potential arises, the system is grounded for safety purposes and stray currents come into the stage. The connection or disconnection of the grounding system can be done using contactors or semiconductor devices. In Paul (2002) a deep explanation of this kind of grounding systems can be found.

Figure 4. Typical AC/DC conversion systems for DC traction systems

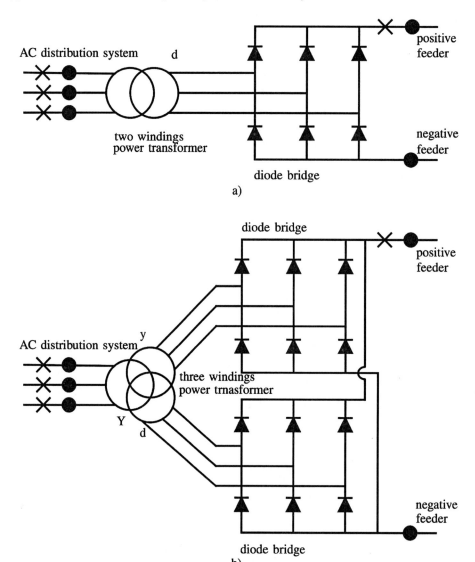

TRAIN MODELING AND INTERACTION WITH THE TRACTION SYSTEM

The analysis of the electrical trains over the traction networks is crucial in the design stage of network infrastructures but also in the operation of these infrastructures. For these reasons, many researchers proposed several techniques for analyze the interaction between trains and the electrical networks.

There some assumptions that are generally made and accepted for most of the authors, those are:

1. The trains have no overcurrent protection.
2. The trains have not squeeze control.
3. There are no voltage catenary constraints.

Following these assumptions, the train model and the network model can be decoupled, this means that we would be able to simulate the trains separately from the network that is what most of the researchers do until the date, see for instance (Coto, Arboleya, & Gonzalez-Moran, 2013) and (Cai, Irving & Case, 1995).

When the train modeling and the network modeling are decoupled, it is possible to estimate some effects that the train behavior produce in the network such us voltage increase in the catenary during braking, reversible power flows in substations equipped with bidirectional devices, voltage drops in the catenaries when the trains are in traction mode, etc. (Ho et al., 2012), (Mariscotty, Pozzobon, & Vanti, 2007) and (Pires, Nabeta, & Cardoso, 2007, 2009). However the effect of the network over the train is not considered at all. This problem can be neglected in the case of the braking mode, because if the train is not able to inject in the network all the regenerated energy, the surplus would be burned using the rheostatic braking system, and thus, the required braking force would be applied to the train even under network saturation conditions, or even if the whole electric braking system fails, the braking force could be provided by the back-up mechanical braking system. This means that the deviation between the required electrical braking force and the real provided braking force could be neglected. The same behavior cannot be assumed when the trains are in traction mode. Under these circumstances, if the demanded electrical power is too high, the voltage drop could trigger the train overcurrent protection and thus, the real absorbed power from the grid would be lower than the demanded. In this particular scenario, the acceleration of the train will also differs from the scheduled and also the speed and the location of the trains in the successive instants, creating a mismatch between the scheduled positions and the real ones. In the work of Chymera, Renfrew, Barnes and Holden (2010), the interaction between the train and the electrical network is considered.

To study the effect of the network congestion over the schedule deviation, the train model and the network model must be coupled and the above-mentioned assumptions must be withdrawn considering:

1. The trains torque limitation constraints: The maximum traction force reference that the train control can demand from the electric motor will be limited.
2. The trains overcurrent protection: When the voltage in the network is lower than a given value, the demanded power reference will be linearly diminished reaching zero if the voltage is lower than the minimum catenary voltage.
3. The trains squeeze control: If during the braking process the regenerated power rise the catenary voltage over a given value, the power is linearly diminished until the maximum catenary voltage is reached. Above this point the injected power will be zero and all regenerated power will be burned using the rheostatic system.
4. The catenary voltage constraints: This last constraint is implicit in the squeeze control constraint and the overcurrent protection, because when the train is in traction mode and the squeeze control is implemented, the regenerated power injected into the grid is strangulated before the maximum catenary voltage constraint is violated. A similar situation occurs during the traction mode if the overcurrent protection is implemented.

Train Mechanical Model

Under the above-mentioned assumptions, the train will be modeled as follows (Pilo, 2003), (Hong, Lee, & Kwak 2007):

$$m\frac{ds(t)}{dt} = F_t - F_w - F_c - F_a \tag{1}$$

where:

m is the mass of the train in kg.
$s(t)$ is the speed of the train in m/s.
F_t is the traction, braking or regulation force in N.
F_w is the force caused by the weight of the train in N.
F_c is the force caused by the track curvature in N.
F_a is the Aerodynamic force in N.

The different forces involved in the Equation (1) can be calculated as:

$$F_w = mg \sin(\theta) \tag{2}$$

$$F_c = m\frac{D}{1000r} \tag{3}$$

$$F_a = k\alpha s2(t) + \beta s(t) + \gamma \tag{4}$$

where:

g is the gravity force in N/ms².
θ is the slope of the track in degrees.
D is an empiric constant that usually varies from 500 to 1200.
r is the curvature radius of the track in m.
α, β, γ and k are aerodynamic coefficients; k is used for incrementing the aerodynamic force inside the tunnels.

With the previous expressions, a compact train model considering the speed and the location can be expressed as:

$$\frac{d}{dt}d(t) = s(t) \tag{5}$$

$$\frac{d}{dt}s(t) = \alpha' s^2(t) + \beta' s(t) + \gamma' + \frac{F_t}{m} \tag{6}$$

where $d(t)$ represents the location of the train and α', β' and γ' and are location dependent parameters.

Train Electrical Equipment Model

The electrical machines and the train power electronics considering the squeeze and overcurrent control is going to be called train electrical equipment and it can be formulated as follows.

$$p^*_{mech}(t) = F^*_t(t) s(t) \tag{7}$$

$$p^*_{elec}(t) = \mu(p^*_{mech}(t), s(t)) p^*_{mech}(t) \tag{8}$$

$$p_{elec}(t) = f(V(t), p^*_{elec}(t)) \tag{9}$$

$$p_{mech}(t) = g(p^*_{mech}(t), p_{elec}(t)) \tag{10}$$

$$F_t(t) = p_{mech}(t)/s(t) \tag{11}$$

where p_{elec} and p_{mech} are respectively the electrical and the mechanical power. The symbol * next to a variable represents a reference. μ and the function f represent the efficiency of the electromechanical conversion and the simulation of the squeeze and overcurrent controls. The function f is represented in Figure 5.

From the train control model described in the previous subsection, the traction force reference is going to be extracted. Using this reference and the actual speed of the train, the mechanical power reference using expression (7) can be obtained. Using the function μ that calculates the efficiency of the electromechanical conversion, the electrical power reference will be calculated using the expression (8). Obviously when the train is in traction mode ($p^*_{mech}>0$), this function will give us values higher than one, meaning than the electrical power requested by the electrical equipment will be higher than

Figure 5. Function f representing the overcurrent and the squeeze control of the train

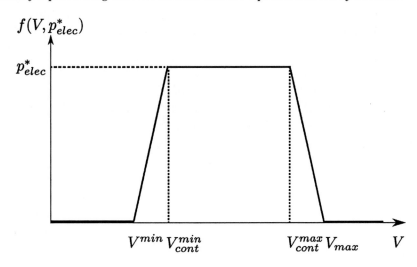

the mechanical power demanded by the train, the difference between them will be the losses of the electromechanical conversion. In the opposite way, if the train is in braking mode, the μ function will return values lower than one. It means that the electrical equipment will try to inject in the network the available mechanical power for regenerating purposes minus the losses of the electromechanical system.

Once the electrical power reference is obtained, the actual electrical power can be calculated using the expression (9). Analyzing such expression and the Figure 5, it can be observed that when the catenary voltage is within the limits, in the interval (V_{cont}^{min}, V_{cont}^{max}), the electrical power will be equal to the reference. However, if the train tries to inject an electrical power in the catenary making the voltage increase above V_{cont}^{max} the squeeze control diminish the injected voltage making it zero if the voltage reaches the maximum catenary voltage V_{max}. If the catenary voltage is too low, the requested electrical power will be reduced, this overcurrent protection works in both traction and braking mode.

The function f can be analytically expressed as follows:

$$f(V(t), p_{elec}^*(t)) = \begin{cases} 0 & V < V_{min} \\ p_{elec}^* \dfrac{V - V_{min}}{V_{cont}^{min} - V_{min}} & V_{min} < V < V_{cont}^{min} \\ p_{elec}^* & V_{cont}^{min} < V < V_{cont}^{max} \\ p_{elec}^* \dfrac{V_{max} - V}{V_{max} - V_{cont}^{max}} & V_{cont}^{max} < V < V_{max} \\ 0 & V > V_{max} \end{cases} \qquad (12)$$

The next step will be the calculation of the actual mechanical power as a function of the electrical power. This part is complicated because, if the train is in traction mode, the mechanical power can be calculated using the actual electrical power, but if the train is in braking mode and the squeeze control is active, the train will reduce the injected electrical power into the network. However there exist back up systems like the rheostatic system or the pneumatic system that provides always the necessary mechanical braking power. This can be modeled using the expression (10). The function g that calculates the actual mechanical power using the above-explained model can be expressed as:

$$g(p_{mech}^*(t), p_{elec}(t)) = \begin{cases} \mu^{-1} p_{elec}(t) & p_{mech}^*(t) > 0 \\ p_{mech}^*(t) & p_{mech}^*(t) < 0 \end{cases} \qquad (13)$$

Finally, the actual traction force can be calculated using the expression (11).

Train Control Model

In this subsection a very basic control of the train position will be described. Basically, the reference position of the train is given by the railway system schedule, any deviation will be corrected increasing or decreasing the speed of the train. For this first approach, two control loops will be implemented, and outer position loop and an inner speed loop as it can be observed in Figure 6. The output of the speed loop is considered as the traction force reference. A more complex control could be considered, adding

Figure 6. Train control model diagram

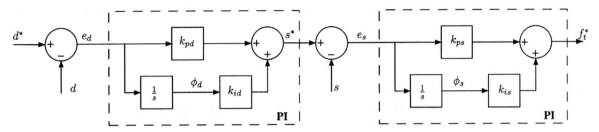

for instance two more inner loops, one based in acceleration and another one based on jerk. With this control, the correction of the train position after the effect of the overcurrent protection could be modeled. If the decoupled approach is applied, the train and the electrical network are modeled in a separate way; the references and the actual magnitudes would always be the same so this module wouldn't be necessary.

This control can be represented with a set of 4 equations per train:

$$\frac{d}{dt}\phi_d(t) = e_d(t) = d^*(t) - d(t) \tag{14}$$

$$\frac{d}{dt}\phi_s(t) = e_s(t) = s^*(t) - s(t) \tag{15}$$

$$F_t^*(t) = k_{ps}\left(s^*(t) - s(t)\right) + k_{is}\phi_s(t) \tag{16}$$

$$s^*(t) = k_{pd}\left(d^*(t) - d(t)\right) + k_{id}\phi_d(t) \tag{17}$$

where k_{pd}, k_{ps}, k_{id} and k_{is} are respectively the proportional and integral position and speed gains in the basic train control system depicted in Figure 6.

TRAINS ALLOCATION PROCEDURES AND NETWORK DESCRIPTION

The electrical analysis of the traction network can be very complicated, not only due to the complexity of the electrical model itself, but also due to the existence of moving loads. The number of trains can vary during the study time interval, creating a set of equations that vary its dimension during the simulation. Even in the case of constant number of trains during the simulation interval, the changes in their relative positions makes very difficult to deal with the system of equations and traditionally these problems were overcome stating one independent problem at each simulation step. This is not a problem when the system is small (a few number of substations and trains), but for huge systems and large time intervals, the problem can become unmanageable. Traditionally, a procedure for determining which trains are in the system and their position must be developed and a new topology must be defined at each simulation step. At each time instant, the relative position of the nodes and the connections between them can vary.

Because of this, if the same enumeration criteria are applied to two different instants, the same line, substation or train can be labeled with different indices. Thus, to compare the same element at different instant, a tracking subroutine must be developed and implemented. The solution vectors at two different instants can have different size, and the position of a specific variable (current or voltage) can vary from one instant to the other.

Some authors worked specifically in this problem, for instance Abrahamson & Söder (2012), modeled the train movement using a discretized model having a fix number of stationary nodes. A special routine was developed to distribute the train load to the two adjacent power system fix load nodes. In this case the real system is substituted by an equivalent fix nodes system. The solution that is going to be explained in this section is the one proposed by Arboleya, Diaz & Coto (2012) based on the use of graph theory.

Using the graph theory approach, the system can be considered as a system with fixed number of trains and fixed topology even when the real number of trains in the system varies and their relative position too. This method considers the substation and the trains as vertices of a graph and the contact lines connecting the nodes as edges. During the considered simulation time, the number of trains will be considered constant. If a train comes out of the stage, it will be deactivated but it will be still considered in the topology, as it will be explained in the next paragraphs. Regarding the enumeration process, the trains will be enumerated first and then the rest of the substations (this is an arbitrary criteria, and enumerating first the substations and then the trains will produce the same result). Using this criterion a given node (train or substation) will always have the same index independently of the simulation step. Nevertheless, the train motion still causes changes in the system topology. To construct an invariant system, in which all the edges connecting trains or trains and substations are invariant too, the graph representing the whole system will be generated as the union of three different sub-graphs.

Figure 7 represents these three sub-graphs set for a traction system with three trains and four DC nodes, three of them substations. The first sub-graph is depicted in Figure 7 a), it represents the real traction network topology with the real connections between the traction network nodes which are not trains nodes 4 to 7. The second sub-graph is represented in Figure 7 b), in this sub-graph all possible connections between trains and the rest of the traction network nodes (from now on substation nodes) are considered, the number of edges of this graph can be obtained multiplying the number of trains n_t by the number of substations n_s. Finally the third sub-graph, represented on Figure 7 c) represents all possible connections between trains, the number of edges in this case can be calculated with the expression $n_t(n_t-1)/2$. The enumeration criterion for the edges will first enumerate the edges whose tail (lower indexed node) is vertex 1 following an ascending order as a function of their head (higher indexed node), then edges whose tail is vertex 2 and so on. With this method a direction is also assigned to each edge. Once all the edges and nodes are enumerated, the node incident matrix Γ, containing all the information about the graph can be built as follows:

$\Gamma ij = 1$ when the tail of the edge i, is vertex j.
$\Gamma ij = -1$ when the head of the edge i, is vertex j.
$\Gamma ij = 0$ otherwise.

Using this definition, the Γ matrix will contain as many rows as edges and as many columns as nodes. Each row will represent an edge and each column will represent a node. Each row representing an edge will have a 1 in the column representing its tail node and a -1 in the column representing its head node; the rest of the positions will be zero.

Figure 7. Sub-graphs describing the whole traction system

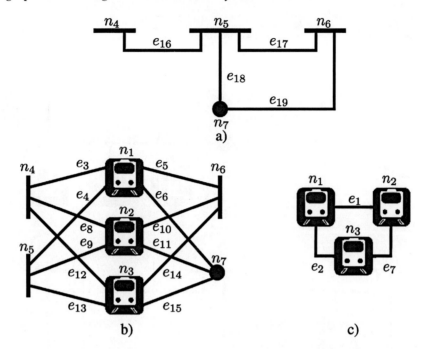

The different scenarios representing different relative positions between the nodes will be achieved activating or deactivating the edges. For instance in Figure 8, two different scenarios are depicted. In the first one Figure 8 a), the three trains are located between substations 6 and 7, this means that the real edge connecting these substations e_{19} (Figure 8 a)) is deactivated and substituted by edges e_6, e_2, e_7 and e_{10}, the other edges belonging to the real topology e_{16}, e_{17} and e_{18} are active edges because there are no trains in such lines. In the second scenario (Figure 8 b)), there are trains in all lines except the one connecting nodes 5 and 6 (e_{17}). In this case, as it can be observed in the figure the edge e_{19} is deactivated and substituted by e_{10} and e_{11}, the edge e_{18} is deactivated too and substituted by the edges e_{13} and e_{15}. Finally, the edge e_{16} is deactivated and substituted by the edges e_3 and e_4. In successive sections it will be explained how and edge is activated or deactivated depending on its assigned impedance. Regarding the non-active trains, they will be placed next to the substation with the lower index and the power assigned to these trains will be zero.

TRACTION SYSTEMS DEVICES MODELS

In this section, different useful models of the devices embedded in the traction networks will be presented, these devices will be added to the traction network models presented in the next sections. Before start with the description of the models, the *per unit* system of calculation will be explained.

Figure 8. Two different scenarios in the same traction system

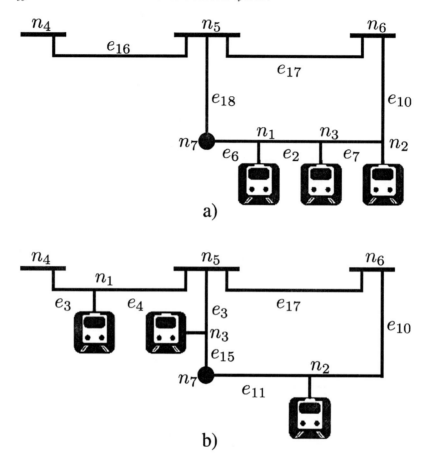

a)

b)

Per Unit System

This method is a very common method based on the definition of a base magnitude of each voltage, current, impedance and power. Then, the real values are expressed as a percentage or per unit with respect to the base, as it can be observed in the expression (18).

$$X_{pu} = \frac{X_{real}}{X_{base}} \tag{18}$$

where X represents a given variable, and the sub-indices *pu*, *real* and *base* represent respectively the per unit value, the real value and the base value. For each voltage level, a voltage base will be chosen. In case of the three-phase system the most common choice is the line-to-line voltage. In case of single-phase systems or DC systems the voltage between overhead line and the return path will be the base voltage. The base power will be common for all voltage levels and in combined AC/DC system will be common also for all AC and DC subsystems. The base current can be calculated for three-phase systems as follows:

$$I_{base} = \frac{S_{base}}{\sqrt{3}U_{base}} \tag{19}$$

where the variables S, V, and I represents respectively the apparent power, the voltage and the current. In case of single phase or DC systems the base current can be computed using expression (20).

$$I_{base} = \frac{S_{base}}{U_{base}} \tag{20}$$

with the expression (21), the base impedance can be calculated for three-phase systems and also for single-phase and DC systems.

$$Z_{base} = \frac{U_{base}^2}{S_{base}} \tag{21}$$

with the use of the per unit system, the comparison of voltage drops and load level between different parts of the system is easier and the values are equalized, so the probability of obtaining ill-conditioned matrices during the calculation process is reduced.

Single-Phase Line Model

From now on, each sine wave $x(t)$, with time invariant amplitude and angular frequency ω, and its time derivative $dx(t)/dt$ will be represented by two phasors $\bar{X}$ and $\dot{\bar{X}}$ respectively as shown in (22) and (23):

$$\bar{X} = X_r + jX_i \tag{22}$$

$$\dot{\bar{X}} = -\omega X_i + j\omega X_r = j\omega \bar{X} \tag{23}$$

where the sub-index i and r represent the real and the imaginary part of the phasor. The equation that will be used to describe a short AC line will be:

$$\Delta v(t) = Ri(t) + L\frac{di(t)}{dt} \tag{24}$$

where Δv represents the voltage drop along the line and R and L are the resistance and the inductance of the line. Assuming steady state and using phasor theory, the Equation (24) can be replaced by:

$$\Delta \bar{v} = R\bar{i} + j\omega L\bar{i} = (R + jX_L)\bar{i} \tag{25}$$

In case of a DC line the model is simplified using the expression (26):

$$\Delta v = Ri \tag{26}$$

Three-Phase Line Model

For three phase lines and unbalanced power flows, the same model as the one presented in previous section is going to be used, adding one equation like the one presented in (24) for each phase. In case of balanced three phase lines, an orthogonal synchronous reference frame with a direct and quadrature axis called d and q respectively is going to be employed. Using this reference frame the line model in steady state can be represented by the next vector equation:

$$\Delta \overline{v}_{dq} = R\overline{i}_{dq} + j\omega L \overline{i}_{dq} \tag{27}$$

Where the sub-index dq represents that the coordinates of the vector are referred to the direct and quadrature axis of the orthogonal synchronous reference frame. Each vector can be separated in its components according the expression (28) and (29):

$$\overline{v}_{dq} = v_d + jv_q \tag{28}$$

$$\overline{i}_{dq} = i_d + ji_q \tag{29}$$

The Equation (27) can be expressed as follows:

$$\Delta v_d = Ri_d - j\omega L i_q \tag{30}$$

$$\Delta v_q = j\omega L i_d + Ri_q \tag{31}$$

In the expressions (30) and (31), the coupling between the d and q axis can be observed. The authors Briz, Degner and Lorentz (2000) and Pogaku, Prodanovic and Green (2007) made a comprehensive analysis of the transformation from the conventional three-phase reference to the orthogonal synchronous reference frame.

Single-Phase Power Transformer

The single-phase power transformer model will be employed for modeling AC traction systems, as it was mentioned in previous sections. One of the most used methods for feeding this kind of systems is by means of single phase transformers connected to the distribution network. In the Figure 9 a), the exact two-winding power transformer model is represented. In this circuit, the primary and the secondary impedances, representing the primary and the secondary leakage inductances and winding resistances are depicted, and also the parallel branch representing the magnetizing inductance and the iron losses. The primary and the secondary impedances will be defined as follows:

Figure 9. Two-winding single-phase transformer equivalent circuits

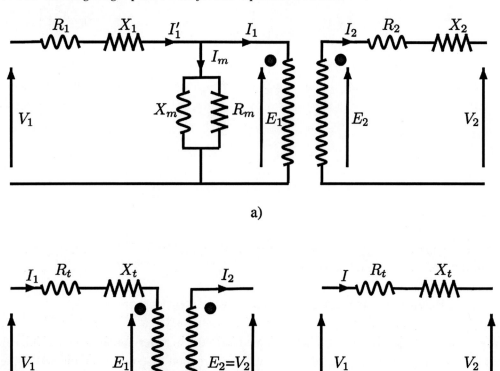

$$Z_1 = R_1 + jX_1 \tag{32}$$

$$Z_2 = R_2 + jX_2 \tag{33}$$

For power flow purposes this exact circuit will be replaced by the one represented in Figure 9 b), where the parallel branch is removed. The secondary impedance is referred to the primary side and added to the primary side impedance, obtaining the power transformer short circuit impedance according to the expression (34):

$$Z_t = R_t + jX_t = Z_1 + \frac{Z_2}{r_t} \tag{34}$$

where r_t is the turn ration and it is defined as:

$$r_t = \frac{N_2}{N_1} \tag{35}$$

Being N_1 and N_2 the number of primary and secondary winding turns respectively.

Using this simplified equivalent circuit the power transformer model is completely defined with the Equations (36) to (38).

$$E_2 = r_t E_1 \tag{36}$$

$$I_1 = r_t I_2 \tag{37}$$

$$E_1 = V_1 - Z_t I_1 \tag{38}$$

Figure 9 c) represents the *per unit* model of the single-phase transformer, where $Z_t = R_t + jX_t$ is the short circuit impedance expressed in per unit.

Three-Phase Power Transformer AC/DC Power Converter

The three-phase power transformer will be combined with the rectifier model to obtain the substation model in DC traction systems. Figure 10 represents a DC traction system substation, where the power transformer is connected to the distribution network and the AC/DC converter to the DC traction system. Considering the per unit system and using the synchronous reference frame, the two elements can be modeled as a whole using the next expressions:

$$V_d^{AC} I_d^{AC} + V_q^{AC} I_q^{AC} - V^{DC} I^{DC} = 0 \tag{39}$$

$$V_q^{AC} I_d^{AC} - V_d^{AC} I_q^{AC} = 0 \tag{40}$$

$$k\sqrt{\left(V_d^{AC}\right)^2 + \left(V_q^{AC}\right)^2} - V^{DC} - R_{eq} I^{DC} = 0 \tag{41}$$

The first two Equations (39) and (40), represents an ideal lossless converter, and (41) represent the voltage drop derived from the commutation and conduction losses. For a six pulses non-controlled rectifier, the factor k is 1.35 and R_{eq} is the equivalent resistance of the power converter in the regular commutation range. This is a very simple non-controlled rectifier model. More complex thyristor based rectifier models are described by Milano (2010). A comprehensive study of how the equivalent resistance is calculated is also presented by Pozzobon (1998). In this study the author presents very detailed models of different kind of rectifiers used in DC traction networks. The models presented in this work are not only for steady state and power flow purposes, but also for short-circuit analysis.

Figure 10. Three-phase power transformer and AC/DC power converter models embedded in a DC substation model

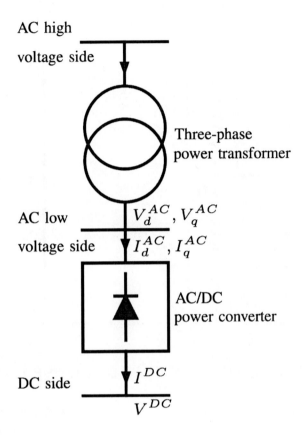

Autotransformer

The autotransformer model will be employed in bi-level AC systems, as it was described in previous sections. A simplified equivalent circuit of this device is presented in Figure 11, as it can be observed the parallel branch representing the magnetizing circuit is omitted. The primary and secondary impedances are referred to the secondary side, considering a single impedance Z_t, so $V_1=E_1$. The turn ratio is defined in the same way that it was previously defined for the single-phase transformer model (see Equation (35)).

Under these assumptions the autotransformer model is defined by:

$$E_2 = r_t E_1 \tag{42}$$

$$I_1 = r_t I_2 \tag{43}$$

$$I'_1 = I_1 + I_2 \tag{44}$$

$$V_2 = E_1 + E_2 - I_2 Z_t \tag{45}$$

Figure 11. Autotransformer simplified equivalent circuit

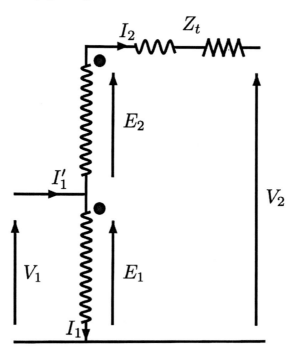

DC SYSTEMS MODELS AND COMBINED AC/DC POWER FLOW

In this section a complete DC traction model will be presented. The AC network feeding the DC traction grid will be also considered, so the model will be a hybrid AC/DC model. Some of the models described in the previous section, like the three-phase power transformer and the rectifier models, and also the line models, will be embedded in the AC/DC grid model. The train allocation procedure previously described will be also used. For the sake of simplicity, a reduced but general model of the whole system like the one represented in Figure 12 will be used in this study. This small system will be used as a support, but the formulation will be expressed in a general form, so it could be applied to any hybrid AC/DC traction system. Figure 12 a) represent the electrical infrastructure without the trains. In Figures 12 b) and c) the connections between trains and the connections between trains and substations are also represented. For this case of study, only two trains were considered, so the graph representing the trains connections (Figure 12 a)) only has one edge (e_1), that will be activated when the two trains are in the same line. The DC traction network, it is composed by two lines (e_8 and e_9) and three nodes (n_3, n_4 and n_5). Two of these substations are connected to the AC feeding system. The two trains and the DC lines and nodes generate the so-called DC subsystem. The AC nodes and the lines connecting these nodes form the AC subsystem. The power transformers are considered as AC lines too, and even the nodes at the power transformers secondary side will be part of the AC subsystem. A special subsystem denoted as link subsystem will also be defined. This system does not contain any node, just the edges representing the power converters that are connected in one side to the AC subsystem and in the other side to the DC subsystem.

Figure 12. General representation of a hybrid AC/DC traction system

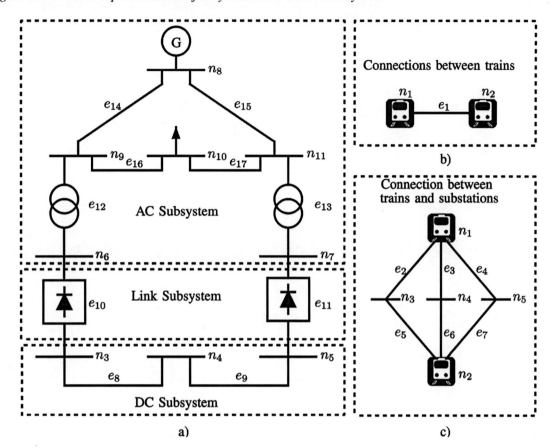

a)   b)   c)

In Figure 12, the node and edge enumeration criteria is also presented. As it was explained in a previous section, first the trains are enumerated and then the DC subsystem nodes. Then the AC nodes corresponding to the secondary of the power transformers are enumerated, and finally the rest of the AC nodes. For obtaining a compact matrix formulation, it is important to enumerate first all DC subsystem edges, then the link subsystem edges and finally, the AC subsystem edges.

The formulation of the problem will be divided into two sets of equations. In the first set, the linear behavior of the model will be described, and the second set will be a package with all the non-linear equations of the system. The linear package is composed by all the Kirchhoff Current and Voltage laws (KCL and KVL) in all lines of the three subsystems. Starting with the AC subsystem, and for a given line, the Equations (30) and (31) can be expressed in a matrix form as follows:

$$\begin{pmatrix} v_{nid} \\ v_{niq} \end{pmatrix} - \begin{pmatrix} v_{njd} \\ v_{njq} \end{pmatrix} = \begin{pmatrix} R_{eij} & -\omega L_{eij} \\ -\omega L_{eij} & R_{eij} \end{pmatrix} \begin{pmatrix} i_{eijd} \\ i_{eijq} \end{pmatrix} \qquad (46)$$

Where the sub-indices *ni* and *nj* represent the nodes *i* and *j* respectively, *eij* represents the edge connecting the nodes *i* and *j*, and *d* and *q* are the direct and quadrature components of the currents and voltages. The Node incident matrix will be used for obtaining a general compact matrix formulation. Three different node incident matrices will be defined, each one is associated with one of the subsystems (AC, DC and Links). In this case the AC node incident matrix (Γ^{AC}) will be defined as follows:

$$\Gamma^{AC} = \begin{pmatrix} \overset{n_6}{+1} & \overset{n_7}{0} & \overset{n_8}{0} & \overset{n_9}{-1} & \overset{n_{10}}{0} & \overset{n_{11}}{0} \\ 0 & +1 & 0 & 0 & 0 & -1 \\ 0 & 0 & +1 & -1 & 0 & 0 \\ 0 & 0 & +1 & 0 & 0 & -1 \\ 0 & 0 & 0 & +1 & -1 & 0 \\ 0 & 0 & 0 & 0 & +1 & -1 \end{pmatrix} \begin{matrix} e_{12} \\ e_{13} \\ e_{14} \\ e_{15} \\ e_{16} \\ e_{17} \end{matrix} \qquad (47)$$

with the use of this matrix, the KVL represented in equation (46) for a single line, can be expressed for the whole AC sub-system in the next way:

$$\Gamma^{AC}\left(v_{Nd}^{AC}\right)^T = R_E^{AC}\left(i_{Ed}^{AC}\right)^T - X_E^{AC}\left(i_{Eq}^{AC}\right)^T \qquad (48)$$

$$\Gamma^{AC}\left(v_{Nq}^{AC}\right)^T = X_E^{AC}\left(i_{Ed}^{AC}\right)^T - R_E^{AC}\left(i_{Eq}^{AC}\right)^T \qquad (49)$$

where the vectors v_{Nd}^{AC}, v_{Nq}^{AC}, i_{Ed}^{AC} and i_{Eq}^{AC} contains the direct and quadrature components of the node voltages and edge currents respectively. The reference for these currents, considers a positive current the one flowing from the tail of the edge to the head. In the particular case presented in this section these vectors can be defined as:

$$v_{Nd}^{AC} = \left[v_{n6d}, v_{n7d}, v_{n8d}, v_{n9d}, v_{n10d}, v_{n11d}\right] \qquad (50)$$

$$v_{Nq}^{AC} = \left[v_{n6q}, v_{n7q}, v_{n8q}, v_{n9q}, v_{n10q}, v_{n11q}\right] \qquad (51)$$

$$i_{Ed}^{AC} = \left[i_{e12d}, i_{e13d}, i_{e14d}, i_{e15d}, i_{e16d}, i_{e17d}\right] \qquad (52)$$

$$i_{Eq}^{AC} = \left[i_{e12q}, i_{e13q}, i_{e14q}, i_{e15q}, i_{e16q}, i_{e17q}\right] \qquad (53)$$

The matrices R_E^{AC} and X_E^{AC} are the resistance and the reactance matrices representing the impedance between the AC nodes, and also the short circuit impedance of the transformers. They are diagonal matrices. Their elements r(i,i) and x(i,i) are respectively the resistance and the inductance of the line or power transformer represented by the edge i.

The KCL for the whole AC subsystem can be easily expressed by means of the node incidence matrix using the next expressions, where "I" represents the identity matrix:

$$\left(\Gamma^{AC}\right)^T \left(i_{Ed}^L\right)^T = I\left[i_{Ed}^L, i_{Nd}^{AC}\right]^T \tag{54}$$

$$\left(\Gamma^{AC}\right)^T \left(i_{Eq}^L\right)^T = I\left[i_{Eq}^L, i_{Nq}^{AC}\right]^T \tag{55}$$

i_{Ed}^L and i_{Eq}^L are vectors containing the d and q components of the Link Subsystem edge currents. Like in the previous case the reference for these currents considers a positive current the one flowing from the tail of the edge to the head. i_{Nd}^{AC} and i_{Nq}^{AC} are vectors containing the currents injected in the AC subsystem nodes, except in the power transformers secondary. The inclusion of these currents in the model, allows the simulation of a slack bus, power plant or load in any of the AC subsystem nodes. A positive nodal current will be injected into the node.

For our particular scenario, these vectors can be defined as:

$$i_{Nd}^{AC} = \left[i_{n8d}, i_{n9d}, i_{n10d}, i_{n10d}\right] \tag{56}$$

$$i_{Nq}^{AC} = \left[i_{n8q}, i_{n9q}, i_{n10q}, i_{n10q}\right] \tag{57}$$

$$i_{Ed}^L = \left[i_{e10d}, i_{e11d}\right] \tag{58}$$

$$i_{Eq}^L = \left[i_{e10q}, i_{e11q}\right] \tag{59}$$

In a similar way, the node incident matrix of the DC subsystem will be used to state all the KVL and KCL of this subsystem, for this particular case of study the Γ^{DC} is defined as:

$$\Gamma^{DC} = \begin{pmatrix} & n_1 & n_2 & n_3 & n_4 & n_5 \\ +1 & -1 & 0 & 0 & 0 \\ +1 & 0 & -1 & 0 & 0 \\ +1 & 0 & 0 & -1 & 0 \\ +1 & 0 & 0 & 0 & -1 \\ 0 & +1 & -1 & 0 & 0 \\ 0 & +1 & 0 & -1 & 0 \\ 0 & +1 & 0 & 0 & -1 \\ 0 & 0 & +1 & -1 & 0 \\ 0 & 0 & 0 & +1 & -1 \end{pmatrix} \begin{matrix} e_1 \\ e_2 \\ e_3 \\ e_4 \\ e_5 \\ e_6 \\ e_7 \\ e_8 \\ e_9 \end{matrix} \tag{60}$$

Analyzing the expression (60) it can be observed that only edges e8 and e9 represent real lines of the DC traction infrastructure, the rest of the edges represent the connections between trains (e1), or between trains and substations (e2 to e7). The KVL's for the whole DC subsystem are represented in the expression (61).

$$\left(\Gamma^{DC}\right)\left(v_N^{DC}\right)^T = R_E^{DC}\left(i_E^{DC}\right)^T \tag{61}$$

v_N^{DC} and i_E^{DC} are vectors with the voltages in the DC nodes and the currents in the DC lines and R_E^{DC} is a diagonal matrix representing the resistance of all DC lines. R_E^{DC} is a diagonal matrix with dimensions (nnDC, nnDC), being nnDC the number of DC nodes considering trains and substations. This matrix will be used to activate or deactivate the DC lines. Thus, the trains' allocations can be changed without varying the system dimension or topology as it was previously explained. For instance, if the train 1 is between nodes n4 and n5 and the train 2 is between nodes n3 and n4 (see Figure 12), the edges e8 and e9 will be deactivated. For deactivating edges e8 and e9, the elements R_E^{DC} (8,8) and R_E^{DC} (9,9) will be set to "infinite" (a value high enough to generate negligible currents). Edge e8 is substituted by edges e5 and e6, which are the edges that connect train 2 with substations n3 and n4; the resistance assigned to each edge will depend on the distance from the train to each substation. In the same manner edge e9 will be substituted by edges e3 and e4. Edge e1 representing the connection between the two trains and edges e2 and e7 will be deactivated too. The values of the resistance matrix will be updated in each iteration, and they will be the only variable part of the problem. The rest of the system will remain constant even when the trains are moving.

The expression (62) represents the KCL's in the whole DC subsystem:

$$\left(\Gamma^{DC}\right)^T \left(i_t^{DC}\right)^T = I\left[i_t^{DC}, i_E^L\right] \tag{62}$$

where i_t^{DC} is a vector with the current of the trains. The train current is considered positive if it is a consumed current. i_E^L is a vector with the currents of the links (in the DC part of the converter). It is important to remark that each DC substation node that it is not connected to the AC sub-system, will add a current labeled as ghost current (igc) that will be set to zero. Under these assumptions, the previous vectors for this particular case of study are defined as:

$$i_t^{DC} = [i_{t1}, i_{t2}] \tag{63}$$

$$i_E^L = [i_{e10}, i_{gc1}, i_{e11}] \tag{64}$$

with all the above stated, the whole system linear behavior can be described in a compact matrix form like the one presented in expression (65). The matrix M and the vector z are generated to fulfill the equations (48), (49), (54), (55), (61) and (62).

$$M \cdot z^T = 0 \tag{65}$$

The detailed structure of matrix M and vector z can be studied in Figure 13. The number of rows of matrix M represent the number of linear equations, and the number of columns the number of unknowns. In the left and bottom part of Figure 13, the total number of equations and unknowns are enumerated. The variable n means number, the sub-indices B and N are branches/edges and nodes respectively, and AC and DC super-indices are used for referring to the AC subsystem and the DC subsystem.

The total number of equations is $2n_N^{AC} + 2n_B^{AC} + n_N^{DC} + n_B^{DC}$ and the total number of unknowns is $4n_N^{AC} + 2n_B^{AC} + 2n_N^{DC} + n_B^{DC}$. The difference between the number of unknowns and the number of equations is $2n_N^{AC} + n_N^{DC}$, but each train will add a new equation. According with the train model that was previously presented this new equation is in the form of:

$$P_{ti}(v_{ni}) - v_{ni}i_{ti} = 0 \tag{66}$$

Being $P_{ti}(v_{ni})$ the power demanded or regenerated by the train. Positive power means demanded power. As it was explained in a previous section, this power is a piecewise defined function depending on the catenary voltage.

Adding the train equations, the unbalance between the number of unknowns and equations is still $2n_N^{AC} + n_L$, (being nL the number of links), but it must be pointed out that each AC node, except the power transformer secondary nodes, can be defined as a slack node, generator node or load node. If an AC node is defined as load node, it will add two expressions like the ones stated in (67) and (68). A generator node will add other two expressions (see (69) and (70)) and the slack node will add equations (71) and (72).

Figure 13. Detailed structure of matrix M, representing the linear behavior of the whole electrical system

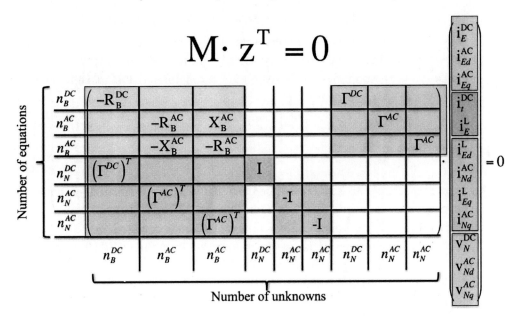

$$P_i = v_{nid}i_{nid} + v_{niq}i_{niq} \tag{67}$$

$$Q_i = v_{niq}i_{nid} - v_{nid}i_{niq} \tag{68}$$

$$P_i = v_{nid}i_{nid} + v_{niq}i_{niq} \tag{69}$$

$$|v_{ni}| = \sqrt{v_{nid}^2 + v_{niq}^2} \tag{70}$$

$$v_{nid} = v_i^{specified} \tag{71}$$

$$v_{niq} = 0 \tag{72}$$

The unbalance between unknowns and equation is now $3n_L$, to solve the problem, three equations must by defined for each of the AC/DC links, that are the ones defined in the substation model.

$$v_{ndj}i_{edk} + v_{nqj}i_{eqk} - v_{nj}i_{ek} = 0 \tag{73}$$

$$v_{nqj}i_{edk} - v_{ndj}i_{eqk} = 0 \tag{74}$$

$$i_{eqk}R_{eq} - v_{ni} + k\sqrt{(v_{ndj})^2 + (v_{nqj})^2} = 0 \tag{75}$$

Modeling non-reversibility in the substations requires that $i_{Ek}^L < 0$. In the Figure 14, the currents used in the expressions (73), (74) and (75) are represented. As it can be observed, each link adds three currents, the two components d and q of the current in the AC side, and the current in the DC side. In the Figure 14, the DC node i is connected to the AC node j, through the link k. and all voltages and currents in the link subsystem are represented.

AC SYSTEMS MODELS

The method presented in a previous section will be here extended to AC traction systems; the case of study will be a conventional bi-level 2x25kV system. Regular AC systems can be modeled using the same technique. In the Figure 15, a typical 2x25kV bi-level system is depicted. In the left side, the feeding high voltage network is connected to the AC traction network through a power transformer with a central tap in the secondary side connected to the rails. The other two poles of the secondary winding are connected to the positive feeder (overhead conductor) and the negative feeder or return line. For referring the variables in the positive feeder or overhead line, a sub-index *0* is used. The sub-index *1*

Figure 14. Detailed representation of a link with its voltages and currents

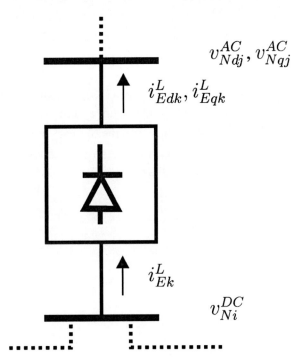

is used for the rails and 2 for the negative feeder. In this case all variables are complex magnitudes. Regarding the node enumeration a different criteria have been used. The first node is the slack and the second one is the primary of the power transformer, the third is the secondary of the power transformer, then the autotransformers are enumerated and finally the trains. In the previous section, the trains were enumerated first, and in this section a different criteria have been followed. However, the two enumeration criteria are equivalent. In this case, it has been considered that the slack feeds directly the traction network, but a more complex AC distribution network could be modeled using the same technique that was explained in the previous section.

Three different node incident matrices will be presented; the first one Γ_f defines the distribution network topology. As it can be observed in the expression (76), this matrix has only one row in this

Figure 15. Representation of a typical 2x25kV bi-level system

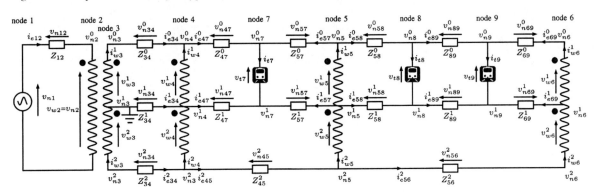

specific scenario because the slack feeds directly the power transformer. Γ_t represents the real connections between the autotransformers and between the autotransformers and the power transformer. With the network depicted in Figure 15, three lines are created. Finally, in Γ_t^*, the real connections, but also all possible connections between trains, and between trains and autotransformers or the power transformer are considered.

As it can be observed in Figure 15 and also expressions (76), (77) and (78) the edges are labeled with two numbers, the first one represents the "from" node or tail and the second one the "to" node or head. The matrix Γ_t^* can be split in three (see expression 78), the top part represent the real connections between autotransformers and transformers. In the middle part, all possible connections between the power transformer and the autotransformers and the trains are represented. The bottom part contains all the possible connections between trains.

$$\Gamma_f = \begin{matrix} n_1 & n_2 \\ [+1 & -1] \end{matrix} \; e_{12} \tag{76}$$

$$\Gamma_t = \begin{matrix} n_3 & n_4 & n_5 & n_6 \\ \begin{bmatrix} +1 & -1 & 0 & 0 \\ 0 & +1 & -1 & 0 \\ 0 & 0 & +1 & -1 \end{bmatrix} \end{matrix} \begin{matrix} e_{34} \\ e_{45} \\ e_{56} \end{matrix} \tag{77}$$

$$\Gamma_t^* = \begin{matrix} n_3 & n_4 & n_5 & n_6 & n_7 & n_8 & n_9 \\ \begin{bmatrix} +1 & -1 & & & & & \\ & +1 & -1 & & & & \\ & & +1 & -1 & & & \\ \hline 1 & & & & -1 & & \\ 1 & & & & & -1 & \\ 1 & & & & & & -1 \\ & 1 & & & -1 & & \\ & 1 & & & & -1 & \\ & 1 & & & & & -1 \\ & & 1 & & -1 & & \\ & & 1 & & & -1 & \\ & & 1 & & & & -1 \\ & & & 1 & -1 & & \\ & & & 1 & & -1 & \\ & & & 1 & & & -1 \\ \hline & & & & 1 & -1 & \\ & & & & 1 & & -1 \\ & & & & & 1 & -1 \end{bmatrix} \end{matrix} \begin{matrix} e_{34} \\ e_{45} \\ e_{56} \\ e_{37} \\ e_{38} \\ e_{39} \\ e_{47} \\ e_{48} \\ e_{49} \\ e_{57} \\ e_{58} \\ e_{59} \\ e_{67} \\ e_{68} \\ e_{69} \\ e_{78} \\ e_{79} \\ e_{89} \end{matrix} \tag{78}$$

Using the above defined node incident matrices all the KVL's and KCL's can be stated in a general form using the next expressions. Equations (79) to (81) represent respectively the KVL in the overhead conductor, rails and negative feeder. Expression (82) represent the KVL in the feeding system.

$$\Gamma_t^* \left(v_N^0\right)^T = Z^0 \left(i_E^0\right)^T \tag{79}$$

$$\Gamma_t^* \left(v_N^1\right)^T = Z^1 \left(i_E^1\right)^T \tag{80}$$

$$\Gamma_t \left(v_N^2\right)^T = Z^2 \left(i_E^2\right)^T \tag{81}$$

$$\Gamma_f \left(v_N\right)^T = Z \left(i_E\right)^T \tag{82}$$

The vectors v_N^0 and v_N^1 contain the voltages in the traction network nodes (nodes 3 to 9) in the overhead conductor and rails circuit respectively. In the case of vector v_N^2 the nodes representing the trains are not considered, so it only contains nodes 3 to 6 negative circuit voltages. The vector v_N, contains the voltages of the feeding system (nodes 1 to 2). Regarding the edge current vectors, they can be defined in the same way that voltage vectors were defined; the super-index indicates the positive, rail or negative circuit, and the absence of super-index means feeding circuit current. In this particular case, for the sake of simplicity, the edge currents were labeled with two numbers in the sub-index indicating the tail and head nodes. However, the same enumeration criteria explained in the previous section could be applied. First, the edges connecting the slack with the power transformer are enumerated, then the edges connecting the power transformer and the autotransformers, then the edges connecting the autotransformers and the trains and finally the edges connecting the trains. In our particular case of study, the current vectors contain the following elements:

$$i_E^0 = \left[i_{e34}^0, i_{e45}^0, i_{e56}^0, i_{e37}^0, i_{e38}^0, i_{e39}^0, i_{e47}^0, i_{e48}^0, i_{e49}^0, i_{e57}^0, i_{e58}^0, i_{e59}^0, i_{e67}^0, i_{e68}^0, i_{e69}^0, i_{e78}^0, i_{e79}^0, i_{e89}^0\right] \tag{83}$$

$$i_E^1 = \left[i_{e34}^1, i_{e45}^1, i_{e56}^1, i_{e37}^1, i_{e38}^1, i_{e39}^1, i_{e47}^1, i_{e48}^1, i_{e49}^1, i_{e57}^1, i_{e58}^1, i_{e59}^1, i_{e67}^1, i_{e68}^1, i_{e69}^1, i_{e78}^1, i_{e79}^1, i_{e89}^1\right] \tag{84}$$

$$i_E^2 = \left[i_{e34}^2, i_{e45}^2, i_{e56}^2\right] \tag{85}$$

$$i_E = \left[i_{e12}\right] \tag{86}$$

The impedance matrices contain the impedance of the lines and also the impedance of the power transformer and autotransformer. Z and Z2 are static impedances, they represent the real topology of the network and they are not affected by the allocation of the trains. Z0 and Z1 are the impedance matrices representing the overhead conductor and the rails impedances. Their definition is similar to the definition of the DC subsystem resistance matrix in the previous section. An impedance value high enough for generating negligible currents will be defined as "infinite" impedance. This infinite impedance will be assigned to non-active edges. The dynamic impedance matrices must be updated at every instant of simulation.

The KCL in all traction network nodes can be obtained using the expressions (83) to (85).

$$\left(\Gamma_t^*\right)^T \left(i_E^0\right)^T = I\left[i_w^1, -i_t\right]^T \tag{87}$$

$$\left(\Gamma_t^*\right)^T \left(i_E^1\right)^T = I\left[\left(i_w^2 - i_w^1\right), i_t\right]^T \tag{88}$$

$$\left(\Gamma_t^*\right)^T \left(i_E^2\right)^T = I\left(-i_w^2\right)^T \tag{89}$$

The vectors i_w^1 i_w^2 contain the currents through the power transformer secondary windings and through the autotransformer windings. Super-index 1 indicates that the winding is connected between the overhead conductor and the rails and super-index 2 represents currents through windings connected between the rails and negative feeder. i_t is the current demanded by the trains.

For the complete definition of the problem, the mathematical models of the power transformer, the autotransformer and the trains must be added to the system of equations. The power transformer will add the next three equations:

$$v_{w2} = \frac{N_2}{N_3^1} v_{w3}^1 \tag{90}$$

$$v_{w2} = \frac{N_2}{N_3^2} v_{w3}^2 \tag{91}$$

$$i_{e12} v_{w2} = i_{w3}^1 v_{w3}^1 + i_{w3}^2 v_{w3}^2 \tag{92}$$

where N2, N31 and N32 are the number of turns of the power transformer primary winding, secondary winding connected between the overhead conductor and the rails and secondary winding between the rails and the negative circuit. Each autotransformer is defined by two equations (89) and (90):

$$v_{wi}^1 = \frac{N_i^2}{N_i^1} v_{wi}^2 \tag{93}$$

$$i_{wi}^1 v_{wi}^1 + i_{wi}^2 v_{wi}^2 = 0 \tag{94}$$

The train model is similar to the one described in the previous section but in this case a complex equation must be defined:

$$S_{ti} = (v_{ni}^0 - v_{ni}^0) i_{ti} \tag{95}$$

where Sti(vti) is the apparent power of the train that is depending on the train voltage $v_{ti} = v_{ni}^0 - v_{ni}^0$.

CONCLUSION

The power system analysis in railway power supply systems is a quite complex task due to many reasons. The first one is that all loads are moving loads that vary their demanded or injected power, but also their relative position. In addition, the value of the loads depends on the train behavior, but also on the congestion of the network, so a very complex non-linear load model must be used. Finally, in some cases, like in the DC traction networks, if the feeding AC system is considered, a hybrid AC/DC power flow problem must be stated. These features make this kind of power systems, at least, very different from the conventional power systems. In this chapter, the basic ideas for understanding the complexity of these systems were presented. The models of the main devices embedded in the systems were explained and finally, the basic tools for modeling DC, AC and hybrid AC/DC traction systems were proposed by means of two cases of study. The first one was a DC subsystem fed through an AC distribution network and the second one an AC bi-level traction system. All the formulation was proposed in a very general way so it could be adapted to whatever kind of traction system.

REFERENCES

Abrahamsson, L., & Söder, L. (2012) An SOS2-based moving trains, fixed nodes, railway power system simulator, *Paper presented at13th International Conference on Design and Operation in Railway Engineering (Comprail '12)*, New Forest, UK.

Arboleya, P., Diaz, G., & Coto, M. (2012). Unified AC/DC power flow for traction systems: A new concept. *IEEE Transactions on Vehicular Technology, 61*(6), 2421–2430. doi:10.1109/TVT.2012.2196298

Briz, F., Degner, M., & Lorenz, L. (2000). Analysis and design of current regulators using complex vectors. *IEEE Transactions on Industry Applications, 36*(3), 817–825. doi:10.1109/28.845057

Cai, Y., Irving, M., & Case, M. (1995). Modeling and numerical solution of multi-branched DC rail traction power systems. *IEE Proceedings. Electric Power Applications, 142*(5), 323–328. doi:10.1049/ip-epa:19952118

Chymera, M. Z., Renfrew, A. C., Barnes, M., & Holden, J. (2010). Modeling Electrified Transit Systems. *IEEE Transactions on Vehicular Technology, 59*(6), 2748–2756. doi:10.1109/TVT.2010.2050220

Coto, M., Arboleya, P., & Gonzalez-Moran, C. (2013). Optimization approach to unified AC/DC power flow applied to traction systems with catenary voltage constraints. *International Journal of Electrical Power & Energy Systems, 53*, 434–441. doi:10.1016/j.ijepes.2013.04.012

Day, L., & McNeil, I. (1966). *Biographical dictionary of the history of technology*. London, United Kingdom: Routledge.

Falvo, M. C., Lamedica, R., Bartoni, R., & Maranzano, G. (2011). Energy management in metro transit systems: An innovative proposal toward an integrated and sustainable urban mobility system including plug-in electric vehicles. *Electric Power Systems Research, 81*(12), 2127–2138. doi:10.1016/j.epsr.2011.08.004

Frey, S. (2012). *Railway electrification systems & engineering*. Delhi, India: White Word Publications.

Hill, R. J. (1994). Electric railway traction. Part 3 Traction power supplies. *Power Engineering Journal, 8*(6), 275–286. doi:10.1049/pe:19940604

Ho, T. K., Chi, Y. L., Wang, J., Leung, K. K., Siu, L. K., & Tse, C. T. (2005). Probabilistic load flow in AC electrified railways. *IEE Proceedings. Electric Power Applications, 152*(4), 1003–1013. doi:10.1049/ip-epa:20045091

Hong, D., Lee, H., & Kwak, J. (2007). Development of a mathematical model of a train in the energy point of view, *Paper presented at international conference on control, automation and systems (ICCAS 2007)*, Seoul, Korea.

Kiessling, F., Puschmann, R., Schmieder, A., & Schneider, E. (2009). *Contact lines for electric railways. Planning, design implementation, Maintenance*. Erlangen, Germany: Publicis Publishing.

Kneschke, T. A. (1985). Control of utility system unbalance caused by single-phase electric traction. *IEEE Transactions on Industry Applications, 21*(6), 1559–1570. doi:10.1109/TIA.1985.349618

Mariscotti, A., Pozzobon, P., & Vanti, M. (2007). Simplified modeling of 2x25kV AT railway system for the solution of low frequency and large-scale problems. *IEEE Transactions on Power Delivery, 22*, 296–301. doi:10.1109/TPWRD.2006.883020

Milano, F. (2010). *Power System Modelling and Scripting*. London, United Kingdom: Springer-Verlag. doi:10.1007/978-3-642-13669-6

Paul, D. (2002). DC traction power system grounding. *IEEE Transactions on Industry Applications, 38*(3), 818–824. doi:10.1109/TIA.2002.1003435

Pilo, E. (2003) Diseño optimo de la electrificación de ferrocarriles de alta velocidad [Doctoral dissertation in Spanish]. Retrieved from https://www.iit.upcomillas.es/personas/eduardo

Pilo, E., Rouco, L., Fernandez, A., & Abrahamsson, L. (2012). A Monovoltage equivalent model of bi-voltage autotransformer-based electrical systems in railways. *IEEE Transactions on Power Delivery, 27*(2), 699–708. doi:10.1109/TPWRD.2011.2179814

Pires, C., Nabeta, S., & Cardoso, J. (2007). ICCG method applied to solve DC traction load flow including earthing models. *IET Electric Power Applications, 1*(2), 193–198. doi:10.1049/iet-epa:20060174

Pires, C. L., Nabeta, S. I., & Cardoso, J. R. (2009). DC traction load flow including AC distribution network. *IET Electric Power Applications, 3*(4), 289–297. doi:10.1049/iet-epa.2008.0147

Pogaku, N., Prodanovic, M., & Green, T. C. (2007). Modeling, analysis and testing of autonomous operation of an inverter-based microgrid. *IEEE Transactions on Power Electronics, 22*(2), 613–625. doi:10.1109/TPEL.2006.890003

Pozzobon, P. (1998). Transient and steady state short-circuit currents in rectifiers for DC traction supply. *IEEE Transactions on Vehicular Technology, 47*(4), 1390–1404. doi:10.1109/25.728534

Raygani, S. V., Tahavorgar, A., Fazel, S. S., & Moaveni, B. (2012). Load flow analysis and future development study for an AC electric railway. *IET Electrical Systems in Transportation, 2*(3), 139–147. doi:10.1049/iet-est.2011.0052

Tzeng, Y., Chen, N., & Wu, R. (1995). A detailed R-L fed bridge converter model for power flow studies in industrial AC/DC power systems. *IEEE Transactions on Industrial Electronics, 42*(5), 531–538. doi:10.1109/41.464617

Tzeng, Y. S., Wu, R. N., & Chen, N. (1998). Electric network solutions of DC transit systems with inverting substations. *IEEE Transactions on Vehicular Technology, 47*(4), 1405–1412. doi:10.1109/25.728537

Tzeng, Y. S., Wu, R. N., & Chen, N. (1998). Unified AC/DC power flow for system simulation in DC. *IET Electric Power Applications, 142*(6), 345–354. doi:10.1049/ip-epa:19952159

KEY TERMS AND DEFINITIONS

Catenary: It is the wire that supports the overhead contact line; it is called catenary because its shape is similar to this geometrical curve. The catenary is linked to the structures by means of clamps. At regular intervals, wires called droppers connect the overhead contact line with the catenary. The droppers have different length to keep constant the height of the overhead conductor. In some countries the catenary is also called the messenger wire.

Cross Coupling Feeding: When a traction system has more than one parallel track connecting two nodes, there is one overhead contact line for each track. Usually all these overheads lines are connected to the same substation. When more connections between the overhead contact lines of the two tracks are added at regular intervals, it can be stated that the overhead lines are cross-coupled.

Double-End Feeding: It is a traction network electrical scheme where two adjacent electrical sections are connected, so the two substations at the same time feed any train in whichever section. This scheme produce lower voltage drops that the single-end feeding scheme but the voltage between the overhead conductors and the rails in the two sections must be in phase.

Electric Locomotive: It is the vehicle that transforms electrical power into mechanical power for moving a train. The electrical power can be obtained from an overhead wire, a third rail or a storage system. The locomotives that use a combustion engine coupled with an electrical generator to produce their own electric power are also considered electric locomotives.

Electrical Section: The whole traction network can be divided into different electrical sections. A substation or a feeding branch can supply energy to one or more electrical sections. Depending on the adopted distribution scheme, the electrical sections can be insulated or electrically connected.

Longitudinal Coupling: This kind of coupling is the connection between two overhead lines of the same track that belongs to different electrical sections. It must be used when the feeding scheme is the double-end feeding. In such cases, even when during the normal operation of the system, the two sections are connected, the longitudinal coupling system must be able to disconnect the sections for isolating possible faults in the network.

Neutral Section: In some occasions, electrical substations connected to different phases feed two adjacent electrical sections. In such cases, the electrical sections cannot be connected and a section that is not electrically supplied separates them. This dead zone is called neutral section. Depending on the kind of trains and their current collector systems, these neutral sections can have different lengths and they must be placed in a location where the train can go drifting (not in traction or braking mode). A safety system must be able to feed the neutral section in case a train is stuck in it.

Overhead Contact Line: It is an electrical line situated over the tracks fed by an electrical substation. The overhead contact line is one of the systems employed to provide electrical power to electric locomotives that are connected to the overhead contact line using a structure called pantograph.

Single-End Feeding: It is a traction network electrical scheme where each electrical section is fed by one substation; two electrical sections fed by different substations cannot be electrically connected. This scheme is usually employed when the two adjacent substations are connected to different phases of the distribution system.

Third Rail: It is a rigid conductor parallel to the tracks used to supply electric energy to the electric locomotives; this method is only employed in DC traction systems like DC metro systems. Ceramic electrical insulators placed over sleepers usually support the third rail. The train is connected to the third rail through current collectors called contact shoes.

Chapter 20
Online Condition Monitoring of Traction Motor

Anik Kumar Samanta
Indian Institute of Technology Kharagpur, India

Devasish Basu
Indian Railways, India

Arunava Naha
Indian Institute of Technology Kharagpur, India

Aurobinda Routray
Indian Institute of Technology Kharagpur, India

Alok Kanti Deb
Indian Institute of Technology Kharagpur, India

ABSTRACT

Squirrel Cage Induction Motors (SCIMs) are major workhorse of Indian Railways. Continuous online condition monitoring of the SCIMs like Traction Motor (TM) are essential to prevent unnecessary stoppage time in case of a complete failure. Before a complete failure, the TMs generally develop incipient or weak faults. Weak faults have minute influence on the motor performance but eventually leads to complete failure of the motor. If these weak faults are identified at the earliest then, a scheduled maintenance can be planned which will prevent any unplanned stoppage. The signals used for SCIM fault detection are motor current, voltage, vibration, temperature, voltage induced in search coil, etc. The most popular fault detection technology is based on Motor Current Signature Analysis (MCSA). MCSA based online and onboard TM condition monitoring system can be very useful for Indian railways to reduce the cost of operation and unplanned delay by shifting from unnecessary scheduled maintenance to condition-based maintenance of TM and other auxiliary SCIMs.

INTRODUCTION

Indian Railway uses both electric and diesel locomotives that derive their tractive power, as well as power for maintaining auxiliary activities in the locomotives from electricity. Power for most of the auxiliary activities in locomotives is mostly generated from the induction motors. Auxiliary activities in locomotives are mostly related to cooling of different sub-systems and equipment of the locomotive. Earlier tractive power was generated by D.C. motors. The advantages offered by A.C. drives and induction motors have

DOI: 10.4018/978-1-5225-0084-1.ch020

changed the scenario with Indian Railways adopting A.C. induction motors in a big way for generating their tractive power as well. Advantages of A.C. drives have led to the induction of WAP5, WAP7 and WAG9 class of electric locomotives and WDP4 and WDG5 class of diesel locomotives. These locomotives also provide higher tractive power compared to DC traction motor based locomotives due to their higher torque to weight ratio.

The maintenance practice of these locomotives is entirely periodic wherein different equipment and motors are checked after a defined period. Shift to induction motors has been able to increase time interval between successive maintenance calls for a locomotive. The practice includes carrying certain replacement activities that may not be necessary as per their condition at that time. Periodic maintenance results in unnecessary expenses and running equipment and components for suboptimal periods. A way out to obviate this drawback is by shifting to condition based maintenance that would require residual life assessment of equipment.

Due to restricted availability of space in the locomotives redundancies are nonexistent. Because of this any outage of any of the machine, drastically reduces the capability of the locomotive and can even completely disable it. The cooling mechanisms like the different blower fans come with no redundancies that makes their functioning extremely critical in the working of the locomotive. While a locomotive comes with four or six traction motors depending on the class of locomotive, yet, outage of any traction motor affects the reliability of the locomotive. Faults in the bearings, mechanical defects in fans, electrical faults like shorting in windings, rotor bar and end ring breakages all occur in the motors in the locomotive and affect its performance.

It, therefore, becomes necessary to have monitoring systems of the motors available in a locomotive that would be able to diagnose the health and fault levels of the motors. In this chapter our focus will be on fault detection of Squirrel Cage Induction Motors (SCIMs) with application to Traction Motor(TM) of A.C. electric locomotives.

BACKGROUND

Fault detection of Inductor motor requires extensive study about the different types of faults and how they are detected. To provide a better understanding, two sections have been dedicated about the faults that occurs and how they are detected. This chapter summarizes the existing techniques and gives an insight into recent technologies that are available for fault detection and diagnosis of SCIMs.

1. Induction Motor Faults and It's Classification

Fault is defined as unpermitted deviation of at least one characteristic property of the system from the acceptable, usual, standard condition, Isermann (2006). Faults are incipient in nature, so even if there is a fault in the system, the system may operate as a normal system with subtle deviation in its states. Fault diagnosis consists of three different steps, 1. *Fault detection*, 2. *Fault Isolation* - localization or classification of the fault, and 3. *Fault identification* - determination of type, magnitude and cause of the fault. Failure is defined as permanent interruption of the system's ability to perform the required functions, Isermann (2006). If faults are not detected and proper maintenance has not been taken, the faulty system leads to complete failure resulting in loss of productivity. Failure prognosis consists of early detection of incipient faults and predicting the remaining useful life before failure and is required for predictive maintenance. *Figure 1* illustrates the different steps involved in fault diagnosis and failure prognosis. Faults in SCIMs can be broadly classified into stator faults, rotor faults and bearing faults. Each fault class can further be classified as given below.

- Rotor Faults
  - Broken rotor bar
  - Broken end ring
  - Eccentricity
    - Static Eccentricity
    - Dynamic Eccentricity
    - Mixed Eccentricity
- Bearing Faults
  - Inner raceway
  - Outer raceway
  - Rolling element
- Stator Faults
  - Inter turn short circuit
  - Phase to phase short
  - Phase to ground short

Brief description about these faults are discussed below

2. Broken Rotor Bar (BRB) and Broken End Ring (BER) Faults

These faults are said to occur when due to stress or mechanical defect, the rotor bars or the end ring, which holds together the bars are damaged. Photographs to illustrate the occurrence of BRB of a 750 KW Trac-

Figure 1. Illustration of fault diagnosis and failure prognosis

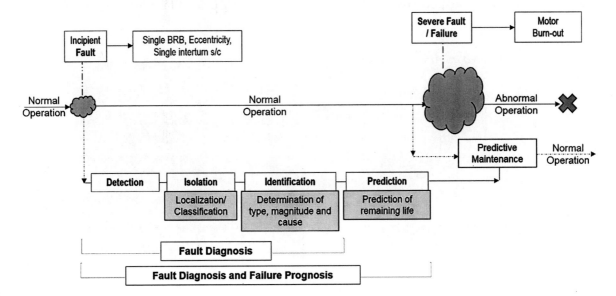

tion Motor are shown in Figure 2. These damaged motors were found during the periodic maintenance in their respective locoshed. The motor on continual run with single BRB fault, develops multiple bar cracks, as found in Figure 2(b). Main causes behind these faults are attributed to manufacturing defects, unbalanced magnetic pull due to load transients and asymmetrical loading patterns, etc., Nandi et al. (2005). This fault results in unbalance of the rotor which causes unbalance in the motor magnetic field. The unbalanced magnetic field is decomposed into negative and positive sequence components. These sequence components of the unbalanced magnetic field induces two fault frequency components on both side of the fundamental frequency given by

$$f_{brb} = f_{ber} = (1 \pm 2ks)f \qquad (1)$$

where s is the slip, k is any integer, and f is the supply frequency. Initially, small cracks may develop in the structure, but with prolonged usage, these may develop into more severe defects and may eventually lead to multiple faults due to uneven thermal stress because of localized heating. If left undetected, this can result in stator-rotor rub. It is, therefore, of utmost importance, that these faults are detected in its inception when the fault is weak. The multiple BRB fault considerably distorts the stator current and hence become conspicuously visible in the spectral signature. However, modulation of the stator current due to single broken bar is weak as mentioned by Xu et al. (2010) and its location is also very close to the fundamental frequency in low load operations. This escalates the difficulty of its detection. Onset of this fault degrades the performance of the motor, and results in modulated stator currents, torque pulsation and reduced average torque, rotor speed pulsation and reduced speed, increased losses, decreased efficiency and excessive heating, etc. Detection of BRB in low load operation was presented through Hilbert modulus with FFT in Puche-Panadero et al. (2009) and Hilbert modulus with ESPRIT in Xu et al. (2013). Low frequency load-torque oscillations introduces BRB-like components in the stator current spectra. Indicators of BRB, independent of load-torque oscillations can be found in Bruzzese (2008); Kim et al. (2015) and Yang et al. (2014). Detection of BRB in low load condition as well as with load torque effect can be found in Pons-Llinares et al. (2015). Knowledge about the fault severity is available in the amplitude of fault specific frequency components as was derived by Xu et al. (2010). A mathematical equations to derive the number of broken bars from the amplitude of the fault component

Figure 2. Reported BRB in different locosheds. Photo Courtesy: Indian Railways

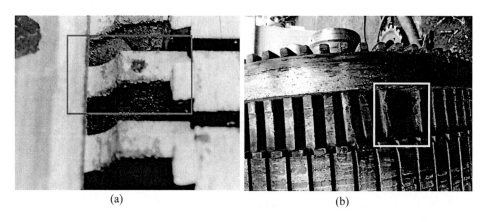

is also provided here. Use of Simulated Annealing algorithm in conjunction with ESPRIT to determine the fault severity was proposed by Xu et al. (2012). Kim et al. (2013) have used least square amplitude estimation with the frequency estimation by ESPRIT to find the fault frequency amplitude. Startup transients are non-stationary in nature. Analysis with wavelet decomposition for non-stationary signal analysis can be found in Antonino-daviu et al. (2012). Effect of simultaneous occurrence of static eccentricity, BRB and speed ripples were studied analytically and experimentally in Kaikaa and Hadjami (2014). Recent advances in detection of BRB can be found in Kang et al. (2015) and Sapena-bano et al. (2015).

3. Eccentricity Faults

Eccentricity related faults occur when there is an uneven airgap the stator and the rotor as defined by Nandi et al. (2005). Larger value of eccentricity results in unbalanced magnetic pull. Which in turn may result in stator-rotor rub or even BRB fault. Similar to the BRB fault, eccentricity fault also creates unbalance in the magnetic field and as a result, fault frequency components get induced in the motor armature current. The Frequency components due to these faults are given by

$$f_{ecc} = \left[(kR \pm n_d) \frac{(1-s)}{p} \pm v \right] f \qquad (2)$$

where R is the number of rotor slots, v is the order of the stator time harmonics that are present in the power supply driving the motor ($v = \pm 1, \pm 3, ...$), p is the number of pole pairs, and n_d is the variable depending on which the eccentricity related faults can be classified as given below.

1. **Static Eccentricity**: The position of the minimum radial air gap remains constant. This can be caused due to ovality of stator or imperfect positioning of the rotor. Mathematically $n_d = 0$ in (2). Static eccentricity fault illustrated in Figure 3(a). It is observed that in this case, the axis of rotation does not coincides with stator axis, but is same as the rotor axis. As a result, a non-uniform stationary air-gap is created, which does not rotate with the rotor.
2. **Dynamic Eccentricity**: In case of dynamic eccentricity the axis of rotation doesn't coincide with the axis of the rotor, but axis of rotation and stator axis are same and the position of the minimal air gap rotates with the rotor. Mathematically, frequency components can be modelled by putting $n_d = 1, 2, 3, ...$ in (2). This phenomenon can be visualized in Figure 3(b). This misalignment is caused by several factors such as bearing wear, misalignment, bent rotor shaft, mechanical resonance due to oscillation of shaft speed, etc.
3. **Mixed Eccentricity**: In reality, both static and dynamic eccentricities tend to coexist. When they exist together, the condition is known as mixed eccentricity. In this case none of the three centers coincides with each other as illustrated by Figure 3(c)

With increasing eccentricity, the resulting unbalanced forces (also known as UMP) can cause stator-to-rotor rub, which may damage the stator and the rotor as confirmed by Faiz et al. (2003). At times the eccentricities in the rotor are tolerated to some extent as a natural manufacturing defect as it cannot be

Figure 3. Illustration of eccentricity faults ((a) Static Eccentricity, (b) Dynamic Eccentricity, and (c) Mixed Eccentricity)

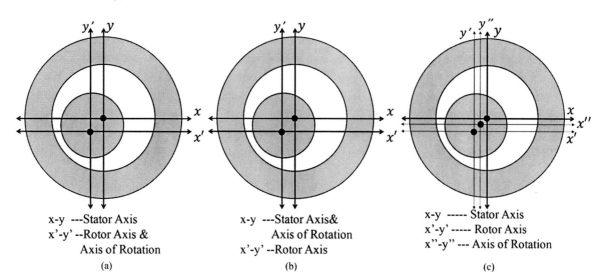

entirely be done away with, especially for heavy rotors. This causes a steady UMP in one direction. This might result in the evolution of some degree of dynamic eccentricity. Unless detected early, these effects may slowly develop into stator or rotor rub, causing a major breakdown of the motor. As a standard practice in the industrial use, air gap eccentricity up to about 10% is permissible according to Nandi and Toliyat (1999). However, manufacturers normally keep the total eccentricity level even lower to minimize UMP as well as to reduce vibration and noise. The occurrence of air gap eccentricity results in unsymmetrical stator and rotor circuit inductances as well as degradation of motor performances. Literature suggests that eccentricity faults account for around 40-50% of total induction motor failures as given in Nandi and Toliyat (1999). Diagnostic methods employed to identify the eccentricity faults can be classified into:

1. Monitoring of electromagnetic field using search coils, as in Chow and Tan (2000),
2. Monitoring of mechanical vibration by Chow and Tan (2000)
3. Model based techniques, Loparo et al. (2000)
4. Artificial intelligence, neural network and fuzzy models by Romero-Troncoso et al. (2011); and Seera et al. (2012) and
5. MOTOR Current Signature Analysis (MCSA) for detection of eccentricity fault can be found in Ayhan et al. (2008); Garcia-perez et al. (2011); and Riera-Guasp et al. (2012).

Nandi et al. (2001) and Gyftakis and Kappatou (2013) established that rotor slot harmonic and eccentricity related faults are available only for a certain combination of number of rotor slots and the number of fundamental pole pair. Drif and Cardoso (2012) and Liu et al. (2004) have used instantaneous power to detect low amplitude fault components in the presence of the fundamental supply frequency. Experimental setups to emulate the eccentricity faults are difficult to construct. Design of experimental setup were elaborately discussed by Knight and Bertani (2005) and Nandi et al. (2011). Experiments with drive connected IMs can be found in Georgakopoulos et al. (2011); Huang et al. (2007); and Hyun et al. (2011). Artificial neural networks (ANN) for fault classification of eccentricity related faults is shown in Huang et al. (2007). Other method based on ANN can be found in Ghate and Dudul (2011). Fault classification with 'Fuzzy Min Max with Classification and Regression Tree' (FMM-CART) can be found in Seera et al. (2012). Detection of eccentricity with terminal voltage at switch off was proposed by Nandi et al. (2011) and under standstill with pulsating input from the drive by Hyun et al. (2011). Condition monitoring of IM with eccentricity fault and its online implementation is shown in Choi et al. (2011). Practical quantification for safe operation of motor under various load with eccentricity fault was discussed in Concari et al. (2011). Using a Rogowski coil for detection of eccentricity fault was shown by Ceban et al. (2012). Effect of slotting on mixed eccentricity was carried out in by Andriamalala et al. (2008) on a dual-stator winding induction motor. Dorrell (2011) have studied the effect of rotor flux, number of bars and saturation on source and the characteristics of Unbalanced Magnetic Pull (UMP) for static and dynamic eccentricity. Time-frequency analysis can be found with wavelet packet decomposition in Ye et al. (2003) and Gabor wavelets in Riera-Guasp et al. (2012). Finally, Dynamic eccentricity components were distinguished from load-torque oscillation using Wigner distribution by Blodt et al. (2009) and with instantaneous active and reactive power by Drif and Cardoso (2012).

4. Bearing Faults

Induction motors in general use ball, taper and rolling element bearings. As reported in literature, specifically by Nandi et al. (2005), bearing faults constitute about 40%-50% of all motor failures. The prevailing practice of detecting bearing failures in IR is based on chemical composition test of the bearings. Most of the reasons behind bearing defect can be owed to:

1. Fatigue due to heavy load and continuous run
2. Contamination with foreign particles due to inadequate sealing and corrosion due to moisture
3. Ineffective lubrication which causes heating
4. Misalignment, presence of eccentricity faults and manufacturing and fabrication defects

Bearing consists of rolling or rotating elements placed closely inside two concentric rings. Defect in any of these parts gives rise to specific signatures. These signatures are mainly manifested in vibrations but can also be sensed in the stator current. Bearing fault signature is given by

$$f_{bear} = f \pm m f_v \tag{3}$$

where f_{bear} are fault frequency component due to different faults in bearing given by their characteristic frequency f_v, $m = 1, 2, 3...$ and f is the supply frequency. Different faults along with their specific frequencies (f_v) are given below

1. Any damage to the outer raceway gives rise fault frequency components according to (4). A pictorial illustration of this fault is shown in Figure 4(c).

$$f_v = (N/2)f_r[1 - b_d \cos(\beta)/d_p] \quad (4)$$

2. Inner raceway fault occurs when there is a damage to the inner raceway as shown in Figure 4(b). The fault specific frequency components are given below.

$$f_v = (N/2)f_r[1 + b_d \cos(\beta)/d_p] \quad (5)$$

3. Ball defect is caused due to any damage to the balls due to wear and tear. This fault also gives rise to specific frequency components given by (6)

$$f_v = d_p f_r / 2b_d \{1 - [b_d \cos(\beta)/d_p]^2\} \quad (6)$$

where $f_r = (1-s)f/p$ is the rotational frequency, N is the number of balls, b_d is the ball diameter, d_p is the ball pitch diameter and β is the contact angle of the ball to the races. A diagram to illustrate the dimension of a bearing is shown in Figure 5.

In Ocak and Loparo (2001), Hidden Markov Models (HMMs) are trained on features extracted from the amplitude demodulated vibration signals for both healthy and faulty states of the induction motor. When a new signal comes in for testing, the feature vectors extracted from it are projected onto the HMMs. The HMM corresponding to the highest probability defines the current state of the machine. Bearing fault result in variations in the induced magnetic field which induces harmonics in the sidebands of the fundamental frequency of the stator current. Akin et al. (2008) describes a bearing fault detection scheme which performs MCSA at the harmonics of the fundamental frequencies of the stator current as well as harmonic frequencies of the inverter current. Fourier decomposition is used by Frosini and Bassi (2010) to detect the frequency components that identify the fault states in the induction machines during bearing faults and along with the motor efficiency are used as indicators of faults is the system. Stator current signature is also analyzed in Ibrahim et al. (2008) to detect bearing faults by monitoring

Figure 4. Bearing Faults ((a) Healthy Bearing, (b) Bearing with inner raceway fault, (c) Bearing with outer raceway fault)

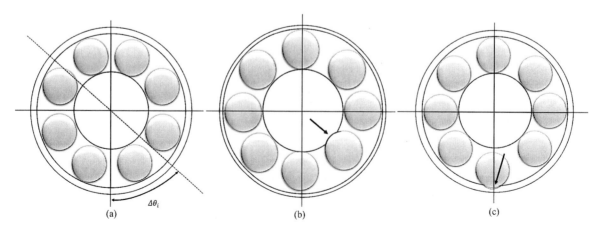

Figure 5. Dimension of a bearing

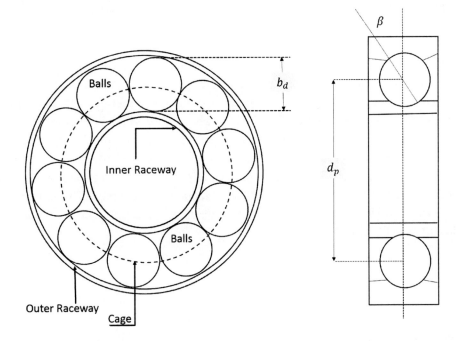

the instantaneous power factor variation which reflects the change in the radial torque in the motor due to bearing defects. Wavelet decomposition of the stator current is used to develop a bearing damage detection method in Eren and Devaney (2004). Apart from the motor current signatures, the vibration signals from the machines have also been reported to be useful for fault detection. In Kankar et al. (2011), features are extracted from the vibration signals which are then used to train neural network and support vector machine classifiers to detect fault states in induction motors due to bearing defects. MFCC features extracted from the vibration signals are used in HMMs that represent normal and fault states in motors. These HMMs are then used to detect faults, Purushotham et al. (2005). In Tandon et al. (2007), four different fault detection mechanisms namely Vibration Measurement, Shock Pulse measurement, Acoustic measurements and Motors current analysis are compared to test their performance in detecting bearing faults in the induction motors. Detection of bearing faults for speed adjustable drives with wavelet packet decomposition was carried out by Teotrakool et al. (2009). Detection of non-conventional faults of the bearing like deformation of seal and corrosion was presented by Frosini and Bassi (2010) using MCSA and efficiency statistics. Using a combination of current, vibration and acoustic sensor, a wireless system was created for detection of eccentricity and bearing faults by Esfahani et al. (2013). Use of trace-ratio linear discriminant analysis by Jin et al. (2014) and spectral kurtosis and envelop analysis of the stator current by Leite et al. (2015) have significant contributions. Detection of bearing faults with stray flux was achieved by Frosini et al. (2015).

5. Stator Faults

Stator winding related faults accounts for a large share of all the failures related to SCIMs. Stator faults such as phase to phase short circuit and phase to ground short circuit starts with incipient stator inter-turn short circuit (ITSC) fault, Gandhi et al. (2011); Tallam et al. (2007). Stator ITSC fault is the result of insulation failure between two adjacent stator turns. Ageing of insulation, mechanical stress, overloading of the machine, voltage spikes from the power electronic controllers are the various possible reasons behind the insulation failures. It does not affect the machine performance on their onset. If the ITSC is not detected at the earliest then, it causes localized heating that leads to more insulation failure and subsequently phase to phase short circuit and/or phase to ground short circuit. Phase to phase or phase to ground faults are detected by over current relay, ground fault relay, etc. The machine, which has developed such severe faults, might have already damaged its core which is irreversible. If stator ITSC fault is detected at the earliest then the machine can be reused only after replacing its windings. Figure 6, taken from a loco shed of Indian Railways, shows various photographs of the stator faults. These motors were damaged permanently and had to be replaced. Due to the absence of a condition monitoring, these motors may have hampered the normal operation of the locomotive severely. ITSC generates unbalance in the machine, which induces fault frequency side bands in the motor stator current. The fault frequencies are evaluated as in Gandhi et al. (2011)

$$f_{itsc} = f\left[\frac{n}{p}(1-s) \pm k\right] \tag{7}$$

where f is the supply frequency, p is the number of pole pairs, s is the slip of the machine. $n = 1, 2, 3, ...$ and $k = 1, 2, 3, ...$. Detecting the fault frequency components in the stator current is one of the popular methods to detect inter-turn faults. ITSC is also detected by measuring vibration, flux etc., Gandhi et al. (2011); Tallam et al. (2007). Effects of different stator faults on the motor can be seen in Figure 7. Joksimovic and Penman (2000) developed a winding function based method for modeling a poly-phase cage induction motor with ITSC. Using the mathematical model, it is demonstrated that under short circuit condition there is a considerable rise in the rotor slot harmonic component. Although this cannot be solely accounted for ITSC, as the rise in slot harmonic can also be due to unbalanced supply voltage,

Figure 6. Reported stator faults at a railway locoshed with stator winding burn out fault. Photo Courtesy: Indian Railways

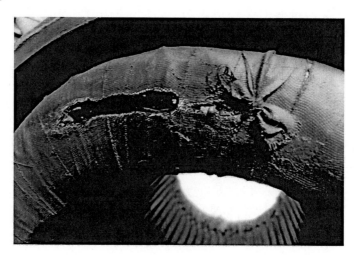

static eccentricity or other abnormal machine condition. Stavrou et al. (2001) have used the stator winding itself as a sensor for detecting abnormalities in the stator winding. This paper suggests that the lower side band of the field rotational frequency with respect the fundamental along with some other components related to slotting are the most reliable indicators of ITSC. Henao et al. (2003) have demonstrated the employment of an external stray flux sensor as an alternative to conventional MSCA methods. It is reported that this method is also effective in the presence of power supply harmonics therefore tracking and filtering the power supply frequency is not required. This method is suitable for low resolution applications. In Da Silva et al. (2008), reconstructed phase space transform of the three-phase stator current envelopes of induction motor is considered for detection of ITSC faults. Das et al. (2014) have used 'Support Vector Machine based Recursive Feature Elimination' (SVMRFE) algorithm to determine the most relevant feature from a list of features obtained from time, frequency, time-frequency and nonlinear analysis of the stator current for ITSC detection. Drif and Cardoso (2014) have used FFT to analyze the instantaneous active and reactive power signature to identify ITSC faults in line fed or inverter fed induction motors.

SIGNALS AND SENSING TECHNIQUES FOR FAULT DETECTION

Fault detection of SCIM demands detection of faults well in advance while the motor is still operational to avoid unscheduled maintenance and prolonged down time. Fault detection can be accomplished by analyzing a variety of signals of the motor like

1. Vibration [Kim et al. (2014); Seshadrinath et al. (2014)] signal is one of the most popularly sensed signals for fault detection of mechanical subsystems. Almost all the motor faults can be detected using vibration. One of the major drawback is sensor placement. Additionally, it requires an elaborate instrumentation arrangement for proper working and is very costly and fragile.

Figure 7. Possible location for current and voltage tapping for auxiliary motors

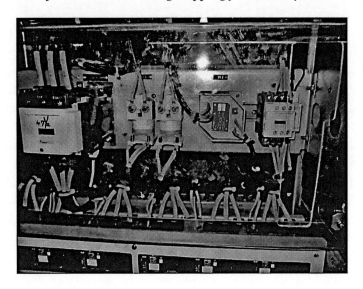

2. Magnetic fields [Cabanas et al. (2011); Ceban et al. (2012)] for fault detection of SCIMs deals with the acquisition of magnetic flux with a search coil wound around one of the stator tooth. This method needs complex sensor arrangement and fixations for each motor. Search coil based fault diagnosis needs installation of pick-up coils at appropriate places in order to capture the flux pattern near stator slots or on the frame of the motor. It is not always feasible to use search coils in all machines in order to monitor the fault modulated flux signal. It is also difficult to locate the best position of the search coils to capture the signal modulated by the faults effectively. Moreover, static and dynamic characteristics of the search coil interfere with the detection procedure.
3. Supply voltage modulation [Nemec et al. (2010)] can also be used effectively for detection of faults. But the major disadvantage lies with the fact that the sensor has to be attached to either of the supply or motor terminals. In case of locomotive systems most of the motor terminals are out of reach and acquiring high voltage signals can pose safety issue. For a 750kW traction motor the terminal voltages may reach to 2.5kV at rated speed. Also for a motors driven by drives it is very difficult to detect the voltage modulation as is the case with traction motor.
4. Active-reactive power analysis [Cruz (2012); Drif and Cardoso (2012)] was popularized due to its effectiveness for analyzing the motor under time varying loads and its ability to distinguish between rotor fault and load-torque oscillations. But again acquiring voltage signals for this method is challenging.
5. Acoustic signals were the earliest known signals that were utilized by human to detect fault in a system. In case of motor, this method was utilized by Kim et al. (2014) to detect rotor eccentricity. Though effective for single motor, the problem is escalated when multiple motors are operating at tandem in an enclosed environment like a locomotive. In this case isolation of the faulty motor can be demanding.
6. Thermal field analysis [Ying (2010)] and Thermal imaging [Picazo-Rodenas et al. (2012)] although used widely used for fault detection in various heavy industries are not very popular in case of induction motor. As extracting fault significant information from thermal images is difficult due to the outer metallic cover of motors it is only be used for stator faults and bearing fault detections.
7. Current [Ayhan et al. (2008); Garcia-perez et al. (2011); Kim et al. (2013); Riera-Guasp et al. (2012); Xu et al. (2010)] signals are most widely used for motor fault detection due to multiple reasons.

Most of the motor faults can be detected from current, the methods are independent of the motor parameters and acquiring current signal is non-invasive as hall sensors are generally used. These sensors are clamped around the cable for sensing and need not be placed on the motor or on the terminals. Fault detection methods concerned with stator current came to be known as Motor Current Signature Analysis (MCSA). A marked disadvantage of MCSA is that it is only valid for the motor operating under steady state condition or the acquired signals are statistically stationary. This problem can be circumvented with time-frequency analysis tools although, spectral analysis of the stator current is the most popular technique, Nandi et al. (2005).

The prevalent practices of fault diagnosis in Indian Railway other than visual inspection involves calculating inductances and resistance of the motor. These methods cannot predict weaker faults and are viable only when the motor is severely damaged. In this study, the main focus is laid on MCSA due to its non-invasive nature and relatively easy of use for online implementation schemes. For this purpose, once the stator current is acquired, with different features of the stator current are derived to carryout the analysis.

ACQUIRING SIGNALS FROM A.C. LOCOMOTIVE

Fault detection of induction motors require the signals mentioned above to be acquired using specific sensors. Now, in a constricted space like that of the locomotive engine, Sensor placement is a challenging task. More so due to the fact that many of the motors are covered. Some of the initial studies about the structural organization of the locomotive has resulted in finding available space for sensor placement. It is also very important to understand the availability of existing sensors which can be used for fault detection.

1. Available Instrumentation

1. **For Auxiliary Motors:** There is no instrumentation directly associated to the auxiliary motors. Current sensors are connected after the auxiliary power converters and each converter supplies current to multiple auxiliary motors.
2. **For Traction Motors:** Each traction motor has one temperature sensor and one speed sensor. Current sensors are connected after the traction motor power converters and each such converter supplies power to three traction motors. Temperature sensors are not very reliable and have a substantial failure rate. In the existing framework, monitoring individual traction motor can only be accomplished with external sensors.

2. Scope of additional Instrumentation

1. **For Auxiliary Motors:** For all motors there are individual Miniature Circuit Breakers (MCBs) and all the MCBs are placed in a single box (Figure 7), so taking the current signal by connecting current sensor for all the motors is possible from a single point.
2. **Traction Motors:** For traction motors current and voltage signals can be tapped from two different places:
    a. After the power converter (located inside the loco) (Figure 8)
    b. From the junction box (located under the bogie) (Figure 9)

Figure 8. Possible location - 1 for current and voltage tapping for TM

Figure 9. Possible location - 2 for current and voltage tapping for TM

TECHNOLOGIES INVOLVED

Specific faults in SCIM induces specific fault frequency components in the stator current. MCSA of SCIM involves accurate identification of these components. Presence of noise especially during light load conditions make this detection through traditional spectral estimation methods such as Fast Fourier Transform (FFT), Periodograms and Welch's method very difficult. Although their RT implementation with the current state of the art embedded system is possible, but accuracy and detectability are major concerns. The main disadvantages of these techniques are the impact of the sidelobe leakage due to inherent windowing of finite data sets and also detection in the presence of a dominating fundamental close to some of the fault specific frequency components. Subspace based methods like Multiple Signal Classification (MUSIC) Garcia-perez et al. (2011), Estimation of Signal Parameters via Rotational Invariance Techniques (ESPRIT) Kim et al. (2013), etc. are very popular for detection of these faults. Other methods available are based on Fuzzy Decisions by Zidani et al. (2008), Neural Networks by Ghate and Dudul (2011), and Hilbert Transforms by Puche-Panadero et al. (2009). Artificial intelligence, neural network and fuzzy model based methods are decision procedure based on the symptomatic behavior of the motor under healthy as well as faulty conditions by Nandi and Toliyat (1999). The later methods require detailed knowledge of motor behavior captured in large amounts of response data followed by rigorous training or tuning. Additionally, Wavelet-based methods by Romero-Troncoso et al. (2012); Seshadrinath et al. (2014) are also popular for variable loading and non-stationary signal analysis. Model based diagnosis methods are based on the hypothesis that tests the change in values of some of the model parameters during faults Thomson et al. (1999).

SPECTRAL ESTIMATORS

As established form the previous sections, it is clear that the spectral analysis of the stator current is advantageous to other methods in terms of sensor placement, non-invasiveness and ease of use. This can be accomplished by FFT, PSD and high-resolution spectral estimators like MUSIC. In this part of the chapter, we will discuss some of the spectral estimation techniques along with some result.

1. Discrete Fourier Transform

Any finite energy signal can be represented by a linear combination of complex exponentials. The representation of the signal in terms of spectrum of coefficients is accomplished with Discrete Time Fourier Transform of a discrete time-series $x[n]$ and is given by

$$X(\omega) = \sum_{n=0}^{N-1} x[n] e^{-j\omega n} \qquad (8)$$

Now, in this method the frequency components are distributed over a continuous spectral band which is not suitable for digital implementation and computation. As a result, Discrete Fourier Transform (DFT) was formulated. In this method, the frequency is represented as a function of an integer k. For a finite time-series of $x[n]$ of length N, the N-point DFT is given by

$$X(k) = \sum_{n=0}^{N-1} x[n] e^{-j\frac{2\pi kn}{N}} \qquad (9)$$

DFT has a computational complexity given by $O(N^2)$. Different algorithms to reduce the computational complexity of DFT have been developed. These algorithms came to be known as Fast Fourier Transform (FFT), and the complexity was reduced to $O(N log_2(N))$. An experiment was conducted with synthetic signal to demonstrate the resolving power of each estimator. Result of this is shown in Figure 10(a).

2. Power Spectral Density

Power spectral density is mathematically represented as the Fourier transform of the autocorrelation sequence r_{xx}. It is given by

$$P_x(\omega) = \sum_{k=-\infty}^{\infty} r_{xx}(k) e^{-jk\omega} \qquad (10)$$

where the autocorrelation sequence, $r_{xx}(k) = \lim_{N \to \infty} \frac{1}{N} \sum_{n=0}^{N-1} x(n+k) x^*(n)$.

There are different ways of estimating power spectrum. One such popular method is Welch's method of averaging periodograms for better accuracy Figure 10(b). The readers are advised to go through chapter 8 of Hayes (2010) for further details.

3. Subspace Based Spectral Estimators

The methods discussed above are not suitable for signals where the sinusoids are closely spaced, or there is considerable amount of noise. In these scenarios, subspace-based methods are used. These methods utilize the property of orthogonality between noise and signal eigenvectors of the signal autocorrelation matrix. These methods can be classified as signal and noise subspace-based methods. Signal subspace based methods or principal component-based methods use vectors that lie in the signal subspace for their computations. Whereas, the noise subspace-based methods or minor component methods are based on the noise eigenvectors. Most of the recent studies on fault detection have mainly concentrated on the noise subspace based methods. The popular methods are MUSIC and ESPRIT. Out of these, we will discuss MUSIC with some details.

MUSIC was developed to find the direction of arrival (DOA) of two closely spaced sources in radar technology and array signal processing and was found to be very efficient in resolving closely spaced sinusoids for fault detection of induction motors. In recent years most of the research in this domain was carried out with signal sources impinging on an array of sensors and the problem is to find the DOA of the signals arising from these sources. Output of the array sensors is considered as the input signal for the spectral analysis. The signal model is represented by

Figure 10. Response of different spectral estimators to a synthetic signal((a) 1024 point FFT, (b) 1024 point PSD via Welch's method, (c)MUSIC pseudo-spectrum with the $N = 1024$)

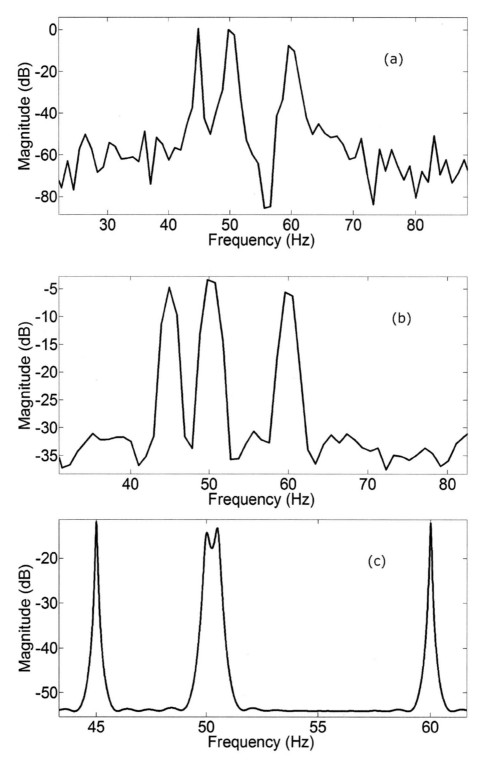

Online Condition Monitoring of Traction Motor

$$x[n] = \sum_{i=1}^{P} s_i e^{j(\omega_i n + \varphi_i)} + v[n]; n = 0, 1, \ldots, (N-1) \tag{11}$$

Where $x[n]$ is the n th instant of the input signal out of N samples and is considered to be complex exponential with additive white Gaussian noise $v[n]$. P is the number of sinusoids, and is considered to be known. Each sinusoid is having an amplitude, s_i, with complex and random phase, φ_i and normalized frequency, ω_i, given by $2\pi f_i T_s$. Here $i = 1, 2, \ldots, P$ and T_s is the sampling period. For mathematical simplicity and to maintain a common structure with the existing works, this model is represented in a vector-matrix form. Where x represents $x[n]$ in vector form, and is given by

$$x = As + v \tag{12}$$

where $\mathbf{s} \in \mathbb{C}^{P \times 1}$ is the vector of P complex amplitudes with random phase

$$s = [s_1 e^{j\varphi_1}, s_2 e^{j\varphi_2}, \ldots, s_p e^{j\varphi_p}]^T \tag{13}$$

and $\mathbf{A} \in \mathbb{C}^{N \times p}$ is the complex source array matrix given by

$$\begin{aligned} A &= [a(\omega_1), a(\omega_2), \ldots, a(\omega_p)], \text{where,} \\ a(\omega_i) &= [1, e^{j w_i}, \ldots, e^{j \omega_i (N-1)}]^T \end{aligned} \tag{14}$$

The noise vector, v is given by

$$v = [v(1), v(2), \ldots, v(n)] \tag{15}$$

The covariance of the input signal is given by

$$\mathbf{R}_x = E[\mathbf{xx}^H] + \sigma^2 \mathbf{I} = \mathbf{A} E[\mathbf{ss}^H] \mathbf{A}^H + \sigma^2 \mathbf{I} = \mathbf{A} \mathbf{R}_s \mathbf{A}^H + \sigma^2 \mathbf{I} \tag{16}$$

Where, σ^2 is the variance of the noise and I is the ($N \times N$) identity matrix. Here, N is the size of the autocorrelation matrix. An analytical derivation to find N is reported by Naha et al. (2015). The largest P eigenvalues of $\mathbf{R}_x$ represents the P sinusoidal components present in the signal, while the other ($m = N - P$) eigenvalues represents the noise. Similarly, the eigenvectors $\mathbf{e}$ corresponding to the P eigenvalues belongs to the signal subspace, whereas the rest, represented by v belongs to the noise subspace. The spectral peaks of MUSIC is computed by

$$h(\omega) = \frac{1}{\sum_{i=1}^{m} |\mathbf{e}^H v_i|^2} \tag{17}$$

where $\mathbf{e} = \left[1, e^{j\omega}, e^{2j\omega}, \cdots e^{j(N-1)\omega}\right]^T$, $\omega = 2\pi f / F_s$. with f being the search space defined by $f \in [0, F_s/2]$, and therefore, $\omega \in [0, \pi]$. It is observed from Figure 10(c) that MUSIC was able to resolve the frequency components close to the fundamental, whereas FFT and PSD were not successful in doing so. For this experiment four sinusoidal signals with frequencies [45, 50.5, 50, 60]Hz each with unity amplitude was used. Noise variance of 0.5 units was used. Sampling frequency was kept at 1000 Samples/s. It is also observed that MUSIC has a very smooth spectrum. As a result false alarm due to spurious peaks is very low as compared to the other methods.

METHODOLOGY

Our main focus for fault detection is based on MCSA, as this method is non-invasive, efficient and can be easily implemented on an embedded hardware. Presence of faults cannot be judged by mere spectral estimation of stator current. The method of fault detection involves pre-processing of the stator current and post processing of the stator spectrum for accurate detection. By now it must be clear that location of the fault specific frequency components depend upon the fundamental supply frequency and rotor speed. So, it is imperative to estimate these quantities before spectral estimation. Presence of fundamental component in the stator current renders the other frequency components invisible and insignificant. It is, therefore, crucial to estimate the fundamental component, store its magnitude for normalization and filter it out to make the signal well-conditioned. A general block diagram used for implementation of this scheme is given below (Figure 11)

1. Estimation of Fundamental Frequency

Fundamental frequency of motor has the highest amplitude in the current spectra. As a result, detection and removal of this component can be accomplished in a variety of ways. In practice, it can either be detected as the maximum component in the current spectra or can also be estimated and tracked using Kalman filter based approaches as presented by Routray et al. (2002). In both the cases, a sharp notch filter can be used for removing this component.

2. Spectral Estimation

Once the stator current is free from the fundamental frequency component, different spectral estimators can be utilized for spectral estimation. It is recommended to use high-resolution spectral estimators based on subspace-based methods for better resolution and accuracy. These methods, although difficult to implement in an online system can be very robust in their performance. In general they are used when the frequency components are deemed to be very close each other. Otherwise, FFT and PSD can also be used. Some results with experiments performed on two motors of 1.5KW and 22KW in lab environment for BRB. Figure 12 shows the FFT, PSD and MUSIC for detection of BRB fault with the motor. In this case the load was high. As a result FFT and PSD could resolve the fault components. Although peaks with MUSIC are more prominent. It is also observed, that the peak frequency information with MUSIC is more accurate than FFT and PSD. Still there are issues related to accurate estimation of amplitude and computation time. In low load applications, FFT and PSD require more number of time samples

Figure 11. Fault detection scheme

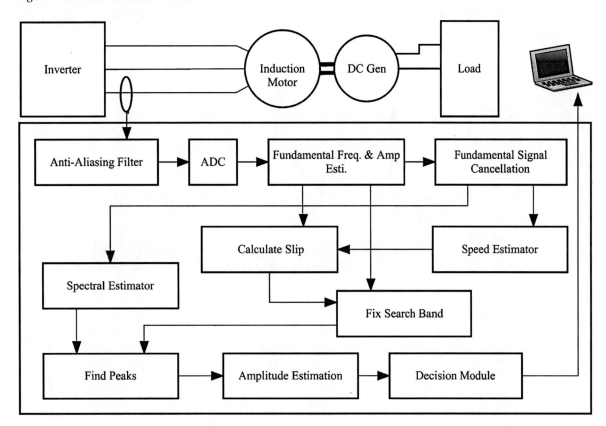

compared to MUSIC for resolving the fault components. Further results with MUSIC for a motor running with different supply frequency and load can be seen in Figure 13. For the experiments on BRB, a 1.5 kW, 4-pole, 220 V, 6.2 Amps, 3-phase squirrel cage induction motor with 45 rotor bars has been used. The motor was mechanically coupled to a dc generator for rheostatic loading. The motor is fed from a V/f drive and its 3-phase supply voltage and 3-phase stator current are recorded. For experiments with the eccentricity fault, a 22kW motor coupled to a 25kW separately excited DC generator was used. The eccentricity condition was found to be inherently present in one of the motor. When compared to an healthy motor, the magnitude of the fault component characteristic to the mixed eccentricity component was found to be more. A maximum deviation of 0.4 mm was found in the motor shaft when it was rotated for a single rotation for the faulty motor. Figure 14 shows the spectral signature of this motor with FFT, PSD and MUSIC. Figure 15 shows the MUSIC pseudospectrum for the healthy as well as the faulty motor running with 4% load. In both the cases, the stator current carried the inherent eccentricity component. But for the faulty motor the magnitude was quite high as compared to the healthy motor. Voltage and current signals for both the motors were recorded with 1 kHz sampling frequency using a Yokogawa DL-750 data logger. The appearance of characteristic fault frequency components can be verified from the equation provided for these two faults.

Figure 12. Spectrum of stator current of a 1.5 KW induction motor for BRB fault obtained with different spectral estimators ((a)4096 point FFT (b)4096 point PSD via Welch's method and (c) MUSIC pseudo-spectrum)

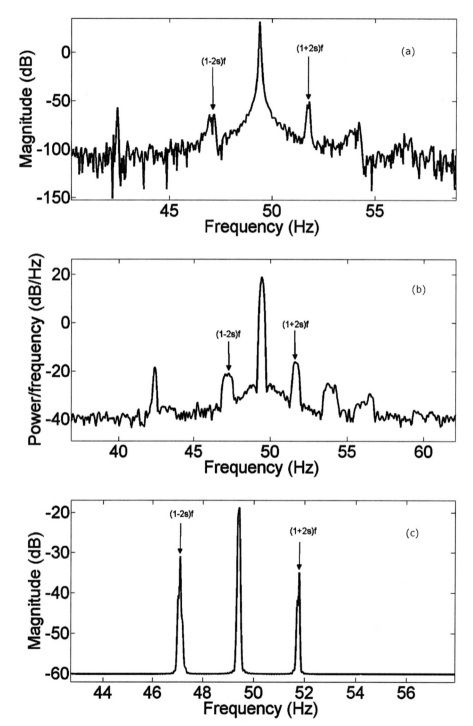

Figure 13. MUSIC pseudo-spectrum of stator current of a 1.5 KW induction motor for BRB fault obtained with load and supply frequency ((a)30 Hz, 9% load and 1.6% slip (b) 40 Hz, 16% load and 1% slip (c) 50Hz, 20% load and 0.8% slip)

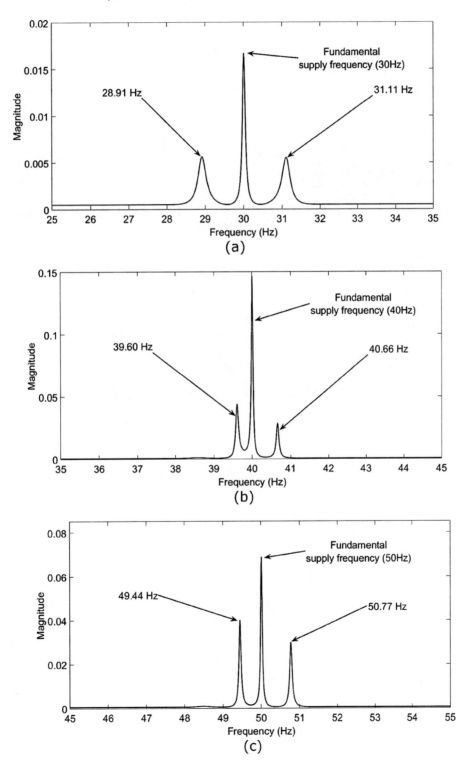

Figure 14. Spectrum of stator current of a 22 KW induction motor with 29% load for mixed eccentricity fault obtained with different spectral estimators((a)4096 point FFT (b)4096 point PSD via Welch's method and (c) MUSIC pseudo-spectrum)

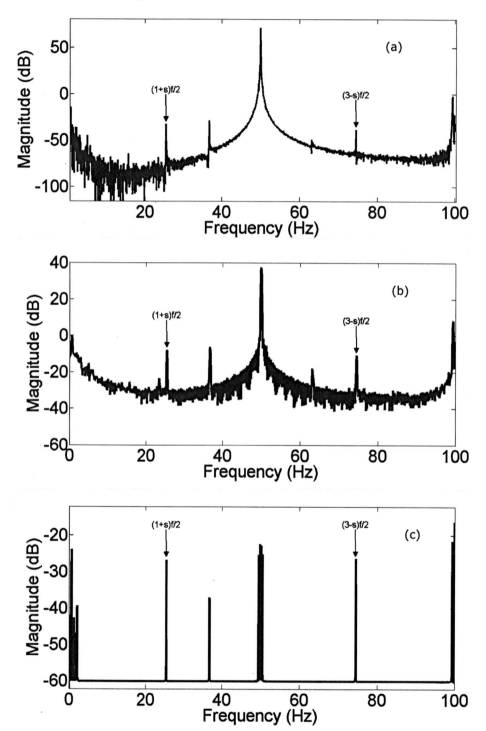

Figure 15. MUSIC pseudo-spectrum of stator current of a 22 KW induction motor with 4% load for mixed eccentricity fault obtained with ((a) Healthy motor (b)Faulty Motor)

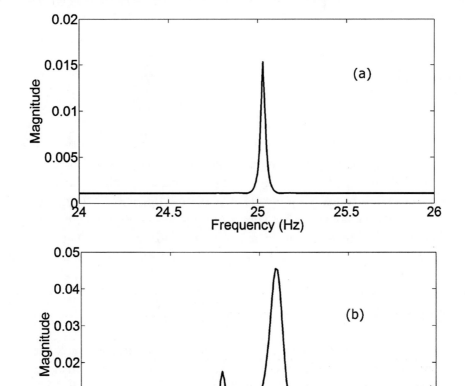

3. Estimation of Rotor Speed

Estimation of rotor speed is very essential as this information is used for searching specific bands in the current spectra for fault detection. Induction motor speed is normally measured using shaft mounted tachometer or by optical speed sensor. Alternative to the physical speed sensors are the soft sensors. Soft sensors measure the speed using the machine current and voltage signals. This soft sensing of speed is also referred to as sensor-less speed estimation method. Physical speed sensors are costly, fragile, and difficult to install in some applications Gao et al. (2011); Li et al. (2006). Also in case of physical speed sensor failure, soft sensing of speed is required for designing fault-tolerant systems, Li et al. (2006). The sensorless speed estimation schemes found in the literature can be classified broadly into two groups as shown in Nandi (2004) as follows

1. Model-observer based schemes: In this scheme, using the current and voltage information the speed is estimated from the induction motor model, Li et al. (2006); Liu et al. (2011);

2. Rotor Slot Harmonic (RSH) based schemes: Slot harmonics in the stator current spectrum are correlated with the speed of the motor. First the slot harmonic frequency is estimated, then speed is calculated from the frequency as shown by Gao et al. (2011). Slot harmonic frequency and the speed are related as by putting $n_d = 0$, $v = 1$, $k = 1$ in (2). The equation is given by

$$f_{PSH} = \left[R\frac{(1-s)}{p} \pm 1 \right] f \tag{18}$$

where f_{PSH} is the slot harmonic frequency, f is the supply frequency, s is slip of the machine, p is the number of pole pairs and should not be confused with the number of sinusoids represented by P.

RSH based methods require spectrum analysis which is time consuming Zaky et al. (2009). These methods do not give satisfactory results for light load, or no-load operation of the machine and for some specific configuration slot harmonics do not appear in the stator current spectrum Gao et al. (2011). Hence, the observer based methods are considered to be better for various loading conditions and also under no-load operations. Observer based methods need exact information about the machine electrical and mechanical parameters

4. Search Bands and Decision

Once the spectrum is estimated, it is required to find the whether particular fault specific frequency components are present. These components are again slip and supply frequency dependent. As a result, dynamic search bands are constructed based on these information for each fault. It is observed that mere presence of these fault components doesn't imply the presence of faults. Moreover, the degree of fault also needs to be quantified. As a result, magnitude of the fault specific components is found out and then are normalized by the magnitude of the fundamental component. A threshold value for each fault is decided a-priori. The normalized magnitude is then compared with this threshold and it is decided whether there is a fault or not. Magnitude of the normalized value can also be used to quantify the amount of fault that has occurred.

ONLINE IMPLEMENTATION SCHEMES

It is vital for condition monitoring systems to actually monitor the signals from an operational motor. Due to the amount of computations involved, the operations are in general carried out in a non-realtime frame-processing based method. For a constricted environment as in a locomotive cabin, the design of the ystem should small, portable and robust to temperature, vibrations and other environmental effects. Khan et al. (2007) implemented an online wavelet based induction motor fault detection system. The online implementation is done using Texas Instrument's board with 32bit floating point processor, TMS320C31. A schematic diagram of the system is given in Figure 16(a). Ordaz-Moreno et al. (2008) have developed low cost FPGA-based online system for BRB detection. The detection scheme is based on the Discrete Wavelet Transform (DWT) on the startup current to identify fault during motor startup. XC3S200 Spartan-3 FPGA from Xilinx is used for online implementation. Rangel-Magdaleno et al. (2009) have reported the development of a FAGA based special purpose System-on-a-Chip (SoC) solu-

Online Condition Monitoring of Traction Motor

tion for online detection of half BRB fault in IM. This method used both current and 3-axis vibration signal to increase the detectability under various loading conditions. XC3S500E Spartan-3 FPGA from Xilinx is used for online implementation. A schematic diagram of the system is given in Figure 16(b). Romero-Troncoso et al. (2011) have implemented another FPGA based online system to detect induction machine faults. The system works based on the Entropy and Fuzzy Inference to detect multiple combined IM faults. FPGA Cyclone-II EP2C35F672C6 from Altera is used for online implementation. A schematic diagram of the system is given in Figure 17(a). Pineda-Sanchez et al. (2013) have detected broken bar in an induction motor by online implementation of detection algorithm in a DSP commercial board, TMS320F28335. A schematic diagram of the system is given in Figure 17(b).

More recently, some of the high performance low-cost ARM-based boards have broadened the choice of hardware. Still there are issues with the computational ability and analog input interfaces. MATLAB

Figure 16. Online IM fault diagnosis system proposed by (a) Khan et al. (2007) (b) Rangel-Magdaleno et al. (2009)

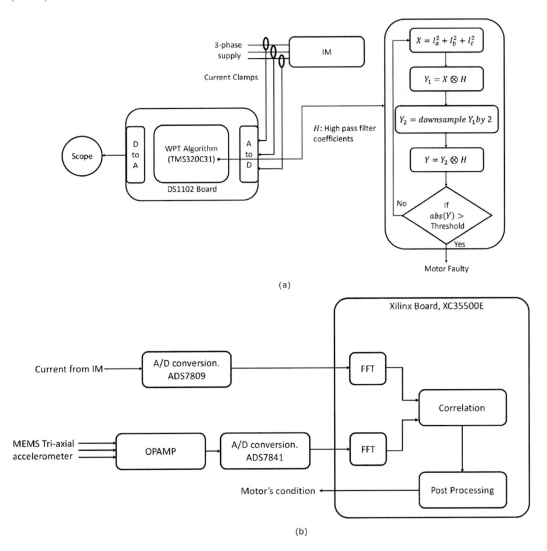

Figure 17. Online IM fault diagnosis system proposed by (a) Romero-Troncoso et al. (2011) and (b) Pineda-Sanchez et al. (2013)

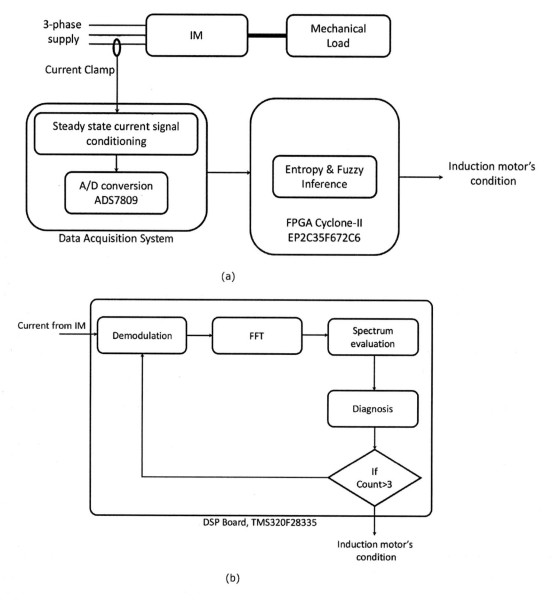

in recent years has developed some interesting tools for implementing SIMULINK/MATLAB models and codes in hardware. The hardware choices are also multiple in numbers, and developers can select them according to their requirements. One such very generic tool is Simulink Realtime. Use of this tool for arc fault detection can be found in Mukherjee et al. (2015). This tool is supported by a realtime operating system with the same name and formerly known as xPC Target. Once software system is developed in SIMULINK in the host computer, with the necessary configurations as required by the tool, it can be dumped into an Intel-based target system for real time execution. The host-target interface is established through Ethernet. A hard disk is used for logging of data, which can be accessed from the host computer. The target computer can also run in standalone mode with no intervention from the host computer. A model was designed to implement an online fault detection system. This system takes current and voltage inputs from the motor and then process them to derive various parameters. The top level schematic of this system is provided in Figure 18. The current and voltages are fed to the 'speed' estimation block for estimation of the speed. An Extended Kalman based 'ECKF1' block is used for estimation of the fundamental frequency. Additionally, this block is used for the input signal conditioning of the stator current by using a notch filter. Once the signal is conditioned, it is sent to the 'Spectral Estimation' block. The spectral estimation block estimates the spectrum in a specified fixed band. Mean values of speed and fundamental frequency are used to calculate the slip. Once the slip is calculated, it is used to fix the search bands. Presence of peak, and their relative magnitude are searched in the bands to confirm the presence of the fault, and also quantify it.

FUTURE RESEARCH DIRECTIONS

In the case of induction motor, the fault specific signatures are well known, but the challenge is to detect those frequencies in offline or online mode. Even though the literature for the detection of fault frequency components in armature current (MCSA) in very rich, there are still many open areas where more research needs to be carried out. MCSA is very popular for detecting faults in IM because of its noninvasiveness, and it requires very less information about the parameters of the motor. The major problems with the traditional MCSA based method is that they assume the condition that the motor is at steady state. So, these traditional methods will not work under transient conditions such as motor starting or braking operation or if the load is time varying in nature or during supply voltage fluctuations. In practice, the loads are time varying in nature for many industry applications including the tractive application in Indian Railways. There are some research works reported in the literature proposing the fault detection in the transient regime using wavelet transforms. The induction motors driven by power electronic drives cannot be diagnosed by those transient methods because the slip remains almost same during startup or breaking operations. Another major challenge is to apply the fault detection methods online. For example, developing an onboard system inside the locomotive to detect weak faults in the

Figure 18. Top level schematic diagram in SIMULINK for fault detection.

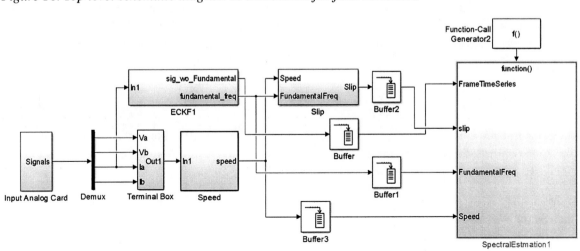

traction motors or other auxiliary motors at very light load and also to predict the remaining useful life. Furthermore, there are conditions when the motor is running under very light load. In this case detecting BRB is very challenging, as the fault components are very close to the fundamental supply frequency. An elegant signal conditioning unit is essential to reduce the effect of the dominant fundamental frequency. This system should track the fundamental, and the cutoff bandwidth should adapt to changes in the slip so as not to suppress the fault components. Even though there are many future scopes in this area, we can narrow down to three major areas which can be useful for Indian Railways,

1. Detection of faults when load is time varying and with supply voltage fluctuations
2. Detection of faults when motor in transient states, when the motor is driven by power electronic drives
3. Detection under light load with major emphasis on efficient signal pre-conditioning
4. Online implementation of fault detection algorithms and prediction of remaining useful life

CONCLUSION

The maintenance strategy followed by Indian Railways is mainly based on scheduled maintenance, which is more costly and time consuming than the condition based maintenance. Traction motors are protected online overcurrent relay and temperature sensors. Currently there is now way to detect incipient fault (like single broken rotor bar, small amount of eccentricity, small problem in bearings etc.) in Traction motors and other auxiliary induction motors. Through this chapter, the current trend in research and the state of the art for fault diagnosis of a induction motor is presented. It is imperative to note that most of the research in this domain has been carried out with low to medium power motors. The fault components arising in large and small SCIMs are similar to each other. As a result, same signatures and their analysis can be carried out for the traction motors and small motors alike. It is also prerogative to mention that MCSA is the most popular method for detection of the faults. In some recent studies, even bearing faults are detected using the stator current instead of vibrations. This increases the reliability and robustness of an online system. Implementations of the algorithms in an online system to detect weak faults and predict remaining useful life of traction motors and other auxiliary induction motors in the Indian Railways can be efficient and economical. But this implementations needs to take care of the harsh environment that is prevalent inside the locomotive. The plots of Figure 11, Figure 13, Figure 14 is the outcome of an original research carried out at IIT Kharagpur.

REFERENCES

Akin, B., Orguner, U., Toliyat, H. A., & Rayner, M. (2008). Low order PWM inverter harmonics contributions to the inverter-fed induction machine fault diagnosis. *IEEE Transactions on Industrial Electronics*, 55(2), 610–619. doi:10.1109/TIE.2007.911954

Andriamalala, R. N., Razik, H., Baghli, L., & Sargos, F.-. (2008). Eccentricity Fault Diagnosis of a Dual-Stator Winding Induction Machine Drive Considering the Slotting Effects. *IEEE Transactions on Industrial Electronics*, 55(12), 4238–4251. doi:10.1109/TIE.2008.2004664

Antonino-daviu, J., Riera-guasp, M., Pons-llinares, J., Park, J., Lee, S. B., Yoo, J., & Kral, C. (2012). Detection of Broken Outer-Cage Bars for Double-Cage Induction Motors under the Startup Transient. *IEEE Transactions on Industry Applications, 48*(5), 1539–1548. doi:10.1109/TIA.2012.2210173

Ayhan, B., Trussell, H. J., Chow, M.-Y., & Song, M.-H. (2008). On the use of a lower sampling rate for broken rotor bar detection with DTFT and AR-based spectrum methods. *IEEE Transactions on Industrial Electronics, 55*(3), 1421–1434. doi:10.1109/TIE.2007.896522

Blodt, M., Regnier, J., & Faucher, J. (2009). Distinguishing Load Torque Oscillations and Eccentricity Faults in Induction Motors Using Stator Current Wigner Distributions. *IEEE Transactions on Industry Applications, 45*(6), 1991–2000. doi:10.1109/TIA.2009.2031888

Bruzzese, C. (2008). Analysis and Application of Particular Current Signatures (Symptoms) for Cage Monitoring in Non-sinusoidally Fed Motors with High Rejection to Drive Load, Inertia, and Frequency Variations. *IEEE Transactions on Industrial Electronics, 55*(12), 4137–4155. doi:10.1109/TIE.2008.2004669

Cabanas, M. F., Pedrayes, F., & Melero, M. G., Garcia, Rojas C. H., Cano, J. M., Orcajo, G. A., & Norniella, J. G. (2011). Unambiguous Detection of Broken Bars in Asynchronous Motors by Means of a Flux Measurement-Based Procedure. *IEEE Transactions on Instrumentation and Measurement, 60*(3), 891–899. doi:10.1109/TIM.2010.2062711

Ceban, A., Pusca, R., & Romary, R. (2012). Study of Rotor Faults in Induction Motors Using External Magnetic Field Analysis. *IEEE Transactions on Industrial Electronics, 59*(5), 2082–2093. doi:10.1109/TIE.2011.2163285

Choi, S., Akin, B., Rahimian, M. M., & Toliyat, H. A. (2011). Implementation of a Fault-Diagnosis Algorithm for Induction Machines Based on Advanced. *IEEE Transactions on Industrial Electronics, 58*(3), 937–948. doi:10.1109/TIE.2010.2048837

Chow, T. W. S., & Tan, H.-Z. (2000). HOS-based nonparametric and parametric methodologies for machine fault detection. *IEEE Transactions on Industrial Electronics, 47*(5), 1051–1059. doi:10.1109/41.873213

Concari, C., Franceschini, G., & Tassoni, C. (2011). Toward Practical Quantification of Induction Drive Mixed Eccentricity. *IEEE Transactions on Industry Applications, 47*(3), 1232–1239. doi:10.1109/TIA.2011.2124434

Cruz, M. A. (2012). An Active Reactive Power Method for the Diagnosis of Rotor Faults in Three-Phase Induction Motors Operating Under Time-Varying Load Conditions. *IEEE Transactions on Energy Conversion, 27*(1), 71–84. doi:10.1109/TEC.2011.2178027

Da Silva, A. M., Povinelli, R. J., & Demerdash, N. A. O. (2008). Induction machine broken bar and stator short-circuit fault diagnostics based on three-phase stator current envelopes. *IEEE Transactions on Industrial Electronics, 55*(3), 1310–1318. doi:10.1109/TIE.2007.909060

Das, S., Purkait, P., Koley, C., & Chakravorti, S. (2014). Performance of a load-immune classifier for robust identification of minor faults in induction motor stator winding. Dielectrics and Electrical Insulation. *IEEE Transactions on, 21*(1), 33–44.

Dorrell, D. G. (2011). Sources and Characteristics of Unbalanced Magnetic Pull in Three-Phase Cage Induction Motors With Axial-Varying Rotor Eccentricity. *IEEE Transactions on Industry Applications, 47*(1), 12–24. doi:10.1109/TIA.2010.2090845

Drif, M., & Cardoso, A. J. M. (2012). Discriminating the Simultaneous Occurrence of Three-Phase Induction Motor Rotor Faults and Mechanical Load Oscillations by the Instantaneous Active and Reactive Power Media Signature Analyses. *IEEE Transactions on Industrial Electronics, 59*(3), 1630–1639. doi:10.1109/TIE.2011.2161252

Drif, M., & Cardoso, A. J. M. (2014). Stator fault diagnostics in squirrel cage three-phase induction motor drives using the instantaneous active and reactive power signature analyses. Industrial Informatics. *IEEE Transactions on, 10*(2), 1348–1360.

Eren, L., & Devaney, M. J. (2004). Bearing damage detection via wavelet packet decomposition of the stator current. *IEEE Transactions on Instrumentation and Measurement, 53*(2), 431–436. doi:10.1109/TIM.2004.823323

Esfahani, E. T., Wang, S., & Sundararajan, V. (2013). Multisensor Wireless System for Eccentricity and Bearing Fault Detection in Induction Motors. IEEE/ASME Trans. *Mechatronics, 19*(3), 818–826. doi:10.1109/TMECH.2013.2260865

Faiz, J., Ardekanei, I. T., & Toliyat, H. A. (2003). An Evaluation of Inductances of a Squirrel-Cage Induction Motor Under Mixed Eccentric Conditions. Energy Conversion. *IEEE Trans., 18*(2), 252–258.

Frosini, L., & Bassi, E. (2010). Stator current and motor efficiency as indicators for different types of bearing faults in induction motors. *IEEE Transactions on Industrial Electronics, 57*(1), 244–251. doi:10.1109/TIE.2009.2026770

Frosini, L., Harlisca, C., & Szabo, L. (2015). Induction Machine Bearing Fault Detection by Means of Statistical Processing of the Stray Flux Measurement. *IEEE Transactions on Industrial Electronics, 62*(3), 1846–1854. doi:10.1109/TIE.2014.2361115

Gandhi, A., Corrigan, T., & Parsa, L. (2011). Recent advances in modeling and online detection of stator interturn faults in electrical motors. *Ind. Electron. IEEE Trans., 58*(5), 1564–1575. doi:10.1109/TIE.2010.2089937

Gao, Z., Turner, L., Colby, R. S., & Leprettre, B. (2011). A Frequency Demodulation Approach to Induction Motor Speed Detection. *IEEE Transactions on Industry Applications, 47*(4), 1632–1642. doi:10.1109/TIA.2011.2153813

Garcia-perez, A., Romero-troncoso, R. D. J., Cabal-yepez, E., & Osornio-rios, R. A. (2011). The Application of High-Resolution Spectral Analysis for Identifying Multiple Combined Faults in Induction Motors. *IEEE Transactions on Industrial Electronics, 58*(5), 2002–2010. doi:10.1109/TIE.2010.2051398

Georgakopoulos, I. P., Mitronikas, E. D., & Safacas, A. N. (2011). Detection of Induction Motor Faults in Inverter Drives Using Inverter Input Current Analysis. *IEEE Transactions on Industrial Electronics, 58*(9), 4365–4373. doi:10.1109/TIE.2010.2093476

Ghate, V. N., & Dudul, S. V. (2011). Cascade Neural-Network-Based Fault Classifier for Three-Phase Induction Motor. *IEEE Transactions on Industrial Electronics*, *58*(5), 1555–1563. doi:10.1109/TIE.2010.2053337

Gyftakis, K. N., & Kappatou, J. C. (2013). A Novel and Effective Method of Static Eccentricity Diagnosis in Three-Phase PSH Induction Motors. *IEEE Transactions on Energy Conversion*, *28*(2), 405–412. doi:10.1109/TEC.2013.2246867

Hayes, M. H. (2010). *Statistical Digital Signal Processing and Modeling*. John Wiley & Sons.

Henao, H., Demian, C., & Capolino, G.-A. (2003). A frequency-domain detection of stator winding faults in induction machines using an external flux sensor. *IEEE Transactions on Industry Applications*, *39*(5), 1272–1279. doi:10.1109/TIA.2003.816531

Huang, X., Habetler, T. G., Harley, R. G., & Wiedenbrug, E. J. (2007). Using a Surge Tester to Detect Rotor Eccentricity Faults in Induction Motors. *IEEE Transactions on Industry Applications*, *43*(5), 1183–1190. doi:10.1109/TIA.2007.904389

Hyun, D., Hong, J., Lee, S. B., Kim, K., Wiedenbrug, E. J., Teska, M., & Nandi, S. (2011). Automated Monitoring of Airgap Eccentricity for Inverter-Fed Induction Motors Under Standstill Conditions. *IEEE Transactions on Industry Applications*, *47*(3), 1257–1266. doi:10.1109/TIA.2011.2126010

Ibrahim, A., El Badaoui, M., Guillet, F., & Bonnardot, F. (2008). A new bearing fault detection method in induction machines based on instantaneous power factor. *IEEE Transactions on Industrial Electronics*, *55*(12), 4252–4259. doi:10.1109/TIE.2008.2003211

Isermann, R. (2006). *Fault-diagnosis systems*. Springer. doi:10.1007/3-540-30368-5

Jin, X., Zhao, M., Chow, T. W. S., & Pecht, M. (2014). Motor bearing fault diagnosis using trace ratio linear discriminant analysis. *IEEE Transactions on Industrial Electronics*, *61*(5), 2441–2451. doi:10.1109/TIE.2013.2273471

Joksimovic, G. M., & Penman, J. (2000). The detection of inter-turn short circuits in the stator windings of operating motors. *IEEE Transactions on Industrial Electronics*, *47*(5), 1078–1084. doi:10.1109/41.873216

Kaikaa, M. Y., Hadjami, M., & Khezzar, A. (2014). Effects of the Simultaneous Presence of Static Eccentricity and Broken Rotor Bars on the Stator Current of Induction Machine. *IEEE Transactions on Industrial Electronics*, *61*(5), 2452–2463. doi:10.1109/TIE.2013.2270216

Kang, T.-, Kim, J., Lee, S. B., & Yung, C. (2015). Experimental Evaluation of Low-Voltage Offline Testing for Induction Motor Rotor Fault Diagnostics. *IEEE Transactions on Industry Applications*, *51*(2), 1375–1384. doi:10.1109/TIA.2014.2344504

Kankar, P. K., Sharma, S. C., & Harsha, S. P. (2011). Fault diagnosis of ball bearings using machine learning methods. *Expert Systems with Applications*, *38*(3), 1876–1886. doi:10.1016/j.eswa.2010.07.119

Khan, M., Radwan, T. S., & Rahman, M. A. (2007). Real-time implementation of wavelet packet transform-based diagnosis and protection of three-phase induction motors. *IEEE Transactions on Energy Conversion*, *22*(3), 647–655. doi:10.1109/TEC.2006.882417

Kim, D.-J., Kim, H.-J., Hong, J.-P., & Park, C.-J. (2014). Estimation of Acoustic Noise and Vibration in an Induction Machine Considering Rotor Eccentricity. *IEEE Transactions on Magnetics*, *50*(2), 857–860. doi:10.1109/TMAG.2013.2285391

Kim, J., Member, S., Shin, S., Member, S., Lee, S. B., Member, S., & Member, S. et al. (2015). *Power Spectrum-Based Detection of Induction Motor Rotor Faults for Immunity to False Alarms. IEEE Trans* (pp. 1–10). Energy Convers.

Kim, Y.-H., Youn, Y.-W., Hwang, D.-H., Sun, J.-H., & Kang, D.-S. (2013). High-Resolution Parameter Estimation Method to Identify Broken Rotor Bar Faults in Induction Motors. *IEEE Transactions on Industrial Electronics*, *60*(9), 4103–4117. doi:10.1109/TIE.2012.2227912

Knight, A. M., & Bertani, S. P. (2005). Mechanical Fault Detection in a Medium-Sized Induction Motor Using Stator Current Monitoring. *IEEE Transactions on Energy Conversion*, *20*(4), 753–760. doi:10.1109/TEC.2005.853731

Leite, V. C. M. N., Borges da Silva, J. G., Veloso, G. F. C., Borges da Silva, L. E., Lambert-Torres, G., Bonaldi, E. L., & de Oliveira, L. E. D. L. (2015). Detection of Localized Bearing Faults in Induction Machines by Spectral Kurtosis and Envelope Analysis of Stator Current. *IEEE Transactions on Industrial Electronics*, *62*(3), 1855–1865. doi:10.1109/TIE.2014.2345330

Li, M., Chiasson, J., Bodson, M., & Tolbert, L. M. (2006). A Differential-Algebraic Approach to Speed Estimation in an Induction Motor. *IEEE Transactions on Automatic Control*, *51*(7), 1172–1177. doi:10.1109/TAC.2006.878775

Liu, P., Hung, C.-Y., Chiu, C.-S., & Lian, K.-Y. (2011). Sensorless linear induction motor speed tracking using fuzzy observers. *IET Electr. Power Appl.*, *5*(4), 325. doi:10.1049/iet-epa.2010.0099

Liu, Z., Yin, X., Zhang, Z., Chen, D., & Chen, W. (2004). Online rotor mixed fault diagnosis way based on spectrum analysis of instantaneous power in squirrel cage induction motors. *IEEE Transactions on Energy Conversion*, *19*(3), 485–490. doi:10.1109/TEC.2004.832052

Loparo, K. A., Adams, M. L., Lin, W., Abdel-Magied, M. F., & Afshari, N. (2000). Fault detection and diagnosis of rotating machinery. *IEEE Transactions on Industrial Electronics*, *47*(5), 1005–1014. doi:10.1109/41.873208

Mukherjee, A., Routray, A., and Samanta, A. (2015). Method for On-line Detection of Arcing in Low Voltage Distribution Systems. Power Delivery, IEEE Trans. Power Del., PP(99):1.

Naha, A., Samanta, A. K., Routray, A., & Deb, A. K. (2015). Determining Autocorrelation Matrix Size and Sampling Frequency for MUSIC Algorithm. *IEEE Signal Processing Letters*, *22*(8), 1016–1020. doi:10.1109/LSP.2014.2366638

Nandi, S. (2004). Modeling of Induction Machines Including Stator and Rotor Slot Effects. *IEEE Transactions on Industry Applications*, *40*(4), 1058–1065. doi:10.1109/TIA.2004.830764

Nandi, S., Ahmed, S., & Toliyat, H. A. (2001). Detection of rotor slot and other eccentricity related harmonics in a three phase induction motor with different rotor cages. *IEEE Transactions on Energy Conversion*, *16*(3), 253–260. doi:10.1109/60.937205

Nandi, S., Ilamparithi, T. C., Lee, S. B., & Hyun, D. (2011). Detection of Eccentricity Faults in Induction Machines Based on Nameplate Parameters. *IEEE Transactions on Industrial Electronics, 58*(5), 1673–1683. doi:10.1109/TIE.2010.2055772

Nandi, S., & Toliyat, H. A. (1999). Condition monitoring and fault diagnosis of electrical machines-a review. Conference Record of the 1999 IEEE Industry Applications Conference, Thirty-Fourth IAS Annual Meeting (Vol. 1, pp. 197–204). IEEE. doi:10.1109/IAS.1999.799956

Nandi, S., Toliyat, H. A., & Li, X. (2005). Condition Monitoring and Fault Diagnosis of Electrical Motors A Review. *IEEE Transactions on Energy Conversion, 20*(4), 719–729. doi:10.1109/TEC.2005.847955

Nemec, M., Drobnic, K., Nedeljkovic, D., & Rastko, F. (2010). Detection of Broken Bars in Induction Motor Through the Analysis of Supply Voltage Modulation. *IEEE Transactions on Industrial Electronics, 57*(8), 2879–2888. doi:10.1109/TIE.2009.2035991

Ocak, H., & Loparo, K. A. (2001). A new bearing fault detection and diagnosis scheme based on hidden Markov modeling of vibration signals. Proceedings of the 2001 IEEE Int. Conf. on Acoust. Speech, Signal Process ICASSP '01 (Vol. 5, pp. 3141–3144). IEEE. doi:10.1109/ICASSP.2001.940324

Ordaz-Moreno, A., de Jesus Romero-Troncoso, R., Vite-Frias, J. A., Rivera-Gillen, J. R., & Garcia-Perez, A. (2008). Automatic online diagnosis algorithm for broken-bar detection on induction motors based on discrete wavelet transform for fpga implementation.. *IEEE Transactions on* Industrial Electronics, 55(5), 2193–2202.

Picazo-Rodenas, M. J., Royo, R., Antonino-Daviu, J., & Roger-Folch, J. (2012). Use of infrared thermography for computation of heating curves and preliminary failure detection in induction motors. *Proceedings of the 2012 20th Int. Conf. Electr. Mach.* (pp. 525–531).

Pineda-Sanchez, M., Perez-Cruz, J., Roger-Folch, J., Riera-Guasp, M., Sapena-Bano, A., & Puche-Panadero, R. (2013). Diagnosis of Induction Motor Faults using a DSP and Advanced Demodulation Techniques. *Proceedings of the 2013 9th IEEE International Symposium SDEMPED* (pp. 609–76. IEEE.

Pons-Llinares, J., Antonino-daviu, J. A., Riera-guasp, M., Lee, S. B., Kang, T.-J., & Yang, C. (2015). Advanced Induction Motor Rotor Fault Diagnosis via Continuous and Discrete Time-Frequency Tools. *IEEE Transactions on Industrial Electronics, 62*(3), 1791–1802. doi:10.1109/TIE.2014.2355816

Puche-Panadero, R., Pineda-Sanchez, M., Riera-Guasp, M., Roger-Folch, J., Hurtado-Perez, E., & Perez-Cruz, J. (2009). Improved Resolution of the MCSA Method Via Hilbert Transform, Enabling the Diagnosis of Rotor Asymmetries at Very Low Slip. *IEEE Transactions on Energy Conversion, 24*(1), 52–59. doi:10.1109/TEC.2008.2003207

Purushotham, V., Narayanan, S., & Prasad, S. A. N. (2005). Multi-fault diagnosis of rolling bearing elements using wavelet analysis and hidden Markov model based fault recognition. *NDT & E International, 38*(8), 654–664. doi:10.1016/j.ndteint.2005.04.003

Rangel-Magdaleno, J., Romero-Troncoso, R., Osornio-Rios, R. A., Cabal-Yepez, E., & Contreras-Medina, L. M. (2009). Novel methodology for online half-broken-bar detection on induction motors. *IEEE Transactions on Instrumentation and Measurement, 58*(5), 1690–1698. doi:10.1109/TIM.2009.2012932

Riera-Guasp, M., Pineda-Sanchez, M., Perez-Cruz, J., Puche-Panadero, R., Roger-Folch, J., & Antonino-Daviu, J. A. (2012). Diagnosis of induction motor faults via gabor analysis of the current in transient regime. *IEEE Transactions on Instrumentation and Measurement, 61*(6), 1583–1596. doi:10.1109/TIM.2012.2186650

Romero-Troncoso, R. J., Pena-Anaya, M., Cabal-Yepez, E., Garcia-Perez, A., & Osornio-Rios, R. (2012). Reconfigurable SoC-Based Smart Sensor for Wavelet and Wavelet Packet Analysis. *IEEE Transactions on Instrumentation and Measurement, 61*(9), 2458–2468. doi:10.1109/TIM.2012.2190340

Romero-Troncoso, R. J., Saucedo-Gallaga, R., Cabal-Yepez, E., Garcia-Perez, A., Osornio-Rios, R., Alvarez-Salas, R., & Huber, N. et al. (2011). FPGA-Based Online Detection of Multiple Combined Faults in Induction Motors through Information Entropy and Fuzzy Inference. *IEEE Transactions on Industrial Electronics, 58*(11), 5263–5270. doi:10.1109/TIE.2011.2123858

Routray, A., Pradhan, A. K., & Rao, K. P. (2002). A novel Kalman filter for frequency estimation of distorted signals in power systems. *IEEE Transactions on Instrumentation and Measurement, 51*(3), 469–479. doi:10.1109/TIM.2002.1017717

Sapena-bano, A., Pineda-sanchez, M., Puche-panadero, R., Perez-cruz, J., Roger-folch, J., Riera-guasp, M., and Martinezroman, J. (2015). Harmonic Order Tracking Analysis: A Novel Method for Fault Diagnosis in Induction Machines. *IEEE Trans. Energy Convers., 1*, 1–9.

Seera, M., Lim, C. P., Ishak, D., & Singh, H. (2012). Fault Detection and Diagnosis of Induction Motors Using Motor Current Signature Analysis and a Hybrid FMM CART Model. *IEEE Trans. Neural Networks Learn. Syst., 23*(1), 97–108. doi:10.1109/TNNLS.2011.2178443 PMID:24808459

Seshadrinath, J., Singh, B., & Panigrahi, B. K. (2014). Investigation of Vibration Signatures for Multiple Fault Diagnosis in Variable Frequency Drives Using Complex Wavelets. *IEEE Transactions on Power Electronics, 29*(2), 936–945. doi:10.1109/TPEL.2013.2257869

Stavrou, A., Sedding, H. G., & Penman, J. (2001). Current monitoring for detecting inter-turn short circuits in induction motors. *IEEE Transactions on Energy Conversion, 16*(1), 32–37. doi:10.1109/60.911400

Tallam, R. M., Lee, S. B., Stone, G. C., Kliman, G. B., Yoo, J.-Y., Habetler, T. G., & Harley, R. G. (2007). A survey of methods for detection of stator-related faults in induction machines. *Ind. Appl. IEEE Trans., 43*(4), 920–933. doi:10.1109/TIA.2007.900448

Tandon, N., Yadava, G. S., & Ramakrishna, K. M. (2007). A comparison of some condition monitoring techniques for the detection of defect in induction motor ball bearings. *Mechanical Systems and Signal Processing, 21*(1), 244–256. doi:10.1016/j.ymssp.2005.08.005

Teotrakool, K., Devaney, M. J., & Eren, L. (2009). Adjustable-speed drive bearing-fault detection via wavelet packet decomposition. *IEEE Transactions on Instrumentation and Measurement, 58*(8), 2747–2754. doi:10.1109/TIM.2009.2016292

Thomson, W. T., & Fenger, M. (2001). Current signature analysis to detect induction motor faults. *Ind. Appl. Mag. IEEE, 7*(4), 26–34. doi:10.1109/2943.930988

Thomson, W. T., Rankin, D., & Dorrell, D. G. (1999). On-line current monitoring to diagnose airgap eccentricity in large three-phase induction motors-industrial case histories verify the predictions. *IEEE Transactions on Energy Conversion, 14*(4), 1372–1378. doi:10.1109/60.815075

Xu, B., Sun, L., & Ren, H. (2010). A New Criterion for the Quantification of Broken Rotor Bars in Induction Motors. *IEEE Transactions on Energy Conversion, 25*(1), 100–106. doi:10.1109/TEC.2009.2032626

Xu, B., Sun, L., Xu, L., & Xu, G. (2012). An ESPRIT-SAA-Based Detection Method for Broken Rotor Bar Fault in Induction Motors. *IEEE Transactions on Energy Conversion, 27*(3), 654–660. doi:10.1109/TEC.2012.2194148

Xu, B., Sun, L., Xu, L., & Xu, G. (2013). Improvement of the Hilbert Method via ESPRIT for Detecting Rotor Fault in Induction Motors at Low Slip. *IEEE Transactions on Energy Conversion, 28*(1), 225–233. doi:10.1109/TEC.2012.2236557

Yang, C., Kang, T.-j., Lee, S. B., Yoo, J.-y., Bellini, A., Zarri, L., & Filippetti, F. (2014). Screening of False Induction Motor Fault Alarms Produced by Axial Air Ducts based on the Space Harmonic-Induced Current Components. *IEEE Transactions on Industrial Electronics, 0046*(c), 1–1.

Ye, Z., Wu, B., & Sadeghian, A. (2003). Current signature analysis of induction motor mechanical faults by wavelet packet decomposition. *IEEE Transactions on Industrial Electronics, 50*(6), 1217–1228. doi:10.1109/TIE.2003.819682

Ying, X. (2010). Performance Evaluation and Thermal Fields Analysis of Induction Motor. *IEEE Transactions on Magnetics, 46*(5), 1243–1250. doi:10.1109/TMAG.2009.2039221

Zaky, M. S., Khater, M. M., Shokralla, S. S., & Yasin, H. (2009). Wide-Speed-Range Estimation With Online Parameter Identification Schemes of Sensorless Induction Motor Drives. *IEEE Transactions on Industrial Electronics, 56*(5), 1699–1707. doi:10.1109/TIE.2008.2009519

Zidani, F., Diallo, D., Member, S., El, M., Benbouzid, H., & Nait-Said, R. (2008). A Fuzzy-Based Approach for the Diagnosis of Fault Modes in a Voltage-Fed PWM Inverter Induction Motor Drive. *IEEE Transactions on Industrial Electronics, 55*(2), 586–593. doi:10.1109/TIE.2007.911951

Chapter 21
Dynamic Analysis of Steering Bogies

Arun K. Samantaray
Indian Institute of Technology Kharagpur, India

Smitirupa Pradhan
Indian Institute of Technology Kharagpur, India

ABSTRACT

Running times of high-speed rolling stock can be reduced by increasing running speed on curved portions of the track. During curving, flange contact causes large lateral force, high frequency noises, flange wears and wheel load fluctuation at transition curves. To avoid derailment and hunting, and to improve ride comfort, i.e., to improve the curving performances at high speed, forced/active steering bogie design is studied in this chapter. The actively steered bogie is able to negotiate cant excess and deficiency. The bogie performance is studied on flexible irregular track with various levels of cant and wheel wear. The bogie and coach assembly models are developed in Adams VI-Rail software. This design can achieve operating speed up to 360 km/h on standard gauge ballasted track with 150mm super-elevation, 4km turning radius and 460m clothoid type entry curve design. The key features of the designed bogie are the graded circular wheel profiles, air-spring secondary suspension, chevron springs in the primary suspension, anti-yaw and lateral dampers, and the steering linkages.

1. INTRODUCTION

Now-a-days, railway is one of the important transportation systems in most of the countries. To increase the popularity of railway transport for further, improving speed, comfort and safety are some of the important issues. However, heavy investment in infrastructure development, design and fabrication, research, maintenance and operations, etc. is required to commission high-speed rails. When operating speed increases up to 300-350 km/h, regular maintenance requirements and related economic issues influence the decisions taken by railway operators. To minimize the wear in the wheels and rails, track quality is the essential factor. The dynamic performance as well as ride comfort can be increased in the straight track as well as curved track by taking care of bogie design as well as track design. To enhance the performance in the straight track, optimization of bogie parameters is sufficient whereas; in curved track, optimization of track design parameters as well as bogie parameters are necessary. Some of the major parameters that influence performance (stability, comfort, etc.) and wheel/track wear in a curved track are curve radius, cant given to the track, wheel-rail geometry, bogie parameters, axle load and the tractive force between rails and the wheel.

For high speed rolling stock, running times can be reduced by increasing running speed on curved portion of track. During curving, flange contact occurs on the gauge corner of the outer rail which causes large lateral force, high frequency noises, wear of flanges and significant change of the wheel load at transition curves. In extreme case, there is a chance of derailment. To avoid derailment and hunting, and to improve ride comfort of the passengers, in the overall, to improve the curving performances at high speed on curved path, two types of bogies have been implemented along with optimization of track parameters: they are tilting bogie and steering bogie.

Running speed can be increased without reducing the ride comfort of passengers by using tilting bogies. When a train enters a curve at high-speed, centripetal force can cause loss of balance of passengers and luggage objects. Tilting is used to compensate this. When tilting is caused due to deformation of suspensions only then it is called passive tilting. When actuators are using to force titling, it is called active tilting. The amount of tilt including that from track super-elevation (much like banking of roads) is usually restricted 6 or 8 degrees. Beyond this, on long distance travel, passengers may suffer from a form of nausea resembling seasickness. This restriction does not allow for speed increase beyond a certain range. Therefore, titling technology if often combined with other technologies.

With the help of steering bogie, speed increases on the curves by minimizing lateral force. Due to high speed on the curve, wheel exerts large lateral force on the rails which causes wear of wheel flanges and rails. Worn wheels or tracks affect the dynamic behavior of bogies. To reduce lateral force as well as lateral displacement, steering mechanisms can be implemented. In fact, active steering mechanism allows for a great deal of speed increase.

Primary steering of rail vehicles is achieved due to conical wheel profile. However, there is an inherent conflict between curving performance (steering) and stability. Stability and curving performance mostly depends on wheel-set guidance and bending stiffness and shear stiffness of primary suspension in the horizontal direction. Generally hunting in the straight track occurs due to self-excited vibration of wheel-sets. The critical speed of railway vehicles depends on the suspension parameters and equivalent conicity of the wheel tread. To overcome the problems of the stability and curving performance (steering), several developments have taken place in the form of implementation of passive steering, semi-active controlled steering and active controlled steering.

This chapter discusses various steering bogies in details and their dynamic behavior is analyzed. The steered bogie and its assembly (car body, front bogie and rear bogie) are modeled using multi-body dynamics framework and run on a designed measured track. The dynamic performance of steered bogie is investigated by changing the design parameters. The creep forces, wheel displacement, contact geometry, accelerations and ride comfort are estimated against standards to ensure safe operation. The input parameters are design parameters of bogie and different track irregularities in vertical as well as lateral directions. Stability, curving behavior and comfort analysis is performed at various speeds. The dynamics of the steered bogie is analyzed using multi-body system (MBS) simulation software ADAMS (VI-Rail).

2. COMMON TERMINOLOGIES

The common terminologies which are used in the rail-vehicle dynamics are given below (The American Public Transportation Association, 2007). Some of them are illustrated in Figure 1 to 3.

- **Angle of Contact:** The angle of a tangent line at the point of contact between the rail and wheel with respect to axis of the wheel-set, i.e., the angle between the contact plane and the axle center line.
- **Angle of Attack:** Angle between the leading outer wheels and outside rail (see Figure 3).
- **Cant Deficiency:** The difference between applied cant and a higher equilibrium cant.
- **Cant Excess:** The difference between applied cant and a lower equilibrium cant.
- **Equilibrium Cant:** The cant needed for the track to neutralize the horizontal acceleration due to curving is called equilibrium cant.
- **Flange Clearance:** The maximum lateral distance by which the wheel-set can shift from its centered position between the rails at which the rail-wheel contact angle becomes 25^0 with respect to the wheel axis.
- **Grade:** Slope of the track in longitudinal direction.
- **Gauge:** The minimum lateral distance between inner faces of the rails.
- **Hunting:** A self-excited lateral oscillation which is produced by forward speed of the vehicle and wheel-rail interaction.
- **Hunting Critical Speed:** The speed above which persistent hunting typically occurs for a given bogie or truck.
- **Rolling Radius:** perpendicular (radial) distance between the wheel/ axle center line and the point of contact with the rail. Rolling radius may vary with respect to lateral location of the point of rolling contact.
- **Rolling Radius Difference:** Different between rolling radius of left wheel and rolling radius of right wheel of a wheel-set. As wheel-set is shifted laterally from its centered position between the rails, rolling radius will vary with respect to lateral location of the point of rolling contact on each wheel.
- **Track Cant:** Amount by which one running rail is raised above the other running rail in a curved track. Track cant is positive when outer rail is raised above the inner rail. The angle made by the track with horizontal plane is called track angle, which is expressed in radians.
- **Super-Elevation:** The vertical distance between left and right rail, same as cant.
- **Tilt Angle:** The angle between the car body floor plane and track plane.
- **Tilting Train:** The train that has the capacity to tilt the car body inward in the curve of the track to reduce the lateral acceleration.
- **Wheel Tread:** The contact portion between the outer rim of the wheel and the rail.

Local or body-fixed reference system (see Figure 2):

- **Roll:** Rotation around the longitudinal axis of the car body.
- **Pitch:** Rotation around the lateral axis of the car body.
- **Yaw:** Rotation around the vertical axis of the car body.

- **Sway:** Combination of the lateral displacement and rotation about the longitudinal axis of car body as well as bogie frame. It occurs due to asymmetrical loading.
- **Bounce:** Vertical motion of rolling stock.

Figure 1. Wheel-rail geometry

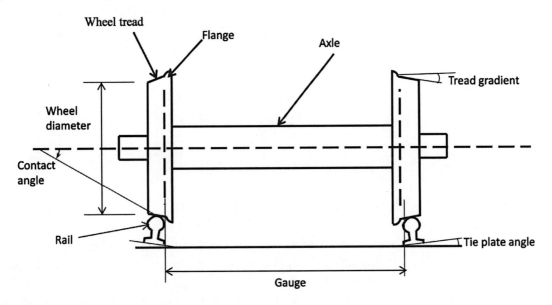

Figure 2. Different types of motion

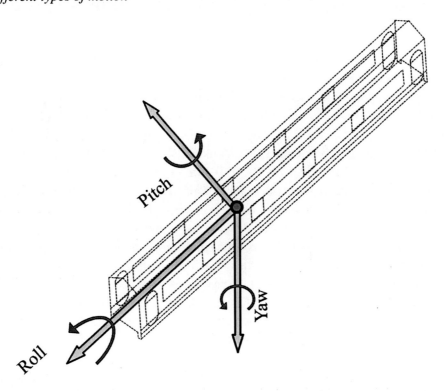

Figure 3. Angle of attack of wheel-set

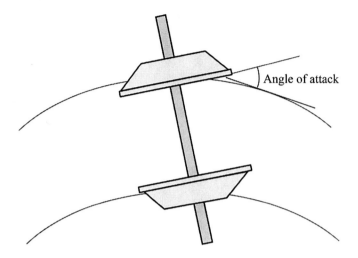

3. SOFTWARE FOR RAIL VEHICLE DYNAMICS STUDY

For railway vehicle dynamics study, several of software are available depending on the application areas such as track design, rail-wheel interaction, vehicle dynamics and structural mechanics (Dukkipati & Amyot, 1988). CONTACT, USETAB and FASTSIM are the software used to study rail-wheel interaction (Polach et al., 2006). CONTACT is a program based on the complete theory of elasticity by which tangential stresses and internal stresses are calculated. Based on simplified theory, Kalker developed a program named as FASTSIM which is a FORTAN sub-routine. USETAB is a program designed by Kalker for a general wheel-rail contact problem to find out creep forces.

We are concentrating on studying the dynamic behavior of the vehicles. Depending on the complexity of the problem like linear or non-linear analysis, degrees of freedom etc., different types of methods are used. In frequency domain, eigenvalue analysis is often used to evaluate critical speed, natural frequencies, damping factors and mode shapes. But for most of the cases, non-linear models are deployed and time-domain solution techniques are used to analyze the dynamic behavior of the vehicle. Dynamic behavior studies, such as modal analysis, ride comfort, hunting stability and curving performances, etc., are performed by multi-body software (MBS) like ADAMS, MEDYNA, SIMPACK, NUCARS, VOCO, VOCOLIN, VAMPIRE and GENSYS. ADAMS/VI-rail provides a graphical interface which allows developing a model similar to that in computer aided design (CAD) (Polach et al., 2006). This software includes stability analysis, dynamic analysis as well wheel-rail contact model and animation. A large number of computer codes have been developed by railway organizations to design as well as optimize suspension systems, tracks parameters and vehicle parameters. The equations of motion for the system are developed by multi-body dynamics theory, which are then processed by a solver. MATLAB / Simulink can be used to design the control part of the railway vehicles whereas the mechanical part can be modeled using ADAMS/ Rail. Finally MATLAB / Simulink and ADAMS/ VI-rail can be coupled (called co-simulation) for mechatronic simulation. MEDYNA (Mehrkorper- Dynamika) was developed by German Aerospace Research Organization (DLR) in collaboration with Technical University of Berlin. SIMPACK was developed by the same team at DLR for analysis of road vehicles as well as rail vehicles having nonlinear kinematics. In SIMPACK, equations of motion are formulated in terms of relative co-

Dynamic Analysis of Steering Bogies

ordinates. This software package uses standard elements for primary and secondary suspensions having linear spring and damper characteristics. In USA, Association of American Railroads (AAR) funded to the development of a program to simulate the behavior of the railway vehicle negotiating a curve. To analyze running behavior of vehicle, Schupp et al. (2004) built a mathematical model by using multi-body system software, SIMPACK. NUCARS is another simulation package developed by Transportation Technology Center, Inc. (TTCI). It is capable of predicting the response of different rail vehicles on different types of track geometry. VOCO (Voiture en Courbe) was developed by French National Transport Research Institute (INRETS) to simulate multi-body dynamics with a reference frame on a long curve. Initially it was used to simulate Y25 bogie. VOCODYM is the second version of VOCO. VOCOLIN is used to simulate the wheel-rail contact with multi-Hertzian approach. VAMPIRE was developed by British Rail Research for analyzing different dynamic behaviors as well as to predict the behaviors of the worn wheels of railway vehicle (Pombo, 2012). Track data can be fed directly to the software from the track recording coach and the behavior of that portion of the track can be predicted in the real time. Derailment risk, passengers' ride comfort and track creep forces can be calculated for different vehicle models for track engineers. GENSYS is a new three dimensional railway vehicle analysis tool which is used in Sweden for simulating track-vehicle interaction with all pre and post processing programs. DIASTARS (Dynamic Interaction Analysis for Shinkansen train And Railway Structure) is a finite element program which is developed for the simulation of Shinkansen train (Tanabe et al., 2003). Along with DIASTARS, Visualization program (VIS) is used to generate animations.

For optimization of vehicle parameters such as those of suspension, sometimes it is easier to develop models based on equations of motion (Cheng & Hsu, 2011). These equations of motion can be simulated using different solvers such as Runge-Kutta method and used in optimization loops. These models are free from software specific constraints and thus can be employed for various other studies. One such study presented in Lee and Cheng (2005) evaluates the influences of several physical parameters on critical speed by using Lyapunov's indirect method. In Cheng et al. (2009), a heuristic nonlinear creep model is used to derive nonlinear coupled differential equations of motion on the curved track.

In this chapter, the performance of different bogie models will be demonstrated through simulation in Adams/ VI-Rail multi-body dynamics software. These studies can be as well performed with other software like SIMPACK or NUCARS.

4. ANATOMY OF RAILWAY VEHICLES

Rolling stock along with track is a complex system having many degrees of freedom. In addition, in rail-wheel interaction involves complex geometry. Forces between the rail and the wheel, inertia forces and forces exerted by suspension system affect the dynamics of the railway vehicle. The dynamic behavior of the rail and the wheel cannot be controlled but suspension design can be optimized to control the motion of the railway vehicles for achieving good ride quality as well as to reduce the tendency of derailment. Railway vehicles consists of wheel-sets, bogie frame, car-body and suspension systems (springs and dampers), all of which affect dynamics of railway vehicles in a complex way. There is no definite way of decoupling the dynamics of each subsystem of a railway vehicle. The main causes of critical hunting and derailment are improper design or damaged suspension system and wheel. The coach is mounted on several bogies. Bogies may be articulated, i.e., shared between adjacent coaches. The main components of a bogie are given below:

1. **Wheel-Sets:** It consists of two wheels with a common axle. The axle and wheels rotate with common angular velocity. The contact portion between wheel and rail is called wheel tread (see Figure 4.). Lower wheel tread gradient on a straight track increases the hunting stability and allows operating the train at higher speed. However, curving requires difference in rolling radii at contact patches between two wheels and rails. Thus, curving performance improves when the wheel tread gradient is high. As a trade-off between these two contrasting requirements, wheel tread is usually conical with 1:40 tread gradient/configuration (Okamato, 1998). To avoid flange contact, lateral displacement is limited to ± 7 to ± 10 mm. For high speed bogie, graded circular wheel tread configurations are mostly used. In such wheel tread configuration, large numbers of arcs are connected next to each other.
2. **Axle Boxes:** An axle box permits the wheel-set to rotate relative to it. Primary suspension system is mounted on the axle box which is attached to the bogie frame. The main function of an axle box is to transmit the longitudinal, lateral and vertical forces from wheel-set to other bogie elements. The bearing supporting the axle in the axle box can be cylindrical roller bearing, taper roller bearing, conical bearing or ball bearings with cylindrical bearing (Okamato, 1999).
3. **Suspension System:** It consists of elastic elements, dampers and associated components (Orlova & Boronenko, 2006). Generally, suspension system has two stages: A primary suspension which connects the axle box to the bogie frame and a secondary suspension which connects the bogie frame to the bolster or car body.
   a. **Primary Suspension:** The main function of the primary suspension is to guide the wheel-sets on straight as well as curved track and to isolate the dynamic loads produced due to track irregularities from the bogie frame. To achieve high speeds, the longitudinal stiffness should be high and lateral stiffness should be low. In curved track, high longitudinal stiffness leads to increase in contact forces between wheel and rail which causes rapid wear and high lateral stiffness leads to increase in dynamic force when negotiating lateral track irregularities. Generally in a passenger bogie, different stiffness is provided in vertical, lateral, as well as longitudinal directions. Coil springs with cylindrical rubber, traction links with resilient bushes, and chevron (rubber-interleaved) springs are the different types of primary suspensions used in different high speed bogies such as in Shinkansen, ETR-460 and X-2000, respectively.
   b. **Secondary Suspension:** The main function of the secondary suspension is to reduce the dynamic accelerations acting on the car-body which is responsible for ride comfort of passengers. The sources of these accelerations are excitation from track irregularities/ roughness and natural oscillations of the bogie frame and car body. As passengers are more sensitive to lateral oscillations, the stiffness in lateral direction should be as small as possible. Different types of secondary suspensions used in different high speed trains are flexi-coil suspension, air spring suspension and full active suspension (FSA).
4. **Dampers:** In bolster bogie (e.g., DT-200), side bearers (Okamato, 1998) give friction damping and prevent hunting. Normally, yaw rotation of bogie relative to car body is controlled by longitudinal yaw dampers. Yaw dampers are mounted between car body and outside of side beam of bogie frame. Yaw damper is a part of the secondary suspension system. Redundant type yaw dampers (called lateral dampers) are used to stabilize the running behavior in ICF3 high speed bogies. Lateral dampers are also useful in reducing lateral disturbances.

5. **Bogie Frame:** Different types bogie frames are there such as Z- shaped and H-shaped. Most of the high speed bogies (Shinkansen, BT-41 and SF 400) are H-shaped. H-shaped bogies consist of two side beams and two cross beams.

5. RAILWAY WHEEL-SET

The conical shaped wheel of the rolling stock helps to reduce rubbing of flange of the wheel on the rail and improve the curving performance. The presence of flange on the wheel prevents derailment. In the straight track, flanges are not in contact with the rail. In curved track, flange contact provides the required guidance to the wheels. Though different types of wheel profiles are there, they have some common features (Orlova & Boronenko, 2006): width of profile is 125-135 mm, the flange height is 28-30 mm, flange inclination angle is between 65 and 70^0 and the conicity is 1:10 or 1:20. But the conicity for higher speed rolling stock is 1:40 for preventing hunting in straight track. If inside conicity is positive and the flange moves towards the rail due to strong steering action, the wheel-set tries to return to the center of the track. In this condition, the combination of the lateral force applied by the rail to the outer wheel of the leading wheel-set and increased vertical load reduces the risk of derailment. When outside conicity is negative, in the absence of steering action, flange is in contact with rail throughout the track. The lateral force applied by the rail to the inner wheel reduces the amount of the vertical load and increases the risk of derailment.

Figure 4. Different components of a Sinkansen 300 series motor bogie

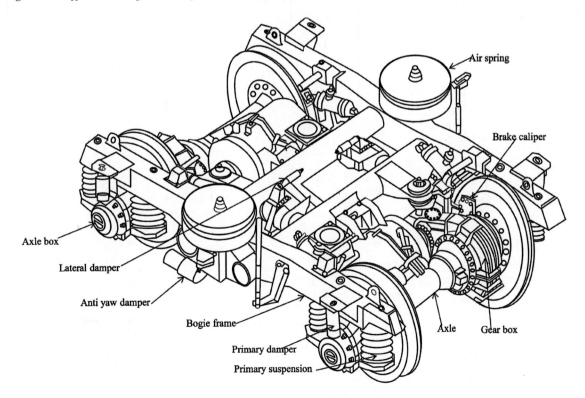

5.1. Hunting of Wheel-Set

In conventional wheel-set, wheels connect to each other through a rigid axle and angular speed of both inner and outer wheels is same. There needs to be a difference in rolling radius between outer and inner wheels on the curved track. The variation of rolling radius can be expressed by using lateral displacement of a wheel in conventional wheel-set. For pure rolling motion, conical wheel-set occupy radial position in the curved track. In the year 1855, Redenbacher derived the first theoretical formulation from the simple geometrical relationship between lateral displacements of the wheel-set on the curve (y), radius of curve (R), wheel radius (r_0), conicity of wheel-set and lateral distance between two contact points on the wheel set ($2l$).

By applying simple geometrical relationship (see Figure 5).

$$\Delta OAB \cong \Delta OCD \Rightarrow \frac{(r_0 - \lambda y)}{(R - l)} = \frac{(r_0 + \lambda y)}{(R + l)}$$
$$\therefore y = \frac{r_0 l}{R \lambda} \qquad (1)$$

While a wheel-set moves on a track, if there is a slight displacement to one side then wheel on one side runs on larger radius as compared to the other side. As two wheels are mounted on the common axle, one wheel will move faster than other than other because of larger instantaneous rolling radius and the wheel set would try to turn. This produces a centripetal force which forces the wheel set to the center of the track and beyond it (due to inertia). The result is a kinematic oscillation called wheel hunting. In the year 1883, Klingel derived the formula for kinematic oscillation with the parameters as wavelength of hunting oscillation (Λ), wheel conicity (λ), wheel radius (r_0) and lateral distance between two rail-wheel contacts ($2l$) (Ayasse & Chollet, 2006). The distance along the track $s=Vt$, where V the forward speed and t is time. Klingel's formula (Ayasse & Chollet, 2006) considers Redtenbacher's formulation of pure rolling of a coned wheel-set on a curve (see Figure 6.):

$$\frac{1}{R} = \frac{d^2 y}{ds^2} = \frac{\omega^2 y}{V^2} = \frac{\omega^2 r_0 l}{V^2 R \lambda} \qquad (2)$$

Figure 5. Rolling of a coned wheel-set on a curve

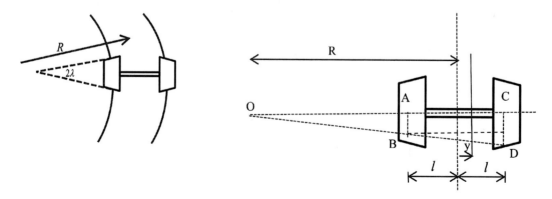

Dynamic Analysis of Steering Bogies

Figure 6. Kinematic oscillation of a wheel-set

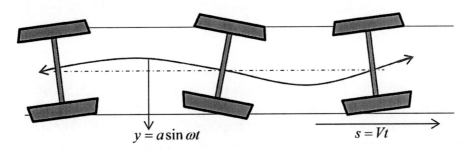

$$\Lambda = 2\pi\sqrt{\frac{r_0 l}{\lambda}} \qquad (3)$$

By increasing the speed, the frequency of the kinematic oscillation also increases. This oscillation damps out below a certain speed called the critical speed. Above the critical speed, this oscillation increases and causes repeated flange contact on two sides of the wheel set which results in an uncomfortable ride and may lead to derailment. This simplified kinematic formulation ignores the inertial and contact forces and the actual behavior is obtained through multi-body dynamics studies. In fact, simple conical wheel sets are not used in modern trains.

Figure 7. Rail-wheel contact at different positions: in (a) normal operation; (b) flange contact (in curved track); (c) tending to derailment; and (d) derailment or wheel jump

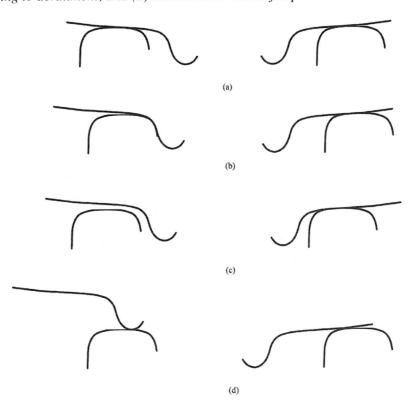

5.2. Flange Climb and Derailment

Dynamic forces act in the frequency range of 0-20 Hz for sprung mass, 20-125 Hz for un-sprung mass and 0-2000 Hz for corrugated, welded and flat wheels (Miura, 1998). The vertical forces in lower frequency range are produced due to track geometry and irregularities. The forces in the higher frequency range are due to discontinuities like rail joints, crossings, and rail and wheel surface irregularities. The track is distorted due to lateral forces by wheel-set, which causes derailment in extreme cases. The lateral forces consist of flange reaction, tread friction and lateral creep force on the rail. The combination of net lateral and vertical forces acting on rail-wheel contact leads to mounting/climbing of the flange on the rail which causes derailment. The derailment coefficient is defined as the ratio of instantaneous lateral force and vertical force at the rail-wheel contact.

Several flange climb and derailment criteria have been proposed as guidelines for railway vehicle engineers. The simplest criterion is Nadal's single wheel L/V limit criterion which is related to instantaneous lateral and vertical forces (Ayasse & Chollet, 2006). Based on the static equilibrium of forces between rail and wheel by considering single point contact at flange (see Figure 8.), the following equations can be derived.

$$F_3 = V\cos\delta + L\sin\delta = V(\cos\delta + \tfrac{L}{V}\sin\delta)$$
$$\begin{cases} F_2 = V\sin\delta - L\cos\delta = V(\sin\delta - \tfrac{L}{V}\cos\delta) & \text{when } V\sin\delta - L\cos\delta < \mu \times F_3 \\ F_2 = \mu \times F_3 & \text{when } V(\sin\delta - \tfrac{L}{V}\cos\delta) \geq \mu \times F_3 \end{cases} \quad (4)$$

The L/V ratio can be expressed as

$$\frac{L}{V} = \frac{\tan\delta - \dfrac{F_2}{F_3}}{1 + \dfrac{F_2}{F_3}\tan\delta} \quad (5)$$

Figure 8. Forces at flange contact location

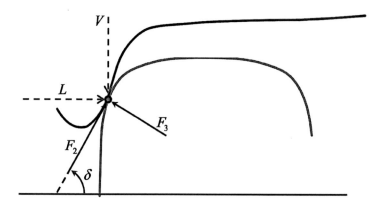

Dynamic Analysis of Steering Bogies

For the saturated condition, $F_2/F_3=\mu$ and the L/V ratio limiting criterion is given by

$$\frac{L}{V} = \frac{\tan\delta - \mu}{1 + \mu\tan\delta} \qquad (6)$$

According to Nadal's formula, derailment coefficient is 1.22 for friction coefficient $\mu=0.35$ and flange angle $\delta=70^0$. Beyond that limiting value, derailment may occur. However, the time duration or the distance for which the value of the derailment coefficient remains at critical value is another governing factor. In Japanese design, the limiting value of derailment coefficient is $0.04/t$ if t is less than 1/20 second and 0.8 if t is more than 1/20 second.

6. RAILWAY TRACKS: DESIGN AND DYNAMICS OF THE TRACK

6.1. Track Design

Track design parameters are important for the dynamic stability of vehicles. The most common parameters are track gauge, track cant / super-elevation, transition length of the curve, radius of curvature (horizontal and vertical), and track irregularities. These parameters are detailed in the following sections.

6.1.1. Track Gauge

In railway track, gauge is the least distance between the inner face of rails on the railway track (Figure 9). Different countries use different types of gauges. They are classified into

- **Standard Gauge:** The lateral width of standard gauge is 1435mm.
- **Broad Gauge:** Track gauge is greater than standard gauge. The values of broad gauge are different in different countries (Russia - 1520mm, United states - 1575mm, 1581mm, 1588mm, 1600mm, 1638mm etc., England- 1727mm, India - 1676mm, France - 1750mm).
- **Narrow Gauge:** Track gauge is smaller than standard gauge. The values of narrow gauge are different in different countries (United States - 508mm, 578mm, United Kingdom - 508mm, 578mm, 660mm, etc., Germany - 520mm,560mm, 860mm,1100mm, Japan - 838mm and 1067mm).
- **Meter Gauge:** Meter gauge is 1000mm.

Figure 9. Track gauge

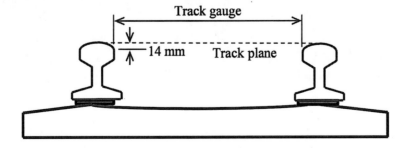

For high speed rolling stock, standard track gauge of 1435mm is universally adopted. This allows for interchangeability between operators and standardization of such a safety critical system.

6.1.2. Track Cant

Track cant or super-elevation is defined as the difference between the levels of two rails (Figure 10) in a curved path which is useful to compensate the unbalanced lateral acceleration. The cant angle can be calculated as follows.

$$\varphi_t = \sin^{-1}\left(\frac{h_t}{2b_0}\right) \tag{7}$$

where $2b_0 = 1435$mm on the standard gauge and h_t is cant. The variation of track cant angle along a track is shown in the Figure 11 (b).

Figure 10. Track cant

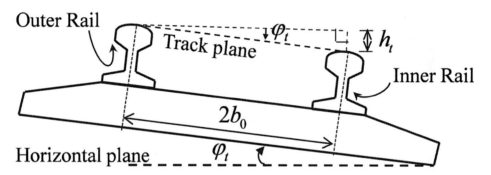

Figure 11. (a) Track curvature, (b) Track cant angle

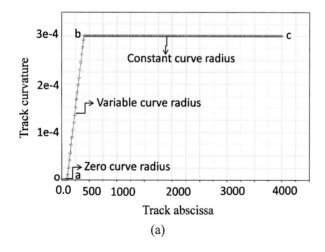
(a)

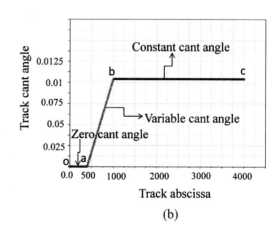
(b)

Dynamic Analysis of Steering Bogies

The track cant is designed based on the radius of curvature and the maximum running speed. More track cant allows higher running speed. However, there is a chance of derailment due to large lateral force and low vertical load on the outer track on a curve when a train runs at low speed on a track designed for high speed. Therefore, there is a limit on the maximum value of cant. When a train runs on a curve at a speed other than the designed speed, there can be cant deficiency or cant excess. These are discussed later in this chapter.

6.1.3. Transition Length of the Curve

The radius of curvature (ρ) is defined as

$$\rho = 1/R \tag{8}$$

where, R is the curve radius. The variation of curvature at the entry and exit of a curved path is called transition curve. Usually, it is of a clothoid type. It is used to join straight track to circular one or two adjacent curves to allow a gradual change in curvature. In a clothoid (also called Euler spiral) the curvature changes linearly with the curve length. The clothoid type of the transition curve (Figure 11 (a)) is given by Lindahl (2001)

$$\rho(s) = \rho_0 + \frac{s}{A^2} \tag{9}$$

where, $s = s_0 = 0$ at the start of the transition curve, ρ is the curvature at a distance s, and A is the clothiod parameter. If a clothoid starts from straight line ($\rho_0 = 0$) and ends in a circle with radius R with the total transition length L_t then

$$A^2 = L_t R \tag{10}$$

Figure 11. (a) shows different portions of a curved track where sections oa, ab and bc, respectively, represent straight portion with infinite radius of curvature, transition portion with variable radius and circular portion with constant radius. With the change in track curvature, the track cant is also varied in proportion (see Figure 11(b)) where the proportionality constant depends on the designed operation speed at which equilibrium cant is desired.

6.1.4. Tracks around the World

There is wide variation in track and speed parameters across railway networks in the world. This is primarily because of different safety standards, guidelines, and demographics and geographic situations. Table 1 gives an overview of the parameters across different railway networks.

Table 1. Speed and track operational parameters in some railway companies throughout the world (Lindahl, 2001)

Organization	TSI/CEN	JR	JR	JR	DB	DB	SNCF	SNCF	BV
Parameter		Tokaido Shinkansen	Sanyo Shinkasen		Hannover-Wuzburg	Koln-RheinMann	TGV Paris Sud Est	TGV Atlantique	Botniaban (partly)
Maximum design speed (kmph)					280	300	300	350	250
Maximum service speed (kmph)		270	300	275	250		270	300	$200^a/250^b$
Max. cant (mm)	180	200	180	180	65	160	180	180	150
Max. cant deficiency (mm)	100	100	100	100	80	150	85	60	100/220
Max. cant excess (mm)	110				50				100
Minimum curve radius (m)		2500	4000	4000	7000	3350	4000	6250	3200
Minimum vertical curve radius (m)		10000	15000	15000	22000		12000 14000		11000

a: Conventional vehicles with older running gear and freight trains, b: Vehicles with improved running gear and car-body tilt system

6.2. Dynamics of the Railway Track

Irregularities of the track lead to oscillations or vibrations which cause human discomfort. Long wavelength track irregularities give rise to low frequency oscillations of train which cause passenger discomfort and can lead to derailment through resonance with different bending modes of the track. Short wavelength irregularities cause vibration and noise which create unpleasant environment for passengers (Miura et al., 1998). Before we proceed further, let us have a look at the track properties.

Figure 12. Different components of the track

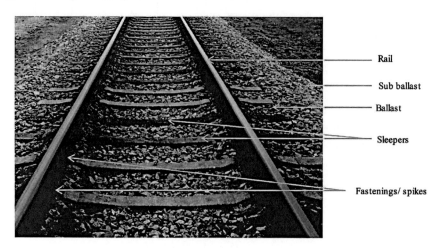

Dynamic Analysis of Steering Bogies

6.2.1. Track and Its Components

A railway track consists of rails, sleepers, rail pads, fastenings, ballast, sub-ballast and subgrade (Dahlberg, 2006). Super structure and sub-structure (sub-ground/ sub-grade) are two sub-systems of a ballasted track. Super structure consists of rails, sleepers, ballast and sub-ballast and sub-structure is composed of the formation layer and the ground. The main function of the track is to guide the train and to carry load of the train and distribute the load over the sleepers and the ballast bed.

1. **Rails:** Modern steel rail has I-section with a flat bottom. The commonly used rail profile is UIC-60, where 60 refers to mass of the rail in kg per meter. Standard rail length is 25m. To improve ride quality and reduce noise as well as vibration, continuous welded rails (CWR) are used in high-speed tracks. CWR carry vertical load (compression) of train, lateral forces from wheel-set, longitudinal forces due to traction and breaking of train. A badly laid track may buckle causing sharp lateral displacement. The above problem can be prevented by using sleepers, ballast and fastening (Miura et al., 1998). Modern dedicated tracks for high-speed rails use fully concrete slab tracks. Note that rail profile UIC 60 with inclination 1:40 has less guiding force as compared to UIC60 with inclination 1:20.
2. **Rail-Pads:** Rail-pads are mainly used in railway track with concrete sleepers. They are placed between steel rails and sleepers to protect the rails from wear and impact damage and also act as electrical insulator. Soft rail pad permits larger deflection of rails, transmits axle load from train to sleepers and also isolates high frequency vibration from rail to sleepers.
3. **Sleepers:** Sleepers provide support to rail. They transmit vertical, lateral and longitudinal forces from the rail to the ballast. Generally wood and concrete sleepers are used. Due to good elastic properties, light weight and ease of handling, timber sleepers are preferable to concrete sleepers. But in case of wood, the service life is very short, especially in tropical and sub-tropical regions and regions affected by termite and other pests. To overcome the drawback of wood, synthetic sleepers made from polyurethane and glass fibers are used in Japan (Miura et al., 1998). Synthetic sleepers have same physical properties as woods/timbers and give better life service. These sleepers are generally used in the places such as steel girder bridge, switches and other places where replacement as well as maintenance is difficult.
4. **Rail Fastenings:** Generally dog spikes are used to fasten timber sleepers to rails. In case of concrete sleepers, different types of springs are used for fastening purposes by using rubber pads. Leaf springs are used in Japan and France, whereas wire springs are used in Germany for their better adjustability, load bearing ability and fastening force.
5. **Ballast:** Coarse stones are used as ballast to support sleepers as well as rail and to transmit vertical and lateral forces from trains to the sub-ballast. Standard depth of ballast is 0.3m and it is packed to 0.5m around sleeper ends to ensure lateral stability. Ballast dissipates the energy transmitted to it through friction between the stones/pebbles. Loosely packed ballast can cause large rail deflection and thus, derailment. Ballast is often covered with wire-mesh to stop the pebbles flying off due to aerodynamic forces (referred to as the flying-ballast problem) as a train passes over it.
6. **Sub-Ballast:** Sub- ballast layer is the transition layer between ballast layers to lower layer of fine graded sub grade. Any sand or gravel may serve as sub-ballast material.

7. **Subgrade:** It is a portion of earth used for foundation of track bed. Sub-ballast and ballast layers rest on subgrade portion. Track failure and poor track quality is mainly dependent on the subgrade quality. Heavy rain or flood can cause damage to the sub-grade.

6.2.2. Dynamic Properties of the Track

For high-speed rails, ballasted tracks or concrete slab tracks are used. Under frequent use, ballast in ballasted track loosens and deforms, causing minor track irregularities. Long wave track irregularity mainly affects vertical and lateral body vibrations, and is the major source of ride discomfort. Rail surface irregularities/short wave irregularities are due to rail welds and rail wear. Short wave irregularities cause wheel vibration, fluctuation of the axle load, high frequency noise and increased dynamic/impact load on track (Miura et al., 1998). Receptance is the ratio of the track deflection and force put on the track. Receptance is the inverse of the track stiffness. The track stiffness is non-linear and depends on the frequency of load. The receptance also depends on the preload on the track.

For soft subgrade, a resonance may occur at the frequency 20-40 Hz. When rail and sleepers vibrate on the ballast bed, the frequency range is 50-300 Hz (Dahlberg, 2006). Here rail and sleepers provide mass. Ballast acts as a spring and also provides large amount of damping. Another resonance can be often found in the frequency range 200 to 600 Hz which is due to bouncing of the rail on rail pads. Sleepers and rail provide masses and ballast provides most of the damping. Highest resonance frequency which is also called pinned-pinned resonance frequency is approximately at 1000Hz. This frequency can be excited when the wavelength of the bending waves of the rails is twice that of the sleepers' spacing. The nodes of the bending vibration of rails are at the support (sleeper) (Dahlberg, 2006).

6.3. Track–Vehicle Interaction

When the vehicle moves in a curved path, horizontal centrifugal acceleration and vertical gravitational accelerations are act on the body as shown in the left side of Figure 13. The resultant acceleration can be split into two components: a_y parallel to the track plane and a_z perpendicular to the track (See right side of Figure 13). The load difference between the inner and outer rails can be neglected at lower velocities.

Figure 13. Acceleration of the car-body during curving

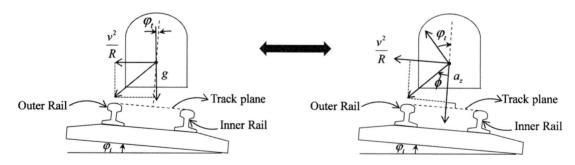

From the Figure 13, static equilibrium conditions give

$$a_y = \frac{v^2}{R}\cos\varphi_t - g\sin\varphi_t = \frac{v^2}{R}\cos\varphi_t - g\cdot\frac{h_t}{2b_0} \tag{11}$$

$$a_z = \frac{v^2}{R}\sin\varphi_t + g\cos\varphi_t \tag{12}$$

By assuming small cant angles ($\varphi_t \leq 0.15$ rad), Equations (11-12) can be written as

$$a_y = \frac{v^2}{R} - g\sin\varphi_t = \frac{v^2}{R} - g\cdot\frac{h_t}{2b_0} \tag{13}$$

$$a_z \approx g \tag{14}$$

The cant for which lateral acceleration is zero for a given radius and given speed is called equilibrium cant (h_{eq}). The equilibrium cant is given by

$$h_{eq} = \frac{2b_0}{g}\cdot\frac{v^2}{R} \tag{15}$$

If speed is expressed in km/h and cant in mm then the equilibrium cant is given by

$$h_{eq} \approx \frac{2b_{0,mm}}{g}\cdot\frac{v^2}{3.6^2 R} \tag{16}$$

For a particular cant, if the speed of the vehicle gives zero acceleration for a given curvature, the corresponding speed is called balanced speed or equilibrium speed. Equilibrium speed is given by

$$v_{eq} = \sqrt{\frac{R\cdot g\cdot h_t}{2b_0}} \tag{17}$$

The net lateral acceleration along the plane of the track is not zero in every case. There is a chance that train moves with low speed or stops in a curved path. Secondly, all the trains do not run at same speed whereas the track remains fixed. So, it is impossible to get zero acceleration for all the trains. Cant deficiency arises when actual cant is less than equilibrium cant. Cant deficiency (h_d) is the difference between equilibrium cant and actual cant (h_t). Cant deficiency is given by

$$h_d = h_{eq} - h_t = \frac{2b_0}{g}\cdot\frac{v^2}{R} - h_t \tag{18}$$

When the actual cant is higher than the equilibrium cant, cant excess is caused. In this case, the vehicle is running at a lower speed than designed speed of the track. Cant excess is the difference between the actual cant and equilibrium cant and is given by

$$h_e = h_t - h_{eq} \tag{19}$$

Cant is somewhat helpful to reduce lateral acceleration as well as the creep force in lateral direction in the curved track. So, the dynamic behavior of car-body is improved with the presence of cant in the curved track. Some results are given at the end of this chapter showing the influence of cant on dynamic behavior of the vehicle. The track parameters are cant 150mm (cant angle is 0.1047 radian), 460 m transition curve and radius 4000m.

7. STEERING BOGIE

There is an inherent conflict between stability on the straight track and good curving behavior in the curved track. Lower conicity helps to achieve high speed on the straight track whereas higher conicity gives better curving performance. For the stability of vehicles, equivalent conicity is usually maintained between 0.1 and 0.4. In low equivalent conicity (in case of new wheel), there is a possibility of creating gravitational stiffness, i.e., wheel-sets cannot be returned back to the center of track when disturbed laterally from central position. To achieve both high speed as well as better curving performance, Shinkansen bogies use conical wheel tread configuration with wheel gradient 1:40 at the nominal contact position on straight track. Bending stiffness and shear stiffness are the two other parameters of primary suspension that influence stability and curving performance. For better curving performance, low bending stiffness in primary suspension is required whereas for better stability, shear stiffness of the primary suspension should be increased. Shear stiffness helps to stabilize the vehicle and bending stiffness improves the curving performance (steering action) of the wheel. The shear stiffness is responsible for critical speed of the vehicle while bending stiffness determines the angle of attack of the wheel-set in curves.

The main purpose of using steering bogie is to reduce noise and wear during curving, improving dynamic performances and minimize the chance of derailment. During curve negotiation, tangential forces are generated due to creepage (longitudinal and lateral) at the point of contact which plays a crucial role in curving and stability of railroad vehicles. The longitudinal and lateral creepages of wheel-sets are determined as (Shabana, et.al. 2008)

$$\left.\begin{array}{l}\zeta_x = \dfrac{\left(\dot{r}_P^w - \dot{r}_P^r\right).t_1^r}{V}, \\[2mm] \zeta_y = \dfrac{\left(\dot{r}_P^w - \dot{r}_P^r\right).t_2^r}{V}, \\[2mm] \varphi = \dfrac{\left(\omega^w - \omega^r\right).n^r}{V},\end{array}\right\} \tag{20}$$

where ξ_x, ξ_y, φ are the longitudinal, lateral and spin creepages, respectively; $\dot{r}_P^w$ and $\dot{r}_P^r$ are time derivative of wheel-rail contact position vector, respectively; t_1^r and t_2^r are orthogonal tangent vectors to the rail at the contact point in longitudinal and lateral directions; n^r is the normal vector unit to the surface at the contact point and V is the wheel velocity in the longitudinal direction. The wheel and rail distort due to compressive force at the contact zone. If two bodies (rail and wheel) move relative to each other, tangential forces (shear traction) are generated at the contact zone due to Coulomb friction. Shear traction between two bodies is $F_t=[F_{tx}, F_{ty}]^T$. The longitudinal and lateral creep forces and the spin creep moment, respectively, can be expressed by integrating local force components and moments over the contact patch in the x-y plane:

$$\left. \begin{array}{l} F_x = \iint F_{tx} \, dxdy \\ F_y = \iint F_{ty} \, dxdy \\ M = \iint \left(xF_{ty} - yF_{tx} \right) dxdy \end{array} \right\} \quad (21)$$

By using elasticity theory based traction-displacement relation, linear relation between creepage, creep forces and moment can be calculated as

$$\begin{bmatrix} F_x \\ F_y \\ M \end{bmatrix} = -Gab \begin{bmatrix} c_{11} & 0 & 0 \\ 0 & c_{22} & \sqrt{ab}c_{23} \\ 0 & -\sqrt{ab}c_{23} & abc_{33} \end{bmatrix} \begin{bmatrix} \xi_x \\ \xi_y \\ \varphi \end{bmatrix}, \quad (22)$$

where ξ_x, ξ_y, and φ are the longitudinal, lateral and spin creepages, respectively; a and b are, respectively, the semi-axis dimensions in the rolling and lateral directions of the elliptical contact patch; G is the modulus of rigidity and c_{ij} ($i,j = 1..3$) are creepage coefficients which depend on Poisson's ratio and ratio of the semi-axes lengths of the contact ellipse (Shabana, et.al., 2008; Kalker, 1990). Kalker introduced the equivalent modulus of rigidity and Poisson's ratio to be used in Equation (22) as

$$G = \frac{1}{2}\left[\frac{1}{G^w} + \frac{1}{G^r} \right], \quad \frac{v}{G} = \left[\frac{v^w}{G^w} + \frac{v^r}{G^r} \right], \quad (23)$$

where G is an average shear modulus of wheel w and the rail r, and v is a combined Poisson's ratio of wheel and the rail. Generally, creepage, creep forces and creep moments are key aspects which affect wear and noise in the curved track. While curving, speed of outer wheel is faster than inner wheel in order to satisfy pure rolling of all wheels.

When the wheel-sets of bogies adapt or are forced to the radial position in the curved track as shown in Figure (16), then the bogies are called radially steered bogies. In the case of steered bogie, the angle of attack is small which reduces track creep forces and the resulting flange wear. Radially steered bogies are classified depending on the control principle used. In case of passive/ self–steering, wheel sets take up radial position due to low bending stiffness of primary suspension. The yaw of a wheel-set is induced by the wheel-rail contact forces or by the relative rotation between bogie frame and vehicle body (either roll or yaw).

In both the above cases, the amount of yaw of the wheel-set depends on the radius of curvature. When wheel-sets are forced to occupy radial position in the curved track, the corresponding steering mechanism is called active steering. In an active steering bogie, electric, hydraulic or pneumatic actuators are used to give yaw motion to the wheel-sets. Links, levers and sliders are used between wheel-sets and the bogie frame to execute different types of forced/ active steering mechanisms.

7.1. Self-Steering

Wheel tread is helpful for the train to negotiate the curve and keep the vehicle in the mean position while rolling on a straight track. Due to presence of conicity in the wheel and rigid connection between wheels of wheel-sets, the bogie is self-steered. However, self-steering may not be sufficient while curving in small radius turn where large lateral forces exerted on wheels and rail cause wear of the both. Self-steering/ passive steering is achieved due to difference in rolling radii between the left and right wheels. If a wheel-set is moved laterally from its centered position, a difference in rolling radii arises (Wickens, 2003). From Figure 14,

$$\text{Equivalent conicity } (\gamma_e) = \frac{\Delta r}{2y} \qquad (24)$$

where Δr is the instantaneous difference in rolling radii between left and right wheels and y is the lateral displacement from the centered position. Rolling radius difference is a function of both the wheel and

Figure 14. Conicity of wheel profile

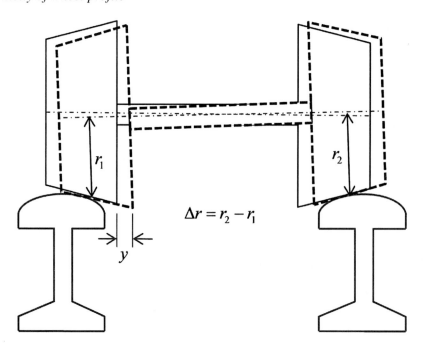

Dynamic Analysis of Steering Bogies

rail shape and equivalent conicity is described by both rail and wheel profiles. The actual profiles are similar to the shapes shown in Figure 15.

As the centrifugal force causes the center of axle to move towards the outside of the curve, the presence of wheel-tread gradient leads to steering action in the curved portion of the track. The vehicle primary suspensions are made more flexible in bending so that they can occupy more and less radial position on the curves to improve curving performance.

Due to presence tread gradient, the effective outer wheel diameter increases and that of inner wheel decreases on the curves (see Figure 16). This difference in the effective diameters leads to self-steering action. A common two-axle conventional bogie tends to turn outward with respect to tangent of the curve. The corresponding angle between the tangent of the curve and direction of the leading outer wheel is called angle of attack (see Figure 17). Due to the angle of attack, lateral creep force acts on the outer wheel towards outer rail in the lateral direction. At the same time, the rear axle experiences opposite forces and tries to turn the opposite way. Longitudinal creep force between rail and wheel creates high lateral force between rail and wheel which creates high lateral force on the front wheel-set towards the outer rail. In steering action, wheelbase on the outer rail is longer than on the inner and axles turn radially in the direction of the curve.

Figure 15. Wheel-rail profile

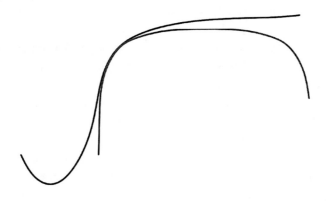

Figure 16. Self-steering characteristics of wheel-sets

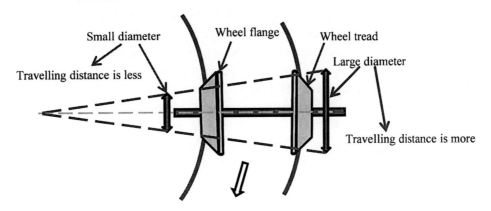

Figure 17. Behavior of self-steering on conventional bogie wheel-set

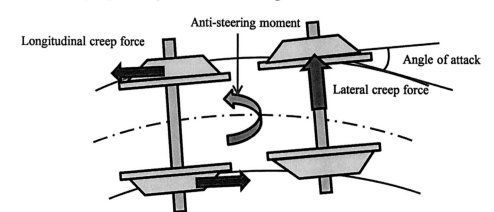

Self-steering / passive steering can be achieved in a number of ways such as

- Reducing the stiffness of the axle box suspension / primary suspension
- Increasing the tread gradient
- Interconnecting two wheel-sets by cross bracing
- Using three-axle vehicles with suitable linkage

By reducing the stiffness of the axle box suspension / primary suspension, wheels and axle can easily move sideways in response to lateral forces. But excess reduction of stiffness of primary suspension can affect running stability and ride comfort.

Increase in tread gradient/ conicity of the wheel set improves the curving behavior but simultaneously compromises stability on the straight track. To achieve both better curving performances as well as stability on the straight track, tread gradient should be optimized. In modern high speed trains such as in almost all bolster-less Shinkansen trains, graded circular wheel tread configurations are used in which large numbers of arcs are joined next to each other (Okamato, 1998).

Circular graded profile reduces the contact area as well as contact bearing forces between wheel tread and rail. Also, the contact points during curving are uniformly distributed over the tread and lead to uniform wear instead of large localized wear. In this chapter, the circular tread profile with various levels of wear has been considered (see Figure 18). In a worn wheel, the tread gradient is larger as compared to a new wheel. As the centrifugal force causes the center of axle to move towards the outside of the curve, the presence of wheel-tread gradient leads to steering action in the curved portion of the track. In a worn wheel, self-steering effect is more pronounced as compared to a new wheel.

Excess tread wear leads may lead to contact with the fishplate bolt. With flange wear, flange angle increases and flange thickness decreases which increases the chance of derailment. Dynamic performance of the vehicle as well as steering effect increases up to certain depth of wear and after that the dynamic behavior (comfort, critical speed, etc.) as well as stability deteriorates. When the center of wheel tread is worn below the level of end of the tread, there can be significant deterioration of the vehicle's dynamic performance, i.e., stability and ride comfort. Different depths of worn profiles considered in this chapter are given in Figure 19.

Figure 18. Tread and flange wear

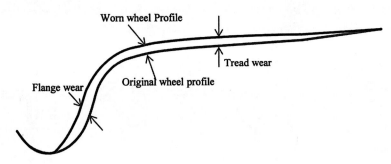

Figure 19. Original wheel profile and two worn profiles with different depths of wear

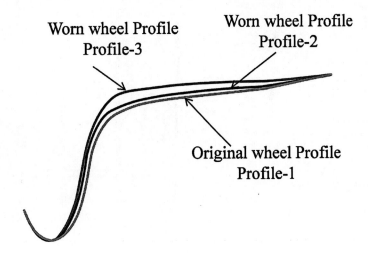

Self-steering or passive steering can also be achieved by interconnecting two wheels by different ways as shown in Figure 20 (Wickens, 2006). Wheel-sets interconnected by cross-bracing (Figure 20(a)) is one of the oldest methods of self-steering. In a three-axle vehicle, central axle steers the outer one through suitable linkage (Figure 20 (b)). Wheel sets of three-axle vehicle with suitable linkage can occupy the radial position in curved track and rearrange themselves on the straight track. In all the above cases, passive/ self-steering is possible only through rigid linkages and pivots.

In conventional suspension arrangement, bending and shear stiffness are dependent to each other. So, when shear stiffness is decreased, the bending stiffness also decreases. Thus, in conventional suspension arrangement, improvement of curving performances reduces the stable operation speed on the straight track. Therefore, in self- steering bogies, shear and bending stiffness are allowed to be independently designed. This is achieved with use of cylindrical or conical laminated rubber springs. Material (rubber) can be removed from certain portions of the spring to give directional properties to the spring's stiffness and damping parameters.

To enhance the curving–stability relationship, Scheffel proposed different types of inter-axle linkage arrangements which are given in the Figure 20 (c), (d) and (e) (Orlova & Boronenko, 2006). These bogie designs are based on the principle of three-piece bogie which consists of two side frames and one bolster. In case of inter connected wheel-sets for self-steering of locomotives, transfer of tractive forces

Figure 20. Passive steering (self-steering): (a) Direct connection between wheels by cross bracing; (b) three axle vehicle; (c) Scheffel HS bogie with diagonal linkage between wheel-sets; (d) inter axle linkage for Scheffel's A-frame bogie; (e) inter axle linkage for Scheffel's radial arm bogie

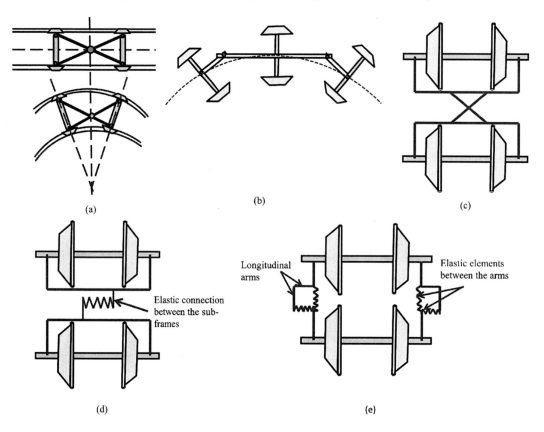

between the wheel-set to bogie frame should not influence the axle guidance. In this case, wheel-sets are steered by inter-axle linkage instead of frame (as in conventional bogie design). To be more effective, steering inter-axle links have low longitudinal as well as lateral primary suspension stiffness. Scheffel's bogies are usually double suspended.

Conicity of the wheels is responsible in natural/ self-steering action but at the same time conicity creates instability. This problem can be overcome by using springs between wheel-sets and the bogie or between the wheel-sets within the bogie. The additional springs interfere with natural steering action. So it is necessary to optimize the mechanical components of the vehicles to give the best stability and steering action. The excess reduction of stiffness of primary suspension can affect running stability and ride comfort.

Generally in case of conventional steering bogies, both the wheel-sets are steered, so the mechanism is bulky. To avoid that problem, a new concept, called single-axle steering, proposes that only one wheel-set should be steered according to the requirement in the curved track. In this case, only rear wheel-set is steered as explained in the Figure 21. When rear axle is steered, angle of attack of rear axle is increased and consequently, large lateral creep force is generated towards the outer rail. By shifting the real axle towards the outer rail, the difference between the diameters of rear wheels increases which helps the bogie to be in radial position. As a result, angle of attack as well as lateral creep force of non-steered

Dynamic Analysis of Steering Bogies

Figure 21. Behavior of single axle steering

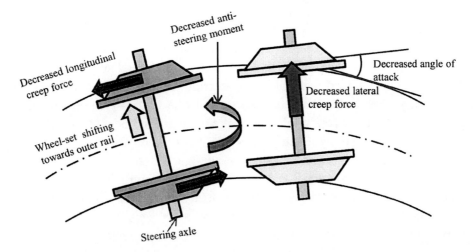

axle (front) decreases. Longitudinal creep force of rear axle and lateral creep force of front wheel lead to reduce creation of an anti-steering moment on the bogie (similar to self-aligning moment in steered road vehicles). As railway vehicles run in both directions, for a coach supported on two bogies with four axles, axles 1 and 4 are non-steered, whereas axles 2 and 3 are steered. When axle-3 (leading axle of rear bogie) is steered, its angle of attack as well as lateral force on the outer rail decreases. This type of steering is used in SC101 bogie. It a link type passive actuation steering in which swing bolster, truck frame and axle boxes are connected to the bogie through levers and links. When vehicle enters into a curve, axle boxes change their position by changing the position of the linkage (Shimokawa & Mizuno, 2013).

To overcome the drawbacks of self-steering/passive steering, assist/ forced/ active steering principle has been adopted in modern bogie designs.

7.2. Active/Assist/Forced Steering

Hunting motion and large centrifugal forces often lead to flange contact of wheel. Active steering is used to control lateral displacement and yaw of wheel-set so that wheel contact patch is constrained to the tread region. Active yaw damping and active lateral damping are two types of active damping which gives stability for solid axle wheel-set. In active yaw damping, an actuator yaw torque is applied through various means to the wheel-set in direct proportion to lateral velocity of the wheel-sets. In addition, unstable modes are stabilized by active lateral damping control technique where the applied lateral force is proportional to the yaw velocity of the wheel-set (Mei & Goodall, 2003[a]). Various sensors, actuators and controllers are used for active stability control. Filters and estimators are used to measure the state variables and local track reference so that natural curving action of the wheel-set is not affected by the stabilization method. Note that it is impossible to realize such modification of the dynamics by purely passive components. For cost effectiveness as well as better reliability, optimized passive components are used for the improving stability of the vehicle. However, these affect steering action (curving performance) of the vehicle and there is no means to improve both stability and curving performance with passive components. Although directional properties can be modified (e.g., different stiffness in longitudinal and lateral directions), the passive design still remains sub-optimal.

7.2.1. Configuration of Active Primary Suspension

Various common forms of active steering configurations in bogie with active primary suspension system are shown in Figure 22. The control system of actuated solid wheel-set (ASW) is designed for increasing stability and enhancing the curving performances. In actuated independently rotating wheels (AIRW) configuration, active control provides guidance and steering. Other types of actuations such as driven independently rotating wheel (DIRW) and directly steered wheel (DSW) are possible through driving and braking torque (Bruni, et.al. 2007). In some designs, control rod is used to enhance curving performances without affecting stability (Shen & Goodall, 1999). To improve steering performances and stability, closed loop control of solid axle wheel-set was designed in (Mei & Goodall, 1999; 2000^b). Two actuators termed lateral actuation and yaw actuation were used to control lateral force and yaw moment or couple, respectively. It was found that less control force (actuator power) is required to achieve optimal stability and ride comfort in yaw actuation as compared to lateral actuation. Actuation is introduced in form of steering torque applied on the axle in AIRW. Control torque required to run DIRW vehicle is less than the torque required for solid wheel-set (Mei & Goodall, 1999). Note that lateral creep force on DIRW nearly reduces to zero. But in straight track, AIRW requires active guidance which may create problems for sensor and controller implementations.

Figure 22. Different types of active steering actuation configurations: (a) yaw actuation of solid wheel-set, (b) lateral actuation of solid wheel-set, (c) independently driven wheels, and (d) steered wheels with separate axles.

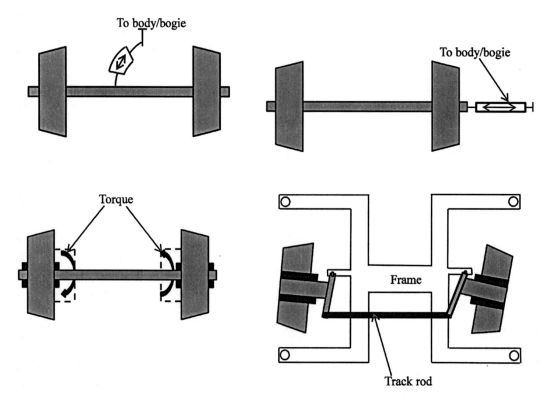

The wheels on the same axle are free to rotate in independently rotating wheel-set (IRW) which is used largely in Spanish Talgo train. For long distance and high speed purposes, appropriate driving torques are applied on IRWs with help of servo-meter which is connected to wheels through differential gear box; this arrangement is named as driven independently rotating wheel (DIRW) as shown in Figure22 (c). Note that separate traction motors may produce differential torque without use of differential (Mei & Goodall, 2003[a]). In IRW, guidance forces are provided by applying a differential torque on the two traction motors of the wheel pair to obtain traction control and active guidance (Powel, 1999). Also, differential braking may be used instead of differential traction on the trailing bogies (Goodall, et.al. 2006[a]). The combined use of AIRW and DIRW strategies have been implemented to improve steering performances which is suitable for high speed and long distance application with conventional bogie features. Such active system consists of one steering actuator and one traction motor for each wheel-set (Perez, et.al. 2004). To obtain compact and light weight design, permanent magnet motors were used inside wheel to implement DIRW control strategy. In Figure22 (d), wheels of IRW are mounted on the wheel frame instead of axle which is actively steered to achieve guidance and curving (Aknin, et.al. 1992). A track rod is used to steer the wheels directly by producing lateral force between wheels and frame. Stabilization, guidance and better curving performance were found in the curving track by using lateral wheel displacement in a feedback loop (Wickens, 1993; 1994). Power steering control mechanism is used to actuate track rod and enhance the curving performance. Power steering through DC motor performs two roles: it acts as a passive magnetic damper (generator mode where power is dissipated) to enhance damping of yaw oscillation in straight track and acts as a powered actuator to negotiate the curve in the transition curves (Michitsuji & Suda, 2006). To improve stability and steering performance, yaw torque is applied on the bogie; this is the basic concept of secondary yaw control. This control has been used in tilting bogie/ high speed train where electro-mechanical actuator provides control force (Diana, 2002). Active control takes place the role of passive yaw damper in straight track and yaw torque is applied on the curved path. The yaw applied torque is computed as a function of cant deficiency and curve radius. A longitudinal actuator may be used to control the amount of torque to the wheel-set in yaw direction. This strategy is helpful to control lateral displacement and yaw motion by using laser signals which are mounted on both ends of the wheel-sets to measure lateral displacement of front and rear axles (Kim, et.al. 2008).

7.2.2. Control Strategies

During curving, both solid axle wheel-set and IRW have their own separate merits and demerits. In case of solid axle wheel-set, passive suspension severely affects the curving performances; whereas IRW has problem in guidance control both in straight and curve track, but no issue with stability.

7.2.2.1. Control Strategies for Stability

Kinematic instability mode of solid axle wheel-sets has been studied by different researchers. Stabilization in case of IRW mounted on common axis can be achieved more easily than wheel-sets with solid axle. The control strategy is called active lateral damping, where applied control force in lateral direction is proportional to wheel-set yaw angular speed. For a two axle vehicle, active yaw damping strategy needs less actuation forces and power and gives better ride comfort as compared to active lateral damping (Mei & Goodall, 1999; Mei & Goodall, 2000[b]). In active yaw damping, control torque is proportional to lateral velocity of the wheel-set with respect to car-body. The active yaw damping is more effective due to its adaptability to speed variation and also due to its stabilizing action as compared to the passive yaw dampers (Mei & Goodall, 2003[b]). To avoid the influence of the passive yaw stiffness on stability, sky-hook/ absolute stiffness control strategy is used, where control torque is directly proportional to yaw of the wheel-set rather than the relative yaw between the bogie and the wheel set. In sky-hook spring control, actuators are attached between the wheel-set and bogie. Measurement or estimate of yaw angle of each wheel-set is fed to the controller to control the corresponding actuator. The concept of sky-hook damper is similar to sky-hook spring (Mei & Goodall, 2006). It is used in secondary suspension to improve ride quality.

Two transfer functions, one between lateral displacement of wheel and control input and other between yaw angle of wheel-set and control torque of single wheel-set were used to explain the stability of solid axle wheel-set in (Mai & Li, 2008). In their study, two unstable and two stable poles were found. They proposed to provide feedback signals of yaw angle and lateral movement in order to provide damping effect on kinematic instability. The feedback method turned out to be equivalent to conventional passive stabilization with yaw stiffness. To ensure better stability, (Mai & Li, 2008) have further proposed a phase lead compensator to compensate the delays.

7.2.2.2. Control Strategies for Guidance

Guidance control is necessary in IRW to ensure that wheel-sets follow the track and do not contact the flange. However, hunting may take place due to track irregularities and cause temporary flange contact. To avoid flange contact, high band width control is necessary. Tracking error is the error of position of the vehicle with respect track centerline. Active guidance is provided by steering the wheels in order to reduce tracking error. Due to active guidance, the vehicle has the ability to follow on the curved path without flange contact and minimize the lateral creep force. Direct guidance control uses wheel-rail deflection measurement (relative displacement between wheel-set and track) in a feedback control system. But direct measurement of wheel–rail deflection is costly and difficult (Wickens, 1994). In another complex approach, differential torque is applied with the help of a PID controller using linear combination of measurements or estimates of leading and trailing wheel-sets' lateral displacements and yaw velocity of leading wheel-set (Gretzschel, 2002). In DIRW control strategy, direct control for guidance is possible by controlling the traction torque (Perez, et.al. 2004).

7.2.2.3. Control Strategies for Steering

The key objective of steering control is to make sure that wheels/ wheel-set follow track and the wheel-rail creep forces are reduced. Longitudinal creep (except during traction/ braking) is undesirable and can cause unwanted yaw motion. However, lateral creep is somewhat essential to provide curving force to compensate the cant deficiency, i.e., the amount of centrifugal force which cannot be balanced through the given amount of track cant. Control strategies are responsible for reduction of unnecessary creep forces and noise/ wear at rail-wheel interface during curving. The basic working principle of forced steering is to control the relative angular displacement between car body and bogie or between bogie and wheel-set. Forced steering control strategy can be perfectly implemented for zero cant deficiency. But in normal operating conditions, zero cant deficiency is usually impossible. Hence, unwanted lateral movement of wheels and creep forces appear at the rail-wheel interface. This problem can overcome by controlling the wheels/ wheel-sets in order to achieve the desired angle of attack. Even distribution of curving force to each wheel is required for minimizing the wheel shifting on the curves. Active steering bogie controls the wheel-set motion when it travels in curved track. In straight track, controller is off and the bogie is equivalent to the conventional bogie provided that the steering mechanism is perfectly locked (acts as a rigid connection). The conditions for perfect steering/ curving are (Goodall & Mei, 2006[h]):

1. The longitudinal creep forces in the wheels on the same axle are same,
2. The lateral creep forces of two wheel-sets are same,
3. The angle of attack of wheel-sets is same so that the entire bogie is in line with the track on the curve.

Equal longitudinal forces on the wheels (perfectly distributed traction/braking force) leads to eliminate wear and damage of the wheels and equal lateral forces are required to balance the centrifugal forces caused by cant deficiency or cant excess. Same angle of attack on different wheel-sets is possible when actuators control the yaw angle with respect to bogie. The required yaw angle is determined from the radius of curvature of track, cant, velocity of vehicle and wheel base. In the other way, yaw torque can be applied for cancelling the effect of longitudinal stiffness of primary suspension. Steering strategy does not affect stability when the effect of longitudinal stiffness of primary suspension is cancelled at frequencies lower than that of kinematic modes. Then relative yaw angle can be measured between the wheel-set to the bogie using various sensors like encoders or geared potentiometers.

Perfect curving requires angle of attack and radial angular position of two wheel-sets to be equal and the bogie to be in-line with the track, which is called yaw relaxation concept (Shen & Goodall, 1997). For this, yaw motions of wheel-sets with respect to bogie frame need to be controlled through actuators. The required yaw angle can be calculated by using track parameter (curve radius (R)) and parameters of wheel-rail interaction (lateral creep force for each wheel-set (F_y) as calculated in Equation (21), creepage coefficient (c_{22}) (Kalker, 1990) and wheel base (l_x)) (Shen et.al. 2004). The small desired yaw angles in leading and trailing wheel sets, respectively, can be expressed as

$$\varphi_{trailing} = \sin^{-1}\left(\frac{F_c}{2f_{22}}\right) + \sin^{-1}\left(\frac{l_x}{R}\right) \approx \frac{F_c}{2f_{22}} + \frac{l_x}{R}$$

$$\varphi_{leading} = \sin^{-1}\left(\frac{F_y}{2c_{22}}\right) - \sin^{-1}\left(\frac{l_x}{R}\right) \approx \frac{F_y}{2c_{22}} - \frac{l_x}{R}, \qquad (25)$$

$$\varphi_{trailing} = \sin^{-1}\left(\frac{F_y}{2c_{22}}\right) + \sin^{-1}\left(\frac{l_x}{R}\right) \approx \frac{F_y}{2c_{22}} + \frac{l_x}{R}.$$

For positioning purpose, the relative yaw angle between individual wheel-set and bogie is measured (Shen & Goodall, 1997; Perez, et.al. 2004). In AIRW control strategy, difference of angular speed between inner and outer wheels are controlled by applied yaw torque in order to run the wheel-set on the center position of the track (Perez, et. al. 2002). The guidance concept used in DIRW control strategy for low speed curve (300 m) is helpful for controlling the wheel-set either in radial position or centered position of the track (Gretzschel, et. al. 2002). In DSW control strategy, feed forward predictive control has been proposed to enhance the performance along transition curves in which a control torque is applied on each steered wheel pair depending upon the variation of track curve with time and track geometry

(Michitsuji & Suda, 2006). In secondary yaw control (SYC) concept, feed forward control strategy is used only for curving in which cant deficiency and bogie yaw rate are measured by using low pass filtered lateral accelerometer signal from the bogie and gyroscope, respectively, and the steering torque applied on the bogie is estimated from track curvature and level of wheel and track wear (Diana, et. al. 2002).

Actuators are mounted on the bogie and car-body to generate yaw torque. Different wheel-sets have different conditions, mostly due to level of wear and local track irregularities. So the individual wheel-sets should be controlled separately to optimize the steering performances (Park, et.al. 2010). In forced steering, the axle needs to yaw relative to the bogie frame. According to Goodall & Mei (2006[b]), the assist/forced steering can be achieved in the following different ways:

1. Axles can be connected by suitable linkage which can be actuated to achieve different radial positions on curved tracks.
2. Control torque can be applied to the wheel-set in the direction perpendicular to the plane of the wheel-set.
3. Actuators can be used in the lateral direction of wheel-set, but ride quality is affected by this arrangement.
4. Wheel-set can be controlled by active torsional coupling between the wheels.

Active primary suspension (actuators with primary suspension) along with passive components can be used in the same way as an active secondary suspension. Passive stiffness can be used to stabilize the kinematic oscillations (hunting instability) and actuator is used to provide steering action in the curved track. In the case of active steering, stability and steering actions are designed separately.

To improve the curving performances by active steering, different types of control strategies are developed. They are (Perez, et. al. 2004):

1. Lateral position of wheel-set may be controlled in such a way that only pure rolling will take place which reduces longitudinal creep force.
2. Yaw moment which is dependent on lateral wheel-rail displacement is to be controlled by using traction motors to run the wheel in central position of the track.
3. The relative angle between wheel-sets is to be controlled by using traction motors to implement differential torque control or yaw actuation for IRW steering control strategy.

By implementing above three strategies, wheel lateral displacement, angle of attack and lateral contact forces can be reduced; thereby reducing material loss/wear of wheel. In some designs, anti-yaw dampers are replaced by actuators to steer the bogie against car body in active steering bogie (Matsumoto, et.al. 2005). Radial steering is the most well-known steering mechanism which consists of mechanical linkages with joints. Park et.al. (2010) proposed a link type steering, which consists of two driving links, two steering links, a transverse link and a linear actuator. The driving links, which are on both sides of the bogie frame, are connected to each other by transverse link. Due to presence of revolute joint between bogie frame and driving links, the links can rotate easily depending on the movement of the linear actuator. Steering links act as the bridge between the axle box and driving links through universal joints. Axle boxes are pulled or pushed according to the movement of the driving links. Left steering link of front wheel-set is connected to the end of left driving links, whereas right steering link is attached to the middle of the right driving link. In addition, there is a transverse link to transmit linear motion and actuator force to driving link. Steering mechanism for rear and front wheel-sets are diagonally symmetrical. In this type of active steering, four PID controllers are used to control the lateral displacement. Control gains are selected in such a way that the generated control forces remain within the allowable actuator force range.

Controller design is an important factor for active steering of railway systems. Linear quadratic optimal control is suitable for both solid axle wheel-set and IRW. To avoid flange contact between wheel and rail, it is necessary to control both the lateral displacement and yaw angle (angle of contact) of wheel with respect to track. Generally controllers severely interfere in natural curving in case of solid axle wheel-set, but it is less critical in IRW due to its extra degree of freedom. Adequate weighting factors are used in optimal control to avoid interference in performance of solid axle wheel-set. In both the cases of solid axle and IRW, optimal controller needs to implement integral action to reduce steady state error (Mei & Goodall, 2003[b]). Due to simplicity and well known design method, classic PID control is often used. But in these approaches, measurements of angle of contact and wheel deflection are required for feedback. However, it is difficult to directly measure these variables in practice. Full state feedback control with ideal condition of the track, i.e. no irregularities, avoids use of such complicated measurements, but requires a well-developed vehicle model. The complete vehicle model has multiple inputs and outputs. Modal control is another approach where lateral and yaw motions of the body/ bogie are decoupled using certain transformations to yield decoupled subsystems. Separate controllers can be then independently designed for these subsystems and inverse transformation is performed to transfer the desired control actuations to the coupled system. For example, outputs from decoupled lateral and yaw controllers are recombined to control two actuators for the wheel-set (Mei & Goodall, 2003[a]).

7.2.3. Sensors and Actuators

Measurement of wheel-set movement with respect to track is used in all the above discussed controllers. But mounting sensors on the wheel-sets is costly and difficult. Model based estimation techniques such as Kalman filters and observers are often used for indirect measurement. The outputs from the sensors are compared with output from a mathematical model (observer model) and the deviation is used to produce corrective action through gain matrix to compensate for the inaccuracy in the model output. Estimated state variables from the observer model are then used as feedback signals for controllers. Inertial sensors on the wheel-sets and bogies have been shown to provide excellent results (Mei & Goodall, 2000[a]). In modern designs, these sensors have been replaced by more accurate bogie based displacement sensors which provide the primary suspension deflection (Pearson, et.al. 2004) and observer models are used to estimate other states from the displacement measurements. Note that application of such type of control strategies needs the track data at the exact position of the vehicle. Usually, track data is stored in a database and the positions of different wheel-sets are computed with respect to a reference on the train. The position of the reference point is obtained through GPS or other trackside tracking and relaying devices. Communication infrastructure is vital to implementation of such type of active steering principle. Another way of obtaining track data is to sense the track through sensors such as through optical sensors and then applying image processing or other techniques to determine the track curvature. Usually, a constant correlation is assumed between the track curvature and cant.

Different types of actuators such as servo-hydraulic, servo-pneumatic, electro-hydraulic and electro-mechanical are used in active suspensions. Though servo-hydraulic actuators are compact and easy to handle, but along with power supply, the whole system tends to be heavy. Pneumatic actuators are basically used in air-spring for many rolling stocks, but compressibility of air causes actuation delay and undesired low frequency oscillations. Electro-mechanical actuators offer quick response and high efficiency. However, they are less compact and careful attention is required for reliability and life of mechanical components. Electro-magnetic actuators give extremely high reliability and better performance, but they are heavy (Goodall & Mei, 2006[b]).

7.2.4. Controller Set-Point Specification

We can increase the curving performance and reduce the lateral creep force by inserting steering mechanism. As a result, wear and tear of wheel (especially, in the flange portion) reduces significantly. Steering effect can be improved in the presence of cant/ super elevation on the curved portion of the track. But excess cant or deficient cant affect the steering mechanism, so optimization of cant angle in the curved portion of the track is required to improve the steering performance at a given operating speed.

Link type forced steering can reduce the lateral force to one-half or one–third than a conventional bogie on conventional tracks (Okamato, 1999). For a given radius of track at a particular position, the steering angle is defined by the angle by which each axle has to be turned to make its wheels' angle of attack with the track zero and align the axle with the radius of the track. Perfect steering can be achieved when the angle of attack for all wheel sets is the same, so that the bogie center line is tangent to the mean track circle. The controller tries to align the wheel sets to a desired angle called a set point. The set point depends on the local curvature of the track.

This steering angle/ yaw movement may be calculated as (see Figure23)

$$\alpha = \sin^{-1}\left(\frac{W_b}{2} \times \frac{1}{R}\right), \tag{26}$$

where W_b is the wheel-base, α is the required steering angle and $1/R = \rho$ is the radius of curvature of the track.

However, the above formulation does not consider the existing cant in the track and hence can result in under or over steering leading to yaw oscillations of the wheel set. Also, the vehicle encounters large centrifugal force in high speed and needs more steering angle. Thus, steering angle in presence of cant angle may be calculated in terms of cant deficiency or cant excess depending on speed of the train. Then using Equation (25), the instantaneous steering angle/ yaw movement at leading and trailing axles, respectively, can be modified to

$$\alpha_l = \sin^{-1}\left(\frac{F_c}{2F_{22}}\right) - \sin^{-1}\left(\frac{W_b}{2} \times \frac{1}{R}\right) \approx \frac{F_c}{2F_{22}} - \frac{W_b}{2R}$$

$$\alpha_t = \sin^{-1}\left(\frac{F_c}{2F_{22}}\right) + \sin^{-1}\left(\frac{W_b}{2} \times \frac{1}{R}\right) \approx \frac{F_c}{2F_{22}} + \frac{W_b}{2R} \tag{27}$$

where $F_c = M(v^2/R - g\sin\phi)$ is the lateral force due to cant deficiency, v is the velocity, g is the acceleration due to gravity, M is the mass corresponding to static axle load, ϕ is cant angle in radians and F_{22} is a creep coefficient. The calculated steering angle depends on track design parameters such as curve radius

Dynamic Analysis of Steering Bogies

Figure 23. Steering angle of wheel set, W_b is distance between two axles of the wheel-set in a bogie

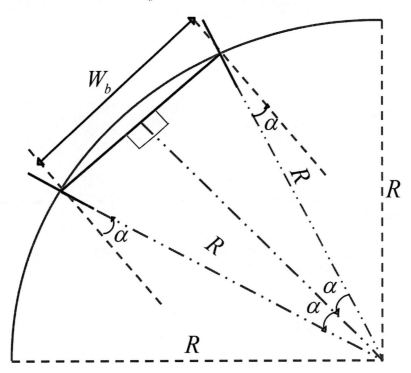

and cant angle and running velocity of the vehicle. However, the computed angle goes out of bounds for large magnitude of cant excess or deficiency, i.e., at very low and very high speeds. Thus, the steering angle is restricted to a specified limit.

In the following, the performance of steering systems would be demonstrated through simulation studies performed with Adams VI-Rail multi-body dynamics simulations.

8. DYNAMIC ANALYSIS

Though a large number of components are present in railway vehicle, we are interested only on the dynamic behavior of the main components such as body components and suspension systems. The main body components are car-body, bogie frame, wheel-set, primary and secondary suspensions, and the axle box. Dynamic performance is analyzed here through multi-body simulation (ADAMS/VI-Rail) software. A bogie template is created as shown in Figure 24. The bogie template is then assembled with car body to produce an integrated coach model (Figure 25). We analyze the dynamic behavior when the train runs at a constant speed. Note that aerodynamic load is not significant on a coach (except for crosswind) and thus a single coach model is sufficient to approximate the behavior of a whole train. Although we have performed analysis with integration of a few coaches, the results are similar and hence only results from the single coach model are presented here. It may however be noted that if variable speed run is to be analyzed, such as during acceleration and braking, then the complete train model with assembly of coaches is required.

Figure 24. Bogie template (without steering mechanism)

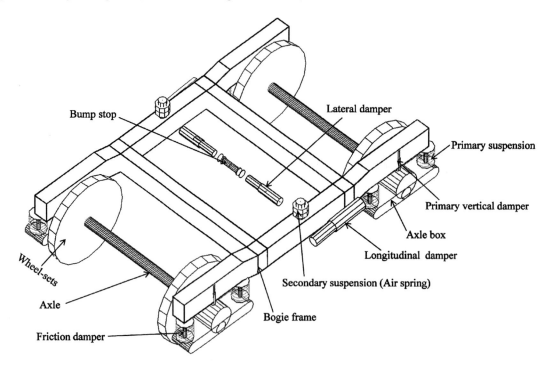

Figure 25. Bogie assembly (with steering)

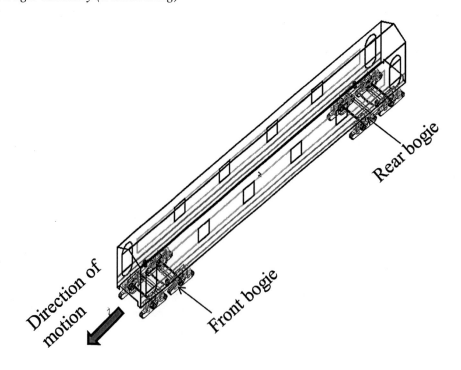

Dynamic Analysis of Steering Bogies

The mass inertia properties are essential for simulation and they are obtained from solid modeling software. Generally track is separately modeled and track parameters are taken as input parameters for simulation. Here, we have considered flexibility of track (lateral, vertical and roll) as well as track irregularities. Track flexibility introduces variable track stiffness as wheels move on a rail (beam) supported at regular intervals on sleepers. Between two sleepers, the track flexibility varies and the frequency of the change in flexibility (a parametric excitation) depends on the sleeper spacing and the train speed. The flexible track also assumes the presence of ballast. Track irregularities are generally given in the lateral and vertical directions in terms of the deviations from the track centerline, track cant and track gauge. The track parameters for the straight track are length of the track, track irregularities, rail inclination and nature of the track (rigid and flexible) and those for the curved track are length of the track, curve radius, cant, track irregularities, rail inclination and nature of the track (rigid and flexible). For the analysis, curve radius is taken as 4000m, cant angle is 0.1047 radian (cant height is 150 mm), rail inclination is 1:40 of rail profile UIC60. At nominal contact, the tread conicity of s1002 wheel profile is 1:20 and accounting for the 1:40 rail inclination/tie plate angle, the equivalent conicity is 1:40. The types of track irregularities are sinus, ramp, DIP, power spectral density (PSD) and measured type. The results shown in this are for measured type track irregularities as shown in Figure 26.

The outputs of the analysis are the critical speed, derailment speed, creep forces between rail and wheel, wheel lateral displacement, acceleration of the wheel-set and the ride comfort.

The bogie template shown in Figure 24 was modified with implementation of various passive steering mechanisms (Figure 27). The simulation results for these passive mechanisms showed improved performance up to 270 km/h speed on curved tracks.

We present the results for the active steering mechanism whose bogie template is shown in Figure 28. The proposed link type active steering consists of steering beam, steering link and steering lever.

Figure 26. Measured type track irregularities

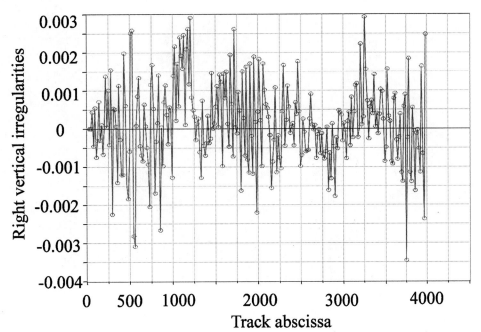

Figure 27. Bogie templates (with passive steering mechanism)

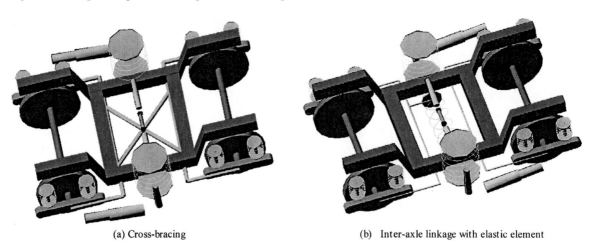

(a) Cross-bracing (b) Inter-axle linkage with elastic element

Steering beam and steering levers are connected by revolute joints, and steering levers and steering links are connected by spherical joints. An actuator is placed in the center of the bogie frame to rotate the steering beam (actuated revolute joint). Generally in this design, yaw torque is applied through the actuator at the center of the steering beam.

Figure 28. Bogie template (Link type active steering mechanism)

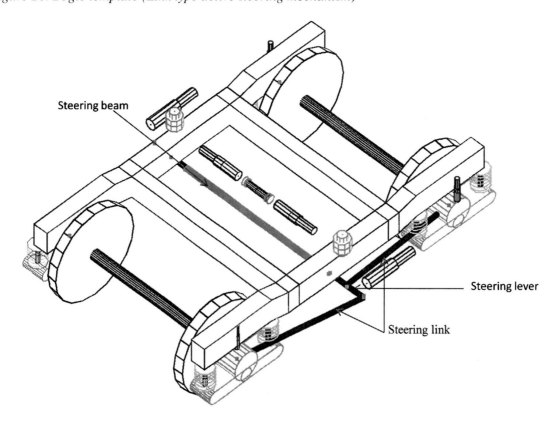

Dynamic Analysis of Steering Bogies

We need to control the steering angle (Equation 26) at center of axle-box which is attached to the axle of wheel-set. The correlation of yaw movement between center of steering beam and at the axle box is calculated from analysis of steering linkage mechanism in ADAMS (Figure 29). This correlation of steering beam and axle box steering angles is shown in Figure 30. For a specific set of chosen link lengths, this correlation can be curve fitted as

$$\alpha_b = 17\alpha_W^3 - 20\alpha_W^2 + 15\alpha_W \tag{28}$$

where α_b and α_W are the desired steering beam angle and desired axle steering angle. The steering linkage applies forces directly on the axle-box and the rear and front axles turn in opposite directions. The angle of steering is in fact very small (less than ±0.01 radian). It is computed based on the operating speed and existing cant in the track at the given location of the bogie. The steering angle varies linearly from 0 to a steady value as the train transits from the straight track to the constant radius of curvature track.

For vehicle stability, track stresses as well as dynamic analysis, it is very essential to understand the behavior of lateral and vertical forces developed between rail and wheel. Basically forces acting on the rail as well as wheel can be classified as static forces, quasi-static forces and dynamic forces. Static forces arise due to static load of wheel on the rail. Several factors are responsible for quasi-static forces: centrifugal forces caused by cant excess or deficiency, crossing of rails and crosswind. Dynamic forces are produced due to stiffness of primary suspension, track geometry and irregularities of tracks as well as discontinuities like rail joints and crossings. Before starting any dynamic simulation, it is necessary to allow the vehicle to settle to its equilibrium position due to static loads. This process develops the necessary forces in suspension elements and contacts to support the static load of the vehicle. These static deflections automatically work as initial conditions for dynamic simulation.

Figure 29. Analysis of steering linkage in ADAMS

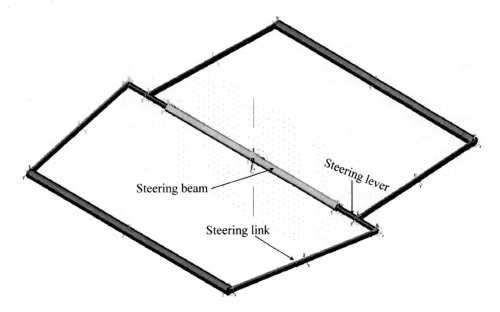

Figure 30. Correlation between yaw angle at axle box and steering beam

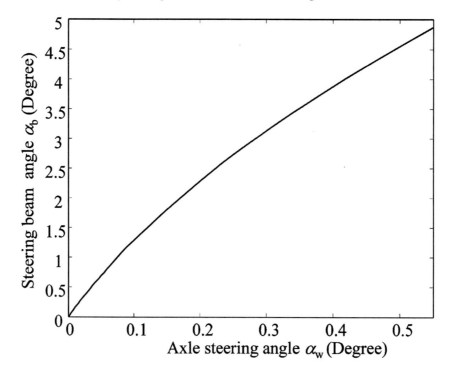

8.1. Wheel-Rail Contact Animation

When wheel is in contact with rail, elastic deformation occurs in the form of contact area which transmits normal load from wheel to rail. The contact area or contact patch is very small as compared to the dimensions of wheel as well as rails. The shape and orientation of contact patch depends on the transverse radius of curvature of wheel tread, radius of curvature of rail head at the point of contact and wheel radius. The contact area is large at the flange throats and corners of the rail.

The wheel-rail interaction is very crucial for dynamic analysis of the vehicle. In the multi-body simulation (MBS) software VI-rail (ADAMS), it is necessary to run time-domain simulations with short time steps (in the order 1ms). The contact parameters are calculated in a preprocessor program. The input parameters are wheel-rail geometry, rail inclination and track gauge. Normally it is assumed that all the parameters are kept constant throughout the simulation. Two wheel-rail elements (one for the left side and the other for the right side) consist of a wheel-set model. Contact angle, conicity and roll angle are the parameters necessary for the dynamic analysis of vehicle and these are computed from specialized algorithms devoted to contact determination.

In tread contact, the radii of curvature are change slowly with position and contact patch is almost elliptical shape depending up on the position of the contact patch, velocity and track properties. If the contact is conformal or the radii are changing suddenly, the contact patch is quite different from elliptical shape. As the rail-wheel contact is a non-linear type, multi-Hertzian method is used to find the nature of contact patches. Some animation contact graphics for s1002 wheel profile and UIC60 rail profile are given in Figure 31. The contact patches are both elliptical and non-elliptical. In Figure 31(a), the train is taking a curve at slow speed and the elliptical contact patches at the tread are stable throughout the

Dynamic Analysis of Steering Bogies

uniform track curvature zone. In Figure 31(b), the speed is close to maximum designed operating speed on curved track. There is a multi-point contact in one wheel at the tread as well as the flange, and the contact patches are still elliptical. Figure 31(c) shows the case when the train is curving at a speed very close to the derailment limit. There is repeated flange contact and rebound at this speed and sometimes one of the wheels loses contact from the rail and then suddenly falls back on the rail. The contact forces are large due to impacts. Figure 31(d) shows the situation when the one of the wheel climbs over the rail and leads to derailment.

8.2. Critical Speed and Derailment

Critical speed is the speed at which vehicle becomes unstable on straight track and the hunting vibrations persists either with or without flange contact. Above critical speed, in addition to lateral vibrations, twist and roll, and resonance type of motions in which car body oscillates about the longitudinal axis take

Figure 31. Different types of contact patches for s1002 wheel profile and UIC60 rail profile

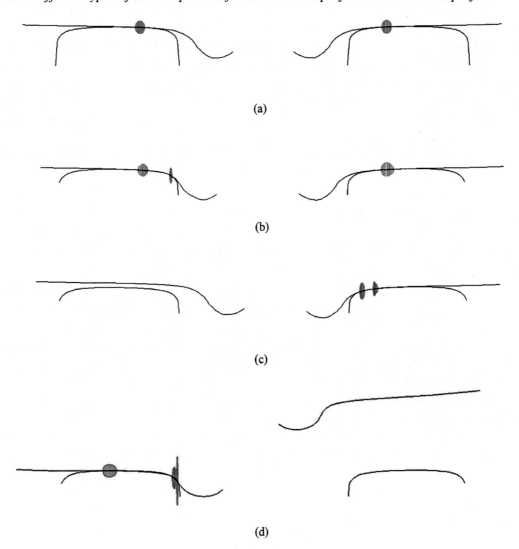

place. Derailment takes place at some speed above the critical speed. For critical speed analysis, inputs are vehicle design parameters, stiffness and damping values of suspension system and moment of inertia (MI). The simplest form of analysis of critical speed is performed using a linearized model about fixed wheel conicity. The minimum speed at which at least one of the eigenvalues of the linearized model has a positive real part is the critical speed. The linearized model employs Kalker's linearized model for rail-wheel contact forces. The critical speeds for the coach with the considered bogie design at various levels of wheel conicity with flexible track and with no rail inclination are shown in Figure 32. The contact angle is taken as 85 λ, where λ is the equivalent conicity. It is found that the critical speed at the conicity of 1:40 (which is equivalent conicity in our case with 1:20 conicity of s1002 wheel profile and 1:40 UIC60 rail inclination) is 102 m/s (367 km/h).

However, the linear analysis only gives a ballpark value and the actual critical speed should be evaluated through full transient simulation of the non-linear model with non-linear contact force formulation. The linear analysis assumes constant conicity throughout the tread. In graded circular wheel, the change in rolling radius due to lateral shift is different for both wheels. Thus, the actual critical speed of trains with graded circular wheels is much higher than the value predicted from linearized model's eigenvalue analysis.

Hunting motion of wheels is a self-excited vibration. The initial excitation can be caused by any form of disturbance such as by track irregularities (Dukkipati & Amyot, 1988). For simulation in Adams VI-Rail, we have used a small amplitude ramp type discontinuity in vertical direction. The same results are obtained for other forms of small disturbance. The results in Figure 33 show the lateral displacement of the wheel-set on a curved track at various operating speeds. As is evident that after the initial

Figure 32. Stability range of vehicle with steered bogies from linear stability analysis with different values of conicity

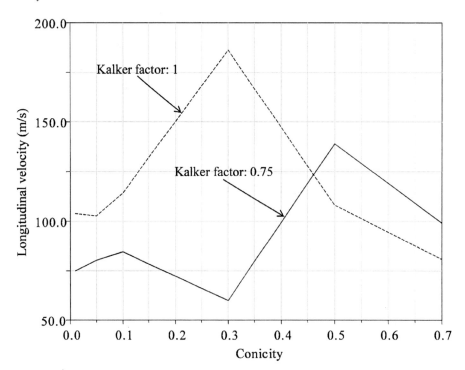

Dynamic Analysis of Steering Bogies

disturbance the hunting vibrations stabilize for a speed below the critical speed, persist at the critical speed and grow (leading to flange contact) above the critical speed. There is no derailment, though, on the straight flexible track.

While a train operates below the critical speed, irregularities of track can still excite other modes of the vehicle. Symmetric track irregularities can give pitch and bounce. Asymmetric irregularities excite roll and yaw modes. Occasionally, these two types of motions are responsible for derailment in which the wheel climbs over the flange. Derailment may be caused at speeds lower than the critical speed due to bad track conditions such as loose ballast, rail waviness, rail discontinuity and any other form of large/abnormal excitation. Even when there is no derailment, temporary wheel separation from one rail causes large vertical impact forces and passenger discomfort. When the speed exceeds the critical speed by a small amount, self-excited yaw and sway motions are sustained and there is a good chance of derailment under sufficiently large disturbance.

Usually, high-speed rail tracks are so well laid out that there is practically no chance of derailment on straight tracks. The derailment can only occur on curved tracks and turnouts. The limiting speed at which derailment occurs in curved track is dependent upon the track radius of curvature, entry curve length, cant/super-elevation, track irregularities and other track parameters, and bogie design parameters. For the considered bogie design with forced steering, the derailment occurs above the speed of 114m/s (410 km/h) on 4km radius track with 460m long entry/transition curve, and flexible track condition with measured irregularities. An animation frame grab from Adams VI-Rail during good curving of a coach on a flexible curved track is shown in Figure 34. The wheel jump and derailment scenarios are shown in Figure 35.

Figure 33. Hunting behavior at various speeds with an initial disturbance

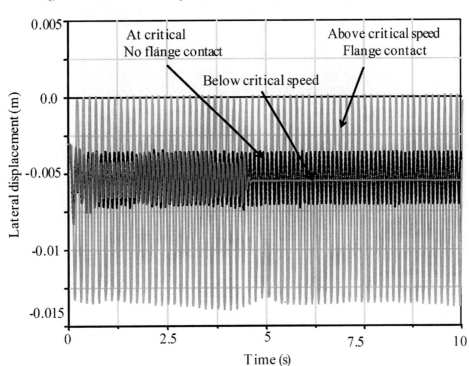

Figure 34. A coach with non-steered bogie approaching the entry curve

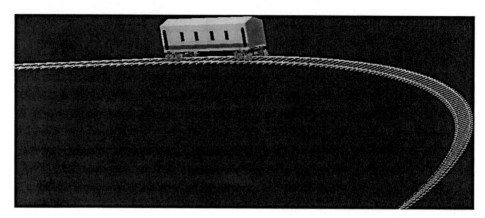

Figure 35. Derailment of coach within the entry curve

8.3. Influence of Worn Wheel

Wheels and rails wear due to the friction between them. Generally, rail wear is less as compared to the wheel wear. The places where wheel wear takes place is determined from the contact patch location. On straight runs, wheels wear in the nominal contact position. On curves, wheels wear near the flanges. In general, some points on the wheel wear out at a much faster rate. A wear evolution model based on forces and locations at the contact patch is used to produce worn wheel profiles (see Figure 19). If a train traverses the same path in both forward and backward directions then both the wheel on the inside of a curve during forward run also remains on the inside of the curve on the reverse run. Thus, wheels wear out at flanges in an asymmetric manner whereas tread wear is faster (due to longitudinal creep forces)

Dynamic Analysis of Steering Bogies

and symmetric. Since the purpose of this chapter is about steering of rail vehicles, we will consider that the train mostly operates on straight track and few curves. Thus, we consider symmetric wheel wear. Two worn wheel profiles and the original new wheel profile as shown in Figure19 are used in this study.

Figures 36 and 37 show the influence of wheel wear on the hunting behavior of the vehicle on a straight flexible track with irregularities. All the simulation results shown here are at the same speed as the critical speed of the vehicle with new wheels. The results reveal that wheels with small wear tend to increase the critical speed and reduce the lateral creep forces. This happens due to flattening of the nominal contact patch (reduction of the conicity) on the wheel tread. Also, the reduction in creep forces indicates that the rate of wear decreases in marginally worn wheels. It is evident that the wheels wear out at a faster rate in the beginning (new wheel) and then start wearing at a smaller rate till a critical level of wear is reached. At this critical level of wear, the critical speed of the vehicle starts to reduce and the contact forces and wheel wear start to increase rapidly.

Small level of wear reduces the wheel conicity near the nominal contact patch. As a result, when the vehicle has to take a turn, wheels have to be displaced more to produce the desired level of difference in radius of curvature on inside and outside rails (see Figure 38). However, small level of wear surprisingly reduces the lateral creep forces during curving as shown in Figure 39. This happens due to stiffer tread gradient near the worn flanges.

When the tread wear level is very high (the rolling radius at nominal tread contact becomes 5mm less than the radius at tread end) then there is a negative cant effect. This causes the bogie to move at a very fast rate towards the flange. Although the worn flanges provide steeper gradient, the lateral momentum is sufficient to cause derailment. Thus, wheel wear levels for safe running are determined from simulations. Those results are out of the scope of this chapter.

Figure 36. Lateral displacement (hunting behavior) of wheels of various levels of wear on straight track

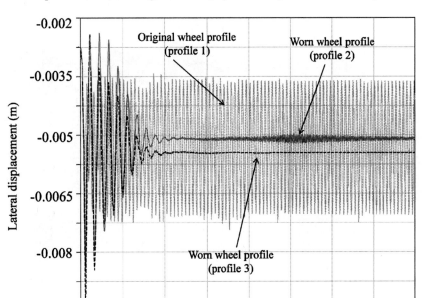

Figure 37. Lateral creep forces of wheels of various levels of wear on straight track

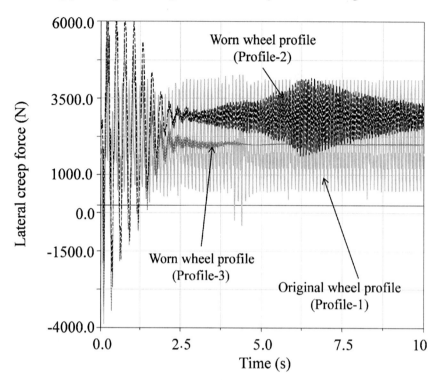

Figure 38. Lateral displacement of wheels of various levels of wear on curved irregular and flexible track of 4km radius

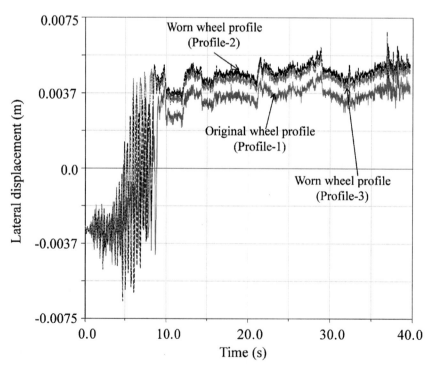

Figure 39. Lateral creep forces of wheels of various levels of wear on curved irregular and flexible track of 4km radius

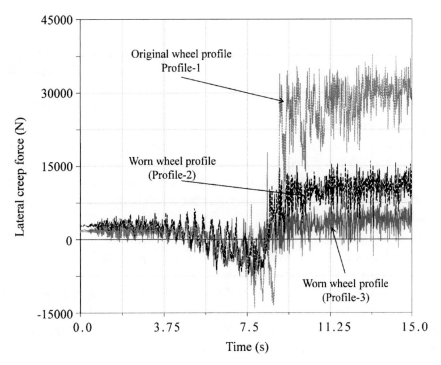

Worn wheels also change the passenger comfort. Comfort level is described in terms of the Sperling ride index, which must be less than 2.5 and preferably less than 2 (see Table 2). Small to moderate wear causes better passenger comfort on straight tracks and slightly lower comfort on straight tracks (Table 3). However, note that the comfort levels in Table 3 are given for very high-speed operation on irregular track (worst case scenario) and the actual comfort level is much better at smaller speed ranges and/or on more regular tracks.

Table 2. Different ranges of ride comfort (The American Public Transportation Association, 2007).

Ride Comfort (W_z)	Subjective Ride Comfort
1	Just noticeable
2	Clearly noticeable
2.5	More pronounced but not unpleasant
3	Strong, irregular but still tolerable
3.25	Very irregular
3.5	Extremely irregular, unpleasant, annoying, prolonged exposure intolerable
4	Extremely unpleasant, prolonged, exposure harmful

Table 3. Ride comfort with worn wheels

Type of Track and Operating Speed	Wheel Profile	Ride Comfort In Y-Direction	Ride Comfort in Z-Direction
Straight track with measured irregularity (100 m/s or 360 km/h)	Profile-1	2.5030	2.0465
	Profile-2	2.4450	2.0457
	Profile-3	2.2900	2.0490
Curved track with measured irregularity (100 m/s or 360 km/h)	Profile-1	2.0740	1.9397
	Profile-2	2.3564	2.2020
	Profile-3	2.2611	2.1130

8.4. Influence of Steering

Proper cant is helpful in reducing lateral accelerations as well as creep forces in lateral direction in the curved track. The cant adds to the self-steering action. If one needs to increase the operating speed around the curves then the track cant has to be increased. However, that causes excess cant for low speed operation on that same track. Therefore, it is not possible to increase the cant beyond a certain limit. Thus, to improve the dynamic behavior of vehicles on curved portions of the track, forced steering will be required. Actuation from the active steering mechanism has to counter the cant deficiency in order to increase the speed on the curves.

The dynamic behavior of the vehicle with forced-steering mechanism is improved in the presence of cant in the curved track. Forced-steering effect is more pronounced in the presence of cant/ super-elevation on the curved portion of the track. Figures 40 and 41 show the influence of cant in a well-designed track when the vehicle with bogies having active steering linkage takes a turn of 4km radius at the recommended speed (270 km/h).

Forced/active steering is effective not only when the train needs to operate with cant deficiency on the track but also when it needs to operate with cant excess. Excess cant is nullified by steering the wheels in opposite direction to the track curvature. Incorrect actuation of the self-steering mechanism can lead to severe problems like derailment due to instability of the feedback control loop. Therefore, precise measurement of location, correct correlation with track geometry at that location, proper selection of controller gains and minimization of controller and actuator delay are some of the important factors in mechatronic implementation of the active steering system.

The control algorithm turns the wheel-sets of each bogie in a coach in opposite directions to minimize the angle of attack. The controller is based on a proportional feedback system where the wheel-set orientation and vehicle speed are measured variables. The local cant deficiency/excess at the given speed is used to compute the desired steering angle and the error with the measured wheel steering angle is

Figure 40. Comparison of the lateral displacement of the left front wheel with track cant (150mm super elevation) and without cant at 75 m/s (270km/h) with forced steering

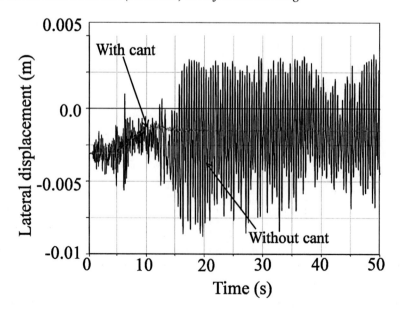

Figure 41. Comparison of the creep force in the left front wheel with track cant (150mm super elevation) and without cant at 75 m/s (270km/h) with forced steering

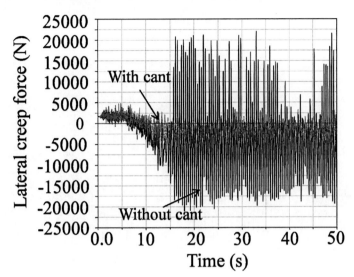

used to provide the additional amount of steering. The controller gains are kept small in order to avoid feedback delay induced instability problems.

Figures 42 and 43 show some results for active steering bogie operating beyond the recommended speed on the curved track (100 m/s as compared to 75 m/s considered in Figs. 40 and 41). Because of steering action, the lateral displacement is reduced and the frequency of lateral accelerations is less. Likewise, creep forces are smaller. With no steering, the wheel on the outer side remains continuously in flange contact whereas with steering action, there is continuous oscillation with no flange contact. The oscillations are result of simple proportional feedback control. With implementation of proportional-integral-derivative (PID) control through co-simulation in Matlab/Simulink and Adams VI-rail software (block diagram as shown in Figure 44), the oscillations in lateral displacement and creep forces become much less. Those results are not presented here.

8.5. Ride Comfort

The steering effect influences the ride comfort to a great extent. The effect of vibration on a person riding the vehicle is measured in terms of ride comfort which depends on carbody acceleration and the person's physical structure. Here, ride comfort for different tracks (with measured irregularities) is calculated by Sperling's ride index method. The weighting function B is different for vertical and horizontal directions. These weighting functions represent the average human body's vibration transmissibility models.

The weighting factor for transmissibility in horizontal (lateral) direction is given by (Chandra & Roy, 2002; Gangadharan et al., 2004)

$$B_h = 0.737 \times \left[\frac{1.911 \times f^2 + (0.25 \times f^2)^2}{(1 - 0.277 \times f^2)^2 + (1.563 \times f - 0.0368 \times f^3)^2} \right]^{1/2} \tag{29}$$

Figure 42. Lateral left wheel displacement of front wheel of front bogie at 100m/s (360 km/h) on an irregular flexible track with 150mm super-elevation

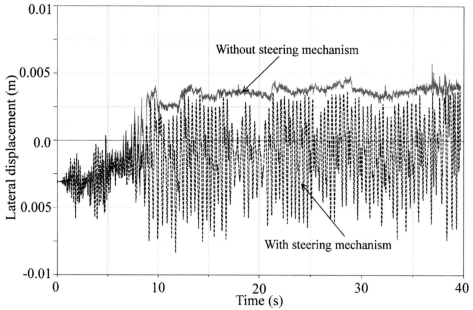

Figure 43. Lateral creep force in left wheel of front wheel of front bogie at 100m/s (360 km/h) on an irregular flexible track with 150mm super-elevation

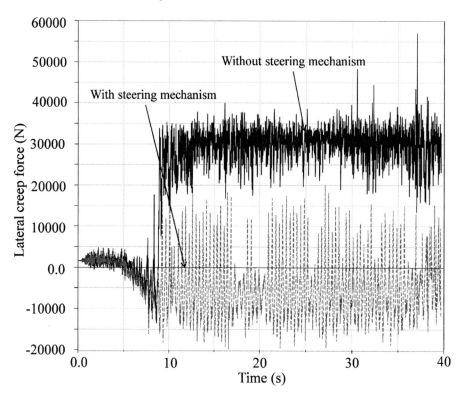

Dynamic Analysis of Steering Bogies

Figure 44. Block diagram of PID control of steering angle calculation.

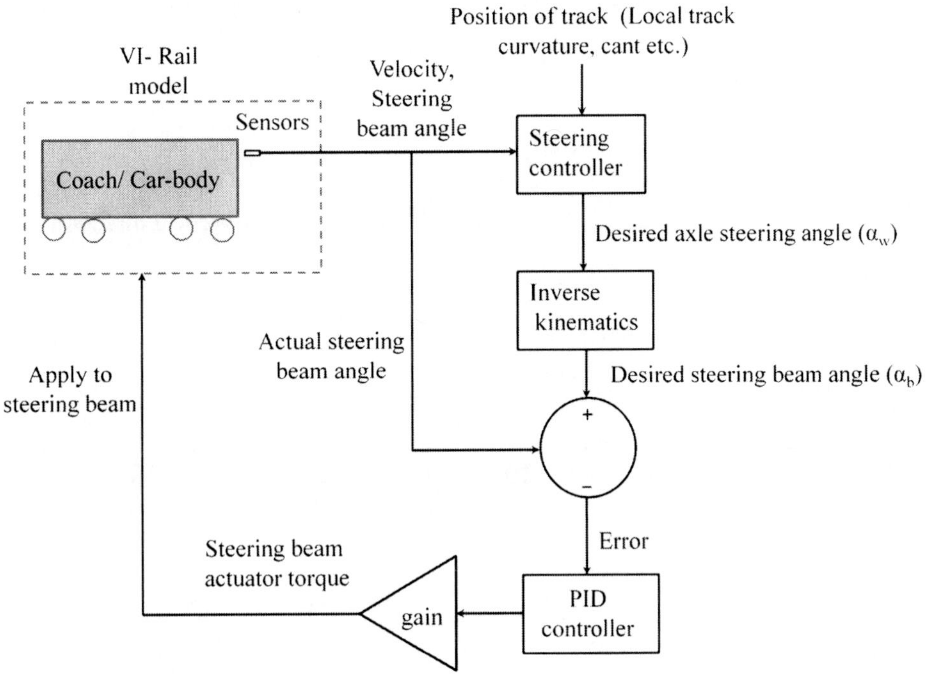

where *f* is the frequency in Hz. Likewise, the weighting factor for transmissibility in vertical direction is given by

$$B_v = B_k/1.25 \tag{30}$$

These weighting functions are graphically shown in Figure 45. It is seen that the low frequency content in the range of 1 to 10 Hz is most uncomfortable to the human beings. The suspension systems should be designed to isolate this frequency range so that they are not transmitted to the car-body.

The ride comfort (Sperling's ride index) in respective directions is expressed as

$$W_z = (a^2 B^2)^{1/6.67} \tag{31}$$

where a is the amplitude of acceleration in cm/s². This amplitude is obtained from Fourier transform (frequency response) of experimental/simulated car-body acceleration data. The ride index factor is determined for each individual frequency, and cumulative ride index (for *n* frequency components) is calculated as

$$W_{z\_total} = (Wz_1^{6.67} + Wz_2^{6.67} + Wz_3^{6.67} + \Lambda + Wz_n^{6.67})^{1/6.67} \tag{32}$$

The continuous vibration spectrum of vehicle can be used to obtain the correct cumulative ride index (ride comfort) by integrating the convoluted transmissibility and acceleration frequency responses over a frequency range of interest:

Figure 45. Weighting factors (frequency weighting curves) in difference directions

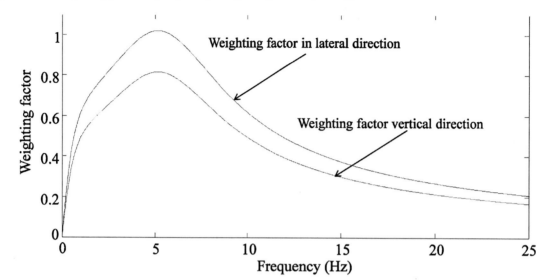

$$W_z = \left(\int_{f_1}^{f_2} a^2 B^2 df \right)^{1/6.67}, \tag{33}$$

where f_1 and f_2 are the lower and upper bounds of the frequency range.

The ride comfort for the forced steering bogie at different speeds and regular and irregular track conditions with various levels of cant are given in Table 4. It is found that at the designed operating speed of 75 m/s (270 km/h) on the curved track, the ride comfort is good for cant values between 0.05 and 0.1047. The steering bogie does not give good ride comfort when there is no cant in the track. Also, the ride becomes very uncomfortable when the operating speed is increased to 100 m/s (360 km/h) on the curved track. Up to 50 m/s (180 km/h) speed on curved track, the ride comfort is exceptional.

9. CONCLUSION

The design of a mechatronic railway bogie with passive and forced/active steering mechanisms is presented in this chapter. Influence of various track parameters on the vehicle performance is evaluated through simulations performed using Adams VI-Rail software. It was shown that the forced steering bogie is able to adapt to cant excess and cant deficiency on curved tracks. In addition, it has been shown that the bogie performance remains good even for small to moderate levels of wheel tread and flange wears. The forced steering bogie also gives exceptional ride comfort up to 270 km/h speed on irregular flexible curved tracks.

Table 4. Ride comfort at different speed, track condition and cant for vehicle with forced steering bogies

Cant	Track Condition	Speed in m/s (km/h)	Ride Comfort	
			Lateral Direction	Vertical Direction
0	Smooth	25 (90)	0.81214	0.3519
		50 (180)	1.10692	0.31017
		75 (270)	2.98507	0.9612
		100 (360)	1.79949	0.61052
	Irregular	25 (90)	1.1047	1.6875
		50 (180)	1.602	1.3874
		75 (270)	**3.3177**	**1.7834**
		100 (360)	2.5026	1.9753
0.05	Smooth	25 (90)	0.8179	0.7643
		50 (180)	1.16873	0.4988
		75 (270)	1.4455	0.688
		100 (360)	2.5566	1.02169
	Irregular	25 (90)	1.113	1.81398
		50 (180)	1.47	1.389
		75 (270)	**1.763**	**1.777**
		100 (360)	2.5569	1.0217
0.075	Smooth	25 (90)	0.822	0.366
		50 (180)	1.13	0.47
		75 (270)	1.44	0.625
		100 (360)	3.199	0.973
	Irregular	25 (90)	1.137	1.68
		50 (180)	1.465	1.389
		75 (270)	**1.788**	**1.67**
		100 (360)	3.116	1.988
0.1047	Smooth	25 (90)	0.8216	0.715
		50 (180)	1.0767	0.5837
		75 (270)	1.2927	0.86023
		100 (360)	1.54	1.0520
	Irregular	25 (90)	1.1744	1.678
		50 (180)	1.4693	1.39
		75 (270)	**1.7886**	**1.78454**
		100 (360)	3.54	3.30866

10. ACKNOWLEDGMENT

The Adams VI-Rail software used in this research has been procured with financial support from Center for Railway Research (CRR), Indian Institute of Technology, Kharagpur. Some validations have been performed with Simpack Rail which has been procured with financial support from Department of Mechanical Engineering, Indian Institute of Technology, Kharagpur. We also acknowledge help from Mr. Sital Singh of Research, Design & Standards Organization (RDSO), Ministry of Railways, Lucknow, for his help during initial stages of software learning.

REFERENCES

Aknin, P., Ayasse, J. B., & Devallez, A. (1991). Active steering of railway wheel sets. *Proceedings of the 12th IAVSD Conference*, Lyon, August 1991.

Ayasse, J.-B., & Chollet, H. (2006). Wheel-Rail Contact. In S. Iwnicki (Ed.), *Handbook of railway vehicle dynamics* (pp. 86–120). Boca Raton, FL: CRC Press.

Bruni, S., & Goodall, R. M., Mei, T.X Tsunashima, H. (2007). Control and monitoring for railway vehicle dynamics. *Vehicle System Dynamics*, *45*(7-8), 743–779. doi:10.1080/00423110701426690

Chandra, K., & Roy, D. G. (2002). *A technical guide on oscillation trial* (p. 334). MT: RDSO.

Cheng, Y. C., & Hsu, C. T. (2012). Hunting stability and derailment analysis of car model of a railway vehicle system. *Rail and Rapid Transit*, *226*(2), 187–202. doi:10.1177/0954409711407658

Cheng, Y. C., Lee, S. Y., & Chen, H. H. (2009). Modeling and nonlinear hunting stability analysis of high speed railway vehicle moving on the curved track. *Journal of Sound and Vibration*, *324*(1-2), 139–160. doi:10.1016/j.jsv.2009.01.053

Dahlberg, T. (2006). Track issues. In S. Iwnicki (Ed.), *Handbook of railway vehicle dynamics* (pp. 143–179). Boca Raton, FL: CRC Press. doi:10.1201/9781420004892.ch6

Diana, G., Bruni, S., Cheli, F., & Resta, F. (2002). Active control of the running behavior railway vehicle: Stability and curving performances. *Vehicle System Dynamics*, *37*(Suppl.), 157–170. doi:10.1080/00423114.2002.11666229

Dukkipati, R. V., & Amyot, J. (1988). *Computer aided simulation in railway dynamics*. New York: Marcel Dekker. Inc.

Gangadharan, K. V., Sujata, C., & Ramamurti, V. (2004), Experimental & analytical ride comfort evaluation of a railway coach. *Proc. IMAC-XXII*, Dearborn, Michigan.

Goodall, R. M., Bruni, S., & Mei, T. X. (2006). Concepts and prospects for actively controlled railway running gear. *Vehicle System Dynamics*, *44*(Suppl.), 60–70. doi:10.1080/00423110600867374

Goodall, R. M., & Mei, T. X. (2006). Active Suspensions. In: S. Iwnicki (Ed.), Handbook of railway vehicle dynamics (pp. 327-357). Boca Raton, FL: CRC Press.

Gretzschel, M., & Bose, L. (1999). A mechatronic approach for active influence on railway vehicle running behaviour. *Vehicle System Dynamics*, *33*(Suppl.), 418–430.

Gretzschel, M., & Bose, L. (2002). A new concept for integrated guidance and drive of railway running gears. *Control Engineering Practice*, *10*(9), 1013–1021. doi:10.1016/S0967-0661(02)00046-1

Kalker, J. J. (1990). *Three Dimensional Elastic Bodies in Rolling Contact.* Dordrecht, Netherlands: Kluwer. doi:10.1007/978-94-015-7889-9

Kim, M.S., Park, J.H. & You, W.H. (2008). Construction of active steering system of scaled railway vehicle. *International journal of systems applications, engineering and development*, 4(2), 217-226.

Lee, S. Y., & Cheng, Y. C. (2005). Hunting stability analysis for high-speed railway vehicle trucks on tangent tracks. *Journal of Sound and Vibration, 282*(3-5), 881–898. doi:10.1016/j.jsv.2004.03.050

Lindahl, M. (2001), Track geometry for high speed railways: A literature survey & simulation of dynamic vehicle response.

Matsumoto, A., Sato, Y., Ohno, H., Suda, Y., Michitsuji, Y., Komiyama, M., & Nakai, T. et al. (2005), Multi-body dynamics simulation and experimental evolution for active steering bogie. *Proc. Int'l symposium on speed up and service technology for railway and maglev systems* (pp. 103-107).

Mei, T. X., & Goodall, R. M. (1999). Wheelset control strategies for a two axle railway vehicle. *Vehicle System Dynamics, 33*(Suppl.), 653–664.

Mei, T. X., & Goodall, R. M. (2000). LQG solution for active steering of solid axle railway vehicles. *IEE Proceedings. Control Theory and Applications, 147*(1), 111–117. doi:10.1049/ip-cta:20000145

Mei, T. X., & Goodall, R. M. (2000). Modal controller for active steering of railway vehicles with solid axle wheel-sets. *Vehicle System Dynamics, 34*(1), 25–41. doi:10.1076/0042-3114(200008)34:1;1-K;FT025

Mei, T. X., & Goodall, R. M. (2003). Recent development in active steering of railway vehicles. *Vehicle System Dynamics, 39*(6), 415–436. doi:10.1076/vesd.39.6.415.14594

Mei, T. X., & Goodall, R. M. (2003). Practical strategies for controlling railway wheel-sets with independently rotating wheels, *Transactions of the ASME Journal of Dynamic Systems, Measurement and Control, 125,* 354–360.

Mei, T. X., & Goodall, R. M. (2006). Stability control of railway bogies using absolute stiffness–skyhook spring approach. *Vehicle System Dynamics, 44*(Suppl.), 83–92. doi:10.1080/00423110600867440

Mei, T. X., & Li, H. (2008), Control design for the active stabilization of rail wheel-sets. *Proceedings of theTransactions of the ASME Journal of Dynamic Systems, Measurement and Control.*

Michitsuji, Y., & Suda, Y. (2006). Running performance of power-steering railway bogie with independently rotating wheels. *Vehicle System Dynamics, 44*(Suppl.), 71–82. doi:10.1080/00423110600867416

Miura, S., Takai, H., Uchida, M., & Fukada, Y. (1998). The mechanism of railway track. In *Japan Railway & Transport Review* (pp. 38-45).

Okamato, I. (1998). How bogies work. *Japan Railway & Transport Review, 18,* 52–61.

Okamato, I. (1999). Shinkansen bogies. *Japan Railway & Transport Review, 19,* 46–52.

Orlova, A., & Boronenko, Y. (2006). The Anatomy of Railway Vehicle Running Gear. In S. Iwnicki (Ed.), *Handbook of railway vehicle dynamics* (pp. 42–83). Boca Raton, FL: CRC Press. doi:10.1201/9781420004892.ch3

Park, J. H., Koh, H. I., Hyun, H. M., Kim, M. S., & You, W. H. (2010). Design and analysis of active steering bogie for urban trains. *Journal of Mechanical Science and Technology, 24*(6), 1353–1362. doi:10.1007/s12206-010-0341-4

Pearson, J. T., Goodall, R. M., Mei, T. X., & Shen, S. (2004). Kalman filter design for high speed bogie active stability system. Proceedings of UKACC control, Bath, UK

Perez, J., Busturia, J. M., & Goodall, R. M. (2002). Control strategies for active steering of bogie based railway vehicles. *Control Engineering Practice, 10*(9), 1005–1012. doi:10.1016/S0967-0661(02)00070-9

Perez, J., Busturia, J. M., Mei, T. X., & Vinolas, J. (2004). Combined active steering and traction for mechatronic bogie vehicles with independently rotating wheels. *Annual Reviews in Control, 28*(2), 207–217. doi:10.1016/j.arcontrol.2004.02.004

Perez, J., Mauer, L., & Busturia, J. M. (2002). Design of active steering systems for bogie-based railway vehicles with independently rotating wheels. *Vehicle System Dynamics, 37*(Suppl.), 209–220. doi:10.1080/00423114.2002.11666233

Polach, O., Berg, M., & Iwnicki, S. (2006). Simulation. In S. Iwnicki (Ed.), *Handbook of railway vehicle dynamics* (pp. 359–421). Boca Raton, FL: CRC Press.

Pombo, J. (2012). Application of computational tool to study influence of worn wheels on railway vehicles. *Journal of software engineering and application, 5*, 51-61.

Powell, A. J. (1999). On the dynamics of actively steered railway vehicles. *Vehicle System Dynamics, 33*(Suppl.), 442–452.

Schupp, G., Weidemann, C., & Mauer, L. (2004). Modeling the contact between wheel and rail within multi-body system, Vehicle System Dynamics. *International Journal of Vehicle Mechanics and Mobility, 41*(5), 349–364.

Shabana, A. A., Zaazaa, K. E., & Sugiyama, H. (2008). *Rail road Vehicle Dynamics: A Computational Approach*. Boca Raton, FL: CRC Press.

Shen, G., & Goodall, R. M. (1997). Active yaw relaxation for improved bogie performance. *Vehicle System Dynamics, 28*(4-5), 273–289. doi:10.1080/00423119708969357

Shen, S., Mei, T. X., Goodall, R. M., Pearson, J., & Himmelstein, G. (2004). A study of active steering strategies for railway bogie. *Vehicle System Dynamics, 41*(Suppl.), 282–291.

Shimokawa, Y., & Mizuno, M. (2013). *Development of new concept steering bogie*. Nippon Steel & Sumitomo Metal Technical Report No. 105, Osaka.

Standard for Definition and Measurement of Wheel Tread taper (Publication No. APTA PR-M-S-017-06). (2007The American Public Transportation Association. Washington, DC: APTA Press.

Tanabe, M., Komiya, S., Wakui, H., Matsumoto, N., & Sogade, M. (2003). Simulation and visualization of high speed Shinkansen train on the railway structure. *Japan J. Indust. Appl. Math, 17*(2), 309–320. doi:10.1007/BF03167350

Wicken, A. H. (1993), Dynamic stability of articulated and steered railway vehicles guided by lateral displacement feedback. *Proceedings of the 13th IAVSD symposium*, Chengdu, China.

Wickens, A. H. (1994). Dynamic Stability of articulated and steered railway vehicles guided by lateral displacement feedback. *Vehicle System Dynamics, 23*(Suppl.), 541–553. doi:10.1080/00423119308969539

Wickens, A. H. (2003). *Fundamentals of rail vehicle dynamics*. Lisse, The Netherlands: Swets & Zeitlinger. doi:10.1201/9780203970997

Wickens, A. H. (2006). A History of Railway Vehicle Dynamics. In S. Iwnicki (Ed.), *Handbook of railway vehicle dynamics* (pp. 5–38). Boca Raton, FL: CRC Press. doi:10.1201/9781420004892.ch2

Chapter 22
Ventilation and Air Conditioning in Tunnels and Underground Stations

Frederic Waymel
Egis Tunnels, France

Christophe Butaud
Egis Tunnels, France

ABSTRACT

This chapter is an overview of the state of the art and advanced principles in the field of ventilation and air conditioning (AC) in tunnels and underground stations. The first part is dedicated to the background which deals with the design objectives that are generally retained for normal and emergency operation of underground rail projects. The second part provides solutions and recommendations of ventilation and AC strategies that can be used in metro and rail projects. Advantages and drawbacks of the proposed solutions are also discussed. The main parameters that can influence the design are introduced in this section. The possibility of using draught relief shaft is detailed. Advantages of Platform Screen doors and heat sink effects are also described. Various cooling technologies of station air conditioning systems are presented. Critical issues when designing longitudinal ventilation system for tunnel emergency situations are also discussed. The last part is a short list of future research directions in the field of cooling / heating production for air conditioning systems.

INTRODUCTION

In order to avoid insertion issues related to dense urban areas, urban rail and metro infrastructures are now-a-days mostly underground. This leads to a particular care to maintain the operation under acceptable environment and safety conditions. Indeed, in normal operating conditions trains and the presence of passengers in the transportation network lead to large amounts of sensible and latent heat released in the underground spaces that may lead to high temperature and humidity levels. Moreover, the underground rail infrastructures are also exposed to a high level of fire risks due to the confined space that increases

DOI: 10.4018/978-1-5225-0084-1.ch022

Ventilation and Air Conditioning in Tunnels and Underground Stations

the consequences of fire on passengers but also more generally on civil and assets. Ventilation and air conditioning systems are part of the main systems which contribute to the achievement of acceptable environment and safety conditions. The following chapter is dedicated to those systems detailing the state of the art and recent researches and developments.

BACKGROUND

This section gives an overview of the state of the art regarding the main objectives that are currently retained for the design of ventilation and air conditioning system in rail and more particularly metro infrastructures.

BACKGROUND AND OBJECTIVES DURING NORMAL OPERATION

1. Thermodynamic Conditions

In normal operation, ventilation and air conditioning systems are dedicated and designed for various purposes.

One of the first targets is to maintain acceptable hygienic and thermal conditions for the passengers using the metro system.

For hygienic conditions, the aim of the ventilation systems is to provide a fresh air renewal into the underground spaces in order to reduce the level of contaminants and carbon dioxide. The volume of fresh air to be provided is generally expressed as a number of volumes to be renewed in a dedicated time but also as a volume of flow rate per passenger. Various standards already exist depending on the country where the infrastructure is built. One of the most common standards used internationally is the ASHRAE 62-1 (ASHRAE, 2007). In this standard, the volume of fresh air for a transportation waiting area is set to 0.3 L/s/m^2 or 4.1 L/s/person.

Thermal comfort of passengers is also becoming one of most critical issues when designing Air conditioning and Ventilation systems in tunnels and underground station areas. Indeed, the quantity of heat released in the station is generally at high level due to the presence of passengers, the functioning of various electrical and mechanical systems but also the large quantity of heat released by trains in the tunnels and stations. The thermal comfort in metro system was very often not properly addressed in the 20th century. Now-a-days, passengers are more on more accustomed to air conditioning which also tends to increase the demand of suitable thermal comfort area in the underground stations. Thermal acceptability is a quite a complicated field of research as it depends on various factors such as the origin of the population, the type of activity, the manner to be clothed, the influence of draught, the subjectivity of passengers regarding the thermal comfort sensation etc. Many research works have already been performed such as those provided by Berglund (1979), Fanger (1988) or Goldman (1978). The results of those researches have been largely detailed in the international standard ASHRAE 55 (ASHRAE, 2004). This standard indicates a comfort level between 19°C and 28°C depending on the humidity ratio and the level of clothing insulation. Those values can be increased up to 4°C depending on the air speed. The upper limit for the absolute humidity is also fixed at 0.012. This standards provides interesting information for the design of ventilation and air conditioning systems in underground stations but should also be used with a particular care. Indeed, most of researches data have been developed for buildings and standing areas where the people are supposed to stay several hours. Although it can be the case inside train coaches for particular long trips, things are different in underground stations as the passengers are supposed to stay only few minutes.

The Relative Warmth Index (RWI) was introduced in the sixties and was largely developed in the Subway Environmental Design Handbook (SEDH, 1976) as a rule to set more appropriate criteria for the design of HVAC systems in subways. This parameter includes the fact that the time response of the human body to a change of thermal conditions is around six minutes. This means that passengers do not feel immediately uncomfortable when a rise of temperature appears along their trip. Moreover, the purpose of using this parameter is mainly for reducing the level of uncomfortable situations along the trip instead of achieving a specific comfort level. The use of this parameter is particularly interesting as it takes into account the fact that passengers generally come from outside where temperature levels can already be high and uncomfortable. During extreme outdoor ambient conditions, it is consequently not necessary to achieve low temperatures in the underground stations.

Another parameter which is generally considered for the thermal comfort of passengers is the air speed. This has especially been introduced in the Beaufort scale in which a value of around 5 m/s should not be exceeded.

Platforms and public spaces in below ground stations can also be regarded as transit areas with some large openings to outside (entrances). For that building category, ASHRAE standards are not completely adapted to determine acceptable thermal conditions.

In summer, for the purpose of designing an air conditioning system devoted to public spaces, an acceptable difference in temperature between outside and inside the station can also be chosen.

In the first place, air conditioning habits are to control that people are not exposed to a difference of temperature of more or less 7/8 °C. Beyond that variation, the difference between two ambiances is awkward for a majority of people.

Considering some large volumes for the public spaces in station, and all obstacles that could obstruct the diffusion of the conditioned air flow into the station (such as staircases and corridors), it is assumed that thermal conditions in spaces are not equals.

To ensure comfortable thermal conditions in standing areas, the conditioning of public spaces can be brought under control in order to follow a threshold of 5°C maximum above outside conditions.

In lower levels, some fresh air is directly blown on platform in normal mode where passengers are waiting for their trains. Especially in urban area, this air can be filtered before being blown in station.

Another major objective of the ventilation systems in underground transportation networks is to maintain air temperatures in consistency with the normal operation of electrical and mechanical equipment and more particularly with the rolling stock. The rolling stock is more and more equipped with air conditioning (AC) systems which are quite sensitive to the air temperature outside of the train. If the temperature exceeds the allowable operating condition of the AC system, the condensing units might not be able to operate properly and might stop running. The state of the art is an air temperature threshold in running tunnels of 40-45°C under normal operation and up to 45-50°C in case of congested mode (train blocked in a running tunnel). Those thresholds can be a particular design key challenge especially in hot climate areas.

Ventilation and Air Conditioning in Tunnels and Underground Stations

2. Attenuation of Trains Piston Effects

When trains are moving into tunnels, the limited cross-section of tunnels can lead to a significant increase of the pressure level in the underground network. Those effects can be interesting as they could be used to provide natural ventilation of tunnels. However, three types of undesirable consequences may appear due to train piston effects:

- High velocities inside underground stations. This is particularly the case for stations that are not equipped with full Platform Screen Doors (PSD)
- High static pressure level along PSD. When stations are equipped with PSD, the ventilation network must consequently be designed in order to reduce the pressure level in consistency with the mechanical resistance of the doors. This mechanical resistance generally corresponds to a pressure level between 200 Pa and 400 Pa but new generation of PSD used in rapid transit system can also resist up to 1,000 Pa.
- Rapid pressure changes that can lead to physiological effects on subway patrons especially for ears. The recommendations regarding the pressure change limits generally vary between 400 Pa/s and 800 Pa/s (Carstens, 1965). It can be noted that those effects are generally perceptible in rapid transit transportation system and are less frequent in normal metro system where the speed of trains is generally below 100 km/h.

BACKGROUND AND OBJECTIVES DURING EMERGENCY OPERATION

Even if during last decades a large amount of research and development has been performed to introduce non-flammable or fire retardant materials in the rolling stock manufacturing, the probability of a fire event still exists and must still be considered as a major safety issue more particularly in confined spaces. The emergency ventilation is one of the most important safety features that contribute to the improvement of the fire safety level by reducing the consequences of these undesirable events.

1. Review of the Main Fire Safety Targets

The first objective of the emergency ventilation is to provide a tenable environment along the path of egress from a fire incident in enclosed stations and enclosed trainways in order to maximize the survivability of occupants intimate with the initial fire development. To achieve this objective, a solution could be the implementation of emergency ventilation systems that are able whatever the fire situation to prevent smoke propagation and keep the air completely free of smoke along paths of egress or at least at passengers head height. However, such an objective might lead to overdesigned ventilation systems whose impacts on civil, on ventilation safety features and on the overall cost could be unreasonable in comparison with the level of fire risks in metro and rail systems (low probability). For that reason, the term of tenable environment is as per the state of the art in transportation system considered as an environment that permits the self-rescue of occupants for a specific period of time. It is then assumed that occupants intimate with the initial fire development can be exposed to smokes. However, consequences of such an exposition should be reduced as low as possible. The evaluation of consequences can be estimated by temperature, toxicity and visibility thresholds that should not be exceeded along paths of egress during the evacuation time. More recently, the principle of Fractional Effective Dose (FED) has been introduced as detailed in the NFPA 130 standards (NFPA 130, 2010). The time of exposure to the smoke temperature or toxicity is taken into account in this parameter. First injuries occur when the FED is closed to 0.3 and the fatality may appear for a FED value of 1. As an example, the maximum exposure time to avoid injuries is 3.8 min. for a temperature of 80°C and 10.1 min. for a temperature of 60°C. For the toxicity (carbon monoxide), an average toxicity of 1150 ppm is acceptable during 6 min. and can grow up to 15 min. for a toxicity level of 450 ppm. For an accurate evaluation of consequences of a fire based on this FED principle, appropriate engineering analysis using fire dynamic simulations and evacuation models are recommended.

Another important target generally retained for the design of emergency ventilation systems is to protect occupants not intimate with the initial fire development. This means that the ventilation system should be able to achieve at least the following purposes:

- Prevent smoke propagation toward other non-incident trains that could be blocked by the train in fire. For that, it is necessary to accommodate the maximum number of trains which could be between ventilation shafts during an emergency. The design of the tunnel ventilation system should consequently be done in consistency with operating objectives and the design of the signaling system.
- In case of fire in tunnel, prevent smoke propagation into adjacent stations where other passengers may wait.
- In case of fire in a station, prevent smoke propagation into other station levels.

2. Consideration Regarding the Design Fire Size

The heat release rate (or HRR) of a train fire is one of most important inputs in the design of emergency ventilation. Unlike the case of fires in road tunnels, there are no standard values of heat release rates for the design of subways. The HRR depends mostly on the rolling stock type and the safety level that have to be achieved for the underground infrastructure. Depending on the size of coaches and the type of materials they are made of, each has a specific fire load, which is the quantity of heat that can be released by the complete combustion of all combustible materials in the train. Since the complete combustion is rarely obtained, the real energy released during an actual fire is a fraction of the fire load, which depends on ventilation conditions, the tunnel geometry and the time which has to be considered (evacuation only, fire brigade intervention…).

2.1 Experimental References

There is not much data available concerning heat release rate in the literature. Most of them come from the EUREKA project EU 499 FIRETUN. It is a series of experiments which took place at the beginning of the 90's in Norway.

Experiments on railways coaches and subway cars have shown an average HRR between 13 and 20 MW. This wide range is explained by the differences between trains.

Results of the experiments also teach that the HRR depends strongly on time. During a first stage (which length depends on the size of the primary fire source), the fire develops inside the vehicle. The HRR is still quite low and increases slowly, since the heat is contained by the metallic exterior shell of

Ventilation and Air Conditioning in Tunnels and Underground Stations

the train. When a critical value of heat release rate is reached (around 10 MW), windows and doors break because of the heat and massive supply of oxygen in fresh air stir up the fire. This is the phenomenon called flash-over. A peak of HRR is then observed, depending on the quantity of available combustible materials. Temperatures at the fire location can rise up to 800 - 900°C. This stage only lasts for a few tens of seconds to a few minutes before the HRR decreases slowly until complete extinction, after a typical duration of about an hour.

As an illustration, one of EUREKA tests was conducted on an aluminium exterior shell subway railcar with internal dimensions of 18.0 m long, 2.8 m high and 3.0 m wide. Results show that the flashover occurs after only 10 minutes and the heat release rate reaches a peak of 35 MW. After another 20 minutes, the HRR drops to 10 MW and then linearly decreases to 1-2 MW during the next 120 minutes. Temperatures at the fire location 40 minutes after the beginning of the fire were around 400 °C and 100 °C after 100 minutes.

Other references include the study of the European work group FIT (Fire in Tunnels). After the analysis of fire scenarios in 210 design references, they recommend to use values between 12 and 47 MW.

2.2 Examples of Design Fire in Recent Metro Projects

The following table presents few examples of heat release rate values chosen for the design of some of the latest subway projects. The range of values is between 9 and 20 MW, depending on rolling stock data, operating conditions (typically headways) and the level of risk considered.

It is particularly recommended to consider in the choice of the design HRR the residual risk which can be assumed by owners, operators and authorities. Indeed, the maximum heat release rate that can be developed by a dedicated rolling stock may appear but with a very low probability and using this value for the design of the ventilation system is not always relevant. Consequently, for exactly the same rolling stock, the design HRR can be different from a project to another depending on the acceptability of this residual risk.

Table 1. Typical train design fire considered in different metro project

Country	City	Heat Release Rate
Denmark	Copenhagen	20 MW
Iran	Tabriz	10 MW
India	Chennai	15 MW
India	Bangalore	15 MW
Ireland	Dublin	18 MW
Algeria	Alger	9 MW
France	Marseilles	15 MW
France	Grand Paris	5 MW
Switzerland	Lausanne	20 MW
Italia	Parma	10 MW

SOLUTIONS AND RECOMMENDATIONS: THE MAIN KEYS FOR THE DESIGN OF VENTILATION AND AIR CONDITIONING SYSTEMS

The first part of this section provides a description of the main parameters that should be considered in modern design methodologies of tunnel ventilation and Heating Ventilation & Air Conditioning (HVAC) systems. Second and third parts present an overview of the state of the art and various strategies that have been developed in recent rail and metro projects for normal and emergency operation. Advantages, drawbacks and key issues are also discussed. The last part of the section is dedicated to a presentation of design approaches and design tools that have been developed and recently improved for better designs.

1. Discussion Regarding the Main Factors Influencing the Design

1.1 Geometry

Geometry of metro project is obviously one of main parameters that influence the tunnel and station ventilation system design. Within the geometry, following parameters can be listed and discussed:

- **Length of Interstations:** This parameter can influence the number of trains running between two stations and consequently the number of trains that could be blocked upstream a train in fire. In some cases, additional interstation ventilation shafts are required to prevent smoke propagation toward following trains.
- **Twin-Tube Unidirectional or Single Bore Bidirectional Tunnels:** In normal operation, the level of train piston effects strongly depends on this parameter. In twin tube tunnels, piston effect is very high due to the limitation of the tunnel cross-section. Additional draught relief shafts can be required to reduce the pressure and air velocities in adjacent stations (see explanation of draught relief shafts in the next sections). Nevertheless, train piston effects combined with an appropriate design of draught relief shafts gives some opportunities to ventilate the tunnel in normal operation without using mechanical ventilation (energy saving). This opportunity is more difficult for single bore tunnels due to the compensation of piston effects between both traffic directions. Generally, mid-tunnel ventilation shafts are required for this type of tunnel to bring adequate amount of fresh air in the middle of the tunnel. In emergency situation, a particular care has to be paid in single bore tunnel regarding the possibility for trains located in the opposite track of the train in fire to exit the tunnel. In twin-tube tunnels, other trains in front of the train in fire have generally the possibility to leave easily the tunnel and smokes can be pushed in the same direction of the traffic direction without any particular issue for blocked trains.
- **Tunnel Cross Section:** The tunnel cross-section also plays a significant role on the level of train piston effects. Indeed, smaller is the cross-section higher is the pressure level generated by the trains movements. In emergency situations, small tunnel cross-sections generally require reversible longitudinal ventilation systems in order to be able to push smokes in one direction or the other. Indeed, small cross-section does not allow smoke stratification above head height. Moreover, the dilution of smoke which helps reducing temperature and toxicity levels is generally limited. For those reasons, when the fire occurs for example at the tail of the train, it is necessary to push the smoke in the opposite traffic direction to avoid large amount of injuries and fatalities. Additional

tunnel ventilation shafts can be required to prevent smoke propagation towards following trains potentially blocked upstream the train in fire. In opposite, greater tunnel cross-section of single bore tunnels can lead to the opportunity of having a strategy which consists of pushing smokes always in the traffic direction or in the direction of the nearest ventilation shaft without considering the location of the fire on the train. Indeed, the smoke stratification leading to a fresh air layer at head height can still appear up to a reasonable distance from the fire in large cross-section. Moreover, the amount of available fresh air for smoke dilution when stratifications effects disappear at a certain distance from the fire may reduce the temperature downstream of the fire where passengers may have to evacuate if the fire occurs for example at the tail of the train. However, it is strictly recommended to assess carefully this type of "unidirectional ventilation strategy" case by case in order to verify that ambient conditions downstream the fire are still acceptable.
- **Civil Constraints:** A particular attention of civil constraints of the project must be paid when designing tunnel ventilation shafts. Indeed, in particular cases the construction of ventilation shafts can be difficult, costly or sometimes unfeasible. Particularly, the presence of dense urban areas, mountain areas, and specific geotechnical properties of the soil must be considered when selecting the location of shafts.
- **Station Architecture:** The design of station smoke extraction system mainly depends on the architecture of the station. It is generally easier to control the smoke propagation in a station whose the different underground level are properly separated. A particular attention has to be paid for stations including large open spaces such as atriums.

1.2 Train Characteristics

Train characteristics can also influence the design of the ventilation and air conditioning system. Among those characteristics, the following can play a significant role in the design:

- **Kinematic Properties:** Maximum speed, acceleration and deceleration which define the time spent by a train in an interstation tunnel. Consequently, those parameters can also influence the number of trains running between two stations and the number of trains that could be blocked upstream a train in fire for which additional ventilation shaft could be required as explained earlier. The speed of the train is also an important key factor regarding the pressure generated by train piston effects.
- **Train Capacity:** This influences the number of passengers to be evacuated in case of fire and the evacuation time to reach a safety place. When the smoke is pushed towards passengers of the train in fire in case of single direction longitudinal ventilation strategy, increase of evacuation time could lead to unacceptable levels of injuries of fatalities which may lead to the need of having reversible ventilation system and additional shafts to protect other following trains.
- **Amount and Properties of Combustible Materials:** Those can lead to higher fire heat release rate. However as mentioned earlier, the design fire is also fixed by the state of the art and the acceptability of the residual risk by owners and authorities.
- **Heat Released During Normal Operation:** The heat released by the train during normal operation can be very high especially when trains are equipped with air conditioning systems. The amount of fresh air and the design of ventilation shaft and air conditioning capacities can consequently be influenced by this parameter. It can however be noted that in last decades, lot of

developments have been done to improve the braking regenerative efficiency of trains and traction power systems. Those developments allow a better recovering of the kinematic energy of the train and reduce consequently the heat release by trains during the braking phase.

1.3 Passengers

The number of passengers using the metro system is also quite important in the design of the ventilation system.

Firstly, sensible and latent heats released by passengers can very often be non-negligible. The amount of the heat released per passengers strongly depends on the type of activity but also on the dry-bulb and wet-bulb temperature of the air. The ASHRAE 55 (ASHRAE, 2004) provides some reference of the total heat released per person depending on the type of activity. For the typical walking speed in station area, the value is set around 115 W/m^2 (average of body surface of 1.8 m^2). The fraction between the sensible and the latent heat depends on enthalpy conditions of the ambient air. The ratio of the latent heat tends to increase when the dry bulb temperature is higher.

Secondly, the number of passengers may affect the sizing of the fresh air flow rate required to achieve hygienic objectives mentioned earlier.

Thirdly, the number of passengers may also affect the evacuation time especially when evacuation means are under designed. More performing smoke extraction systems could be required in particular situations to reduce the Fractional Effective Dose during the evacuation phase.

1.4 Operation

Characteristics of operation systems and implementation of particular operation procedures are also considered in modern ventilation design approaches. Indeed, those can largely influence the design and the optimization of the ventilation system. Following points are more particularly considered:

- **Headways:** As for the interstation length and kinematic properties of the rolling stock, this can also influence the number of trains running between two stations and consequently the number of trains that could be blocked by a train in fire.
- **Signaling Systems:** The design of fixed signaling blocks or specific additional safety functions on a moving block signaling system can be proposed in such a way that the number of trains in a particular tunnel section can be limited to one train per track in each emergency tunnel ventilation section (between two shafts). However, this has to be done in consistency with the design objective related to the desired minimum operating headway.
- **Particular Operating Procedures:** Procedure to remove following trains can also be proposed (automatic or turn-back procedure) in order to reduce requirements of additional tunnel ventilation shafts to protect from the smokes potential non-incident trains blocked by the train in fire. However, this generally needs to be done in a short time before the smoke reaches the following train. Moreover, it is generally necessary to cut very quickly the traction power supply in the incident bore in order to proceed quick evacuation of the incident train. The turn back of a non-incident blocked train should consequently be done quickly before cutting the traction especially if the train is located in the same traction power section of the incident train. The limited fire

resistance of the catenary is also a particular issue that should be considered in such a procedure. More generally, it is largely recommended to include this kind of procedure when adopted in the emergency operating plan.

1.5 Means of Egress

Sizing egress means plays a significant role on the evacuation time in tunnels and stations. When the complete evacuation can be achieved in a reasonable time, it is generally possible to reduce performances of the emergency ventilation system. Indeed, the exposure time to unsuitable temperature and toxic ambient conditions can be minimized reducing the risk of injuries and fatalities (see above for the discussion related to FED parameters). The three main characteristics of egress means which influence the evacuation time in tunnels are the following:

- Distance between safety exits.
- Width of the evacuation pathway along the tunnel
- Width of egress doors.

2. Strategies in Normal Operation

2.1 Tunnels

Maintaining acceptable temperature conditions during operating hours in tunnels is one of main key challenges. The quantity of heat released by the rolling stock can be very high especially when air-conditioning systems are installed on the trains but also when the kinetic energy of moving trains is not necessarily regenerated during the braking phase. Various ventilation strategies are applied to minimize effects on the tunnel temperature. Mechanical ventilation but also natural ventilation using the benefit of train piston effects and efficient draught relief shafts are introduced hereafter explaining also advantages and drawbacks.

2.1.1 Consideration Regarding the Design of Draught Relief Shafts

Normal operation ventilation strategies for twin tube tunnels generally relies on the construction and the use of draught relief shafts as detailed in the Subway Environmental Design Handbook (SEDH, 1976). A draught relief shaft is a shaft with low aerodynamic resistance open to the atmosphere and excluding any use of mechanical ventilation system. Draught relief shafts are in most of the cases located at the interface between tunnels and stations but can also be implemented along the interstation. These shafts are designed more particularly for the two following main purposes:

- To reduce the influence of train piston effects in the station areas. When stations are equipped with platform screen doors, this helps to reduce the static pressure on screen doors and also the leakage through doors. Without platform screen doors, the air speed generated by train piston effects in the station can be reduced by draught relief shafts.
- To reduce the level of rapid pressure changes and attenuates undesirable physiological effects on passengers' ears.

- To create a natural ventilation for the tunnel using the action of train piston effects. The negative static pressure generated by the train at the rear of the train leads to fresh air entering through the shaft located at the tunnel entrance. In opposite, the positive static pressure generated at the front of the train leads to warm air exhausting through the shaft located at the tunnel exit.

The way to design draught relief shafts is one of the key issues to achieve desirable effects especially regarding the natural ventilation of tunnels. Two kinds of designs principles have been proposed in recent metro projects that can be discussed here.

The first principle consists of building one shaft per tunnel end shared between the two bores as shown on the Figure 1. One of the main advantages of this design is to reduce civil works and land acquisition as only one shaft termination is required for this. This single shaft is also generally efficient enough to reduce train piston effects in the station area. Nevertheless, the efficiency of the natural ventilation can be very low because it leads to significant warm air recirculation between the two opposite traffic bores.

The other principle consists of building one shaft per tunnel end and per bore as shown on the Figure 2. This requires additional civil works and land acquisition but may lead to significant efficiency for the natural ventilation. Indeed in such case, the independency between the two bores avoids any flow recirculation.

Figure 1. Single draught relief shaft configuration

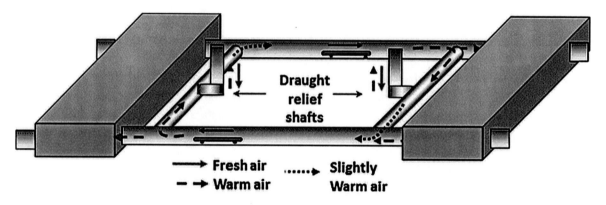

Figure 2. Double draught relief shaft configuration

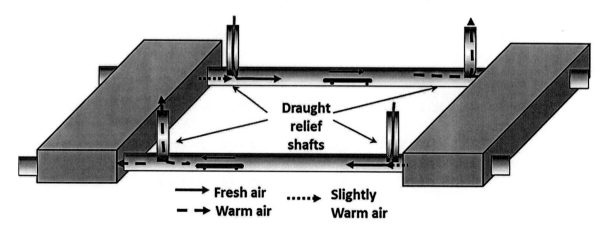

For single bore tunnel with bidirectional trains traffic, performances of natural ventilation based on train piston effects and draught relief shafts are significantly reduced especially during rush hours. Indeed, the warm air pushed by a train in one direction does not generally has the time to exit the shaft and is often reintroduced in the tunnel due to the train moving in the opposite direction which tends to create negative pressure effects in the tunnel when leaving the shaft. Draught relief shafts can still be used to reduce the residual draught in stations or any other undesirable pressure effects. However, the ventilation of the tunnel to achieve temperature level objectives generally requires mechanical ventilation located in a mid-tunnel ventilation shaft. A particular attention has to be paid when using fans during this normal operation. Indeed, the increase of static pressure in running tunnels induced by train piston effects must be considered for the design of fans characteristics. Stalling effects that could lead to the breaking of the fan may appear if this issue is not properly addressed. Fan manufacturers have developed during the last decades fans that are less sensitive to those effects thanks to the implementation of anti-stall rings on the fan casing.

2.1.2 Consideration Regarding Platform Screen Doors (PSD)

At the current time, lot of metro underground stations is not equipped with Platform Screen Doors. Consequently, ventilation and air-conditioning strategies applied in tunnels and in the underground stations are strongly dependent. Indeed, in such a case lot of airflow exchanges can occur between stations and tunnels due to train piston effects. Those effects are particularly high in twin-tubes tunnels because of the limitation of the tunnel cross-section. Moreover, the construction of draught relief shafts is generally not sufficient to remove them completely.

The fact is that generally in tropical weather or any other warm areas underground stations need air conditioning in order to keep ambient conditions under acceptable and comfortable situations for the passengers. The temperature usually set in stations is between 25°C and 30°C with a maximum relative humidity of around 55%-60%. On the other side, the temperature allowed in the tunnel depends on the temperature which remains acceptable for the normal operation of the rolling stocks. This temperature is commonly 40°C corresponding to the one required for a suitable functioning of train air conditioning systems. But when stations are not equipped with PSD, the temperature of 40°C may not be achieved in the tunnels as it would lead to a large amount of unsuitable warm air flow entering into stations. In such case, stations air conditioning system would not be able to cover the corresponding peak heat load and unsuitable temperature fluctuations may occur in the station. For that reason, the tunnel air must generally not exceed a temperature of 32-33°C and the tunnel ventilation system has to operate in closed mode when the outdoor temperature exceeds a value of around 30°C. During this mode of operation, cooled air is dumped into the tunnel by the air conditioning system to compensate the heat released by trains in tunnels.

Some other issues occur when PSDs are not installed in stations. First, a large part of the heat released by trains during dwell time is dissipated in the platform area. Moreover, residual train piston effects in the station tend to generate unsteady heat loads corresponding to the air entering and leaving the station through entrances ((Kumar,2003), (Waymel, 2003), (Waymel,2006)). All of these need to be compensated by the cooling capacity of the station air conditioning system.

However, a metro system without PSD reduce significantly the benefit of station fresh air entering into tunnels (Eckford, 2003) and a special care has to be made for the design of the tunnel ventilation system to achieve the allowable temperature during normal operation.

2.1.3 Consideration Regarding the Heat Exchanges with the Tunnel Walls and the Surrounding Soil

During normal operation, the air speed generated by train piston effects or the mechanical ventilation leads to heat convection effects between the air and the wall surface of the tunnel. When the temperature of the wall surface is less than the air temperature of the tunnel, suitable air cooling effects appear and may contribute to keep the air temperature in the tunnel in desired conditions. In opposite, when the air temperature is less than the wall surface temperature, cooling effects for the wall surface appear which contribute to the increase of the tunnel air temperature.

This physical principle can be used on one hand to reduce the air temperature in the tunnel during operating hours and on the other hand, to control and avoid unsuitable temperature escalation of the wall surface by using for instance night cooling mechanical ventilation. However, cooling effects achieved in the tunnels strongly depend on the tunnel wall surface temperature. This parameter depends on many aspects that need to be considered during the design phase. Indeed, the wall surface temperature is a consequence of the air temperature in the tunnel but also of the heat transfer phenomena which occur through the concrete wall and the surrounding soil. Consequently, the thermal inertia of the system has to be carefully taken into account and the overall process has to be planned considering short term but also long term heat sink effects between the air, the concrete wall and the surrounding soil.

2.2 Underground Stations

2.2.1 Renewal Circulation of Fresh Air into Public Spaces

The total air flow rate blown in stations can be controlled by fans or Air Handling Units to ensure the correct fresh air flow rate for passengers in the station according to the hourly expected frequentation. The fresh air can be distributed in the various public spaces of the station by a set of ducts and dampers. The minimum flow rate to be provided during rush hour should be at least equal to recommendations or standards as detailed earlier. During off-peak, the fresh air flow rate can be minimized in order to save energy but an equivalent to one air - change per hour of the volume of the station is generally recommended to ensure the renewal of fresh air in different spaces of the station.

In the case where the station is fitted with platform screen doors, the distribution of fresh air in the station can consist in a horizontal sweep at each level or in a vertical sweep of the whole public space.

When the station has full height platform screen doors that isolate the volume of station platforms with track area, this flow rate also contribute to create a relative overpressure into the station with the tunnel.

Thus, when platform screen doors are opened, this overpressure reduces the contamination of platforms coming from the tunnel: fines particles produced during mechanical braking of trains fitted with irons bearing (steel wheels) for example. However, this type of action can be limited especially when trains arrived in the station due to the significant rise of the static pressure in the tunnel area induced by the piston effect.

2.2.2 Conditioning of Fresh Air Blown into the Station

Depending on weather outside conditions and the type of pollution present in air intakes (urban pollution, dusty environment, beachfront…), fresh air could be filtered and treated to achieve objectives in term of temperature, humidity, gas concentration, particle and dust level.

Ventilation and Air Conditioning in Tunnels and Underground Stations

In a particularly dusty environment, an inertial filter can be set upstream the air handle unit. In the air handle unit, a charcoal filter is associated with a particulate filter to trap urban gas, odor and particle. If an air-air energy recovery system is selected (wheel type heat exchanger, plate heat exchanger, heat pipe exchanger type…) a particle filter is required on the return air duct.

In winter conditions, a heating coil can also be used to increase the blown air temperature to avoid creating discomfort for people. Also, a minimum temperature in station may be run by the instrumentation system in station relayed by control program standard equipment in auxiliary heating coil of the air handle unit (hydraulic pumps or electrical resistance).

In summer conditions, a cooling coil or an air washer system can be used to cool down the fresh air before blowing it in the station.

The air washer system is appropriate when outside climatic conditions are typical of a dry climate (i.e. a high temperature combined with a low relative humidity). The air washer system has the property to cool the fresh air passing through the system by evaporation process of water droplets produced by spray nozzles. The consequences are the decrease of dry-bulb temperature and an increase of air humidity in an isenthalpic transformation. Such equipment also has filtration properties: while passing across a water curtain, some solid particles carried in the air flow can be caught by water droplets. A water softening treatment should be provided to prevent scaling of sprayers and a demineralization should be done if water is not intended to be recycled. The air washer system can only be used with open mode ventilation strategy (without air recirculation) because the humidity of the air in the station which has already been treated by the air washer is not compatible anymore with the fundamental principle of the air washer that requires dry air. For the same cooling capacity, air washer systems require less energy compared to air handling units. However, the water consumption can be very high which is not always appropriate in some countries where water resources can be limited.

When the local climate is not a dry-type climate with high dry temperatures in summer combined with a low relative humidity, a cooling coil is more appropriate to cool down the temperature and decreasing the humidity level if necessary (depending on the sensible heat ratio of the cooling coil designed for the application by the HVAC engineer) of the fresh air introduced in the station.

Heating and cooling coils are generally connected to a hydraulic distribution network of hot and cold water in the station.

In order to facilitate maintenance operations and to simplify the establishing of all technical related systems (such as electrical power supply, control systems, hydraulic distribution networks…), air handling equipment are gathered in a dedicated room.

2.2.3 Air Extraction

Air extraction can also be done for the ventilation of stations. In some cases, the fresh air can be provided by negative pressure only by extracting the air in the volume of the station. The smoke extraction system can for example be used for this application (optimization). In that case, the air enters through station entrances but can also come from the tunnel. Nevertheless, the control of the pollution or warm air coming from outside or from the tunnel is not really possible and this kind of ventilation strategy is not really recommended in high polluted and warm areas.

Extraction can also be required when the fresh air flow rate is very important and could lead to high velocity in the station especially at station entrances. In that case, the extraction flow rate remains below or at the same level of the fresh air flow rate.

Air extraction is also necessary in closed mode air conditioning strategy in order to recycle partially the air of the station into the air conditioning system. This closed mode strategy is generally recommended when the air enthalpy inside the station is lower than the outside air enthalpy.

Last but not the least, local air extraction points can also be recommended to extract efficiently the heat released by equipment (see also UPE /OTE).

2.2.4 Production and Distribution of Heating and Cooling Energy

On one side, the hydraulic distribution network is connecting some thermodynamic equipment (like chiller, heat pump, multipurpose heat pump) with a hot water tank and a cold water tank. This is the "primary loop circuit". This loop has its own control system to deal with problems of freezing, unexpected start or potential bypass between several machines for example.

On the other side the hydraulic distribution network is connecting the air handle units and other fan coils with the hot and cold water tanks. This is the "secondary loop circuit" as shown on Figure 3. This loop has also its own control system to supply the terminals according to heating and cooling requirements.

To lower the hot coil power installed, a heat recovery air handle unit fitted with a heat exchanger between fresh air and return air could be provided. The aim of that system is to reduce heating costs especially in winter by taking advantage of the heat dissipated in the station. If the air in the station has to be extracted during summer time, the heat exchanger must be bypassed or stopped if the temperature inside the station is higher than the external temperature.

Figure 3. Double Fluxes Air Handling Unit

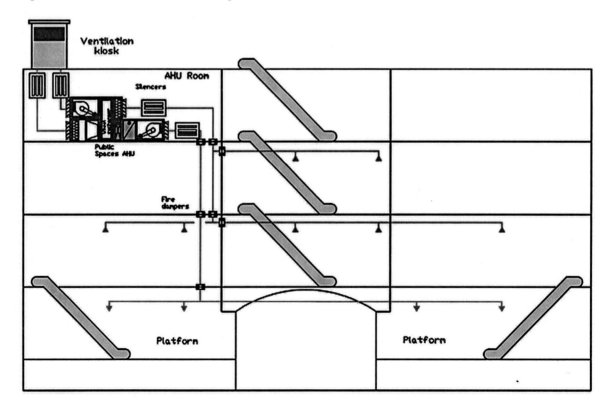

Ventilation and Air Conditioning in Tunnels and Underground Stations

The Figure 4 represents an installation of production able to cover both heating and cooling requirements in station.

A geothermal multi-purpose heat pump connected with pipes buried in the deep foundation in concrete of the station and with the hot and cold water tanks can increase the efficiency of the installation by providing simultaneously hot and cold water. Deep foundations of the station fitted with geothermal tubes are used for discharging excess energetic production (hot or cold as needed in station). The design of the geothermal mutli-purpose heat pump must comply with the annual charging and discharging cycles of the soil so as not to deteriorate thermal properties of the ground.

This installation can provide a low operating cost and improve the profitability of the installation although initially, the geothermal system induces an additional cost.

The installation is completed by a heat pump and a chiller (which are well known and proven) to cover the remaining requirement or to ensure the production of heating and cooling energy in case of maintenance or damage on the geothermal installation. In normal operation, the geothermal multi-purpose heat pump is used in priority because a thermodynamic machine that exchanges on water has better performances than a thermodynamic machine that exchanges on air, even if heating and cooling requirements are not simultaneous.

When heating and cooling requirements in station are simultaneous, production of heating and cooling energy can be achieved by means of multipurpose heat pump only as detailed on Figure 5 (type water-air or water-water). With this installation, the geothermal multi-purpose heat pump is also used in priority to produce heating and cooling energy in order to increase the performance of the system.

Figure 4. Geothermal energy recovering

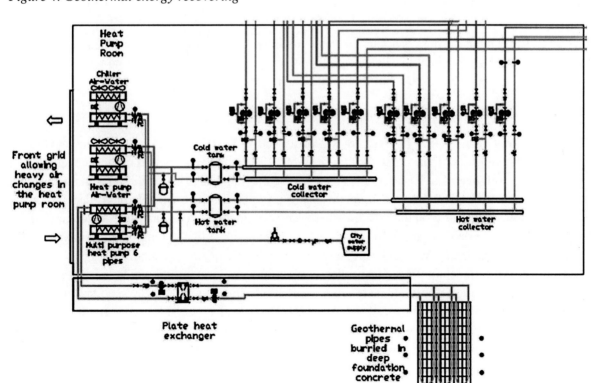

Figure 5. Multipurpose heat pump and geothermic

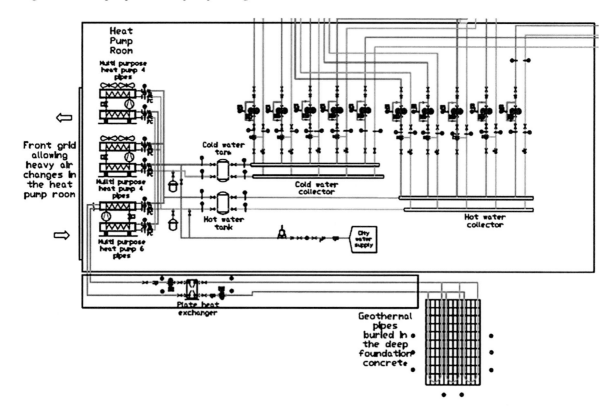

The air to water multi-purpose heat pumps can be used to complete the energy production. In case of simultaneous and balanced calls of hot and cold power, the multi-purpose heat pump will exchange on water otherwise the multi-purpose heat pump will exchange on air.

2.2.5 Under Platform and Over Track Exhaust Systems

Under Platform Exhaust (UPE) or Over Track Exhaust (OTE) Systems can also be implemented in the underground stations. The main purpose of those systems is to extract a large part of the heat released by trains during dwell time. The location of heat sources on the train has to be carefully taken into account when designing this kind of system. Indeed, the efficiency of the heat extraction strongly depends on the extraction flow rate but also on the proximity between extraction points and auxiliary equipment of the train releasing the most important part of the heat. The locations of the braking system (braking disks, braking resistors) and train AC condensing units are particularly the main parameter to be considered.

In case of station air conditioning using closed mode system, it can be recommended to combine the OTE or UPE extraction system with the return air ventilation and to equilibrate the supplied air flow rate with the extraction flow rate of the UPE / OTE. This is generally required for reducing the amount of outside air entering at station entrances. However, it is recommended to study case by case as the most appropriate strategy strongly depends on the enthalpy of the air collected by the UPE / OTE system and the enthalpy of the outside air (Figure 6).

Figure 6. UPE / OTE Ventilation system

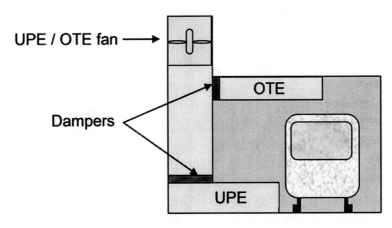

3. Strategies in Emergency Operation

3.1 Tunnels

A longitudinal ventilation principle is generally used in case of a train fire in a tunnel. This consists in pushing the smoke in one direction and keeping the tunnel section upstream the fire free of smoke.

A push-pull strategy is generally applied in metro tunnels which consists of blowing fresh air from one shaft and exhausting smokes through an opposite shaft. Jet fans can also be used for this ventilation principle in order to help the smoke going in the desired direction.

With this longitudinal ventilation principle, the velocity upstream the fire must be strong enough to prevent backlayering in order to leave the upstream section of the tunnel free of smoke to improve the evacuation conditions. This particular requirement means that the ventilation system must be designed to achieve a longitudinal velocity at least equal to the critical velocity. This critical velocity strongly depends on the design fire size, the cross-section and the slope of the tunnel (Danziger & Kennedy, 1982).

Another issue for the ventilation in case of tunnel fire is to prevent smoke propagation into adjacent station trackways where other trains can be stopped. The limitation of the smoke propagation into the tunnel can also be an important issue in order to protect passengers standing in other trains blocked in the incident bore. In several countries, regulations and standards also provide requirements for reducing the length of the smoke propagation in the tunnel not only for safety reasons but also to reduce the consequences on the assets. The role and the location of ventilation shafts are consequently critical to achieve those particular objectives. Moreover, the design of the flow rate capacity available in the shafts is also one of the most critical issues of the design. Indeed, extraction fans must be able to extract the total smoke flow rate generated by the longitudinal ventilation. Moreover, it is generally necessary to create a fresh air flow rate (confinement flow) upstream the shaft to reduce the risk of smoke propagation over the extraction point. It has to be noted that this fresh air flow rate also contributes to the increase of the fan capacity.

Thermal effects induced by the fire also have a significant impact on the overall behavior of the emergency ventilation system and should consequently be taken into account during the design phase. One of these effects is the dilatation of the air when passing through the fire. The volume flow rate generated upstream the fire to achieve the critical velocity can be significantly increase depending on the fire size. If the extraction shaft is closed to the fire, cooling effects of the wall are very limited and the volume flow rate of hot gases should be considered for the design of fans flow rate. The lower density of hot gases may also have a significant impact on fan performances as the total pressure provided by a fan strongly depends on the density of the flow extracted by the fan itself.

An appropriate balance of the longitudinal flow rate upstream and downstream extraction fans can also be a key challenge in the design of emergency ventilation scenarios. Appropriate simulations using one dimensional ventilation network models are necessary to determine how the ventilation system has to be operated for each fire case. It is particularly necessary to verify through those simulations that the proposed ventilation scenario is able to achieve not only the critical velocity upstream the fire but also the confinement velocity downstream the shaft. The difficulty is that the balance of flow rates may vary following various parameters that cannot always be controlled such as the exact location of the fire in the tunnel which may change stack effects or the impact of thermal effects on the fan, the location of other trains in the underground network, adverse wind pressure effects at tunnel portals, etc. To take into account the variability that may happens in the balance of flow rate, it is recommended to include a certain level of margin in the design of the fan capacity in order to compensate the influence of those effects. However, this can sometimes lead to a significant increase of those capacities and have a large impact on civil requirements. Other technologies have been developed recently, largely applied in road tunnels and are currently extended in several recent metro projects. One of the principles consists in implementing a regulation process in the control of the ventilation system in order to adapt the functioning of the emergency ventilation scenario according to the situation. This regulation process is based on air velocity measurements with anemometers in the tunnel and the response of the ventilation system is modified according to those measurements. This kind of process is really interesting to balance properly flow rates whatever the situation of the fire event and can significantly reduce the margin to be considered in the sizing of the fan flow rates. However, a particular care needs to be paid in the reliability of the velocity measurement and in the regulation process implemented in the software of the ventilation control and monitoring system.

Although the principle of longitudinal ventilation is largely expended in rail and metro infrastructures, other emergency ventilation strategies can also be developed. Indeed, longitudinal ventilation systems can lead in some cases to a large amount of longitudinal smoke flow rate to be extracted. This is especially the case when the fire occurs in cross-over where the cross section of the tunnel is very high. In some particular situation, it could also be necessary to achieve smoke stratification objectives to improve evacuation conditions at head height and at both sides of the train. But, smoke stratification is generally very limited with a longitudinal system due to the large amount of turbulent effects generated by the longitudinal velocity. Consequently, one alternative principle is the transverse ventilation. This principle consists in minimizing the longitudinal velocity in the fire zone to improve the smoke stratification and to extract the smoke layer through a ventilation duct. This kind of ventilation is generally not reasonable in term of civil for normal running tunnels due to the construction of the duct but can be proposed in specific and limited areas such as cross-overs.

3.1.2 Underground Stations

The main principles of emergency ventilation systems in underground station are generally fixed by building regulations and standards. Lot of design rules already exist such as the international standards NFPA 92 (NFPA, 2012). In these standards, design rules for the sizing of smoke extraction flow rates and smokes barriers are particularly well detailed and will not be detailed and commented in this Handbook. However, specificities exist in the case of underground stations.

One of these is the case of train fire at platform level. The heat release rate is generally more important than the design HRR used in normal building areas. This can lead to higher smoke extraction flow rates and to additional particular requirements to prevent smoke propagation from the platform level to the other station levels.

Under platform and over track exhaust systems can also be used as a smoke extraction system when they are implemented for the normal operation. In some cases, the use of the tunnel ventilation system which is already design for train fires in tunnel can also be used as a good alternative in order to avoid additional massive extraction system in the station. However, the risk of pushing the smokes into adjacent tunnels must be carefully assessed.

The presence of full height platform screen doors provides also several advantages to control the smoke propagation into the station. However, these advantages can be limited in time as the fire resistance of the screens is generally limited in temperature and in time. Moreover, smoke propagation into the platform area can still appear through open doors and activation of the smoke extraction system dedicated to the platform is generally necessary.

4. Improvement of Design Approaches and Design Tools

4.1 Prescriptive Design Approaches vs. Performance based Design

4.1.1 Prescriptive Approaches based on Regulations and Standards

Ventilation design in underground metro infrastructures is in some cases partially described by the dedicated safety regulations and standards of the country in which the metro is built.

As an example, the French regulation is very prescriptive regarding tunnel ventilation shafts design (French Ministry of transportation, 2005). Indeed, this indicates that smoke shall not propagate at a higher distance than 800 m from the fire. It is also specified that the smoke in case of fire in a tunnel shall never propagate into an adjacent station. Those requirements lead to the following constraints for the location of tunnel ventilation shafts:

- The maximum distance between two ventilation shafts cannot exceed 1600 m. This distance can drop to 800 m when a reversible longitudinal ventilation system is required to be able to push smoke in either traffic or opposite traffic directions especially when the location of the fire on the train is considered as a critical issue for the choice of the evacuation direction in the tunnel.
- When emergency procedures indicate that it is necessary to push smoke toward the station to achieve adequate tunnel ambient conditions for the evacuation of passengers located in the train in fire. In this case, stations ends must be equipped with tunnel ventilation shafts to prevent smoke propagation into the station.

Another interesting example is the Brazilian standard for the fire safety in tunnels, ABNT NBR 2009, which simply mentions that a longitudinal ventilation system pushing smoke in one direction of the tunnel can be applied up to a distance of 3000 m. It is also indicated that longitudinal ventilation with massive extraction can be used especially for longer tunnel. However, there is no specific prescription regarding the location of the shaft used for the massive extraction.

Prescriptive approaches are generally interesting as they give simple rules for the design of the ventilation systems. However, in some cases, the requirements can be inappropriate to the project constraints and may not give clear understanding of the safety objectives to be achieved.

4.1.2 Performance based Approaches

In opposite to prescriptive approaches, performance based approaches do not give detailed technical requirements for the design of the tunnel or underground stations ventilation systems. For example the distance between ventilation shafts may not be specifically detailed. This kind of approach is more focused on objectives to be achieved by the design. For instance, the NFPA 130 standard (NFPA, 2010) reminds the following fire safety objectives:

- Protect occupants not intimate with the initial fire development
- Maximize the survivability of occupants intimate with the initial fire development

Considering those objectives, this leads to the following main targets to be achieved by the design of the emergency ventilation system:

- Prevent smoke propagation towards passengers standing in station and in safe trains. This standard particularly mentions that the ventilation system has to accommodate the maximum number of trains that could be between ventilation shafts during an emergency
- To reduce as much as possible the consequences of the fire on passengers evacuating the train in fire.

In opposite to prescriptive approaches, performance based approaches give some degrees of freedom for engineers to design ventilation systems closer to the particular constraints of the project and lead to a range of opportunities to optimize the ventilation systems.

However, a special care has to be paid when using this kind of approach. Indeed, this particularly requires a clear identification of parameters that could play a significant role in the objectives to be achieved. Moreover, this approach generally needs appropriate evaluation of the proposed systems to demonstrate that safety or comfort levels achieved by the proposed ventilation system fulfill target objectives.

5. Design Numerical Tools

Since last two decades, computer performances and also fluid & thermodynamics numerical tools have seriously evolved. 1D ventilation network numerical software but also 3D Computational Fluid Dynamics tools are consequently more and more frequently used during design studies allowing more accurate assessment and sizing of the proposed design and more generally optimization of ventilation systems. The following subsections give some brief details about the functionalities of these tools and their improvements since the last decades.

Ventilation and Air Conditioning in Tunnels and Underground Stations

5.1 One Dimensional Ventilation Tools

The development of one dimensional ventilation tools started in the sixties firstly for mine applications and was extended to underground transportation infrastructure applications in the seventies. Those tools include a one dimensional representation of the complete ventilation network including the tunnels and the ventilation ducts. Depending on the complexity of the tools, various parameters that can affect the thermodynamic properties of the air and the balance of flow rates in the network can be taken into account such as heat sources, traffic, fans, jet fans, pressure losses, etc. Conduction models are also implemented to take into account the heat exchanges with the concrete and the surrounding soil.

This kind of tool is particularly interesting to evaluate the performances of the ventilation scenarios proposed for normal or emergency operation.

One of the famous one dimensional software that was developed in the last two decades of the 20th century is the Subway Environmental Simulator (SES). This software has been largely used for the design of metro ventilation systems and still remains the reference in this application. However, this software presents some limitations due to some physical models that were simplified to achieve reasonable calculations time in line with the computational resources of the past century. Some of the limitations are the influence of the density of the hot gas and dilatation effects that were not properly addressed on the functioning of the fans. The heat conduction model used for normal operation and emergency operation are also based on analytic models or on a single layer heat conduction model which is not really appropriate to assess properly heat sink effects of the soil in normal operation and tend also to minimize the gas temperature in emergency calculations. Since the last fifteen years, new numerical tools have been developed by engineering companies specialized in tunnel ventilation including more accurate physical and numerical models. This new generation of one dimensional tools allows a better assessment of the performance of the ventilation system for a safer design and also provides more accurate results which can lead in some cases to significant optimizations of ventilation and air conditioning systems.

5.1.1 Three Dimensional Computational Fluid Dynamic (CFD) Tools

The improvement of computational capacities has led in the last ten years to a more frequent use of 3D CFD tools. One of most important advantages of this kind of tools is the 3D representation of the geometry and also the implementation of accurate numerical algorithms to solve the Navier-Stokes turbulent equations of the fluid mechanics. Some particular effects such as the smoke stratification can be addressed in this kind of tools which is not possible with one dimensional software. The modeling of fire sources including combustion models and numerical radiation models have also recently been improved. The use of 3D CFD model is particularly interesting to assess in detail the ambient air conditions in the vicinity of a fire in a tunnel or to evaluate particular emergency ventilation strategies in complex underground stations. When combined with evacuation tools, it is also possible to calculate with a good level of accuracy the Fractional Effective Dose of patrons during evacuation time.

However, despite the improvement of the computational capacities, it is still not possible to model the overall ventilation network with this kind of tools and one dimensional simulations are generally required to determine flow rate boundary conditions of the 3D model.

FUTURE RESEARCH DIRECTIONS

Energy saving is and will be in the future one of the main key challenge for the design and the use of ventilation and air conditioning systems. Future researches should consequently be focalized in this area.

The use of seepage water could be for example interesting (see Figure 7). If the seepage water flow around the station is sufficient, heat pump water-to-water connected to a plate heat exchanger offers theoretically some great energetic performances throughout the year to meet the need of heating and cooling.

A water loop can also be installed in the tunnel to use the heating produced by passing trains to produce heating energy at a lower cost.

Figure 7. Water / Water chiller with seepage water including water loop in tunnels

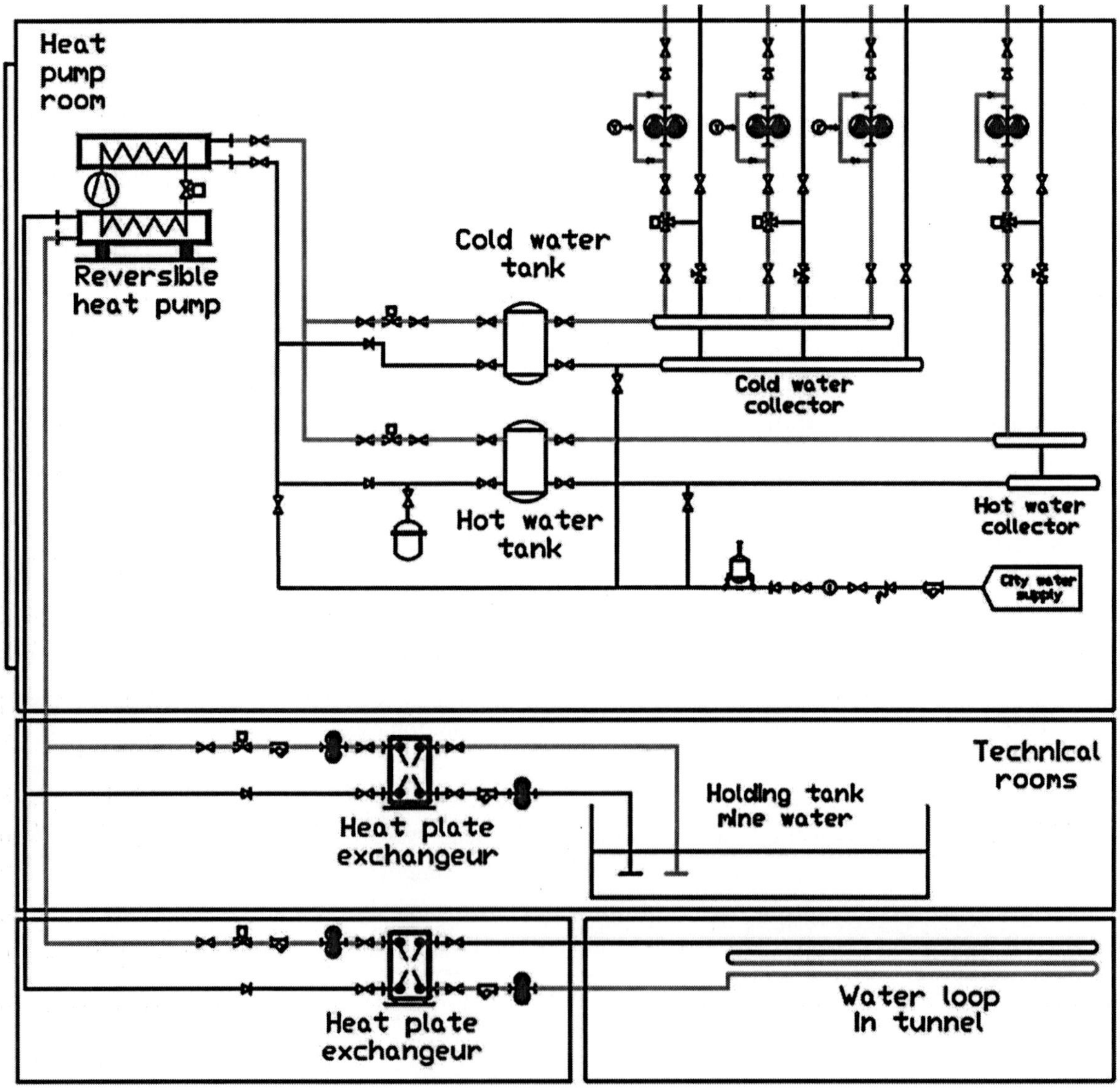

If the seepage water flow is not sufficient to deliver enough cooling energy by the means of a heat pump (see Figure 8), it is possible to add a cooling coil in front of the air intake on a chiller. After filtration, seepage water is flowing through the cooling coil to cool down the external air upstream the evaporator of the chiller in order to improve the energetic efficiency of the machine.

CONCLUSION

Tunnel and station ventilation & air-conditioning systems are major mechanical / electrical systems in underground rail infrastructures including metro lines. Firstly, they play a significant role on the comfort but also on the safety of passengers. Secondly, civil requirements for the construction of the ventilation shafts but also for corresponding technical rooms are generally crucial regarding land acquisitions and all other construction issues and may have significant impacts in the overall capital cost of the project. Thirdly, the power demand and energy consumptions to operate those systems are very often at high level. A particular care is consequently recommended for the design of those systems. As reminded in this chapter, objectives of the design and all parameters that can influence the design should be clearly known and defined at the beginning of a new infrastructure project.

Various principles can be applied for normal and emergency operations, each having advantages and drawbacks as discussed in this chapter. Appropriate analyses are consequently recommended at the design stage in order to determine the most appropriate solution for the infrastructure in terms of performances of the systems, civil / power requirements and energy savings.

Figure 8. Air / water chiller with seepage water

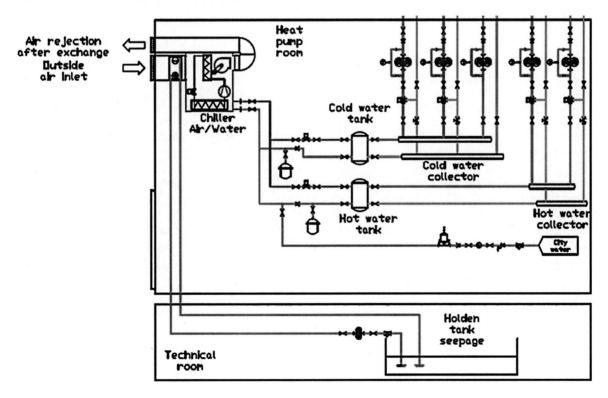

REFERENCES

ABNT NBR 15661. (2009). Proteção contra incêndio em tuneis, *Norma Brasileira*.

ASHRAE Standard 55-2007. Thermal Environmental Conditions for Human Occupancy. (2007).

ASHRAE Standard 62.1-2007, Ventilation for Acceptable Indoor Air Quality. (2007).

Berglund, L. G. (1979). Thermal acceptability. *ASHRAE Transactions*, 85(2), 825–834.

Cartens, J. P. (1965). *Literature Survey of Passenger Comfort Limitations for High Speed Ground Transports*. Hartgforf, Conn.: United Aircraft Corp.

Danziger, N. H., & Kennedy W.D. (1982). Longitudinal ventilation analysis for the glenwood canyon tunnels. *Proc 4th Int. Symposium Aerodynamics & Ventilation of Vehicle Tunnels* (pp. 169-186).

Eckford, D. C., Pope, C. W., Owoeye, A., & Henson, D. H. (2003). The ventilation of metro systems with and without full height platform screen doors. *Proceedings of the 11th International Symposium on Aerodynamics & Ventilation in Vehicle Tunnels*.

Fanger, P. O., & Christensen, N. K. (1986). Perception of draught in ventilated spaces. *Ergonomics*, 29(2), 215–235. doi:10.1080/00140138608968261 PMID:3956473

Arrêté du 22 novembre 2005 relatif à la sécurité dans les tunnels des systèmes de transport public guidés urbains de personnes. (2005). *French Ministry of Transportation. Journal officiel*. p31.

Goldman, R. F. (1978). *The role of clothing in achieving acceptability of environmental temperatures between 65°F and 85°F (18°C and 30°C). Energy conservation strategy in buildings* (J. Stolwijk, Ed.). New Haven: Yale University Press.

Kumar, S., Pahuja, D., Bakre, A., & Saha, S. K. (2003). Prediction of unsteady heatgains using SES analysis in an interchange subway station of the Delhi metro. *Proceedings of the 11th International Symposium on Aerodynamics & Ventilation in Vehicle Tunnels*.

NFPA 130, Standard for Fixed Guideway Transit and Passenger Rail Systems. (2010).

Subway Environmental Design Handbook. (1976). *Principles and Applications* (2nd ed., Vol. I). United States Department of Transportation.

Waymel, F., Monnoyer, F., & William-Louis, M. (2006). Numerical simulation of the unsteady three-dimensional flow in confined domains crossed by moving bodies. *Computers & Fluids*, 35, pp. 525-543.

Waymel, F., William-Louis, M., & Monnoyer, F. (2003). 3D simulation of airflow and heat transfer in a subway station with moving trains. *Proceedings of the 11th International Symposium on Aerodynamics & Ventilation in Vehicle Tunnels*.

Compilation of References

ABNT NBR 15661. (2009). Proteção contra incêndio em tuneis, *Norma Brasileira*.

Abrahamsson, L., & Söder, L. (2012) An SOS2-based moving trains, fixed nodes, railway power system simulator, *Paper presented at 13th International Conference on Design and Operation in Railway Engineering (Comprail '12)*, New Forest, UK.

Adler, N., Pels, E., & Nash, C. (2010). High-speed rail and air transport competition: Game engineering as tool for cost-benefit analysis. *Transportation Research Part B: Methodological*, 44(7), 812–833. doi:10.1016/j.trb.2010.01.001

Ahmadi, S., & Spanias, A. S. (1999). Cepstrum-based pitch detection using a new statistical V/UV classification algorithm. *IEEE Transactions on Speech and Audio Processing*, 7(3), 333–338. doi:10.1109/89.759042

Ahmadlou, M., Adeli, H., & Adeli, A. (2010). New diagnostic EEG markers of the Alzheimer's disease using visibility graph. *Journal of Neural Transmission (Vienna, Austria : 1996)*, 117, 1099–1109. http://doi.org/<ALIGNMENT.qj></ALIGNMENT>10.1007/s00702-010-0450-3

Ahmadlou, M., & Adeli, H. (2012). Visibility graph similarity: A new measure of generalized synchronization in coupled dynamic systems. *Physica D. Nonlinear Phenomena*, 241(4), 326–332. doi:10.1016/j.physd.2011.09.008

Airplane crashes are very often caused by fatigue. (n. d.). *LAFuel.com*. Retrieved from http://www.lafuel.com/2013/04/-3&d=78

Akin, B., Orguner, U., Toliyat, H. A., & Rayner, M. (2008). Low order PWM inverter harmonics contributions to the inverter-fed induction machine fault diagnosis. *IEEE Transactions on Industrial Electronics*, 55(2), 610–619. doi:10.1109/TIE.2007.911954

Aknin, P., Ayasse, J. B., & Devallez, A. (1991). Active steering of railway wheel sets. *Proceedings of the 12th IAVSD Conference*, Lyon, August 1991.

Albrecht, T. (2009). The Influence of Anticipating Train Driving on the Dispatching Process in Railway Conflict Situations. *Networks and Spatial Economics*, 9(1), 85–101. doi:10.1007/s11067-008-9089-0

Alkulaibi, A. (1996). Fast HOS based simultaneous voiced/unvoiced detection and pitch estimation using 3-level binary speech signals. *Proceedings of the 8th IEEE signal processing workshop on Statistical Signal and array processing* (pp. 194-197). Retrieved from http://ieeexplore.ieee.org/xpls/abs_all.jsp?arnumber=534851

Allison, W. (2013). *Signal Design Principles*. NSW, Australia: RailCorp.

Alonso, W. (1964). *Location and Land-use: Towards a General Theory of Land Rent*. Cambridge: Harvard University Press. doi:10.4159/harvard.9780674730854

Álvarez Rueda, R. (2006, November 1). Assessing alertness from EEG power spectral bands. *Bibdigital.epn.edu*. Retrieved from http://bibdigital.epn.edu.ec/handle/15000/9872

Ammar, S., Duncombe, W., Jump, B., & Wright, R. (2004). Constructing a fuzzy-knowledge-based-system: An application for assessing the financial condition of public schools. *Expert Systems with Applications*, 27(3), 349–364. doi:10.1016/j.eswa.2004.05.004

An, M. (2003, July 6-7). Application of a knowledge-based intelligent safety prediction system to railway infrastructure maintenance. *Proceedings of International Railway Engineering Conference*, London. Edinburgh: Engineering Technics Press.

Anderson, R. E., Poon, A., Lustig, C., Brunette, W., Borriello, G., & Kolko, B. E. Building a transportation information system using only GPS and basic SMS infrastructure. Proceedings of ICTD'09. doi:10.1109/ICTD.2009.5426678

Anderson, B. D., & Moore, J. B. (1979). *Optimal Filtering*. Englewood Cliffs: Prentice-Hall.

Anderson, T. K. (2010). Using geodemographics to measure and explain social and environment differences in road traffic accident risk. *Environment & Planning A*, 42(9), 2186–2200. doi:10.1068/a43157

Andriamalala, R. N., Razik, H., Baghli, L., & Sargos, F.-. (2008). Eccentricity Fault Diagnosis of a Dual-Stator Winding Induction Machine Drive Considering the Slotting Effects. *IEEE Transactions on Industrial Electronics*, 55(12), 4238–4251. doi:10.1109/TIE.2008.2004664

Andriantiatsaholiniaina, L. A., Kouikoglou, V. S., & Phillis, Y. A. (2004). Evaluating strategies for sustainable development: Fuzzy logic reasoning and sensitivity analysis. *Ecological Economics*, 48(2), 149–172. doi:10.1016/j.ecolecon.2003.08.009

An, M. (2005). A review of design and maintenance for railway safety – the current status and future aspects in the UK railway industry. *World J. Eng.*, 2(3), 10–23.

An, M., Chen, Y., & Baker, C. J. (2011). A fuzzy reasoning and fuzzy-analytical hierarchy process based approach to the process of railway risk information: A railway risk management System. *Information Sciences*, 181(18), 3946–3966. doi:10.1016/j.ins.2011.04.051

An, M., Huang, S., & Baker, C. J. (2007). Railway risk assessment – the fuzzy reasoning approach and fuzzy analytic hierarchy process approaches: a case study of shunting at Waterloo depot. *Proc. Instn Mech. Engrs, Part F: J. Rail and Rapid Transit*, 221(3), 365–383. doi:10.1243/09544097JRRT106

An, M., Lin, W., & Stirling, A. (2006). Fuzzy-based-approach to qualitative railway risk assessment. *Proc. Instn Mech. Engrs, Part F: J. Rail and Rapid Transit*, 220, 153–167. doi:10.1243/09544097JRRT34

Annual Report 2012-2013. (2013). *Delhi Metro Rail Corporation Ltd.*, New Delhi.

Anselin, L. (1995). The Local Indicators of Spatial Association LISA. *Geographical Analysis*, 27(2), 93–115. doi:10.1111/j.1538-4632.1995.tb00338.x

Antonino-daviu, J., Riera-guasp, M., Pons-llinares, J., Park, J., Lee, S. B., Yoo, J., & Kral, C. (2012). Detection of Broken Outer-Cage Bars for Double-Cage Induction Motors under the Startup Transient. *IEEE Transactions on Industry Applications*, 48(5), 1539–1548. doi:10.1109/TIA.2012.2210173

Arboleya, P., Diaz, G., & Coto, M. (2012). Unified AC/DC power flow for traction systems: A new concept. *IEEE Transactions on Vehicular Technology*, 61(6), 2421–2430. doi:10.1109/TVT.2012.2196298

Compilation of References

Arief, Z., Purwanto, D., Pramadihanto, D., Sato, T., & Minato, K. (2009). Relation between eye movement and fatigue: Classification of morning and afternoon measurement based on Fuzzy rule. *Proceedings of theInternational Conference on Instrumentation, Communication, Information Technology, and Biomedical Engineering 2009, ICICI-BME 2009*. http://doi.org/ doi:10.1109/ICICI-BME.2009.5417286

Arrêté du 22 novembre 2005 relatif à la sécurité dans les tunnels des systèmes de transport public guidés urbains de personnes. (2005). *French Ministry of Transportation. Journal officiel*. p31.

ASHRAE Standard 55-2007. Thermal Environmental Conditions for Human Occupancy. (2007).

ASHRAE Standard 62.1-2007, Ventilation for Acceptable Indoor Air Quality. (2007).

Assad, A.A. (1980). Models for rail transportation. *Transportation Research A*, 14 (A), 205-220.

Assuncao, R., & Reis, E. A. (1999). A new proposal to adjust Moran's I for population density. *Statistics in Medicine*, *18*(16), 2147–2162. doi:10.1002/(SICI)1097-0258(19990830)18:16<2147::AID-SIM179>3.0.CO;2-I PMID:10441770

ATOC. (2005). *Passenger Demand Forecasting Handbook*. London: Association of Train Operating Companies.

Australian Transport Safety Bureau Report RO-2009-009. (2009). *Australian Safety Bureau*.

Ayasse, J.-B., & Chollet, H. (2006). Wheel-Rail Contact. In S. Iwnicki (Ed.), *Handbook of railway vehicle dynamics* (pp. 86–120). Boca Raton, FL: CRC Press.

Ayhan, B., Trussell, H. J., Chow, M.-Y., & Song, M.-H. (2008). On the use of a lower sampling rate for broken rotor bar detection with DTFT and AR-based spectrum methods. *IEEE Transactions on Industrial Electronics*, *55*(3), 1421–1434. doi:10.1109/TIE.2007.896522

Ayoob, E., Steinfeld, A., & Grace, R. (2003). Identification of an "appropriate" drowsy driver detection interface for commercial vehicle operations. *Proceedings of The Human Factors and Ergonomics society annual meeting*, 47(16), 1840-1844. Retrieved from http://pro.sagepub.com/content/47/16/1840.short

Babcock, M., Lu, X., & Norton, J. (1999). Time series forecasting of quarterly railroad grain carloadings. *Transportation Research Part E, Logistics and Transportation Review*, *35*(1), 43–57. doi:10.1016/S1366-5545(98)00024-6

Bahl, R., Linn, J.F., & Wetzel, D.L. (Eds.), (2013). Financing Metropolitan Governments in Developing Countries. Cambridge, MA, USA: Lincoln Institute of Land Policy.

Balan, R., Khoa, N., & Lingxiao, J. (2011, June). Real-time trip information service for a large taxi fleet. *Proceedings of Mobisys*, *11*. doi:10.1109/ITST.2007.4295824

Ball, D., & Sunderland, M. (2001). *An economic history of London, 1800-1914*. New York: Routledge.

Baltard, V., & Callet, F. F. (1863). Monographie des Halles Centrales de Paris, construites sous le règne de Napoléon III et sous l'administration de M. le Baron Haussmann. Paris.

Banister, D., & Goodwin, M. T. (2011). Quantification of the non-transport benefits resulting from rail investment. *Journal of Transport Geography*, *19*(2), 212–223. doi:10.1016/j.jtrangeo.2010.05.001

Bardhan, R. (2013). Simulation of Land Use Consequences of Urban Corridor in Kolkata: An Integrated Spatial and Expert System Model. *Environment and Urbanization Asia*, *4*(2), 267–286. doi:10.1177/0975425313510767

Barker, F., & Hyde, R. (1982). *London. As it might have been*. London: John Murray.

Barker, Th., & Robbins, M. (1963). *A History of London Transport. Passenger travel and the development of the metropolis* (Vol. 1). London: George Allen & Unwin Ltd.

Barlow, P. W. (1864, September 9). Patent No. 2207. London Transport Museum.

Barlow, P. W. (1871). *The Relief of Street Traffic. Advantages of the City and Southwark Subway, with Reasons why the Proposed Connection of Street Tramways from the Elephant and Castle through the City is Unnecessary and Undesirable. Second Pamphlet.* London.

Barlow, P. W. (1867). *On the Relief of London Street Traffic, with a Description of the Tower Subway now Shortly to be Executed.* London.

Barry, J. W. (1885). The City Lines and Extensions (Inner Circle Completion) of the Metropolitan and District Railways. *Minutes of Proceedings of the Institution of Civil Engineers, 81*, 34-51.

Bassett, D. S., & Bullmore, E. (2006). Small-world brain networks. *The Neuroscientist, 12*(6), 512–523. doi:10.1177/1073858406293182 PMID:17079517

Baum-Snow, N., & Kahn, M. E. (2000). The effects of new public projects to expand urban rail transit. *Journal of Public Economics, 77*(2), 241–263. doi:10.1016/S0047-2727(99)00085-7

Beck, K., Jäger, B., & Lemmer, K. (2007). *Optimisation of point life cycle costs through load-dependent maintenance* (M. C. Forde, Ed.). Edinburg, UK: Engineering Technics Press Edinburgh.

Beltran, A. (1988). Une Victoire Commune. L'alimentation en énergie électrique du Métropolitain (1re moitié du XXe siècle). In *Métropolitain. L'autre dimension de ville* (pp. 111–122). Paris: Mairie de Paris.

Ben-Akiva, M. E., & Lerman, S. R. (1985). *Discrete Choice Analysis: Theory and Application to Travel Demand.* MA: MIT Press.

Ben-Akiva, M. E., & Morikawa, T. (1990). Estimation of travel demand models from multiple data sources.*Proceedings 11th International Symposium on Transportation and Traffic Theory*, Yokohama.

Bentham, J. (1789). *An Introduction to the Principles of Morals and Legislation.* Oxford: Clarendon Press.

Bergasa, L. M., & Nuevo, J. (2005). Real-time system for monitoring driver vigilance. *Proceedings of theIEEE International Symposium on Industrial Electronics* (Vol. III, pp. 1303–1308). http://doi.org/ doi:<ALIGNMENT.qj></ALIGNMENT>10.1109/ISIE.2005.1529113

Berglund, L. G. (1979). Thermal acceptability. *ASHRAE Transactions, 85*(2), 825–834.

Bernardeschi, C., Fantechi, S., Gnesi, S., Larosa, S., Mongardi, G., & Romano, D. (1998). A Formal Verification environment for railway signaling system design. *Formal Methods in System Design, 12*(2), 139–161. doi:10.1023/A:1008645826258

Bhoraskar, R., Vankadhara, N., Raman, B., & Kulkarni, P. (2012, January). Wolverine: Traffic and road condition estimation using smartphone sensors. Proceedings of WISARD.

Bhowmick, B., & Kumar, K. S. C. (2009). Detection and classification of eye state in ir camera for driver drowsiness identification. *Proceedingsof the 2009 IEEE International Conference on Signal and Image Processing ApplicationsConferenceICSIPA '09* (pp. 340–345). http://doi.org/ doi:10.1109/ICSIPA.2009.5478674

Bienvenüe, F. (1896). *Chemins de Fer Urbains à traction électrique. Devis descriptif et estimatif.* Paris: Régie Autonome des Transports Parisiens.

Bierlaire, M., Axhausen, K., & Abbay, G. (2001). Acceptance of Model Innovation: The Case of the Swiss metro.*Proceedings of the 1st Swiss Transport Research Conference*.

Compilation of References

Bjorner, D., & Jones, C. B. (1978). *The Vienna Development Method: The Meta-Language, LNCS* (Vol. 61). Springer. doi:10.1007/3-540-08766-4

Blank, L., & Tarquin, A. (2002). *Engineering Economy* (6th ed.). USA: McGraw-Hill Higher Education.

Blank, L., & Tarquin, A. (2008). *Basics of Engineering Economy*. USA: McGraw-Hill Higher Education.

Blodt, M., Regnier, J., & Faucher, J. (2009). Distinguishing Load Torque Oscillations and Eccentricity Faults in Induction Motors Using Stator Current Wigner Distributions. *IEEE Transactions on Industry Applications, 45*(6), 1991–2000. doi:10.1109/TIA.2009.2031888

Böhm, T. (2012). *Accuracy Improvement of Condition Diagnosis of Railway Switches via External Data Integration* (C. Boller, Ed.). Germany.

Böhm, T., & Gutsche, K. (2011). *Diagnosis and Prediction for a Successful Management of Railway Infrastructure* (M. Singh, J. P. Liyanage, & R. B. Rao, Eds.). Kolkata, India: Publishing Services PL.

Bojadziev, G., & Bojacziev, M. (1997). *Fuzzy logic for business, finance, and management*. Singapore: World Scientific.

Bollinger, C. R., & Ihlanfeldt, K. R. (1997). The Impact of Rapid Rail Transit on Economic Development: The Case of Atlanta's MARTA. *Journal of Urban Economics, 42*(2), 179–204. doi:10.1006/juec.1996.2020

Boudon, F. et al. (1977). Système de l'architecture urbaine: Le Quartier des Halles de Paris. Paris.

Bourillon, F. (2001). A propos de la Commission des embellissements. In K. Bowie (Ed.), (textes réunis par) La Modernité avant Haussmann. Formes de l'espace urbain à Paris 1801 – 1853 (pp. 139–151). Paris: Éditions Recherches.

Bowes, D. R., & Ihlanfeldt, K. R. (2001). Identifying the Impacts of Rail Transit Stations on Residential Property Values. *Journal of Urban Economics, 50*(1), 1–25. doi:10.1006/juec.2001.2214

Bowie, K., & Texier, S. (Eds.), (2003). Paris et ses Chemins de Fer. Paris: Action Artistique de la Ville de Paris.

Bowles, J. B., & Pelaez, C. E. (1995a). Fuzzy logic prioritisation of failure in a system failure mode, effects and criticality analysis. *Reliability Engineering & System Safety, 50*(2), 203–213. doi:10.1016/0951-8320(95)00068-D

Bowles, J. B., & Pelaez, C. E. (1995b). Application of fuzzy logic to reliability engineering. *Proceedings of the IEEE, 83*(3), 435–449. doi:10.1109/5.364489

Bowman, D., Schaudt, W., & Hanowski, R. (2012). Advances in drowsy driver assistance systems through data fusion. In *Handbook of Intelligent Vehicles* (pp. 895-912). Retrieved from http://link.springer.com/10.1007/978-0-85729-085-4_34

Bradley, K. (2006). The Development of the London Underground, 1840-1933: The Transformation of the London Metropolis and the Role of Laissez-Faire in Urban Growth [PhD Thesis]. Emory University.

Brännlund, U., Lindberg, P. O., Nou, A., & Nilsson, J.-E. (1998). Railway timetabling using lagrangian relaxation. *Transportation Science, 32*(4), 358–369. doi:10.1287/trsc.32.4.358

Briz, F., Degner, M., & Lorenz, L. (2000). Analysis and design of current regulators using complex vectors. *IEEE Transactions on Industry Applications, 36*(3), 817–825. doi:10.1109/28.845057

Brown, B. (1968). Delphi Process: A Methodology Used for the Elicitation of Opinions of Experts. An earlier paper published by RAND (Document No: P-3925, 1968).

Brucker, P., Heitmann, S., & Knust, S. (2005). *Scheduling Railway Traffic at a Construction site*. Springer Berlin Heidelberg. doi:10.1007/3-540-26686-0_15

Bruni, S., & Goodall, R. M., Mei, T.XTsunashima, H. (2007). Control and monitoring for railway vehicle dynamics. *Vehicle System Dynamics*, *45*(7-8), 743–779. doi:10.1080/00423110701426690

Bruzzese, C. (2008). Analysis and Application of Particular Current Signatures (Symptoms) for Cage Monitoring in Non-sinusoidally Fed Motors with High Rejection to Drive Load, Inertia, and Frequency Variations. *IEEE Transactions on Industrial Electronics*, *55*(12), 4137–4155. doi:10.1109/TIE.2008.2004669

Buckley, J. J. (1985). Fuzzy hierarchical analysis. *Fuzzy Sets and Systems*, *17*(3), 233–247. doi:10.1016/0165-0114(85)90090-9

Budai, G., Maróti, G., Dekker, R., Huisman, D., & Kroon, L. (2009). Rescheduling in Passenger Railways: The Rolling Stock Rebalancing Problem. *Journal of Scheduling*. doi:10.1007/s10951-009-0133-9

Bundele, M., & Banerjee, R. (2009). Detection of fatigue of vehicular driver using skin conductance and oximetry pulse: a neural network approach. *Proceedings of the 11th International Conference on Information Integration and web-based applications & services* (pp. 739-744). Retrieved from http://dl.acm.org/citation.cfm?id=1806478

Burdett, R., & Kozan, E. (2008). A sequencing approach for creating new train timetables. OR Spectrum. doi:10.1007/s00291-008-0143-6

Bussieck, M. (1998). Optimal Lines in Public Rail Transport [PhD thesis]. Technischen Universität Braunschweig, Braunschweig.

Cabanas, M. F., Pedrayes, F., & Melero, M. G., Garcia, Rojas C. H., Cano, J. M., Orcajo, G. A., & Norniella, J. G. (2011). Unambiguous Detection of Broken Bars in Asynchronous Motors by Means of a Flux Measurement-Based Procedure. *IEEE Transactions on Instrumentation and Measurement*, *60*(3), 891–899. doi:10.1109/TIM.2010.2062711

Cacchiani, V. (2008). Models and Algorithms for Combinatorial Optimization Problems Arising in Railway Applications [PhD thesis]. Universit di Bologna.

Cacchiani, V., Caprara, A., & Toth, P. (2008). A column generation approach to train timetabling on a corridor. *4OR: A Quarterly Journal of Operations Research*, *6*(2), 125-142.

Cacchiani, V., & Toth, P. (2012). Nominal and robust train timetabling problems. *European Journal of Operational Research*, *219*(3), 727–737. doi:10.1016/j.ejor.2011.11.003

Cadarso, L., Marín, Á., & Maróti, G. (2013). Recovery of disruptions in rapid transit networks. *Transportation Research Part E, Logistics and Transportation Review*, *53*, 15–33. doi:10.1016/j.tre.2013.01.013

Caffier, P. P., Erdmann, U., & Ullsperger, P. (2003). Experimental evaluation of eye-blink parameters as a drowsiness measure. *European Journal of Applied Physiology*, *89*(3), 319–325. doi:10.1007/s00421-003-0807-5 PMID:12736840

Cai, R. S., Zhu, Y. T., & Guo, Y. M. (2007). Wavelet-based multi-feature voiced/unvoiced speech classification algorithm. *Proceedings of theIET Conference on Wireless, Mobile and Sensor Networks 2007 (CCWMSN07)* (Vol. 2007, pp. 897–900). IEE. http://doi.org/ doi:10.1049/cp:20070294

Caimi, G., Fuchsberger, M., Burkolter, D., Herrmann, T., Wüst, R., & Roos, S. (2009). Conflict-free train scheduling in a compensation zone exploiting the speed profile. *Proceedings of the 3rd international seminar on railway operation modeling and analysis*.

Cai, X., & Goh, C. J. (1994). A Fast Heuristic for the Train Scheduling Problem. *Computers & Operations Research*, *21*(5), 499–510. doi:10.1016/0305-0548(94)90099-X

Cai, Y., Irving, M., & Case, M. (1995). Modeling and numerical solution of multi-branched DC rail traction power systems. *IEE Proceedings. Electric Power Applications*, *142*(5), 323–328. doi:10.1049/ip-epa:19952118

Cajochen, C., Zeitzer, J. M., Czeisler, C. A., & Dijk, D. J. (2000). Dose-response relationship for light intensity and ocular and electroencephalographic correlates of human alertness. *Behavioural Brain Research*, *115*(1), 75–83. doi:10.1016/S0166-4328(00)00236-9 PMID:10996410

Campbell, H. F., & Brown, R. P. (2003). Investment Appraisal: Decision-Rules. In Benefit-Cost Analysis (pp. 36-61). United Kingdom: Cambridge University Press.

Canca, D., Barrena, E., Laporte, G., & Ortega, F. A. (2014). A short-turning policy for the management of demand disruptions in rapid transit systems. *Annals of Operations Research*, 2014, 1–22.

Canca, D., Barrena, E., Zarzo, A., Ortega, F., & Algaba, E. (2012). Optimal train reallocation strategies under service disruptions. *Procedia: Social and Behavioral Sciences*, *54*, 402–413. doi:10.1016/j.sbspro.2012.09.759

Cao, Z., Yuan, Z., & Li, D. (2014). Estimation method for a skip-stop operation strategy for urban rail transit in China. *Journal of Modern Transportation*, *22*(3), 174–182. doi:10.1007/s40534-014-0059-6

Caprara, A., Fischetti, M., Guida, P. L., Monaci, M., Sacco, G., & Toth, P. (2001). Solution of real-world train timetabling problems. *Proceedings of the 34th Hawaii International Conference on System Sciences* (pp. 1-10).

Caprara, A., Fischetti, M., & Toth, P. (2002). Modeling and solving the train timetabling problem. *Operations Research*, *50*(5), 851–861. doi:10.1287/opre.50.5.851.362

Caprara, A., Kroon, L., Monaci, M., Peeters, M., & Toth, P. (2007). Passenger railway optimization. *Handbooks in Operations Research and Management Science*, *14*, 129–187. doi:10.1016/S0927-0507(06)14003-7

Caprara, A., Monaci, M., Toth, P., & Guida, P. L. (2006). A lagrangian heuristic algorithm for a real-world train timetabling problem. *Discrete Applied Mathematics*, *154*(5), 738–753. doi:10.1016/j.dam.2005.05.026

Caprara, L., Kroon, G., Monaci, M., Peeters, M., & Toth, P. (2007). Passenger Railway Optimization. In C. Barnhart & G. Laporte (Eds.), *Handbooks in Operations Research and Management Science* (Vol. 14, pp. 129–187). Elsevier.

Carbone, A., Papa, F., & Sacco, N. (2012). An Optimization Approach for Delay Recovery in Urban Metro Transportation Systems.

Caron, F. (1997). *Histoire de chemins de fer en France* (Vol. I). Paris: Fayard.

Carrel, A., Mishalani, R. G., Wilson, N. H., Attanucci, J. P., & Rahbee, A. B. (2010). Decision Factors in Service Control on High-Frequency Metro Line. *Transportation Research Record: Journal of the Transportation Research Board*, *2146*(1), 52–59. doi:10.3141/2146-07

Cartens, J. P. (1965). *Literature Survey of Passenger Comfort Limitations for High Speed Ground Transports*. Hartgforf, Conn.: United Aircraft Corp.

Casares, P., & Coto-Millan, P. (2011). Passenger transport planning. A Benefit-Cost Analysis of the High Speed Railway: The case of Spain. *Atlantic Review of Economics*, *2*, 1–12.

Castillo, E., Gallego, I., Ureña, J. M., & Coronado, J. M. (2009). Timetabling optimization of a single railway track line with sensitivity analysis. *Top (Madrid)*, *17*(2), 256–287. doi:10.1007/s11750-008-0057-0

Ceban, A., Pusca, R., & Romary, R. (2012). Study of Rotor Faults in Induction Motors Using External Magnetic Field Analysis. *IEEE Transactions on Industrial Electronics*, *59*(5), 2082–2093. doi:10.1109/TIE.2011.2163285

Cervero, R. (1984). Journal Report: Light Rail Transit and Urban Development. *Journal of the American Planning Association, 50*(2), 133–147. doi:10.1080/01944368408977170

Cervero, R. (1994). Rail Transit and Joint Development: Land Market Impacts in Washington, D.C. and Atlanta. *Journal of the American Planning Association, 60*(1), 83–94. doi:10.1080/01944369408975554

Cervero, R., & Duncan, M. (2001). *Rail transit's value added: effects of proximity to light and commuter rail transit on commercial land values in Santa Clara County, California*. Washington, DC, USA: Urban Land Institute, National Association of Realtors.

Cervero, R., & Landis, J. (1997). Twenty years of the Bay Area Rapid Transit system: Land use and development impacts. *Transportation Research Part A, Policy and Practice, 31*(4), 309–333. doi:10.1016/S0965-8564(96)00027-4

Cervero, R., Murakami, J., & Miller, M. (2010). Direct Ridership Model of Bus Rapid Transit in Los Angeles County, California. *Transportation Research Record: Journal of the Transportation Research Board, 2145*, 1–7. doi:10.3141/2145-01

Chandra, K., & Roy, D. G. (2002). *A technical guide on oscillation trial* (p. 334). MT: RDSO.

Chang, S. C., & Chung, Y. C. (2005). From Timetabling to train regulation-a new train operation model. *Information and Software Technology, 47*(9), 575–585. doi:10.1016/j.infsof.2004.10.008

Chang, S. K. J., & Hsu, C. L. (2001). Fare and Service Headway for a High Speed Rail system with private sector involvement. *Journal of the Eastern Asia Society for Transportation Studies, 4*(1), 2001.

Chattopadhyay, G., & Kumar, S. (2009). Parameter Estimation for Rail Degradation Model. *International Journal of Performability Engineering, 5*(2), 119–130.

Chaudhuri, A., Routray, A., & Kar, S. (2012). Effect of sleep deprivation on estimated distributed sources for Scalp EEG signals: A case study on human drivers. *Proceedings of the 2012 4th International Conference on Intelligent Human Computer Interaction (IHCI)* (pp. 1–6). IEEE. http://doi.org/ doi:10.1109/IHCI.2012.6481805

Chen, M., & Niu, H. (2009). Modeling Transit Scheduling Problem with Short-Turn Strategy For A Congested Public Bus Line. *Paper presented at the Logistics@ sThe Emerging Frontiers of Transportation and Development in China*. doi:10.1061/40996(330)631

Chen, B., Cheng, H. H., & Palen, J. (2009, February). Integrating mobile agent technology with multi-agent systems for distributed traffic detection and management systems. *Transportation Research Part C, Emerging Technologies, 17*(1), 1–10. doi:10.1016/j.trc.2008.04.003

Cheng, C. H. (1997). Evaluating naval tactical missile systems by fuzzy AHP based on the grade value of membership function. *European Journal of Operational Research, 96*(2), 343–350. doi:10.1016/S0377-2217(96)00026-4

Cheng, C. H., Yang, K. L., & Hwang, C. L. (1999). Evaluating attack helicopters by AHP based on linguistic variable weight. *European Journal of Operational Research, 116*(2), 423–435. doi:10.1016/S0377-2217(98)00156-8

Cheng, Y. (1996). Optimal Train Traffic Rescheduling Simulation by a Knowledge-Based System Combined with Critical Path Method. *Simulation Practice and Theory, 4*(6), 399–413. doi:10.1016/S0928-4869(96)00034-1

Cheng, Y. (1998, November). Hybrid Simulation for Resolving Resource Conflicts in Train Traffic Rescheduling. *Computers in Industry, 35*(3), 233–246. doi:10.1016/S0166-3615(97)00071-7

Cheng, Y. C., & Hsu, C. T. (2012). Hunting stability and derailment analysis of car model of a railway vehicle system. *Rail and Rapid Transit, 226*(2), 187–202. doi:10.1177/0954409711407658

Cheng, Y. C., Lee, S. Y., & Chen, H. H. (2009). Modeling and nonlinear hunting stability analysis of high speed railway vehicle moving on the curved track. *Journal of Sound and Vibration*, *324*(1-2), 139–160. doi:10.1016/j.jsv.2009.01.053

Chen, X. (2012). Managing transportation financing in an innovative way. *Management Research and Practice*, *4*(3), 5–17.

Chevalier, L. (1973). *Labouring Classes and Dangerous Classes in Paris during the first half of the nineteenth century* (F. Jellinek, Trans.). London: Routledge & Kegan Paul. (Original work published 1958)

Chialvo, D. R. (2004). Critical brain networks. In A. Physica (Ed.), *Statistical Mechanics and its Applications* (Vol. 340, pp. 756–765). Doi:10.1016/j.physa.2004.05.064

Choi, S., Akin, B., Rahimian, M. M., & Toliyat, H. A. (2011). Implementation of a Fault-Diagnosis Algorithm for Induction Machines Based on Advanced. *IEEE Transactions on Industrial Electronics*, *58*(3), 937–948. doi:10.1109/TIE.2010.2048837

Chowdhary, M. A., & Sadek, A. (2008). *Fundamentals of Intelligent Transportation systems planning*. US: Artech House Inc.

Chow, T. W. S., & Tan, H.-Z. (2000). HOS-based nonparametric and parametric methodologies for machine fault detection. *IEEE Transactions on Industrial Electronics*, *47*(5), 1051–1059. doi:10.1109/41.873213

Chrimes, M. (2013). Hawkshaw, Sir John (1811–1891). Oxford Dictionary of National Biography. Retrieved from http://0-www.oxforddnb.com.catalogue.ulrls.lon.ac.uk/view/article/12690

Chymera, M. Z., Renfrew, A. C., Barnes, M., & Holden, J. (2010). Modeling Electrified Transit Systems. *IEEE Transactions on Vehicular Technology*, *59*(6), 2748–2756. doi:10.1109/TVT.2010.2050220

Cimatti, A., Corvino, R., Lazzaro, A., Narasamdya, I., …, Tchaltsev, A. (2012). Formal Verification and Validation of ERTMS Industrial Railway Train Spacing System. Computer Aided Verification, LNCS (Vol. 7358, pp. 378–393). Springer.

Cimatti, A., Giunchiglia, F., Mongardi, G., Romano, D., Torielli, F., & Traverso, P. (1998). Formal Verification of a railway interlocking system using model checking. *Formal Aspects of Computing*, *10*(4), 361–380. doi:10.1007/s001650050022

City of London & Southwark Subway Company. (n. d.). Minute book 1, 1884 – 1889. London Metropolitan Archives.

City of London & Southwark Subway Company. (n. d.). Minute book 2, 1889 – 1892. London Metropolitan Archives.

Clifton, K. J., Burnier, C. V., & Akar, G. (2009). Severity of injury resulting from pedestrian–vehicle crashes: What can we learn from examining the built environment? *Transportation Research Part D, Transport and Environment*, *14*(6), 425–436. doi:10.1016/j.trd.2009.01.001

Concari, C., Franceschini, G., & Tassoni, C. (2011). Toward Practical Quantification of Induction Drive Mixed Eccentricity. *IEEE Transactions on Industry Applications*, *47*(3), 1232–1239. doi:10.1109/TIA.2011.2124434

Cordeau, J. F., Toth, P., & Vigo, D. (1998, November). A Survey of Optimization models for Train Routing and Scheduling. *Transportation Science*, *32*(4), 380–404. doi:10.1287/trsc.32.4.380

Cortés, C. E., Jara-Díaz, S., & Tirachini, A. (2011). Integrating short turning and deadheading in the optimization of transit services. *Transportation Research Part A, Policy and Practice*, *45*(5), 419–434. doi:10.1016/j.tra.2011.02.002

Coto, M., Arboleya, P., & Gonzalez-Moran, C. (2013). Optimization approach to unified AC/DC power flow applied to traction systems with catenary voltage constraints. *International Journal of Electrical Power & Energy Systems*, *53*, 434–441. doi:10.1016/j.ijepes.2013.04.012

Cottrill, C. D., & Thakuriah, P. (2010). Evaluating pedestrian crashes in areas with high low-income or minority populations. *Accident; Analysis and Prevention*, *42*(6), 1718–1728. doi:10.1016/j.aap.2010.04.012 PMID:20728622

Cowie, J. (2010). *The Economics of Transport – A theoretical and applied perspective*. USA: Routledge.

Croft, R. J., & Barry, R. J. (2000). Removal of ocular artifact from the EEG: A review. *Neurophysiologie Clinique*, *30*(1), 5–19. doi:10.1016/S0987-7053(00)00055-1 PMID:10740792

Cruz, M. A. (2012). An Active Reactive Power Method for the Diagnosis of Rotor Faults in Three-Phase Induction Motors Operating Under Time-Varying Load Conditions. *IEEE Transactions on Energy Conversion*, *27*(1), 71–84. doi:10.1109/TEC.2011.2178027

Da Silva, A. M., Povinelli, R. J., & Demerdash, N. A. O. (2008). Induction machine broken bar and stator short-circuit fault diagnostics based on three-phase stator current envelopes. *IEEE Transactions on Industrial Electronics*, *55*(3), 1310–1318. doi:10.1109/TIE.2007.909060

Dahlberg, T. (2006). Track issues. In S. Iwnicki (Ed.), *Handbook of railway vehicle dynamics* (pp. 143–179). Boca Raton, FL: CRC Press. doi:10.1201/9781420004892.ch6

Dailey, D. J., Maclean, S. D., Cathey, F. W., & Wall, Z. R. (2001). Transit vehicle arrival prediction: An algorithm and a large scale implementation. *Transportation Research Record*.

Dalkey, N., & Helmer, O. (1963). An Experimental Application of the Delphi Method to the use of experts. *Management Science*, *9*(3), 458–467. doi:10.1287/mnsc.9.3.458

Damm, W., & Klose, J. (2001). Verification of a radio-based signaling system using the STATEMATE verification environment. *Formal Methods in System Design*, *19*(2), 121–141. doi:10.1023/A:1011279932612

Danziger, N. H., & Kennedy W.D. (1982). Longitudinal ventilation analysis for the glenwood canyon tunnels. *Proc 4th Int. Symposium Aerodynamics & Ventilation of Vehicle Tunnels* (pp. 169-186).

Dasgupta, A., George, A., Happy, S. L., & Routray, A. (2013). A Vision-Based System for Monitoring the Loss of Attention in Automotive Drivers. *IEEE Transactions on Intelligent Transportation Systems*, *15*(1), 1–14. doi:10.1109/TITS.2013.2271052

Dasgupta, A., George, A., Happy, S. L., Routray, A., & Shanker, T. (2013). An on-board vision based system for drowsiness detection in automotive drivers. *International Journal of Advances in Engineering Sciences and Applied Mathematics*, *5*(2-3), 94–103. doi:10.1007/s12572-013-0086-2

Dasgupta, P. (2006). *A Roadmap for Formal Property Verification*. Springer. doi:10.1007/978-1-4020-4758-9_8

Das, S. (2014). Hidden Cost in Public Infrastructure Project: A Case Study of Kolkata East–West Metro. In K. S. Sridhar & G. Wan (Eds.), *Urbanization in Asia: Governance, Infrastructure and the Environment* (pp. 149–164). Springer. doi:10.1007/978-81-322-1638-4_9

Das, S., Purkait, P., Koley, C., & Chakravorti, S. (2014). Performance of a load-immune classifier for robust identification of minor faults in induction motor stator winding. Dielectrics and Electrical Insulation. *IEEE Transactions on*, *21*(1), 33–44.

Day, L., & McNeil, I. (1966). *Biographical dictionary of the history of technology*. London, United Kingdom: Routledge.

de Andrade, L., Vissoci, J. R. N., Rodrigues, C. G., Finato, K., Carvalho, E., Pietrobon, R., & de Barros Carvalho, M. D. et al. (2014). Brazilian road traffic fatalities: A spatial and environmental analysis. *PLoS ONE*, *9*(1), e87244. doi:10.1371/journal.pone.0087244 PMID:24498051

de la Broise, T., & Torres, F. (1996). *Schneider, histoire en force*. Paris: Éditions de Monza.

Compilation of References

de Rus, G. (2012). Economic evaluation of the high speed rail. Expert Group on Environmental Studies. Sweden: Ministry of Finance. Retrieved from http://www.ems.expertgrupp.se/Default.aspx?pageID=3

De Waard, D., & Brookhuis, K. A. (1991). Assessing driver status: A demonstration experiment on the road. *Accident; Analysis and Prevention, 23*(4), 297–307. doi:10.1016/0001-4575(91)90007-R PMID:1883469

Deckoff, A. A. (1990). *The short-turn as a real time transit operating strategy*. Massachusetts Institute of Technology.

Del Signore, E., Giuliano, R., Mazzenga, F., & Petracca, M. (2014). On the suitability of public mobile networks for supporting train control/management systems. *Proceedings of the IEEE WCNC/14 conference* (pp. 3302 – 3307).

Délibérations. (1898, June 27). Conseil Municipal de Paris.

Delle Site, P., & Filippi, F. (1998). Service optimization for bus corridors with short-turn strategies and variable vehicle size. *Transportation Research Part A, Policy and Practice, 32*(1), 19–38. doi:10.1016/S0965-8564(97)00016-5

Dennis, R. (2008). Cities in Modernity Representations and Productions of Metropolitan Space, 1840-1930. Cambridge: Cambridge University Press.

Development Plan for Chennai Metropolitan Area. (2006). CMDA.

Development Plan for Greater Mumbai 2014-2034: Preparatory Studies. (2013). Retrieved from http://www.mcgm.gov.in/

Dhupati, L. S., Kar, S., Rajaguru, A., & Routray, A. (2010). A novel drowsiness detection scheme based on speech analysis with validation using simultaneous EEG recordings. *Proceedings of the 2010 IEEE International Conference on Automation Science and Engineering* (pp. 917–921). IEEE. http://doi.org/ doi:10.1109/COASE.2010.5584246

Diana, G., Bruni, S., Cheli, F., & Resta, F. (2002). Active control of the running behavior railway vehicle: Stability and curving performances. *Vehicle System Dynamics, 37*(Suppl.), 157–170. doi:10.1080/00423114.2002.11666229

Dinges, D., Mallis, M., Maislin, G., & Powell, I. (1998). Evaluation of techniques for ocular measurement as an index of fatigue and the basis for alertness management. Retrieved from http://trid.trb.org/view.aspx?id=647942

Ding, Y., & Chien, S. I. (2001). Improving transit service quality and headway regularity with real-time control. *Transportation Research Record: Journal of the Transportation Research Board, 1760*(1), 161–170. doi:10.3141/1760-21

Doi, M., & Allen, W. B. (1986). A Time Series Analysis of Monthly Ridership for an Urban Rail Rapid Transit Line. *Transportation, 13*(3), 257–269. doi:10.1007/BF00148619

Doll, C. N. H., Dreyfus, M., Ahmad, S., & Balaban, O. (2013). Institutional framework for urban development with co-benefits: The Indian experience. *Journal of Cleaner Production, 58*(0), 121–129. doi:10.1016/j.jclepro.2013.07.029

Dorrell, D. G. (2011). Sources and Characteristics of Unbalanced Magnetic Pull in Three-Phase Cage Induction Motors With Axial-Varying Rotor Eccentricity. *IEEE Transactions on Industry Applications, 47*(1), 12–24. doi:10.1109/TIA.2010.2090845

Draft Concept Note on Smart City Scheme. (2014). MoUD-GoI.

Drif, M., & Cardoso, A. J. M. (2012). Discriminating the Simultaneous Occurrence of Three-Phase Induction Motor Rotor Faults and Mechanical Load Oscillations by the Instantaneous Active and Reactive Power Media Signature Analyses. *IEEE Transactions on Industrial Electronics, 59*(3), 1630–1639. doi:10.1109/TIE.2011.2161252

Drif, M., & Cardoso, A. J. M. (2014). Stator fault diagnostics in squirrel cage three-phase induction motor drives using the instantaneous active and reactive power signature analyses. Industrial Informatics. *IEEE Transactions on, 10*(2), 1348–1360.

Driver fatigue and road accidents. (n. d.). *ROSPA*. Retrieved from http://www.rospa.com/roadsafety/info/fatigue.pdf

Duffy, M. (2003). *Electric Railways 1880-1990*. London: Institution of Electrical Engineers.

Dukkipati, R. V., & Amyot, J. (1988). *Computer aided simulation in railway dynamics*. New York: Marcel Dekker. Inc.

Dupuit, J. (1844). De la mesure de l'utilité des travaux publics. *Annales des Ponts et Chaussées, 8*.

Durbin, J., & Koopman, S. (2001). *Time series analysis by state space methods*. Oxford University Press.

Eberlein, X. J., Wilson, N. H., Barnhart, C., & Bernstein, D. (1998). The real-time deadheading problem in transit operations control. *Transportation Research Part B: Methodological, 32*(2), 77–100. doi:10.1016/S0191-2615(97)00013-1

Eberlein, X. J., Wilson, N. H., & Bernstein, D. (1999). *Modeling real-time control strategies in public transit operations Computer-aided transit scheduling* (pp. 325–346). Springer.

Eberlein, X. J., Wilson, N. H., & Bernstein, D. (2001). The Holding Problem with Real–Time Information Available. *Transportation Science, 35*(1), 1–18. doi:10.1287/trsc.35.1.1.10143

EC. (2012, 01 25). *2012/88/EU. Commission decision of 25 January 2012 on the technical specification for interoperability relating to the control-command and signalling subsystems of the trans-European rail system*. Brussels, Belgium: TSI CCS.

Eckford, D. C., Pope, C. W., Owoeye, A., & Henson, D. H. (2003). The ventilation of metro systems with and without full height platform screen doors. *Proceedings of the 11th International Symposium on Aerodynamics & Ventilation in Vehicle Tunnels*.

Eisner, C. (2002). Using symbolic CTL model checking to verify the railway stations of Hoorn-Kersenboogerd. *Software Tools for Technology Transfer, 4*(1), 107–124. doi:10.1007/s100090100063

Ekelund, R., & Hébert, R. (1985). Consumer's surplus - the first hundred years. *History of Political Economy, 17*(3), 419–454. doi:10.1215/00182702-17-3-419

EN 50128, Railway Applications Communications, Signaling and Processing Systems Software for Railway Control and Protection Systems. (2001European Committee for Electrotechnical Standardization. CENELEC.

EN50126 Railway applications-the specification and demonstration of reliability, availability, maintainability and safety (RAMS). (1999). *European Standard*, CENELEC.

Engle, R. F., & Granger, C. J. (1987). Cointegration and error correction: Representation, estimation and testing. *Econometrica, 55*(2), 251–276. doi:10.2307/1913236

Eoh, H. J., Chung, M. K., & Kim, S. H. (2005). Electroencephalographic study of drowsiness in simulated driving with sleep deprivation. *International Journal of Industrial Ergonomics, 35*(4), 307–320. doi:10.1016/j.ergon.2004.09.006

Eren, L., & Devaney, M. J. (2004). Bearing damage detection via wavelet packet decomposition of the stator current. *IEEE Transactions on Instrumentation and Measurement, 53*(2), 431–436. doi:10.1109/TIM.2004.823323

Eriksson, J., Girod, L., Hull, B., Newton, R., Madden, S., & Balakrishnan, H. (2008). The Pothole Patrol: Using a Mobile Sensor Network for Road Surface Monitoring.*Proceedings of the sixth annual international conference on mobile systems, applications and services MobiSys'08*.

Esfahani, E. T., Wang, S., & Sundararajan, V. (2013). Multisensor Wireless System for Eccentricity and Bearing Fault Detection in Induction Motors. IEEE/ASME Trans. *Mechatronics, 19*(3), 818–826. doi:10.1109/TMECH.2013.2260865

Compilation of References

ETCS / ERTMS Specifications. (n. d.). *European Train Control System.* http://www.era.europa.eu/Core-Activities/ERTMS/Pages/Set-of-specifications-2.aspx

Evolving Perspectives in the Development of Indian Infrastructure (Vol. 2). *Infrastructure Development Finance Company Limited.* (2012). Orient Blackswan Private Limited, Greater Noida.

Ewing, R., Hamidi, S., & Grace, J. B. (2014). Urban sprawl as a risk factor in motor vehicle crashes. *Urban Studies.*

Fahmy, H. M. A. (2001). Reliability evaluation in distributed computing environments using the AHP. *Computer Networks, 36*(5-6), 597–615. doi:10.1016/S1389-1286(01)00175-X

Faiz, J., Ardekanei, I. T., & Toliyat, H. A. (2003). An Evaluation of Inductances of a Squirrel-Cage Induction Motor Under Mixed Eccentric Conditions. Energy Conversion. *IEEE Trans., 18*(2), 252–258.

Falvo, M. C., Lamedica, R., Bartoni, R., & Maranzano, G. (2011). Energy management in metro transit systems: An innovative proposal toward an integrated and sustainable urban mobility system including plug-in electric vehicles. *Electric Power Systems Research, 81*(12), 2127–2138. doi:10.1016/j.epsr.2011.08.004

Fanger, P. O., & Christensen, N. K. (1986). Perception of draught in ventilated spaces. *Ergonomics, 29*(2), 215–235. doi:10.1080/00140138608968261 PMID:3956473

Fantechi, A., Fokkink, W., & Morzenti, A. (2012). Some trends in Formal Methods applications to Railway Signaling. In S. Gnesi, & T. Margaria (Eds.), *Formal Methods for Industrial Critical Systems: A survey of applications.* John Wiley and Sons, Inc. doi:10.1002/9781118459898.ch4

Fatigue is a Major Cause in Truck Crashes. (n. d.). *OPTAlert.* Retrieved from http://www.optalert.com/news/truck-crashes-fatigue

Fay, A., & Schnieder, E. (1999). Knowledge-based decision support system for real-time train traffic control. In N. Wilson (Ed.), Computer-Aided Transit Scheduling (pp. 347-370). Berlin: Springer. doi:10.1007/978-3-642-85970-0_17

Fewstar, M., & Graham, D. (1999). *Software Testing Automation: Effective use of test execution tools.* ACM Press, Addison Wesley.

Flamini, M., & Pacciarelli, D. (2008). Real time management of a metro rail terminus. *European Journal of Operational Research, 189*(3), 746–761. doi:10.1016/j.ejor.2006.09.098

Fleury, M., & Pronteau, J. (1987). *Petit Atlas Pittoresque des quarante-huit Quartiers de la ville de Paris (1834) par A.-M. Perrot, ingénieur.* Paris: Commission des Travaux Historiques.

Fok, D., Frances, H., & Paap, R. (2002). Econometric Analysis of the Market Share Attraction Model. In P. H. Frances & A. L. Montgomery (Eds.), *Econometric Models in Marketing* (pp. 223–256). New York: JAI/Elsevier. doi:10.1016/S0731-9053(02)16010-5

Fokkink, W. (1996). Safety criteria for the vital processor interlocking at Hoorn-Kersenboogerd. *Proceedings of the 5th Conference on Computers in Railways.* Computational Mechanics Publications.

Fouracre, P., Dunkerley, C., & Gardner, G. (2003). Mass rapid transit systems for cities in the developing world. *Transport Reviews, 23*(3), 299–310. doi:10.1080/0144164032000083095

Frank, O. (1966). Two-way traffic on a single line of railway. *Operations Research, 14*(5), 801–811. doi:10.1287/opre.14.5.801

Freeman, M. J., & Aldcroft, D. (1985). *The Atlas of British Railway History.* London: Croom Helm.

Frey, S. (2012). *Railway electrification systems & engineering*. Delhi, India: White Word Publications.

Fringuelli, B., Lamma, E., Mello, P., & Santocchia, G. (1992). Knowledge-based technology for controlling railway stations. *IEEE Intelligent Systems*, 7(6), 45–52.

Frosini, L., & Bassi, E. (2010). Stator current and motor efficiency as indicators for different types of bearing faults in induction motors. *IEEE Transactions on Industrial Electronics*, 57(1), 244–251. doi:10.1109/TIE.2009.2026770

Frosini, L., Harlisca, C., & Szabo, L. (2015). Induction Machine Bearing Fault Detection by Means of Statistical Processing of the Stray Flux Measurement. *IEEE Transactions on Industrial Electronics*, 62(3), 1846–1854. doi:10.1109/TIE.2014.2361115

Furth, P. (1987). Short turning on transit routes. *Transportation Research Record*, 1987, 1108.

Fu, X., Zhang, A., & Lei, Z. (2012). Will China's airline industry survive the entry of high-speed rail? *Research in Transportation Economics*, 35(1), 13–25. doi:10.1016/j.retrec.2011.11.006

Gadd, D. S., Keeley, D. D., & Balmforth, D. H. (2003). Good practice and pitfalls in risk assessment. Health & Safety Laboratory, HMSO, HSE Book, London.

Gadgil, V. B. (2011). *No metro project is financially viable on its own across the world*. Retrieved from http://www.reachouthyderabad.com/newsmaker/hw376.htm

Gandhi, A., Corrigan, T., & Parsa, L. (2011). Recent advances in modeling and online detection of stator interturn faults in electrical motors. *Ind. Electron. IEEE Trans.*, 58(5), 1564–1575. doi:10.1109/TIE.2010.2089937

Gangadharan, K. V., Sujata, C., & Ramamurti, V. (2004), Experimental & analytical ride comfort evaluation of a railway coach. *Proc. IMAC-XXII*, Dearborn, Michigan.

Ganti, R., Pham, N., Ahmadi, H., Nangia, S., & Abdelzaher, T. GreenGPS: a participatory sensing fuel-efficient maps application. Proceedings of the sixth annual international conference on mobile systems, applications and services MobiSys'10. doi:10.1145/1814433.1814450

Gao, Z., Turner, L., Colby, R. S., & Leprettre, B. (2011). A Frequency Demodulation Approach to Induction Motor Speed Detection. *IEEE Transactions on Industry Applications*, 47(4), 1632–1642. doi:10.1109/TIA.2011.2153813

Garcia-perez, A., Romero-troncoso, R. D. J., Cabal-yepez, E., & Osornio-rios, R. A. (2011). The Application of High-Resolution Spectral Analysis for Identifying Multiple Combined Faults in Induction Motors. *IEEE Transactions on Industrial Electronics*, 58(5), 2002–2010. doi:10.1109/TIE.2010.2051398

Garrett, M., & Taylor, B. (1999). Reconsidering Social Equity in Public Transit. *Berkeley Planning Journal, 13*(1).

Georgakopoulos, I. P., Mitronikas, E. D., & Safacas, A. N. (2011). Detection of Induction Motor Faults in Inverter Drives Using Inverter Input Current Analysis. *IEEE Transactions on Industrial Electronics*, 58(9), 4365–4373. doi:10.1109/TIE.2010.2093476

Geske, U. (2006). Railway scheduling with declarative constraint programming. InDeclarative Programming for Knowledge Management, *LNCS* (Vol. *4369*, pp. 117–134).

Getis, A., & Ord, J. K. (1992). The analysis of spatial association by use of distance statistics. *Geographical Analysis*, 24(3), 189–206. doi:10.1111/j.1538-4632.1992.tb00261.x

Ghate, V. N., & Dudul, S. V. (2011). Cascade Neural-Network-Based Fault Classifier for Three-Phase Induction Motor. *IEEE Transactions on Industrial Electronics*, 58(5), 1555–1563. doi:10.1109/TIE.2010.2053337

Compilation of References

Ghiaus, C. (2001). Fuzzy model and control of a fan-coil. *Energy and Building, 33*(6), 545–551. doi:10.1016/S0378-7788(00)00097-9

Ghoneim, N., & Wirasinghe, S. (1986). Optimum zone structure during peak periods for existing urban rail lines. *Transportation Research Part B: Methodological, 20*(1), 7–18. doi:10.1016/0191-2615(86)90032-9

Ghoseiri, K. & Morshedsolouk, F. (2006). ACS-TS: Train Scheduling using Ant Colony System. *Journal of Applied Mathematics and Decision Sciences*. doi:.10.1155/JAMDS/2006/95060

Global Status Report on Road Safety: Time for Action. (2009). World Health Organization, Geneva, Switzerland.

Gluchschenko, O., & Förster, P. (2013). Performance based approach to investigate resilience and robustness of an ATM System. *ATM Seminar.* Chicago, IL, USA.

Goldman, R. F. (1978). *The role of clothing in achieving acceptability of environmental temperatures between 65°F and 85°F (18°C and 30°C). Energy conservation strategy in buildings* (J. Stolwijk, Ed.). New Haven: Yale University Press.

Gomez, V. (2012). SSMMATLAB, a Set of MATLAB Programs for the Statistical Analysis of State-Space Models. Retrieved from http://www.sepg.pap.minhap.gob.es/sitios/sgpg/en-GB/Presupuestos/Documentacion/Paginas/SSM-MATLAB.aspx

Gomez-ibanez, J. A. (1985). A Dark Side to Light Rail? The Experience of Three New Transit Systems. *Journal of the American Planning Association, 51*(3), 337–351. doi:10.1080/01944368508976421

Gonzalez, R.C., Woods, R.E., & Eddins, S.L. (2004). Digital image processing using MATLAB. Retrieved from http://course.sdu.edu.cn/Download/a273bf58-2fc1-4bbf-92a3-fbaa8c54f534.pdf

Goodall, R. M., & Mei, T. X. (2006). Active Suspensions. In: S. Iwnicki (Ed.), Handbook of railway vehicle dynamics (pp. 327-357). Boca Raton, FL: CRC Press.

Goodall, R. M., Bruni, S., & Mei, T. X. (2006). Concepts and prospects for actively controlled railway running gear. *Vehicle System Dynamics, 44*(Suppl.), 60–70. doi:10.1080/00423110600867374

Goossens, J. W. (2004). Models and Algorithms for Railway Line Planning Problems [PhD thesis]. Universiteit Maastricht.

Gorman, M. (1998). An Application of Genetic and Tabu searches to the Freight Railroad Operating Plan Problem. *Annals of Operations Research, 78*(19), 51–69. doi:10.1023/A:1018906301828

Grandjean, E. (1979). Fatigue in industry. *British Journal of Industrial Medicine, 36*, 175–186. doi:10.2105/AJPH.12.3.212 PMID:40999

Greathead, J. (1896). The City and South London Railway; with some Remarks upon Subaqueous Tunnelling by Shield and Compressed Air. London.

Grech, R., Cassar, T., Muscat, J., Camilleri, K. P., Fabri, S. G., Zervakis, M., & Vanrumste, B. et al. (2008). Review on solving the inverse problem in EEG source analysis. *Journal of Neuroengineering and Rehabilitation, 5*(1), 25. doi:10.1186/1743-0003-5-25 PMID:18990257

Greeley, H., & Friets, E. (2006). Detecting fatigue from voice using speech recognition. *Proceedings of the 2006 IEEE International Symposium on Signal Processing and Information technology.* Retrieved from http://ieeexplore.ieee.org/xpls/abs_all.jsp?arnumber=4042307

Gretzschel, M., & Bose, L. (1999). A mechatronic approach for active influence on railway vehicle running behaviour. *Vehicle System Dynamics, 33*(Suppl.), 418–430.

Gretzschel, M., & Bose, L. (2002). A new concept for integrated guidance and drive of railway running gears. *Control Engineering Practice*, *10*(9), 1013–1021. doi:10.1016/S0967-0661(02)00046-1

Grimm, M., & Treibech, C. (2012). Determinants of road traffic crash fatalities across Indian States. *Institute of Social Studies Working Paper Series/General Series*, Working Paper 531. The Hague, Netherlands.

Grube, P., Núñez, F., & Cipriano, A. (2011). An event-driven simulator for multi-line metro systems and its application to Santiago de Chile metropolitan rail network. *Simulation Modelling Practice and Theory*, *19*(1), 393–405. doi:10.1016/j.simpat.2010.07.012

Guha, S., Plarre, K., Lissner, D., Mitra, S., Krishna, B., Dutta, P., & Kumar, S. (2010, November). Autowitness: locating and tracking stolen property while tolerating GPS and radio outages. Proceedings of the 8th ACM conference on embedded networked sensor systems (pp. 29-42). ACM. doi:10.1145/1869983.1869988

Guidance on the preparation of risk assessment within railway safety cases. (2002). Railway Safety Railway Group Guidance Note – GE/GN8561, 1.

Guide to Cost-Benefit Analysis of Investment Projects. (2008). *European Commission*. Retrieved from http://ec.europa.eu/regional_policy/sources/docgener/guides/cost/guide2008_en.pdf

Guo, Y. N., Shi, X. P., & Zhang, X. D. (2010). A study of short term forecasting of the railway freight volume in China using ARIMA and Holt-Winters models. *Proceedings of the 8th International Conference on Supply Chain Management on Information Systems*.

Gupta, S., Kar, S., Gupta, S., & Routray, A. (2010). Fatigue in human drivers: A study using ocular, psychometric, physiological signals. *Proceedings of the 2010 IEEE Students'Technology SymposiumTechSym '10* (pp. 234–240). Doi:<ALIGNMENT.qj></ALIGNMENT>10.1109/TECHSYM.2010.5469152

Gutsche, K., & Böhm, T. (2011). e-Maintenance of Railway Assets Based on a Reliable Condition Prediction. *International Journal of Performability Engineering*, *7*(6), 573–582.

Gyftakis, K. N., & Kappatou, J. C. (2013). A Novel and Effective Method of Static Eccentricity Diagnosis in Three-Phase PSH Induction Motors. *IEEE Transactions on Energy Conversion*, *28*(2), 405–412. doi:10.1109/TEC.2013.2246867

Halliday, S. (2001). *Underground to Everywhere. London's underground railway in the life of the capital*. Sutton Publishing London Transport Museum.

Hammoud, R., Witt, G., Dufour, R., Wilhelm, A., & Newman, T. (2008). On driver eye closure recognition for commercial vehicles. Retrieved from http://papers.sae.org/2008-01-2691/

Hanowski, R., & Bowman, D. (2008). PERCLOS+: Development of a robust field measure of driver drowsiness. *Proceedings of the 15th World Congress on Intelligent transport systems and ITS America's 2008 Annual meeting*. Retrieved from http://trid.trb.org/view.aspx?id=904975

Hansen, I., & Pachl, J. (2014). Railway Timetabling & Operations: Analysis, Modelling, Optimisation, Simulation, Performance Evaluation (2nd ed.). Hamburg: Eurailpress/DVV Media Group.

Hansen, I. A., & Pachl, J. (2014). *Railway Timetabling & Operations. Analysis - Modelling - Optimisation - Simulation - Performance Evaluation*. Germany: Eurailpress.

Hansen, I., & Pachl, J. (2008). *Railway timetable and traffic: analysis, modelling, simulation*. Hamburg, Germany: Eurailpress.

Compilation of References

Hansen, M. (1990). Airline competition in a hub-dominated environment: An application of non-cooperative game theory. *Transportation Research Part B: Methodological*, *24*(1), 27–43. doi:10.1016/0191-2615(90)90030-3

Hansen, P., & Mladenović, N. (2001). Variable neighborhood search: Principles and applications. *European Journal of Operational Research*, *130*(3), 449–467. doi:10.1016/S0377-2217(00)00100-4

Hansen, P., Mladenović, N., & Pérez, J. A. M. (2010). Variable neighbourhood search: Methods and applications. *Annals of Operations Research*, *175*(1), 367–407. doi:10.1007/s10479-009-0657-6

Hanson, C. S., Noland, R. B., & Brown, C. (2013). The severity of pedestrian crashes: An analysis using Google Street View imagery. *Journal of Transport Geography*, *33*, 42–53. doi:10.1016/j.jtrangeo.2013.09.002

Harold, L., & Turoff, M. (1975). *The Delphi Method: Techniques and Applications*. Reading, Mass.: Addison-Wesley.

Harris, R. G. (1977). Economies of traffic density in the rail freight industry. *The Bell Journal of Economics*, *8*(2), 556–564. doi:10.2307/3003304

Harvey, C., & Aultman-Hall, L. (2015). Urban Streetscape Design and Crash Severity. *Proceedings of the Transportation Research Board 94th Annual Meeting* (No. 15-2942).

Hasannayebi, E., Sajedinejad, A., Mardani, S., & Mohammadi, K. (2012). *An integrated simulation model and evolutionary algorithm for train timetabling problem with considering train stops for praying*. Paper presented at the Simulation Conference (WSC). *Proceedings of the*, *2012*(Winter).

Hassannayebi, E., Sajedinejad, A., & Mardani, S. (2014). Urban rail transit planning using a two-stage simulation-based optimization approach. *Simulation Modelling Practice and Theory*, *49*, 151–166. doi:10.1016/j.simpat.2014.09.004

Hauser, D., & Tadikamalla, P. (1996). The analytic hierarchy process in an uncertain environment: A simulation approach. *European Journal of Operational Research*, *91*(1), 27–37. doi:10.1016/0377-2217(95)00002-X

Hayami, T., Matsunaga, K., Shidoji, K., & Matsuki, Y. (2002). Detecting drowsiness while driving by measuring eye movement - a pilot study. *Proceedings of the IEEE 5th International Conference on Intelligent Transportation Systems*. Doi:<ALIGNMENT.qj></ALIGNMENT>10.1109/ITSC.2002.1041206

Hayes, M. H. (2010). *Statistical Digital Signal Processing and Modeling*. John Wiley & Sons.

Heilporn, L., De Giovanni, L., & Labbé, M. (2008). De Giovanni, and M. Labbè. Optimization Models for the Single Delay Management Problem in Public Transportation. *European Journal of Operational Research*, *189*(3), 762–774. doi:10.1016/j.ejor.2006.10.065

Henao, H., Demian, C., & Capolino, G.-A. (2003). A frequency-domain detection of stator winding faults in induction machines using an external flux sensor. *IEEE Transactions on Industry Applications*, *39*(5), 1272–1279. doi:10.1109/TIA.2003.816531

Hicks, J. (1941). The rehabilitation of consumers' surplus. *The Review of Economic Studies*, *8*(2), 108–116. doi:10.2307/2967467

Higgins, A., Kozan, E., & Ferreira, L. (1997). Heuristic techniques for single line train scheduling. *Journal of Heuristics*, *3*(1), 43–62. doi:10.1023/A:1009672832658

Hill, R. J. (1994). Electric railway traction. Part 3 Traction power supplies. *Power Engineering Journal*, *8*(6), 275–286. doi:10.1049/pe:19940604

Hindrayanto, I., Koopman, S. J., & Ooms, M. (2010). Exact maximum likelihood estimation for non-stationary periodic time series models. *Computational Statistics & Data Analysis*, *54*(11), 2641–2654. doi:10.1016/j.csda.2010.04.010

Hoh, B., Gruteser, M., Herring, R., Ban, J., Work, D., Herrera, J., ..., Jacobson, Q. (2008). Virtual trip lines for distributed privacy-preserving traffic monitoring. Proceedings of the 6th international conference on mobile systems, applications and services Mobisys '08.

Höjer, M. (2000). What is the Point of IT? Backcasting Urban Transport and Land-use Futures. Royal Institute of Technology, Stockholm (Trita-IP. FR; 00-72)

Homnan, B., & Benjapolakul, W. (2004). Application of fuzzy inference to CDMA soft handoff in mobile communication systems. *Fuzzy Sets and Systems*, *144*(2), 345–363. doi:10.1016/S0165-0114(03)00206-9

Hong, D., Lee, H., & Kwak, J. (2007). Development of a mathematical model of a train in the energy point of view, *Paper presented at international conference on control, automation and systems (ICCAS 2007)*, Seoul, Korea.

Hong, T., & Qin, H. (2007). Drivers drowsiness detection in embedded system. *Proceedings of the IEEE International conference on Vehicular Electronics and Safety ICVES '07*. Retrieved from http://ieeexplore.ieee.org/xpls/abs_all.jsp?arnumber=4456381

Horeau, H. (1845). *Examen critique du projet d'agrandissement et de construction des Halles Centrales d'Approvisionnement pour la Ville de Paris*. Paris.

Horeau, H. (1846). *Nouvelles observations sur le projet d'agrandissement et de construction des Halles Centrales d'Approvisionnement pour la Ville de Paris*. Paris.

Hossain, M., & Rahman, M. S. (2007). Economic feasibility of Dhaka-Laksam direct railway link. *Journal of Civil Engineering*, *35*(1), 47–58.

Ho, T. K., Chi, Y. L., Wang, J., Leung, K. K., Siu, L. K., & Tse, C. T. (2005). Probabilistic load flow in AC electrified railways. *IEE Proceedings. Electric Power Applications*, *152*(4), 1003–1013. doi:10.1049/ip-epa:20045091

Houghton, K., Miller, E., & Foth, M. (2013). Integrating ICT into the planning process: Impacts, opportunities and challenges. *Australian Planner*, *51*(1), 24–33. doi:10.1080/07293682.2013.770771

Howlett, P. G., & Pudney, P. (1995). *Energy-Efficient Train Control*. Berlin, Heidelberg, New York: Springer. doi:10.1007/978-1-4471-3084-0

Hsiao, M., Malhotra, A., Thakur, J. S., Sheth, J. K., Nathens, A. B., Dhingra, N., & Jha, P. Million Death Study Collaborators. (2013). Road traffic injury mortality and its mechanisms in India: Nationally representative mortality survey of 1.1 million homes. *BMJ Open*, *8*, 1–9. PMID:23959748

HSRC. (n. d.). Retrieved from http://hsrc.in/backgroundBudgetSpeech.html

Huang, S., An, M., & Baker, C. J. (2005). A fuzzy based approach to risk assessment for track maintenance incorporated with AHP. Proceedings of International Railway Engineering Conference, London, Edinburgh: Engineering Technical Press.

Huang, R. S., Jung, T. P., Delorme, A., & Makeig, S. (2008). Tonic and phasic electroencephalographic dynamics during continuous compensatory tracking. *NeuroImage*, *39*(4), 1896–1909. doi:10.1016/j.neuroimage.2007.10.036 PMID:18083601

Huang, R. S., Tsai, L. L., & Kuo, C. J. (2001). Selection of valid and reliable EEG features for predicting auditory and visual alertness levels. *Proceedings of the National Science Council, Republic of China. Part B, Life Sciences*, *25*, 17–25. PMID:11254168

Huang, X., Habetler, T. G., Harley, R. G., & Wiedenbrug, E. J. (2007). Using a Surge Tester to Detect Rotor Eccentricity Faults in Induction Motors. *IEEE Transactions on Industry Applications*, *43*(5), 1183–1190. doi:10.1109/TIA.2007.904389

Hueper, J., Dervisoglu, G., Muralidharan, A., Gomes, G., Horowitz, R., & Varaiya, P. (2009). Macroscopic Modeling and Simulation of Freeway Traffic Flow. *Proceedings of the 12th IFAC Symposium on Control in Transportation Systems* (pp. 112-116).

Huet, E. (1896). Métropolitain urbain à traction électrique: Rapport du Directeur Administratif des Travaux. Paris, 2 March 1896, plus Annexe au rapport, Paris, 23 March 1896. Régie Autonome des Transports Parisiens.

Huet, O. (1878). *Les Chemins de Fer Métropolitains de Londres. Étude d'un réseau de chemins de fer métropolitains pour la ville de Paris. Mission à Londres en mai, 1876.* Paris.

Hüisman, D., Kroon, L. G., Lentink, R. M., & Vromans, M. C. J. M. (2005). *Operations Research in Passenger Railway Transportation. Technical report, Erasmus Research Institute of Management (ERIM).* Rotterdam School of Management.

Hullinger, D. R. (1999). Taylor enterprise dynamics. *Paper presented at theSimulation Conference.*

Hunt, D. (2005). *Return on Investment on Freight Rail Capacity Improvement.* United Kingdom: Cambridge Systematics. Inc.

Hwang, J.-G., & Jo, H.-J. (2010). Deduction of Coding Rules and Implementation of Automatic Inspection Tool for Railway Signaling Software. *Proceedings of theInternational Congress on Railtransport Technology* (pp. 81-85).

Hwang, J.-G., Cho, H.-J., & Kim, B.-H. (2013, May). Results of coding rules testing of train control system software. *Int'l Journal of Software Engineering and Its Applications, 7,* 249–259.

Hwang, J.-G., & Jo, H.-J. (2009). Development of automated testing tool for coding standard to check the safety of train control system software. *Journal of the Korean Society for Railway, 12,* 81–87.

Hwang, J.-G., & Jo, H.-J. (2009). Development of the automated metric analysis tool for train control system software. *Journal of the Korean Society for Railway, 12,* 450–456.

Hwang, J.-G., & Jo, H.-J. (June 2008). Development of automatic testing tool for design & coding standard for railway signal system software.*Proceedings of ICCMSE*2008.

Hwang, J.-G., Jo, H.-J., Kim, B.-H., & Baek, J.-H. (2013, April). Applicability analysis of software testing for actual operating railway software. *Proceedings of SoftTech* (Vol. *2013,* pp. 257–278).

Hwang, J.-G., Jo, H.-J., & Kim, H.-S. (2008). Design of the safety assessment tool for train control system software. *Journal of the Korean Society for Railway, 11,* 139–144.

Hyun, D., Hong, J., Lee, S. B., Kim, K., Wiedenbrug, E. J., Teska, M., & Nandi, S. (2011). Automated Monitoring of Airgap Eccentricity for Inverter-Fed Induction Motors Under Standstill Conditions. *IEEE Transactions on Industry Applications, 47*(3), 1257–1266. doi:10.1109/TIA.2011.2126010

Ibraheem, A. Th. (2011). Evaluating light-rail transit alternatives using the rating and ranking method. *Journal of Engineering and Applied Sciences, 6*(10), 93–104.

Ibrahim, A., El Badaoui, M., Guillet, F., & Bonnardot, F. (2008). A new bearing fault detection method in induction machines based on instantaneous power factor. *IEEE Transactions on Industrial Electronics, 55*(12), 4252–4259. doi:10.1109/TIE.2008.2003211

IEC 61508-3. (1998). IEC.

IEC 62279. (2002). Railway Applications – software for Railway Control and Protection Systems.

IEC 62425. (2005). *Railway Application: Communications, signaling and processing systems - Safety related electronic system for signaling*.

Inchingolo, P., & Spanio, M. (1985). On the identification and analysis of saccadic eye movements-A quantitative study of the processing procedures. *IEEE Transactions on Biomedical Engineering*, BME-32(9), 683-695. Retrieved from http://ieeexplore.ieee.org/xpls/abs_all.jsp?arnumber=4122143

Isermann, R. (2006). *Fault-diagnosis systems*. Springer. doi:10.1007/3-540-30368-5

ISO. (03 2012). EN 50128. *Railway applications - Communication, signalling and processing systems - Software for railway control and protection systems*.

ISO. (2000). *EN 50126. Railway applications - The specification and demonstration of reliability, availability, maintainability and safety*. RAMS.

Iyer, R.V.; Ghosh, S. (1995, February). DARYN-a distributed decision-making algorithm for railway networks: modeling and simulation. *IEEE Transactions on Vehicular Technology*, 44(1), 180-191.

Jackson, A. (1986). *London's Metropolitan Railway*. Newton Abbot: David & Charles.

Jackson, C. A., & Earl, L. (2006). Prevalence of fatigue among commercial pilots. *Occupational Medicine (Oxford, England)*, 56(4), 263–268. doi:10.1093/occmed/kql021 PMID:16733255

Jahn, M. (1982). Suburban development in Outer West London, 1850-1900. In F.M.L. Thompson (Ed.), *The rise of suburbia*. Leicester: Leicester University Press.

Jamili, A., Ghannadpour, S., & Ghorshinezhad, M. (2014). The optimization of train timetable stop-skipping patterns in urban railway operations.

Jap, B. T., Lal, S., Fischer, P., & Bekiaris, E. (2007). Using spectral analysis to extract frequency components from electroencephalography: Application for fatigue countermeasure in train drivers. *Proceedings of the 2nd International Conference on Wireless Broadband and Ultra Wideband Communications, AusWireless '07*. Doi:<ALIGNMENT.qj></ALIGNMENT>10.1109/AUSWIRELESS.2007.83

Jawaharlal Nehru National Urban Renewal Mission. (2005). New Delhi: Overview.

Jespersen-Groth, J., Pottho, D., Clausen, J., Huisman, D., Kroon, L. G., Maroti, G., & Nielsen, M. N. (2009). Disruption Management in Passenger Railway Transportation. Robust and Online Large-Scale Optimization. Springer Berlin / Heidelberg.

Jia, W., Mao, B., Liu, H., Chen, S., & Ding, Y. (2008). Service robustness analysis of trains by a simulation method. *Paper presented at the Traffic and Transportation Studies*. doi:10.1061/40995(322)70

Jillella, S., Newman, P., & Matan, A. (2014). Participatory Sustainability Approach to Value Capture Based Urban Rail Financing in India Through Deliberated Stakeholder Engagement.*Proceedings of the 4th World Sustainability Forum* (Vol. 4). doi:10.3390/wsf-4-b002

Jin, X., Zhao, M., Chow, T. W. S., & Pecht, M. (2014). Motor bearing fault diagnosis using trace ratio linear discriminant analysis. *IEEE Transactions on Industrial Electronics*, 61(5), 2441–2451. doi:10.1109/TIE.2013.2273471

Ji, Q., & Yang, X. (2002). Real-Time Eye, Gaze, and Face Pose Tracking for Monitoring Driver Vigilance. *Real-Time Imaging*, 8(5), 357–377. doi:10.1006/rtim.2002.0279

Jiuran, H., & Bingfeng, S. (2013). The application of ARIMA-RBF model in urban rail traffic volume forecast.*Proceedings of the 2nd International Conference on Computer Science and Electronic Engineering* (pp. 1662-1665).

Joksimovic, G. M., & Penman, J. (2000). The detection of inter-turn short circuits in the stator windings of operating motors. *IEEE Transactions on Industrial Electronics*, *47*(5), 1078–1084. doi:10.1109/41.873216

Jones, C. (2004). *Paris Biography of a City*. Penguin Books.

Jong, G., Daly, A., Pieters, M., & Hoorn, T. (2006). The logsum as an evaluation measure: Review of the literature and new results. *Transportation Research Part A: Policy and Practice*, 41(9), 874–889.

Jovanović, P. (2005). Upravljanje investicijama. Faculty of Organizational Sciences, Serbia.

Jovanovic, D., & Harker, P. (1991). Tactical Scheduling of Rail Operations: SCAN I system. *Transportation Science*, *25*(1), 46–64. doi:10.1287/trsc.25.1.46

Jung, T. P., Makeig, S., Stensmo, M., & Sejnowski, T. J. (1997). Estimating alertness from the EEG power spectrum. *IEEE Transactions on Bio-Medical Engineering*, *44*(1), 60–69. doi:10.1109/10.553713 PMID:9214784

Jurdak, R., Corke, P., Dharman, D., & Salagnac, G. Adaptive GPS duty cycling and radio ranging for energy-efficient localization. *Proceedings of the 8th ACM conference on embedded networked sensor systems* (pp. 57-70). ACM. doi:10.1145/1869983.1869990

Kadry, S. (2011). A New Proposed Technique to Improve Software Regression Testing Cost. *International Journal of Security and Its Applications*, *5*, 45–48.

Kaikaa, M. Y., Hadjami, M., & Khezzar, A. (2014). Effects of the Simultaneous Presence of Static Eccentricity and Broken Rotor Bars on the Stator Current of Induction Machine. *IEEE Transactions on Industrial Electronics*, *61*(5), 2452–2463. doi:10.1109/TIE.2013.2270216

Kalker, J. J. (1990). *Three Dimensional Elastic Bodies in Rolling Contact*. Dordrecht, Netherlands: Kluwer. doi:10.1007/978-94-015-7889-9

Kalman, R. (1960). A new approach to linear filtering and prediction problems. *Journal of Basic Engineering*, *82*(1), 35–45. doi:10.1115/1.3662552

Kang, T.-, Kim, J., Lee, S. B., & Yung, C. (2015). Experimental Evaluation of Low-Voltage Offline Testing for Induction Motor Rotor Fault Diagnostics. *IEEE Transactions on Industry Applications*, *51*(2), 1375–1384. doi:10.1109/TIA.2014.2344504

Kankar, P. K., Sharma, S. C., & Harsha, S. P. (2011). Fault diagnosis of ball bearings using machine learning methods. *Expert Systems with Applications*, *38*(3), 1876–1886. doi:10.1016/j.eswa.2010.07.119

Kar, S., Bhagat, M., & Routray, A. (2010). EEG signal analysis for the assessment and quantification of driver's fatigue. *Transportation Research Part F: Traffic Psychology and Behaviour*, *13*(5), 297–306. doi:10.1016/j.trf.2010.06.006

Kar, S., & Routray, A. (2013). Effect of Sleep Deprivation on Functional Connectivity of EEG Channels. *IEEE Transactions on Systems, Man, and Cybernetics Systems*, *43*(3), 666–672. doi:10.1109/TSMCA.2012.2207103

Keaton, M. H. (1989). Designing optimal railroad operating plans: Lagrangian relaxation and Heuristic approaches. *Transportation research Part B: Methodological*, 23 B(6), 415-431.

Keeler, T. E. (1974). Railroad costs, returns to scale, and excess capacity. *The Review of Economics and Statistics*, *56*(2), 201–208. doi:10.2307/1924440

Kellett, J. R. (1979). Railways and Victorian Cities. London et al. Routledge.

Kelton, W., Sadowski, R., & Swets, N. (2009). *Simulation with Arena*. USA: McGraw-Hill Higher Education.

Kennedy, A. (1997). Risk management and assessment for rolling stock safety cases. *Proc. Instn Mech. Engrs, Part F: J. Rail and Rapid Transit, 211*(1), 67–72. doi:10.1243/0954409971530914

Khan, M. B., & Zhou, X. (2009). Stochastic Optimization Model and Solution Algorithm for Robust Double-Track Train-Timetabling Problem. *IEEE Transactions on Intelligent Transportation Systems, 10*(3). doi:10.1109/TITS.2009.2030588

Khan, M., Radwan, T. S., & Rahman, M. A. (2007). Real-time implementation of wavelet packet transform-based diagnosis and protection of three-phase induction motors. *IEEE Transactions on Energy Conversion, 22*(3), 647–655. doi:10.1109/TEC.2006.882417

Kharas, H. (2010). The emerging middle class in developing countries *Global Development Outlook* (Working Paper No. 285). OECD Development Centre.

Kiessling, F., Puschmann, R., Schmieder, A., & Schneider, E. (2009). *Contact lines for electric railways. Planning, design implementation, Maintenance*. Erlangen, Germany: Publicis Publishing.

Kim, M.S., Park, J.H. & You, W.H. (2008). Construction of active steering system of scaled railway vehicle. *International journal of systems applications, engineering and development, 4*(2), 217-226.

Kim, D.-J., Kim, H.-J., Hong, J.-P., & Park, C.-J. (2014). Estimation of Acoustic Noise and Vibration in an Induction Machine Considering Rotor Eccentricity. *IEEE Transactions on Magnetics, 50*(2), 857–860. doi:10.1109/TMAG.2013.2285391

Kim, J., Member, S., Shin, S., Member, S., Lee, S. B., Member, S., & Member, S. et al. (2015). *Power Spectrum-Based Detection of Induction Motor Rotor Faults for Immunity to False Alarms. IEEE Trans* (pp. 1–10). Energy Convers.

Kim, Y.-H., Youn, Y.-W., Hwang, D.-H., Sun, J.-H., & Kang, D.-S. (2013). High-Resolution Parameter Estimation Method to Identify Broken Rotor Bar Faults in Induction Motors. *IEEE Transactions on Industrial Electronics, 60*(9), 4103–4117. doi:10.1109/TIE.2012.2227912

Kiymik, M. K., Akin, M., & Subasi, A. (2004). Automatic recognition of alertness level by using wavelet transform and artificial neural network. *Journal of Neuroscience Methods, 139*(2), 231–240. doi:10.1016/j.jneumeth.2004.04.027 PMID:15488236

Kleijnen, J. P., & Wan, J. (2007). Optimization of simulated systems: OptQuest and alternatives. *Simulation Modelling Practice and Theory, 15*(3), 354–362. doi:10.1016/j.simpat.2006.11.001

Klimesch, W. (1999). EEG alpha and theta oscillations reflect cognitive and memory performance: A review and analysis. *Brain Research. Brain Research Reviews, 29*(2-3), 169–195. doi:10.1016/S0165-0173(98)00056-3 PMID:10209231

Kneschke, T. A. (1985). Control of utility system unbalance caused by single-phase electric traction. *IEEE Transactions on Industry Applications, 21*(6), 1559–1570. doi:10.1109/TIA.1985.349618

Knight, A. M., & Bertani, S. P. (2005). Mechanical Fault Detection in a Medium-Sized Induction Motor Using Stator Current Monitoring. *IEEE Transactions on Energy Conversion, 20*(4), 753–760. doi:10.1109/TEC.2005.853731

Kolonko, M., & Engelhardt-Funke, O. (2001). Cost-Benefit Analysis of Investments into Railway Networks with Periodically Timed Schedules. *Computer-Aided Scheduling of Public Transport Lecture Notes in Economics and Mathematical Systems, 505*, 443–459. doi:10.1007/978-3-642-56423-9_25

Konstadinos, G.G. (2007, June). *Travel Behavior and Demand Analysis and Prediction* (Working Paper). Department of Geography, University of California Santa Barbara.

Koselleck, R. (2004). Futures Past on the Semantics of Historical Time. Translated and with an Introduction by Keith Tribe. New York: Columbia University Press. originally published in German in 1979

Compilation of References

Koukoumidis, E., Peh, L., & Martonosi, M. (2011, June). Signalguru: Leveraging mobile phones for collaborative traffic signal schedule advisory. Proceedings of Mobisys.

Koutsopoulos, H. N., & Wang, Z. (2007). Simulation of urban rail operations: Application framework. *Transportation Research Record: Journal of the Transportation Research Board, 2006*(1), 84–91. doi:10.3141/2006-10

Krajewski, J., Wieland, R., & Batliner, A. (2008). An acoustic framework for detecting fatigue in speech based human-computer-interaction. In Computers Helping People with Special Needs, LNCS (Vol. 5105, pp. 54–61). doi:10.1007/978-3-540-70540-6_7

Krajewski, J., Batliner, A., & Golz, M. (2009). Acoustic sleepiness detection: Framework and validation of a speech-adapted pattern recognition approach. *Behavior Research Methods, 41*(3), 795–804. doi:10.3758/BRM.41.3.795 PMID:19587194

Krajewski, J., Schnieder, S., Sommer, D., Batliner, A., & Schuller, B. (2012). Applying multiple classifiers and non-linear dynamics features for detecting sleepiness from speech. *Neurocomputing, 84*, 65–75. doi:10.1016/j.neucom.2011.12.021

Krishnaveni, V., Jayaraman, S., Aravind, S., Hariharasudhan, V., & Ramadoss, K. (2006). Automatic identification and Removal of ocular artifacts from EEG using Wavelet transform. *Measurement Science Review, 6*, 45–57. Retrieved from http://www.freewebs.com/biomedical-eng/scprs/2.pdf

Krishnaveni, V., Jayaraman, S., Anitha, L., & Ramadoss, K. (2006). Removal of ocular artifacts from EEG using adaptive thresholding of wavelet coefficients. *Journal of Neural Engineering, 3*(4), 338–346. doi:10.1088/1741-2560/3/4/011 PMID:17124338

Krumholz, N. (1982). A Retrospective View of Equity Planning Cleveland 1969–1979. *Journal of the American Planning Association, 48*(2), 163–174. doi:10.1080/01944368208976535

Kulshreshtha, M., Nag, B., & Kulshreshtha, M. (2001). A Multivariate Cointegrating Vector Auto Regressive Model of Freight Transport Demand: Evidence from Indian Railways. *Transport Research Part A, 35*, 29–45.

Kumar, S., Pahuja, D., Bakre, A., & Saha, S. K. (2003). Prediction of unsteady heatgains using SES analysis in an interchange subway station of the Delhi metro. *Proceedings of the11th International Symposium on Aerodynamics & Ventilation in Vehicle Tunnels.*

Kuo, R. J., Chi, S. C., & Kao, S. S. (1999). A decision support system for locating convenience store through fuzzy AHP. *Computers & Industrial Engineering, 37*(1-2), 323–326. doi:10.1016/S0360-8352(99)00084-4

Kuo, R. J., Chi, S. C., & Kao, S. S. (2002). A decision support system for selecting convenience store location through integration of fuzzy AHP and artificial neural network. *Computers in Industry, 47*(2), 199–214. doi:10.1016/S0166-3615(01)00147-6

Kusakabe, T., Iryo, T., & Asakura, Y. (2010). Estimation method for railway passengers' train choice behavior with smart card transaction data. *Transportation, 37*(5), 731–749. doi:10.1007/s11116-010-9290-0

Laguna, M., & Marti, R. (2002). *The OptQuest callable library Optimization software class libraries* (pp. 193–218). Springer.

Lal, S. K. L., & Craig, A. (2001). A critical review of the psychophysiology of driver fatigue. *Biological Psychology, 55*(3), 173–194. doi:10.1016/S0301-0511(00)00085-5 PMID:11240213

Lal, S. K. L., Craig, A., Boord, P., Kirkup, L., & Nguyen, H. (2003). Development of an algorithm for an EEG-based driver fatigue countermeasure. *Journal of Safety Research, 34*(3), 321–328. doi:10.1016/S0022-4375(03)00027-6 PMID:12963079

Lambert, A. J. (1984). *Nineteenth Century Railway History through the Illustrated London News*. London: David and Charles.

Landis, J., Guathakurta, S., & Zhang, M. (1994). *Capitalization of transportation investments into single-family home prices* (Working paper 619). Institute of Urban and Regional Development, University of California, Berkeley, USA.

Lano, K. (1996). *The B Language and Method: A Guide to Practical Formal Development*. Springer. doi:10.1007/978-1-4471-1494-9

Larroque, D. (2002). Le Métropolitain: histoire d'un projet. In D. Larroque, M. Margairaz, P. Zembri, Paris et ses Transports XIXe – XXe siècles. Deux siècles de décisions pour la ville et sa région. Paris: Éditions Recherches, 41-94.

Lascelles, T. S. (1987). *The City & South London Railway*. Oxford: The Oakwood Press.

Lavedan, P. (1969). *La question du déplacement de Paris et du transfert des Halles au Conseil municipal sous la monarchie de Juillet*. Paris: Commission des Travaux Historiques.

Lawrence, J. D. (2000). Software qualification in safety applications. *Reliability Engineering & System Safety*, 70(2), 167–184. doi:10.1016/S0951-8320(00)00055-7

Leander, P. & Lukaszewicz, P. (2008). EETROP: Energy Efficient Train Operation-State of the art of train ECO-operation (Technical Report 20).

Lee, A. H. I., Chen, W. C., & Chang, C. J. (2008). A fuzzy AHP and BSC approach for evaluating performance of IT department in the manufacturing industry in Taiwan. *Expert Systems with Applications*, 34(1), 96–107. doi:10.1016/j.eswa.2006.08.022

Lee, C. E. (1973). *The Northern Line. A Brief History*. London: London Transport.

Lee, J.-D., Jung, J.-I., Lee, J.-H., Hwang, J.-G., Hwang, J.-H., & Kim, S.-U. (2007). Verification and conformance test generation of communication protocol for railway signaling systems. *Computer Standards & Interfaces*, 29(2), 143–151. doi:10.1016/j.csi.2006.03.001

Lee, J.-H., Hwang, J.-G., & Park, G.-T. (2005). Performance evaluation and verification of communication protocol for railway signaling systems. *Computer Standards & Interfaces*, 27(3), 207–219. doi:10.1016/S0920-5489(04)00097-2

Lee, S. Y., & Cheng, Y. C. (2005). Hunting stability analysis for high-speed railway vehicle trucks on tangent tracks. *Journal of Sound and Vibration*, 282(3-5), 881–898. doi:10.1016/j.jsv.2004.03.050

Lee, T., & Ghosh, S. (1998, November). RYNSORD: A novel, decentralized algorithm for railway networks with "soft reservation." *IEEE Transactions on Vehicular Technology*, 47(4), 201–222. doi:10.1109/25.728526

Lee, Y., & Chen, C.-Y. (2009). A heuristic for the train pathing and timetabling problem. *Transportation Research Part B: Methodological*, 43(8-9), 837–851. doi:10.1016/j.trb.2009.01.009

LeGoff, G. (1996). Using synchronous languages for interlocking. *First Int. Conf. on Computer Application in Transportation Systems*.

Leite, V. C. M. N., Borges da Silva, J. G., Veloso, G. F. C., Borges da Silva, L. E., Lambert-Torres, G., Bonaldi, E. L., & de Oliveira, L. E. D. L. (2015). Detection of Localized Bearing Faults in Induction Machines by Spectral Kurtosis and Envelope Analysis of Stator Current. *IEEE Transactions on Industrial Electronics*, 62(3), 1855–1865. doi:10.1109/TIE.2014.2345330

Lemoine, B. (1980). Les Halles de Paris. L'histoire d'un lieu, les péripéties d'une reconstruction, la succession des projets, l'architecture d'un monument, l'enjeu d'une "Cité". Paris: L'Equerre.

Compilation of References

Lenné, M. G., Triggs, T. J., & Redman, J. R. (1997). Time of day variations in driving performance. *Accident; Analysis and Prevention, 29*(4), 431–437. doi:10.1016/S0001-4575(97)00022-5 PMID:9248501

Leonard Janer, L., Bonet, J. J., & Lleida-Solano, E. (1996). Pitch detection and voiced/unvoiced decision algorithm based on wavelet transforms. *Proceedings of theFourth International Conference on Spoken Language Processing* (Vol. 2, p. 1209). doi:10.1109/ICSLP.1996.607825

Levinson, H., Zimmerman, S., Clinger, J., Rutherford, S., Smith, R. L., Cracknell, J., & Soberman, R. (2003). Case Studies in Bus Rapid Transit Bus Rapid Transit (TCRP Report 90). Washington, D.C.

Levinson, D. M., & Istrate, E. (2011). *Access for value: financing transportation through land value capture*. The Brookings Institute, Transportation Research Board.

Levinson, D., Mathieu, J. M., Gillen, D., & Kanafani, A. (1997). The full cost of high speed rail: An engineering approach. *The Annals of Regional Science, 31*(2), 189–215. doi:10.1007/s001680050045

Lewin, H. G. (1936). *Railway mania and its aftermath: 1845-1852*. London: Railway Gazette.

Li Zhu, Yu F.R., Bing Ning, & Tao Tang. (2014). Communication-based train control (CBTC) systems with cooperative relaying: design and performance analysis. *IEEE Vehicular Technology, 63*, 2162-2172.

Li, Y., & Gendreau, M. (1991). Real time scheduling on a transit bus route: a 0-1 stochastic programming model.

Liebchen & Stiller. S. (2009) Delay resistant timetabling. *Public transport*, 1(1):55-72. doi:10.1007/s12469-008-0004-3

Liebchen, C. (2008). The first optimized railway timetable in practice. *Transportation Science, 42*(4), 420–435. doi:10.1287/trsc.1080.0240

Li, M., Chiasson, J., Bodson, M., & Tolbert, L. M. (2006). A Differential-Algebraic Approach to Speed Estimation in an Induction Motor. *IEEE Transactions on Automatic Control, 51*(7), 1172–1177. doi:10.1109/TAC.2006.878775

Lin, W., & Zeng, J. (1999). Experimental study of real-time bus arrival time prediction with GPS data. *Transportation Research Record*, 1666, 101–109.

Lindahl, M. (2001), Track geometry for high speed railways: A literature survey & simulation of dynamic vehicle response.

Lindner, T. & Zimmermann, U. (2005). Cost Optimal Periodic Train Scheduling. *Mathematical Methods of Operations Research*, 62(2), 281-295.

Lin, W., & Sheu, J. (2010). Automatic train regulation for metro lines using dual heuristic dynamic programming. *Proceedings of the Institution of Mechanical Engineers. Part F, Journal of Rail and Rapid Transit, 224*(1), 15–23. doi:10.1243/09544097JRRT283

Li, S., & Wang, C. (2013). Evaluation of Postponement Strategies in Benetton using OptQuest Simulation. *International Journal of Advancements in Computing Technology*, 5(7).

Litman, T. (2007). Evaluating rail transit benefits: A comment. *Transport Policy, 14*(1), 94–97. doi:10.1016/j.tranpol.2006.09.003

Liu, P., Hung, C.-Y., Chiu, C.-S., & Lian, K.-Y. (2011). Sensorless linear induction motor speed tracking using fuzzy observers. *IET Electr. Power Appl., 5*(4), 325. doi:10.1049/iet-epa.2010.0099

Liu, S. Q., & Kozan, E. (2009). Scheduling trains as a blocking parallel-machine job shop scheduling problem. *Computers & Operations Research, 36*(10), 2840–2852. doi:10.1016/j.cor.2008.12.012

Liu, Z., Yin, X., Zhang, Z., Chen, D., & Chen, W. (2004). Online rotor mixed fault diagnosis way based on spectrum analysis of instantaneous power in squirrel cage induction motors. *IEEE Transactions on Energy Conversion*, *19*(3), 485–490. doi:10.1109/TEC.2004.832052

Lobo, A., & Loizou, P. (2003). Voiced/unvoiced speech discrimination in noise using gabor atomic decomposition. *Acoustics, Speech, and Signal* .... Retrieved from http://ieeexplore.ieee.org/xpls/abs_all.jsp?arnumber=1198907

Lohia, S. K. (2011). Sustainable Urban Transport Sustainable Urban Transport; Initiatives by Govt. of India. *Paper presented at theThird Biennial Conference of the Indian Heritage Cities Network Karnataka*, India.

Lootsma, F. A. (1996). A model for the relative importance of the criteria in the multiplicative AHP and SMART. *European Journal of Operational Research*, *94*(3), 467–476. doi:10.1016/0377-2217(95)00129-8

Loparo, K. A., Adams, M. L., Lin, W., Abdel-Magied, M. F., & Afshari, N. (2000). Fault detection and diagnosis of rotating machinery. *IEEE Transactions on Industrial Electronics*, *47*(5), 1005–1014. doi:10.1109/41.873208

López Galviz, C. (2012). Converging Lines Dissecting Circles: Railways and the Socialist Ideal in London and Paris at the Turn of the Twentieth Century. In M. Davies & J. Galloway (Eds.), *London and Beyond: Essays in honour of Derek Keene* (pp. 317–337). London: Institute of Historical Research Series.

López Galviz, C. (2013a). Mobilities at a standstill: Regulating circulation in London c.1863-1870. *Journal of Historical Geography*, *42*, 62–76. doi:10.1016/j.jhg.2013.04.019

López Galviz, C. (2013b). Metropolitan Railways: Urban Form and the Public Benefit in London and Paris 1850-1880. *The London Journal*, *38*(2), 184–202. doi:10.1179/0305803413Z.00000000030

Luèthi, M. (2009). Improving the Efficiency of Heavily Used Railway Networks through Integrated Real-Time Rescheduling [PhD thesis]. Swiss Federal Institute Of Technology, ETH Zurich.

LUL. (2001). London Underground Limited Quantified Risk. *Assessment Update*, *2001*, 1.

Lutz, R.R. & Woodhouse, R.M. (1999). Bi-directional Analysis for Certification of Safety-Critical Software. Proceedings of 1st International Software Assurance Certification Conference. *Proceedings of the International Software Assurance Certification Conference* (pp. 1-9).

Macdonald, W. A. (1985). *Human Factors & Road Crashes - A Review of Their Relationship*.

Makeig, S., & Jung, T. P. (1995). Changes in alertness are a principal component of variance in the EEG spectrum. *Neuroreport*, *7*(1), 213–216. doi:10.1097/00001756-199512000-00051 PMID:8742454

Makitalo, M., & Hilmola, O. P. (2010). Analysing the future of railway freight competition: A Delphi study in Finland. *Foresight*, *12*(6), 20–37. doi:10.1108/14636681011089961

Mandel, B., Gaudry, M., & Rothengatter, W. (1997). A disaggregate Box-Cox logit mode choice model of intercity passenger travel in Germany and its implications for high-speed rail demand forecasts. *The Annals of Regional Science*, *31*(2), 99–120. doi:10.1007/s001680050041

Mannino, C., & Mascis, A. (2009). Optimal real-time traffic control in metro stations. *Operations Research*, *57*(4), 1026–1039. doi:10.1287/opre.1080.0642

Marchand, B. (1993). *Paris, histoire d'une ville XIXe – XXe siècles*. Paris: Éditions du Seil.

Mariscotti, A., Pozzobon, P., & Vanti, M. (2007). Simplified modeling of 2x25kV AT railway system for the solution of low frequency and large-scale problems. *IEEE Transactions on Power Delivery*, *22*, 296–301. doi:10.1109/TPWRD.2006.883020

Maròti. (2006). *Operations Research Models for Railway Rolling Stock Planning* [PhD thesis]. Technische Universiteit, Eindhoven, Amsterdam.

Marshall, A. (1898). *Principles of Economics* (4th ed.). London: Macmillan.

Martin, A. (1894). *Étude historique et statistique sur les moyens de transport dans Paris avec plans, diagrammes et cartogrammes*. Paris: Ministère de l'Instruction Publique et des Beaux-Arts.

Martinez, L.M.G., & Viegas, J.M. (2012). The value capture potential of the Lisbon subway. *The journal of transport and land use*, 5(1), 65-82.

Martin, G. (2004). *Past Futures: The Impossible Necessity of History*. Toronto, Buffalo, London: Toronto University Press.

Mascis, A., & Pacciarelli, D. (2002). Job-shop scheduling with blocking and no-wait constraints. *European Journal of Operational Research*, 143(3), 498–517. doi:10.1016/S0377-2217(01)00338-1

Maskeliunaite, L., Sivilevičius, H., & Podvezko, V. (2009). Research on the quality of passenger transportation by railway. *Transport*, 24(2), 100–112. doi:10.3846/1648-4142.2009.24.100-112

Matousek, M., & Petersen, I. (1983). A method for assessing alertness fluctuations from EEG spectra. *Electroencephalography and Clinical Neurophysiology*, 55(1), 108–113. doi:10.1016/0013-4694(83)90154-2 PMID:6185295

Matsumoto, A., Sato, Y., Ohno, H., Suda, Y., Michitsuji, Y., Komiyama, M., & Nakai, T. et al. (2005), Multi-body dynamics simulation and experimental evolution for active steering bogie. *Proc. Int'l symposium on speed up and service technology for railway and maglev systems* (pp. 103-107).

Matthews, G., Davies, D. R., Westerman, S. J., & Stammers, R. B. (2000). *Human performance: Cognition, stress, and individual differences*. East Sussex: Psychology Press.

Maycock, G. (1997). Sleepiness and driving: The experience of U.K. car drivers. *Accident; Analysis and Prevention*, 29(4), 453–462. doi:10.1016/S0001-4575(97)00024-9 PMID:9248503

Mazzarello, M., & Ottaviani, E. (2007). A Traffic Management System for Real-time Traffic Optimisation in Railways. *Transportation Research Part B: Methodological*, 41(2), 246–274. doi:10.1016/j.trb.2006.02.005

Mcfadden, D. (1974). Conditional logit analysis of qualitative choice behavior. In Frontiers of econometrics (pp. 105-142).

McIntosh, J. (2014). *Framework to capture the value created by urban transit in car dependent cities* [Unpublished doctoral dissertation]. Curtin University, Perth, Australia.

McIntosh, J., Newman, P., Crane, T., & Mouritz, M. (2011). *Alternative Funding Mechanisms for Public Transport in Perth: the Potential Role of Value Capture, Committee for Perth*. Retrieved from http://www.committeeforperth.com.au/pdf/Advocacy/Report%20-%20AlternativeFundingforPublicTransportinPerthDecember2011.pdf

McIntosh, J., Trubka, R., & Newman, P. (2014a). Can Value Capture Work in Car Dependent Cities? Willingness to pay for transit access in Perth, Western Australia. *Transport Research – Part A*, 67(September), 320–339.

McIntosh, J., Trubka, R., & Newman, P. (2014b). Tax Increment Financing framework for integrated transit and urban renewal projects in car dependent cities. *Urban Planning and Research Journal On-line*, 3(December). doi:10.1080/08111146.2014.968246

Medda, F. R. (2012). Land value capture finance for transport accessibility: A review. *Journal of Transport Geography*, 25, 154–161. doi:10.1016/j.jtrangeo.2012.07.013

Mees, A. I. (1991). Railway scheduling by network optimization. *Mathematical and Computer Modelling*, *15*(1), 33–42. doi:10.1016/0895-7177(91)90014-X

Mei, T. X., & Li, H. (2008), Control design for the active stabilization of rail wheel-sets. *Proceedings of theTransactions of the ASME Journal of Dynamic Systems, Measurement and Control*.

Mei, T. X., & Goodall, R. M. (1999). Wheelset control strategies for a two axle railway vehicle. *Vehicle System Dynamics*, *33*(Suppl.), 653–664.

Mei, T. X., & Goodall, R. M. (2000). LQG solution for active steering of solid axle railway vehicles. *IEE Proceedings. Control Theory and Applications*, *147*(1), 111–117. doi:10.1049/ip-cta:20000145

Mei, T. X., & Goodall, R. M. (2000). Modal controller for active steering of railway vehicles with solid axle wheel-sets. *Vehicle System Dynamics*, *34*(1), 25–41. doi:10.1076/0042-3114(200008)34:1;1-K;FT025

Mei, T. X., & Goodall, R. M. (2003). Practical strategies for controlling railway wheel-sets with independently rotating wheels, *Transactions of the ASME Journal of Dynamic Systems, Measurement and Control*, 125, 354–360.

Mei, T. X., & Goodall, R. M. (2003). Recent development in active steering of railway vehicles. *Vehicle System Dynamics*, *39*(6), 415–436. doi:10.1076/vesd.39.6.415.14594

Mei, T. X., & Goodall, R. M. (2006). Stability control of railway bogies using absolute stiffness–skyhook spring approach. *Vehicle System Dynamics*, *44*(Suppl.), 83–92. doi:10.1080/00423110600867440

Metronet SSL (2005b). *Framework for the assessment of HS&E risks* (Second Metronet SSL Interim Report).

Metronet SSL, (2005a). *Development of Metronet staff risk model* (First Metronet SSL Interim Report).

Meyer zu Hörste, M. (2004). *Methodische Analyse and generische Modellierung von Eisenbahnleit- und –sicherungssystemen [Dissertation]*. VDI Fortschritt Berichte, Dusseldorf, Germany. (In German).

Meyer zu Horste, M., & Schnieder, E. (1999). Formal Modeling and Simulation of Train Control Systems using Petri nets. *Proceedings of theWorld Congress on Formal Methods in the development of Computing Systems*. Toulouse, LNCS (Vol. 1709). Springer.

Meyer zu Hörste, M., Jaschke, K., & Lemmer, K. (2003). A test facility for ERTMS/ETCS conformity. In E. Schnieder, & G. Tarnai (Eds.), FORMS 2003.

Meynadier, H. (1843). *Paris sous le point de vue pittoresque et monumental ou éléments d'un plan général d'ensemble de ses travaux d'art et d'utilité publique*. Paris.

Michaelis, M., & Schöbel, A. (2009). Integrating line planning, timetabling, and vehicle scheduling: a customer-oriented heuristic. *Public Transport*, *1*(3). doi:10.1007/s12469-009-0014-9

Michitsuji, Y., & Suda, Y. (2006). Running performance of power-steering railway bogie with independently rotating wheels. *Vehicle System Dynamics*, *44*(Suppl.), 71–82. doi:10.1080/00423110600867416

Mikhailov, L. (2004). A fuzzy approach to deriving priorities from interval pairwise comparison judgements. *European Journal of Operational Research*, *159*(3), 687–707. doi:10.1016/S0377-2217(03)00432-6

Milan, J. (1993). A model of competition between high speed rail and air transport. *Transportation Planning and Technology*, *17*(1), 1–23. doi:10.1080/03081069308717496

Milano, F. (2010). *Power System Modelling and Scripting*. London, United Kingdom: Springer-Verlag. doi:10.1007/978-3-642-13669-6

Milenković, M., Bojović, N., & Nuhodžić, R. (2012). A Comparative Analysis of Neuro-Fuzzy and Arima Models for Urban Rail Passenger Demand Forecasting. *Proceedings of International Conference on Traffic and Transport Engineering*, Belgrade, Serbia (pp. 569-577).

Milenković, M., Bojović, N., Glisovic, N., & Nuhodžić, R. (2013). Use of SARIMA models to assess rail passenger flows: a case study of Serbian Railways. *Proceedings of the 2nd International Symposium & 24th National Conference on Operational Research*, Athens, Greece (pp. 296-302).

Milenković, M., Bojović, N., Glišović, N., & Nuhodžić, R. (2014). Comparison of Sarima-Ann and Sarima-Kalman Methods for Railway Passenger Flow Forecasting. *Proceedings of the Second International Conference on Railway Technology: Research, Development and Maintenance*, Ajaccio, France. Stirlingshire, UK: Civil-Comp Press. doi:10.4203/ccp.104.193

Milenković, M., & Bojović, N. (2014). A Recursive Kalman Filter Approach to Forecasting Railway Passenger Flows. *International Journal of Railway Technology*, *3*(2), 39–57. doi:10.4203/ijrt.3.2.3

MIL-STD-882D Standard practice for system safety. (2000). USA Department of Defense.

MISRA Coding Standard. (2004). MISRA (Motor Industry Software Reliability Association).

Miura, S., Takai, H., Uchida, M., & Fukada, Y. (1998). The mechanism of railway track. In *Japan Railway & Transport Review* (pp. 38-45).

Mladenović, N., & Hansen, P. (1997). Variable neighborhood search. *Computers & Operations Research*, *24*(11), 1097–1100. doi:10.1016/S0305-0548(97)00031-2

Mohammed, S. I., Graham, D. J., Melo, P. C., & Anderson, R. J. (2013). A meta-analysis of the impact of rail projects on land and property values. *Journal of Transport Research Part A*, *50*, 158–170.

Mohan, D., Tsimhoni, O., Sivak, M., & Flannagan, M. J. (2009). Road safety in India: challenges and opportunities (Report UMTRI-2009-1). University of Michigan, Transportation Research Institute.

Mohan, P., Padmanabhan, V., & Ramjee, R. (2008). Nericell: Rich monitoring of road and traffic conditions using mobile smartphones. In *SenSys*. ACM. doi:10.1145/1460412.1460444

Möller, D. P. (2014). Simulation Tools in Transportation *Introduction to Transportation Analysis. Modelling and Simulation (Anaheim)*, *2014*, 195–228.

Mon, D. L., Cheng, C. H., & Lin, J. C. (1994). Evaluating weapon system using fuzzy analytic hierarchy process based on entropy weight. *Fuzzy Sets and Systems*, *62*(2), 127–134. doi:10.1016/0165-0114(94)90052-3

Mönsters, M., Linder, C., & Lackhove, C. (2013). Betriebliche Ansätze zur Minderung von Schienenverkehrslärm: Untersuchung der Auswirkungen einer Verringerung der Fahrgeschwindigkeit von Güterzügen auf die Leistungsfähigkeit verschiedener Streckenstandards. *Der Eisenbahningenieur (EI)* (pp. 16-19).

Montez, T., Linkenkaer-Hansen, K., van Dijk, B. W., & Stam, C. J. (2006). Synchronization likelihood with explicit time-frequency priors. *NeuroImage*, *33*(4), 1117–1125. doi:10.1016/j.neuroimage.2006.06.066 PMID:17023181

Montgomery, D. C., Jennings, C. L., & Kulahci, M. (2008). *Introduction to Time Series Analysis and Forecasting*. Wiley-Interscience.

Morley, M. (1993). Safety in Railway Signaling Data: A behavioral analysis. *Proceedings of the 6th Workshop on Higher Order Logic Theorem Proving and its Applications*, *LNCS*(Vol. 740). Springer.

Motraghi, A., & Marinov, M. V. (2012). Analysis of urban freight by rail using event based simulation. *Simulation Modelling Practice and Theory*, *25*, 73–89. doi:10.1016/j.simpat.2012.02.009

Mukaidono, M. (2002). *Fuzzy Logic for Beginners*. Singapore, New Jersey, London, Hong Kong: World Scientific Publishing Co.

Mukherjee, A., Routray, A., and Samanta, A. (2015). Method for On-line Detection of Arcing in Low Voltage Distribution Systems. Power Delivery, IEEE Trans. Power Del., PP(99):1.

Müller-Hannemann, M., Schulz, F., Wagner, D., & Zaroliagis, C. D. (2004). Timetable Information: Models and Algorithms. In ATMOS (pp. 67-90).

Mumbai City Development Plan 2005-2025. (2005). *MCGM*. Retrieved from http://www.mcgm.gov.in/irj/go/km/docs/documents/MCGM%20Department%20List/City%20Engineer/Deputy%20City%20Engineer%20(Planning%20and%20Design)/City%20Development%20Plan/Strategy%20for%20transportation.pdf

Murphy, K. M., Shleifer, A., & Vishny, R. W. (1989). Industrialization and the Big Push. *Journal of Political Economy*, 97(5), 1003–1026. doi:10.1086/261641

Muthukannan, M., & Thirumurthy, A. M. (2008). Modeling for optimization of urban transit system utility: A case study. *Journal of Engineering and Applied Sciences (Asian Research Publishing Network)*, 3, 71–74.

Nachtigall, K., & Voget, S. (1996). A Genetic Algorithm Approach to Periodic Railway Synchronization. *Computers & Operations Research*, 23(5), 453–463. doi:10.1016/0305-0548(95)00032-1

Naha, A., Samanta, A. K., Routray, A., & Deb, A. K. (2015). Determining Autocorrelation Matrix Size and Sampling Frequency for MUSIC Algorithm. *IEEE Signal Processing Letters*, 22(8), 1016–1020. doi:10.1109/LSP.2014.2366638

Nandi, S., & Toliyat, H. A. (1999). Condition monitoring and fault diagnosis of electrical machines-a review. Conference Record of the 1999 IEEE Industry Applications Conference, Thirty-Fourth IAS Annual Meeting (Vol. 1, pp. 197–204). IEEE. doi:10.1109/IAS.1999.799956

Nandi, S. (2004). Modeling of Induction Machines Including Stator and Rotor Slot Effects. *IEEE Transactions on Industry Applications*, 40(4), 1058–1065. doi:10.1109/TIA.2004.830764

Nandi, S., Ahmed, S., & Toliyat, H. A. (2001). Detection of rotor slot and other eccentricity related harmonics in a three phase induction motor with different rotor cages. *IEEE Transactions on Energy Conversion*, 16(3), 253–260. doi:10.1109/60.937205

Nandi, S., Ilamparithi, T. C., Lee, S. B., & Hyun, D. (2011). Detection of Eccentricity Faults in Induction Machines Based on Nameplate Parameters. *IEEE Transactions on Industrial Electronics*, 58(5), 1673–1683. doi:10.1109/TIE.2010.2055772

Nandi, S., Toliyat, H. A., & Li, X. (2005). Condition Monitoring and Fault Diagnosis of Electrical Motors A Review. *IEEE Transactions on Energy Conversion*, 20(4), 719–729. doi:10.1109/TEC.2005.847955

Narayanaswami, S., & Mohan, S. (2013). The roles of ICT in driverless, automated railway operations. *International Journal of Logistics System and Management*, 14(4), 490–503. doi:10.1504/IJLSM.2013.052749

Naumann, A., Grippenkoven, J., Giesemann, S., Stein, J., & Dietsch, S. (2013). Rail Human Factors: Human-centred design for railway systems. *Proceedings of the 12th IFAC Symposium on Analysis, Design and Evaluation of Human-Machine Systems,* Las Vegas, NV, USA. IFAC.

Nelson, P., Baglino, A., Harrington, W., Safirova, E., & Lipman, A. (2007). Transit in Washington, DC: Current benefits and optimal level of provision. *Journal of Urban Economics*, 62(2), 231–251. doi:10.1016/j.jue.2007.02.001

Compilation of References

Nemec, M., Drobnic, K., Nedeljkovic, D., & Rastko, F. (2010). Detection of Broken Bars in Induction Motor Through the Analysis of Supply Voltage Modulation. *IEEE Transactions on Industrial Electronics*, *57*(8), 2879–2888. doi:10.1109/TIE.2009.2035991

Newman, A. M., Nozick, L. K., & Yano, C. A. (2002). Optimization in the rail industry. In Handbook of Applied Optimization (pp. 704-719).

Newman, P., Glazebrook, G., & Kenworthy, J. (2013). Peak car use and the rise of global rail: Why this is happening and what it means for large and small cities. *Journal of Transportation Technologies*, *3*(04), 272–287. doi:10.4236/jtts.2013.34029

Newman, P., & Kenworthy, J. (1999). *Sustainability and cities overcoming automobile dependence*. USA: Island Press.

Newman, P., & Kenworthy, J. (2015). *The End of Automobile Dependence: How Cities are Moving Beyond Car-based Planning*. Washington, DC: Island Press. doi:10.5822/978-1-61091-613-4

NFPA 130, Standard for Fixed Guideway Transit and Passenger Rail Systems. (2010).

Ngai, E.W.T., & Wat, F.K.T. (2003). Design and development of a fuzzy expert system for hotel selection. *International Journal of Management Sciences*, *31*(4), 275–286.

Nie, L., & Hansen, I. A. (2005). System analysis of train operations and track occupancy at railway stations. *European Journal of Transport and Infrastructure Research*, *5*(1), 31–54.

Ning, B. (2010). *Advanced Train Control. Wessex*. WIT Press.

Noland, R. B., & Quddus, M. A. (2004). A spatially disaggregate analysis of road casualties in England. *Accident; Analysis and Prevention*, *36*(6), 973–984. doi:10.1016/j.aap.2003.11.001 PMID:15350875

NUTP. (2006). *National Urban Transport Policy*. New Delhi.

O'Dell, S. W., & Wilson, N. H. (1999). *Optimal real-time control strategies for rail transit operations during disruptions Computer-aided transit scheduling* (pp. 299–323). Springer.

Ocak, H., & Loparo, K. A. (2001). A new bearing fault detection and diagnosis scheme based on hidden Markov modeling of vibration signals. Proceedings of the 2001 IEEE Int. Conf. on Acoust. Speech, Signal Process ICASSP '01 (Vol. 5, pp. 3141–3144). IEEE. doi:10.1109/ICASSP.2001.940324

Odgers, J. F., & Schijndel, A. V. (2011). Forecasting Annual Train Boardings in Melbourne Using Time Series Data. *Proceedings of the 34th Australian Transport Research Forum*. Retrieved from http://www.atrf11.unisa.edu.au/Assets/Papers/ATRF11_0109_final.pdf

Odijk, M. A. (1996). A constraint generation algorithm for construction of periodic railway timetables. *Transportation Research Part B: Methodological*, *30*(6), 455–464. doi:10.1016/0191-2615(96)00005-7

Okamato, I. (1998). How bogies work. *Japan Railway & Transport Review*, *18*, 52–61.

Okamato, I. (1999). Shinkansen bogies. *Japan Railway & Transport Review*, *19*, 46–52.

Olsson, N., Okland, A., & Halvorsen, S. (2012). Consequences of differences in cost-benefit methodology in railway infrastructure appraisal - A comparison between selected countries. *Transport Policy*, *22*, 29–35. doi:10.1016/j.tranpol.2012.03.005

openETCS Project-Team. (2014, 12 12). *openETCS*. Retrieved from www.openetcs.org

Ordaz-Moreno, A., de Jesus Romero-Troncoso, R., Vite-Frias, J. A., Rivera-Gillen, J. R., & Garcia-Perez, A. (2008). Automatic online diagnosis algorithm for broken-bar detection on induction motors based on discrete wavelet transform for fpga implementation. . *IEEE Transactions on* Industrial Electronics, *55*(5), 2193–2202.

O'Regan, G. (2012). Z Formal Specification Language. In *Mathematics in Computing* (pp. 109–122). Springer.

Orlova, A., & Boronenko, Y. (2006). The Anatomy of Railway Vehicle Running Gear. In S. Iwnicki (Ed.), *Handbook of railway vehicle dynamics* (pp. 42–83). Boca Raton, FL: CRC Press. doi:10.1201/9781420004892.ch3

Osbourne, M.J. (2003). *An Introduction to Game Theory*. Toronto, Canada: Oxford University Press.

Pachl, J. (2011). Deadlock avoidance in railroad operations simulations. In *Transportation Research Board 90th Annual Meeting* (Paper No: 11-0175). Washington DC, USA.

Pachl, J. (2009). *Railway Operation and Control*. USA: VTD Rail Publishing.

Padam, S., & Singh, S. K. (2001). Urbanization and Urban transport in India: the sketch for a policy. *Paper presented at theTransport Asia project workshop*, Pune, India.

Paolucci, M., & Pesenti, R. (1999). An object-oriented approach to discrete-event simulation applied to underground railway systems. *Simulation*, *72*(6), 372–383. doi:10.1177/003754979907200601

Papadelis, C., Kourtidou-Papadeli, C., Bamidis, P. D., Chouvarda, I., Koufogiannis, D., Bekiaris, E., & Maglaveras, N. (2006). Indicators of sleepiness in an ambulatory EEG study of night driving. *Proceedings of the Annual International Conference of the IEEE Engineering in Medicine and Biology Society* (Vol. 1, pp. 6201–6204). Doi:<ALIGNMENT.qj></ALIGNMENT>10.1109/IEMBS.2006.259614

Papadelis, C., Chen, Z., Kourtidou-Papadeli, C., Bamidis, P. D., Chouvarda, I., Bekiaris, E., & Maglaveras, N. (2007). Monitoring sleepiness with on-board electrophysiological recordings for preventing sleep-deprived traffic accidents. *Clinical Neurophysiology*, *118*(9), 1906–1922. doi:10.1016/j.clinph.2007.04.031 PMID:17652020

Papageorgiou, G., Damianou, P., & Pitsilides, A. (2006, July 6-7). A Microscopic Traffic Simulation Model for Transportation Planning in Cyprus, Ayia Napa, Cyprus.

Papayanis, N. (2004). *Planning Paris before Haussmann*. Baltimore, London: The Johns Hopkins University Press.

Papyrus SysML editor. (2014). Retrieved http://eclipse.org/papyrus

Park, J. H., Koh, H. I., Hyun, H. M., Kim, M. S., & You, W. H. (2010). Design and analysis of active steering bogie for urban trains. *Journal of Mechanical Science and Technology*, *24*(6), 1353–1362. doi:10.1007/s12206-010-0341-4

Pascual-Marqui, R. D., Michel, C. M., & Lehmann, D. (1994). Low resolution electromagnetic tomography: A new method for localizing electrical activity in the brain. *International Journal of Psychophysiology*, *18*(1), 49–65. doi:10.1016/0167-8760(84)90014-X PMID:7876038

Patel, V., Chatterji, S., Chisholm, D., Ebrahim, S., Gopalakrishna, G., Mathers, C., & Reddy, K. S. et al. (2011). Chronic diseases and injuries in India. *Lancet*, *377*(9763), 413–428. doi:10.1016/S0140-6736(10)61188-9 PMID:21227486

Pattara-Atikom, W., & Peachavanish, R. Estimating road traffic congestion from cell dwell time using neural network. *Proceedings of the 7th International Conference on ITS ITST '07*.

Paul, D. (2002). DC traction power system grounding. *IEEE Transactions on Industry Applications*, *38*(3), 818–824. doi:10.1109/TIA.2002.1003435

Compilation of References

Pearson, J. T., Goodall, R. M., Mei, T. X., & Shen, S. (2004). Kalman filter design for high speed bogie active stability system. Proceedings of UKACC control, Bath, UK

Peleska, J. (2013). Industrial-Strength Model-Based Testing - State of the Art and Current Challenges. In A. K. Petrenko, & H. Schlingloff, (Eeds.), *Proceedings Eighth Workshop on Model-Based Testing, Rome, Italy, 17th March 2013. Electronic Proceedings in Theoretical Computer Science* (S. 3-28).

Pender, B., Currie, G., Delbosc, A., & Shiwakoti, N. (2013). Disruption Recovery in Passenger Railways. *Transportation Research Record: Journal of the Transportation Research Board, 2353*(1), 22–32. doi:10.3141/2353-03

Penetar, D., McCann, U., & Thorne, D. (1993). Caffeine reversal of sleep deprivation effects on alertness and mood. *Psychopharmacology, 112*(2), 359-365. Retrieved from http://link.springer.com/article/10.1007/BF02244933

Peng, J., & Aston, J. (2006). The State Space Models Toolbox for MATLAB. *Journal of Statistical Software, 41*(6), 1–26.

Perez, J., Busturia, J. M., & Goodall, R. M. (2002). Control strategies for active steering of bogie based railway vehicles. *Control Engineering Practice, 10*(9), 1005–1012. doi:10.1016/S0967-0661(02)00070-9

Perez, J., Busturia, J. M., Mei, T. X., & Vinolas, J. (2004). Combined active steering and traction for mechatronic bogie vehicles with independently rotating wheels. *Annual Reviews in Control, 28*(2), 207–217. doi:10.1016/j.arcontrol.2004.02.004

Perez, J., Mauer, L., & Busturia, J. M. (2002). Design of active steering systems for bogie-based railway vehicles with independently rotating wheels. *Vehicle System Dynamics, 37*(Suppl.), 209–220. doi:10.1080/00423114.2002.11666233

Perreymond. (1844). *Revue Générale d'Architecture et des Travaux Publics*, col. 184-188 and 232-235.

Petersen, E. R., & Taylor, A. J. (1982). A structured model for rail line simulation and optimization. *Transportation Science, 16*(2), 192–206. doi:10.1287/trsc.16.2.192

Peters, G.A., & Peters, B.J. (2006). *Human error: causes and control.* CRC Press. doi:10.1201/9781420008111

Picazo-Rodenas, M. J., Royo, R., Antonino-Daviu, J., & Roger-Folch, J. (2012). Use of infrared thermography for computation of heating curves and preliminary failure detection in induction motors. *Proceedings of the 2012 20th Int. Conf. Electr. Mach.* (pp. 525–531).

Pickrell, D. H. (1992). A Desire Named Streetcar Fantasy and Fact in Rail Transit Planning. *Journal of the American Planning Association, 58*(2), 158–176. doi:10.1080/01944369208975791

Picon, A. (1994). Les Fortifications de Paris. In B. Belhoste, F. Masson, & A. Picon (Eds.), Le Paris des Polytechniciens. Des ingénieurs dans la ville 1794 – 1994 (pp. 213 – 221). Paris: Délégation à l'action artistique de la ville de Paris.

Pilo, E. (2003) Diseño optimo de la electrificación de ferrocarriles de alta velocidad [Doctoral dissertation in Spanish]. Retrieved from https://www.iit.upcomillas.es/personas/eduardo

Pilo, E., Rouco, L., Fernandez, A., & Abrahamsson, L. (2012). A Monovoltage equivalent model of bi-voltage autotransformer-based electrical systems in railways. *IEEE Transactions on Power Delivery, 27*(2), 699–708. doi:10.1109/TPWRD.2011.2179814

Pineda-Sanchez, M., Perez-Cruz, J., Roger-Folch, J., Riera-Guasp, M., Sapena-Bano, A., & Puche-Panadero, R. (2013). Diagnosis of Induction Motor Faults using a DSP and Advanced Demodulation Techniques. *Proceedings of the 2013 9th IEEE International Symposium SDEMPED* (pp. 609–76. IEEE.

Pires, C. L., Nabeta, S. I., & Cardoso, J. R. (2009). DC traction load flow including AC distribution network. *IET Electric Power Applications, 3*(4), 289–297. doi:10.1049/iet-epa.2008.0147

Pires, C., Nabeta, S., & Cardoso, J. (2007). ICCG method applied to solve DC traction load flow including earthing models. *IET Electric Power Applications*, *1*(2), 193–198. doi:10.1049/iet-epa:20060174

Platzer, A., & Quesel, J.-D. (2009). European Train Control System: A Case Study in Formal Verification. *Proceedings of the11th International Conference on Formal Engineering Methods ICFEM 2009* (pp. 246-265). Springer. doi:10.1007/978-3-642-10373-5_13

Pnueli, A. (1977). The Temporal Logic of Programs. In Foundations of Computer Science (pp. 46-57).

Pogaku, N., Prodanovic, M., & Green, T. C. (2007). Modeling, analysis and testing of autonomous operation of an inverter-based microgrid. *IEEE Transactions on Power Electronics*, *22*(2), 613–625. doi:10.1109/TPEL.2006.890003

Polach, O., Berg, M., & Iwnicki, S. (2006). Simulation. In S. Iwnicki (Ed.), *Handbook of railway vehicle dynamics* (pp. 359–421). Boca Raton, FL: CRC Press.

Pombo, J. (2012). Application of computational tool to study influence of worn wheels on railway vehicles. *Journal of software engineering and application*, *5*, 51-61.

Pons-Llinares, J., Antonino-daviu, J. A., Riera-guasp, M., Lee, S. B., Kang, T.-J., & Yang, C. (2015). Advanced Induction Motor Rotor Fault Diagnosis via Continuous and Discrete Time-Frequency Tools. *IEEE Transactions on Industrial Electronics*, *62*(3), 1791–1802. doi:10.1109/TIE.2014.2355816

Porter, R. (2000). *London: A Social History*. London: Penguin Books.

Powell, A. J. (1999). On the dynamics of actively steered railway vehicles. *Vehicle System Dynamics*, *33*(Suppl.), 442–452.

Pozzobon, P. (1998). Transient and steady state short-circuit currents in rectifiers for DC traction supply. *IEEE Transactions on Vehicular Technology*, *47*(4), 1390–1404. doi:10.1109/25.728534

Principles of Interlocking. (2009). Indian Railways Institute of Signal Engineering and Telecommunications.

Procès verbal. (1898, June 27). Conseil Municipal de Paris.

Profile of safety risk on the UK mainline railway. (2003). Railway Safety SP-RSK-3.1.3.11, 3.

Profillidis, V. A. (2006). *Railway Management and Engineering*. Ashgate Publishing Limited.

Profillidis, V. A., & Botzoris, G. N. (2006). Econometric models for the forecast of passenger demand in Greece. *Journal of Statistics & Management Systems*, *9*(1), 37–54. doi:10.1080/09720510.2006.10701192

Puche-Panadero, R., Pineda-Sanchez, M., Riera-Guasp, M., Roger-Folch, J., Hurtado-Perez, E., & Perez-Cruz, J. (2009). Improved Resolution of the MCSA Method Via Hilbert Transform, Enabling the Diagnosis of Rotor Asymmetries at Very Low Slip. *IEEE Transactions on Energy Conversion*, *24*(1), 52–59. doi:10.1109/TEC.2008.2003207

Pucher, J., Korattyswaropam, N., Mittal, N., & Ittyerah, N. (2005). Urban transport crisis in India. *Transport Policy*, *12*(3), 185–198. doi:10.1016/j.tranpol.2005.02.008

Puong, A. (2001). *A Real-Time Train Holding Model for Rail Transit Systems*. Massachusetts Institute of Technology, Department of Civil and Environmental Engineering.

Purushotham, V., Narayanan, S., & Prasad, S. A. N. (2005). Multi-fault diagnosis of rolling bearing elements using wavelet analysis and hidden Markov model based fault recognition. *NDT & E International*, *38*(8), 654–664. doi:10.1016/j.ndteint.2005.04.003

Quaglietta, E. (2013). A simulation-based approach for the optimal design of signaling block layout in railway networks. *Simulation Modelling Practice and Theory*.

Compilation of References

Quatieri, T. F. (2002). *Discrete-Time Speech Signal Processing: Principles and Practice*. Pearson Education. Retrieved from http://books.google.com/books?hl=en&lr=&id=UMR9ByupVy8C&pgis=1

Raban, M. Z., Dandona, L., & Dandona, R. (2014). The quality of police data on RTC fatalities in India. *Injury Prevention*, *20*(5), 293–301. doi:10.1136/injuryprev-2013-041011 PMID:24737796

Railway Safety Act. (2006). *Act of the Ministry of Land, Transport Affairs*.

Railways (safety case) regulations 2000. (2000). Health & Safety Executive (HSE).

Raju, S. (2008). Project NPV, positive externalities, social cost-benefit analysis—the Kansas City light rail project. *Journal of Public Transportation*, *11*(4), 59–88. doi:10.5038/2375-0901.11.4.4

Ramanathan, R., & Ganesh, L. S. (1995). Using AHP for resource allocation problems. *European Journal of Operational Research*, *80*(2), 410–417. doi:10.1016/0377-2217(93)E0240-X

Rana, R., Chou, C., Kanhere, S., Bulusu, N., & Hu, W. (2010, April). Earphone: an end-to-end participatory urban noise mapping system. Proceedings of IPSN.

Rangel-Magdaleno, J., Romero-Troncoso, R., Osornio-Rios, R. A., Cabal-Yepez, E., & Contreras-Medina, L. M. (2009). Novel methodology for online half-broken-bar detection on induction motors. *IEEE Transactions on Instrumentation and Measurement*, *58*(5), 1690–1698. doi:10.1109/TIM.2009.2012932

Rao; D.S., Disha Handa, Gaurav Bagga, Ajay Kumar Rangra, & Nandini Nayar. (2010). Extra-Organizational Systems: A Challenge to the Software Engineering Paradigm. *International Journal of Advanced Science and Technology*, *20*, 25–42.

Raturi, V., Srinivasan, K., Narulkar, G., Chandrashekharaiah, A., & Gupta, A. (2013). Analyzing inter-modal competition between high speed rail and conventional transport systems: A game theoretic approach. *Proceedings of the 2nd Conference of Transportation Research Group of India. Procedia: Social and Behavioral Sciences*, *104*, 904–913. doi:10.1016/j.sbspro.2013.11.185

Raygani, S. V., Tahavorgar, A., Fazel, S. S., & Moaveni, B. (2012). Load flow analysis and future development study for an AC electric railway. *IET Electrical Systems in Transportation*, *2*(3), 139–147. doi:10.1049/iet-est.2011.0052

Real-time 3D graphics and virtual set solutions. (n. d.). *Brainstorm*. Retrieved from http://neuroimage.usc.edu/brainstorm/

Reed, M. E. (1996). *The London & North Western Railway*. Penryn: Atlantic Transport Publishers.

Rees, R. (1984). *Public Enterprise Economics* (2nd ed.). Oxford: Weidenfeld and Nicolson.

Rényi, A. (1961). On Measures of Entropy and Information. *Proceedings of the Fourth Berkeley Symposium on Mathematical Statistics and Probability* (Vol. 1, pp. 547–561). The Regents of the University of California.

Report of the Commissioners appointed to investigate the Various Projects for establishing Railway Termini Within or in the immediate Vicinity of the Metropolis. (1846). Parliamentary Papers, Houses of Commons and Lords, London.

Report, R. A. I. B. (2010). *RAIB review of the railway industry's investigation of an irregular signal sequence at Milton Keynes, 29 Dec 2008. Railway Accident Investigation Branch*, Department for Transport, UK.

Riera-Guasp, M., Pineda-Sanchez, M., Perez-Cruz, J., Puche-Panadero, R., Roger-Folch, J., & Antonino-Daviu, J. A. (2012). Diagnosis of induction motor faults via gabor analysis of the current in transient regime. *IEEE Transactions on Instrumentation and Measurement*, *61*(6), 1583–1596. doi:10.1109/TIM.2012.2186650

Riggs, J., & West, T. (1986). *Engineering Economics*. USA: McGraw-Hill.

Road Accidents in India. (2010). *Ministry of Roads Transport and Highways (MoRTH)*. Retrieved from http://morth.nic.in/writereaddata/mainlinkFile/File761.pdf

Road transport Year Book (2011-12). (2013). Transport Research Wing. New Delhi.

Robert, J. (1967). *Notre Métro*. Paris.

Rolling stock asset strategic safety risk model. (2004). *Tube Lines*, Interim Report.

Roman, C., Espino, R., & Martin, J. (2007). Competition of high-speed train with air transport: The case of Madrid–Barcelona. *Journal of Air Transport Management, 13*(5), 277–284. doi:10.1016/j.jairtraman.2007.04.009

Romero-Troncoso, R. J., Pena-Anaya, M., Cabal-Yepez, E., Garcia-Perez, A., & Osornio-Rios, R. (2012). Reconfigurable SoC-Based Smart Sensor for Wavelet and Wavelet Packet Analysis. *IEEE Transactions on Instrumentation and Measurement, 61*(9), 2458–2468. doi:10.1109/TIM.2012.2190340

Romero-Troncoso, R. J., Saucedo-Gallaga, R., Cabal-Yepez, E., Garcia-Perez, A., Osornio-Rios, R., Alvarez-Salas, R., & Huber, N. et al. (2011). FPGA-Based Online Detection of Multiple Combined Faults in Induction Motors through Information Entropy and Fuzzy Inference. *IEEE Transactions on Industrial Electronics, 58*(11), 5263–5270. doi:10.1109/TIE.2011.2123858

Rosenblum, M., Pikovsky, A., & Kurths, J. (1996). Phase synchronization of chaotic oscillators. *Physical Review Letters, 76*(11), 1804–1807. Retrieved from http://www.ncbi.nlm.nih.gov/pubmed/10060525 doi:10.1103/PhysRevLett.76.1804 PMID:10060525

Routray, A., & Kar, S. (2012). Classification of brain states using principal components analysis of cortical EEG synchronization and HMM. *Proceedings of the 2012 IEEE International Conference on Acoustics, Speech and Signal Processing (ICASSP)* (pp. 641–644). IEEE. http://doi.org/ doi:10.1109/ICASSP.2012.6287965

Routray, A., Pradhan, A. K., & Rao, K. P. (2002). A novel Kalman filter for frequency estimation of distorted signals in power systems. *IEEE Transactions on Instrumentation and Measurement, 51*(3), 469–479. doi:10.1109/TIM.2002.1017717

Rubinov, M., & Sporns, O. (2010). Complex network measures of brain connectivity: Uses and interpretations. *NeuroImage, 52*(3), 1059–1069. doi:10.1016/j.neuroimage.2009.10.003 PMID:19819337

Rules regarding safety standards of railway facilities. (June 2011). Ordinance of the Ministry of Land, Transport Affairs.

Rus, G., & Gustavo, N. (2007). Is investment in High Speed Rail socially profitable? *Journal of Transport Economics and Policy, 41*(1), 3–23.

Saaty, T. L. (1980). *Analytical Hierarchy Process*. New York: McGraw-Hill.

Saaty, T. L. (1990). *The analytic hierarchy process*. Pittsburgh: RWS Publications.

Saboya, M. (1991). *Presse et Architecture au XIXe siècle. César Daly et la Revue Générale de l'Architecture et des Travaux Publics*. Paris: Picard Éditeur.

Sackman, H. (1974). Delphi Assessment: Expert Opinion, Forecasting and Group Process. Brown, Thomas, An Experiment in Probabilistic Forecasting. (R-944-ARPA, 1972).

Sadiq, R., & Husain, T. (2005). A fuzzy-based methodology for an aggregative environmental risk assessment: A case study of drilling waste. *Environmental Modelling & Software, 20*(1), 33–46. doi:10.1016/j.envsoft.2003.12.007

Sahin, I. (1999). Railway Traffic Control and Train Scheduling based on Inter-train Conflict Management. *Transportation Research Part B: Methodological, 33*(7), 511–534. doi:10.1016/S0191-2615(99)00004-1

Compilation of References

Sajedinejad, A., Mardani, S., Hasannayebi, E., & Mir Mohammadi, K., S., & Kabirian, A. (2011). SIMARAIL: simulation based optimization software for scheduling railway network. *Paper presented at the Simulation Conference (WSC)*. doi:10.1109/WSC.2011.6148066

Salim, V., & Cai, X. (1997). A Genetic Algorithm for Railway Scheduling with Environmental Considerations. *Environmental Modelling & Software*, *12*(4), 301–310. doi:10.1016/S1364-8152(97)00026-1

Sánchez, Á. G., Ortega-Mier, M., & Arranz, R. (2011). Discrete-Event Simulation Models for Assessing Incidents in Railway Systems. *International Journal of Information Systems and Supply Chain Management*, *4*(2), 1–14. doi:10.4018/jisscm.2011040101

Sanchez, T. W. (1999). The Connection Between Public Transit and Employment. *Journal of the American Planning Association*, *65*(3), 284–296. doi:10.1080/01944369908976058

Sapena-bano, A., Pineda-sanchez, M., Puche-panadero, R., Perez-cruz, J., Roger-folch, J., Riera-guasp, M., and Martinezroman, J. (2015). Harmonic Order Tracking Analysis: A Novel Method for Fault Diagnosis in Induction Machines. *IEEE Trans. Energy Convers.*, *1*, 1–9.

Scheier, B., Schumann, T., Meyer zu Hörste, M., Dittus, H., & Winter, J. (2014). Wissenschaftliche Ansätze für einen energieoptimierten Eisenbahnbetrieb (Scientific Approaches for an energy-efficient railway operation). In Eisenbahn Ingenieur Kalender [Railway Engineer Calendar] (pp. 265-278). Hamburg: Media Group | Eurailpress (In German).

Schimmelpfennig, J. (2011). The South Eastern and Chatham Railways Managing Committee: A case for vertically-integrated regional duopolies? *International Journal of Strategic Decision Sciences*, *2*, 95–103. doi:10.4018/jsds.2011040106

Schleicher, R., Galley, N., Briest, S., & Galley, L. (2008). Blinks and saccades as indicators of fatigue in sleepiness warnings: Looking tired? *Ergonomics*, *51*(7), 982–1010. doi:10.1080/00140130701817062 PMID:18568959

Schnieder, E. (2001). Section 6.43.36.5: Train and Railway Operations Control. In *UNESCO, Encyclopedia of Life Support Systems*. Paris: UNESCO.

Schön, W., Larraufie, G., Moens, G., & Poré, J. (2014). Railway Signalling and Automation (Vol. 1). Paris: La Vie du Rail.

Schön, W., Larraufie, G., Moens, G., & Poré, J. (2013). *Railway Signalling and Automation* (Vol. 2). Paris: La Vie du Rail.

Schupp, G., Weidemann, C., & Mauer, L. (2004). Modeling the contact between wheel and rail within multi-body system, Vehicle System Dynamics. *International Journal of Vehicle Mechanics and Mobility*, *41*(5), 349–364.

Seera, M., Lim, C. P., Ishak, D., & Singh, H. (2012). Fault Detection and Diagnosis of Induction Motors Using Motor Current Signature Analysis and a Hybrid FMM CART Model. *IEEE Trans. Neural Networks Learn. Syst.*, *23*(1), 97–108. doi:10.1109/TNNLS.2011.2178443 PMID:24808459

Sekon, G. A. (1899). Illustrated Interviews, Mr. Thomas Chellew Jenkin. The Railway Magazine (Vol. 5, pp. 1–16). July to December.

Sengupta, A., Routray, A., & Kar, S. (2013). Complex brain networks using Visibility Graph synchronization. *Proceedings of the 2013 Annual IEEE India Conference (INDICON)* (pp. 1–4). IEEE. http://doi.org/doi:10.1109/INDCON.2013.6726126

Seshadrinath, J., Singh, B., & Panigrahi, B. K. (2014). Investigation of Vibration Signatures for Multiple Fault Diagnosis in Variable Frequency Drives Using Complex Wavelets. *IEEE Transactions on Power Electronics*, *29*(2), 936–945. doi:10.1109/TPEL.2013.2257869

Shabana, A. A., Zaazaa, K. E., & Sugiyama, H. (2008). *Rail road Vehicle Dynamics: A Computational Approach*. Boca Raton, FL: CRC Press.

Shah, J., & Iyer, A. (2004). Robust voiced/unvoiced classification using novel features and gaussian mixture model. *Proceedings of the IEEE Conference on Acoustics, Speech and Signal Processing.*

Shannon, C. E. (1948). A Mathematical Theory of Communication. *ACM SIGMOBILE mobile computing and communications review, 5*(1), 3–55.

Sharma, D., & Tomar, S. (2010). Mainstreaming climate change adaptation in Indian cities. *Environment and Urbanization, 22*(2), 451–465. doi:10.1177/0956247810377390

Shen, G., & Goodall, R. M. (1997). Active yaw relaxation for improved bogie performance. *Vehicle System Dynamics, 28*(4-5), 273–289. doi:10.1080/00423119708969357

Shen, S. (2000). *Integrated real-time disruption recovery strategies: a model for rail transit systems.* Massachusetts Institute of Technology.

Shen, S., Fowkes, T., Whiteing, T., & Johnson, D. (2009). Econometric modelling and forecasting of freight transport demand in Great Britain.*Proceedings of the European Transport Conference*, Noordwijkerhout, The Netherlands.

Shen, S., Mei, T. X., Goodall, R. M., Pearson, J., & Himmelstein, G. (2004). A study of active steering strategies for railway bogie. *Vehicle System Dynamics, 41*(Suppl.), 282–291.

Shen, S., & Wilson, N. H. (2001). *An optimal integrated real-time disruption control model for rail transit systems Computer-aided scheduling of public transport* (pp. 335–363). Springer. doi:10.1007/978-3-642-56423-9_19

Shimokawa, Y., & Mizuno, M. (2013). *Development of new concept steering bogie.* Nippon Steel & Sumitomo Metal Technical Report No. 105, Osaka.

Shimomura, Y., Yoda, T., & Sugiura, K. (2008). Use of frequency domain analysis of skin conductance for evaluation of mental workload. *The Journal of Physiology.* PMID:18832780

Shyr, O., & Hung, M. (2010). Intermodal Competition with High Speed Rail-A Game Theory Approach. *Journal of Marine Science and Technology, 18*(1), 32–40.

Siemiatycki, M. (2006). Message in a Metro: Building Urban Rail Infrastructure and Image in Delhi, India. *International Journal of Urban and Regional Research, 30*(2), 277–292. doi:10.1111/j.1468-2427.2006.00664.x

Sii, H. S., Ruxton, T., & Wang, J. (2001). A fuzzy-logic-based approach to qualitative safety modelling for marine systems. *Reliability Engineering & System Safety, 73*(1), 19–34. doi:10.1016/S0951-8320(01)00023-0

Sikdar, P. K., & Bhavsar, J. N. (2009). *Road Safety Scenario in India and Proposed Action Plan. Transport and Communications Bulletin for Asia, 79.* Road Safety.

Simmons, J. (1995). *The Victorian Railway.* London: Thames & Hudson.

Singh, A., Singh, P., Yadav, K., Naik, V., & Chandra, U. (2012, July). Low energy and sufficiently accurate localization for non-smartphones. Proceedings of MDM.

Singh, S. K. (2005). Review of urban transportation in India. *Journal of Public Transportation, 8*(1), 79–97. doi:10.5038/2375-0901.8.1.5

Small, K. A., & Rosen, H. S. (1981). Applied welfare economics with discrete choice models. *Econometrica, 49*(1), 105–129. doi:10.2307/1911129

Smith, P., Shah, M., & da Vitoria Lobo, N. (2003). Determining driver visual attention with one camera. *IEEE Transactions on Intelligent Transportation Systems, 4*(4), 205–218. doi:10.1109/TITS.2003.821342

Compilation of References

Smolka, M. O. (2012). A New Look at Value Capture in Latin America. *Land Lines, 24*(3), 10–15.

Spoerri, A., Egger, M., & von Elm, E. (2011). Mortality from road traffic accidents in Switzerland: Longitudinal and spatial analyses. *Accident; Analysis and Prevention, 43*(1), 40–48. doi:10.1016/j.aap.2010.06.009 PMID:21094295

Sporns, O., & Zwi, J. D. (2004). The small world of the cerebral cortex. *Neuroinformatics, 2*(2), 145–162. doi:10.1385/NI:2:2:145 PMID:15319512

Sprague, J. C., & Whittaker, J. D. (1986). *Economic Analysis for Engineers and Managers*. USA: Prentice Hall.

Stam, C. J. (2004). Functional connectivity patterns of human magnetoencephalographic recordings: A "small-world" network? *Neuroscience Letters, 355*(1-2), 25–28. doi:10.1016/j.neulet.2003.10.063 PMID:14729226

Stam, C. J. (2005). Nonlinear dynamical analysis of EEG and MEG: Review of an emerging field. *Clinical Neurophysiology, 116*(10), 2266–2301. doi:10.1016/j.clinph.2005.06.011 PMID:16115797

Stam, C. J., de Haan, W., Daffertshofer, A., Jones, B. F., Manshanden, I., van Cappellen van Walsum, A. M., & Scheltens, P. et al. (2009). Graph theoretical analysis of magnetoencephalographic functional connectivity in Alzheimer's disease. *Brain. Journal of Neurology, 132*, 213–224. doi:10.1093/brain/awn262 PMID:18952674

Stam, C. J., Jones, B. F., Nolte, G., Breakspear, M., & Scheltens, P. (2007). Small-world networks and functional connectivity in Alzheimer's disease. *Cerebral Cortex, 17*(1), 92–99. doi:10.1093/cercor/bhj127 PMID:16452642

Stam, C. J., & Reijneveld, J. C. (2007). Graph theoretical analysis of complex networks in the brain. *Nonlinear Biomedical Physics, 1*(1), 3. doi:10.1186/1753-4631-1-3 PMID:17908336

Stam, C. J., & Van Dijk, B. W. (2002). Synchronization likelihood: An unbiased measure of generalized synchronization in multivariate data sets. *Physica D. Nonlinear Phenomena, 163*(3-4), 236–251. doi:10.1016/S0167-2789(01)00386-4

Standard for Definition and Measurement of Wheel Tread taper (Publication No. APTA PR-M-S-017-06). (2007The American Public Transportation Association. Washington, DC: APTA Press.

Stanley, P., & IRSE (Eds.). (2011). *ETCS for Engineers*. Hamburg, Germany: Eurailpress / DVV Media Group.

Stavrou, A., Sedding, H. G., & Penman, J. (2001). Current monitoring for detecting inter-turn short circuits in induction motors. *IEEE Transactions on Energy Conversion, 16*(1), 32–37. doi:10.1109/60.911400

Stefanis, V., Profillidis, V., Papadopoulos, B., & Botzoris, G. (2001). Analysis and Forecasting of Intercity Rail Passenger Demand by Econometric and Fuzzy Regression Models. *Proceedings of the 8th SIGEF Congress*, Italy.

Strogatz, S. H. (2001). Exploring complex networks. *Nature, 410*(6825), 268–276. doi:10.1038/35065725 PMID:11258382

Subway Environmental Design Handbook. (1976). *Principles and Applications* (2nd ed., Vol. I). United States Department of Transportation.

Sugita, Y., Okamoto, S., Kuwabara, T., Hiraishi, M., Goda, K., & Ito, A. (2005). *New Solution for Urban Traffic: Small-Type Monorail System Automated People Movers 2005*. Orlando, Florida, United States: American Society of Civil Engineers.

Suhl, L., Mellouli, T., Biederbick, C., & Goecke, J. (2001). *Managing and preventing delays in railway traffic by simulation and optimization Mathematical methods on optimization in transportation systems* (pp. 3–16). Springer. doi:10.1007/978-1-4757-3357-0_1

Sundaravalli Narayanaswami. (2010). *Dynamic and Realtime Rescheduling Models: An Empirical Analysis from Railway Transportation*. Saarbuchen, Germany: Lambart Academic Publishers.

Sutcliffe, A. (1983). London and Paris: Capitals of the Nineteenth Century. *The Fifth H.J. Dyos Memorial Lecture.* London, Paris: Victorian Studies Centre, University of Leicester.

Suzuki, H., Cervero, R., & Iuchi, K. (2013). *Transforming Cities with Transit.* Washington, DC: The World Bank. doi:10.1596/978-0-8213-9745-9

Szpigel, B. (1973). Optimal train scheduling on a single track railway. *Operations Research, 72*, 343–352.

Taghizadeh, A. O., Soleimani, S. V., & Ardalan, A. (2013). Lessons from a flash flood in Tehran subway, Iran. *PLoS Currents, 2013*, 5.

Tallam, R. M., Lee, S. B., Stone, G. C., Kliman, G. B., Yoo, J.-Y., Habetler, T. G., & Harley, R. G. (2007). A survey of methods for detection of stator-related faults in induction machines. *Ind. Appl. IEEE Trans., 43*(4), 920–933. doi:10.1109/TIA.2007.900448

Tamil Nadu State Transport Authority. (2010). Retrieved from http://www.tn.gov.in/sta/ra2.pdf

Tanabe, M., Komiya, S., Wakui, H., Matsumoto, N., & Sogade, M. (2003). Simulation and visualization of high speed Shinkansen train on the railway structure. *Japan J. Indust. Appl. Math, 17*(2), 309–320. doi:10.1007/BF03167350

Tanaka, S., Kumazawa, K., & Koseki, T. (2009). *Passenger flow analysis for train rescheduling and its evaluation.Paper presented at theInternational Symposium on Speed-up, Safety and Service Technology for Railway and Maglev Systems.*

Tandon, N., Yadava, G. S., & Ramakrishna, K. M. (2007). A comparison of some condition monitoring techniques for the detection of defect in induction motor ball bearings. *Mechanical Systems and Signal Processing, 21*(1), 244–256. doi:10.1016/j.ymssp.2005.08.005

Tang, M. T., Tzeng, G. H., & Wang, S. W. (2000). A hierarchy fuzzy MCDM method for studying electronic marketing strategies in the information service industry. *J. Inter. Info. Man., 8*(1), 1–22.

Tao, R., Liu, S., Huang, C., & Tam, C. M. (2011). Cost-Benefit Analysis of High-Speed Rail Link between Hong Kong and Mainland China. *Journal of Engineering, 1*(1), 36–45.

Taplin, J. H. E. (1999). Simulation Models of Traffic Flow. *Proceedings of the 34th Annual Conference "OR in the New Millennium" ORSNZ '99*, Hamilton, New Zealand (pp. 175-184). Operational Research Society of New Zealand.

Teotrakool, K., Devaney, M. J., & Eren, L. (2009). Adjustable-speed drive bearing-fault detection via wavelet packet decomposition. *IEEE Transactions on Instrumentation and Measurement, 58*(8), 2747–2754. doi:10.1109/TIM.2009.2016292

Thiagarajan, A., Ravindranath, L., Balakrishnan, H., Madden, S., & Girod, L. (2011). Accurate, low-energy trajectory mapping for mobile devices. Proceedings of NSDI.

Thiagarajan, A., Ravindranath, L., LaCurts, K., Madden, S., Balakrishnan, H., Toledo, S., & Eriksson, J. (2009, November). Vtrack: Accurate, energy-aware road traffic delay estimation using mobile phones. Proceedings of Sensys.

Thies W., Ratan, A.L., & Davis, J. (2011). Paid crowdsourcing as a vehicle for global development. *Proceedings of theCHI Workshop on Crowd-sourcing and Human Computation.*

Thomson, W. T., & Fenger, M. (2001). Current signature analysis to detect induction motor faults. *Ind. Appl. Mag. IEEE, 7*(4), 26–34. doi:10.1109/2943.930988

Thomson, W. T., Rankin, D., & Dorrell, D. G. (1999). On-line current monitoring to diagnose airgap eccentricity in large three-phase induction motors-industrial case histories verify the predictions. *IEEE Transactions on Energy Conversion, 14*(4), 1372–1378. doi:10.1109/60.815075

Thornbury, W. (1878). The Tower Subway and London Docks. In *Old and New London:* (Vol. 2, pp. 122-128). London: Cassell, Petter, & Galpin. Retrieved from http://www.british-history.ac.uk/report.aspx?compid=45081

Timmermans, H. J. P., & Zhang, J. (2009). Modeling household activity travel behavior: Examples of state of the art modeling approaches and research agenda. *Transportation Research Part B: Methodological*, 43(2), 187–190. doi:10.1016/j.trb.2008.06.004

Tirachini, A., Cortés, C. E., & Jara-Díaz, S. R. (2011). Optimal design and benefits of a short turning strategy for a bus corridor. *Transportation*, 38(1), 169–189. doi:10.1007/s11116-010-9287-8

Tiwari, G. (2011). Key Mobility Challenges in Indian Cities. International Transport Forum Discussion Papers, Discussion Paper No. 2011-18.

Tiwari, G. (2011). Key Mobility Challenges in Indian Cities. *International Transport Forum*, Germany. doi:10.1787/5kg9mq4m1gwl-en

Tiwari, G., and D. Jain. (2012). Accessibility and safety indicators for all road users: case study Delhi BRT. *Journal of Transport Geography*, 22, 87-95.

Tormos, P., Lova, A., Barber, F., Ingolotti, L., Abril, M., & Salido, M. A. (2008). A genetic algorithm for railway scheduling problems. *Studies in Computational Intelligence*, 128, 255–276. doi:10.1007/978-3-540-78985-7_10

Törnquist, J. (2012). Design of an effective algorithm for fast response to the rescheduling of railway traffic during disturbances. *Transportation Research Part C, Emerging Technologies*, 20(1), 62–78. doi:10.1016/j.trc.2010.12.004

Train, K. (2003). *Discrete Choice Methods with Simulation*. Cambridge: Cambridge University Press. doi:10.1017/CBO9780511753930

Transforming City Bus Transport in India through Financial Assistance for Bus Procurement under JnNURM. (2012). MoUD. New Delhi: Retrieved from http://jnnurm.nic.in/wp-content/uploads/2012/02/booklet-on-transforming-City-Bus-Transport-in-India.pdf

Transportation's Role in Reducing US Greenhouse Gas Emissions Volume 1: Synthesis Report. (2010, April). *US Department of Transportation*. Retrieved from http://ntl.bts.gov/lib/32000/32700/32779/DOT_Climate_Change_Report_-_April_2010_-_Volume_1_and_2.pdf

Tzeng, Y. S., Wu, R. N., & Chen, N. (1998). Electric network solutions of DC transit systems with inverting substations. *IEEE Transactions on Vehicular Technology*, 47(4), 1405–1412. doi:10.1109/25.728537

Tzeng, Y. S., Wu, R. N., & Chen, N. (1998). Unified AC/DC power flow for system simulation in DC. *IET Electric Power Applications*, 142(6), 345–354. doi:10.1049/ip-epa:19952159

Tzeng, Y., Chen, N., & Wu, R. (1995). A detailed R-L fed bridge converter model for power flow studies in industrial AC/DC power systems. *IEEE Transactions on Industrial Electronics*, 42(5), 531–538. doi:10.1109/41.464617

Ueno, A., & Uchikawa, Y. (2004). Relation between human alertness, velocity wave profile of saccade, and performance of visual activities. *Proceedings of the Annual International Conference of the IEEE Engineering in Medicine and Biology Society* (Vol. 2, 933–935). Doi:<ALIGNMENT.qj></ALIGNMENT>10.1109/IEMBS.2004.1403313

Ulusoy, Y. Y., Chien, S. I.-J., & Wei, C.-H. (2010). Optimal all-stop, short-turn, and express transit services under heterogeneous demand. *Transportation Research Record: Journal of the Transportation Research Board*, 2197(1), 8–18. doi:10.3141/2197-02

UNISIG. (2013). ETCS Subset 026. ETCS System Requirement Specification (SRS) (Issue 3.3.0). (ERA, Ed.) Lille, Frankreich: ERA.

UNISIG. (2013). ETCS Subset 076-0. Test Plan (Issue 2.3.3). (ERA, Ed.) Lille, France: ERA.

UNISIG. (2013, 11 22). ETCS Subset 076-3. Methodology of Testing (Issue 2.3.11). (ERA, Ed.) Lille, France: ERA.

UNISIG. (2015). ETCS Subset 076-6-3. Test Sequences(Issue 3.0.0). (ERA, Ed.) Lille, Frankreich: ERA.

UNISIG. (2015). ETCS Subset 076-7. Test Sequence Validation and Evaluation / Scope of the Test. (ERA, Ed.) Lille, France: ERA.

UNISIG. (2015). ETCS Subset 094. Functional Requirements for an On-Board reference Test Facilitiy(Issue 3.0.0), 3.0.0. (ERA, Ed.) Lille, France: ERA.

Van Dongen, H. P. A., & Dinges, D. F. (2000). Circadian Rhythms in Fatigue, Alertness and Performance. In *Principles and Practice of Sleep Medicine* (pp. 391–399). Retrieved from http://www.nps.navy.mil/orfacpag/resumepages/projects/fatigue/dongen.pdf

vanLaarhoven, P.J.M., & Pedrycz, W. (1983). A fuzzy extension of Saaty's priority theory. *Fuzzy Sets and Systems*, *11*(1), 229–241. doi:10.1016/S0165-0114(83)80082-7

Vázquez-Abad, F. J., & Zubieta, L. (2005). Ghost simulation model for the optimization of an urban subway system. *Discrete Event Dynamic Systems*, *15*(3), 207–235. doi:10.1007/s10626-005-2865-9

Veelenturf, L. P., Nielsen, L. K., Maroti, G., & Kroon, L. G. (2011). Passenger oriented disruption management by adapting stopping patterns and rolling stock schedules. *Proceedings of the 4th International Seminar on Railway Operations Modelling and Analysis RailRome '11*.

Veelenturf, L. P., Potthoff, D., Huisman, D., Kroon, L. G., Maróti, G., & Wagelmans, A. P. (2012). *A recoverable robust solution approach for real-time railway crew rescheduling: Technical report*. Erasmus University Rotterdam.

Verma, A., Sudhira, H., Rathi, S., King, R., & Dash, N. (2010). Sustainable urbanization using high speed rail (HSR) in Karnataka, India. *Research in Transportation Economics*, 38(1), 67–77.

Viola, P., & Jones, M. (2001). Rapid object detection using a boosted cascade of simple features. *Proceedings of the 2001 IEEE Computer Society Conference on Computer Vision and Pattern Recognition. CVPR '01* (Vol. 1, pp. I–511–I–518). http://doi.org/ doi:10.1109/CVPR.2001.990517

Vision 2020 Board. Government of India - New Delhi: Vision 2020. (2009). *Indian Railways*. Retrieved from http://www.nwr.indianrailways.gov.in/uploads/files/1299058054467-englishvision.pdf

Voith, R. (1993). Changing capitalization of CBD-oriented transportation systems: Evidence from Philadelphia 1970-1988. *Journal of Urban Economics*, *33*(3), 361–376. doi:10.1006/juec.1993.1021

Vuchic, V. R. (1973). Skip-stop operation as a method for transit speed increase. *Traffic Quarterly*, *27*(2), 307–327.

Vuckovic, A., Radivojevic, V., Chen, A. C. N., & Popovic, D. (2002). Automatic recognition of alertness and drowsiness from EEG by an artificial neural network. *Medical Engineering & Physics*, *24*(5), 349–360. doi:10.1016/S1350-4533(02)00030-9 PMID:12052362

Walker, G., Snowdon, J. N., & Ryan, D. M. (2005). Simultaneous Disruption Recovery of a Train Timetable and Crew Roster in Real Time. *Computers & Operations Research*, *32*(8), 2077–2094. doi:10.1016/j.cor.2004.02.001

Wang, Y., De Schutter, B., van den Boom, T., Ning, B., & Tang, T. (2013). Real-time scheduling for single lines in urban rail transit systems. *Paper presented at the 2013 IEEE International Conference on Intelligent Rail Transportation (ICIRT)*. doi:10.1109/ICIRT.2013.6696258

Wang, J. Y., & Yang, H. (2005). A game-theoretic analysis of competition in a deregulated bus market. *Transportation Research Part E, Logistics and Transportation Review*, *41*(4), 329–355. doi:10.1016/j.tre.2004.06.001

Wang, L., Chu, J., & Wu, J. (2007). Selection of optimum maintenance strategies based on a fuzzy analytic hierarchy process. *International Journal of Production Economics*, *107*(1), 151–163. doi:10.1016/j.ijpe.2006.08.005

Wang, R., Kudrot-E-Khuda, M., Nakamura, F., & Tanaka, S. (2014). A Cost-Benefit Analysis of Commuter Train Improvement in the Dhaka Metropolitan Area. Bangladesh. *Procedia: Social and Behavioral Sciences*, *138*, 819–829. doi:10.1016/j.sbspro.2014.07.231

Waters, D. (2008). *Quantitative methods for business*. Prentice Hall.

Watts, D. J., & Strogatz, S. H. (1998). Collective dynamics of "small-world" networks. *Nature*, *393*(6684), 440–442. doi:10.1038/30918 PMID:9623998

Waymel, F., Monnoyer, F., & William-Louis, M. (2006). Numerical simulation of the unsteady three-dimensional flow in confined domains crossed by moving bodies. *Computers & Fluids, 35*, pp. 525-543.

Waymel, F., William-Louis, M., & Monnoyer, F. (2003). 3D simulation of airflow and heat transfer in a subway station with moving trains. *Proceedings of the 11th International Symposium on Aerodynamics & Ventilation in Vehicle Tunnels*.

Wedgwood, R. L. (1909). Statistics of railway costs. *The Economic Journal*, *19*(73), 13–31. doi:10.2307/2220505

Wegele, S. (2008). Automatic dispatching of train operations using a hybrid optimisation method. *Proceedings of the 8th World Conrgress on Rail Research (WCRR)(CD-Rom)*. Seoul, South Korea: Korea Railroad Corporation; Korea Rail Network Authority; Korea Railroad Research Institute.

Weißleder, S., & Schlingloff, H. (2011). Automatic Model-Based Test Generation from UML State Machines. In J. Zander, I. Schieferdecker, & P. J. Mosterman (Eds.), *Model-Based Testing for Embedded Systems* (pp. 77–109). Oxon, UK: CRC Press. doi:10.1201/b11321-5

White, J. (2008). *London in the Nineteenth Century: A Human Awful Wonder of God*. London: Vintage.

Whitmore, J., & Fisher, S. (1996). Speech during sustained operations. *Speech Communication*, *20*(1-2), 55–70. doi:10.1016/S0167-6393(96)00044-1

WHO. (2011). Burden of disease from environmental noise - Quantification of healthy life years lost in Europe. New York: World Health Organisation (WHO).

Wicken, A. H. (1993), Dynamic stability of articulated and steered railway vehicles guided by lateral displacement feedback. *Proceedings of the 13th IAVSD symposium*, Chengdu, China.

Wickens, A. H. (1994). Dynamic Stability of articulated and steered railway vehicles guided by lateral displacement feedback. *Vehicle System Dynamics*, *23*(Suppl.), 541–553. doi:10.1080/00423119308969539

Wickens, A. H. (2003). *Fundamentals of rail vehicle dynamics*. Lisse, The Netherlands: Swets & Zeitlinger. doi:10.1201/9780203970997

Wickens, A. H. (2006). A History of Railway Vehicle Dynamics. In S. Iwnicki (Ed.), *Handbook of railway vehicle dynamics* (pp. 5–38). Boca Raton, FL: CRC Press. doi:10.1201/9781420004892.ch2

Wiegmans, B. W. (2008). *The economics of a new rail freight line: The case of the Betuweline in the Netherlands.* Association for European Transport and contributors.

Wierwille, W. W., Wreggit, S. S., Kirn, C. L., Ellsworth, L. A., & Fairbanks, R. J. (n. d.). Research on vehicle-based driver status/performance monitoring; development, validation, and refinement of algorithms for detection of driver drowsiness (Final report). Retrieved from http://trid.trb.org/view.aspx?id=448128

Wijeweera, A., To, H., & Charles, M. B. (2013). An Econometrics Analysis of Freight Rail Demand Growth in Australia. *Proceedings of the 42nd Australian Conference of Economists Conference Proceedings beyond the Frontiers: New Directions in Economics*, Murdoch University, Perth, Western Australia.

Wijeweera, A., To, H., Charles, M. B., & Sloan, K. (2014). A time series analysis of passenger rail demand in major Australian cities. *Economic Analysis and Policy, 44*(3), 301–309. doi:10.1016/j.eap.2014.08.003

Williams, B. (2008). Intelligent transportation systems standards. London: Artech House.

Williams, G. W. (1963). Highway hypnosis: An hypothesis. *The International Journal of Clinical and Experimental Hypnosis, 11*(3), 143–151. doi:10.1080/00207146308409239 PMID:14050133

Wilson, J.R. (2007). *People and Rail Systems: Human Factors at the Heart of the Railway.* Ashgate Publishing, Ltd. Retrieved from https://books.google.com/books?hl=en&lr=&id=RLOrqGhSMOsC&pgis=1

Winston, C., & Maheshri, V. (2007). On the social desirability of urban rail transit systems. *Journal of Urban Economics, 62*(2), 362–382. doi:10.1016/j.jue.2006.07.002

Wolmar, C. (2004). *The Subterranean Railway. How the London underground railway was built and how it changed the city forever.* London: Atlantic Books.

Wong, O., & Rosser, M. (1978). Improving system performance for a single line railway with passing loops. *New Zealand Operational Research, 6*, 137–155.

Woolford, P. (2003). *Interlocking Principles.* London, UK: Railway Safety and Standards Board Ltd.

Wu, Q., Sun, B., Xie, B., & Zhao, J. (2010). A PERCLOS-based driver fatigue recognition application for smart vehicle space. *Proceedings of the 3rd International Symposium on Information Processing ISIP '10* (pp. 437–441). Doi:<ALIGNMENT.qj></ALIGNMENT>10.1109/ISIP.2010.116

Wüest, R., Laube, F., Roos, S., & Caimi, G. (2008). Sustainable Global Service Intention as objective for Controlling Railway Network Operations in Real Time.*Proceedings of the 8th World Congress of Railway Research (WCRR)*, Seoul, Korea.

Wu, K. Y. (2011). Applying the fuzzy Delphi method to analyze the evaluation indexes for service quality after railway re-opening: using the old mountain line railway as an example.*Proceedings of the 15th WSEAS international conference on Systems* (pp. 474-479).

Xu, B., Sun, L., & Ren, H. (2010). A New Criterion for the Quantification of Broken Rotor Bars in Induction Motors. *IEEE Transactions on Energy Conversion, 25*(1), 100–106. doi:10.1109/TEC.2009.2032626

Xu, B., Sun, L., Xu, L., & Xu, G. (2012). An ESPRIT-SAA-Based Detection Method for Broken Rotor Bar Fault in Induction Motors. *IEEE Transactions on Energy Conversion, 27*(3), 654–660. doi:10.1109/TEC.2012.2194148

Xu, B., Sun, L., Xu, L., & Xu, G. (2013). Improvement of the Hilbert Method via ESPRIT for Detecting Rotor Fault in Induction Motors at Low Slip. *IEEE Transactions on Energy Conversion, 28*(1), 225–233. doi:10.1109/TEC.2012.2236557

Compilation of References

Yadaiah, N., Kumar, A. G. D., & Bhattacharya, J. L. (2004). Fuzzy based coordinated controller for power system stability and voltage regulation. *Electric Power Systems Research, 69*(2-3), 169–177. doi:10.1016/j.epsr.2003.08.008

Yalçınkaya, Ö. (2010). *A feasible timetable generator simulation modelling framework and simulation integrated genetic and hybrid genetic algorithms for train scheduling problem* [Unpublished doctoral dissertation]. Dokuz Eylül University, İzmir, Turkey.

Yalçınkaya, Ö., & Bayhan, G. M. (2012). A feasible timetable generator simulation modelling framework for train scheduling problem. *Simulation Modelling Practice and Theory, 20*(1), 124–141. doi:10.1016/j.simpat.2011.09.005

Yang, J., & Zhang, D. (2004). Two-dimensional PCA: a new approach to appearance-based face representation and recognition. *IEEE Transactions on Pattern Analysis and Machine intelligence, 26*(1). Retrieved from http://ieeexplore.ieee.org/xpls/abs_all.jsp?arnumber=1261097

Yang, C., Kang, T.-j., Lee, S. B., Yoo, J.-y., Bellini, A., Zarri, L., & Filippetti, F. (2014). Screening of False Induction Motor Fault Alarms Produced by Axial Air Ducts based on the Space Harmonic-Induced Current Components. *IEEE Transactions on Industrial Electronics, 0046*(c), 1–1.

Yao, J.-S., Chen, M.-S., & Lin, H.-W. (2005). Valuation by using a fuzzy discounted cash flow model. *Expert Systems with Applications, 28*(2), 209–222. doi:10.1016/j.eswa.2004.10.003

Yeo, M.V.M., Li, X., Shen, K., & Wilder-Smith, E.P.V. (2009). Can SVM be used for automatic EEG detection of drowsiness during car driving? *Safety Science, 47*(1), 115–124. doi:10.1016/j.ssci.2008.01.007

Yeom, H.-G., & Hwang, S.-M. (2009). A Study on Tool for supporting the Software Process Improvement based on ISO /IEC 15504. *International Journal of Software Engineering and Its Applications, 3*, 1–8.

Ye, Z., Wu, B., & Sadeghian, A. (2003). Current signature analysis of induction motor mechanical faults by wavelet packet decomposition. *IEEE Transactions on Industrial Electronics, 50*(6), 1217–1228. doi:10.1109/TIE.2003.819682

Yin, X., Han, J., & Yu, P. S. Truth discovery with multiple conflicting information providers on the web. Proceedings of SIGKDD'07. doi:10.1145/1281192.1281309

Ying, X. (2010). Performance Evaluation and Thermal Fields Analysis of Induction Motor. *IEEE Transactions on Magnetics, 46*(5), 1243–1250. doi:10.1109/TMAG.2009.2039221

Yoko, T., & Norio, T. (2005). Robustness Indices for Train Rescheduling. *Proceedings of the 1st International Seminar on Railway Operations Modelling and Analysis*, Delft, The Netherlands.

Yu, C. S. (2002). A GP-AHP method for solving group decision-making fuzzy AHP problems. *Computers & Operations Research, 29*(14), 1969–2001. doi:10.1016/S0305-0548(01)00068-5

Yu, G., & Qi, X. (2004). *Disruption management*. World Scientific. doi:10.1142/5632

Zahran, E., & Abbas, A. (2009). High performance face recognition using PCA and ZM on fused LWIR and VISIBLE images on the wavelet domain. *Proceedings of the international conference on computer engineering & systems ICCES '09*. Retrieved from http://ieeexplore.ieee.org/xpls/abs_all.jsp?arnumber=5383223

Zaky, M. S., Khater, M. M., Shokralla, S. S., & Yasin, H. (2009). Wide-Speed-Range Estimation With Online Parameter Identification Schemes of Sensorless Induction Motor Drives. *IEEE Transactions on Industrial Electronics, 56*(5), 1699–1707. doi:10.1109/TIE.2008.2009519

Zeileis, A., Kleiber, C., & Jackman, S. (2007). Regression models for count data in R. Research report series. *Vienna Univ. of Economics and Business Administration*.

Zhang, X., Liu, Z., & Wang, H. (2013). Lessons of Bus Rapid Transit from Nine Cities in China. *Transportation Research Record: Journal of the Transportation Research Board*, *2394*, 45–54. doi:10.3141/2394-06

Zhou, Q., Prasanna, V. K., Wang, Y., & Chang, H. (2009, September). A Semantic Framework for Integrated Modeling and Simulation of Transportation Systems. *Proceedings of theIFAC Symposium on Control in Transportation Systems (CTS '09)*.

Zhou, X., & Zhong, M. (2005). Bicriteria train scheduling for high-speed passenger railroad planning applications. *European Journal of Operational Research*, *167*(3), 752–771. doi:10.1016/j.ejor.2004.07.019

Zhou, X., & Zhong, M. (2007). Single-track train timetabling with guaranteed optimality: Branch-and-bound algorithms with enhanced lower bounds. *Transportation Research Part B: Methodological*, *43*(3), 320–341. doi:10.1016/j.trb.2006.05.003

Zhu, G., Li, Y., & Wen, P. P. (2014). Analysis and Classification of Sleep Stages Based on Difference Visibility Graphs from a Single Channel EEG Signal. *IEEE Journal of Biomedical and Health Informatics*, *18*(6), 1813-1821. Doi:<ALIGNMENT.qj></ALIGNMENT>10.1109/JBHI.2014.2303991

Zhu, Z.Z.Z., & Ji, Q.J.Q. (2004). Real time and non-intrusive driver fatigue monitoring. *Proceedings of the 7th International IEEE Conference on Intelligent Transportation Systems (IEEE Cat. No.04TH8749)*. doi:<ALIGNMENT.qj></ALIGNMENT>10.1109/ITSC.2004.1398979

Zidani, F., Diallo, D., Member, S., El, M., Benbouzid, H., & Nait-Said, R. (2008). A Fuzzy-Based Approach for the Diagnosis of Fault Modes in a Voltage-Fed PWM Inverter Induction Motor Drive. *IEEE Transactions on Industrial Electronics*, *55*(2), 586–593. doi:10.1109/TIE.2007.911951

Zimmermann, A., & Hommel, G. (2005). Towards modeling and evaluation of ETCS real-time communication and operation. *Journal of Systems and Software*, *77*(1), 47–54. doi:10.1016/j.jss.2003.12.039

Zito, P., & Salvo, G., & LaFranca, L. (2011). Modelling Airlines Competition on Fares and Frequencies of Service by Bi-Level Optimization. *Procedia: Social and Behavioral Sciences*, *20*, 1080–1089. doi:10.1016/j.sbspro.2011.08.117

About the Contributors

Min An (BEng, MSc, PhD, CEng, MCICE, MIMechE, MIEngD) is a Professor (Reader) of Transport Risk Management in the University of Birmingham and Beijing Jiaotong University. He is the Director of MSc in Civil Engineering and MSc in Civil Engineering and Management, Leader of Safety, Risk ands Reliability Engineering, and Head of JBM Accreditation Response Committee on H&S. He is an Editor/ Associate Editor/ Member of Editorial Boards for 11 International Journals, a member of international Assessor Boards for UK EPSRC, Hong Kong Engineering and Sciences Research Council Hong Kong City University Research Committee. He has been involved in developing and applying more rational and sustainable safety, risk, reliability and decision-making techniques and methods to facilitate engineering safety and reliability analyses. This work has been sustained over the past twenty-six years and has resulted in over one hundred technical papers related to railway and transportation safety and reliability. He is member of Organisation/Scientific Committees for many International Conferences, and has delivered many keynote lectures at conferences and seminars. Min An has also delivered many workshops to transfer his research results to the industry. He has been awarded a special prize, two first-prizes and two second-prizes in advanced science and technology for his scientific research.

Pablo Arboleya: Received the M.Eng. and Ph.D. (with distinction) degrees from the University of Oviedo, Gijon, Spain, in 2002 and 2005, respectively, both in electrical engineering. He is Senior member of the IEEE Power and Energy Society since 2013 and was a recipient of the University of Oviedo Outstanding Ph.D. Thesis Award in 2008. Nowadays, he works as an Associate Professor in the Department of Electrical Engineering at the University of Oviedo (with tenure since 2010). In 2013, he was a visitor professor at the University of Rome (La Sapienza) and In 2014, he was a visitor professor at the University of Illinois at Urbana-Champaign. He worked several years in the field of electrical machines design and faults detection, participating in more than 20 research projects with companies like ABB, Arcelor, EdP, REE, Hidrocantabrico, CAF. Presently his main research interests are focused in the electrical networks modeling and analysis techniques, he worked with conventional transmission network analysis but he also developed models for DC traction networks and special distribution networks like microgrids with embedded distributed generators. Pablo Arboleya is the author of a large number of publications that have appeared in the leading journals in the field of Electric Machines and Power systems and have been presented at a wide array of international conferences. Regarding his teaching activities, he is the coordinator in a International Master Course in "Electrical Energy Conversion and Power Systems" (EECPS).

Lennart Asbach studied at Dresden University of Technology in Germany. He holds a Master in Mechatronics and has joined the German Aerospace Centre (DLR) in 2009. Since then his main field of research is conformity and interoperability lab-testing of complex systems (e.g. ETCS). Since 2012 he is the leader of the Railway Technology team and the head of the laboratory RailSiTe (Railway Simulation and Testing), which is accredited under the ISO 17025 for ETCS Conformity Tests. He has several publications in this field and is leading different national and international projects in the field of testing.

Ronita Bardhan is an Assistant Professor at Centre for Urban Science and Engineering, Indian Institute of Technology Bombay, India. She holds a PhD in Urban Engineering from The University of Tokyo, Japan, a Masters in City & Regional Planning from Indian Institute of Technology Kharagpur and a Gold Medal for Bachelor's in Architecture from Indian Institute of Engineering Science and Technology, India. Dr. Bardhan was nominated by Ministry of Human Resource Development, Government of India as a government scholar and is recipient of the prestigious Japanese Government Monbukagakusho Scholarships. She was awarded the Young Researcher Award-2012, Global Center of Excellence for Sustainable Urban Regeneration, The University of Tokyo, Japan, for her contribution towards effective and innovative research which contributes to sustainable urban regeneration.

Devasish Basu has a B. Tech. (Elect. Engg.) from the Institute of Technology- BHU (now IIT-BHU), Varanasi in 1993, a M. Tech. (Power Systems/ Elect. Engg.) from the Institute of Technology- BHU (now IIT-BHU), Varanasi in 1999, and a MBA (PGDM) from Indian Institute of Management- Calcutta in 2010. He is currently pursuing a Ph.D. from IIT, Kharagpur.

Thomas Böhm studied at University of Magdeburg in Germany and the EPFL Lausanne in Switzerland. He holds a Master in Computer Science in Engineering and has joined the German Aerospace Centre (DLR) in 2007. Since then his main field of research is condition based maintenance and health management using statistics and data mining. Since 2010 he is leader of the Life Cycle Management team with whom he further advanced the topic of efficient, predictive maintenance of railway infrastructure. He has several publications in this field and is session chairman of the International Congress of Condition Monitoring and Diagnostic Engineering Management. He worked in and led several international and national Projects as well as in development projects for diagnostic and prognostic algorithms for Deutsche Bahn.

Nebojša Bojović received the M.Sc. degree in Railway Transport Engineering, Magisterial degree. and the Ph.D. degree in Traffic and Transport area, all from the Faculty of Transport and Traffic Engineering, University of Belgrade, Serbia. At present, he is full time professor at the Department of Management in Railway, Rolling stock and Traction. His current areas of research are in the fields of operations research with special emphasis to the area of transport research.

Christophe Butaud Ventilation engineer in underground spaces at egis tunnels since 2004.

Aritra Chaudhuri has received his B.E. degree from Bengal Engg. and Science University, Shibpur, in Electrical Engineering. In 2010 he received his M. Tech in Instrumentation from IIT Kharagpur in 2012. He is currently pursuing his Ph.D. in the Advanced Technology Development Center, IIT Kharagpur, India.

About the Contributors

Anirban Dasgupta has received his B.Tech. (Hons.) degree from National Institute of Technology (NIT), Rourkela, India in 2010 and his M.S. degree from the Indian Institute of Technology (IIT)Kharagpur, in 2014. He is currently pursuing his Ph.D at the Department of Electrical Engineering, IIT Kharagpur, India. His research interests include computer vision, embedded systems, signal and image processing.

Pallab Dasgupta holds a B.Tech, a M.Tech and a Ph.D in Computer Science & Engineering from the Indian Institute of Technology Kharagpur. He is a professor in Computer Science and Engineering at IIT Kharagpur. His research interests include Formal Verification, Artificial Intelligence and VLSI. He has over 160 research papers and 3 books in these areas. He has collaborated with several industries as a consultant in formal verification, including Intel, Synopsys, General Motors, Freescale, Semiconductor Research Corporation and Indian Railways. He is a Fellow of the Indian National Academy of Engineering, Fellow of the Indian Academy of Sciences, and a Fellow of the Institution of Electronics and Telecommunication Engineers, India.

Carlos Lopez Galviz is a Lecturer in the Theories and Methods of Social Futures at Lancaster University. He has been a consultant in several planning initiatives in Europe and South America and was in 2014 a visiting scholar of the Shanghai Academy of Social Sciences with the project 'Past Futures: 19th-century London and Paris and the future of Chinese cities.' His contribution to this volume is partly based on this project. Currently, he leads Re-configuring Ruins (www.reconfigruins.com), a collaborative and inter-disciplinary project funded by the Arts and Humanities Research Council in the UK which examines the relationship between the arts, heritage and urban regeneration. Carlos has published widely on the history of London and Paris, including *Going Underground: New Perspectives* (co-editor; London, 2013) and, more recently, the co-edited volume *Global Undergrounds: Exploring Cities Within* (Reaktion Books, 2016).

Anjith George has received his B.Tech. (Hons.) degree from Calicut University, India in 2010. and a M-Tech degree from the Indian Institute of Technology (IIT) Kharagpur, India in 2012. Presently, he is pursuing a Ph.D. from IIT Kharagpur. His current research interests include real-time computer vision and its applications.

Michael Meyer zu Hörste was born in 1969. He studied mechanical engineering at the Technical University of Braunschweig, Braunschweig, Germany and graduated in 1995. From 1995 to 2001 he was working as a scientific staff member at the Institute of Control and Automation Engineering of the Technical University of Braunschweig, Germany, in the field of railway safety. In 2004, he received his PHD in mechanical engineering for his PHD-thesis about modeling and simulation of the generic behavior of train control systems. Since 2001, he has been working as scientific staff at the German Aerospace Center (DLR) in the Department of Railway Systems of the Institute of Transportation Systems in Braunschweig, Germany. His main research focuses on the European Train Control System (ETCS), the European Rail Traffic Management System (ERTMS), interlocking and train localization.

Hardi Hungar Since 2012, he has been a researcher at the German Aerospace Center (DLR), Institute of Transportation Systems, Dept. Rail Systems, concerned with test methods and technology for signaling and safety equipment. Before that, he worked for more than 20 years at several research organizations and universities and in the industry on formal methods and their applications in the development of embedded systems. His academic qualifications include a habilitation (1998, Carl von Ossietzky University Oldenburg, Germany) and a PhD (1991, Christian-Albrechts University Kiel).

Jong-Gyu Hwang received the B.S and M.S. degrees in Electrical Engineering from Konkuk University, Korea in 1994 and 1996, respectively, and a PH. D in Electrical and Computer Engineering, Hanyang University in 2005. As of 1995, he has been a Principal Researcher and a Professional Engineer on the Railway Signaling System with the Korea Railroad Research Institute. He was a visiting scholar at Virginia Commonwealth Univ. from 2011 to 2012. His research interests are in the areas of railway signaling systems, computer network technology, PRT (Personal Rapid Transit) systems, and software testing of embedded systems.

Arnab Jana is an Assistant Professor at the Centre for Urban Science and Engineering, Indian Institute of Technology (IIT) Bombay, India. He works in the area of urban infrastructure policy and planning, primarily focusing on public health policy, application of ICT in urban and regional planning, sustainability & environmental issues. He holds a PhD in Urban Engineering from The University of Tokyo, Japan. He was awarded the prestigious MEXT Scholarship by the Ministry of Education, Culture, Sports, Science, and Technology, Govt. of Japan. He is also an alumnus of the Asian Program of Incubation of Environmental Leadership (APIEL), the University of Tokyo. Dr. Jana holds a Master's degree in City Planning from IIT Kharagpur, where he was awarded the Institute Medal. He has also worked in consulting firms, and has professional and academic experience of working in several countries.

Hyun-Jeong Jo received a B.S. degree from the Hankuk Aviation University, Goyang, Gyonggi-do, Korea, in 2003. She worked towards a M.S. degree at the Gwangju Institute of Science and Technology (GIST), Gwangju, Korea. Since 2005, she has been with the Train Control System Research Team of the Korea Railroad Research Institute (KRRI). His research interests are the areas of railway signaling, software safety, and communication application technology.

BIbek Kabi obtained his M.S. (by research) degree from IIT Kharagpur in 2015. His research interests include Fixed-point Arithmetic and speech processing.

Jillella Satya Sai Kumar is an accomplished senior international expert in urban planning, sustainable transportation, logistics and ITS areas with over 24 years of professional experience in research, consultancy, policy advice, capability building, technical assistance and providing domain based technical solutions worldwide. He served in various leadership roles and got associated with reputed organizations namely; United Nations ESCAP, CiSTUP-Indian Institute of Science, Infosys, ITC InfoTech, to name a few. Currently, he is pursuing his doctoral research on value capture-based innovative transit financing strategies at Curtin University, Australia. In addition, he holds two complimenting master's degrees: firstly in City Planning from School of Architecture and Planning, Anna University and the second master's in Transportation Engineering from the Asian Institute of Technology (AIT), respectively. He holds a bachelor's degree in Civil Engineering from Jawaharlal Nehru Technological University, India and also is an MDP alumni of IIM-Ahmedabad. He is committed to the principles of green cities and sustainable mobility planning with technical aspects and deliberative consultation approaches.

About the Contributors

Christoph Lackhove is Leader of the Railway Operations team at the Railway Systems department of the Institute of Transportation Systems has joined the German Aerospace Centre (DLR) in 2007. His main field of research are the migration of new technologies in the railway domain as well as low noise railway operations. He has several publications in his field and gained his PhD about the optimized migration of ETCS in Europe. He worked in several national and international projects.

Mahesh Mangal holds a BE degree in Electronics and Communication from MBM Engineering college, Jodhpur and a M.Tech degree in Reliability Engineering from Indian Institute of Technology, Kharagpur. He has more than 37 years of experience in various fields of Indian Railways which includes Administration, research, design, project, maintenance and operation of signaling and telecommunication. He had an extensive tenure in Research Design and Standards Organization, Lucknow where he was involved in devolvement of new signaling technologies including testing and evaluation of advance signaling system for use on Indian Railways as cross approved product. He was also involved in development of "Train collision Avoidance System" a Train Control system which is presently undergoing field evaluation. He has also worked as Director, Network Planning and marketing in RailTel Corporation of India Limited, an Indian Railways PSU, also as Divisional Railway Manager, Bangalore and Additional Member, Telecom in Railway Board, New Delhi. Mr. Mangal is currently working as General Manger, Railway Electrification and is involved in Railway Electrification projects on Indian Railways.

Soheil Mardani graduated in Industrial Engineering from Tarbiat Modares University. An expert in using simulation solutions and engines including Enterprise Dynamics, Aimsun and EMME. He has directed simulation projects at the national level, including an optimized time table generator for the Iranian Railway Company.

Anne (Annie) Matan is a researcher and a lecturer at Curtin University Sustainability Policy (CUSP) Institute in Australia, interested in creating sustainable, vibrant and people-focused urban places. Her research focus is on walkability, pedestrian planning, urban design and transportation planning. She has worked in both the State and Local Government, and joined CUSP in 2011 after finishing her PhD at Curtin University.

Vlastimil Melichar was born in 1950. He received magisterial degree in 1973 in Railway Transport Operation and Economics and the Ph.D. degree in Sectoral and Cross-Economy in 1984 all from the Faculty of Transport Operation and Economics University of Transport and Communication of Žilina. Professor of Technology and Management in Transport was appointed in 2002 at the University of Pardubice. At present he is professor at the Department of Transport Management, Marketing and Logistics at the Jan Perner Transport Faculty of University of Pardubice. His current areas of research are in the fields of analysis and modeling of factors affecting supply and demand in the transport sector, cost-benefit analysis, which acts on the efficiency of transport, logistics reengineering processes in transport and evaluating the effectiveness of investment in transport.

Miloš Milenković received the M.Sc. degree in Railway Transport Engineering, Magisterial degree. and the Ph.D. degree in Traffic and Transport area in 2013., all from the Faculty of Transport and Traffic Engineering, University of Belgrade, Serbia. At present, he is a research fellow at Zaragoza Logistic

Center (Spain) and assistant professor at the Faculty of Transport and Traffic Engineering (University of Belgrade, Serbia). His current areas of research are in the fields of time series analysis, combinatorial optimization, operations research with special emphasis to the area of railway transport.

Arunava Naha received B.E. degree in Electrical Engineering from Bengal Engineering and Science University, Shibpur, West Bengal in 2006. He recently submitted his thesis for MS (by Research) to the Electrical Engineering Dept. of Indian Institute of Technology, Kharagpur in 2013. He is currently working towards his PhD degree in Electrical Engineering at IIT Kharagpur. His research interests cover induction motor fault modeling and diagnosis under light loading conditions using statistical and intelligent signal processing.

Sundaravalli Narayanaswami earned her PhD in Industrial Engineering and Operations Research from IIT Bombay, after a Master's in Computer Science. As part of her PhD thesis, she developed different models for transportation rescheduling in presence of disruptions. Her current research spans automation of transportation operations. She is also exploring certain topics in the sustainability of urban transportation and emerging ICT applications in few domains. Her teaching interests are in Intelligent Transportation systems, Operations Research, Operations Management, Urban transportation, in addition to Management Information systems and Knowledge Management. She was earlier affiliated with Mumbai University and Federal education institutes in UAE and she has also taught in many Executive development programs. She publishes and reviews regularly and presents her research work in International forums. Prof Narayanaswami's early career was in IT services marketing and electronics equipment design and production. She is a life member of several professional associations and holds a Fellowship from the British Computer Society. She also serves on the Editorial board of the Annals of Management Science.

Erfan Hassan Nayebi is a PhD student in Industrial Engineering at Tarbiat Modares University, Iran. He received his M.Sc. for industrial engineering at Sharif university of Technology in 2012. During his career at SIMARON he has contributed to a wide variety of simulation-based optimization studies for railway companies. His research interests are in the area of mathematical programming, simulation and meta-heuristic algorithms.

Peter Newman is the Professor of Sustainability at Curtin University and Director of CUSP. His books include 'Green Urbanism in Asia' (2013), 'Resilient Cities: Responding to Peak Oil and Climate Change' (2009), 'Green Urbanism Down Under' (2009) and 'Sustainability and Cities: Overcoming Automobile Dependence' with Jeff Kenworthy which was launched in the White House in 1999. In 2001-3, Peter directed the production of Western Australia's Sustainability Strategy in the Department of the Premier and Cabinet. In 2004-5, he was a Sustainability Commissioner in Sydney advising the government on planning and transport issues. In 2006/7 he was a Fulbright Senior Scholar at the University of Virginia Charlottesville. Peter was on the Board of Infrastructure Australia 2008-14 and was a Lead Author for Transport on the IPCC for their 5th Assessment Report. In 2011, Peter was awarded the Sidney Luker medal by the Planning Institute of Australia (NSW) for his contribution to the science and practice of town planning in Australia and in 2014 he was awarded by the Order of Australia for his contributions

About the Contributors

to urban design and sustainable transport, particularly related to the saving and rebuilding of Perth's rail system. He was an elected Fremantle City Councilor from 1976-80 where he still lives. He is working in Bangalore and Pune on a project funded by the Australian Government called 'Deliberative Democracy and Sustainable Transport in India.'

Smitirupa Pradhan is currently pursuing research on railway vehicle dynamics at the Center for Railway Research, Department of Mechanical Engineering, IIT Khragpur. She has a master's degree in Machine Design from IIT Kharagpur. She has worked as a teacher for 5 years.

Yong Qin is a Professor of Rail Traffic Control and Safety in State Key Laboratory of Rail Traffic Control and Safety at Beijing Jiaotong University. He is also the Deputy Dean of the Beijing Research Center of Urban Traffic Information Intelligent Sensing and Service Technologies, the Deputy Dean and the General Secretary of the Rail Transportation Electrotechnical Committee of China, Electrotechnical Society, a member of the Intelligent Automation Committee of Chinese Automation Association, and a member of Fuzzy Mathematics and Systems Committee of Systems Engineering Society of China. Professor Yong Qin's expertise is mainly in the areas of intelligent transportation systems, railway safety and emergency management, rail network management and traffic models. He is the author/co-author of more than one hundred journal and conference papers and has published 5 books. He has been awarded 7 prizes in science and technology by the Chinese Ministry of Transportation and the Chinese Ministry of Education for his research achievements.

Varun Raturi completed his B-Tech at NIT Nagpur and his M-Tech at IISc Bangalore. Currently, he is a research scholar in the Dept. of Civil Engg., IISc Bangalore, working under Dr. Ashish Verma.

Aurobinda Routray is a professor in the Department of Electrical Engineering, Indian Institute of Technology, Kharagpur. His research interest includes non-linear and statistical signal processing, signal based fault detection and diagnosis, real time and embedded signal processing, numerical linear algebra, and data driven diagnostics.

Anik Kumar Samanta received his B. Tech. degree from Dr. B. C. Roy Engineering College, Durgapur, India, in 2011. He is currently working towards his M.S. degree from Indian Institute of Technology Kharagpur. He is also associated with the Centre for Railway Research for designing diagnostics of induction motors for the Indian Railways. His research interests include high-resolution spectral estimation, signal based fault detection and diagnosis of induction motors, real-time and embedded signal processing.

Arun K. Samantaray is a full professor at IIT Kharagpur. His research areas are vehicle dynamics, systems and control, non-linear mechanics and rotor dynamics.

Anwesha Sengupta was born in Kolkata, West Bengal, India in 1982. She received B.E. and M.S. degrees in Electrical Engineering from Jadavpur University and the Indian Institute of Technology in 2005 and 2011, respectively. She is currently pursuing her Ph.D. in the Department of Electrical Engineering at the Indian Institute of Technology since 2011. Her research interests include signal processing and EEG analysis.

Jörg Schimmelpfennig is Professor of Theoretical and Applied Microeconomics at Ruhr University Bochum, Germany. His main research areas are economic regulation, rail transport economics and defense economics. He is a regular contributor at international conferences and has published in leading academic journals. He is a member of, inter alia, the Institute for Defense and Government Advancement, the U.S. Naval Institute, the Naval Historical Foundation, the Air Force Association, the Army Historical Foundation, the Royal United Services Institute and the Army Records Society. He served as an advisor to renowned institutions and companies as well as regulatory authorities. He is also a contributor and reviewer on the arts. Dr Schimmelpfennig studied mathematics, physics and economics at the University of Bielefeld, Germany and obtained his Doctorate in Economics from the University of Osnabrück, Germany.

T. G. Sitharam has a Master's from the Indian Institute of Science, Bangalore in 1986 and a Ph.D. from the University of Waterloo, Waterloo, Ontario, Canada in 1991. Further he was a research engineer at University of Texas at Austin, Texas, USA until 1994. He has served in many organizations like Govt of India, University of Waterloo, Canada, University of Texas at Austin, USA, and Yamaguchi University, Japan. Since 1994, he has been at the Indian Institute of Science and presently he is a professor in the department of Civil Engineering. He was the founder Chairman of the Centre for infrastructure, Sustainable Transportation and Urban Planning (CiSTUP) at the Indian Institute of Science from Jan, 2009 to March 2014, Bangalore. He is the Chief Editor of the International journal of Geotechnical Earthquake Engineering published by IGI Global, PA, USA. He was also an Associate Editor (AE) for the ASCE Journal of Materials in Civil Engineering, USA for the period 2006-2009; an Editorial Board member of Geotechnique letters, ICE, UK (for 3 years), editorial board member of EJGE and also a Member, Committee on Soils and Rock Instrumentation (AFS20), Transportation Research Board of the National Academies, Division of National research Council (NRC), USA for the period 2007-2009. Professor Sitharam has successfully guided 25 Ph.D students and has published more than 150 papers in international and national journals and 250 publications in conferences of repute. He has an H-index of 25 and i-10 index of 50. He has written six textbooks and has also guest edited several volumes for Current Science, the Journal of earth system science and other journals. He is a recipient of the "Sir C.V. Raman State Award for Young Scientists," from the Government of Karnataka in the year 2004, in recognition and appreciation of exceptional contributions made to Engineering Sciences. He is also the recipient of the 1998 S.P. Research award (SAARC) and the 2014 Prof. Gopal ranjan award from IIT Roorkee. He was a visiting professor in the Department of Civil Engineering at Yamaguchi University, Ube, Japan, for one year (1999-2000). He was a visiting professor at the Technical University of Nova Scotia, Halifax, Canada and University of Waterloo, Waterloo, Ontario. He was a William Mong fellow at the University of Hong Kong, Hong Kong during 2011. Prof. Sitharam is a life member of several professional organizations including ISSMGE, ISRM, JGS, IGS. He is also a member of many committees formed by the Government of Karnataka, India and advises on infrastructure and urban infrastructure-related topics.

Sumeeta Srinivasan is a lecturer in the Department of Urban and Environmental Policy and Planning at Tufts University, Medford, MA, USA. She teaches courses in Geographic Information Systems. She is an affiliate of the China project at the School of Engineering and Applied Sciences at Harvard University. Her research interests are in Transportation and land use planning. She has a PhD from the Massachusetts Institute of Technology, Cambridge MA, USA.

About the Contributors

Libor Švadlenka graduated the University of Pardubice, Jan Perner Transport Faculty, the course of Transport Management, Marketing and Logistics in 2001. He vindicated the dissertation „Specific aspects of management in the postal services "in 2004. In 2004 he worked at the Board of management of Czech Post as an Expert of postal transportation. Since 2005 he has been working as a lecturer of department of Transport Management, Marketing and Logistics at the department of Transport Management, Marketing and Logistics of University of Pardubice. From 2007 he has been a head of the section "Economy and Management of Communications" at this department. In 2010 he has obtained degree Assoc. Prof at the University of Zilina with the paper called "Ensuring Universal Postal Service in the Conditions of Fully Liberalised Postal Market". In his revolting activities are focused on management and marketing in postal services, next on transport and communication system of the Czech Republic and the problem of electronic business. He was solver of Transformation and development programme of Ministry of Education, Youth and Sports "Introduction of the new field of study Management, Marketing and Logistics in communications" and currently he is guarantee of this field of study. He is the supervisor of students of doctor study programme. He participates on solving of grant projects of Czech Science Foundation and on projects of institutional research. Currently he lectures on international and national conferences and publishes in revolting papers.

Ashish Verma is a Ph.D. from IIT Bombay and currently serving as an Asst. Prof. of Transportation Engg. in the Dept. of Civil Engg. and Centre for infrastructure, Sustainable Transportation, and Urban Planning (CiSTUP) at the Indian Institute of Science (IISc), Bangalore, India. Before joining IISc, he has served at IIT Guwahati, and for the Mumbai Metropolitan Region Development Authority (MMRDA). His research interest are in; sustainable transportation planning, public transport planning and management, modeling and optimization of transportation systems, application of geoinformatics in transportation, driver behaviour and road safety, intelligent transportation system (ITS) and traffic management. He has authored more than 70 research publications in the areas of sustainable transportation and road safety. He is an editorial board member of several leading international journals of Elsevier, Springer, the American Society of Civil Engineers (ASCE) and other publishers. He has also been a guest editor of special issues of leading journals for Elsevier, ASCE, and Current Science, focused on sustainable transportation in India and developing countries. He has received DAAD and CONNECT fellowships from the Humbolt foundation in the past, for collaborative research with Germany. He has visited and delivered guest lectures at several international universities such as, the Univ. of Texas at Austin, the Imperial College, London, the Swiss Federal Institute of Technology, Leeds University, UK. He has also delivered Prof. SVC Aiya memorial and World Habitat Day themed lectures for the Institutes of Electronics and Telecommunication Engineers (IETE) and Institution of Engineers (India), respectively. He has been involved in many national and internationally funded research projects related to sustainable transportation and road safety. He is the founding President of the Transportation Research Group of India (TRG, www.trgindia.org), which is a registered society whose mission is to aid India's overall growth through focused transportation research, education, and policies in the country. He is also presently serving as a Country Representative from India and as a Scientific Committee Member of the World Conference on Transport Research Society (WCTRS) based in Lyon, France.

Frederic Waymel Engineer and PhD in Fluid mechanics and Energetics focused on the study of thermal and ventilation effects in subway stations at the University of Valenciennes Frederic has 15 years of experience in tunnel ventilation. Frederic has started his career at the French National Institute of Environment and Industrial Risks in the fire and ventilation department. He was in charge of research activities and project manager in the field of ventilation and fire in underground spaces. Frederic works for Egis Tunnels since 2007 as a tunnel ventilation expert and has been involved in the design and commissioning of ventilation systems in several road, rail and metro projects. Frederic is also an active member of the PIARC world road association.

Özgür Yalçınkaya was born in 1978, in Turkey. He completed his BSc degree in 2000 from the Industrial Engineering Department of Dokuz Eylül University (Turkey), ranked first in the department. He completed his MSc degree in 2004 with a thesis entitled "An optimization study for urban public transport system," and a PhD degree in 2010 with a dissertation entitled "A feasible timetable generator simulation modelling framework and simulation integrated genetic and hybrid genetic algorithms for train scheduling problem" from the same university. He spent a year (from November 2013 to October 2014) at the Technical University of Braunschweig (Germany) for his Postdoc research at the Institute of Transport, Railway Construction and Operation and the Institute of Railway Systems Engineering and Traffic Safety. His academic research interests are: simulation and simulation optimization, train scheduling problem, response surface methodology, metamodelling, evolutionary algorithms, genetic algorithms, strategic management, strategic planning and the organization of the industrial engineering profession. He is married and has a son.

Index

A

AC Traction Systems 456, 470, 480
Accidents 23-29, 31-32, 35, 37, 40-41, 46, 55, 60, 175, 179, 197, 228, 273-276, 300, 322, 389, 391-392, 395, 438
Adjacency Matrix 293-294
Air Handling Units 592-593
Alertness 273-279, 290, 295-297, 300
American Public Transportation Association (APTA) 526
appraisal techniques 67-68, 83, 97
Arima 119-120, 126
Artificial intelligence 502
Asset and Maintenance Management System (AMM) 419
Automatic Control 228, 438
average waiting time 420, 422, 440-444, 446

B

Bearing Faults 491, 495-497, 516
Block Section 223, 450
Broken Rotor Bar 491, 516
Built environment 23-25, 35, 37

C

cash flow 68-70, 72-73, 76-80, 83
Catenary 461, 464, 479, 487
central parts 26, 35
Characteristic Path Length 293-294, 296, 310
Clustering Coefficient 293-294, 296, 310
Coding rules 234-236, 238-241, 248
coding standards 233-235, 238-239, 242-243, 246, 248
Cointegration 110-112
Command-Control and Signaling System (CCS) 419
Complex Network 273, 279, 293
component failure event 177, 179-181, 193

compound interest 71, 78
Condition Monitoring 414, 417, 489, 495, 498, 512
Condition-Based Maintenance 419, 489
conditioning systems 580-581, 586, 591, 601-602
Conformity 252, 254-256, 258-259, 266, 271
constraints 43-44, 60, 161-162, 170, 228, 316, 318, 322-324, 326, 330-331, 335, 337, 383, 386, 395, 409, 412, 425-426, 445, 529, 599-600
consumer surplus 148, 150, 155, 158, 161-165
continuous development 67
Control Table 218
conventional programming 398
correlation coefficients 29
Cost Benefit 67-68, 80, 85, 94, 169
Cost-Benefit Analysis 80, 84, 168
Cross Coupling Feeding 487

D

DC Traction Systems 452-453, 458-460, 472, 485, 488
Deadlock 323-324, 336, 364-365, 383, 386
demand frequency 318
Derailment 174-175, 180, 202, 205, 207, 524-525, 529, 531, 533-535, 537-538, 542, 546, 559, 564-567, 570
design ventilation 600
discount rate 69-71, 73-75, 77-78, 81, 85, 91
Discounted cash 69-70, 73
discrete choice model 146, 149, 151
Discrete Event System 450
Dispatching 316, 323, 335, 383, 411, 417, 424
Disruption Management 420-422, 424, 426, 436, 440, 445, 450
Disruption recovery 422, 424, 426, 445-446
Disturbance 119-120, 123, 371, 374, 408, 413, 417, 419-420, 422, 440-441, 444, 565
DLR 259, 261-263, 270, 411, 528
Double-End Feeding 459, 487-488
Dupuit 161-167

Dwell Time 386, 427, 438, 591, 596
dynamic behavior 150, 525, 528-529, 542, 546, 557, 570

E

Eccentricity Faults 493-494
Economic Affluence 41, 45, 60
Economic Welfare 161, 163, 165, 169-170
EEG 273-275, 277-280, 287, 290-291, 299-300, 311
Electric Locomotive 452-453, 487
Electrical Section 454, 458, 487-488
electromagnetic communication 392
Electronic Interlocking 212-213
embedded software 234-235, 238, 242
Energy-Efficiency 411, 419
Entropy 278, 290, 311, 513
ERTMS 252-254, 256, 261, 264
ETCS 215, 223, 250-257, 262-266, 271-272
European Railway Agency (ERA) 271
External Effects 169

F

Fans 490, 591-592, 597-598, 601
Fatigue 273-280, 290, 294-297, 300, 311
Fault Detection 489-490, 496-497, 499-501, 503, 506-507, 511-512, 514-515
Feasible Solution 371, 386
Fire Safety 583, 599-600
first-best pricing 161-162, 166-168
Fluid Mechanics 601
Forced/Active Steering 524-525, 544, 549-550, 552-556, 559-560, 565, 570-571, 574
forecasting methods 86, 100-102, 104, 114, 119
Formal Methods 212, 214-215, 220-221, 223, 228, 258, 264
Formal Verification 212, 220-224, 227-228, 230
frequency components 278, 492-493, 496, 498, 502, 506-507, 512, 515, 573
fundamental frequency 296, 492, 496, 506, 515-516

G

game theory 149-151, 158
GIS 26

H

hazard identification 174, 177, 179
Headway 84, 327, 373, 386, 395, 421-422, 425-426, 428, 450
Hicks 161, 165

I

income elasticity 111, 113, 162, 164
Integrated Land use-Transport 58
integrated traffic 405, 408, 410, 414, 417
Intelligent Transportation 387, 395, 401
Interlocking 208, 212-215, 223, 231, 250, 259, 265-266, 406, 419
intermodal competition 146, 150, 158
Internal Rate 67-68, 77, 84, 91
Internal Rate of Return 67-68, 77, 84, 91
international standards 232-236, 242, 246, 598
Internet and Communication Technology 44
Interoperability 252-254, 256, 271-272, 397-400
investment alternative 73
Investments 51, 67-68, 80, 85, 89-91, 94, 97, 113, 132-135, 137, 141-143, 148, 150, 223, 229, 314, 330

J

JnNURM 52-53

K

Kalman filtering 119

L

Land Use 37, 40, 43-44, 53, 57-58, 60, 100, 131-134, 139-141, 143, 330, 390
Longitudinal Coupling 454, 456, 487-488
Long-Run Capacity 161, 168

M

marginal costs 161-164, 166-170
Metro 25, 35, 37, 44, 46-49, 54-55, 130-131, 134-135, 149, 215, 223, 421-422, 425, 436, 438-440, 446, 488, 580-581, 583, 585-586, 588, 590-591, 597-599, 601, 603
Million Plus 41-43, 52
Minimum Headway Time 426, 450
mobility management 406, 417
mode choice 146, 158
Model Checking 221, 224, 231
Model-based Testing 265-266, 270
Modeling 25, 28, 84, 100-101, 104-105, 110, 113, 122, 126, 150, 223, 265-266, 290, 313-315, 331, 389, 391, 394-397, 399, 401, 420, 425, 428-430, 445, 452-453, 460-461, 470, 480, 485, 498, 559, 601
monitoring systems 387, 490, 512
Motor Current Signature Analysis 489, 500

MRTS 25, 35, 43, 45-48, 52, 60
Multi-Agent Systems 397-398, 401
multi-body dynamics 525, 528-529, 533, 557

N

Nash equilibrium 146, 149, 157
natural monopoly 149, 164, 166-168
network structure 126, 294, 406, 426-427
Neural networks 119, 124-125, 495, 502
Neutral Section 488
Nineteenth century 2-3, 6, 8, 10, 13, 15-16, 18
NUTP 50, 53

O

online implementation 495, 500, 512-513
operational control 317, 321, 422
Operational Interoperability 252, 271
optimal pricing 161-162
Overhead Contact Line 487-488

P

passenger service 102
Past Futures 1, 5-6, 18
PCA 273, 284-285
Pearson correlation 29
PERCLOS 273-274, 276-277, 280, 282, 284-286, 297, 300, 311
political arguments 84
Power Flow 452-454, 459, 471-472, 474, 485
Prediction 100, 118, 124, 158, 277, 300, 408-410, 414-416, 419
present value 67, 69-70, 72-73, 75, 77, 84-85
Present Worth 68, 71, 73-75, 77, 84
primary suspension 524-525, 542-543, 546, 548, 550, 553-555, 561
profit maximization 148, 150, 156, 162-163
public transport 41, 45-47, 49-51, 53, 55, 57, 60, 134, 137, 148, 406

Q

qualitative descriptors 177, 179, 182-183, 186-190, 192, 194, 198-199, 208

R

radial position 532, 543-545, 547-548, 553
RailSiTe® 259

Railway Infrastructure 84, 101, 323, 338, 340-341, 344, 386, 450
Railway Power Supply Systems 452-455, 485
railway safety 173-175, 177, 179, 192, 199, 208, 233, 274
Railway Signaling 212, 214-215, 218, 221-224, 226, 228, 231-236, 238-239, 242-243, 248
Railway Signaling Principles 215, 218, 221, 226, 231
Railway Signaling system software 232-234, 236, 238-239, 242-243, 248
Railway Yard 213, 216, 231
Railways 1-2, 4-11, 14, 16-18, 54, 112, 114, 118, 120, 137, 147-148, 173-174, 212-215, 218, 223, 226-227, 250, 253, 265, 279, 314, 317, 387, 389, 393, 401, 405, 411, 417, 419, 425, 453, 489-490, 492, 498, 515-516, 576, 584
real-time planning 400, 450
Recovery Task 450
Regression 23-24, 28, 32, 35, 105, 107, 109, 112-114, 119, 126, 135, 495
Relay Ladder Logic 218, 231
Relay Logic 220, 222
Rescheduling 323, 335, 383, 425, 442
resource capacity 321-322
Ride Comfort 524-525, 528-529, 546, 548, 550-551, 559, 571, 573-574
Route Locking 218, 227

S

Saccades 311
Saccadic Ratio 311
Safety 23-24, 37, 50, 58, 60, 152, 173-177, 179-181, 192, 199, 202-203, 208, 212, 214-215, 218, 221-222, 227-228, 232-236, 238-239, 243, 248, 250-251, 253-254, 258-259, 271, 274, 321, 323-324, 337, 383, 392, 394, 416, 426, 438, 450, 459, 488, 524, 536-537, 580-581, 583-584, 597, 599-600, 603
Schedule 158, 168, 273, 315-316, 318-319, 321-324, 386, 407, 409, 412, 414, 416, 419, 424-425, 461, 464
Scheduling 313, 315-317, 320-323, 330, 335, 337-338, 383, 390, 395, 400-401, 414, 421-422, 425-426
second-best pricing 161-162, 166-167, 169-170
Serviceability 272
Short Turn 424-425, 427-429, 438, 440, 450
short turning 425, 428, 443-444
Short-Run Capacity 162, 168
signal conditioning 515-516

signaling system software 232-234, 236, 238-239, 242-243, 248
Simulation Model 335-337, 340, 342, 344, 346, 349, 355, 363, 366-367, 369, 371, 373, 383, 386, 422, 425, 429, 431, 441-442, 444
Single-End Feeding 459, 487-488
Slutzki Equation 165
Small-World Network 294
social concerns 1-2, 18
social costs 80, 85
social welfare 146, 148, 150, 154-158, 161
spectral estimation 502, 506, 515
Squirrel Cage Induction Motor 507
station skipping 444
Steering Linkage 561, 570
steering mechanisms 525, 544, 559, 574
Stop Skipping 445, 450
SVM 273, 285-286
Synchronization 273, 279-280, 287, 290-294

T

tactical 176, 314, 317, 319, 330-331, 421
Tamil Nadu 24, 37
Technical Interoperability 272
Temporal Logic 220-222
Testing 110, 213-215, 220, 222, 225-228, 232-235, 238-240, 242-244, 247-248, 250-260, 264-266, 268-270, 286, 400, 421, 496
Thermal Comfort 581-582
Thermodynamics 600
Third Rail 452, 487-488
Time series methods 114, 119-120
Timetable 85, 275, 315, 320-321, 323, 335-338, 340, 364-369, 371, 374, 383, 386, 405-407, 411, 417, 419, 422, 426, 442, 445
Timetabling 317, 319-321, 335-338, 383, 422, 426, 429
Traction Motor 489-491, 551
Traction Network 456, 465-467, 474, 480-481, 483-484, 487-488
traffic management 250, 252, 338, 393, 398-400, 405-410, 414, 416-417, 419, 429, 440

Traffic Management System (TMS) 252, 400, 405-406, 410, 419
Traffic safety 23-24, 383
traffic systems 396, 410
Train Control System 215, 223, 228, 250, 252-253, 271-272, 419, 450, 465
Transit Funding and Financing 130-131, 138, 143
transition region 174, 179
transport systems 53, 57, 161-162, 167, 335, 389, 406
transportation systems 259, 274, 314, 336, 383, 387, 392-399, 401, 406, 420, 422, 524
Travel Demand 60

U

uncertainty 86, 98, 100, 173, 175, 177, 179, 182, 185, 190, 194, 208, 311, 398, 400, 420, 422, 442, 444-446
underground spaces 8, 580-581
Urban Land Values 131-132
Urban Rail 112, 114, 119, 130-132, 134-135, 137-138, 143, 161, 170, 420-422, 580
Urbanization 32, 35, 37, 40, 42-44, 49, 60, 130, 147

V

Validation 140, 154, 215, 221, 229, 232-235, 238, 242, 248, 253, 258, 270, 272, 400
Value Capture 130-132, 135, 137-138, 141, 143
VC Mechanisms 130-132, 137-138, 140-143
Verification 119, 212-213, 215, 220-224, 227-228, 230, 233, 238, 253, 265, 269, 272, 366, 394, 400
Visibility Graph 279, 290-293, 311

W

Walkability 54, 58, 140
Waterloo depot 177, 197, 207-208
Wheel Wear 524, 566-567
wheel-rail contact 528-529, 543-544, 562

Information Resources Management Association

Become an IRMA Member

Members of the **Information Resources Management Association (IRMA)** understand the importance of community within their field of study. The Information Resources Management Association is an ideal venue through which professionals, students, and academicians can convene and share the latest industry innovations and scholarly research that is changing the field of information science and technology. Become a member today and enjoy the benefits of membership as well as the opportunity to collaborate and network with fellow experts in the field.

IRMA Membership Benefits:

- **One FREE Journal Subscription**
- **30% Off Additional Journal Subscriptions**
- **20% Off Book Purchases**
- Updates on the latest events and research on Information Resources Management through the IRMA-L listserv.
- Updates on new open access and downloadable content added to Research IRM.
- A copy of the Information Technology Management Newsletter twice a year.
- A certificate of membership.

IRMA Membership $195

Scan code to visit irma-international.org and begin by selecting your free journal subscription.

Membership is good for one full year.

www.irma-international.org